Bioidentical Hormones 101

2nd Edition

Menopausal Hormone Replacement

Jeffrey Dach, MD

Bioidentical Hormones 101: Menopausal Hormone Replacement, 2nd Edition

Published by Medical Muse Press

ISBN 978-1-7324210-6-6 (Hardback)
ISBN 978-1-7324210-5-9 (Paperback)

Library of Congress Control Number: 2025909319 1

Author Contact information:
Jeffrey Dach, MD
7450 Griffin Road, Suite 180/190
Davie, Florida 33314
Office Telephone 954–792–4663
Website: www.jeffreydachmd.com

Disclaimer: The reader is advised to discuss this information found in these pages with his or her personal physician and to act only on the advice of his/her personal physician. This book is not intended as a substitute for the medical advice of a physician. The reader should regularly consult a physician in matters relating to his or her health, particularly in respect to any symptoms that may require diagnosis or medical attention. The author and publisher disclaim responsibility for any adverse effects resulting directly or indirectly from the information or advice contained in this book. Regarding nutritional supplements mentioned in this book, they have not been reviewed or approved by the Food and Drug Administration (FDA). Any statements in this book about nutritional supplements are informational of a general nature and do not claim to prevent or cure any medical condition or disease.

Cover Design: Yesna99 @ 99designs.com
Interior Design: Deborah Stocco, modernbookdesign.com

Table of Contents

Foreword by Lindsey Berkson MA, DC, CNS, DACBN, ACN

THE FACT IS HORMONES ARE the most important cell signaling system in our body. Hormones signal the gut, the esophagus, the heart, the kidneys, the lungs, and the brain which contains the all-mighty hippocampus, the site of brain regeneration. The hippocampus stores and retrieves memories that define who we are. Without all these hormone signals, we cannot live a full, healthy, and meaningful life.

When we think of doctors who specialize in hormone replacement, we think of gynecologists and urologists. Truth be told, most OB/GYNS and urologists know very little about hormones. A survey in 2023 revealed that only 31% of gynecologists had any training in hormones, usually just one short class in medical school. For the past 22 years, hormones have not been taught in medical schools, osteopathic schools, naturopathic schools or nursing schools. If you are not "taught" hormones, you won't "think" hormones. Yet, OB/GYN, urologists, gastroenterologists, cardiologists, nephrologists, pulmonologists, and especially gerontologists are not trained in hormones, and are not thinking about hormones during your office visit or while they examine you. They do not include hormone levels as part of your yearly health evaluation. You might ask why did this happen? It all goes back to the 2002 Women's Health Initiative study which was wrongly interpreted. The original authors reported to the world that hormones cause breast cancer and heart disease, two medical conditions hormones were originally thought to prevent. Because of this 2002 study, hormones became feared, vilified, and no longer taught in medical schools. However, within months of the publication of the WHI study, re-analyses began unveiling deeper facts that went unnoticed by the mass media. Nineteen years later, the original authors republished their findings admitting that the original "Be scared of estrogen" was wrong. In fact, re-analysis found, that

estrogen "protects" healthy breasts more than any other agent ever investigated.

Estrogen therapies taken for an average of 5 years reduce the risk of getting breast cancer by 23%. If you do get breast cancer and were lucky enough to have been on estrogen therapies for an average of 5 years, your risk of dying from breast cancer was reduced by 44%.

Nothing has ever been so breast-protective as estrogen. The new 10 million study published in April 2024 by our own NIH is the largest study ever run on hormone replacement. This study of 10 million Medicare-age women tracked hormone use and health outcomes, demonstrating that estrogen therapies statistically reduce the risk of getting breast cancer. There is no debate. The data is clear that estrogen is breast-protective. They also found estrogen statistically (beyond chance) prevents lung cancer and colorectal cancer, prevents dementia, and helps older women live 19% longer with better quality of life.

Today, doctors and patients alike are frightened of hormones. Mass misinformation has created mass fear. The fear and avoidance of hormones have led to mass downstream poor health. Doctors are not testing our hormone levels or providing hormone treatments. Your doctor is more apt to write a script for birth control pills than test and treat your hormones with natural, safer, bioidentical hormones. Instead, we are bombarded with the message that hormones cause cancer and should be feared. Doctors tell us we do not need hormones, and we should age gracefully without hormone replacement. When this happens, YOU lose out and miss out on great hormonal medical care.

Enter center stage Dr. Jeffrey Dach with this second edition of *Bioidentical Hormones 101*. Step by step, Dr. Dach explains the studies that show the "fear" of estrogen is "irrational" and "unfounded." It is widely believed estrogen causes breast cancer. This belief is false, based on old studies which are wrong. In this book, Dr. Dach dives deep into the science, obliterating the myths and misgivings about hormones. He examines the safety of bioidentical hormones versus synthetic hormones which have been "pushed" on doctors by Big Pharma. When the money is followed, the corrupting influence of Big Pharma becomes obvious. What about high-risk women with a history of breast cancer? Dr. Dach even addresses hormone safety and efficacy in breast cancer patients. Did you know that testosterone is breast cancer protective? Testosterone up-regulates estrogen receptor beta (ER-beta), the anti-cancer receptor. There are two main estrogen receptors, ER-alpha and ER-beta. The ER-beta receptor is a "tumor suppressor." The ER-beta receptor also tamps down chronic excessive inflammation. So, it is also an "inflammation suppressor." Science now recognizes that many diseases start with chronic excessive inflammation.

The male hormone testosterone's final metabolite (break-down product) is called 3-beta-adiol, which activates estrogen receptor beta (ER-beta). This inhibits (blocks) breast cell proliferation (or growth-out-of-control). You will learn in this book that breast cancer patients can be given testosterone. Women with adequate testosterone blood levels have "less" breast cancer in the first place. And Dr. Dach shares how testosterone, can even be given to ER+ breast cancer patients on aromatase inhibitors as well as to triple-negative breast cancer patients. These cell types have no hormone receptors at all, so why deprive them of hormones? Most doctors do not know this unless they read Dr. Dach's book! We think that doctors know the science. But in reality, very few do. They are too busy working. In reading this book, you will learn about hormones from A to Z. Dr. Dach goes into striking detail about how the fear of breast cancer associated with hormone replacement is no longer valid. Instead, the increased breast cancer risk applies only to the combination of estrogen with a synthetic version of progesterone, called a "pro-gestin" (medroxyprogesterone, MPA) and not to estrogen alone. And there is no increased risk of breast cancer with the combination of estrogen and natural progesterone. Dr. Dach demonstrates that avoiding hormone replacement puts women in harm's way. Avoiding hormones means a dysfunctional immune system, less breast protection, less quality of life, and aging before your time.

I met Dr. Dach in April 2024, when Dr. David Brownstein and I put on our own continuing medical education course for MDs, NPs, nurses, PAs and NDs called "Everything Hormones" offering 16 hours of CME, continuing medical education credits. We had almost 300 doctors and nurse practitioners from 7 countries. Many were OB/GYNs and urologists. They had feared hormones. Most admitted to us they knew very little about hormones and knew little about how to test, track, and prescribe hormones. Honestly, it was quite shocking. Both you and your providers need education in hormones. Healthy hormones and balanced natural hormone replacement leads to:

Less breast cancer.
Less prostate cancer.
Less dementia.
Less hip fractures.
Less heart and renal disease.
More energy.
More immune resilience.

And oh, lest we forget, less diabetes. Hormones help our blood sugar stay healthy and stable.

My entire life focus has been on hormones. I have written many books on hormones. I have been a distinguished hormone scholar at an environmental estrogen think tank, the Center for Bioenvironmental Research at Tulane and Xavier Universities. My life had been ruined by hormones. I was a DES daughter. This means my mother was pregnant with me, she was given the most powerful synthetic estrogen ever invented, a nasty drug, that became the model compound for endocrine disruption, diethylstilbestrol – DES. This drug was given to 38 million pregnant women before DES was banned in 1971 as the most cancer-causing compound ever invented. I had breast cancer, kidney cancer, and battled eye disease. Once my hormones got balanced on bioidentical hormone replacement therapies, my fragile health returned. I have lived the truth that "strong hormones mean a stronger life". Do not let your life be ruined by "hormonal imbalance" or "hormonal fear." Dr. Dach is an agile thinker. He has focused his science-detailed cognition on hormones, for you to get unafraid. For you to get and stay strong. This *Bioidentical Hormones 101*, second edition book is worth its weight in hormonal gold. Dr. Dach did the heavy hormonal lifting, so you can keep on trucking. Dr. Dach is sharing how to keep our hormones resilient and strong, based on science that most doctors do not know. But now you can know, thanks to Dr. Dach for a job well done.

Lindsey Berkson, MA, DC, CNS, DACBN, ACN
Website: https://drlindseyberkson.com/
Agile Thinking Substack: https://drlindseyberkson.substack.com/
Everything Hormones online with 16 hrs of CME: https://drlindseyberkson.com/everythinghormones/

Dr. Lindsey Berkson is the author of *Hormone Deception*, Contemporary Books., 2000 and *Safe Hormones Smart Women.* iUniverse, 2010. Stay tuned for *Oxytocin Medicine* coming out in 2025. Dr. Berkson is a chiropractic physician, best-selling author, nutritionist, and integrative nutritional, gastrointestinal, and endocrine specialist. Dr. Berkson consults all over the world working remotely as a consultant as part of your personalized healthcare team. Dr. Berkson specializes in complex cases, high-risk hormonal patients, and severe gastroenterology cases trying to avoid surgery. Dr. Berkson spent 35 years studying hormones from every perspective, as a scientist, clinician, health advocate, author, inventor, and most importantly, as a patient. Dr. Berkson has worked with and trained by Dr. Jonathan Wright the pioneer in bioidentical hormones. Dr. Berkson was a distinguished hormone scholar for years at the Center for Bioenvironmental Research at Tulane University studying with Elwood Jensen PhD and Jan-Ake Gustafsson PhD who discovered the first estrogen receptors, ER Alpha and ER Beta. Dr. Berkson wrote one of the first breakthrough books on endocrine disruption (*Hormone Deception*, McGraw-Hill 2000, Awakened Medicine Press 2016). This book is mandatory reading in many environmental science university programs. She has published another 21 books. Dr. Berkson has led grand rounds at the University of Minnesota, and the University of Southern Florida. Dr. Berkson's book *Healthy Digestion the Natural Way* (Wiley & Sons 2000) sold over a million copies. In the 1970's Dr. Berkson lived with Hopi Indians in Arizona and 1992 lived with Aboriginals in Australia teaching nutrition and learning Indigenous cultures.

Foreword by David Brownstein, MD

I STARTED USING BIOIDENTICAL, NATURAL hormones over three decades ago. I was not trained in their use in medical school, but the subject fascinated me. Instead, my medical school training was the application of synthetic, foreign hormone molecules that lack receptors in the human body. Instead, synthetic hormones have unpredictable off-target effects, making them endocrine-disrupting chemicals. Why was I not made aware of the benefits and common sense of using bioidentical, natural hormones? The reason is clear: Bioidentical, natural hormones cannot be patented, just as any natural substance cannot be patented. The big pharma cartel has no interest in using non-patentable products as their profit margin can be greatly exaggerated by using synthetic items that can be patented. When I learned of the difference between a natural hormone, where the human body had receptors available to bind the hormone, utilize, and then detoxify it from the body, and a synthetic hormone, where there were no receptors and a markedly reduced ability to detoxify the molecule, the choice was clear to me: There simply is no justification for using a synthetic hormone when a bioidentical, natural version is available.

This epiphany made biochemical sense as well as common sense. Unfortunately, in medicine, common sense is not so common, and the majority of physicians inappropriately prescribe toxic, synthetic hormones. Synthetic hormones are associated with a plethora of adverse effects, while bioidentical, natural hormones are a much safer option. This is predictable because the human body is designed to function with natural hormones.

Dr. Jeffrey Dach has written another round of excellent books. *Bioidentical Hormones 101*, Second Edition, provides the reader with a comprehensive analysis of the clinical use of hormones. This collection is a masterpiece of information, integral to any physician interested in the use of natural, bioidentical hormones. It should be required reading for all medical students. I have read many medical books during my career. Jeffrey Dach's *Bioidentical Hormone 101*, Second Edition, rivals any of my favorite books. I will treasure these and utilize them for many years. Read Dr. Dach's *Bioidentical Hormones 101,* Second Edition and become educated on how and why to utilize natural, bioidentical hormones.

David Brownstein, M.D.
Medical Director of the Center for Holistic Medicine
6089 West Maple Road, Suite 200
West Bloomfield, MI 48322
Telephone: 248-851-1600
www.drbrownstein.com
Author of 17 books, including:
The Miracle of Natural Hormones
Iodine: Why You Need It Why You Can't Live Without It.

Foreword by Carol Petersen, RPh, CNP

THIS MORNING, I SENT A text to someone inquiring about breast cancer risk, recurrence, and the use of bioidentical hormones. The best advice I could give was to refer her to *Bioidentical Hormones 101*, second edition. I simply did not have the words—or the time—to untangle the wealth of information that someone with these questions deserves. To say that information about hormone therapies is entangled would be an understatement. The medical literature, organized medical groups like the Menopause Society, the Endocrine Society, and the American College of Obstetrics and Gynecology, pharmaceutical companies, and media outlets are rife with mixed messages. The advent of quasi-hormones—synthetic compounds created to secure exclusive sales through patented formulations—has only added to the confusion. Both medical professionals and their patients are left grappling with the challenge of discerning truth from the tangles.

Dr. Jeffrey Dach has taken on this challenge with kindness, gentleness, and care. In *Bioidentical Hormones 101*, second edition, he untangles the medical literature, prevailing attitudes, and media propaganda to remind us that bioidentical hormones consistently address hormone deficiencies with precision—the right molecule in the right amount. He weaves this message throughout the book, ensuring readers fully grasp its importance. We live in a time when many individuals who desperately need hormone support are fearful—haunted by concerns over breast cancer, other cancers, or blood clots like deep vein thrombosis. Meanwhile, there's significant financial incentive in withholding hormone support; instead, patients are offered a broad palette of drugs—benzodiazepines for anxiety, sleep aids, antipsychotics, antidepressants, antihypertensives for blood pressure management, or medications for bone loss that paralyze bone remodeling but ensure favorable bone density test results.

Enter Dr. Dach. In a thorough and organized fashion, he navigates through the medical literature to reveal how studies—and their professional or media interpretations—can mislead us. Carefully and methodically, he clears a pathway through this tangled jungle of information. By the time you finish a chapter, you will find your decisions remain your own—but now you will have the comfort of understanding the backstory and knowing your choices are sound. *Bioidentical Hormones 101*, second edition fills a critical gap in our understanding of the endocrine system. Medical professionals will appreciate its concise discussion of issues most relevant to their patients. Women who have felt disempowered by poor healthcare choices will find themselves armed with knowledge—and empowered to take control of their health again. As I conclude this forward, I am reminded of the hero's journey: where once there was a sword, then a pen—and now a keyboard. The story is not over yet; there are briar patches to clear of tangles and dragons to slay. *Bioidentical Hormones 101*, second edition is a remarkable gift to us all—and I eagerly await Bioidentical Hormones 201, 301, and beyond.

Carol Petersen, RPh, CNP
The Wellness by Design Project
Madison, Wisconsin
Phone: 608-469-8821
carol@thewellnessbydesignproject.com
https://thewellnessbydesignproject.com

With her **Wellness by Design Project**, Carol provides Bioidentical Hormone Consultation and Coaching Support. You are invited to join her to explore aging gently with vitality and exuberance at https://thewellnessbydesignproject.com/meet-carol/

Carol Petersen is an accomplished compounding pharmacist with decades of experience helping patients improve their quality of life through bio-identical hormone replacement therapy. She graduated from the University of Wisconsin School of Pharmacy and is a Certified Nutritional Practitioner. Her passion for optimizing health and commitment to compounding is evident in her involvement with organizations including the International College of Integrated Medicine and the American College of Apothecaries, the American Pharmacists Association, and the Alliance for Pharmacy Compounding She was also the founder and first chair of the Compounding Special Interest Group with the American Pharmacists Association. She is the chair of the Integrated Medicine Consortium. She co-hosts a radio program "Take Charge of Your Health" in the greater New York area. She is on the Medical Advisory Board for the Centre for Menstrual Cycle and Ovulation Research (CeMCOR.ca).

Foreword by Donna White

IT IS MY GREAT HONOR to write this Foreword on behalf of Dr. Jeffrey Dach, MD—a pioneer, a scholar, and one of the most respected authorities in the field of Bioidentical Hormone Replacement Therapy. For years, I have followed Dr. Dach's work with admiration—especially his newsletters, which consistently distill complex medical literature into accessible, evidence-based insights that providers and patients alike can trust. Now, with the release of *Bioidentical Hormones 101* Second Edition, Dr. Dach offers us the most comprehensive, thoroughly researched resource on bioidentical hormones ever published. This extraordinary three-volume series is more than a book—it is the encyclopedia of bioidentical hormones. It should be required reading for every medical provider who prescribes hormone therapy, and it will undoubtedly become an essential reference for patients who are seeking to reclaim their health, their vitality, and their future through informed, science-based care. The depth of scholarship in these pages is matched only by the urgency of the message. For far too long, hormone therapy has been clouded by fear, misinformation, and outdated conclusions—particularly those stemming from misinterpreted studies like the early arms of the Women's Health Initiative. Dr. Dach unflinchingly corrects these misconceptions, highlighting the protective benefits of estrogen, progesterone, testosterone, and estriol when prescribed appropriately. He clearly differentiates between synthetic and bioidentical hormones, and he supports every statement with meticulously referenced data from the medical literature. Having been in the field of bioidentical hormone replacement therapy for over 30 years, I have seen firsthand the transformation that occurs when patients receive proper hormone care—and when providers are equipped with the right tools and knowledge. In my work as Founder of the BHRT Training Academy, we have trained providers from 4 countries on how to safely and effectively prescribe bioidentical hormones. Many of those practitioners now consider Dr. Dach's research an indispensable part of their learning journey. His work empowers providers with clarity and conviction—and most importantly, it equips them to serve women more confidently and compassionately. Women deserve to enter perimenopause and menopause with their questions answered and their concerns addressed by well-trained, evidence-informed medical providers. With this second edition, Dr. Dach not only arms providers with that knowledge, but also helps restore hope for women who have been told that suffering through hormonal decline is inevitable. It is not. Dr. Dach is not just a physician—he is a truth-teller, a teacher, and a trailblazer. His commitment to scientific integrity and patient-centered care shines through every chapter of this series. I am deeply grateful for the work he has done to advance this field and proud to share this space as a fellow voice in the Bioidentical Hormone Revolution. If you are a medical professional, may these volumes strengthen your practice. If you are a patient, may they embolden your journey. Either way, you are holding in your hands the gold standard of hormone education. Let the revolution continue.

Donna White, Author of *The Hormone Makeover and The Bioidentical Hormone Revolution*
Founder and President of the BHRT Training Academy
www.BHRTTrainingAcademy.com
donna@bhrttrainingacademy.com
704-396-5677
9935-D Rea Road #415
Charlotte, NC 28277

Donna White is a leading authority in Bioidentical Hormone Replacement Therapy (BHRT) with over 30 years of dedicated experience. As the Founder and Executive Director of Education at BHRT Training Academy, Donna is a pioneer in developing BHRT training programs, significantly advancing the hormone health movement. Donna expresses genuine care and a readiness to engage personally with both patients and practitioners. Drawing on her journey with hormonal imbalance, she offers an empathetic and expert perspective and has earned the trust of her colleagues and in the field. She excels in making complex medical topics accessible and practical, greatly benefiting both patient care and practitioner knowledge. Under her leadership, BHRT Training Academy has transformed hormone therapy practice, substantially improved patient outcomes, and fostered growth in medical practices. A sought-after speaker on BHRT, Donna has been named to the Board of Advisors for the Menopause Association, underscoring her influence and leadership in women's health. A staunch advocate for evidence-based treatments, she tirelessly works to ensure women have the information needed to make informed healthcare decisions. Her commitment has not only raised the standard of care but has also empowered a generation of medical professionals with the tools to drive positive change in hormone health.

Foreword by Amy Brenner, MD

I FIRST ENCOUNTERED *BIOIDENTICAL HORMONES 101* through its original edition, and it quickly became one of the most valued resources on my shelf. As a healthcare provider, I continually seek materials that are not only evidence-based but also clear, accessible, and empowering for patients. Dr. Jeffrey Dach's work checked every box. With this newly updated second edition, *Bioidentical Hormones 101* builds on its strengths, offering deeper insights, current research, and an even more compelling case for bioidentical hormone therapy as a safe, effective, and patient-centered approach. It remains a book I turn to regularly and one I confidently recommend to patients seeking to better understand their hormonal health. Dr. Dach distinguishes himself by doing what few in the medical community are willing to do: challenging outdated paradigms with data, compassion, and clinical expertise. His work presents a credible, science-based alternative to conventional hormone replacement therapy—one that honors both physiology and the body's innate design. This book is far more than a primer on bioidentical hormones—it is a roadmap for reclaiming control over one's health. It replaces fear with facts, dispels long-standing myths, and equips readers with the knowledge to make informed decisions about their care. Whether you're a patient navigating hormone imbalance or a practitioner searching for reliable, trustworthy resources, this second edition is essential reading. I am deeply appreciative of Dr. Dach's dedication to truth in medicine and proud to offer this book as a foundational reference within my practice.

Amy Brenner, MD, FACOG, ABAARM
Amy Brenner MD and Associates, LLC
6413 Thornberry Ct, Mason, OH 45040
(513) 770-0787
https://www.spamedicca.com/www.dramy-brenner.com

Foreword by Phyllis Bronson, PhD

IN THIS COMPLETELY REVISED SECOND edition of *Bioidentical Hormones 101*, you will explore the profound impact bioidentical hormones have on physical, emotional, and mental well-being. This updated edition not only revisits foundational concepts but also incorporates the latest advancements in medical research on the science of hormone replacement. Throughout the book, Dr. Dach unravels the complexities of menopausal hormone replacement, dispelling myths and providing clarity on the safe and effective use of bioidentical hormones.

In 2007, I was a keynote speaker at the International Society of Orthomolecular Medicine's (ISOM) annual meeting. That is when I first met Dr. Jeffrey Dach, who was in the audience for my presentation on premenstrual syndrome (PMS) and the effective treatment with bioidentical progesterone. Dr. Dach is a brilliant, scholarly doctor with a long career in conventional and integrative medicine. In 2007, Dr. Dach was just getting started in the field of bioidentical hormone replacement and has since become a close colleague and leading advocate for bioidentical hormones. My original work on progesterone revealed the critical distinctions between bioidentical progesterone and synthetic versions of progesterone called progestins. Bioidentical progesterone shares the same chemical structure as progesterone produced by the human ovary and adrenal glands. However, progestins are chemically altered versions of progesterone. Progestins are endocrine-disrupting chemicals found in birth control pills, intrauterine devices (IUDs), and older versions of *Hormone Replacement Therapy* (HRT). As explained in this book, progestin-containing hormone formulas should be declined. Bioidentical hormone formulas are preferred. For many years, I have been fighting an uphill battle against pharmaceutical companies who have promoted synthetic progestins as equivalent to bioidentical progesterone. They are not equivalent, as clearly explained throughout this book.

Dr. Dach has been a staunch ally and has done much to advance the discourse on bioidentical hormone therapy for women of all ages. Over the years, Dr. Dach has written extensively in his monthly newsletter about bioidentical hormone topics, relying on my research and that of others. Much of this information has been condensed into this second edition of *Bioidentical Hormones 101*. As you turn these pages, you will find personal anecdotes, patient stories, and expert commentary that will resonate with your own experiences. Whether you are a healthcare practitioner, a patient, or simply curious about the subject, this book aims to empower you with knowledge and understanding. In the ever-evolving landscape of health and wellness, bioidentical hormones have emerged as a beacon of hope for many seeking balance and vitality in their lives. I invite you to join Dr. Jeffrey Dach on this enlightening journey as he explores the transformative potential of bioidentical hormones to reclaim our health, one chapter at a time.

Phyllis Bronson, PhD
https://phyllisbronsonphd.com/
Author of *Moods, Emotions, and Aging: Hormones and the Mind-body Connection*, Rowman Littlefield, New York, 2013.

Phyllis Bronson, Ph.D. holds a doctorate in biochemistry. Her ongoing research involves studying the biological impact of molecules on mood and emotion. Dr. Bronson works with women who have hormone-based mood disorders utilizing her original research on human identical hormones. She lives in Aspen, Colorado, and is President of Biochemical Consulting and The Biochemical Research Foundation.

Foreword by Daved Rosensweet, M.D.

JEFFREY DACH, MD'S *BIOIDENTICAL HORMONES 101*, Second Edition is a truly remarkable work. Far more advanced than the title suggests—more of a "505" than a "101"—this book offers a scholarly and comprehensive exploration of the medical literature underlying menopause care. Dr. Dach's deep understanding allows him to present complex clinical insights in a clear, accessible way. This makes the book an essential read for any medical professional treating women in menopause. It thoughtfully guides the reader through the common hurdles and pitfalls, offering a roadmap to optimal clinical approaches. The book also provides an in-depth analysis of breast cancer risk, prevention, and treatment, making it especially valuable for women with a history of breast cancer. I am deeply grateful for the knowledge, dedication, and effort that went into creating this important resource. Thank you, Dr. Dach.

Daved Rosensweet M.D., Founder of The Menopause Method and The Institute of Bioidentical Medicine and author of *Menopause and Natural Hormones* and *Happy Healthy Hormones: How to Thrive in Menopause*
Medical Director
1058 N. Tamiami Trail, Suite 108
Sarasota, FL 34236
941-220-5444
www.iobim.org
www.brite.live

Introduction by Jeffrey Dach MD

THIS BOOK IS A COMPLETELY updated and revised second edition of *Bioidentical Hormones 101* originally published in 2011. This authoritative book represents 20 years of experience prescribing bioidentical hormone replacement in an outpatient setting and is intended for both menopausal women as well as health care professionals. Bioidentical hormones used at appropriate dosages are safe, and effective for relieving menopausal symptoms. Health benefits extend to all organ systems. Bioidentical hormone treatments are safest when used in the correct combination, much like a symphony with multiple musical instruments making beautiful music. Why get your information about hormone replacement from the newspapers and magazines? This book gives you the true story directly from the medical literature, translated into plain English.

What are the benefits and adverse effects of HRT on the cardiovascular system, on the brain, on the risk for breast cancer, and the risk for blood clotting? All forms of oral estrogen are associated with blood clotting. Transdermal forms of estrogen are not associated with blood clots. Thus, transdermal estrogen is preferable to oral estrogen. The safest estrogens reduce the risk of breast cancer. They target ER beta, the tumor suppressor receptor, such as unsaturated B-ring steroids found in Premarin (CEE), estriol (E3), and testosterone metabolite 3-beta-diol. Exposure to pregnancy levels of hormones confers protection from breast cancer. Long-term estrogen deprivation (LTED) transforms estrogen from a growth factor to a death factor in breast cancer cells.

The exact hormone formula is important for optimal results. For example, a formula using Premarin alone is not safe because of the increased risk of endometrial cancer. Synthetic progestins such as MPA are endocrine-disrupting chemicals that increase the risk of breast cancer. In 2011, Dr. Cynthia L. Bethea writes:

While the WHI trial [2002, first arm] made a valuable contribution in revealing the risks associated with conjugated equine estrogens treatment in postmenopausal women, it unfortunately generated considerable controversy in the field because it was interpreted as an indictment of postmenopausal hormone replacement, **when in fact, it did not study hormone replacement at all: that would have required use of the natural hormones, estradiol and progesterone**. The actions of the natural hormones are significantly different from those of Premarin and MPA [medroxyprogesterone]. (1)

The 2004 second arm Women's Health Initiative 18 year follow-up shows that estrogen-treated women (Premarin-alone) enjoy a **45 percent reduction in mortality from breast cancer**. This is a paradigm shift in our thinking about hormones and breast cancer. Estrogen is breast cancer protective. In 2022, Dr. Isaac Manyonda writes:

The evidence now compels a paradigm shift from the traditional thinking that estrogen could cause breast cancer to a recognition that it actually prevents the disease, and that when the disease does occur (no preventative intervention achieves a 100% preventative effect), it is often picked up early and mortality is reduced by up to 44%. (2)

There is no further need for menopausal women to fear hormone replacement and suffer needlessly with menopausal symptoms of hot flashes night sweats, brain fog, insomnia, osteoarthritis, osteoporosis, coronary artery disease and dementia. We will show you the routine bioidentical HRT formulas and dosages we have been using for 20 years. These are safe and effective containing estradiol, estriol, progesterone, and testosterone in the correct dosage, and route of delivery. This book shows you the medical studies that prove the safety and efficacy of bioidentical hormone replacement formulas.

In this book, I argue that menopausal estrogen deficiency is the harbinger of chronic disease, and the manufactured fear of estrogen is contributing to the chronic disease epidemic. Step by step, I explain the studies showing that the fear of estrogen is irrational and unfounded. I also explain the exact hormone formulation matters. Synthetic chemically altered hormones are endocrine-disrupting chemicals. These should be avoided. Synthetic progestins such as medroxy-progesterone (MPA) are known to increase the risk of breast cancer and should be avoided.

What is the safest and most effective bioidentical hormone formula? Why has our medical system failed to provide hormone replacement for menopausal women? What causes breast cancer and how do we prevent it? These questions are answered in painstaking detail with copious references to the medical literature.

Enjoy the book and sign up for my free monthly newsletter at: https://jeffreydachmd.com/ where you can also download a pdf FREE copy of the first edition of *Bioidentical Hormones 101* (2011). Or use this shortened URL: https://shorturl.at/gXG8W

Or the longer URL version:

https://jeffreydachmd.com/bioidentical-hormones-101-first-edition-2011-by-jeffrey-dach-md/

♦ **References for Introduction**

1) Bethea, Cynthia L. "MPA: Medroxy-Progesterone Acetate Contributes to Much Poor Advice for Women." (2011): 343-345.

2) Manyonda, Isaac, et al. "Could Perimenopausal Estrogen Prevent Breast Cancer? Exploring The Differential Effects of Estrogen-Only Versus Combined Hormone Replacement Therapy." Journal of Clinical Medicine Research 14.1 (2022): 1-7.

Chapter 1

Menopausal Hormone Replacement Health Benefits

SUSAN IS A 54-YEAR-OLD ACCOUNTANT suffering from many diverse health problems, vague aches and pains in the muscles and joints, difficulty remembering things or finding words, insomnia, dry skin, chronic fatigue, mood disturbance, night sweats, and hot flashes. Instead of offering Susan menopausal hormone therapy (MHT), her primary care doctor ordered a DEXA bone density scan which showed osteopenia (low bone density) and then offered Susan a drug called Fosamax ®, a bisphosphonate drug FDA approved in 2005 for the treatment of postmenopausal osteoporosis. I explained to Susan all her symptoms are due to menopausal estrogen deficiency, which is very treatable with bioidentical hormone replacement therapy (HRT). I advised against the Fosamax drug, as estrogen is more effective for improving bone density while avoiding the adverse side effects of Fosamax, namely osteonecrosis of the jaw and atypical femur fractures.

Estrogen Prevents Bone Loss

Postmenopausal bone loss is caused by estrogen deficiency. The bone loss is progressive with a 10-12 percent loss of bone density over the first two years after menopause, and then 0.5 percent every year thereafter. The progressive loss of bone density can be prevented and even reversed with menopausal hormone therapy (MHT) reducing the risk for fracture by 20-40 percent. In 2021 Dr. Anna Gosset from France writes:

> Postmenopausal osteoporosis is a frequent clinical condition which affects nearly 1 in 3 women. Estrogen deficiency leads to rapid bone loss which is maximal within the first 2-3 years after the menopause transition and can be prevented by menopause hormone therapy (MHT). Not only, MHT prevents bone loss and the degradation of the bone microarchitecture but it significantly reduces the risk of fracture at all bone sites by 20-40%. It is the only anti-osteoporotic therapy that has a proven efficacy regardless of basal level of risk, even in low-risk women for fracture... ...In the absence of contraindication, use of MHT should be considered as a first option for the maintenance of bone health in those women where specific bone-active medications [bisphosphonates] are not warranted. (1-3)

FDA Safety Warning on Bisphosphonates

In 1998, Dr. Steven R. Cummings published the FIT trial data on Fosamax (Alendronate) which failed to show any benefit for the majority of the worried well, the osteopenia group defined as T score greater than -2.5. **Note:** DEXA scan (dual-energy X-ray absorptiometry) is an X-ray imaging test for bone density commonly ordered by primary care doctors. The DEXA scan provides a number called the T-score. Osteoporosis is defined as a T score more negative than -2.5. Osteopenia is defined as T-score between -1 and -2.5. Getting back to the FIT study, the osteopenia group treated with alendronate had higher fracture rates than placebo. This is exactly the group that was targeted for treatment by Sally Fields in Boniva (Ibandronate) drug ads. Bisphosphonate drugs, Boniva (ibandronate) Fosamax (alendronate), Actonel (risedronate) and Reclast (zoledronic acid) have severe adverse side effects of jaw necrosis (OJN), spontaneous mid-femur fracture, heart rhythm disturbances, and severe bone and joint pain. The spontaneous mid-femur fractures are especially troubling since these occur without any trauma. Subtrochanteric fractures are pathological fractures with underlying abnormal bone matrix directly caused by the bisphosphonate drug. The bottom line is that bisphosphonates are **BAD** drugs that make the bones weaker not stronger. For this reason, the FDA has issued multiple safety warnings on bisphosphonate drugs. In 2024, Dr. Gregory Curfman discusses the fate of 1,000 victims of atypical femoral fractures on alendronate who took Merck to court in 2011, claiming injury from failure to warn of adverse side effects. Thirteen years later, their case is still unresolved, writing:

> In 2011, a group of approximately 1000 plaintiffs who had been affected by atypical femoral fractures while receiving alendronate filed a lawsuit against Merck, the drug manufacturer... After 13 years in multiple courts at all levels of the federal judiciary, the plaintiffs in this case have not yet been provided a legal remedy for their injury. It is time for justice to be served. (4-15)

Failing to Mention Hormone Replacement

Susan wanted to know why her primary care doctor failed to mention hormone replacement, the first-line treatment for the prevention and treatment of osteoporosis, as discussed in 2023 by Dr. Silva. The error and tragedy of modern medicine is failure to prescribe bioidentical hormone replacement to maintain bone density. Instead, the menopausal patient is given a dangerous bisphosphonate drug that causes spontaneous fractures. Also, Susan is worried that hormone replacement therapy might cause breast cancer because her sister was treated for breast cancer. These types of concerns are quite common. I then spent the time to explain to Susan the health benefits of menopausal hormone replacement. I also explained to Susan that an increased risk for breast cancer is associated with the use of carcinogenic synthetic hormones progestins, such as MPA (medroxy-progesterone). We avoid these and use human bioidentical hormones found to be safe in

multiple studies, as discussed below. In addition to the hormone replacement program, we use a separate breast cancer prevention program which includes iodine supplementation, testosterone, selenium, and DIM (di-indole methane). For more on Iodine and DIM, see Chapters 21-23. For more on Testosterone as Breast Cancer Preventive, see Chapter 18. In 2023, Dr. Barbara Campolina Silva, an academic endocrinologist from Brazil, discusses estrogen therapy as the first-line treatment for post-menopausal women at risk for osteoporosis, writing:

> estrogen therapy should be used not only for the prevention of postmenopausal osteoporosis but also as the **first line in the treatment** of early postmenopausal women at high risk of osteoporotic fracture. (16-17)

Eighty Percent Decline in Women Using HRT

Susan's story is not unusual. Since 2002, there has been an 83 percent decline in the use of menopausal hormone replacement, causing needless suffering, a result of the negative publicity following the 2002 WHI (Women's Health Initiative Study) convincing menopausal women and their doctors to abandon hormone replacement. Worse, since 2002, medical schools have dropped training programs in hormone replacement creating a lack of core competency in health professionals to prescribe menopausal HRT. This is unfortunate, since bioidentical hormone replacement is the single most effective medical intervention for menopausal women. Thankfully, this insanity is ending. Menopausal hormone replacement therapy (MHT) is now recognized by mainstream medicine as absolutely necessary for maintaining health after menopause. In 2024, Dr. Lin Yang reported a decline in MHT use among menopausal women of 27 percent in 1999 to 4.7 percent in 2020. This is an 83 percent decline! More than 1 million women in the United States enter menopause each year. Yet only 47,000 are given hormone replacement. This leaves 953,000 women untreated annually. Dr. Lin Yang writes:

> In this serial cross-sectional study that included 13,048 postmenopausal women, the estimated prevalence of MHT use declined from 26.9% to 4.7% over 2 decades, with the greatest declines observed among women aged 52 years to younger than 65 years. Women of racial and ethnic minority groups had lower prevalence of MHT use compared to non-Hispanic White women... MHT use declined among postmenopausal women in the US from 1999 to 2020, with considerable variation in patterns of use and correlates across age and racial and ethnic groups. (18) (106)

In 2019, Dr. Santiago Palacios bemoans the reluctance to treat menopausal women with HRT has created a large and unnecessary burden of suffering, writing:

> Reluctance to treat menopausal symptoms [with HRT] has derailed and fragmented the clinical care of midlife women, **creating a large and unnecessary burden of suffering.** Clinicians who stay current regarding hormonal and non-hormonal treatments can put menopause management back on track by helping women make informed treatment choices. In addition, we must train and equip the next generation of healthcare providers with the skills to address the current and future needs of this patient population. **Note: All bold letters in quotations throughout this book are my emphasis.** (19-20)

A Situation Detrimental to Women's Health

In 2014 Dr. Richard Santen laments that internal medicine specialists fail to provide HRT to their menopausal patients, lacking the core competency and experience to do so. This situation is detrimental to women's health. Internists and general practitioners should obtain the training to prescribe hormone replacement for every one of their menopausal patients, yet they fail to do so. Instead, there

is a generalized fear of estrogen. This is the great tragedy and error of modern medicine. Dr. Richard Santen writes:

> Most internists currently lack the core competencies and experience necessary to address menopausal issues and meet the needs of women who have completed their reproductive years. We believe that this situation is detrimental to women's health, leads to fragmented care, and should change. Note: for more on the Fear of Estrogen, see Chapter 7. (21)

HRT Was the Standard of Care Until 2002

In 2021, Dr. Robert D. Langer, Professor Emeritus in Family Medicine and Public Health at the University of California, says menopausal hormone replacement therapy was the standard of care until the release of the 2002 WHI study (Women's Health Initiative Study) when HRT use plummeted because of unfounded fears of cancer and heart disease, not supported by the data. The actual data shows HRT reduces all-cause mortality and reduces the risk of coronary disease, osteoporosis, and dementia. Drs. Robert D. Langer and Howard Hodis write:

> HRT use plummeted following the WHI in 2002 and has remained low...Unfortunately, among many women and clinicians, the perception of HRT benefit/risk is distorted, and its use avoided, leading to unnecessary distress. Following the WHI, many clinicians have not received adequate training to feel comfortable prescribing HRT. When initiated within 10 years of menopause, HRT reduces all-cause mortality and risks of coronary disease, osteoporosis, and dementias. (22-24)

Hormone Replacement Therapy (HRT) Reduces All-Cause Mortality

There are considerable health benefits of menopausal hormone replacement involving all organ systems, the bones, joints, heart, and brain. Hormone Replacement Therapy (HRT) with bioidentical hormones reduces all-cause mortality, reduces the risk of cardiovascular disease, reduces bone loss, prevents osteoporosis, and reduces the risk of dementia. In 2023, Dr. Guilherme Renke writes:

> Recent evidence suggests that, when initiated within 10 years of menopause, HRT reduces all-cause mortality and risks of coronary disease, osteoporosis, and dementia. (44) (25-27)

Estrogen: The Only Effective Medication for Low Bone Density

In 1988, the FDA approved estrogen for the prevention of osteoporosis as an indicated use. Estrogen therapy prevents the annual loss of bone density caused by estrogen deficiency, making stronger bones, and thus preventing fractures. In 2005, Dr. Janelle Guirguis-Blake says estrogen is the only effective medication FDA-approved for the prevention of fractures in women with low bone density, writing:

> Estrogen, either unopposed [alone] or in combination with progestin [progesterone or MPA], **currently is the only effective medication approved by the U.S. Food and Drug Administration (FDA) for the prevention of fractures in women with low bone mineral density**. Note: although the medical literature mistakenly categorized progesterone as a progestin, in this book, we will define progestin as synthetic version of progesterone. Thus, I will exclude natural progesterone from the definition of progestins. (28-30)

WHI and DOP Shows HRT Reduces Fractures

The Women's Health Initiative (WHI) was the first randomized trial to show postmenopausal estrogen therapy provides a 34% reduction in osteoporotic fractures of the hip and vertebral bodies. These same numbers were independently confirmed by the Danish Osteoporosis Study (DOP) conducted by Dr. Leif Mosekilde (2000). Estrogen replacement therapy is now accepted by mainstream med-

icine as the first line for the prevention and treatment of osteoporosis. In my opinion, anti-resorptive drugs such as bisphosphonates (Fosamax®) and the newer osteoporosis drugs should be avoided. Instead, estrogen hormone replacement should be offered as the preferred treatment for all post-menopausal women concerned about bone density. For more on estrogen for the prevention of osteoporosis, see Chapter 8. In 2018, Dr. Veronika Alexa Levin, an OB-Gyne surgeon affiliated with Johns Hopkins Hospital says estrogen is the primary prevention and treatment of osteoporosis. Dr. Levin also says that transdermal estrogen is safer than oral estrogen pills, as discussed more completely in Chapters 4 and 5 of this book. Dr. Levin writes:

> Menopause predisposes women to osteoporosis due to declining estrogen levels. This results in a decrease in bone mineral density (BMD) and an increase in fractures. Osteoporotic fractures lead to substantial morbidity and mortality...The Women's Health Initiative (WHI) randomized controlled trial first proved hormonal therapy (HT) [with estrogen] reduces the incidence of all osteoporosis-related fractures in postmenopausal women... These studies support that HT [hormone therapy with estrogen] improves BMD [bone mineral density] and reduces fracture risk in women with and without osteoporosis... **we propose that HT should be considered for the primary prevention and treatment of osteoporosis** in appropriate candidates.... **low-dose and transdermal HT that have been shown to be safer than oral standard-dose HT**. (31-39)

In 2023, Dr. Leslie Cho, a cardiologist and Director of the Cleveland Clinic's Women's Cardiovascular Center says four major U.S. medical societies currently recommend hormone therapy (HT) for menopausal women, writing:

> four major North American medical societies, the **American College of Obstetricians and Gynecologists (ACOG), American**

Association of Clinical Endocrinology (AACE), The Endocrine Society (ENDO), and The North American Menopause Society (NAMS), now recommend HT in appropriate patients for the management of menopausal symptoms. (40)

Hormone Replacement Therapy (HRT): 52 Percent Reduction in Cardiovascular Disease

Menopausal HRT using estradiol and progesterone reduces the risk of coronary artery disease by 52 percent as discussed in 2012 by Dr. Louise Lind Schierbeck from Denmark. One must avoid MPA medroxyprogesterone and other synthetic forms which increase heart disease and breast cancer risk. In 2022, Dr. Howard Hodis says that when initiated within 10 years of menopause, menopausal hormone replacement **significantly reduces all-cause mortality and cardiovascular disease, while other therapies such as statin drugs for lipid-lowering fail to do so**, writing:

> Initiated in women <60 years of age and/or at or near menopause, **HRT significantly reduces all-cause mortality and cardiovascular disease (CVD) whereas other primary CVD prevention therapies such as lipid-lowering [with statin drugs] fail to do so**. Magnitude and type of HRT-associated risks, including breast cancer, stroke and venous thromboembolism are rare (<10 events/10,000 women), not unique to HRT and comparable with other medications. Note: For more on this, see Chapter 9 on Bioidentical Hormone Prevent Coronary Artery Disease, and Chapter 27 on The Failure of Cholesterol Lowering Drugs. (41)

In 2012, Dr. Louise Lind Schierbeck reported on the Danish Osteoporosis Study, a randomized controlled trial (RCT) of 1006 healthy postmenopausal women. Half were treated with triphasic estradiol and norethisterone acetate (a progestin) for 10 years. The other half served as controls. The DOP trial showed a **52 percent reduction in cardiovascular disease after 10 years of hormone replacement therapy (HRT)**, with estrogen and with or without progestogen, relative to no HRT, writing:

> Survival curve from the Danish Osteoporosis Study [DOP] showing a statistically significant reduction of cardiovascular disease by 52% (**HR, 0.48**; 95% CI, 0.27–0.89) after 10 years of randomized hormone replacement therapy (HRT) relative to no HRT...After 10 years of randomised treatment, women receiving hormone replacement therapy early after menopause had a significantly reduced risk of mortality, heart failure, or myocardial infarction, without any apparent increase in risk of cancer, venous thromboembolism, or stroke. (42)

Start HRT As Soon As Possible

In 2020, Dr. Izaäk Schipper says menopausal hormone replacement eliminates the increased cardiovascular mortality associated with premature menopause, also called premature ovarian insufficiency (POI). For the greatest benefit, Dr. Schipper advises women should start HRT immediately after menopause, writing:

> It has been substantiated that hormone replacement therapy (HRT) decreases the risk for ischemic heart disease and eliminates the increased cardiovascular disease mortality. It is therefore crucial to start HRT as soon as possible, particularly in women with premature ovarian insufficiency [premature menopause]. (43)

HRT Increases Mitochondrial Biogenesis

In 2023, Dr. Guilherme Renke discusses the mechanism of estrogen's cardioprotective properties, proposing a novel insight that estrogen promotes mitochondrial biogenesis and modulates the renin-angiotenisn-aldosterone system (RAAS). Dr. Guilherme Renke advises using human-identical hormone replacement to receive these health benefits, writing:

> E2 [estradiol] mediates its cardioprotective actions by increasing mitochondrial biogenesis, angiogenesis, and vasodilation,

decreasing reactive oxygen species (ROS) and oxidative stress, and modulating the renin-angiotensin-aldosterone system (RAAS). In this review, we assess whether it is prudent to develop an approach to managing postmenopausal women based on modifying the patient's CV [cardiovascular] risk that includes **human-identical hormone replacement therapy (HRT),** modulation of RAAS, and stimulating mitochondrial biogenesis. Note: mitochondrial biogenesis is the regeneration of new mitochondria, the organelles responsible for cellular energy production. Angiogenesis is the creation of new blood vessels, such as collateral vessels to tissues in need of more blood flow. Mitochondrial biogenesis is the body's ability to increase size and number of mitochondria, thus increasing cellular energy production required by the high energy requirements of heart muscle and brain. (44)

HRT Reduces Risk for Breast Cancer

The Women's Health Initiative Study (WHI) 2004, second arm, using Premarin-alone reduced breast cancer by 22 percent. The 18-year follow-up of this 2004 study showed estrogen-alone (Premarin, CEE) reduces mortality from breast cancer by 40 percent as discussed in 2020 by Dr. Rowan Chlebowski. However, in the first arm (2002) using the combination of Premarin and medroxyprogesterone (MPA), there was a 28 percent increase in breast cancer. A synthetic progestin called medroxyprogesterone (MPA), a known carcinogen, was the culprit responsible for the increased breast cancer in the first arm study (2002). Synthetic progestins should be avoided. Natural progesterone is breast cancer preventive and should be used in place of synthetic progestins for endometrial protection. Dr. Rowan Chlebowski writes:

> In long-term follow-up of 2 placebo-controlled randomized clinical trials involving 27, 347 postmenopausal women, prior randomized use of conjugated equine estrogen (CEE), compared with placebo, among women with prior hysterectomy was significantly associated with lower risk of breast cancer (annualized incidence, 0.30% vs 0.37%; **hazard ratio [HR], 0.78**); and breast cancer mortality (annualized mortality, 0.031% vs 0.046%; **HR, 0.60**), whereas prior randomized use of CEE plus medroxyprogesterone acetate (MPA) [first arm WHI 2002] , compared with placebo, among women with an intact uterus, was significantly associated with higher risk of breast cancer (annualized incidence, 0.45% vs 0.36%; **HR, 1.28**) and no significant difference in breast cancer mortality (annualized mortality, 0.045% vs 0.035%; **HR, 1.35).** (45-55)

Avoiding HRT Puts Women in Harm's Way

In 2022, Dr. Delfin A. Tan, Ob Gyne Surgeon and Reproductive Endocrinologist in Manila, Philippines agrees that the carcinogenic culprit is MPA, a synthetic progestin, and should be avoided by using natural bioidentical progesterone. Dr. Tan suggests women should no longer fear breast cancer risk from hormone replacement. Fear and failure to use menopausal hormone replacement puts women in harm's way. Dr. Delfin A. Tan writes:

> The threat that women may develop breast cancer is the major reason why both physicians and women are afraid to use menopausal hormone therapy (MHT). The fear pertains to **estrogen-progestin** replacement therapy (EPRT) as **estrogen-alone replacement therapy has no, or even a reduced, breast cancer risk**... **avoiding MHT use when indicated puts a woman in harm's way**. (56)

Synthetic Progestins Are Not the Same as Progesterone

In 2021, Dr. Irene Lambrinoudaki, Professor of Endocrinology in the Department of Obstetrics and Gynecology, University of Athens Medical School, reviewed breast cancer risk associated with menopausal hormone therapy. Dr. Irene Lambrinoudaki cited two large nested, case-controlled studies from the UK of 98,611 women with breast cancer diagnosis, matched to 457,498 female controls as reported in 2020 by Dr. Yana Vinogradova. This study showed the excess risk of breast cancer **arises from the use of a synthetic progestin such as medroxyprogesterone (MPA), levonorgestrel or norethisterone.** Synthetic progestins are known carcinogens, avoided by using natural progesterone instead. Dr. Irene Lambrinoudaki writes:

> Estrogen-only [estradiol or Premarin] therapy for up to 9 years increased only marginally the risk of breast cancer (**OR 1.14**, CI 1.08–1.21), whereas **estrogen–progestin combination** therapy for the same duration was associated with **a more pronounced increase in breast cancer risk** (OR **1.70**, CI 1.64 to 1.76). **The risk differed according to the progestin used, being higher with medroxyprogesterone acetate, levonorgestrel and norethisterone (OR 1.87** CI 1.71 to 2.05, **1.79** CI 1.68 to 1.90 and **1.88** CI 1.79 to 1.99 respectively) and lower with dydrogesterone (OR 1.24 CI 1.03 to 1.48) for more than 5 years of therapy. The excess risk dissipated in past users. **The excess risk of breast cancer associated with MHT [Menopausal Hormone Therapy] is mainly conferred by the progestin.** Note: OR is the Odds Ratio. (57-59)

Natural Progesterone is the Safest

In 2020, Dr. John Stevenson discussed the selection of the safest progestogen agent for menopausal hormone replacement, stating micronized progesterone is the safest option, advising against the use of synthetic progestins, writing:

> **Micronized progesterone and dydrogesterone appear to be the safest options**, with lower associated cardiovascular, thromboembolic, and breast cancer risks compared with other progestogens, and are the first-choice options for use in 'special situations,' such as in women with high-density breast

tissue, diabetes, obesity, smoking, and risk factors for venous thromboembolism, among others. Note: In my opinion, natural bioidentical progesterone should be the first choice for all women using HRT. Dydrogesterone is a synthetic oral progesterone, a stereoisomer of natural progesterone meaning it is very close to being identical to natural progesterone. It was withdrawn from the US market in 1997 by Solvay because natural progesterone replaced it. However, dydrogesterone is still widely used outside the US. (60)

Natural Progesterone Prevents Breast Cancer

In 2017, Dr. Allan Lieberman reviewed the literature on natural progesterone, stating it prevents breast cancer, as well as other cancers, and should be used instead of synthetic progestins, writing:

> A meta-analysis of 3 studies involving 86,881 postmenopausal women reported that the use of **natural progesterone was associated with a significantly lower risk of breast cancer compared with synthetic progestins.** Anovulation and low levels of serum progesterone have been associated with a significantly higher risk of breast cancer in premenopausal women. Use of progesterone has been linked to lower rates of uterine and colon cancers and may also be useful in treating other cancers such as ovarian, melanoma, mesothelioma, and prostate. Progesterone may also be helpful in preventing cardiovascular disease and preventing and treating neurodegenerative conditions such a stroke and traumatic brain injury...Physicians should have no hesitation prescribing natural progesterone. **The evidence is clear that progesterone does not cause breast cancer. Indeed, progesterone is protective and preventative of breast cancer.** (61)

HRT for Breast Cancer Survivors

Breast cancer survivors who suffer from menopausal symptoms and seek HRT are denied treatment by the mainstream medical system over misplaced concerns for cancer recurrence. Quite to the contrary, studies show breast cancer survivors treated with menopausal hormone replacement enjoy reduced all-cause mortality and reduced breast cancer-related mortality. Again, avoiding the use of carcinogenic progestins such as norethisterone is advisable, since the only study to demonstrate increased breast cancer recurrence, the HABITS trial from Sweden, used norethisterone, a known carcinogen. By using natural progesterone, cancer risk associated with norethisterone can be avoided. In 2022, Dr. Avrum Zvi Bluming reviewed hormone replacement after breast cancer suggesting "it is time" to change mainstream dogma related to hormone replacement in menopausal breast cancer survivors. Of 25 studies using hormone replacement after breast cancer, only one, the HABITS trial in Sweden, showed increased cancer recurrence. This HABITS trial used estradiol combined with a synthetic progestin called norethisterone, known to be carcinogenic. The 2008 French Cohort study by Dr. Agnès Fournier showed the use of the carcinogenic progestin, norethisterone, doubles the risk for breast cancer. Dr. Fournier's findings were confirmed in 2010 by Dr. Heli Lyytinen from Finland. Next, Dr. Bluming looked at only those studies including estrogen receptor (ER) positive breast cancer cells. Of 17 studies of hormone replacement after breast cancer with positive estrogen receptors in the cancer cells, **NONE** of 17 showed an increased risk for breast cancer recurrence. In 2022, Dr. Avrum Zvi Bluming writes:

[I found] twenty-five studies of HRT after a breast cancer diagnosis, published between 1980 and 2013...Only 1 of the 25 studies, the HABITS trial, demonstrated an increased risk of recurrence, which was limited to local or contralateral, and not distant, recurrence. None of the studies, including HABITS, reported increased breast cancer mortality associated with HRT... Of the 25 studies reporting the risk of HRT

administered to breast cancer survivors, 17, including HABITS, listed those with positive estrogen receptor assays...None of those reports identified an increased risk of breast cancer recurrence associated with a positive receptor assay...For more on this, see the Chapters 6 and 19, on Hormone Replacement for Breast Cancer Survivors. Note: For more on HRT for breast cancer survivors, see Chapters 6 and 19. (47) (62-68)

39% Reduction in All-Cause Mortality with HRT

In 2017, Dr. JoAnn E. Manson reviewed all-cause mortality for younger menopausal women aged 50-59 using estrogen alone (Premarin®, CEE) in the second arm of the WHI study (2004) finding a **thirty-nine percent reduction in all-cause mortality**. Dr. Manson writes:

When examined by 10-year age groups comparing younger women (aged 50-59 years) to older women (aged 70-79 years) in the pooled cohort, the ratio of nominal HRs [Hazard Ratio] for **all-cause mortality was 0.61** (95% CI, 0.43-0.87). Note: all-cause mortality is reduced when HRT is started early, within 10 years of menopausal transition. (69)

Estrogen Prevents Osteoarthritis

Pre-clinical animal studies show rendering the animal estrogen-deficient induces cartilage degeneration and arthritis. However, restoring estrogen to the animal stimulates cartilage regeneration. In my clinical practice, I have found this to be true in menopausal women as well. Estrogen hormone replacement prevents joint degeneration. Post-menopausal women who already have osteoarthritis when starting HRT have considerable improvement in joint pain after starting estrogen applied topically over the joint. For more on this, see Chapter 12 Estrogen Prevents Osteoarthritis. (70-82)

HRT Improves Cognitive Function and Prevents Dementia

Menopausal HRT improves cognitive function (concentration and focus) and reduces long-term risk for dementia. In 2018, Dr. Robert Speth reviewed the literature on estrogen replacement therapy for prevention of dementia citing the 2002 Cache County, Utah study by Dr. Peter Zandi which found a **59 percent decrease in risk of AD [Alzheimer's Disease]** for the over 10 years of use of Menopausal HRT [Menopausal Hormone Therapy], writing:

> The Cache County, UT, study of more than 1,800 women showed that MHT [Menopausal Hormone Therapy] reduced the risk of AD [Alzheimer's Disease], with the longest usage (>10 yr) having the lowest risk (**HR: 0.41**, 95% CI: 0.17–0.86). (83-84)

In 2021, Dr. Yu Jin Kim conducted a retrospective analysis of insurance claim records of 379,352 women aged 45 years or older in the Humana dataset, Louisville, Kentucky, a U.S. health insurance company, from 2007 to 2016. Dr. Yu Jin Kim found HRT use was associated with **58 percent decrease in all neurodegenerative diseases, AD (Alzheimer's Disease), and dementia** with greater duration of therapy, and greater efficacy with natural bioidentical hormone preparations such as 17-beta-estradiol (E2) combined with natural progesterone. Dr. Yu Jin Kim writes:

> HT [Hormone Therapy] was associated with **58% reduced risk** of all NDDs [neurodegenerative diseases] including AD [Alzheimer's Dementia] and dementia, with greater duration of therapy and natural steroid formulations (17β-estradiol and/ or progesterone) associated with greater efficacy. (85)

In 2023, Dr. Matilde Nerattini reviewed the medical literature on menopause hormone therapy for preventing Alzheimer's disease and dementia, finding six RCTs (Randomized Controlled Trials) and 45 observational reports.

Both the RCTs and observational studies showed an increased risk of dementia with the use of synthetic progestins (MPA) which should be avoided. The observational studies showed a 19-22% reduction in risk of dementia with estrogen use, but not with the estrogen plus synthetic progestin combination. Again, synthetic progestins are to be avoided. (86-101)

Vasomotor Symptoms, Hot Flashes, Night Sweats

Estrogen is the most effective treatment for vasomotor symptoms of menopausal estrogen deficiency, namely hot flashes and night sweats. This statement is accepted by mainstream medicine. (102-106)

Vaginal Atrophy

Menopausal HRT with estrogen is the most effective treatment for vaginal atrophy, vaginal dryness, pain, vaginal itching, and discomfort. **Note:** vaginal atrophy is thinning of the lining caused by menopausal estrogen deficiency. Mainstream OB/GYN doctors will commonly prescribe estradiol vaginal inserts for dryness and atrophy. (107-112)

HRT Relieves Insomnia, Improves Sleep Quality

Insomnia is caused by estrogen deficiency, a common menopausal symptom related to me every day in my office. Both estrogen and progesterone components of menopausal hormone replacement therapy (HRT) improve sleep quality. Part of menopausal insomnia is due to sleep interruption by hot flashes and nights sweats. However, another part is related to the brain neurotransmitter roles of estrogen and progesterone. 2021, Dr. B.J. Nolan reviewed nine randomized controlled trials with 388 post-menopausal women using progesterone for sleep, acting through the GABA receptors in the brain, writing:

> Micronized progesterone improves various sleep outcomes in randomized controlled

trials, predominantly in studies enrolling postmenopausal women...Preclinical data has shown progesterone metabolites improve sleep parameters through positive allosteric modulation of the gamma-aminobutyric acid type A receptor [GABA Type A Receptor]. (113-121)

In 2022, Dr. Annika Haufe reviewed the medical literature on estrogen improving sleep quality finding estrogen significantly improves insomnia, writing:

> The majority of studies that relied on self-reported data found that oral **estrogen therapy significantly improved insomnia**. Relief persisted throughout 24 months of therapy. Transdermal estrogen therapy likewise alleviated sleep disturbances and both oral and transdermal application improved sleep quality, latency and satisfaction. Difficulties in getting to sleep and difficulties of maintaining sleep significantly improved with intranasal estradiol application...(122-124)

In 2019, Dr. Jinju Lee from Korea reviewed HRT for menopausal sleep disorders, finding both estrogen and progesterone improve sleep quality and relieve insomnia of menopause through a wide range of effects, writing:

> Progesterone has both sedative and anxiolytic features. It stimulates the production of the NREM [Non-Rapid Eye Movement sleep] associated gamma-aminobutyric acid receptors by stimulating benzodiazepine receptors. In addition, progesterone also acts as a respiratory stimulant and has been used to treat mild obstructive sleep apnea (OSA). The effect of estrogen on sleep structure is complex as estrogen has a wide range of effects that potentially affects sleep structure. First, it is associated with metabolism of norepinephrine, serotonin, and acetylcholine-neurotransmitters that affect sleep pattern. Estrogen has been proven to decrease sleep latency, the number of awakening after sleep occurs, and cyclic spontaneous arousals; and increase total sleep time. Second, estrogen has a regulating effect on body temperature. During the night, estrogen plays a role in keeping the central body temperature low. In mammals, estrogen is a hormone that regulates the temperature of the lowest body temperature during the night. When decreased estrogen, this time shift forward and the depth of the temperature drop changes. Estrogen has a direct effect on mood by affecting the norepinephrine activity and serotonin response and uptake in the brain. All of these effects mean that estrogen would have an antidepressant effect. If we treat menopausal symptoms earlier in the menopause period with estrogen or estrogen-progesterone therapy, it will have a more beneficial effect in improving menopausal symptoms. Estrogen therapy is very effective in treating vasomotor symptoms, which improves sleep quality, and hormone replacement therapy is one of the main therapies for osteoporosis, mood disorder, and depression. Studies have shown that estrogen replacement therapy improves sleep quality, enables falling asleep, decreases nighttime wakefulness and also reduces vasomotor symptoms. Therefore, hormone replacement therapy is recommended for menopausal insomnia to improve the quality of sleep and life. For those who are starting hormone replacement therapy, use of low-dose estradiol rather than conjugated estrogen is more suitable. (125)

Testosterone Improves Libido and Well-Being

Testosterone improves libido for menopausal women. Currently, testosterone is indicated for the diagnosis of Hypoactive Sexual Desire Disorder (HSDD) i.e. loss of libido. As discussed in 2024 by Dr. Omoniyi Y. Adebisi, medical studies show testosterone treatment is effective in menopausal women with decreased libido. In 2020, Drs. Alice Scott and Louise Newson, two menopause specialists in the UK reviewed testosterone for menopausal women, suggesting primary care doctors should be prescribing testosterone, writing:

Numerous studies have shown that adding testosterone to hormonal therapy can improve sexual function and general well-being among women during their menopause. A recent systematic review and meta-analysis of testosterone treatment in women has provided robust support for a trial of testosterone in women when clinically indicated. In postmenopausal women, **testosterone supplementation improved several domains of sexual response, including sexual desire, pleasure, arousal, orgasm, and self-image**...It has also been shown to have additional benefits including the improvement of urogenital, psychological, and somatic symptoms, an increase in bone density, and enhancement of cognitive performance when combined with estrogen as part of HRT. Many women notice that taking **testosterone improves their mood, concentration, motivation, and energy levels.** Note: for more on testosterone for breast cancer prevention, see Chapter 18. (126-131)

HRT Improves Urinary Tract Symptoms Genito-Urinary Syndrome

The genito-urinary syndrome consists of repeated urinary tract infections and other genito-urinary symptoms caused by post-menopausal estrogen deficiency. Both estrogen and testosterone strengthen the pelvic floor and bladder muscles, preventing recurrent urinary tract infections, and incontinence, treating the genitourinary syndrome of menopause. Estrogen (estradiol) vaginal inserts are commonly prescribed by conventional OB/GYN doctors. My routine office formula vaginal capsule includes estriol (E3) and testosterone for mild symptoms, and a stronger formula containing higher dosage of estradiol (E2) and progesterone for more severe symptoms. The testosterone component may be given as a vaginal insert or as a topical cream. (132-134)

HRT Improves Bioenergetics and Weight

Many women mistakenly think that hormone replacement will cause weight gain. The reality is just the opposite. Menopausal estrogen deficiency causes weight gain and visceral fat accumulation. Postmenopausal hormone replacement prevents weight gain, and women on HRT report less accumulation of visceral fat. The WHI study (2002) showed that in postmenopausal women taking HRT (Premarin and MPA), the body mass index (BMI) and waist circumference decreased as discussed in 2021 by Dr. Anna Fenton. This effect is thought to be due to estrogen's ability to increase bioenergetics, stimulate mitochondrial biogenesis, and increase the metabolic rate. This increased bioenergetics is especially important for the brain which has a high energy demand. HRT benefits for the brain include improved memory and cognitive function. In 2014, Dr. Jamaica Rettberg reviewed estrogen as the master regulator of energy production for the brain and body, thus explaining protection against weight gain and cognitive dysfunction, writing:

> After menopause, when estrogen levels drop, women experience a general increase in weight as well as a redistribution of adipose tissue leading to increased abdominal fat deposition. Importantly, the increased abdominal fat in postmenopausal women tends to be visceral and not subcutaneous fat (Lovejoy et al., 2008).... meta-analysis conducted by the Endocrine Society reported that **HT [Hormone Therapy] was associated with less accumulation of weight, fat mass, and/or centrally located fat mass ...postmenopausal HT protects against weight gain, and also promotes less adipose tissue deposition in visceral fat stores**....Results from several imaging studies support the idea that postmenopausal **HT can modulate brain bioenergetics**, likely leading to the maintenance **of cognitive function and reduced risk of AD [Alzheimer's Dementia]**...HT users showed increased rCBF [regional cerebral blood flow with PET imaging] over time

compared to nonusers in the hippocampus, parahippocampal gyrus, and temporal lobe, regions that are critical for memory formation and are also vulnerable to decreased glucose metabolism in preclinical AD [Alzheimer's disease] (Maki and Resnick, 2000). As before, the HT users scored higher on a battery of **memory tests** than nonusers (Maki and Resnick, 2000). (135-137)

In 2023, Dr. Jing Zhu reviewed estrogen's regulation of menopausal energy production, reporting estrogen deficiency of menopause triggers decreased energy production in the brain and other organs leading to type 2 diabetes mellitus, hypertension, cardiovascular disease, and dementia, writing:

> Within the brain, central estrogen via ER [estrogen receptor] regulates appetite and energy expenditure and maintains cell glucose metabolism, including glucose transport, aerobic glycolysis, and mitochondrial function. In the whole body, **estrogen has shown beneficial effects on weight control, fat distribution, glucose and insulin resistance, and adipokine secretion**. As demonstrated by multiple in vitro and in vivo studies, menopause-related decline of circulating estrogen may induce the disturbance of metabolic signals and a significant decrease in bioenergetics, which could trigger an increased incidence of late-onset Alzheimer's disease, type 2 diabetes mellitus, hypertension, and cardiovascular diseases in postmenopausal women. (138)

Mood Swings, Depression

Menopausal HRT treats mood swings and depression commonly reported during menopause. Both testosterone and estrogen serve as effective antidepressants. Progesterone has anti-anxiety effects. In 2019, Dr. Giulia Gava from Bologna, Italy discusses the beneficial effects of estrogen on mood and cognition as a neurosteroid in the brain, writing:

> During this [menopausal] transition, women can suffer symptoms related to menopause (such as hot flushes, sleep disturbance, mood changes, memory complaints and vaginal dryness). Neurological symptoms such as sleep disturbance, "brain fog" and mood changes are a major complaint of women transitioning menopause, with a significant impact on their quality of life, productivity and physical health…During menopausal transition, women are at higher risk of developing depression, stress, anxiety and emotional distress…Estrogen activity found in regions known to be involved in mood and cognitive regulation (e.g., prefrontal cortex, hippocampus) is contributing evidence of the concept of mediating effects (and possible therapeutic effects) of this hormone on mood. In general, the effects of E2 on serotonin (5-hydroxytryptamine 5-HT) may be described as **favorable to mood**, with an increase in 5-HT synthesis and availability…. Estrogens also improve mitochondrial respiratory efficiency, helping to prevent the formation of oxygen free radicals that are known to negatively affect mitochondria energetics in depression. Estrogen effects also promote noradrenaline (NE) availability through a reduction of monoamine oxidases (MAOs) and an increase in the activity of tyrosine hydroxylase, the rate limiting enzyme in the synthesis of catecholamine. Acute E2 administration stimulates dopamine b-hydroxylase (DBH) gene transcription, catalyzing the hydroxylation of dopamine to form NE. Finally, estrogen may also play a role as an anti-depressant, because of its stimulating effect on then brain-derived neurotrophic factor (BDNF), an important neuroprotective and growth factor agent, found to be deficient in depression. (139-142)

Menopausal Aging of the Skin

Menopausal HRT improves the skin by increasing skin collagen content, thickness, elasticity, and hydration, serving as an anti-aging treatment for skin with fewer wrinkles and sagging. In 2022, Dr. Alexandra Kendall reviewed menopausal skin changes such as skin thinning, reduced elasticity, and impaired wound healing, all prevented or reversed by hormone replacement, writing:

> The menopause occurs when a decline in ovarian follicular activity leads to the cessation of menstruation and a loss of fertility. The associated reduction in levels of reproductive hormones, particularly estrogen, causes additional non-reproductive symptoms that include changes to the metabolism, central nervous system, and skin. Cutaneous changes include skin thinning, reduced elasticity, and impaired wound healing, all of which can be linked to dermal changes such as age-related reductions in collagen production. However, other skin changes, such as skin sensitivity, reduced stratum corneum (SC) cohesion, and changes to epidermal structure and biomechanical function point to a dysfunctional epidermal barrier. (145-153)

Testosterone Relieves Dry Eye Syndrome

Lubrication of the eye and skin is through the pilo-sebaceous unit controlled by testosterone. Menopausal dry eye and dry skin are the result of testosterone and estrogen deficiency. Treatment with testosterone may be curative for dry eye syndrome if treated early enough in the course of the disease as discussed in 2003 and 2005 by Dr. Charles Conner, who has a PhD in Cell Biology and is a Professor at the Rosenberg School of Optometry in San Antonio Texas. Dry eye syndrome can be caused by or aggravated by estrogen-blocking drugs such as aromatase inhibitors. (154-159)

Menopausal Hair Loss

About 50% of post-menopausal women report alterations in hair quality, loss of hair, and hair thinning, attributed to female pattern hair loss (FPHL). Another common type of hair loss is called Telogen Effluvium, typically seen in post-menopausal women **starting hormone replacement**, sometimes mistaken for FPHL as discussed in 2012 and 2013 by Dr.

Rebecca Glaser from Dayton Ohio who finds improvement in hair quality with testosterone treatment. In Telogen effluvium there is rapid hair shedding after some metabolic stress, **hormonal changes**, or medication trigger. It is usually temporary and hair often grows back without any treatment after 3 to 6 months. In 2022, Dr. Bob Kronemyer writes:

> More than half of postmenopausal women experience female pattern hair loss (FPHL), according to results of a cross-sectional study published in Menopause. The authors attributed their findings to the pathophysiological changes in FPHL involving progressive miniaturization of hair follicles and subsequent conversion of terminal follicles into vellus-like follicles. Furthermore, alterations in growth, diameter, and pigmentation of the hair progressively increase, along with scalp hair thinning, thus indicating that FPHL strongly correlates with menopause. (160)

In 2012, Dr. Rebecca Glaser studied the effect of testosterone pellet treatment on hair quality in 285 patients, writing:

> Subcutaneous testosterone therapy was found to have a **beneficial effect on scalp hair growth** in female patients treated for symptoms of androgen deficiency. We propose this is due to an anabolic effect of testosterone on hair growth. The fact that no subject complained of hair loss as a result of treatment casts doubt on the presumed role of testosterone in driving female scalp hair loss. No patient in this cohort reported scalp hair loss on testosterone therapy. A total of 262 women (92%) reported some increase in facial hair growth. (161-165)

63% Reduction in Colorectal Cancer

The self-reported use of menopausal hormone replacement reduces risk for colon cancer by 63 percent. In 2016, Dr. Cecilia Williams reviewed colorectal cancer prevention by targeting the estrogen receptor beta (ER-beta). Natural compounds that bind to and activate ER-beta are Premarin (CEE), estriol (E3), testosterone (3-beta-diol), genistein, coumestrol, daidzein (particularly when metabolized to equol), apigenin, naringenin, and kaempferol. Estradiol (E2) binds to ER-alpha and ER-beta equally. Dr. Cecilia Williams writes:

A large body of evidence from preclinical studies indicates that expression of the estrogen receptor beta (ER-beta/ESR2) demonstrates an inverse relationship with the presence of colorectal polyps and stage of tumors, and can mediate a protective response. **Natural compounds, including phytoestrogens, or synthetic ERβ selective agonists, can activate or upregulate ERβ** in the colon and promote apoptosis in preclinical models and in clinical experience. Importantly, this activity has been associated with a reduction in polyp formation and, in rodent models of CRC, has been shown to lower incidence of colon adenocarcinoma. (166-169)

HRT Reduces Diabetes by 30 Percent

Menopausal HRT decreases the risk of developing new-onset diabetes by 30 percent and improves diabetic control. Human clinical studies show menopausal HRT improves insulin secretion by beta cells of the pancreas, glucose effectiveness, and insulin sensitivity. **Note:** Beta-cells are the insulin-secreting cells in the pancreas. In 2017, Dr. Mauvais-Jarvis reviewed the medical literature on hormone replacement and diabetes prevention, writing:

In a meta-analysis by Salpeter et al. (2006) of 107 randomized trials comparing MHT to placebo or no treatment in women without diabetes, MHT was associated with a reduction in fasting glucose and fasting insulin that led to a 13% drop in insulin resistance...This was associated with an estimated reduction of 30% in new-onset diabetes...MHT [Menopausal Hormone Therapy] improves beta-cell insulin secretion, glucose effectiveness, and insulin

sensitivity, as measured in clinical settings. (170-175)

Wound Healing: Menopausal HRT enhances wound healing, and reduces the incidence of wound complications. (176)

Musculoskeletal System: Menopausal HRT with estrogen and testosterone maintains muscle mass, and muscle strength, and eliminates muscle aches and pains. (177-180)

Migraines: Estrogen helps with migraine symptoms. Triptan drugs may be needed as well. We have found the nasal spray version of Sumatriptan more effective than the oral tablets. (181-190)

Generalized Inflammation: Menopausal estrogen deficiency causes low-grade inflammation in the body and brain, relieved by estrogen therapy. Inflammation within the hypothalamus may cause cognitive dysfunction, hypothalamic dysfunction, metabolic syndrome, and obesity, all relieved with estrogen HRT. (191-195)

Macular Degeneration / Retinopathy: Retinopathy is an adverse effect of estrogen-blocking drugs such as clomiphene, tamoxifen, and aromatase inhibitors such as letrozole, anastrozole and exemestane. The retina is a hormonal organ, having estrogen receptors, and manufacturing its own estrogen needed for mitochondrial energy production. Estrogen deficiency leads to retinopathy and macular degeneration, prevented by estrogen hormone replacement. A nutritional supplement program with Lutein, Meso-Zeanthin, Zeaxanthin can help preserve retinal function. (196-212)

HRT for the Elderly Over 65?

In 2022, Dr. Seo Baik studied Medicare records for 7 million elderly menopausal women in the U.S. using estrogen hormone therapy (ERT), showing compared to non-users, ERT users enjoy a 20% reduction in mortality and a reduction in 5 different cancer cell types, less heart disease, and less dementias, writing:

The 7 million NIH study, the largest study ever run on women, with 1.5 million American women on ERT, shows that women 65 and older on estrogen therapies have statistically "less" of all 5 cancers studied (breast, ovarian, uterine, lung, and colon). As well as live almost 20% longer, healthier lives, with less heart disease (unless on oral estrogens) and less dementias. (213)

HRT for Premature Menopause

In 2023, Dr. Lawrence Nelson described menopause as a "waning of ovarian function" with low estrogen levels. The sudden decline in estrogen levels associated with premature menopause is associated with significant morbidity and early mortality, Dr. Lawrence Nelson writes:

> The essential biology of this physiologic midlife transition in women is a waning of ovarian endocrine function. The physiologic midlife transition to so-called menopause is a state of low serum E2 [estradiol], and early E2 deficiency is associated with significant morbidity and early mortality…A few prospective population-based cohort studies provide convincing evidence that women with early onset menopause, and the associated E2 deficiency, have 1) a shorter life expectancy, 2) increased risk of type II diabetes, 3) adverse effects on cognitive function, 4) significant correlation between age at menopause and age at diagnosis of dementia, and 5) a significant correlation between age at menopause and age at death…Evidence is clear, extremely low E2 [estradiol] levels increase the risk for some women. For example, there is a 2.5-fold increase in hip and vertebral fractures in older women with total E2 levels less than 5.0 pg/ml. …Intriguingly, even minimal increases in E2 serum concentrations have a proven beneficial effect on bone mineral density in menopausal women, with little effect on endometrial proliferation…This same therapeutic window of low doses of E2, proven to improve bone mineral density, could theoretically also improve health for menopausal women regarding their cardiovascular health, central nervous system health, mood, and related cosmetic benefits to skin and hair. (112)

Summary by Dr. Jane Yang

In 2024, Dr. Jane Yang reviewed the medical literature from 1972 to 2023 on menopausal estrogen deficiency and the role of hormone replacement, finding menopausal estrogen deficiency negatively impacts almost all organ systems in a profound way, most of which can be alleviated with estrogen replacement therapy, writing:

> Menopause is a natural and inevitable part of aging for women. As women approach perimenopause and menopause, the dramatic fluctuations and eventual **substantial decline in circulating estrogen levels affect almost all organ systems, including urogenital, reproductive, cardiovascular, neurologic, psychiatric, skeletal, dermatologic, immune, and digestive systems. These symptoms, first acute then chronic, have a profound and overall negative impact on the quality of life of women, and most symptoms can be alleviated with estrogen therapy.** It is clear from basic science literature that ERs [estrogen receptors] are found throughout the body and that they are associated with uncountable beneficial actions in both women and men, even as they age. (20)

Summary by Dr. Avrum Bluming

In 2022, Dr. Avrum Bluming, a retired oncologist from California with a special interest in hormone replacement therapy (HRT), concluded it was time to reject the prevailing dogma of withholding HRT from breast cancer survivors for fear of recurrence, and advocates treating menopausal breast cancer survivors with hormone replacement. Dr. Bluming also comments on the importance of HRT for improving the quality of life for menopausal women, and the unfortunate dissemination of misinformation and fear-mongering by the mass media regarding the 2002 WHI study which triggered a stampede away from HRT, writing:

> Estrogen therapy/HRT has been reproducibly found to improve quality of life in 80% of women who experience perimenopausal and postmenopausal symptoms, which could include hot flushes, night sweats, insomnia, difficulty concentrating, decreasing recent memory, bladder/urinary discomfort, frequent urinary tract infections, mood swings, arthralgias, and palpitations and which will last a median of 7.4 years. Estrogen therapy/ HRT is the most effective treatment for these symptoms, relieving most of them in the great majority of treated patients. Nothing else comes close. And yet, as a result of widely circulated misinformation stemming largely from the Women's Health Initiative (WHI) in 2002, these treatments are not widely used even among eligible women with no history of treated breast cancer. Indeed, in 2020, the British Medical Association published a report showing that a third of female general practitioners were considering cutting back their working hours or retiring prematurely due to untreated menopausal symptoms. (62)

Conclusion: Hormone replacement is the single most important medical intervention for the menopausal woman. The failure of the medical system to provide hormone replacement is one of the great errors of modern medicine. How is this error possible? Follow the money. The fear of estrogen and the avoidance of menopausal hormone replacement has created a windfall for the drug industry which sells a drug for each menopausal symptom. Deceptive drug industry advertising and the capture of medical research, medical meetings, and medical journals is directly responsible for this atmosphere of fear of estrogen. We now have proof from the WHI second arm (2004) that estrogen (Premarin, CEE) prevents breast cancer and reduces mortality from breast cancer by 40 percent. We have 25 of 26 studies showing hormone replacement in breast cancer survivors does not increase breast cancer recurrence. The hormone formula makes a difference. Hormone replacement formulas containing synthetic progestins are carcinogenic and should be avoided. Use natural progesterone instead. The goal of this chapter is to reverse the brainwashing and erase the fear of estrogen, thus liberating the thinking of the menopausal woman. I have included extensive references to the medical literature to make this more authoritative and convincing. As discussed above, the health benefits of menopausal hormone replacement extend beyond the alleviation of vasomotor symptoms of hot flashes and night sweats, to include improved all-cause mortality, bone density, sleep quality, cognition, depression, mood, body weight, diabetic control, skin quality, wound healing, preservation of retina, eye and skin lubrication, muscle mass and strength, prevention of breast cancer, cardiovascular disease, osteoporosis, osteoarthritis, and dementia. Relieving vasomotor symptoms with a non-hormonal drug, Veozah, is not a good idea since all the health benefits of estrogen listed above are lost.

♦ **References for Chapter 1, Menopausal Hormone Replacement Health Benefits**

1) Gosset, Anna, et al. "Menopausal Hormone Therapy for The Management of Osteoporosis." Best Practice & Research Clinical Endocrinology & Metabolism 35.6 (2021): 101551.

2) Elsheikh, Arwa, et al. "Postmenopausal osteoporosis." Contemporary OB/GYN 68.5 (2023): 7-12.

3) McClung, Michael Roy, et al. "Management of Osteoporosis in Postmenopausal Women: The 2021 Position Statement of The North American Menopause Society." Menopause 28.9 (2021): 973-997.

4) Curfman, Gregory. "Fosamax Fractures—Justice Has Not Been Served." JAMA 331.22 (2024): 1887-1888.

5) FDA Safety Announcement [10-13-2010] FDA Drug Safety Communication: Safety update for osteoporosis drugs, bisphosphonates, and atypical fractures. https://www.fda.gov/drugs/drug-safety-and-availability/fda-drug-safety-communication-safety-update-osteoporosis-drugs-bisphosphonates-and-atypical

6) Boniva: What If Sally Field Told the Truth? By Vivian Goldschmidt, MA. Save Our Bones. https://saveourbones.com/boniva-what-if-sally-field-told-the-truth/

7) Cummings, Steven R., et al. "Effect of Alendronate on Risk of Fracture in Women With Low Bone Density But Without Vertebral Fractures: Results From the Fracture Intervention Trial." Jama 280.24 (1998): 2077-2082.

8) Cummings, Steven R., et al. "History of Alendronate." Bone 137 (2020): 115411.

9) Bottai, Vanna, et al. "Prevalence of Atypical Femoral Fractures, A Clinical Update: A Comparative Retrospective Study 7 Years Later." Injury 55 (2024): 111346.

10) Xiao, Yao, et al. "Atypical Femur Fracture Associated with Common Anti-Osteoporosis Drugs In FDA Adverse Event Reporting System." Scientific Reports 13.1 (2023): 10892.

11) Isaacs, Joseph D., et al. "Femoral Insufficiency Fractures Associated with Prolonged Bisphosphonate Therapy." Clinical Orthopaedics and Related Research® 468.12 (2010): 3384-3392.

12) Odvina, Clarita V., et al. "Unusual Mid-Shaft Fractures During Long-Term Bisphosphonate Therapy." Clinical Endocrinology 72.2 (2010): 161-168.

13) Lenart, Brett A., Dean G. Lorich, and Joseph M. Lane. "Atypical Fractures of The Femoral Diaphysis in Postmenopausal Women Taking Alendronate." New England Journal of Medicine 358.12 (2008): 1304-1306.

14) Dimitrakopoulos, I., C. Magopoulos, and D. Karakasis. "Bisphosphonate-Induced Avascular Osteonecrosis Of The Jaws: A Clinical Report Of 11 Cases." International journal of oral and maxillofacial surgery 35.7 (2006): 588-593.

15) Merigo, E., et al. "Jaw Bone Necrosis Without Previous Dental Extractions Associated With The Use Of Bisphosphonates (Pamidronate And Zoledronate): A Four-Case Report." Journal of oral pathology & medicine 34.10 (2005): 613-617.

16) Silva, Barbara Campolina, and Marcio Alexandre Hipólito Rodrigues. "Estrogen Hormone Therapy and Postmenopausal Osteoporosis: Does It Really Take Two to Tango?" Women & Health 63.10 (2023): 770-773.

17) Manjer, Jonas, et al. "Serum Iodine and Breast Cancer Risk: A Prospective Nested Case-Control Study Stratified for Selenium Levels." Cancer Epidemiology, Biomarkers & Prevention 29.7 (2020): 1335-1340.

18) Yang, Lin, and Adetunji T. Toriola. "Menopausal Hormone Therapy Use Among Postmenopausal Women." JAMA Health Forum. Vol. 5. No. 9. American Medical Association, 2024.

19) Palacios, Santiago, et al. "Hormone Therapy for First-Line Management of Menopausal Symptoms: Practical Recommendations." Women's Health 15 (2019): 1745506519864009.

20) Yang, Jane L., et al. "Estrogen Deficiency in the Menopause and the Role of Hormone Therapy: Integrating the Findings of Basic Science Research with Clinical Trials." Menopause (2024): 10-1097.

21) Santen, Richard J., et al. "Competency in Menopause Management: Whither Goest the Internist?" Journal of Women's Health 23.4 (2014): 281-285.

22) Langer, R. D., et al. "Hormone Replacement Therapy–Where Are We Now?" Climacteric 24.1 (2021): 3-10.

23) Lobo, Roger A., et al. "Back to the Future: Hormone Replacement Therapy as Part of a Prevention Strategy for Women at the Onset of Menopause." Atherosclerosis 254 (2016): 282-290.

24) Akter, N., et al. "The Effect of Hormone Replacement Therapy on The Survival of UK Women: A Retrospective Cohort Study 1984– 2017." BJOG 129.6 (2022): 994.

25) Hodis, Howard N., and Wendy J. Mack. "Menopausal Hormone Replacement Therapy and Reduction of All-Cause Mortality and Cardiovascular Disease: It Is About Time and Timing." The Cancer Journal 28.3 (2022): 208-223.

26) Stute, Petra, et al. "The Impact of Menopausal Hormone Therapy on Overall Mortality–A Comprehensive Review." Climacteric 23.5 (2020): 447-459.

27) Kim, Yu Jin, et al. "Association Between Menopausal Hormone Therapy and Risk of Neurodegenerative Diseases: Implications for Precision Hormone Therapy." Alzheimer's & Dementia: Translational Research & Clinical Interventions 7.1 (2021): e12174.

28) Guirguis-Blake, Janelle. "Hormone Therapy for The Prevention of Chronic Conditions in Postmenopausal Women." American Family Physician 72.12 (2005): 2520-2522.

29) Zaheer, Sarah, and Meryl S. LeBoff. "Osteoporosis: Prevention and Treatment." Endotext [Internet] (2022).

30) Cagnacci, Angelo, and Martina Venier. "The Controversial History of Hormone Replacement Therapy." Medicina 55.9 (2019): 602

31) Levin, Veronika Alexa, X. Jiang, and R. Kagan. "Estrogen Therapy for Osteoporosis in The Modern Era." Osteoporosis International 29 (2018): 1049-1055.

32) Na, Zhao, et al. "Role of Menopausal Hormone Therapy in The Prevention of Postmenopausal Osteoporosis." Open Life Sciences 18.1 (2023): 20220759.

33) Anagnostis, Panagiotis, et al. "Estrogen and Bones After Menopause: A Reappraisal of Data and Future Perspectives." Hormones 20 (2021): 13-21.

34) Rozenberg, Serge, et al. "Is There a Role for Menopausal Hormone Therapy in The Management Of Postmenopausal Osteoporosis?" Osteoporosis International 31 (2020): 2271-2286.

35) Chang, Cherry Yin-Yi, et al. "Timing and Dosage of And Adherence to Hormone Replacement Therapy and Fracture Risk In Women With Menopausal Syndrome In Taiwan: A Nested Case-Control Study." Maturitas 146 (2021): 1-8.

36) Newson, Louise, et al. "Letter to Editor: Our Concerns About HRT Not Having a Priority As A Treatment For Osteoporosis In The NOGG Guidelines." Osteoporosis International 34.4 (2023): 815-816.

37) Stepan, Jan J., et al. "Update on Menopausal Hormone Therapy for Fracture Prevention." Current Osteoporosis Reports 17 (2019): 465-473.

38) Gamsjaeger, Sonja, et al. "Effect of Hormone Replacement Therapy on Bone Formation Quality And Mineralization Regulation Mechanisms In Early Postmenopausal Women." Bone reports 14 (2021): 101055.

39) Mosekilde, Leif, et al. "Hormonal Replacement Therapy Reduces Forearm Fracture Incidence in Recent Postmenopausal Women—Results of the Danish Osteoporosis Prevention Study." Maturitas 36.3 (2000): 181-193.

40) Cho, Leslie, et al. "Rethinking Menopausal Hormone Therapy: For Whom, What, When, And How Long?" Circulation 147.7 (2023): 597-610.

41) Hodis, Howard N., and Wendy J. Mack. "Menopausal Hormone Replacement Therapy and Reduction of All-Cause Mortality and Cardiovascular Disease: It Is About Time And Timing." The Cancer Journal 28.3 (2022): 208-223.

42) Schierbeck, Louise Lind, et al. "Effect of Hormone Replacement Therapy on Cardiovascular Events In Recently Postmenopausal Women: Randomised Trial." BMJ 345 (2012).

43) Schipper, I., and Y. V. Louwers. "Premature and Early Menopause in Relation to Cardiovascular Disease." Seminars in Reproductive Medicine. Vol. 38. No. 4-05. 2020.

44) Renke, Guilherme, et al. "Cardio-Metabolic Health and HRT In Menopause: Novel Insights in Mitochondrial Biogenesis and RAAS." Current Cardiology Reviews 19.4 (2023): 1-5.

45) Chlebowski, Rowan T., et al. "Association of Menopausal Hormone Therapy with Breast Cancer Incidence and Mortality During Long-Term Follow-Up of The Women's Health Initiative Randomized Clinical Trials." Jama 324.4 (2020): 369-380.

46) De Lignieres, B., et al. "Combined Hormone Replacement Therapy and Risk of Breast Cancer in A French Cohort Study of 3175 Women." Climacteric 5.4 (2002): 332-340.

47) Fournier, Agnès, et al. "Unequal Risks for Breast Cancer Associated with Different Hormone Replacement Therapies: Results from The E3N Cohort Study." Breast Cancer Research and Treatment 107 (2008): 103-111.

48) Campagnoli, Carlo, et al. "Progestins and Progesterone in Hormone Replacement Therapy and The Risk of Breast Cancer." The Journal of Steroid Biochemistry and Molecular Biology 96.2 (2005): 95-108.

49) Tan, D. A., and A. R. B. Dayu. "Menopausal Hormone Therapy: Why We Should No Longer Be Afraid of The Breast Cancer Risk." Climacteric 25.4 (2022): 362-368.

50) Climént-Palmer, María, and David Spiegelhalter. "Hormone Replacement Therapy and The Risk of Breast Cancer: How Much Should Women Worry About It?" Post reproductive health 25.4 (2019): 175-178.

51) Hipolito Rodrigues, et al. "Micronized Progesterone, Progestins, And Menopause Hormone Therapy." Women & Health 61.1 (2021): 3-14.

52) Stevenson, John C., et al. "Progestogens as a Component of Menopausal Hormone Therapy: The Right Molecule Makes the Difference." Drugs In Context 9 (2020).

53) Ferretti, Gianluigi, et al. "The Protective Side of Progesterone." Breast Cancer Research 9 (2007): 1-1.

54) Allan Lieberman, M. D., and M. D. Luke Curtis. "In Defense of Progesterone: A Review of the Literature." Alternative Therapies in Health and Medicine 23.6 (2017): 24-32.

55) Kohn, Grace E., et al. "The History of Estrogen Therapy." Sexual Medicine Reviews 7.3 (2019): 416-421.

56) Tan, Delfin. A., and A. R. B. Dayu. "Menopausal Hormone Therapy: Why We Should No Longer Be Afraid of The Breast Cancer Risk." Climacteric 25.4 (2022): 362-368.

57) Lambrinoudaki, Irene. "Menopausal Hormone Therapy and Breast Cancer Risk: All Progestogens Are Not the Same." Case Reports in Women's Health 29 (2020): e00270.

59) Vinogradova, Yana, Carol Coupland, and Julia Hippisley-Cox. "Use of Hormone Replacement Therapy and Risk of Breast Cancer: Nested Case-Control Studies Using the QResearch and CPRD databases." BMJ 371 (2020).

60) Stevenson, John C., et al. "Progestogens as A Component of Menopausal Hormone Therapy: The Right Molecule Makes the Difference." Drugs in Context 9 (2020).

61) Allan Lieberman, M. D., and M. D. Luke Curtis. "In Defense of Progesterone: A Review of the Literature." Alternative Therapies in Health and Medicine 23.6 (2017): 24-32.

62) Bluming, Avrum Zvi. "Hormone Replacement Therapy After Breast Cancer: It Is Time." The Cancer Journal 28.3 (2022): 183-190.

63) Bluming, Avrum Z. "Safety of Systemic Hormone Replacement Therapy in Breast Cancer Survivors." Breast Cancer Research and Treatment 191.3 (2022): 685-686.

64) Menopause, Breast Cancer, and What Comes Next-a Conversation with Dr. Avrum Bluming By: Alloy Staff, September 5, 2024. https://www.myalloy.com/blog/menopause-breast-cancer-and-what-comes-next

65) Holmberg, L. "Increased Risk of Recurrence After Hormone Replacement Therapy In Breast Cancer Survivors." J Natl Cancer Inst 100 (2008): 475-482.

66) Ugras, Stacy K., and Rakhshanda Layeequr Rahman. "Hormone Replacement Therapy After Breast Cancer: Yes, No or Maybe?" Molecular and Cellular Endocrinology 525 (2021): 111180.

67) Deli, Tamás, et al. "Hormone Replacement Therapy in Cancer Survivors–Review of The Literature." Pathology & Oncology Research 26 (2020): 63-78.

68) Lyytinen, Heli, et al. "Do The Dose or Route of Administration of Norethisterone Acetate as A Part Of Hormone Therapy Play A Role In Risk Of Breast Cancer: National-Wide Case-Control Study From Finland." International Journal of Cancer 127.1 (2010): 185-189.

69) Manson JE, et al. Menopausal Hormone Therapy and Long-Term All-Cause and Cause-Specific Mortality: The Women's Health Initiative Randomized Trials. JAMA 2017; 318:927–38.

70) Roman-Blas, Jorge A., et al. "Osteoarthritis Associated with Estrogen Deficiency." Arthritis research & therapy 11 (2009): 1-14.

71) Gilmer, Gabrielle, et al. "A Network Medicine Approach to Elucidate Mechanisms Underlying Menopause-Induced Knee Osteoarthritis." bioRxiv (2023): 2023-03.

72) Gulati, Malvika, et al. "The Influence of Sex Hormones on Musculoskeletal Pain And Osteoarthritis." The Lancet Rheumatology 5.4 (2023): e225-e238.

73) Castañeda, Santos, and Esther F. Vicente-Rabaneda. "Disentangling the Molecular Interplays Between Subchondral Bone and Articular Cartilage in Estrogen Deficiency-Induced Osteoarthritis." Osteoarthritis and Cartilage 31.1 (2023): 6-8.

74) Burkard, Theresa, et al. "Risk of Hand Osteoarthritis in New Users of Hormone Replacement Therapy: A Nested Case-Control Analysis." Maturitas 132 (2020): 17-23.

75) Mei, Yixue, et al. "Roles of Hormone Replacement Therapy and Menopause on Osteoarthritis and Cardiovascular Disease Outcomes: A Narrative Review." Frontiers in Rehabilitation Sciences (2022): 45.

76) Felson, David T., and Michael C. Nevitt. "The Effects of Estrogen on Osteoarthritis." Current Opinion In Rheumatology 10.3 (1998): 269-272.

77) Pang, Huiwen, et al. "Low Back Pain and Osteoarthritis Pain: A Perspective of Estrogen." Bone Research 11.1 (2023): 42.

78) Tang, Jinshuo, et al. "Estrogen-Related Receptors: Novel Potential Regulators of Osteoarthritis Pathogenesis." Molecular Medicine 27.1 (2021): 1-12.

79) Yang, Xiaohui, et al. "Meta-Analysis of Estrogen in Osteoarthritis: Clinical Status and Protective Effects." Alternative Therapies in Health and Medicine 29.1 (2023): 224-230.

80) Ge, Yuxiang, et al. "Estrogen Prevents Articular Cartilage Destruction in A Mouse Model of AMPK Deficiency Via ERK-Mtor Pathway." Annals of translational medicine 7.14 (2019).

81) Dreier, Rita, et al. "Estradiol Inhibits ER Stress-Induced Apoptosis in Chondrocytes and Contributes To A Reduced Osteoarthritic Cartilage Degeneration In Female Mice." Frontiers in Cell and Developmental Biology 10 (2022): 913118.

82) Sniekers, Yvonne H., et al. "Estrogen is Important for Maintenance of Cartilage and Subchondral Bone In A Murine Model Of Knee Osteoarthritis." Arthritis research & therapy 12 (2010): 1-12.

83) Speth, Robert C., et al. "A Heartfelt Message, Estrogen Replacement Therapy: Use It or Lose It." American Journal of Physiology-Heart and Circulatory Physiology 315.6 (2018): H1765-H1778.

84) Zandi, Peter P., et al. "Hormone Replacement Therapy and Incidence of Alzheimer Disease in Older Women: the Cache County Study." Jama 288.17 (2002): 2123-2129.

85) Kim, Yu Jin, et al. "Association Between Menopausal Hormone Therapy and Risk of Neurodegenerative Diseases: Implications for Precision Hormone Therapy." Alzheimer's & Dementia: Translational Research & Clinical Interventions 7.1 (2021): e12174.

86) Nerattini, Matilde, et al. "Systematic Review and Meta-Analysis of The Effects of Menopause Hormone Therapy on Risk of Alzheimer's Disease and Dementia." Frontiers in Aging Neuroscience 15 (2023): 1260427.

87) Russell, Jason K., et al. "The Role of Estrogen in Brain and Cognitive Aging." Neurotherapeutics 16 (2019): 649-665.

88) Torromino, Giulia, et al. "Estrogen-Dependent Hippocampal Wiring as A Risk Factor for Age-Related Dementia in Women." Progress in Neurobiology 197 (2021): 101895.

89) Georgakis, Marios K., et al. "Surgical Menopause in Association with Cognitive Function And Risk Of Dementia: A Systematic Review And Meta-Analysis." Psychoneuroendocrinology 106 (2019): 9-19.

90) Gava, Giulia, et al. "Cognition, Mood and Sleep in Menopausal Transition: The Role Of Menopause Hormone Therapy." Medicina 55.10 (2019): 668.

91) Stute, Petra, et al. "Cognitive Health After Menopause: Does Menopausal Hormone Therapy Affect It?" Best Practice & Research Clinical Endocrinology & Metabolism 35.6 (2021): 101565.

92) Saleh, Rasha NM, et al. "Hormone Replacement Therapy Is Associated with Improved Cognition And Larger Brain Volumes In At-Risk APOE4 Women: Results From The European Prevention Of Alzheimer's Disease (EPAD) Cohort." Alzheimer's Research & Therapy 15.1 (2023): 10.

93) McCarthy, Micheline, and Ami P. Raval. "The Peri-Menopause in A Woman's Life: A Systemic Inflammatory Phase That Enables Later Neurodegenerative Disease." Journal of Neuroinflammation 17 (2020): 1-14.

94) Zimmerman, Benjamin, et al. "Longitudinal Effects of Immediate and Delayed Estradiol On Cognitive Performance In A Spatial Maze And Hippocampal Volume In Menopausal Macaques Under An Obesogenic Diet." Frontiers in Neurology 11 (2020): 539.

95) Klinge, Carolyn M. "Estrogenic Control of Mitochondrial Function." Redox Biology 31 (2020): 101435.

96) Zhao, Wei, et al. "Estrogen Deficiency Induces Mitochondrial Damage Prior To Emergence of Cognitive Deficits in A Postmenopausal Mouse Model." Frontiers in Aging Neuroscience 13 (2021): 713819.

97) Guo, Hang, et al. "The Critical Period for Neuroprotection by Estrogen Replacement Therapy and The Potential Underlying Mechanisms." Current Neuropharmacology 18.6 (2020): 485-500.

98) Marchant, Ivanny Carolina, et al. "Estrogen, Cognitive Performance, and Functional Imaging Studies: What Are We Missing About Neuroprotection?" Frontiers in Cellular Neuroscience 16 (2022): 866122.

99) Russell, Jason K., Carrie K. Jones, and Paul A. Newhouse. "The Role of Estrogen in Brain and Cognitive Aging." Neurotherapeutics 16 (2019): 649-665.

100) Ali, Noor, et al. "The Role of Estrogen Therapy as A Protective Factor for Alzheimer's Disease and Dementia In Postmenopausal Women: A Comprehensive Review Of The Literature." Cureus 15.8 (2023).

101) Salinero, Abigail E., et al. "Brain Specific Estrogen Ameliorates Cognitive Effects of Surgical Menopause in Mice." bioRxiv (2023).

102) Archer, D. F., et al. "Menopausal Hot Flushes and Night Sweats: Where Are We Now?" Climacteric 14 (2011): 515-528.

103) Steingold, Kenneth A., et al. "Treatment of Hot Flashes with Transdermal Estradiol Administration." The Journal of Clinical Endocrinology & Metabolism 61.4 (1985): 627-632.

104) Liu James, et al. "SAT-239 Bioidentical Estradiol and Progesterone Improved Hot Flushes, Night Sweats and Sweating." Journal of the Endocrine Society 3. Supplement_1 (2019): SAT-239.

105) Nelson, Heidi D. "Commonly Used Types of Postmenopausal Estrogen for Treatment of Hot Flashes: Scientific Review." JAMA 291.13 (2004): 1610-1620.

106) Kling, Juliana M., Cynthia A. Stuenkel, et al. "Management of the Vasomotor Symptoms of Menopause: Twofers In Your Clinical Toolbox." Mayo Clinic Proceedings. Vol. 99. No. 7. Elsevier, 2024.

107) Tanmahasamut, Prasong, et al. "Effect of Estradiol Vaginal Gel on Vaginal Atrophy in Postmenopausal Women: A Randomized Double-Blind Controlled Trial." Journal of Obstetrics and Gynaecology Research 46.8 (2020): 1425-1435.

108) Villa, Paola, et al. "Local Ultra-Low-Dose Estriol Gel Treatment of Vulvo-Vaginal Atrophy: Efficacy and Safety of Long-Term Treatment." Gynecological Endocrinology 36.6 (2020): 535-539.

109) Donders, Gilbert GG, et al. "Pharmacotherapy for The Treatment of Vaginal Atrophy." Expert Opinion on Pharmacotherapy 20.7 (2019): 821-835.

110) Simon, James, et al. "Effective Treatment of Vaginal Atrophy with An Ultra–Low-Dose Estradiol Vaginal Tablet." Obstetrics & Gynecology 112.5 (2008): 1053-1060.

111) Nothnagle, Melissa, and Julie Scott Taylor. "Vaginal Estrogen Preparations for Relief Of Atrophic Vaginitis." American Family Physician 69.9 (2004): 2111-2112.

112) Nelson, Lawrence M. "The Truth About 17-Beta Estradiol: Menopause Beyond "Old Wives' Tales"." Frontiers in Endocrinology 14 (2023): 1229804.

113) Nolan, B. J., B. Liang, and A. S. Cheung. "Efficacy of Micronized Progesterone for Sleep: A Systematic Review and Meta-analysis of Randomized Controlled Trial Data." The Journal of Clinical Endocrinology and Metabolism 106.4 (2021): 942-951.

114) Hachul, Helena, et al. "Sleep During Menopause." Sleep Medicine Clinics 18.4 (2023): 423-433.

115) Bashir, Kiran, et al. "Prevalence of Insomnia in Menopausal Women: Prevalence of Insomnia." Pakistan Journal of Health Sciences (2023): 43-46.

116) Li, Caixia, et al. "Analysis of the Long-Term Beneficial Effects of Menopausal Hormone Therapy On Sleep Quality And Menopausal Symptoms." Experimental and Therapeutic Medicine 18.5 (2019): 3905-3912.

117) Andenæs, Randi, et al. "Associations Between Menopausal Hormone Therapy and Sleep Disturbance in Women During the Menopausal Transition and Post-Menopause: Data from The Norwegian Prescription Database and The HUNT Study." BMC Women's Health 20.1 (2020): 1-9.

118) Pan, Zhuo, et al. "Different Regimens of Menopausal Hormone Therapy for Improving Sleep Quality: A Systematic Review and Meta-Analysis." Menopause (New York, NY) 29.5 (2022): 627.

119) Dias, Rejane Camila Alvarenga, et al. "Fibromyalgia, Sleep Disturbance, and Menopause: Is There a Relationship? A Literature Review." International Journal of Rheumatic Diseases 22.11 (2019): 1961-1971.

120) Zhou, Qian, et al. "Investigation of the Relationship Between Hot Flashes, Sweating and Sleep Quality in Perimenopausal and Postmenopausal Women: The Mediating Effect of Anxiety and Depression." BMC Women's Health 21.1 (2021): 1-8.

121) Polo-Kantola, Päivi, et al. "When Does Estrogen Replacement Therapy Improve Sleep Quality?" American Journal of Obstetrics and Gynecology 178.5 (1998): 1002-1009.

122) Haufe, Annika, Fiona C. Baker, and Brigitte Leeners. "The Role of Ovarian Hormones In The Pathophysiology Of Perimenopausal Sleep Disturbances: A Systematic Review." Sleep Medicine Reviews 66 (2022): 101710.

123) Kagan, Risa, et al. "Improvement in Sleep Outcomes with A 17β-Estradiol–Progesterone Oral Capsule (TX-001HR) For Postmenopausal Women." Menopause (New York, NY) 26.6 (2019): 622.

124) Geiger, Paul J., et al. "Effects of Perimenopausal Transdermal Estradiol On Self-Reported Sleep, Independent Of Its Effect On Vasomotor Symptom Bother And Depressive Symptoms." Menopause (New York, NY) 26.11 (2019): 1318.

125) Lee, Jinju, et al. "Sleep Disorders and Menopause." Journal of Menopausal Medicine 25.2 (2019): 83-87.

126) Scott, Alice, and Louise Newson. "Should We Be Prescribing Testosterone to Perimenopausal and Menopausal Women? A Guide to Prescribing Testosterone for Women in Primary Care." British Journal of General Practice 70.693 (2020): 203

127) Adebisi, Omoniyi Y., and Karen Carlson. "Hypoactive Sexual Desire Disorder in Women." StatPearls [Internet]. StatPearls Publishing, 2024.

128) Islam, Rakibul M., et al. "Safety and Efficacy of Testosterone for Women: A Systematic Review and Meta-Analysis of Randomised Controlled Trial Data." The Lancet Diabetes & Endocrinology 7.10 (2019): 754-766.-204.

129) Parish, Sharon J., and Juliana M. Kling. "Testosterone Use for Hypoactive Sexual Desire Disorder in Postmenopausal Women." Menopause 30.7 (2023): 781-783.

130) Davis, Susan R., and Jane Tran. "Testosterone Influences Libido and Well-Being In Women." Trends in Endocrinology & Metabolism 12.1 (2001): 33-37.

131) Basson, Rosemary. "Testosterone Therapy for Reduced Libido In Women." Therapeutic advances in endocrinology and metabolism 1.4 (2010): 155-164.

132) Rahn, David D., et al. "Vaginal Estrogen for Genitourinary Syndrome of Menopause: A Systematic Review." Obstetrics & Gynecology 124.6 (2014): 1147-1156.

133) Ferrante, Kimberly L., et al. "Vaginal Estrogen for The Prevention of Recurrent Urinary Tract Infection in Postmenopausal Women: A Randomized Clinical Trial." Female Pelvic Medicine & Reconstructive Surgery 27.2 (2021): 112-117.

134) Buck, Emory S., et al. "Effective Prevention of Recurrent UTIS With Vaginal Estrogen: Pearls for A Urological Approach to Genitourinary Syndrome of Menopause." Urology 151 (2021): 31-36.

135) Rettberg, Jamaica R., Jia Yao, and Roberta Diaz Brinton. "Estrogen: A Master Regulator of Bioenergetic Systems in The Brain and Body." Frontiers In Neuroendocrinology 35.1 (2014): 8-30.

136) Chopra, Sakshi, et al. "Weight Management Module for Perimenopausal Women: A Practical Guide for Gynecologists." Journal Of Mid-Life Health 10.4 (2019): 165.

137) Fenton, Anna. "Weight, Shape, And Body Composition Changes at Menopause." Journal of Mid-life Health 12.3 (2021): 187.

138) Zhu, Jing, et al. "Role of Estrogen in The Regulation of Central and Peripheral Energy Homeostasis: From a Menopausal Perspective." Therapeutic Advances in Endocrinology and Metabolism 14 (2023): 20420188231199359.

139) Gava, Giulia, et al. "Cognition, Mood and Sleep in Menopausal Transition: The Role Of Menopause Hormone Therapy." Medicina 55.10 (2019): 668.

140) Halbreich, Uriel, and Linda S. Kahn. "Role of Estrogen in the Etiology and Treatment of Mood Disorders." CNS Drugs 15 (2001): 797-817.

141) Wieland, Scott, et al. "Anxiolytic Activity of The Progesterone Metabolite 5α-Pregnan-3α-Ol-20-One." Brain research 565.2 (1991): 263-268.

142) Picazo, O., and A. Ferna. "Anti-Anxiety Effects of Progesterone and Some of Its Reduced Metabolites: An Evaluation Using the Burying Behavior Test." Brain research 680.1-2 (1995): 135-141.

145) Kendall, Alexandra C., et al. "Menopause Induces Changes to The Stratum Corneum Ceramide Profile, Which Are Prevented by Hormone Replacement Therapy." Scientific Reports 12.1 (2022): 21715.

146) Lephart, Edwin D., and Frederick Naftolin. "Menopause and the Skin: Old Favorites and New Innovations in Cosmeceuticals for Estrogen-Deficient Skin." Dermatology and Therapy 11 (2021): 53-69.

147) Lephart ED. A Review of the Role of Estrogen in Dermal Aging and Facial Attractiveness In Women. J Cosmet Dermatol (2018) 17(3):282–8.

148) Wilkinson, Holly N., and Matthew J. Hardman. "The Role of Estrogen in Cutaneous Ageing and Repair." Maturitas 103 (2017): 60-64.

149) Rzepecki, Alexandra K., et al. "Estrogen-Deficient Skin: The Role of Topical Therapy." International Journal of Women's Dermatology 5.2 (2019): 85.

150) Hall, Glenda, and Tania J. Phillips. "Estrogen and Skin: The Effects of Estrogen, Menopause, And Hormone Replacement Therapy on The Skin." Journal of the American Academy of Dermatology 53.4 (2005): 555-568.

151) Zouboulis, C. C., et al. "Skin, Hair and Beyond: The Impact of Menopause." Climacteric 25.5 (2022): 434-442.

152) Borda, Luis J., et al. "Bioidentical Hormone Therapy in Menopause: Relevance in Dermatology." Dermatology online journal 25.1 (2019).

153) Reus, Thamile Luciane, et al. "Revisiting the Effects of Menopause on The Skin: Functional Changes, Clinical Studies, In Vitro Models and Therapeutic Alternatives." Mechanisms of Ageing and Development 185 (2020): 111193.

154) Connor, Charles Gerald. "Treatment of Dry Eye with A Transdermal 3% Testosterone Cream." Investigative Ophthalmology & Visual Science 44.13 (2003): 2450-2450.

155) Connor, Charles Gerald. "Symptomatic Relief of Dry Eye Assessed with The OSDI In Patients Using 5% Testosterone Cream." Investigative Ophthalmology & Visual Science 46.13 (2005): 2032-2032.

156) Feng, Zhao Xun, et al. "Risk of Ocular Adverse Events with Aromatase Inhibitors." Canadian Journal of Ophthalmology (2023).

157) Almafreji, Ibrahim, et al. "Review of the Literature on Ocular Complications Associated with Aromatase Inhibitor Use." Cureus 13.8 (2021).

158) Supalaset, Sumet, et al. "A Randomized Controlled Double-Masked Study of Transdermal Androgen In Dry Eye Patients Associated With Androgen Deficiency." American Journal of Ophthalmology 197 (2019): 136-144.

159) Aryani, Inda Astri, et al. "Role of Androgen on Physiological Function of Pilosebaceous Unit." Bioscientia Medicina: Journal of Biomedicine and Translational Research 5.6 (2021): 545-551.

160) Kronemyer, Bob. "Female Pattern Hair Loss in Postmenopausal Women." Contemporary OB/GYN 67.4 (2022): 26-26.

161) Glaser RL, et al. Improvement in Scalp Hair Growth in Androgen-Deficient Women Treated with Testosterone: A Questionnaire Study. Br J Dermatolo. 2012 Feb; 166(2): 274-278.

162) Glaser R, Dimitrakakis C. Testosterone Therapy in Women: Myths and Misconceptions. Maturitas. 2013; 74: 230-234.

163) Rinaldi, Fabio, et al. "The Menopausal Transition: Is the Hair Follicle "Going through Menopause"?" Biomedicines 11.11 (2023): 3041.

164) Chaikittisilpa, Sukanya, et al. "Prevalence of Female Pattern Hair Loss In Postmenopausal Women: A Cross-Sectional Study." Menopause 29.4 (2022): 415-420.

165) Hughes, Elizabeth C., et al. "Telogen Effluvium." StatPearls [Internet]. StatPearls Publishing, 2024.

166) Williams, Cecilia, et al. "Estrogen Receptor Beta as Target for Colorectal Cancer Prevention." Cancer Letters 372.1 (2016): 48-56.

167) Rennert, Gad, et al. "Use of Hormone Replacement Therapy and The Risk of Colorectal Cancer." Journal Of Clinical Oncology 27.27 (2009): 4542.

168) Botteri, Edoardo, et al. "Menopausal Hormone Therapy and Colorectal Cancer: A Linkage Between Nationwide Registries in Norway." BMJ open 7.11 (2017).

169) Lin, Kueiyu Joshua, et al. "The Effect of Estrogen vs. Combined Estrogen-Progestogen Therapy on The Risk of Colorectal Cancer." International Journal of Cancer 130.2 (2012): 419-430.

170) Mauvais-Jarvis, Franck, et al. "Menopausal Hormone Therapy and Type 2 Diabetes Prevention: Evidence, Mechanisms, And Clinical Implications." Endocrine Reviews 38.3 (2017): 173-188.

171) Salpeter, S. R., et al. "Meta-Analysis: Effect of Hormone-Replacement Therapy On Components Of The Metabolic Syndrome In Postmenopausal Women." Diabetes, Obesity and Metabolism 8.5 (2006): 538-554.

172) Xu, Youhua, et al. "Combined Estrogen Replacement Therapy on Metabolic Control In Postmenopausal Women With Diabetes Mellitus." The Kaohsiung journal of medical sciences 30.7 (2014): 350-361.

173) Anagnostis, Panagiotis, et al. "Early Menopause and Premature Ovarian Insufficiency Are Associated with Increased Risk Of Type 2 Diabetes: A Systematic Review And Meta-Analysis." European journal of endocrinology 180.1 (2019): 41-50.

174) Mauvais-Jarvis, Franck. "Is Estradiol A Biomarker of Type 2 Diabetes Risk in Postmenopausal Women?" Diabetes 66.3 (2017): 568.

175) Pu, D., et al. "Metabolic syndrome in menopause and associated factors: a meta-analysis." Climacteric 20.6 (2017): 583-591.

176) El Mohtadi, Mohamed, et al. "Estrogen Deficiency—A Central Paradigm In Age-Related Impaired Healing?" EXCLI journal 20 (2021): 99.

177) Collins, Brittany C., et al. "Aging of the Musculoskeletal System: How the Loss of Estrogen Impacts Muscle Strength." Bone 123 (2019): 137-144.

178) Pellegrino, Andrea, et al. "Mechanisms of Estrogen Influence on Skeletal Muscle: Mass, Regeneration, And Mitochondrial Function." Sports Medicine 52.12 (2022): 2853-2869.

179) Dam, Tine Vrist, et al. "Transdermal Estrogen Therapy Improves Gains in Skeletal Muscle Mass After 12 Weeks of Resistance Training in Early Postmenopausal Women." Frontiers in Physiology 11 (2021): 596130.

180) Tian, Xu, et al. "From Mitochondria to Sarcopenia: Role Of 17β-Estradiol and Testosterone." Frontiers in Endocrinology 14 (2023): 1156583.

181) Raffaelli, Bianca, et al. "Menstrual migraine is Caused By Estrogen Withdrawal: Revisiting The Evidence." The Journal of Headache and Pain 24.1 (2023): 131.

182) Nappi, Rossella E., et al. "Role of Estrogens in Menstrual Migraine." Cells 11.8 (2022): 1355.

183) Nappi, Rossella E., et al. "Hormonal management of migraine at menopause." Menopause international 15.2 (2009): 82-86

184) Reddy, Nihaal, et al. "The Complex Relationship Between Estrogen and Migraines: A Scoping Review." Systematic Reviews 10 (2021): 1-13.

185) Ornello, Raffaele, et al. "Acute and Preventive Management of Migraine During Menstruation and Menopause." Journal of Clinical Medicine 10.11 (2021): 2263.

186) Gilmore, Katherine Louise, and Diana Mansour. "A Case Vignette Describing Management of Menopausal Symptoms and Migraine in The Perimenopause." BMJ Sexual & Reproductive Health (2021).

187) MacGregor, E. Anne. "Migraine, Menopause and Hormone Replacement Therapy." Post Reproductive Health 24.1 (2018): 11-18.

188) Silberstein, S. D., and B. De Lignières. "Migraine, Menopause and Hormonal Replacement Therapy." Cephalalgia 20.3 (2000): 214-221.

189) Loder, Elizabeth, Paul Rizzoli, and Joan Golub. "Hormonal Management of Migraine Associated With Menses and the Menopause: A Clinical Review: CME." Headache: The Journal of Head and Face Pain 47.2 (2007): 329-340.

190) Ibrahimi, Khatera, et al. "Migraine and Perimenopause." Maturitas 78.4 (2014): 277-280.

191) McCarthy, Micheline, and Ami P. Raval. "The Peri-Menopause in A Woman's Life: A Systemic Inflammatory Phase That Enables Later Neurodegenerative Disease." Journal Of Neuroinflammation 17 (2020): 1-14.

192) Abu-Taha, May, et al. "Menopause and Ovariectomy Cause a Low Grade of Systemic Inflammation That May Be Prevented by Chronic Treatment with Low Doses of Estrogen Or Losartan." The Journal Of Immunology 183.2 (2009): 1393-1402.

193) Yang, Hye Rim, et al. "Obesity induced by estrogen deficiency is associated with hypothalamic inflammation." Biochemistry and Biophysics Reports 23 (2020): 100794.

194) Vegeto, Elisabetta, Valeria Benedusi, and Adriana Maggi. "Estrogen Anti-Inflammatory Activity In Brain: A Therapeutic Opportunity For Menopause And Neurodegenerative Diseases." Frontiers in neuroendocrinology 29.4 (2008): 507-519.

195) Au, April, et al. "Estrogens, Inflammation and Cognition." Frontiers In Neuroendocrinology 40 (2016): 87-100.

196) Wergenthaler, Nousal, et al. "Etiology of Idiopathic Macular Holes in the Light of Estrogen Hormone." Current Issues in Molecular Biology 45.8 (2023): 6339-6351.

197) Korpole, Nilay Reddy, et al. "Gender Difference in Ocular Diseases, Risk Factors And Management With Specific Reference To Role Of Sex Steroid Hormones." Journal of Mid-life Health 13.1 (2022): 20.

198) Patnaik, J. L., et al. "Hormone Therapy as a Protective Factor for Age-Related Macular Degeneration." Ophthalmic Epidemiology 27.2 (2019): 148-154.

199) Wei, Qingquan, et al. "17β-estradiol Ameliorates Oxidative Stress and Blue Light-Emitting Diode-Induced Retinal Degeneration By Decreasing Apoptosis And Enhancing Autophagy." Drug Design, Development and Therapy 12 (2018): 2715.

200) Cascio, Caterina, et al. "The Estrogenic Retina: The Potential Contribution to Healthy Aging and Age-Related Neurodegenerative Diseases of The Retina." Steroids 103 (2015): 31-41.

201) Bazvand, Fatemeh, et al. "Tamoxifen Retinopathy." Survey of Ophthalmology: S0039-6257.

202) Tenney, Stephen, et al. "Tamoxifen Retinopathy: A Comprehensive Review." Survey of Ophthalmology (2023).

203) Vinding, Troels, et al. "Retinopathy Caused by Treatment With Tamoxifen In Low Dosage." Acta Ophthalmologica 61.1 (1983): 45-50.

204) Mckeown, Craig A., et al. "Tamoxifen retinopathy." The British Journal of Ophthalmology 65.3 (1981): 177.

205) Tunc, M. "Maculopathy Following Extended Usage of Clomiphene Citrate." Eye 28.9 (2014): 1144-1146.

207) Russom, Mulugeta, et al. "Blindness and Retinal Disorder Associated with Clomifene Citrate: Cases Series Assessment." Clin Med Invest 2.3 (2017): 1-4.

208) Feng, Zhao Xun, et al. "Risk of Ocular Adverse Events with Aromatase Inhibitors." Canadian Journal of Ophthalmology (2023).

209) Azeem, Sitara, et al. "A case Report on Letrozole-Related Maculopathy." Oman Journal of Ophthalmology 16.2 (2023): 322-325.

210) Almafreji, Ibrahim, et al. "Review of the Literature on Ocular Complications Associated with Aromatase Inhibitor Use." Cureus 13.8 (2021).

211) Keenan, Tiarnán DL, et al. "Oral Antioxidant and Lutein/Zeaxanthin Supplements Slow Geographic Atrophy Progression to the Fovea in Age-Related Macular Degeneration." Ophthalmology (2024).

212) Mrowicka, Małgorzata, et al. "Lutein and Zeaxanthin and Their Roles in Age-Related Macular Degeneration—Neurodegenerative Disease." Nutrients 14.4 (2022): 827

213) Baik, Seo H., et. Al. "Effects of Hormone Therapy on Survival, Cancer, Cardiovascular and Dementia Risks In 7 Million Menopausal Women Over Age 65: A Retrospective Observational Study." medRxiv (2022): 2022-05.

Chapter 2

Russell Marker and the Origins of Bioidentical Hormones

Where Do Bioidentical Hormones Come From? Who was Russell Marker?

A COUPLE OF TIMES A week, I get the question, "Where do hormones come from?" The story starts in 1938 with an obscure chemistry professor at Pennsylvania State College by the name of Russel Earl Marker who was working on plant steroid chemistry. Around this time, Marker found a plant steroid from the Dioscorea family called diosgenin which could be easily converted into the human bioidentical hormone, progesterone. Dr. Russell E. Marker then invented the Marker Degradation Process, a practical way to mass produce human hormones using diosgenin, a plant steroid found in Mexico. Next, Marker needed an economical source of plant material to isolate diosgenin. (1-4)

Finding the Plant Material in Mexico

In November 1941, Russell Marker found what he had been searching for in a botany textbook describing a Dioscorea plant indigenous to Veracruz in Mexico, called Cabeza de Negro. In 1942, Marker traveled to Mexico where he purchased Dioscorea plants in bulk and began mass production of the bioidentical hormone, progesterone. In 1943, Marker resigned from Penn State University and moved to Mexico to run his factory for the mass production of progesterone and other bioidentical hormones. Although progesterone was available, it was extremely expensive. Russell Marker made mass-produced progesterone available to the market at an affordable price. Dr. Marker decided not to register his invention at the patent office, donating his Marker Degradation Process to the world. In 2024, Hannah Kleinschmit wrote her doctoral thesis, an overview of women's hormone replacement, revealing that Marker's production method was so successful, it remains the production process for hormones in modern-day pharmaceuticals, writing:

> In 1938, American chemist Russell Earl Marker [was the first to] manufacture a bioidentical version of progesterone from the sarsasapogenin sterol of the plant Sarsaparilla by utilizing the resemblances in structure between cholesterol and progesterone and a chemical technique dubbed 'Marker's degradation'... Marker's extensive research and future experiments would later lead to the discovery and use of diosgenin sterols of the Dioscorea species, a wild Mexican yam, to produce progesterone instead...Marker's production method and use of the Dioscorea species resulted in such success for the generation of steroidal hormones that it remains the standard progesterone production process for modern-day pharmaceutics. (5-9)

Many Different Factories in Mexico

In early 1944, Syntex was formed in Mexico to manufacture progesterone from the Diascorea Diosgenin. In May 1945, Russell Marker left Syntex over a dispute and started a new company, Botanica-mex, near Mexico City which then made several kilograms of progesterone. However, Botanica-mex folded in March 1946, and was restarted as a new company called Hormonosynth. During this time, the cabeza de negro plant source was replaced by another yam called barbasco, containing 500% more diosgenin. After Marker's retirement, the company was again reorganized as Diosynth. In 1949, at the age of 47, Russell Marker retired from chemistry altogether and started a company offering Mexican reproductions of objects made by three great 18th-century silversmiths such as teapots, coffee urns, tureens, dinner service sets, and candlesticks. Russell Marker passed away on March 23, 1995, age 93. (1-9)

Dr. Marker Leaves Syntex

After Dr. Russell Marker left Syntex, the company recruited another chemist George Rosenkranz, who began in October 1945, and Syntex was again selling progesterone. Rosenkranz extended the process to the production of testosterone and other bioidentical hormones. Rosenkranz built a research program at Syntex, and recruited other Ph.D. chemists including Carl Djerassi and Alejandro Zaffaroni. (10)

Cortisone and the "Pill"

Further research at Syntex led to the production of the powerful anti-inflammatory steroid called cortisone, using the same plant steroid diosgenin as a precursor. In 1951, Syntex was first to make a synthetic birth control pill, norethistrone (norethindrone). By the 1950s, Syntex and its competitors were the major suppliers of bioidentical and synthetic hormones to the United States. (11)

Health Benefits of Diosgenin

While using diosgenin as a precursor to progesterone, Russel Marker had no way of knowing that diosgenin had many health benefits of its own. For example, diosgenin is a natural anti-cancer and anti-inflammatory drug. Diosgenin is non-toxic and has benefits for an array of chronic diseases as described in 2022 by Dr. P. Semwal, writing:

> A steroidal saponin, diosgenin, and its variants are useful in the treatment of a great variety of chronic conditions, including cardiovascular disease, **several forms of lethal malignancies,** nervous system problems, and autoimmune diseases... Diosgenin and its derivatives have been shown to have pharmacological benefits

against cancer, diabetes, osteoporosis, Alzheimer's disease, and stroke in several investigations...Furthermore, investigations demonstrating its **nontoxic nature** significantly promote the inclusion of this medicine in additional clinical studies or trials in the forthcoming days. (12-17)

Diosgenin is an ER-alpha Inhibitor

The anticancer benefit of Diosgenin for breast cancer prevention involves the suppression of ER-alpha activity. Estrogen receptor alpha (ER-alpha) is the proliferative receptor, and ER-beta is the tumor suppressor receptor. Estrogen receptors and their importance are discussed in later chapters. One would think Diosgenin would be a good candidate for wide-scale use in the prevention of breast cancer. However, since it is a natural substance, it is not eligible for patent protection, and therefore will never receive the funding needed for FDA approval. Come to think of it, many other natural substances suppress ER-alpha and bind to and activate ER-beta, thus breast cancer preventive. These are phytoestrogens like genistein, daidzein, equol (found in soy products), silibinin (from milk thistle), and certain flavonoids. Natural dietary supplements such as resveratrol, pterostilbene, quercetin, isoflavones, tea extracts (EGCG), and curcumin are promising candidates to be studied for breast cancer prevention and treatment. These are safe and non-toxic. Instead, most of the research on breast cancer has been focused on patented estrogen-blocking drugs. Tamoxifen and aromatase inhibitors are both extensively studied and found to reduce the incidence of breast cancer by 50-65 percent among high-risk women. Yet, only ten percent of eligible high-risk women use these patented drugs. What is the reason for this low uptake? The lack of mortality benefit and lack of improvement in quality of life due to toxicity makes doctors and women hesitant to use these drugs for primary prevention of breast cancer. In 2017, Dr. Katherine D. Crew reviewed anti-estrogen drugs for breast cancer prevention, and asked the question: How do we increase uptake of tamoxifen and other anti-estrogens for breast cancer prevention?"

The reason why **none of the prevention studies [of Tamoxifen and aromatase inhibitors] have shown an improvement in overall survival** is because the primary end point of the prevention studies has been breast cancer incidence. **Without demonstrating an improvement in overall survival**, some physicians may be reluctant to prescribe a long-term medication with **potential for toxicity.** Additionally, none of the prevention studies have demonstrated an improvement in quality of life (QOL) for women who choose chemoprevention. (18-24)

Conclusion: It is mind-boggling to think that in 1938, the fate of bioidentical hormone production rested on one eccentric chemist, Russell Earl Marker, who singlehandedly invented a production process so successful, that it is still used today to produce modern steroidal drugs. (1-14)

◆ References Chapter 2 Russell Marker and the Origins of Bioidentical Hormones

1) Seeman, Jeffrey I. "Russell Earl Marker and the Beginning of The Steroidal Pharmaceutical Industry." The Chemical Record 23.4 (2023): e202300048.

2) Marker, Russell E., and Josefina Lopez. "Steroidal Sapogenins. No. 166. The Neosapogenins." Journal of the American Chemical Society 69.10 (1947): 2393-2395.

3) Kovács, Lajos. "The Campfire Stories of Russell Marker, A Pioneer of Chemistry." Notes and Records 78.3 (2024): 467-492.

4) Marker, Russell E., et al. "Steroidal Sapogenins." Journal of the American Chemical Society 69.9 (1947): 2167-2230.

5) Kleinschmit, Hannah. "An Overview of Women's Hormone Replacement Therapy (HRT)." (2024).

6) Marker RE, Krueger J (1940). "Sterols. CXII. Sapogenins. XLI. The Preparation of Trillin and its Conversion to Progesterone." Journal of the American Chemical Society. 62 (12): 3349–3350.

7) Road to Hope for Female Infertility: Progesterone, By Stephen T. Spagnol, Spring 2010

8) Loriaux, D. Lynn. "Russell Earl Marker (1902–1995)." The Endocrinologist 18.3 (2008): 107-108.

9) Piette, P. "The History of Natural Progesterone, The Never-Ending Story." Climacteric 21.4 (2018): 308-314.

10) Russell Marker and the Mexican Steroid Hormone Industry, December 2, 1999, the American Chemical Society. https://www.acs.org/education/whatischemistry/landmarks/progesteronesynthesis.html

11) Mandy, Redig. "Yams of Fortune: The (Uncontrolled) Birth of Oral Contraceptives." Journal of Young Investigators 6.7 (2005).

12) Semwal, Prabhakar, et al. "Diosgenin: an updated pharmacological review and therapeutic Perspectives." Oxidative Medicine and Cellular Longevity 2022.1 (2022): 1035441.

13) Karami-Mohajeri, Somayyeh, et al. "Diosgenin: mechanistic insights on its anti-inflammatory effects." Anti-Inflammatory & Anti-Allergy Agents in Medicinal Chemistry (Formerly Current Medicinal Chemistry-Anti-Inflammatory and Anti-Allergy Agents) 21.1 (2022): 2-9.

14) Son, In Suk, et al. "Antioxidative and hypolipidemic effects of diosgenin, a steroidal saponin of yam (Dioscorea spp.), on high-cholesterol fed rats." Bioscience, biotechnology, and biochemistry 71.12 (2007): 3063-3071.

15) Mondal, Sadhan. "Anticancer Potential of Biologically Active Diosgenin and its Derivatives: An Update." Current Traditional Medicine 10.1 (2024): 67-80.

16) Khanal, Pukar, et al. "Systems and in vitro pharmacology profiling of diosgenin against breast cancer." Frontiers in Pharmacology 13 (2023): 1052849.

17) Arya, Prajya, and Pradyuman Kumar. "Diosgenin a steroidal compound: An emerging way to cancer management." Journal of Food Biochemistry 45.12 (2021): e14005.

18) Chun, Jaemoo, et al. "The Induction of Apoptosis by A Newly Synthesized Diosgenyl Saponin Through the Suppression of Estrogen Receptor-Alpha In MCF-7 Human Breast Cancer Cells." Archives of pharmacal research 37 (2014): 1477-1486.

19) Hwang, Sun Tae, et al. "Genistin Attenuates Cellular Growth and Promotes Apoptotic Cell Death Breast Cancer Cells Through Modulation of ER-Alpha Signaling Pathway." Life Sciences 263 (2020): 118594.

20) Crew, Katherine D., et al. "How Do We Increase Uptake Of Tamoxifen And Other Anti-Estrogens For Breast Cancer Prevention?" NPJ Breast Cancer 3.1 (2017): 20.

21) Ávila-Gálvez, María Ángeles, et al. "Dietary Phenolics Against Breast Cancer. A Critical Evidence-Based Review and Future Perspectives." International journal of molecular sciences 21.16 (2020): 5718.

22) Torrens-Mas, Margalida, and Pilar Roca. "Phytoestrogens for Cancer Prevention and Treatment." Biology 9.12 (2020): 427.

23) Sudhakaran, Meenakshi, Sagar Sardesai, and Andrea I. Doseff. "Flavonoids: New Frontier for Immuno-Regulation And Breast Cancer Control." Antioxidants 8.4 (2019): 103.

24) Obakan-Yerlikaya, Pinar, et al. "Breast Cancer and Flavonoids as Treatment Strategy." Breast Cancer Biol. Med 5 (2017): 305-326.

Chapter 3

Do Not Monkey with My Hormones!

Why Are Monkeys a Good Model to Study Hormone Replacement?

HORMONE STUDIES FROM THE PRIMATE Center at Winston-Salem allow us to answer a few questions about different hormone formulations. There are several reasons why non-human primates (Macaque monkeys) are a good model for studying hormone replacement therapy. Macaque monkeys have 93.54% identity with the human genome. Both humans and Macaque monkeys have similar steroidal response to hormone administration and breast cancer pathophysiology. Both have similar menstrual cycles and experience a distinct menopause with a decline in estrogen levels. Both have similar endometrial and menstrual physiology. In 2004, Dr. Jay Kaplan explains why macaque monkeys are a good model, writing:

> Researchers first made systematic use of macaque monkeys to elucidate the reproductive biology of women. Much of this research was organized around study of the menstrual cycle, a phenomenon shared uniquely by women and the Old World anthropoid primates...Monkey studies also demonstrate that **estrogen replacement effectively reduces bone loss**...Old World monkeys are among the most appropriate models for studying cancers of the reproductive tract because **they have a similar endometrial and menstrual physiology**...large brain size, high intelligence, and a reproductive physiology similar to that of women make Old World monkeys among the most salient models for evaluating hormonal influences on emotional states and cognitive abilities...the greatest contribution of animal models, especially nonhuman primates, is that they offer the ability to conduct controlled experiments that would be logistically or ethically proscribed [forbidden] in women. (1)

The 2008 French Cohort Human Observational Study

One of the human observational studies I frequently cite is the 2008 French Cohort Study by Dr. Agnes Fournier who compared various hormone preparations demonstrating the safety of the estradiol/progesterone combination for menopausal women. This study showed that bioidentical hormone users in France using the combination of estradiol with natural progesterone had **no increased risk of breast cancer**. The hazard ratio (HR) was equal to 1.0, meaning the hormone-treated group had the same risk for breast cancer risk as non-hormone users in the general population. (21)

Here is the Data from Table 3 of the French Cohort Study (Fournier, 2008):

Hormone Combination	Relative Risk of Invasive Breast Cancer
Non-User	1.0
estrogen (estradiol) alone	1.29 (p=0.73 non-significant)
estrogen (estradiol) / progesterone (bioidentical)	1.0
estrogen (estradiol) / medroxyprogesterone	1.48

Adding Medroxyprogesterone (MPA)

The relative risk of breast cancer with estradiol alone was increased by 28 percent (RR= 1.28). When medroxyprogesterone (MPA, a progestin) is added to estradiol, the findings are quite different, showing increased breast cancer risk (RR=1.48). However, there is no increased risk if progesterone is used instead of MPA (RR=1.0). Notice the relative risk of 1.28 with estradiol alone is reduced to 1.0 when progesterone is added to it in combination. Progesterone has remarkable anti-cancer effects against breast cancer, as well as many other cancers, as discussed in 2017 by Dr. Allan Lieberman. **Note:** RR is a relative risk. (2)

Conflicting Studies with Estradiol Alone

I would add here that some would claim there is no increased breast cancer risk with estradiol alone based on the 2017 systemic review and meta-analysis by Dr. Shilan Yang of randomized controlled trials (RCT). This study is flawed because of relatively few cases in each study with relatively short follow-up time. For example, in the 2001 study by Dr. Catherine Viscoli in which 664 women were treated with estradiol and followed for 2.8 years. There were 5 cases of breast cancer in the estradiol arm and 5 cases in the placebo arm. In the 2003 study by Dr. Howard Hodis, 226 postmenopausal women were treated with estradiol for 3.3 years follow-up. There was only one case of breast cancer in the control group. Perhaps the largest study with the longest follow-up was done in 2014 by Dr. Nicola Cherry in which 1017 women were randomized to estradiol treatment for 2 years or placebo and observed for 14 years. There were seven cases of breast cancer in the estradiol arm, and 15 in the placebo arm, indicating a 53 percent reduction in breast cancer in those taking estradiol alone. This study by Dr. Cherry suggests that estradiol alone is breast cancer protective. Thus, we have two conflicting studies, the 2008 French cohort (observational) (RR=1.28) and the 2014 Nicola Cherry (RCT) study (RR=0.47). However, for the sake of discussion, let us accept the French cohort observational study of RR=1.28 for breast cancer with estradiol alone, meaning estradiol alone increases the risk for breast cancer by 28 percent. (3-6)

Medroxyprogesterone (MPA) is Carcinogenic

The 2002 WHI study (the first arm using Premarin/MPA) confirmed the carcinogenicity of MPA, the synthetic progestin. This study showed increased invasive breast cancer (HR=1.25) with the combination of Premarin and Medroxyprogesterone (MPA), also called Prempro. This hazard ratio (HR) of 1.25 for the added MPA roughly matches the increased relative risk of HR = 1.48 in the French Cohort study (Fournier, 2008) for the estradiol/MPA combination. MPA is a known carcinogen, and is routinely used to induce breast cancer in the MPA animal model using BALB-c mice as demonstrated in 2009 by Dr. Claudia Lanari. BALB-c mice are immunocompetent albino mice widely used for research. (7-8) (21)

The DMBA Mouse Model of Breast Cancer

Although there are many models of breast cancer in mice using carcinogenic chemicals such as DMBA and nitrosomethylurea (NMU), genetic engineered mouse models (GEMMs), and xenografting tumors into mice. All attempts to induce cancer by giving 17-beta-estradiol (human estrogen) to wild-type mice have failed. Quite to the contrary, pregnancy levels of estrogen and progesterone protect mice from breast cancer tumorigenesis in multiple strains, as discussed in 2006 by Dr. Collin M. Blakely who found a common gene signature thought to contribute to pregnancy-induced protection from breast cancer. This finding was confirmed in 2007 by Dr. Lakshmanaswamy Rajkumar. Dr. Collin M. Blakely writes:

Our findings show that hormone-induced protection against mammary tumorigenesis is widely conserved among divergent rat strains and define a gene expression signature that is tightly correlated with reduced mammary tumor susceptibility as a consequence of a normal developmental event. Given the conservation of this signature, these pathways may contribute to pregnancy-induced protection against breast cancer. (9-10)

Yes, estrogen is a growth factor for breast cancer and is required to sustain breast cancer xenografts in many mouse models. One such model is the ACI rat, an inbred strain susceptible to breast cancer, and other cancers. The ACI model is the **only** rodent model in which doses of estradiol at luteal phase levels induce breast cancer in 100% of the animals as discussed in 1997 and 2018 by Dr. James D. Shull. (11-12)

However, estrogen does not induce breast cancer in wild-type mice. If it did, estrogen would be used instead of MPA, an endocrine disrupting chemical routinely used to induce breast cancer in mice. This is the well-known MPA mouse model of breast cancer as discussed in 2009 by Claudia Lanari. In 2006, Dr. Jose Russo proclaimed estrogen causes breast cancer in mice, but his experiment used a SCID mouse (severe combined immunodeficient mouse) injected with MCF-10F breast cancer cells pretreated with a high dose 17-beta-estradiol and then injected into SCID mice. According to Drs. Hanahan and Weinberg one of the 10 hallmarks of cancer is immortality. MCF-10F cells are immortal cells. Thus, the MCF-10F cells already demonstrate a feature of cancer even before treatment with high-dose estrogen. Wild-type mice treated with estradiol have been studied, finding estrogen does not reliably induce breast cancer. As mentioned above, if estradiol could induce breast cancer reliably in wild-type mice, this would be a commonly used mouse breast cancer model for study. There is no such model, and has never been found to be the case. In the later stage of his career, Dr. Jose Russo abandoned his attempts to prove estrogen causes breast cancer. Instead, he took up the study of high estrogen levels of pregnancy as breast cancer protective. (13-19)

Estrogen Alone (Premarin) Decreases Breast Cancer by 23 Percent

Now that we have the 11-year follow-up data of the 2004 Second Arm, Premarin-Alone (CEE) study of the Women's Health Initiative,

we can ask another question. The Premarin-alone arm (CEE) showed a 20-27% decrease in breast cancer compared to placebo. As mentioned above, the 18-year follow-up showed the estrogen-alone group (Premarin) enjoyed a 40 percent reduction in mortality from breast cancer. This finding of estrogen (Premarin) as breast cancer protective falsifies the mainstream medical dogma that estrogen causes breast cancer. (20)

Why Less Cancer with Premarin-Alone, and More Cancer with Estradiol Alone?

One might ask the obvious question: Why is Premarin-alone associated with **decreased** (23 percent) breast cancer risk compared to non-hormone users, while **estradiol-alone** (bioidentical) is associated with 29% increase in invasive breast cancer in the 2008 French Cohort study (HR=1.29, p=0.73)? What could explain the difference between estradiol and Premarin? In 2008, Dr. Agnes Fournier asks this question, writing:

> Compared with HRT never-use, use of **estrogen alone [estradiol] was associated with a significant 1.29-fold increased risk** …our finding of a 1.3-fold increased breast cancer risk associated with the use of estrogen alone (almost exclusively estradiol compounds, and mostly administered through the skin) **differs with that of the WHI estrogen-alone trial** which found a decreased risk [23 percent] when **oral conjugated equine estrogens [Premarin]** were used in a population of older and often overweight women. Note: Premarin is CEE, conjugated equine estrogen from pregnant horses. (21)

Monkey Studies Explain the Findings

The answer comes from Dr. Charles E. Wood at the primate center in Winston Salem in a 2008 study treating menopausal macaque monkeys with hormone replacement. Dr. Wood compared Premarin (CEE) to 17-beta estradiol (human estrogen) as hormone replacement in a non-human primate model and found a highly significant 259-330% increase in breast cell proliferation in estradiol-treated monkeys compared to controls. However, the Premarin (CEE) treated monkeys had far less cell proliferation, only a 75% increase compared to controls. This was determined with the KI-67 antigen test of cell proliferation. This difference in cell proliferation, 330 percent for estradiol vs. 75 percent for Premarin explains why estradiol-alone, used in the French Cohort study, is associated with a slightly increased breast cancer risk, while Premarin-alone as used in the WHI second arm, has NO increased risk. The risk was decreased by 23% less breast cancer in the Premarin-alone users in the 2004 WHI second arm. (22)

It is the Estrogen Receptors!

The next logical question is: how can we explain this difference in proliferation for the two estrogens, 17-beta-estradiol vs. Premarin? The answer lies in the estrogen receptor targeting. While ER-alpha is a pro-carcinogenic growth promoter receptor, ER-beta is a tumor suppressor receptor. Estradiol binds to and activates both ER-alpha and ER-beta equally. Premarin preferentially binds to and activates ER-beta by virtue of the unsaturated B-ring steroids in Premarin (CEE, equine estrogen). Dr. Barbara Levy reviewed the role of Premarin (CEE) as an ER-beta ligand. This is due to the unsaturated B-ring steroids contained in Premarin (CEE), writing:

> We now know that there are two unique estrogen receptors, a [alpha] and b [beta], which were cloned in 1996. Although estradiol binds equally to the a and b estrogen receptors, **conjugated equine estrogen binds predominantly to the b [beta] receptors, leading** to overall more potent clinical effects. (23)

In 2008, Dr. Bhagu R. Bhavnani studied equine estrogens (CEE) finding they contain "ring B unsaturated equine estrogens" which selectively bind to and activate ER-beta, writing:

> Our data indicate that some natural estrogens such as the **ring B unsaturated equine estrogens of the type present in the drug CEE [Premarin]** have characteristics that can be useful as **selective ER-beta ligands**. Note: a ligand is an agent that binds to a receptor. (24-25)

Hormones, endocrine disrupting chemicals (EDCs), and plant substances that activate ER-alpha stimulate growth and proliferation, increasing the risk for carcinogenesis. For example, in 2021, the endocrine-disrupting chemical Bisphenol-C was studied by Dr. Xiaohui Liu, finding Bisphenol C the "strongest bifunctional ER-alpha-agonist and ER-beta-antagonist due to magnified halogen bonding". Bisphenol is an endocrine-disrupting chemical (EDC) that causes breast cancer by binding to and activating ER-alpha while inhibiting ER-beta. This is a very bad combination leading to more pronounced carcinogenic effects. (26)

ER-Beta Acts as a Tumor Suppressor

On the other hand, hormones and plant substances that bind to and activate ER-beta are tumor suppressors and can be used in cancer prevention and treatment. Two examples of hormones that preferentially bind to and activate ER-beta are estriol (E3), and the testosterone metabolite called 3-Beta-Diol. This is the basis for ER-beta based breast cancer prevention and a good reason to make sure every post-menopausal hormone replacement program includes estriol and testosterone. In 2020, Dr. Rahul Mal studied ER-beta as a tumor suppressor, downregulating the proliferative effects of estrogen via downregulation of the expression of the oncogene, cyclin D1. Dr. Rahul Mal suggested hormones, drugs or plant extracts that activate ER-beta could serve as a targeted therapy for breast cancer, thus replacing chemotherapy, writing:

High ER-beta expression is associated with improved overall survival in women with breast cancer...The promise of ER-beta activation, as a potential targeted therapy, is based on concurrent activation of multiple tumor suppressor pathways with few side effects compared to chemotherapy. Thus, **ER-beta is a nuclear receptor with broad-spectrum tumor suppressor activity**, which could serve as a potential treatment target in a variety of human cancers including breast cancer...Relative to ER alpha, **ER beta binds estriol and ring B unsaturated estrogens [Premarin, CEE] with higher affinity, while the reverse is true of 17β-estradiol and estrone**. On the other hand, the dihydrotestosterone metabolites 5-androstenediol and 3beta androstanediol **[3Beta-Diol] are relatively selective (3-fold) for ER-beta over ER-alpha**...The cell division protein **cyclin D1** (CCND1), one target of AP-1 and SP1 mediated transcription, is upregulated by ER alpha and induces estrogen-mediated proliferation...Opposing actions and dominance of ER-beta over ER-alpha with respect to activation of **cyclin D1 gene expression** may explain why ER-beta is a negative regulator of the proliferative effects of estrogen... Thus, ER-beta and ER-alpha have shown opposing effects on proliferation and the expression of various oncogenes and tumor suppressors in breast cancer cell lines in the presence of estradiol...ER-beta is unique in that it functions as a tumor suppressor in diverse biologic contexts. **ER-beta has been implicated in various cancer types, including breast, prostate, lung, glioblastoma, thyroid, and ovarian cancer...Thus, ER-beta is a nuclear receptor with broad-spectrum tumor suppressor activity that could serve as a potential treatment target in a variety of human cancers**. (27)

As you might expect, researchers are now gearing up for the next gold rush, the search for drugs and plant extracts that preferentially activate ER-beta. For example, in Chapter 2 on Russell Marker and the Origin of Bioidentical Hormones, we discussed how the plant ste-roid diosgenin inhibits ER-alpha, thus reducing breast cancer risk. Similarly, activation of ER-beta confers anticancer benefits. In 2023, Dr. Sukhbir Singh discussed the soy isoflavone, daidzein, which preferentially activates ER-beta. Preclinical studies show daidzein targets cancer stem cells, having anticancer activity against breast, prostate, and lung cancer, and melanoma. (28-33)

What is the Estrogen Paradox?

The "Estrogen Paradox" was explained in 2008 by Dr. V. Craig Jordan, PhD, a hormone researcher whose name is synonymous with tamoxifen. Estrogen is a growth factor for breast cancer, and this is the rationale for anti-estrogen drugs such as tamoxifen and aromatase inhibitors. Breast cancer cells recruit surrounding fibroblasts in the micro-environment to feed them more estrogen. Quite to the contrary, in women or animals that undergo long-term estrogen deprivation (LTED) and then start estrogen replacement, the estrogen induces apoptosis and programmed cell death in breast cancer cells. This is counterintuitive to the dogma that estrogen causes breast cancer. The LTED can be either five years of menopausal estrogen deficiency or 5 years of an estrogen-blocking drug such as tamoxifen or aromatase inhibitors. Since all the women in the WHI second arm 2004 were estrogen deprived for more than 5 years, this could be another factor to explain the 23 percent reduced breast cancer in the estrogen-alone arm of the WHI study, and the 40-45 percent reduction in mortality from breast cancer in the 18-year follow-up of the WHI second arm study (HR: 0.55), as discussed by Dr. JoAnn Manson in 2017 and Dr. Howard Hodis in 2018. LTED switches estrogen from a survival signal to a death signal. In 2008, Dr. V. Craig Jordan writes:

> An estrogen deprivation gap of 5 years after menopause is required for high-dose estrogen to be an effective treatment for breast cancer. In addition, the same applies to 5 years of adjuvant tamoxifen therapy

when recurrence and mortality continue to decrease after adjuvant tamoxifen treatment is stopped …the paradox, which is maintained throughout the WHI [Women's Health Initiative] evaluation of more than 12 years, is **estrogen causes a decrease in mortality and a decrease in the incidence of new breast cancers. This is counterintuitive to the scientific and medical community** unless one embraces and understands the known clinical evidence that governs safe estrogen use for the treatment of breast cancer after menopause. These were established 70 years ago. (34-39)

Studies Supporting the Estrogen Paradox - The Million Woman Study 2011

In 2011, Dr. Valerie Beral published the Million Woman Study which supports the above Estrogen Paradox described by Dr. V. Craig Jordan. In this study, women who had been long-term estrogen deprived (LTED) for at least 5 years after menopause, and then started estrogen-alone, had no increased risk of breast cancer (RR=1.05). However, if the estrogen-alone therapy had been started immediately after menopause without LTED, there was an increased risk of breast cancer (RR=1.43). Users of estrogen and progestin combinations had even greater risk (RR=1.53). (40)

In 2009, Agnès Fournier's French E3N cohort study also supported the Estrogen Paradox theory of V. Craig Jordan described above. Her data showed the initiation of hormone replacement (HRT) close to menopause without LTED was associated with an increased risk of breast cancer. However, when initiating HRT much later after LTED, there is reduced risk. Additionally, natural progesterone combined with estrogen has no increased breast cancer risk, even without LTED. In 2009, Dr. Agnès Fournier writes:

In conclusion, our results indicate that contrary to what is currently hypothesized regarding the risk of heart disease, initiation of MHT [menopausal hormone therapy] close to menopause onset rather than later may not limit the increased risk of breast

cancer. Instead, even short durations of use of some EP-MHTs [estrogen-progestins] were associated with substantially elevated risks of breast cancer when treatment was initiated close to menopause. Finally, our finding that, for short durations of use around menopause, **progesterone** in EP-MHT **may be safer regarding breast cancer risk than other progestagens** needs to be confirmed in other settings… Our results suggest that, for some EP-MHT, **the timing of treatment initiation transiently modulates the risk of breast cancer; and that, when initiated close to menopause, even short durations of use are associated with an increased breast cancer risk. Estrogen-progesterone [natural progesterone] combinations might be an exception in this regard**. (41)

How to Make the Estradiol Less Proliferative?

What if we are starting hormone replacement soon after menopause and there is no estrogen deprivation gap? Without LTED, estrogen is a growth signal, not a death signal, and will not paradoxically induce apoptosis in breast cancer cells. Now, we have the obvious question: how can we make menopausal hormone replacement safer? How can we make estradiol less proliferative in breast tissue? This brings us to the importance of adding progesterone to an estradiol-alone HRT regimen. For example, commercially available estradiol patches such as Vivelle-Dot and Climara Patch contain estradiol-alone. Simply add progesterone to the estradiol. Oral progesterone is usually taken separately as a capsule or gelcap around bedtime. For women with surgically induced menopause after hysterectomy, the mainstream OB/GYN doctors will routinely give an estrogen-alone regimen without progesterone. These women have undergone hysterectomy and have no uterus, so a progestin drug is not needed for endometrial protection. **Note:** In the 1960's, Premarin alone was found to cause endometrial hyperplasia and endometrial cancer. After this was discovered, a synthetic progestin, medroxyprogesterone (MPA) was added to Premarin in the early 1980s to prevent endometrial cancer. This combination pill is called Prempro. (42-44)

After hysterectomy, estradiol-alone may induce proliferative effects in breast tissue. Reducing breast cell proliferation is the key. This can be done by combining natural progesterone with estradiol. It can also be done by switching from estradiol-alone to the Bi-est/ progesterone combination. My office preferentially uses a Bi-est/ progesterone combination in a compounded topical cream to ensure both hormones are taken together in the proper ratio to prevent endometrial hyperplasia and reduce breast proliferation. In addition, every post-menopausal woman is routinely given an oral micronized progesterone capsule (100-200 mg) nightly. Micronized progesterone is FDA-approved (1988, Solvay) for the prevention of endometrial hyperplasia induced by estrogen therapy, thus preventing endometrial cancer as described in 1993 by Dr. Dean Moyer. **Note:** Bi-est is a compounded estrogen with the formula of 20% estradiol (E2) and 80% estriol E3, first developed by Dr. Jonathan V. Wright from Tahoma Washington. (45-46)

Progesterone: Mechanism of Breast Cancer Prevention

In 2015, Dr. Hisham Mohammed used in vitro and in vivo xenograft studies of ER-alpha positive breast cancer cells to elucidate the protective mechanism by which progesterone acts as a proliferative brake on the ER-alpha receptor, thus downregulating ER-alpha proliferation, writing:

> We now show that PR [progesterone receptor] is not merely an ER-alpha-induced gene target, but is also an **ER-alpha-associated protein that modulates its behavior**. In the presence of agonist ligands, **PR associates with ER-alpha** to direct ER-alpha chromatin binding events within breast cancer cells, resulting in a unique

gene expression program that is associated with good clinical outcomes. Progesterone inhibited estrogen-mediated growth of ER-alpha+ cell line xenografts and primary ER-alpha+ breast tumor explants and had **increased anti-proliferative effects when coupled with an ER-alpha antagonist**. Copy number loss of PgR [progesterone receptor] is a common feature in ER-alpha+ breast cancers, explaining lower PR levels in a subset of cases. Our findings indicate that **PR functions as a molecular rheostat to control ER-alpha chromatin binding and transcriptional activity**, which has important implications for prognosis and therapeutic interventions…We conclude that activation of PR [progesterone receptor] results in a robust association between PR and the ER-alpha complex…**Progesterone blocks ER-alpha+ tumour growth**…PR is a critical determinant of ER-alpha function due to crosstalk between PR and ER-alpha. In this scenario, under estrogenic conditions, **an activated PR functions as a proliferative brake in ER-alpha+ breast tumors** by re-directing ER-alpha chromatin binding and altering the expression of target genes that induce a switch from a proliferative to a more differentiated state. (47)

Breast Cancer Prevention Program

Our breast cancer prevention program includes: A formula that includes Bi-est (80 percent estriol, 20 percent estradiol), testosterone (metabolized to 3-Beta-Diol), and both topical and oral progesterone. See the appendix. All patients are given the breast cancer prevention program: Iodine testing and supplementation when found low. Compounded testosterone topical cream. I3C and DIM (Indole-3-carbinol I3C, and Di-Indole-Methane), and methyl-folate to cover those patients harboring a MTHFR mutation which impairs the functioning of the COMT enzyme. We also do Vitamin D3 and Selenium testing and supplementation when found low. Our target for Vitamin D3 is above 50 ng/mL, and for selenium is above 135 mcg/L. (48)

Typical Compounded Dosing Schedules

In 2011, Dr. Andres D Ruiz provided typical Menopausal HRT Dosing Schedules in Table 1 of her publication. In her table, the words "topical estrogen" is referring to estradiol (E2). I prefer to use Bi-est, a formula devised by Dr. Jonathan Wright, a pioneer in the use of bioidentical hormones. Bi-est contains 80 percent estriol (E3) and 20 percent estradiol (E2). E3 preferentially binds to and activates ER-beta, thus conferring breast cancer protective properties. (48-51)

Typical Compounded HRT Dosing Courtesy of Dr. Andres D. Ruiz

Dose Classification Dose Range

Topical Estrogen:
Low Dose ≤0.5 mg
Moderate Dose 0.51-1 mg
High Dose >1 mg

Oral Estrogen:
Low Dose ≤1 mg
Moderate Dose 1.1-2 mg
High Dose >2 mg

Topical Progesterone:
Low Dose <20 mg
Moderate Dose 21-50 mg
High Dose >50 mg

Oral Progesterone:
Low Dose <100 mg
Moderate Dose 101-200 mg
High Dose >200 mg

Why is Bi-Est Superior to Estradiol Alone?

Bi-Est is a combination of estradiol (E2) and estriol (E3), in a ratio of 20% estradiol, and 80% estriol. Studies show that estriol (E3) preferentially binds to and activates ER-beta, is associated with less breast proliferation, and is thought to be breast cancer preventive. Estriol preferentially binds to and activates ER-beta which is a tumor breast suppressor. Much of the early work on estriol as a breast cancer preventive was done by Dr. Henry Lemon. (52-57)

How Does Estriol (E3) Prevent Breast Cancer?

Estriol and the Estrogen Receptor Beta.

As it turns out, basic science has given us important answers here. As mentioned above, we have two estrogen receptors, ER-alpha cloned in 1986, and ER-beta, cloned in 1996. ER-alpha is associated with breast cell proliferation, while ER-beta prevents proliferation. ER-alpha is pro-carcinogenic, while ER-beta is a tumor suppressor. Estradiol (E2) binds equally to both receptors, whereas estriol (E3) binds preferentially to ER-beta, explaining its protective effect. Notice estrone (E1) is not included in the formula for two reasons. E1 is reversibly converted to estradiol (E2). Estrone (E1) preferentially binds to ER-alpha. Testosterone also has breast cancer protective effects because its metabolite, 3-Beta-Diol, preferentially binds to and activates ER-beta. Synthetic progestins block the androgenic receptor disrupting the beneficial ER-beta cancer suppressive effects of testosterone, thus explaining the carcinogenicity of synthetic progestins. At high doses, however, MPA becomes like an androgen and can be used to treat metastatic breast cancer with a 30 percent response rate, about the same response rate for metastatic breast cancer as DES and tamoxifen. (58-61)

In 2008, Dr. Bao Ting Zhu asks the question: is it necessary to control ER-alpha and ER-beta activation in post-menopausal hormone therapy to achieve optimal outcome? I would say yes, it is necessary to preferentially activate ER-beta while suppressing ER-alpha. Premarin and estriol [E3] are similar in their strong binding to ER-beta. The estradiol metabolite, 2-methoxyestradiol (2MeEO) has strong anti-cancer activity, discussed further in Chapter 21. Dr. Zhu writes:

> All together, it is evident that many of the equine estrogens contained in Premarin have a strong differential binding affinity for human ER-beta over ER-alpha, which is very similar to the human pregnancy estrogen [estriol] E3...The inclusion of methoxyestrogen sulfates in HRT may be beneficial **because of 2-methoxyestradiol's strong anti-tumorigenic activity**... Our recent study showed that endogenous estrogens (such as E1 and 2-OH-E1) present in non-pregnant women **mainly activate**

the ER-alpha system, whereas estrogens (such as E3 and epi-E3) present in pregnant women **predominantly activate the ER-beta system**... It is believed that an optimally-adjusted activation of the ER-alpha and ER-beta signaling systems would help maximize the beneficial effects of HRT, and additionally minimize its untoward effects. (62-64)

Progestins Activate ER-Alpha

In 2012, Dr. Sebastián Giulianelli studied a mouse model, finding medroxyprogesterone (MPA) activates ER-alpha which binds to the PR (progesterone receptor) protein forming an ER-Alpha/PR dimer, which then co-localizes in the breast cell nucleus to the Cyclin D1 and MYC promoter regions. This is very bad thing, because both Cyclin D1 and MYC are oncogenes. These are genes that increase proliferation involved in carcinogenesis. Dr. Sebastián Giulianelli writes:

> We found that treatment with **the progestin medroxyprogesterone acetate (MPA) induced the expression and activation of ER alpha,** as well as rapid nuclear colocalization of activated ER alpha with PR [Progesterone Receptor]...Chromatin immunoprecipitation studies showed that **MPA triggered binding of ER alpha and PR to the CCND1 [Cyclin D1] and MYC promoters**...We confirmed that nuclear colocalization of both receptors also occurred in human breast cancer samples... Together, our findings argued that ER alpha–PR association on target gene promoters is essential for **progestin-induced cell proliferation**. (65-67)

The above 2012 findings of Dr. Giulianelli were confirmed in 2024 by Dr. Meghan S. Perkins with in-vitro studies of MCF-7 breast cancer cells. Dr. Meghan S. Perkins found that all three progestins, medroxyprogesterone acetate (MPA), norethisterone (NET), and levonorgestrel (LNG) had carcinogenic properties. All three progestins are used widely for oral birth control pills and menopausal hormone therapy, writing:

> all progestogens [synthetic progestins] promoted the association of the PR [progesterone receptor] and ER alpha [estrogen receptor alpha] on the promoter of the PR target gene, MYC [a proliferative oncogene], thereby increasing its expression under non-estrogenic and estrogenic conditions...**These progestins are used globally in both contraception and menopausal hormone therapy (MHT)**...MHT containing progestins such as first generation medroxyprogesterone acetate (MPA) or norethisterone (NET), or second generation levonorgestrel (LNG) have been **associated with a higher risk [of breast cancer] than estrogen-only MHT**...Considering that the expression of MYC is often upregulated in breast cancer, and that it plays a role in promoting proliferation, **these results suggest that the progestogens evaluated in this study all promote breast cancer cell proliferation, albeit to different extents, via a mechanism requiring an association of the PR and ER alpha on the MYC promoter**. Similarly, many other studies find synthetic progestins increase breast cell proliferation. (68-76)

Bi-Est is Safer than Estradiol Alone

Now that we know that estriol (E3) preferentially binds to and activates ER-beta, we can confidently recommend the Bi-est (20% estradiol/ 80% estriol). Bi-est combined with both progesterone and testosterone is perhaps the safest HRT regimen currently available. However, we want even more breast cancer protection by adding supplements: iodine (Iodoral), I3C/DIM (Di-Indole-Methane), methyl-folate, Vitamin D3, and selenium supplementation. These supplements are discussed further in Chapter 21 on Iodine, Estrogen Metabolism and 2MeEO.

Protective Effects of Progesterone Compared with Harmful Effects of Medroxyprogesterone – Breast Cell Proliferation

Let us now take a moment to look at a few primate studies comparing the efficacy and adverse effects of natural progesterone to its synthetic progestin counterpart medroxyprogesterone (MPA). How can we explain why synthetic progestins such as medroxyprogesterone (MPA) increase breast cancer risk while natural progesterone is breast cancer protective? In 2007, Dr. Charles E. Wood was interested in how to attenuate breast cell proliferation induced by estradiol in a primate model. Dr. Wood compared the effect of medroxyprogesterone (MPA) to that of progesterone in a primate model of hormone replacement. The estradiol/MPA combination increased breast cell proliferation, **while the estradiol/progesterone combination did not increase proliferation**. Dr Wood writes:

> **Estradiol plus medroxyprogesterone (MPA) significantly increased breast cell proliferation using Ki67 markers.** However, **estradiol with progesterone did not increase cellular proliferation**... These findings suggest that oral micronized progesterone has a more favorable effect on risk biomarkers for postmenopausal breast cancer than medroxyprogesterone acetate. Note: the Ki-67 marker is routinely used as a measure of breast cell proliferation. (77)

Previous primate studies by Dr. Cline in 1998 showed the same increased breast cell proliferation with the addition of medroxyprogesterone (MPA) to Premarin treated monkeys. In 1998, Dr Cline concluded:

> These findings indicate that progestogens (MPA) may exacerbate, not antagonize mammary gland proliferation induced by estrogen (Premarin) replacement therapy. (78)

Protective Effect of Progesterone in Human Studies

The anti-proliferative, protective effect of natural progesterone was confirmed in two placebo controlled randomized double-blind human trials in 1995 by Dr. King-Jen Chang and in 1998 by Dr. Jean-Michel Foidart. Dr. Chang studied 40 post-menopausal women undergoing lumpectomy for a breast mass. A hormone gel was applied topically to the breast daily for two weeks preceding lumpectomy. This hormone gel contained either placebo, estradiol (E2), natural progesterone (P), or a combination of estradiol/progesterone (E2/P). Breast biopsies revealed the women who had topical progesterone (P) gel applied to the breast had decreased proliferation of breast cells. (79)

In 1998, Dr. Foidart studied forty untreated postmenopausal women who were planning cosmetic breast surgery. Natural progesterone was applied to the breast as a topical gel for 14 days before breast surgery, or excision of a benign breast lesion. Dr. Foidart found the progesterone topical gel reduced the proliferative effects of estrogen, thereby explaining its cancer-preventive effect. Breast epithelial cell proliferation is the underlying factor for increasing cancer risk. When proliferation is increased, this increases the risk of breast cancer. When proliferation is suppressed, this decreases the risk of breast cancer. (80)

Natural Progesterone Prevents Breast Cell Hyperplasia

In 2011, Dr. Andres D. Ruiz studied the effectiveness of natural (bioidentical) hormone replacement in menopausal women in a cohort study. Progesterone not only reduces breast epithelial cell hyperplasia, it also acts as a neurosteroid reducing emotional lability, irritability and anxiety, writing:

> Two RCTs [randomized controlled trials] by Chang and colleagues and Foidart and colleagues demonstrated that women receiving topical P4 [natural progesterone]

experienced a reduction in breast epithelial proliferative markers via reductions in mitotic divisions and proliferating cell nuclear antigen (PCNA) labeling index % compared to placebo. **These studies support the belief that P4 [natural progesterone] may prevent breast epithelial hyperplasia**... Women [treated with progesterone] experienced a **25% decrease in emotional lability** (p < 0.01), a **25% decrease in irritability** (p < 0.01), and a **22% reduction in anxiety** (p = 0.01) within 3 to 6 months. (49)

The Invention of Premarin

In the 1940's Premarin was invented. This is an estrogen drug isolated from the urine of pregnant horses called conjugated equine estrogen. Premarin-alone was FDA approved in 1942 without understanding that it caused endometrial hyperplasia and is carcinogenic to the uterine endometrium. Premarin alone was thought safe and handed out freely to millions of women. By 1975, it was discovered that Premarin caused an estimated 15,000 cases of endometrial cancer, representing the largest epidemic of serious iatrogenic disease in the history of medicine. (81-84)

One might think this would be the end of the drug. However, Premarin was promptly rehabilitated with the addition of a synthetic progestin, MPA to prevent endometrial cancer. Thus, in 1995, Prempro was born, a synthetic hormone pill containing both Premarin (CEE, the horse estrogen) and the synthetic progestin medroxyprogesterone (MPA), also called Provera. MPA was FDA-approved on June 18, 1959, and Prempro was FDA-approved on June 4, 2003. (85-86)

More Dr. Charles Wood Monkey Studies on MPA

In 2007, Dr. Charles Wood was interested in what would happen to risk markers for breast cancer when progesterone is substituted for MPA in his primate model, so he studied 25 menopausal macaque primates comparing hormone replacement (HRT) using two different combinations: estradiol E2+MPA compared to estradiol E2+natural progesterone. Breast biopsies were studied in both monkey groups after 2 months of hormone treatment. Dr. Wood found the E2+MPA combination increased proliferative activity not seen with E2+P4 (estradiol with natural progesterone). The E2+MPA combination increased the KI-67 proliferation marker and cyclin B1, but the E2+P4 treatment did not increase proliferation. The E2+ MPA combination increased the oncogene c-Myc expression 2.5-fold when compared to controls. In addition, E2+ MPA increased expression of proliferation genes, epidermal growth factor (EGF), and transforming growth factor alpha (TGFa). In 2007, Dr. Charles Wood writes:

> Estrogen plus progestin [MPA] hormone therapy has been associated with increased breast proliferation, breast density, and breast cancer risk in postmenopausal women, beyond that seen with estrogen alone. The goal of this study was to evaluate progestogen [MPA] effects on gene expression profiles in the breast contributing to this promotional effect. (87)

In 2015, Dr. Valerie Flores reviewed this Charles Wood primate study (above), writing:

> The postmenopausal animals [macaque primates] received one of four treatment regimens, with doses reflecting commonly prescribed doses in MHT [menopausal hormone therapy] for postmenopausal women—placebo, E2 [estradiol] daily, E2+P4 [estradiol plus natural progesterone] daily, or E2+MPA [estradiol plus medroxyprogesterone] daily. After two months of treatment macaques treated with E2+MPA demonstrated **a significant increase in proliferation of breast lobular and ductal cells, compared to placebo; this proliferative activity was not seen with E2+P4 treatment.** There was also increased expression of proliferation markers **Ki67 and cyclin B1 in the E2+MPA treated monkeys, but not in the E2+P4 treatment** group. In a follow-up study using this same animal

model, Wood et al also demonstrated differences in gene expression profiles for a given progestogen treatment...Breast biopsies were collected after two months of treatment, and analyzed for differences in gene expression. It was found that **breast tissue exposed to E2+MPA demonstrated increased expression of genes in the ErbB proliferative pathway—epidermal growth factor (EGF) and transforming growth factor alpha (TGFa)**. Genes of the Jak/Stat signal transduction pathway, including **c-MYC** gene expression were also differentially expressed, with a **2.5 fold change in the E2+MPA treatment when compared to control** (P< 0.01). **cMYC induces signals for cell proliferation, and is known to be involved in tumorigenesis...**Thus... providing further support for MPA's role in promoting breast cell proliferation through growth factor signaling mechanisms. Note: ErbB = erythroblastic oncogene B, is an oncogene, frequently referred to as HER2 (human epidermal growth factor receptor 2). ErbB is a cell surface receptor involved in a signaling cascade which triggers cell proliferation via MAPK and PI3K/Akt. cMyc is also an oncogene. (88)

Progesterone Benefits vs. MPA Adverse Effects

Health benefits of progesterone extend to the cardiovascular system, brain, and neurologic system. In 2011, Dr. Cynthia L. Bethea studied the effects of hormone replacement on mood, finding that improved mood with the estradiol/progesterone combination is abrogated by the estradiol/MPA combination. All the beneficial health outcomes seen with progesterone on the brain, breast, and cardiovascular system are abrogated and reversed by synthetic progestin, MPA, writing:

> The actions of the natural hormones are significantly different from those of Premarin and MPA...**progesterone has potent neuroprotective effects and MPA does not**...We found that in nonhuman primates, estradiol or equine estrogens increase

the expression of tryptophan hydroxylase (TPH), the rate-limiting enzyme in serotonin synthesis...In contrast, supplementation with MPA completely blocked the effect of equine estrogens. Thus, **MPA endangers brain cells and potentially facilitates stress and depression**...MPA reduces the dilatory effect of estrogens on coronary arteries, increases the progression of coronary artery atherosclerosis, accelerates low-density lipoprotein uptake in plaque, increases the thrombogenic potential of atherosclerotic plaques, and promotes insulin resistance and its consequent hyperglycemia in primates, whereas progesterone does not...Most of the research papers showing significantly better outcomes in brain, breast, and cardiovascular parameters with estradiol plus progesterone instead of MPA end with rational statements to the effect that hopefully the data will lead to better hormone therapy. (89)

The Protective Effect of Testosterone in Rhesus Monkeys

In 2000, Dr. Jian Zhou studied ovariectomized rhesus monkeys deficient in estrogen, progesterone, and testosterone treated with a placebo, estradiol (E2) alone, or in combination with progesterone (E2/P), estradiol in combination with testosterone (E2/T), or tamoxifen for 3 days. Dr. Jian Zhou found estradiol alone increased breast epithelial cell proliferation as expected. However, **co-administration of testosterone with estradiol reduced breast epithelial cell proliferation by 40 percen**t indicating testosterone is breast cancer protective. Remember ER-alpha is the proliferative, pro-cancer receptor, while ER-beta is the cancer suppressor receptor. **The co-administration of testosterone completely abolished the estradiol (E2) augmented ER-alpha expression.** This is justification for the addition of testosterone to our hormone replacement program. Dr. Jian Zhou writes:

> Progesterone did not alter E2's proliferative effects, but testosterone reduced E2-induced

proliferation by 40% (P< 0.002) and **entirely abolished E2-induced augmentation of ER-alpha expression**...These observations showing androgen-induced down-regulation of mammary epithelial proliferation and ER expression suggest that **combined estrogen/androgen hormone [testosterone] replacement therapy might reduce the risk of breast cancer associated with estrogen replacement.** (90-91)

Twenty-four years ago, Dr. Jian Zhou (above) made the startling discovery that combining estrogen with testosterone is the optimal formula for menopausal hormone replacement. For this reason, all postmenopausal hormone replacement programs should include testosterone. For more on this, see Chapter 18 on Testosterone for Prevention and Treatment of Breast Cancer.

Benefits of Testosterone in Monkey Study, Dr. Constantine Dimitrakakis

In 2003, Dr. Constantine Dimitrakakis studied mammary epithelial proliferation (MEP) in female ovariectomized monkeys. Blocking the androgen receptor resulted in a two-fold increase in breast cell proliferation, however, adding a small amount of testosterone to the estrogen replacement completely attenuated estrogen-induced stimulation of breast cells, with inhibition of ER (estrogen receptor) signaling to the oncogenic C-Myc pathway, writing:

> We show that androgen receptor blockade in normal female monkeys results in a more than twofold increase in MEP [Mammary Epithelial Proliferation], indicating that endogenous androgens normally inhibit MEP. Moreover, we show that addition of a small, physiological dose of T [testosterone] to standard estrogen therapy almost completely attenuates estrogen-induced increases in MEP in the ovariectomized monkey, suggesting that the increased breast cancer risk associated with estrogen treatment could be reduced by

T supplementation. **Testosterone reduces mammary epithelial estrogen receptor ER-alpha and increases ER-beta expression**, resulting in a marked reversal of the ER-alpha/beta ratio found in the estrogen-treated monkey. Moreover, **T treatment is associated with a significant reduction in mammary epithelial MYC expression**, suggesting that T's antiestrogenic effects at the mammary gland involve alterations in ER signaling to MYC. Conclusions: These findings suggest that treatment with a balanced formulation including all ovarian hormones may prevent or reduce estrogenic cancer risk in the treatment of girls and women with ovarian failure. (92-93)

Conclusion: Studies of hormone-treated monkeys are illuminating, showing synthetic progestins such as MPA are carcinogenic and should be avoided. On the other hand, estriol, progesterone, and testosterone reduce breast epithelial cell proliferation caused by estradiol. This was confirmed in two human RCTs (randomized controlled trials) as mentioned above. Estriol (E3) is breast cancer protective, preferentially binding and activating ER-beta. Considering all this, the optimal menopausal hormone replacement program should include Bi-est (20% estradiol/ 80% estriol), progesterone, and testosterone. The exact hormone formulation matters, as discussed in 2022 by Dr. Haim Abenhaim, Associate Professor of Obstetrics and Gynecology at McGill U. in Montreal. Dr. Abenhaim says hormone formulations containing synthetic progestins increase the risk for breast cancer. Instead, use natural progesterone, writing:

> Although menopausal HT [hormone therapy] use appears to be associated with an overall increased risk of breast cancer, this risk appears **predominantly mediated through formulations containing synthetic progestins.** When prescribing menopausal HT, **micronized progesterone may be the safer progestogen to be used.** (94)

Routine HRT Starting Formula

Our routine HRT starting formula is a compounded topical cream containing:

5 mg/gram Bi-est (20% estradiol / 80% estriol) and 50 mg/gram progesterone. Half a gram of cream is applied topically twice a day providing one gram of cream daily.

A separate topical cream containing: Testosterone 12 mg per gram, a quarter gram of cream is applied topically daily (3 mg testosterone)

The dosage is titrated up or down depending on the patient's response to treatment, so the final maintenance dosage may differ from the starting dosage. (**Note:** For post-menopausal women, progesterone is commonly added to the testosterone cream. However, for young cycling women, progesterone is taken separately on a 12–26-day monthly schedule to avoid suppression of ovulation.

♦ **References for Chapter 3 Do Not Monkey with My Hormones!**

1) Kaplan, Jay R. "Modeling Women's Health with Nonhuman Primates and Other Animals." ILAR journal 45.2 (2004): 83-88.

2) Lieberman, Allan, and Luke Curtis. "In Defense of Progesterone: A Review of The Literature." Alternative Therapies in Health & Medicine 23.7 (2017).

3) Yang, Zhilan, et al. "Estradiol Therapy and Breast Cancer Risk in Perimenopausal And Postmenopausal Women: A Systematic Review And Meta-Analysis." Gynecological Endocrinology 33.2 (2017): 87-92.

4) Viscoli, Catherine M., et al. "A Clinical Trial of Estrogen-Replacement Therapy After Ischemic Stroke." New England Journal of Medicine 345.17 (2001): 1243-1249.

5) Hodis, Howard N., et al. "Hormone Therapy and The Progression of Coronary-Artery Atherosclerosis in Postmenopausal Women." New England Journal of Medicine 349.6 (2003): 535-545.

6) Cherry, Nicola, et al. "Long-Term Safety of Unopposed Estrogen Used by Women Surviving Myocardial Infarction: 14-Year Follow-Up of The ESPRIT Randomised Controlled Trial." BJOG: An International Journal of Obstetrics & Gynaecology 121.6 (2014): 700-705.

7) Writing Group for the Women's Health Initiative Investigators. Risks and benefits of estrogen plus progestin in healthy postmenopausal women: Principal results from the Women's Health Initiative randomized controlled trial. JAMA 2002 Jul 17; 288:321-33.

8) Lanari, Claudia and Lee Malvina, et al. "The MPA Mouse Breast Cancer Model: Evidence for a Role of Progesterone Receptors in Breast Cancer." (2009).

9) Blakely, Collin M., et al. "Hormone-Induced Protection Against Mammary Tumorigenesis Is Conserved in Multiple Rat Strains and Identifies a Core Gene Expression Signature Induced by Pregnancy." Cancer research 66.12 (2006): 6421-6431.

10) Rajkumar, Lakshmanaswamy, et al. "Hormone-Induced Protection of Mammary Tumorigenesis in Genetically Engineered Mouse Models." Breast Cancer Research 9 (2007): 1-11.

11) Shull, James D., et al. "Ovary-intact, But Not Ovariectomized Female ACI Rats Treated With 17beta-Estradiol Rapidly Develop Mammary Carcinoma." Carcinogenesis 18.8 (1997): 1595-1601.

12) Shull, James D., et al. "Rat Models Of 17β-Estradiol-Induced Mammary Cancer Reveal Novel Insights Into Breast Cancer Etiology And Prevention." Physiological Genomics 50.3 (2018): 215-234.

13) Russo, Jose, et al. "17-Beta-estradiol Induces Transformation and Tumorigenesis in Human Breast Epithelial Cells." The FASEB journal 20.10 (2006): 1622-1634.

14) Russo, Jose, et al. "The Protective Role of Pregnancy in Breast Cancer." Breast Cancer Research 7 (2005): 1-12.

15) Russo, Jose, and Jose Russo. "The Physiological Basis of Breast Cancer Prevention." The Future of Prevention and Treatment of Breast Cancer (2021): 129-161.

16) Hanahan, Douglas, and Robert A. Weinberg. "Hallmarks of Cancer: The Next Generation." cell 144.5 (2011): 646-674.

17) Liu, Chong, et al. "Advances in Rodent Models for Breast Cancer Formation, Progression, And Therapeutic Testing." Frontiers in Oncology 11 (2021): 593337.

18) Jerry, D. Joseph. "Roles for Estrogen and Progesterone in Breast Cancer Prevention." Breast Cancer Research 9 (2007): 1-3.

19) Lanari, Claudia and Lee Malvina, et al. "The MPA Mouse Breast Cancer Model: Evidence for A Role Of Progesterone Receptors In Breast Cancer." (2009).

20) Chlebowski, Rowan T., et al. "Association of Menopausal Hormone Therapy with Breast Cancer Incidence and Mortality During Long-Term Follow-Up of The Women's Health Initiative Randomized Clinical Trials." JAMA 324.4 (2020): 369-380.

21) Fournier, Agnès, et al. "Unequal Risks for Breast Cancer Associated with Different Hormone Replacement Therapies: Results from The E3N Cohort Study." Breast Cancer Research and Treatment 107 (2008): 103-111.

22) Wood, Charles E., et al. "Comparative Effects of Oral Conjugated Equine Estrogens and Micronized 17β-Estradiol on Breast Proliferation: A Retrospective Analysis." Menopause 15.5 (2008): 978-983.

23) Levy, Barbara, and James A. Simon. "A Contemporary View of Menopausal Hormone Therapy." Obstetrics & Gynecology 144.1 (2024): 12-23)

24) Bhavnani, Bhagu R., and Frank Z. Stanczyk. "Pharmacology Of Conjugated Equine Estrogens: Efficacy, Safety and Mechanism of Action." The Journal of steroid biochemistry and molecular biology 142 (2014): 16-29.

25) Bhavnani, Bhagu R., et al. "Structure Activity Relationships and Differential Interactions And Functional Activity Of Various Equine Estrogens Mediated Via Estrogen Receptors (ERs) ERα and ERβ." Endocrinology 149.10 (2008): 4857-4870.

26) Liu, Xiaohui, et al. "Bisphenol-C Is the Strongest Bifunctional ER-alpha-agonist and ER-beta-antagonist Due To Magnified Halogen Bonding." PloS one 16.2 (2021): e0246583.

27) Mal, Rahul, et al. "Estrogen Receptor Beta (ER-beta): A Ligand Activated Tumor Suppressor." Frontiers In Oncology 10 (2020): 587386.

28) Singh, Sukhbir, et al. "Unveiling the pharmacological and nanotechnological facets of daidzein: Present state-of-the-art and future perspectives." Molecules 28.4 (2023): 1765.

29) Ranjithkumar, R., et al. "Novel daidzein molecules exhibited anti-prostate cancer activity through nuclear receptor ERβ modulation, in vitro and in vivo studies." Journal of Chemotherapy 33.8 (2021): 582-594.

30) Kumar, Vinod, and Shyam S. Chauhan. "Daidzein Induces Intrinsic Pathway Of Apoptosis Along With ER A/B Ratio Alteration And ROS Production." Asian Pacific Journal of Cancer Prevention: APJCP 22.2 (2021): 603.

31) Rajabi, Sadegh, et al. "Anti-Breast Cancer Activities Of 8-Hydroxydaidzein by Targeting Breast Cancer Stem-Like Cells." Journal of Pharmacy & Pharmaceutical Sciences 23 (2020): 47-57.

32) Jin, S., et al. "Daidzein Induces MCF-7 Breast Cancer Cell Apoptosis Via the Mitochondrial Pathway." Annals of Oncology 21.2 (2010): 263-268.

33) Choi, Eun Jeong, and Gun-Hee Kim. "Daidzein causes cell cycle arrest at the G1 and G2/M phases in human breast cancer MCF-7 and MDA-MB-453 cells." Phytomedicine 15.9 (2008): 683-690.

34) Jordan, V. Craig. "Molecular mechanism for breast cancer incidence in the Women's Health Initiative." Cancer Prevention Research 13.10 (2020): 807-816.

35) Jordan, V. Craig. "The 38th David A. Karnofsky lecture: the paradoxical actions of estrogen in breast cancer—survival or death?" Journal of Clinical Oncology 26.18 (2008): 3073-3082.

36) Jordan, V. Craig. "The new biology of estrogen-induced apoptosis applied to treat and prevent breast cancer." Endocrine-related cancer 22.1 (2015): R1-R31.

37) Mehta, Jaya, Juliana M. Kling, and JoAnn E. Manson. "Risks, benefits, and treatment modalities of menopausal hormone therapy: current concepts." Frontiers in endocrinology 12 (2021): 564781.

38) Hodis, Howard N., and P. M. Sarrel. "Menopausal hormone therapy and breast cancer: what is the evidence from randomized trials?" Climacteric 21.6 (2018): 521-528.

39) Manson, JoAnn E., et al. "Menopausal hormone therapy and long-term all-cause and cause-specific mortality: the Women's Health Initiative randomized trials." Jama 318.10 (2017): 927-938.

40) Beral, Valerie, et al. "Breast Cancer Risk In Relation To The Interval Between Menopause And Starting Hormone Therapy." Journal of the National Cancer Institute 103.4 (2011): 296-305.

41) Fournier, Agnès, et al. "Estrogen-Progestagen Menopausal Hormone Therapy And Breast Cancer: Does Delay From Menopause Onset To Treatment Initiation Influence Risks?." Journal of Clinical Oncology 27.31 (2009): 5138-5143.

42) Weiderpass, Elisabete, et al. "Risk of endometrial cancer following estrogen replacement with and without progestins." Journal of the National Cancer Institute 91.13 (1999): 1131-1137.

43) Rodriguez, Adriana C., et al. "Estrogen signaling in endometrial cancer: a key oncogenic pathway with several open questions." Hormones and Cancer 10 (2019): 51-63.

44) Razavi, Pedram, et al. "Long-term postmenopausal hormone therapy and endometrial cancer." Cancer epidemiology, biomarkers & prevention 19.2 (2010): 475-483.

45) Moyer, Dean L., et al. "Prevention of Endometrial Hyperplasia by Progesterone During Long-Term Estradiol Replacement: Influence Of Bleeding Pattern And Secretory Changes." Fertility and sterility 59.5 (1993): 992-997.

46) Ye, Amanda. "Bi-Est 80/20 Cream." US Pharm 48.9 (2023): 59-60.-382.

47) Mohammed, Hisham, et al. "Progesterone Receptor Modulates ER-alpha Action In Breast Cancer." Nature 523.7560 (2015): 313-317.

48) Marsden, Tracy. "Bioidentical Hormone Replacement: Guiding Principles for Practice." Nat Med J 2.3 (2010): 2010-03. https://www.naturalmedicinejournal.com/journal/bioidentical-hormone-replacement-guiding-principles-practice

49) Ruiz, Andres D., et al. "Effectiveness of Compounded Bioidentical Hormone Replacement Therapy: An Observational Cohort Study." BMC women's health 11 (2011): 1-10.

50) Wright, Jonathan V. "Bio-Identical Steroid Hormone Replacement: Selected Observations from 23 Years of Clinical and Laboratory Practice." Annals of the New York Academy of Sciences 1057.1 (2005): 506-524.

51) Friel, Patrick N., Christa Hinchcliffe, and Jonathan V. Wright. "Hormone Replacement With Estradiol: Conventional Oral Doses Result In Excessive Exposure To Estrone." Altern Med Rev 10.1 (2005): 36-41.

52) Head, Kathleen A. "Estriol: Safety And Efficacy." Alternative Medicine Review: A Journal Of Clinical Therapeutic 3.2 (1998): 101-113.

53) Lemon, Henry M. "Antimammary Carcinogenic Activity Of 17-Alpha-Ethinyl Estriol." Cancer 60.12 (1987): 2873-2881.

54) Lemon, Henry M. "Pathophysiologic Considerations in The Treatment Of Menopausal Patients With Estrogens; The Role Of Estriol In The Prevention Of Mammary Carcinoma." Acta endocrinologica. Supplementum 233 (1980): 17-27.

55) Lemon, H. M. "Clinical and Experimental Aspects of The Anti-Mammary Carcinogenic Activity Of Estriol." Frontiers of Hormone Research 5 (1977): 155-173.

56) Lemon, Henry M. "Estriol Prevention of Mammary Carcinoma Induced By 7, 12-Dimethylbenzanthracene and Procarbazine." Cancer Research 35.5 (1975): 1341-1353.

57) Tzingounis, V. A., et al. "The Significance of Ostriol In The Management Of The Post-Menopause." Acta endocrinologica. Supplementum 233 (1980): 45-50.

58) Bentel, Jacqueline M., et al. "Androgen Receptor Agonist Activity of The Synthetic Progestin, Medroxyprogesterone Acetate, In Human Breast Cancer Cells." Molecular and cellular endocrinology 154.1-2 (1999): 11-20.

59) Paruthiyil, Sreenivasan, et al. "Estrogen Receptor Beta Inhibits Human Breast Cancer Cell Proliferation and Tumor Formation By Causing A G2 Cell Cycle Arrest." Cancer research 64.1 (2004): 423-428.

60) Warner, Margaret, and Jan-Åke Gustafsson. "The Role of Estrogen Receptor B (ER-Beta) In Malignant Diseases—A New Potential Target for Antiproliferative Drugs in Prevention and Treatment of Cancer." Biochemical And Biophysical Research Communications 396.1 (2010): 63-66.

61) Muss, H. B., and J. M. Cruz. "High-Dose Progestin Therapy for Metastatic Breast Cancer." Annals of Oncology 3 (1992): S15-S20.

62) Zhu, Bao Ting. "Is It Necessary to Control the Level of Estrogen Receptor alpha and beta Activation In Postmenopausal Hormone Replacement Therapy To Achieve The Optimal Outcome?" Molecular Medicine Reports 1.1 (2008): 15-20.

63) Zhu, Bao Ting, et al. "Quantitative Structure-Activity Relationship of Various Endogenous Estrogen Metabolites for Human Estrogen Receptor A and B Subtypes: Insights into the Structural Determinants Favoring a Differential Subtype Binding." Endocrinology 147.9 (2006): 4132-4150.

64) Zhou, Jian, et al. "Testosterone Inhibits Estrogen-Induced Mammary Epithelial Proliferation And Suppresses Estrogen Receptor Expression." The FASEB Journal 14.12 (2000): 1725-1730.

65) Giulianelli, Sebastián, et al. "Estrogen Receptor Alpha Mediates Progestin-Induced Mammary Tumor Growth by Interacting with Progesterone Receptors at The Cyclin D1/MYC Promoters." Cancer research 72.9 (2012): 2416-2427.

66) Dhanasekaran, Renumathy, et al. "The MYC Oncogene—The Grand Orchestrator of Cancer Growth and Immune Evasion." Nature Reviews Clinical Oncology 19.1 (2022): 23-36.

67) Kim, Jong Kyong, and J. Alan Diehl. "Nuclear Cyclin D1: An Oncogenic Driver in Human Cancer." Journal Of Cellular Physiology 220.2 (2009): 292-296.

68) Perkins, Meghan S., et al. "Upregulation of an estrogen receptor-regulated gene by first generation progestins requires both the progesterone receptor and estrogen receptor alpha." Frontiers in Endocrinology 13 (2022): 959396.

69) de Lignières B. Effects of progestogens on the postmenopausal breast. Climacteric 2002; 5:229–35.

70) Campagnoli C, Clavel-Chapelon F, Kaaks R, et al. Progestins and progesterone in hormone replacement therapy and the risk of breast cancer. J Steroid Biochem Mol Biol 2005; 96:95–108.

71) Ory K, Lebeau J, Levalois C, et al. Apoptosis inhibition mediated by medroxyprogesterone acetate treatment of breast cancer cell lines. Breast Cancer Res Treat 2001; 68:187–98. 554

72) Hofseth LJ, et al. Hormone Replacement Therapy with Estrogen or Estrogen Plus Medroxyprogesterone Acetate Is Associated with Increased Epithelial Proliferation in The Normal Postmenopausal Breast. J Clin Endocrinol Metab 1999; 84:4559–65.

73) Jeng MH, Parker CJ, Jordan VC. Estrogenic Potential of Progestins in Oral Contraceptives Stimulate Human Breast Cancer Cell Proliferation. Cancer Res 1992; 52:6539–46.

74) Kalkhoven E, et al. Synthetic Progestins Induce Proliferation of Breast Tumor Cell Lines Via the Progesterone or Estrogen Receptor. Mol Cell Endocrinol 1994; 102:45–52.

75) Asi, Noor, et al. "Progesterone vs. Synthetic Progestins and The Risk of Breast Cancer: A Systematic Review and Meta-Analysis." Systematic Reviews 5 (2016): 1-8.

76) Mastorakos, G., et al. "Progestins and The Risk of Breast Cancer." Acta Endocrinologica (Bucharest) 17.1 (2021): 90.

77) Wood, Charles E., et al. "Effects of estradiol with micronized progesterone or medroxyprogesterone acetate on risk markers for breast cancer in post-menopausal monkeys." Breast cancer research and treatment 101 (2007): 125-134.

78) Cline, J. Mark, et al. "Effects of Conjugated Estrogens, Medroxyprogesterone Acetate, And Tamoxifen on The Mammary Glands of Macaques." Breast cancer research and treatment 48 (1998): 221-229.

79) Chang, King-Jen, et al. "Influences of percutaneous administration of estradiol and progesterone on human breast epithelial cell cycle in vivo." Fertility and sterility 63.4 (1995): 785-791.

80) Jean-Michel Foidart, M. D., et al. "Estradiol and progesterone regulate the proliferation of human breast epithelial cells." Fertility and Sterility 69.5 (1998): 963-969.

81) Jick, Hershel, et al. "The Epidemic of Endometrial Cancer: A Commentary." American Journal of Public Health 70.3 (1980): 264-267.

82) Ziel, Harry K., and William D. Finkle. "Increased Risk of Endometrial Carcinoma Among Users of Conjugated Estrogens." New England journal of medicine 293.23 (1975): 1167-1170.

83) Antunes, Carlos MF, et al. "Endometrial Cancer and Estrogen Use: Report of A Large Case-Control Study." New England Journal of Medicine 300.1 (1979): 9-13.

84) Key, T. J. A., and M. C. Pike. "The Dose-Effect Relationship Between 'Unopposed' Estrogens And Endometrial Mitotic Rate: Its Central Role In Explaining And Predicting Endometrial Cancer Risk." British journal of cancer 57.2 (1988): 205-212.

85) Cagnacci, Angelo, and Martina Venier. "The Controversial History Of Hormone Replacement Therapy." Medicina 55.9 (2019): 602.

86) Cho, Leslie, et al. "Rethinking Menopausal Hormone Therapy: For Whom, What, When, And How Long?" Circulation 147.7 (2023): 597-610.

87) Wood, Charles E., et al. "Effects of estradiol with micronized progesterone or medroxyprogesterone acetate on risk markers for breast cancer in post-menopausal monkeys." Breast cancer research and treatment 101 (2007): 125-134

88) Flores, Valerie A., and Hugh S. Taylor. "The effect of menopausal hormone therapies on breast cancer: avoiding the risk." Endocrinology and Metabolism Clinics 44.3 (2015): 587-602.

89) Bethea, Cynthia L. "MPA: Medroxy-Progesterone Acetate Contributes to Much Poor Advice for Women." (2011): 343-345.

90) Zhou, Jian, et al. "Testosterone Inhibits Estrogen-Induced Mammary Epithelial Proliferation and Suppresses Estrogen Receptor Expression." The FASEB Journal 14.12 (2000): 1725-1730.

91) Paruthiyil, Sreenivasan, et al. "Estrogen Receptor B Inhibits Human Breast Cancer Cell Proliferation and Tumor Formation By Causing A G2 Cell Cycle Arrest." Cancer research 64.1 (2004): 423-428.

92) Dimitrakakis, Constantine, et al. "A Physiologic Role for Testosterone in Limiting Estrogenic Stimulation Of The Breast." Menopause 10.4 (2003): 292-298.

93) Zhou J, Ng S, et al. Testosterone Inhibits Estrogen-Induced Mammary Epithelial Proliferation and Suppresses Estrogen Receptor Expression. FASEB J 2000;14(12):1725-1730.

94) Abenhaim, Haim A., et al. "Menopausal Hormone Therapy Formulation and Breast Cancer Risk." Obstetrics & Gynecology 139.6 (2022): 1103-1110.

Chapter 4

The Safety of Transdermal Estrogen, Part One

A Case of Chronic Calf Swelling

REBECCA, A FORTY-YEAR-OLD HOUSEWIFE CAME into the office to see me about various complaints related to mood, poor sleep, and weight gain. Rebecca had been taking birth control pills for the past twenty years because of irregular menstrual cycles. At the price of chemical castration, the pills did indeed restore the semblance of regular menstrual cycles. Oral contraceptive pills (OCPs) prevent ovulation, thus serving as birth control. About two years ago, Rebecca took a long airplane flight to a vacation destination, and the next day while out shopping, Rebecca felt an aching, heavy feeling in her right calf, significantly larger than the left. When she returned home, she went to her doctor who ordered an ultrasound scan of the deep veins of the legs and a CAT scan of the abdomen and pelvis, all of which were negative for acute deep venous thrombosis, and no treatment was offered.

A Case of Healed Deep Vein Thrombosis

I explained to Rebecca her calf swelling and discomfort are symptoms of a healed clot formation, clot lysis in the deep vein that has resolved causing incompetent valves, and chronic venous insufficiency. This is a complication of the birth control pills, the aftermath of an episode of deep venous thrombosis of the deep vein of the right calf vein. The blood clot has dissolved, and the vein has recanalized, explaining why the imaging tests showed normal findings. As bad as this may be, it is a far better outcome than pulmonary embolus and stroke caused by birth control pills, a catastrophic outcome. (1)

OCPs – Ninefold Increase in Stroke

One of the major adverse effects of oral birth control pills is increased coagulation and blood clot formation which may cause DVT (deep venous thrombosis), pulmonary embolus, and stroke. OCP use is associated with a ninefold increased risk of stroke, also called cerebral infarction. These strokes are usually caused by blood clots originating in deep leg veins that break loose and enter the systemic circulation. Usually, the blood clots get stuck in the pulmonary circulation, called pulmonary embolus. However, in some people there is a patent foramen ovale which allows the clot to enter the systemic circulation. **Note:** foramen ovale is an opening from the right atrium to the left atrium. Oral contraceptives alter platelet aggregation, enhance antithrombin III activity, decrease serum antithrombin levels, and increase levels of certain coagulation factors, especially factor VII. Additional risk factors for venous thromboembolism (VTE) are: oral estrogen pills, some oral synthetic progestins, smoking, drinking, obesity, hypertension, and diabetes. (2-3)

Adverse Effect is a Reason to Discontinue OCPs

An adverse effect such as clot formation is certainly a good reason to discontinue oral contraceptive pills and switch to natural alternatives to replace the OCPs. Once stopping OCP's ovulation usually resumes uneventfully. However, some women will continue to have anovulation after discontinuing OCPs. To restore ovulation and regular menstrual cycles, natural cyclic progesterone (used on days 12-26), myoinositol, and chaste berry (vitex) can be helpful. Treatment with cyclic progesterone capsules or creams for days 12-26 of the menstrual cycle may help with mood disorders. Good thyroid function is required for regular ovulatory cycles, and thyroid medication may help. For those who want alternatives to OCPs, non-hormonal methods of birth control commonly used are barrier methods with condoms, cervical diaphragm, cervical cap, contraceptive gel, spermicide, and contraceptive sponge. (4-9)

WHI Study Halted Because of Increased Stroke

As mentioned above, the 2002 Women's Health Initiative Study (WHI), the first arm using the oral Premarin and oral medroxyprogesterone (MPA) found a **41 percent increased incidence of stroke** in hormone users compared to controls. Again, this is caused by increased clot formation induced by oral estrogen pills (Premarin, CEE) and oral MPA. (10)

Making the Switch to Transdermal Estrogen

How can we make hormone replacement safer, and eliminate the risk of blood clots? Oral estrogen pills increase blood clot formation, and risk for DVT and stroke because of the first-pass metabolism through the liver. The portal venous system carries oral estrogen directly from the GI tract to the liver, leading to increased hepatic production of clotting factors, factor VII, factor X, and fibrinogen. A safer form of delivery of estrogen is the transdermal (topical) form. The estrogen is applied topically in a skin cream or skin patch, thus avoiding the first pass through the liver. (11-12)

In 2022, Dr. Shuyuan Huang reviewed the medical literature on the role of hormones in ischemic stroke, finding that transdermal estrogen combined with micronized progesterone was the safest and most preferred route. The use of HRT declined from 26.9% to just 4.7% from 1999 to 2020. In 2023, it declined further to only 1.8% of menopausal women in the U.S. using hormone replacement therapy (HRT). In 2023 there were about 63 million women over the age of 50. If we apply the 1.8 percent to these 63 million post-menopausal

women, then we have 1,134,000 HRT users in 2023. About one-third of these use transdermal estrogen/micronized progesterone, and two-thirds (748,440) use oral Premarin/MPA. Dr. Shuyuan Huang estimates if all 748,440 HRT users switched from oral estrogen/MPA pills to transdermal estrogen/micronized progesterone, this would **prevent about 2,245 ischemic strokes per year**. A reasonable public health measure would be for the Secretary of Health and Human Services (HHS) to instruct the National Institute of Health (NIH) to fund a media campaign to advise all women on HRT and their physicians to make this switch, realizing this will prevent up to 2,245 ischemic strokes per year. Dr. Shuyuan Huang writes:

> Estrogen and progestin with ischemic stroke...The Women's Health Initiative (WHI) showed that estrogen (E) plus progestin (P) increased the risk of ischemic stroke in generally healthy post-menopausal women. However, altering the route of hormone administration and the type of hormone may remedy this drawback. Encouraging hormone therapy users to switch from oral to transdermal estrogen and from synthetic [MPA] to micronized progesterone **reduced the risk of ischemic stroke by ≤ 3000 per million hormone therapy users per year**. Note: MPA=medroxyprogesteone. (13-15)

Transdermal Estrogen and Micronized Progesterone Safest

In 2016, Dr. Marianne Canonico conducted a case-controlled study using the French National Health Insurance database over 3 years to identify 3,144 ischemic strokes and 12,158 controls. Compared to non-user controls, oral estrogen users had a 58 percent increase in ischemic stroke. However, transdermal estrogen users had a 17 percent reduction in ischemic stroke compared to non-user controls. Oral micronized and transdermal natural progesterone have no increase in ischemic stroke, however, the norpregnane derivative synthetic progestins have an increased risk of 225 percent. Dr. Canonico agrees with Dr. Shuyuan Huang (above). She encourages all oral estrogen users to switch to transdermal estrogen and micronized natural progesterone, writing:

> Our findings suggest that **transdermal estrogens might be the safest option** for short-term hormone therapy use…Our study provides new findings on the potential safety of transdermal estrogens in relation to risk of IS [ischemic stroke]. It also suggests for the first time the importance of concomitant progestogens in determining the risk of IS. Taken together with findings for thrombotic risk, **transdermal estrogens alone or combined with micronized progesterone may be the best option** to improve the benefit/risk ratio of HT use and may represent **the safest option with respect to both VTE [Venous Thrombo-Embolism] and stroke risk. Based on these observations, ≤3000 cases of these pathologies per year per million HT users could be avoided by encouraging them to change from oral to transdermal estrogens and from synthetic progestins to micronized progesterone**. (16)

In 2012, Dr. Sian Sweetland from the U.K. studied venous thromboembolism (VTE) risk in relation to different types of post-menopausal hormone therapy, "The Million Women Study." This study included 1,058,259 postmenopausal UK women followed by National Health Service hospital admission and death records. Dr. Sian Sweetland found the greatest risk of VTE with the oral estrogen/medroxyprogesterone (MPA) combination. **Note:** This same formula was used in 2002, WHI study, first arm, explaining the increased venous thrombosis and stroke in the 2002 WHI study. Getting back to Dr. Sweetland's study in the U.K., oral estrogen users' risk for VTE was more than double that of non-HRT users (RR= 2.67). However, transdermal estrogen alone was the safest with no increased risk of VTE (RR=0.82). Dr. Sian Sweetlan writes:

> During 3.3 million years of follow-up, 2200 women had an incident VTE [venous thrombo embolism]…the

risk was significantly greater for oral estrogen-progestin than oral estrogen-only therapy (RR = 2.07), **with no increased risk with transdermal estrogen-only therapy (RR=0.82).** Among users of oral estrogen-progestin, the risk from HT varied by progestin type, with significantly **greater risks for preparations containing medroxyprogesterone acetate** than other progestins (RR=2.67 [2.25–3.17] vs. 1.91 [1.69–2.17]… Over 5 years, 1 in 660 who had never used HT were admitted to hospital for (or died from) pulmonary embolism, compared with 1 in 475 current users of oral estrogen-only HT, 1 in 390 users of estrogen-progestin HT containing norethisterone/norgestrel, and **1 in 250 users of estrogen-progestin HT containing medroxyprogesterone acetate**. (17)

In 2024, Dr. Andrea Genazzani from Pisa, Italy discussed the position paper from the FIGO (International Federation of Gynecology and Obstetrics) on the benefits and risks of menopausal hormone therapy, finding transdermal estrogens do not increase the risk of venous thromboembolism, while oral estradiol and oral Premarin do so. In addition, oral synthetic progestins such as norethisterone and MPA [medroxyprogesterone] do increase the risk of VTE, writing:

> **Transdermal estrogens do not increase the risk of venous thromboembolism (VTE), while oral estradiol, and particularly CEE [Premarin], do**…Another critical determinant of thrombotic risk is the type of progestogens associated with estrogens used by women with an intact uterus. Indeed, **micronized progesterone and dydrogesterone do not increase the risk**, but **norpregnanes, namely nomegestrol acetate and promegestone, norethisterone, as well as MPA [medroxyprogesterone], do increase the risk….Therefore, the use of transdermal estrogens and, where indicated, micronized progesterone or dydrogesterone should be preferred** in women who have an increased baseline thrombotic risk. Note: Premarin is CEE, conjugated equine estrogen.

Dydrogesterone an isomer of natural progesterone, is commonly used in Europe and not used in the U.S. (18)

In 2019, Dr. Yana Vinogradova did a nested case-controlled study using two large UK (England) databases finding 80,396 women with a diagnosis of venous thromboembolism (VTE) between 1998 and 2017, and 391,494 female controls. The use of oral estrogen therapy had a 58 percent increased risk of venous thromboembolic disease compared with non-user controls. Transdermal estrogen had no risk, writing:

> In the present study, **transdermal treatment was the safest type** of hormone replacement therapy when risk of venous thromboembolism was assessed. Transdermal treatment appears to be underused, with the overwhelming preference still for oral preparations. (19)

Inherited Thrombophilia – Oral Estrogen Pills Contra-Indicated

Thrombophilia is a genetic mutation that increases the risk of blood clots. The two most common mutations are the Leiden mutation of factor V and the G20210A mutation of prothrombin. Indeed, many women suffering blood clots and strokes while on birth control pills have an underlying genetic thrombophilia, an inherited tendency to form clots, placing them at greater risk. Additional risk factors for VTE are smoking, drinking, obesity, hypertension, and diabetes. Women with genetic clotting disorders should avoid birth control pills (BCPs), oral estrogen, and all other medications that cause blood clot formation. One might argue the case for routine screening for thrombophilia in all young women before starting oral birth control pills. In addition, women suffering from VTE or blood clot formation while on BCPs or any oral form of HRT (hormone replacement therapy) should have thrombophilia screening testing. Screening is not necessary for transdermal estrogen and oral progesterone users, since they have no increased risk of VTE from the HRT. In 2005, Dr. Joseph Caprini from Northwestern reported on thrombophilia screening in 166 women presenting with venous thrombosis and found one quarter, about 23 percent, had a Factor V Leiden genetic mutation. Dr. Caprini's Thrombophilia Testing Panel includes: Factor V Leiden (FVL), Prothrombin 20210A mutation (P2), methylene tetrahydrofolate reductase deficiency (MTHFR), fasting serum homocysteine (HC), lupus anticoagulant (LA), anticardiolipin antibodies (ACA), antithrombin deficiency (AT), protein S deficiency (PS), and protein C deficiency (PC). (20-23)

Cerebral Vein Thrombosis Associated with Inherited Thrombophilia

Dr. Ida Martinelli reported in the 1998 NEJM on 40 patients with idiopathic cerebral-vein thrombosis. All 40 patients were screened for inherited thrombophilia, finding twenty percent were carriers of the prothrombin-gene mutation, and fifteen percent had the Factor V mutation. Dr. Martinelli advises against the use of oral contraceptives, and oral estrogens in these patients. (24)

Conclusion: Oral estrogen pills are associated with increased coagulation, blood clots, venous thromboembolism (VTE), and stroke. However, the use of transdermal estrogen at standard dosages is safe, and not associated with increased risk. Some synthetic progestins are associated with an increased risk of VTE, while oral micronized and transdermal natural progesterone are safe and do not cause increased clotting. For example, oral medroxyprogesterone (MPA) doubles the risk of VTE. Third-generation synthetic progestins found in BCPs, desogestrel, gestodene, and drospirenone are even worse. They increase the risk of VTE six times more than non-users. A reasonable public health measure would be for the Secretary of Health and Human Services (HHS) and National Institute of Health (NIH) to fund a media campaign advising all women taking oral estrogen/MPA and their physicians to switch to transdermal estrogen and natural progesterone. If all users made this switch, then 2,245 ischemic strokes would be prevented annually. If you are looking for a good public health measure, this is it. (25-27)

♦ References For Chapter 4 The Safety of Transdermal Estrogen Part One

1) Masuda, Elna M., et al. "The Natural History of Calf Vein Thrombosis: Lysis of Thrombi and Development of Reflux." Journal of Vascular Surgery 28.1 (1998): 67-74.

2) Bevan, Heather, Khema Sharma, and Walter Bradley. "Stroke in young adults." Stroke 21.3 (1990): 382-386.

3) Li, Feng, et al. "Oral Contraceptive Use And Increased Risk Of Stroke: A Dose–Response Meta-Analysis Of Observational Studies." Frontiers in Neurology 10 (2019): 993.

4) van Die, M. Diana, et al. "Vitex Agnus-Castus Extracts For Female Reproductive Disorders: A Systematic Review Of Clinical Trials." Planta medica 79.07 (2013): 562-575.

5) Jarry, Hubertus, et al. "Evidence For Estrogen Receptor B-Selective Activity Of Vitex Agnus-Castus And Isolated Flavones." Planta medica 69.10 (2003): 945-947.

6) Reiser, Elisabeth, et al. "Non-Hormonal Treatment Options For Regulation Of Menstrual Cycle In Adolescents with PCOS." Journal of Clinical Medicine 12.1 (2022): 67.

7) Bizzarri, Mariano, et al. "Myo-inositol and D-chiro-inositol as modulators of ovary steroidogenesis: A narrative review." Nutrients 15.8 (2023): 1875.

8) Placidi, Martina, et al. "Myo-Inositol and Its Derivatives: Their Roles in the Challenges of Infertility." Biology 13.11 (2024): 936.

9) Asante, Naana Boadiwaa, et al. "Barrier Methods of Contraception." Conception and Family Planning-New Aspects. IntechOpen, 2024.

10) Wassertheil-Smoller, Sylvia, et al. "Effect of Estrogen Plus Progestin on Stroke In Postmenopausal Women: The Women's Health Initiative: A Randomized Trial." Jama 289.20 (2003): 2673-2684.

11) Trenor III, Cameron C., et al. "Hormonal Contraception and Thrombotic Risk: A Multidisciplinary Approach." Pediatrics 127.2 (2011): 347-357.

12) Mehta, Jaya, Juliana M. Kling, and JoAnn E. Manson. "Risks, Benefits, And Treatment Modalities of Menopausal Hormone Therapy: Current Concepts." Frontiers In Endocrinology 12 (2021): 564781.

13) Huang, Shuyuan, et al. "Research Progress on The Role of Hormones in Ischemic Stroke." Frontiers In Immunology 13 (2022): 1062977.

14) Gass, Margery LS, et al. "Use of compounded hormone therapy in the United States: report of the North American Menopause Society Survey." Menopause 22.12 (2015): 1276-1285.

15) Thompson, Jennifer Jo, et al. "Why women choose compounded bioidentical hormone therapy: lessons from a qualitative study of menopausal decision-making." BMC women's health 17 (2017): 1-18.

16) Canonico, Marianne, et al. Postmenopausal Hormone Therapy and Risk of Stroke: Impact of The Route of Estrogen Administration and Type of Progestogen. Stroke (2016) 47(7):1734–41.

17) Sweetland, Sian, et al. "Venous Thromboembolism Risk In Relation To Use Of Different Types Of Postmenopausal Hormone Therapy In A Large Prospective Study." Journal of Thrombosis and Haemostasis 10.11 (2012): 2277-2286.

18) Genazzani, Andrea R., et al. "Counseling in Menopausal Women: How To Address The Benefits And Risks Of Menopause Hormone Therapy. A FIGO Position Paper." South African General Practitioner 5.1 (2024): 14-28.

19) Vinogradova, Yana, Carol Coupland, and Julia Hippisley-Cox. "Use of Hormone Replacement Therapy and Risk Of Venous Thromboembolism: Nested Case-Control Studies Using the QResearch and CPRD databases." BMJ 364 (2019).

20) Caprini, Joseph A., et al. "Thrombophilia Testing In Patients With Venous Thrombosis." European Journal Of Vascular And Endovascular Surgery 30.5 (2005): 550-555.

21) Dalen, James E. "Should Patients With Venous Thromboembolism Be Screened For Thrombophilia?." The American Journal Of Medicine 121.6 (2008): 458-463.

22) Cosmi, B., and S. Coccheri. "Thrombophilia In Young Women Candidate To The Pill: Reasons For And Against Screening." Pathophysiology of Haemostasis And Thrombosis 32.5-6 (2002): 315-317.

23) Andreassi, Maria Grazia, et al. "Factor V Leiden, Prothrombin G20210A Substitution and Hormone Therapy: Indications for Molecular Screening Testing." Laboratoriums Medizin 30.5 (2006): 317-325.

24) Martinelli, Ida, et al. "High risk of cerebral-vein thrombosis in carriers of a prothrombin-gene mutation and in users of oral contraceptives." New England Journal of Medicine 338.25 (1998): 1793-1797.

25) Lidegaard, Øjvind, et al. "Risk of Venous Thromboembolism from Use Of Oral Contraceptives Containing Different Progestogens and Estrogen doses: Danish Cohort Study, 2001-9." Bmj 343 (2011).

26) Cockrum, Richard H., et al. "Association of Progestogens and Venous Thromboembolism Among Women of Reproductive Age." Obstetrics & Gynecology 140.3 (2022): 477-487.

27) Scarabin, Pierre-Yves, et al. "Effects of Oral and Transdermal Estrogen/Progesterone Regimens On Blood Coagulation And Fibrinolysis In Postmenopausal Women: A Randomized Controlled Trial." Arteriosclerosis, thrombosis, and vascular biology 17.11 (1997): 3071-3078.

Chapter 5

Transdermal vs. Oral Estrogen Part Two

ELLEN IS A 53-YEAR-OLD SCHOOLTEACHER seeking a second opinion on her menopausal hormone therapy. Three years ago, Ellen began to experience typical menopausal vasomotor symptoms of hot flashes and night sweats, and her primary care doctor prescribed oral contraceptive pills (OCPs) for her. Two months ago, her doctor informed Ellen she had high blood pressure and added a blood pressure pill to her medication list. I explained to Ellen that OCPs should not be used for menopausal hormone replacement, as they increase CRP (C-reactive protein), cortisol, blood pressure, and the risk of thromboembolic disease. (1-4)

Using Oral Contraceptive Pills to Manage Menopause

Treatment with oral contraceptive pills (OCPs) is not recommended as hormone therapy (HT) in post-menopausal (PM) women because it causes blood clots and hypertension. In 2021, Dr. Felice Gersh writes:

> Of note, **oral contraceptives contain ethinyl estradiol and various progestins, are thrombophilic [induces blood clots], and predispose to hypertension**...Oral contraceptives are not recommended as HT [Hormone Therapy] in PM [Post Menopausal] women and must be used cautiously, with individualized risk management, during the menopausal transition. (5-11)

The previous chapter 4, explores the studies showing transdermal estrogen delivery is the preferred delivery method, and oral estrogen pills are to be avoided. In Chapter 5, we explore further the studies explaining why oral estrogen pills are associated with hypertension, hypercoagulability, and inflammation. Thus, they cause the observed increase in blood clots, deep venous thrombosis, pulmonary embolus, and stroke in young women on OCPs. The same blood clotting abnormality seen with OCPs applies to oral estrogen and medroxyprogesterone (MPA) pills. On the other hand, transdermal estrogen and micronized progesterone do not cause such effects, making them the safer choice. (12)

Bioidentical HRT- Comparing Oral Vs. Transdermal Estrogen

In 1997, Dr. Pierre-Yves Scarabin from France, studied bioidentical HRT in post-menopausal women. He compared oral to transdermal estrogen delivery in 45 post-menopausal women in Paris, France. The women were given either oral or transdermal estradiol, along with natural progesterone, and various parameters of the coagulation system were studied. Dr. Scarabin writes:

> ...that oral estrogen/progesterone replacement therapy may result in coagulation activation and increased fibrinolytic potential, whereas opposed transdermal estrogen appears without any substantial effects on hemostasis. (13)

In 2018, Dr. Scarabin wrote an updated meta-analysis of the medical literature comparing the oral estrogen route to the transdermal route, finding that oral estrogen carries a 48 percent increased risk of venous thromboembolism (VTE), with a relative risk (RR) of 1.48. The added progestogen agent (progesterone vs MPA) may increase the risk for VTE. The transdermal estradiol combined with progesterone is the safest, with no increased risk of VTE. However, when oral estrogen (CEE) is combined with a synthetic progestin such as MPA (CEE +MPA), the **risk of VTE goes up almost 3-fold**. (RR 2.77). Dr. Scarabin writes:

Among women using estrogen-only preparations, oral but not transdermal preparations increased VTE risk (relative risk (RR) **1.48**, 95% confidence interval (CI) 1.39-1.58; RR 0.97, 95% CI 0.87-1.09, respectively)...In transdermal estrogen users, there was no change in VTE risk in women using micronized progesterone (**RR 0.93**, 95% CI 0.65-1.33), whereas norpregnane derivatives were associated with increased VTE risk (**RR 2.42**, 95% CI 1.84-3.18). Among women using opposed oral estrogen, there was higher VTE risk in women using **medroxyprogesterone acetate (RR 2.77**, 95% CI 2.33-3.30) than in those using other progestins. **These clinical findings, together with consistent biological data, emphasize the safety advantage of transdermal estrogen combined with progesterone**... (14)

In 2020, Dr. Pierre-Yves Scarabin responded to an editorial by Dr. Felice Gersh, reiterating the transdermal route avoids first pass through the liver, is not associated with clot formation, and is the safest way to administer estrogen. Estrogen delivery by oral pills is associated with first pass through the liver, increasing thrombogenicity and clot formation. Dr. Scarabin says women should be encouraged to use estrogen and progesterone as the preferred hormones, and transdermal as the preferred route of delivery, writing:

Oral but not trans-dermal estrogens result in a hepatic first-pass effect that may induce reversible prothrombotic changes in hemostatic variables, including resistance to activated protein C. Thrombotic process plays a critical role in the development of both venous and arterial diseases. European studies have clearly shown the advantage of transdermal estrogens with respect to the risk of venous thromboembolism and probably also stroke. The type of progestogens has also emerged as an important determinant of thrombosis. Progesterone has no effect on blood coagulation, and it is the safest progestogen with respect to thrombosis.

Since thrombosis is one main serious adverse effect of hormone therapy, **women should be encouraged to use transdermal estrogens combined with progesterone**. (15)

Avoiding Hepatic First Pass Metabolism

In a 2009 report, Dr. Kopper speculates that transdermal estrogen is safer and more effective than oral estrogen. Dr. Kopper writes:

Transdermal drug delivery may mitigate some of these effects by avoiding gut and hepatic first-pass metabolism. Advantages of transdermal delivery include the ability to administer unmetabolized estradiol directly to the blood stream, administration of lower doses compared to oral products, and minimal stimulation of hepatic protein production. (16)

Elevated CRP with Oral Estrogen

In 2002, Dr. Andrea Decensi from Milan, Italy studied changes in CRP (C-reactive protein, a non-specific inflammatory marker) in 189 post-menopausal women randomized to either oral CEE (Premarin) or transdermal estradiol (50 mcg/day skin patch). Sequential medroxyprogesterone was added to each group. After 12 months, the CRP had increased by 64 percent in the oral Premarin-treated group. In contrast, transdermal estradiol did not elevate CRP after 12 months of use. Dr. Andrea Decensi writes:

Relative to baseline, CRP increased at 12 months... **64%** for oral CEE... In contrast transdermal E2 **does not elevate CRP levels** up to 12 months of treatment. ...The increase in C-reactive protein (CRP) during oral conjugated equine estrogen (CEE) may explain the initial excess of cardiovascular disease observed in clinical studies. (17)

Oral Estrogen has Fibrinogen Levels Five Times Lower than Transdermal

In 2008, Dr. Matteo Lazzeroni from Milan, Italy studied post-menopausal women given either oral or transdermal estrogen, oral CEE

(Premarin) vs. placebo, and transdermal estradiol (E2, 50 mcg/day) vs. placebo. Sequential MPA (medroxyprogesterone10 mg/day) was also given to each treatment group. 55 women were randomized in each group. The oral estrogen group (CEE + MPA) had a **5-fold decrease in fibrinogen** compared to the transdermal estradiol group, (-5.7% vs, -1.1%). A decrease in fibrinogen levels indicates activation of the clotting system with consumption of clotting factors, and increased clot formation. This is a bad thing, leading to venous thromboembolism (VTE). Dr. M. Lazzeroni concludes that transdermal estrogen is safer and preferable to oral estrogen, writing:

> After 12 months, there was a statistically significant effect of the route of administration of hormone replacement therapy (HRT) on fibrinogen levels; the median percentage change being −5.7% with CEE and −1.1% with E2 (p = 0.012)...**Our data indicate that transdermal E2 [estradiol] may be preferable to oral CEE [Premarin] based on its safer cardiovascular risk profile**. (18)

More on CRP by Dr. Koh

In 2006, Dr. Kwang Kon Koh from Korea studied the effects of oral estrogen on blood clotting compared to transdermal estrogen. Dr. Kwang Kon Koh found oral estrogen increases the CRP (C-reactive protein), a nonspecific inflammatory marker, for up to 3 years after stopping the estrogen pill. However, transdermal estrogen does not increase CRP. The decrease in CRP by transdermal estrogen correlates with increasing estrogen blood levels. The higher the estrogen level, the greater the decline in CRP, indicating the anti-inflammatory properties of estrogen. Dr. Koh writes:

> Orally administered estrogens increase blood CRP levels and maintain sustained increases for up to 3 years. By contrast, transdermal estradiol in healthy postmenopausal women significantly lowers CRP levels or remains

unchanged. In premenopausal women, CRP correlates inversely with blood estradiol concentrations during the menstrual cycle. (19)

Oral Estrogen Increases Markers of Hypercoagulability

In 2001, Dr. Satu Vehkavaara from Helsinki Finland studied 27 post-menopausal women using either oral or transdermal estrogen, finding oral estrogen increased CRP and hypercoagulability while transdermal estrogen had no such effects, writing:

> oral estradiol changed markers of coagulation towards hypercoagulability and increased serum CRP concentrations. Transdermal estradiol or placebo had no effects on any of these parameters. (20)

A Three-Year Study Using Bio-Identical Hormones in 75 Menopausal Women

In 2013, Dr. Kenna Stephenson studied 75 peri/postmenopausal women treated over three years with topical Bi-est (80 percent estriol, 20 percent estradiol) combined with natural progesterone. This is the same formula devised by Dr. Jonathan V. Wright who pioneered the use of Bi-est. Dr. Kenna Stephenson measured cardiovascular biomarkers, hemostatic, inflammatory, immune signaling factors, quality-of-life measures, and health outcomes. Unlike the Premarin/MPA combination used in the WHI trial, Dr. Kenna Stephenson's study used topical Bi-est (80 percent estriol, 20 percent estradiol) and natural progesterone over three years. And, unlike Premarin and MPA used in the WHI study which is known to increase CRP (inflammatory markers), increase blood clotting, and increase venous thromboembolism (VTE), the transdermal bioidentical hormones resulted in the opposite. The inflammatory and coagulation markers were reduced, and there were no adverse side effects over three years in 75 women, writing:

Subjects receiving compounded transdermal bioidentical hormone therapy showed significant favorable changes in: ...fasting glucose, fasting triglycerides, ...C-reactive Protein, fibrinogen, Factor VII, Factor VIII, Insulin-like Growth Factor 1...Antithrombin III levels were significantly decreased at 36 months...Administration of compounded transdermal bioidentical hormone therapy in doses targeted to physiologic reference ranges administered in a daily dose significantly relieved menopausal symptoms in peri/postmenopausal women. Cardiovascular biomarkers, inflammatory factors, immune signaling factors, and health outcomes were favorably impacted... The therapy did not adversely alter the net prothrombotic potential, and there were no associated adverse events...Our study supports the efficacy of compounded transdermal BHRT [Bioidentical Hormone Replacement Therapy] for both symptom reduction and benefits on inflammatory, immune and cardiometabolic pathways without a net increase in thrombotic risks. Although it has been asserted that compounded transdermal BHRT should be thought of as inducing the same adverse health risks as medroxyprogesterone acetate and conjugated equine estrogens [used in the WHI study], **our data does not support this with respect to the hemostatic, inflammatory, immune, and cardiovascular factors examined**. (21-22)

In 2017, Dr. Marc L'Hermite from Brussels, Belgium reviewed menopausal hormone replacement concluding the combination of transdermal estradiol with oral micronized progesterone is safest and most preferred. These are available at the corner drug store, avoiding the compounding pharmacy. However, Dr. Hermite ignores the role of estriol and testosterone which are both ER-beta ligands and therefore breast cancer protective. In my opinion, these should be incorporated into the HRT program, which means using a compounding pharmacy. Dr. Marc L'Hermite writes:

combining oral micronized progesterone with transdermal estradiol can presently be considered as the optimal MHT [Menopausal Hormone Therapy]. It is not only safer than custom-compounded bioidentical hormones but also than oral conventional MHT and has the best breast profile; registered products for such optimal MHT are available around the world and must be preferred.

Note: Estradiol skin patch and oral progesterone is far better than premarin and medroxyprogesterone, however, I would disagree with part of the statement. Using a compounded formulation of estradiol and natural progesterone that includes estriol and testosterone, both ER-beta ligand tumor suppressors is safer than the estradiol skin patch and oral progesterone from the local pharmacy. (23)

Conclusion: It is mind-boggling to think that OCPs are still being prescribed for menopausal hormone replacement. OCPs are synthetic, hormone-disrupting chemicals that should never be used as menopausal hormone replacement. There is now overwhelming evidence in the medical literature that transdermal estrogen combined with natural progesterone is the safest formula with no increase in VTE and no increase in CRP, the inflammatory marker. However, the oral estrogen/MPA combination increases the risk for venous thromboembolism by almost three-fold. The 2013 study by Dr. Kenna Stephenson treating 75 post-menopausal women with transdermal Bi-est and natural progesterone over three years with no adverse effects, revealed dramatic reductions in inflammatory and thrombotic markers confirming the safety and superiority of transdermal bioidentical hormones compared to oral estrogen/MPA combination. **Note:** oral estrogen/MPA was used in the first arm 2002 WHI study. VTE=Venous Thromboembolism, MPA=medroxyprogesterone. For more on OCPs, see Chapter 25. (21)

♦ **References for Chapter 5 Transdermal vs. Oral Estrogen Part Two**

1) de Souza, Ingrid Soares, et al. "Association Between the Use of Oral Contraceptives and The Occurrence of Systemic Hypertension: A Systematic Review With Statistical Comparison Between Randomized Clinical Trial Interventions." European Journal of Obstetrics & Gynecology and Reproductive Biology: X 22 (2024): 100307.

2) Dreon, Darlene M., et al. "Oral Contraceptive Use and Increased Plasma Concentration Of C-Reactive Protein." Life Sciences 73.10 (2003): 1245-1252.

3) Boisseau, Nathalie, et al. "Oral Contraception but Not Menstrual Cycle Phase Is Associated with Increased Free Cortisol Levels and Low Hypothalamo-Pituitary-Adrenal Axis Reactivity." Journal of Endocrinological Investigation 36 (2013): 955-964.

4) Masama, Coleka, et al. "Hormone Contraceptive Use in Young Women: Altered Mood States, Neuroendocrine and Inflammatory Biomarkers." Hormones and Behavior 144 (2022): 105229.

5) Gersh, Felice L., et al. "Postmenopausal Hormone Therapy for Cardiovascular Health: The Evolving Data." Heart 107.14 (2021): 1115-1122.

6) Liu, Hui, et al. "Association Between Duration of Oral Contraceptive Use and Risk Of Hypertension: A Meta-Analysis." The Journal of Clinical Hypertension 19.10 (2017): 1032-1041.

7) Cameron, Natalie A., et al. "Oral contraceptive Pills and Hypertension: A Review Of Current Evidence And Recommendations." Hypertension 80.5 (2023): 924-935.

8) Zuhaira, Ulul Azmi, et al. "Meta-Analysis the Effect of Oral Combination Contraceptive on Hypertension and Stroke." (2022): 520-531.

9) Liu, Hui, et al. "Association Between Duration of Oral Contraceptive Use and Risk Of Hypertension: A Meta-Analysis." The Journal of Clinical Hypertension 19.10 (2017): 1032-1041.

20) Laragh, John H. "Oral Contraceptive Hypertension." Postgraduate Medicine 52.3 (1972): 98-105.

11) Laragh, John H. "Oral Contraceptive-Induced Hypertension—Nine Years Later." American Journal Of Obstetrics And Gynecology 126.1 (1976): 141-147.

12) Writing Group for the Women's Health Initiative Investigators. "Risks and Benefits of Estrogen Plus Progestin in Healthy Postmenopausal Women: Principal Results from The Women's Health Initiative Randomized Controlled Trial." JAMA 288.3 (2002): 321-333.

13) Scarabin, Pierre-Yves, et al. "Effects of Oral and Transdermal Estrogen/Progesterone Regimens on Blood Coagulation and Fibrinolysis In Postmenopausal Women: A Randomized Controlled Trial." Arteriosclerosis, Thrombosis, And Vascular Biology 17.11 (1997): 3071-3078.

14) Scarabin P-Y. Progestogens and Venous Thromboembolism In Menopausal Women: An Updated Oral Versus Transdermal Estrogen Meta-Analysis. Climacteric 2018;21:341–5.

15) Scarabin, Pierre-Yves, et al. "Menopause and Hormone Therapy in the 21st Century: Why Promote Transdermal Estradiol and Progesterone?" Heart 106.16 (2020): 1278-1.

16) Kopper, Nathan W., et al. "Transdermal Hormone Therapy in Postmenopausal Women: A Review of Metabolic Effects and Drug Delivery Technologies." Drug Design, Development and Therapy (2009): 193-202.

17) Decensi, Andrea, et al. "Effect of Transdermal Estradiol and Oral Conjugated Estrogen On C-Reactive Protein in Retinoid-Placebo Trial In Healthy Women." Circulation 106.10 (2002): 1224-1228.

18) Lazzeroni, Matteo, et al. "The Effect of Transdermal Estradiol or Oral Conjugated Estrogen and Fenretinide Versus Placebo on Hemostasis and Cardiovascular Risk Biomarkers In A Randomized Breast Cancer Chemoprevention Trial." Ecancermedicalscience 2 (2008).

19) Koh, Kwang Kon, and Byung-Koo Yoon. "Controversies Regarding Hormone Therapy: Insights From Inflammation and Hemostasis." Cardiovascular Research 70.1 (2006): 22-30.

20) Vehkavaara, Satu, et al. "Effects of Oral and Transdermal Estrogen Replacement Therapy on Markers of Coagulation, Fibrinolysis, Inflammation and Serum Lipids And Lipoproteins In Postmenopausal Women." Thrombosis And Haemostasis 85.04 (2001): 619-625.

21) Stephenson, Kenna, et al. "The Effects of Compounded Bioidentical Transdermal Hormone Therapy on Hemostatic, Inflammatory, Immune Factors; Cardiovascular Biomarkers; Quality-Of-Life Measures; And Health Outcomes In Perimenopausal And Postmenopausal Women." International Journal of Pharmaceutical Compounding 17.1 (2013): 74-85.

22) Wright, Jonathan V., and John Morgenthaler. Natural Hormone Replacement: For Women Over 45. Smart Publications, 1997.

23) L'Hermite, Marc. "Bioidentical Menopausal Hormone Therapy: Registered Hormones (Non-Oral Estradiol±Progesterone) Are Optimal." Climacteric 20.4 (2017): 331-338.

Chapter 6

Bioidentical Hormones for Breast Cancer Survivors Part One

Morning Rounds with Dr. Steven Economou at Rush Hospital Breast Cancer Surgeon

BACK IN 1977 IN CHICAGO, a Rush Hospital breast surgeon named Steven Economou was making rounds with his entourage of interns and residents. I was one of the interns in his group. Wearing long white coats and brandishing stethoscopes, we followed Dr. Economou from room to room. Dr. Economou was chief of surgery specializing in breast cancer surgery, and one of the surgical floors of Rush Hospital was reserved for women having breast cancer surgery. Steven Economou MD had a daily ritual of morning rounds on the surgical floor with his entourage of interns and residents in training. As we bobbed in and out of hospital rooms, examining patients and reviewing charts, Dr. Economou enjoyed stumping the house staff with difficult medical questions, as if they were small darts deftly thrown to an imaginary target on our foreheads. Just as if it was yesterday, I can remember one such question he asked me:

Dr. Dach, does estrogen cause breast cancer?

Note: my last name is pronounced "Dash." I had just finished medical school, and I did not know the answer with any certainty. As was my custom in those days, I offered a plausible answer based on accepted knowledge learned in lectures. I replied:

Of course, estrogen causes breast cancer. Estrogen stimulates the growth of breast tissue, and any growth stimulation will cause cancer.

Lucky for me, this was the acceptable mainstream answer, then, and still is today. The reality is that this question is open to debate.

I assumed Dr. Economou would provide the answer. To my surprise, he was silent and offered no reply. He changed the subject and continued to the next room. Fast forward 47 years and medical science is still looking for the answer, continuing to debate the question. Sadly, Dr. Steven G. Economou passed away at the age of 84 in 2007. (1-3)

Does Estrogen Cause Breast Cancer? It All Depends. NO – Estrogen Does NOT Cause Breast Cancer

For the sake of argument, let us accept the hypothesis estrogen does not cause breast cancer, as suggested in 2023 by Dr. Bin Ke who wrote:

no association was identified between estradiol, progesterone, and the risk of breast cancer. (4)

Is the above statement by Dr. Bin Ke true? What if I told you estrogen reduces mortality from breast cancer by 45 percent? Would you believe that? The proof is data from the 2002 and 2004 Women's Health Initiative (WHI) randomized clinical trials. In 2011, Dr. Andrea Z. LaCroix published the data on the 11-year follow-up of the Women's Health Initiative (WHI), estrogen-only, second-arm. Paradoxically, rather than causing increased breast cancer, the estrogen-only arm of the WHI study found a 23 percent reduction in breast cancer in women using estrogen-alone (CEE) (after hysterectomy). In 2018, Dr. Howard Hodis reviewed the 18-year follow-up of this same WHI estrogen-alone, second arm (2004) showing a **45 percent reduction in breast cancer mortality**. For more, see Chapter 7 on Errors in Modern Medicine: The Fear of Estrogen. (5-8)

Yes, Estrogen Does Cause Breast Cancer in-Vitro and in-Vivo (in Mice)

Let us suppose estrogen **does cause** breast cancer. For example, Dr. Jose Russo devoted most of his science career to the hypothesis that estrogen causes breast cancer. In 2006, Dr. Russo treated breast epithelial cells (ER-alpha negative) with estrogen (17-beta-estradiol) in-vitro, finding the cells transformed into cancer cells. When injected into severe combined immune-deficient mice (SCID), the treated cells grew tumors. Dr. Jose Russo writes:

> Altogether our data indicate that 17-beta-estradiol is able to induce complete neoplastic transformation of human breast epithelial cells, as proven by the formation of tumors in SCID [severe combined immunodeficient] mice...The induction of complete transformation of MCF-10F [breast epithelial cells] cells in vitro confirms the carcinogenicity of E2 [estradiol], **supporting the concept that this hormone [estradiol E2] could act as an initiator of breast cancer in women**. Note: MCF-10F cells are immortalized human breast epithelial cells (HBEC) negative for ER-alpha but positive for ER-beta. Estrogen receptor alpha (ER-alpha) is present in normal breast tissue and breast cancer cells, and is absent in MCF-10F cells and all the estradiol-induced tumors in SCID mice. (9-12)

There are a few problems with Dr. Russo's above study. Number one, the starting breast epithelial cells (MCF-10F) are abnormal. Normal cells have both ER-alpha and ER-beta (receptors). These cells have been genetically manipulated to lack ER-alpha, thus making these cells more prone to malignant transformation. Breast cancer cells that have mutated into an ER-alpha negative cell type are more aggressive and carry a higher mortality rate. In 2020, Dr. Emma Zattarin writes:

> patients with neoplasms undergoing a conversion from ER-alpha+ to ER-alpha– status had a 48% increase in the risk of death when compared with BC [breast cancer] patients with stable ER-alpha+ status. (12)

Number two, these starting breast cells have been "immortalized" through laboratory manipulation. One of Drs. Hanahan and Weinberg's hallmark of cancer is the immortality of the cancer cell, so one may consider these MCF-10F cells as already taking the first step to carcinogenesis. Number three, SCID mice are not normal mice. They suffer from severe combined immunodeficiency. If estrogen causes breast cancer, one would think that treating wild-type mice with estrogen would induce breast cancer. This was studied in 2021 by Dr. Chong Liu who reviewed the various breast cancer models in mice finding the simple administration of estrogen to a wild-type mouse was insufficient to create a viable model of breast cancer. The mouse models for breast cancer require an inbred mouse, a genetically engineered mouse, or the use of carcinogenic chemicals for breast cancer induction. For example, inbred Noble (Nb) and August–Copenhagen–Irish (ACI) rats are uniquely susceptible to mammary cancer induction by estradiol. However, female Copenhagen (COP) and Brown Norway (BN) rats are resistant. The mechanism of carcinogenesis is thought to be the genotoxic effects of catechol estrogens and quinone metabolites leading to DNA adduct formation as described in 1997 by Dr. Ercole L Cavalieri. In 2024, Dr Raquel Nicotra studied rat models of hormone-induced breast cancer, writing:

> hormone-induced rat mammary tumors might be derived from the genotoxic effects of E1/E2 [estrone and estradiol] **quinone metabolites**, previously shown to trigger mutations responsible for initiating various human cancers. (13-16)

Estrogen Protects Against Breast Cancer

Although inbred Noble (Nb) and August–Copenhagen–Irish (ACI) rats are good models for hormone-induced breast cancer, other rodent models show hormones prevent breast cancer. In 2014, Drs. Arumugam and Rajkumar studied ovariectomized mice xenografted with human breast cancer cells, after which the mice were treated with various combinations of natural hormones, estrogen, progesterone, testosterone, and DHEA. Dr. Rajkumar found the **natural hormone-treated mice had maximum reduction in tumor growth, and better outcomes than the AI (aromatase inhibitor) treated mice.** This finding raises questions about the dogmatic use of anti-estrogen drugs for breast cancer prevention. In 2007, Dr. Lakshmanaswamy Rajkumar used genetically modified mice that developed spontaneous breast cancer to show that administering hormones, estrogen and progesterone, confers protection from breast cancer. **Note:** Aromatase inhibitors are estrogen-blocking drugs commonly used by oncologists to treat breast cancer patients. (17-19)

11-Year Follow Up - First Arm 2002 WHI

In 2010 (JAMA), Dr. Rowan T. Chlebowski published the 11-year follow-up data for the WHI 2002, first arm of the Women's Health Initiative (WHI) treating 10,739 postmenopausal women randomized to either placebo or Prempro (Premarin plus medroxyprogesterone, MPA). The hormone-treated group had a 25 percent increase in invasive breast cancer compared to the placebo group. The breast cancer in the Prem-Pro treated group tended to be more aggressive, with 78% more lymph node invasion. Mortality from breast cancer was doubled in the Prempro group compared to placebo. **Note:** Premarin is CEE, conjugated equine estrogen derived from pregnant horses. MPA is medroxyprogesterone, a synthetic progestin. Previous observational studies show synthetic progestins such as MPA increase risk for breast cancer. In 2000, Dr. Catherine Schairer published in JAMA the Breast Cancer Detection Demonstration Project (BCDDP) showing increased risk of breast cancer with progestin use (MPA, HR=1.4). In 1996, Dr. Ingemar Persson published the Swedish Record Review

showing increased breast cancer with a synthetic progestin (HR=1.4). In 2003, Dr. Valerie Beral from the U.K. published the Million Woman Study showing the greatest increase in breast cancer risk with progestin use. (20-23)

Estrogen is OK, Medroxyprogesterone is NOT OK

A critical look at the two WHI studies (2002 and 2004) leads us to the obvious conclusion, that estrogen alone (CEE) does not increase breast cancer risk and may even be protective while adding a progestin, medroxyprogesterone (MPA), a synthetic chemically altered form of progesterone to the estrogen, increases breast cancer risk. Not only does MPA induce breast cancer in laboratory animals, but these cancers tend to be more aggressive and are deadlier. In 2000, Dr. Ronald K. Ross at the Norris Comprehensive Cancer Center in Los Angeles did a population-based, case-control study of 30,1996 women from LA County, of which there were 1,897 breast cancer patients. Dr. Ronald K. Ross found menopausal hormone replacement (HRT) with estrogen alone had no increased risk for breast cancer, while estrogen plus progestin (MPA) has a 24-38 percent increased risk. Dr. Ronald K. Ross writes:

> This study provides strong evidence that the addition of progestin [medroxyprogesterone, MPA] to HRT [Hormone Replacement Therapy] enhances markedly the risk of breast cancer relative to estrogen use alone. Note: Dr. Ross did not specify the exact type of estrogen in his study, however, it is most likely Dr. Ross' patients used oral Premarin (CEE) which was in common use in the U.S. at the time. (24-27)

Dr. Cheblowski - 3 RCTs of Estrogen Alone Reduce Risk of Breast Cancer

In 2015, Dr. Rowan Chlebowski reviewed the 13-year follow-up data from the first arm 2002 WHI study showing combined estrogen (CEE) with MPA increased breast cancer risk (HR, 1.37). On the other hand, the estrogen-alone second arm 2004 WHI study showed reduced breast cancer (HR, 0.79 and HR, 0.55) compared to placebo. Dr. Rowan Chlebowski writes:

> In the estrogen alone trial, the HR for invasive breast cancer risk was lower than 1 throughout the intervention phase (HR, 0.79; 95% CI, 0.61-1.02) and remained lower than 1 in the early postintervention phase (HR, 0.55; 95% CI, 0.34-0.89). (28)

These findings were replicated by two other randomized controlled studies (RCT) showing similar (42-53 percent) reductions in breast cancer with estrogen alone, writing:

> A lower breast cancer risk with estrogen alone has received support from other randomized trials. In the Estrogen for the Prevention of Re-Infarction Trial (ESPRIT) in 1017 women, after 14 years follow-up there were fewer breast cancers in the unopposed estrogen (**estradiol valerate**, 2 mg/d) group (7 vs 15 [**HR, 0.47**; 95% CI, 0.19-1.15]) [Cherry, 2014]. Similarly, in a smaller randomized trial in Denmark, in 192 women there were fewer breast cancers in the estrogen alone (17β-**estradiol**, 2 mg/d) group (10 vs 17 [**HR, 0.58**; 95% CI, 0.27-1.29]) [Schierbeck, 2012]. (28-30)

MPA Disrupts Androgen Signaling

As you will see in Chapter 18 on Testosterone as Breast Cancer Prevention, testosterone is breast cancer preventive primarily because its metabolite 3-Beta-Diol preferentially binds to and activates ER-beta, the tumor suppressor. A second mechanism is testosterone downregulates ER-alpha. In 2007, Dr. Stephen N. Birrell discovered that synthetic progestins such as MPA act as endocrine disruptors to androgen signaling, abrogating the protective effects of androgens for breast tissue, thus explaining the carcinogenicity of MPA, writing:

> There is now considerable evidence that using a combination of synthetic progestins and estrogens in hormone replacement therapy (HRT) increases the risk of breast

cancer compared with estrogen alone...We propose that the observed excess of breast malignancies associated with combined HRT may be explained, in part, by **synthetic progestins such as MPA acting as endocrine disruptors to negate the protective effects of androgen signaling in the breast**. ... MPA binds to the AR [androgen receptor] with an affinity comparable to the native androgenic ligand, 5-alpha-dihydrotestosterone (DHT)... Notably, in France, the majority of women taking combined HRT receive oral micronized progesterone rather than synthetic progestin. In two French studies-the E3N-EPIC cohort of 54,548 women and a smaller study of 3175 women, **no significant increase in breast cancer risk due to HRT use with micronized progesterone was observed compared with untreated women.** (31)

In the above quote, Dr. Stephen N. Birrell mentions two French studies by De Lignieres (2002) and Fournier (2005) in which micronized progesterone was substituted for the synthetic progestin, MPA, finding no increased breast cancer when this is done. **Note:** In this book, progestin is defined as a synthetic chemically altered version of natural progesterone. By definition, natural progesterone is not a progestin, rather it is identical in chemical structure to the progesterone made by the corpus luteum of the human ovary. See Chapter 2, Russell Marker and the Origin of Bioidentical Hormones for how progesterone is produced commercially. (32-33)

In 2009, Dr. Kathryn B. Horowitz hypothesized synthetic progestins such as MPA reactivate dormant cancer cells in the breast tissue, writing:

> We propose that women who develop breast cancer while on estrogens plus progestins harbor undiagnosed nascent disease [dormant breast cancer cells] before the start of therapy. **The progestin component, in a nonproliferative step, reactivates receptor-negative cancer stem cells** within

such germinal, perhaps even dormant tumors. (34)

Medroxyprogesterone (MPA) Model of Breast Cancer in Mice

MPA is a well-known animal model of breast cancer. In 2009, Dr. Claudia Lanari injected MPA into BALB-c mice, finding about 80 percent developed breast cancer over 52 weeks. Dr. Claudia Lanari discusses the MPA mouse breast cancer model, writing:

> the administration of medroxyprogesterone acetate (MPA) to BALB/c female mice induces mammary ductal carcinomas with a mean latency of 52 weeks and an incidence of about 80%. These tumors are hormone-dependent (HD), metastatic, express both ER [Estrogen Receptor] and PR [Progesterone Receptor], and are maintained by syngeneic transplants. [Note: syngeneic transplant means breast cancer cells from one mouse are injected into a second genetically identical mouse] ... Probably, the most important advantages [of the MPA breast cancer model in Balb-C mice] are that the model is established **in immunocompetent animals** and that the ductal mammary carcinomas originated express ER [estrogen receptor]and PR [progesterone receptor], are hormone-responsive, and metastasize to lymph nodes and lungs. (25-27)

> Note: Unlike the MPA model mouse model of breast cancer, there is no estradiol mouse model of breast cancer, as estradiol has never been shown to reliably induce breast cancer in wild-type mice. All the mouse models of breast cancer require genetic manipulations or chemical carcinogenesis for the induction of breast cancer. (35)

Another Carcinogenic Progestin, Norethisterone

Norethisterone is another carcinogenic progestin widely used in Sweden where the HABITS study showed roughly three times the incidence of breast cancer in the norethisterone-treated group compared to the placebo. This is no sur-prise. After all, progestins are carcinogenic. (36-38)

Hormone Replacement for Breast Cancer Survivors?

For the past four decades, your oncologist and most of mainstream medicine have held the dogmatic belief that hormone replacement is contra-indicated in breast cancer survivors because of the belief estrogen causes breast cancer. Surely, if estrogen causes cancer, then we should see increased cancer recurrence in breast cancer survivors treated with HRT. We are now faced with the question, does HRT increase the risk of breast cancer recurrence in breast cancer survivors?

In 2022, Dr. Francesca Poggio from Genova Italy did a meta-analysis of studies of hormone replacement in breast cancer survivors, exclusively using four RCTs (randomized controlled trials) of 2,022 patients randomized to receive HRT (estrogen/progestin combination) and 2,023 in their control groups receiving placebo. These four studies used progestins known to be carcinogenic. Dr. Poggio found breast cancer survivors using HRT had 46 percent increase in breast cancer recurrence (HR=1.46). Women with hormone receptor-positive breast cancers had an even greater 80 percent increased recurrence, while hormone receptor-negative cancers only 19 percent increased recurrence. Dr. Poggio writes:

> The use of HRT [hormone replacement therapy] was associated with a detrimental prognostic effect in BC [breast cancer] survivors, particularly in those with hormone receptor-positive disease. Alternative interventions to mitigate menopause-related symptoms should be proposed...Our meta-analysis reported that the use of systemic HRT cannot currently be recommended for BC patients experiencing symptoms related to estrogen deficiency, particularly in the case of hormone receptor-positive disease... **Therefore, this approach [HRT] remains contraindicated in the BC setting**. (39)

Now you can understand why most oncologists and mainstream doctors believe HRT is contraindicated for the breast cancer survivor. They read the above article by Dr. Poggio (2022). Notice the 4 RCT studies included in Dr. Poggio's analysis used progestins which are known to be carcinogenic. The Dr. Peter Kenemans RCT study (2009), which accounts for almost 80 percent of the patients in Dr. Poggio's meta-nalysis, used **tibolone,** a synthetic hormone rejected by regulators at the FDA, and banned by the World Anti-Doping Agency (WADA) for use in sports because its metabolites are potent anabolic steroids category S1. In 2023, Dr. Jae Kyung Lee reviewed tibolone's association with breast cancer. The Million Women Study in the UK showed increased breast cancer with the use of tibolone (HR=1.45). As mind-boggling as this seems, most doctors do not understand synthetic hormones cause cancer. The medical literature is confusing on what is natural progesterone and what is progestin. Clearing up this confusion is quite simple. Progesterone is identical to the natural hormone made by the corpus luteum in the ovary. A progestin is a patented, chemically altered version of progesterone. Natural progesterone cannot be patented. (40-43)

Dr. Avrum Bluming Criticized Dr. Poggio

In 2022, Dr. Avrum Bluming discussed the above metanalysis by Dr. Poggio, saying their conclusion is misleading and adds nothing to the contentious debate, writing:

In the most recently published review of this subject, Poggio et al. concluded that "use of HRT was associated with a detrimental prognostic effect in breast cancer survivors." …To reach their misleading conclusion, they selected 2 of the previously reported [by me] prospective randomized trials and included a publication that combined a prospective cohort study with a prospective, randomized one. They omitted one previously reported prospective, randomized trial from their analysis. **An increased recurrence was**

identified in only 1 of the 3 selected studies, the frequently cited HABITS trial [Dr. Lars Holmberg, 2004] and upon meta-analysis of these 3 studies, no significant difference in recurrence was noted when HRT patients were compared with controls. Only when the authors added another prospectively randomized trial into their meta-analysis, [Kenemans, 2009] which constituted **78% of the total 3995 patients analyzed from all 4 studies,** were they able to report an observed increased recurrence rate. But that additional study did not investigate the role of **estrogen but of tibolone**, a compound that is not available in the United States and that has no reported estrogenic effect on breast tissue or endometrium. **Their conclusion is thus misleading, and their analysis adds nothing of value to the already contentious discussion of this complex issue**. (44-46)

Denying Women the Health Benefits of HRT

In 2002, Dr. Olavi Ylikorkala from Helsinki says refusal to provide HRT for breast cancer survivors denies them indisputable health benefits of HRT, writing:

The categorical refusal [to prescribe HRT] is a double-edged sword because it also denies these women all the indisputable health benefits HRT provides… This refusal is not, however, supported by the observational data available so far on this question, **because HRT has not increased the risk for breast cancer recurrence**. (47)

What if we reviewed all the data and excluded the studies using carcinogenic progestins and the banned tibolone? As you will see below, this will give an entirely different picture with less recurrence and less mortality in breast cancer survivors using HRT. Before we go into this, we will first discuss the breast cancer prevention program used in my office.

Breast Cancer Prevention Program

1) **Vitamin D:** We routinely test vitamin D3 levels and optimize vitamin D to above 50

ng/mL. In 2020, Dr. William Grant reviewed the recent medical literature finding optimal Vitamin D levels decrease the risk for breast, colorectal, prostate, and all other cancers, writing:

> medical practice should embrace and public health advice should encourage use of vitamin D to reduce cancer risk and increase survival rates after diagnosis. (48)

In 2024, Dr. Antia Torres from Spain analyzed the recent medical literature on vitamin D and breast cancer prevention, finding an inverse relationship between serum vitamin D level and risk for breast cancer, breast cancer recurrence, and breast cancer mortality. A serum vitamin D level above 40 ng/ml is protective, writing:

> From the results analyzed in this study, the deficiency of vitamin D is closely associated with the development of BC [breast cancer]. In general, higher serum levels of vitamin D may exert a protective effect against the development of the disease. In the present study, it has been observed and discussed that **serum levels of vitamin D ≥ 40.6 ng/mL ± 14.19 ng/mL could be considered protective against the risk of developing BC [breast cancer]**. (49-50)

2) **Selenium Supplementation:** We test for selenium levels routinely and supplement when found low, using 135 mcg/L as a cut-off based on the 2008 study by Dr. Joachim Bleys. Selenium is a critical mineral involved in the intracellular anti-oxidant system (i.e. glutathione peroxidase) and the DNA repair mechanism. Selenium deficiency is associated with increased risk for all cancers, including breast cancer. In 2021, Dr. Kamil Demircan studied selenium status biomarkers of mortality from breast cancer finding patients with low selenium status indicated by all three biomarkers had the highest mortality risk, writing:

> Patients with low selenium status according to all three biomarkers (triple deficient) had

the highest mortality risk with an **overall survival probability of 50% after 8 years**... Prediction of mortality based on all three biomarkers outperformed established tumor characteristics like histologic grade, number of involved lymph nodes, or tumor size. An assessment of Se [selenium] status at breast cancer diagnosis identifies patients at exceptionally high risk for a poor prognosis. (51-54)

Selenium is important for patients harboring genetic mutations in the selenoprotein anti-oxidant and DNA repair mechanisms, such as BRCA1 gene carriers. **Note:** BRCA1 gene is involved in cancer prevention. When mutated, this function is lost. Selenium is discussed in more detail in my 2011 book, *Bioidentical Hormones 101*, first edition, available for free at: https://bioidenticalhormones101.com/

3) **Iodine Supplementation:** We routinely test the spot urinary iodine level, and give iodine supplements when found low using the WHO guidelines. Iodine deficiency is a risk factor for breast cancer, and iodine supplementation is not only preventive for breast cancer but also useful as adjuvant treatment of breast cancer. For more see Chapters 21-23 on Iodine for the Prevention of Breast Cancer, Iodine for Treatment of Breast Cancer, and Estrogen Metabolism, Iodine, 2MeOE2. (55)

4) **Use Progesterone, not Progestin:** Our office uses human bioidentical progesterone, in both topical and oral forms, avoiding carcinogenic synthetic progestins such as medroxyprogesterone, norethisterone, and norethindrone. In 2022, Dr. Haim A. Abenhaim reviewed breast cancer risk with various hormone formulations, concluding:

> Although menopausal HT [hormone therapy] use appears to be associated with an overall increased risk of breast cancer, this risk appears predominantly mediated through formulations containing synthetic progestins [medroxyprogesterone, norethisterone, and norethindrone]. **When prescribing menopausal HT**, **micronized progesterone**

may be the safer progestogen to be used. (56-58)

5) **Estrogen Formulation:** Our office uses Bi-est which is 80% estriol (E3) and 20% estradiol (E2). Estriol preferentially activates ER-beta, a tumor suppressor, and is breast cancer protective. We avoid oral estrogen and instead use topical delivery to avoid blood clotting and venous thromboembolism known to be associated with oral estrogen products. For more on this see the Chapters 4 and 5 on The Safety of Transdermal Estrogen Part One and Transdermal vs Oral Estrogen Part Two. (59)

6) **Progesterone:** My office uses oral micronized progesterone, FDA-approved (Schering/Solvay December 1998) to prevent endometrial hyperplasia and endometrial cancer. The usual dosage is 100 mg oral capsule taken at bedtime. In addition to the oral progesterone, each patient receives topical progesterone, 50 mg daily. I disagree with the OB/Gyne practice of withholding progesterone for post-hysterectomy patients. We routinely give progesterone to all post-hysterectomy patients, realizing progesterone has many other health benefits besides endometrial protection. For example, natural progesterone has an antiproliferative effect on breast tissue and is breast cancer preventive. Although estradiol-alone increases the proliferation of breast cells in human and macaque monkey studies, combining progesterone with estradiol opposes and neutralizes this proliferative effect of estrogen. Progesterone has neurologic and neuropsychiatric benefits. Progesterone stimulates neurogenesis and has neuroprotective effects in traumatic brain injury and stroke. Progesterone relieves depression, anxiety, vasomotor symptoms, and somatic symptoms in post-menopausal women. and has beneficial cardiovascular effects. (56) (60-64)

Progesterone for Peri-Menopausal Transition

The menopausal transition, peri-menopause, usually starts at age 48-52 with the onset of irregular menstrual cycles, fluctuating estrogen levels, anovulation with low progesterone levels, and mood disturbance. Progesterone as a topical cream, or as an oral capsule usually provides symptomatic relief during peri-menopause. (65-71)

7) **Testosterone:** My office routine HRT program includes testosterone for all postmenopausal women. A metabolite of testosterone, 3-Beta-Diol, preferentially activates ER-beta, a tumor suppressor. In 2019, Dr. Rebecca Glaser showed testosterone prevents breast cancer, decreasing the risk by 39-40 percent. Dr. Glaser's findings were replicated in 2021 by Dr. Gary Donovitz the founder of the Bio-T pellet company. For more on this see Chapter 18 on Testosterone for the Prevention and Treatment of Breast Cancer. (72-73)

8) **I3C and DIM (Di-Indole-Methane):** All pre- and post-menopausal women are routinely given I3C/DIM (Di-Indole-Methane) serving as a breast cancer preventive. For more on this topic, see Chapter 21 on Estrogen Metabolism, Iodine, DIM, and 2MeOE2. (74-77)

Sulforaphane, Resveratrol, and NAC Preventing Reactive Estrogen Quinones

9) **4-Hydroxy Quinones:** Certain estrogen metabolites called 4-hydroxy-quinones attach directly to DNA causing oxidative damage and depurination, DNA adduct formation leads to mutations and an increased risk for cancer. In 2021, Dr. Ercole Cavalieri studied these reactive metabolites of estrogen, called catechol estrogen 3,4, quinones, finding the nutritional supplements resveratrol, NAC (N-acetyl cysteine) and sulforaphane prevent oxidative damage to DNA associated with catechol-estrogen-quinones. Dr. Cavalieri writes:

> Cancer can be initiated by increased formation of reactive estrogen metabolites called catechol estrogen-3,4-quinones. If estrogen metabolism becomes unbalanced and significant amounts of these quinones arise, depurinating estrogen-DNA adducts are primarily formed, leading to cancer-

causing mutations. Note: Depurination is the removal or loss of a purine base such as adenine or guanine from the DNA double helix, causing a mutation that may lead to cancer if not repaired. (78-80)

4-OH-Estrogens

In 2019, Dr. Suyu Miao studied 4-hydroxy estrogen metabolites in the urine of breast cancer patients and controls, finding they serve as the most important risk factor for breast cancer, writing:

> Among many alterations of sex hormone metabolisms, 4-hydroxy estrogen (4-OH-E) metabolite was found to be significantly increased in the urine samples of patients with breast cancer compared with the normal healthy controls. **This was the most important risk factor for breast cancer.** (81-86)

Several commercially available laboratories offer urine testing for estrogen metabolites, including the 4-hydroxy estrogens. At the time of writing, these include Genova lab, ZRT lab, Doctor's Data lab, and Dutch Test (Precision Analytical Inc.) **Note: I have no financial interest in any laboratory.**

Interventions to Reduce 4-Hydrox-Quinones

Interventions to reduce 4 hydroxy quinones are part of our breast cancer prevention program. These supplements include iodine, I3C/DIM, resveratrol, NAC, and sulforaphane. For more on these supplements, see Chapter 21. (87)

10) Polymorphisms in Estrogen Metabolism Increase Breast Cancer Risk – Genotype Analysis

Normally, carcinogenic estrogen metabolites of the 4-OH-quinone variety are metabolized and rendered harmless by methylation pathways. However, genetic mutations in these metabolic pathways such as mutation in the MTHFR gene, or COMT gene (catechol-O-methyl transferase), create a bottleneck in the estrogen detoxification pathway causing the accumulation of 4-OH-estrogen quinone metabolites, thus increasing cancer risk. Genetic testing for polymorphisms or urinary estrogen metabolite testing can be useful here. We routinely give all patients on HRT a good quality multivitamin containing methyl-folate to cover the possibility of underlying MTHR mutation. Inexpensive cheek swab kits are available online for home testing for MTHFR and COMT mutations which may impair methylation pathways needed to prevent DNA adduct formation. (88)

11) **Low-Fat Diet:** In 2017, Dr. R.T. Chlebowski reviewed the data from the Low-Fat Diet arm of the WHI study, revealing a 35 percent reduction in mortality from breast cancer, suggesting a considerable benefit with a low-fat diet, writing:

> Results In the dietary group, fat intake and body weight decreased (all P < .001). During the 8.5-year dietary intervention [**low-fat-diet,**] with 1,764 incident breast cancers… deaths after breast cancer (n = 134) were significantly reduced (**40 deaths** [0.025% per year] vs. **94 deaths** [0.038% per year]; **HR, 0.65**; 95% CI, 0.45 to 0.94; P = .02) by the dietary intervention. (89-91)

What is the mechanism for a high-fat diet as a risk factor for breast cancer? High-fat diets are associated with gut dysbiosis, low-level endotoxemia, metabolic syndrome, obesity, and post-prandial inflammation, all implicated in increased cancer risk. (92-95)

Drs. William T. Creasman and Philip J. DiSaia

Both Drs. William T. Creasman and Philip J. DiSaia are highly regarded academic professors of Obstetrics and Gynecology, and authors of medical textbooks, *Clinical Gynecologic Oncology and Women's Health Review*, now in its tenth edition. In the 1980s, Drs. Creasman and DiSaia first advised menopausal hormone replacement therapy (HRT) for all women. Over the many decades of their careers, they have written extensively about HRT for breast cancer survivors claiming HRT does not increase the risk of breast cancer recurrence. In 2005 *Current Opinions in Oncology*, Dr Creasman writes:

> Several case-control and cohort studies have noted either no increased risk or less risk of recurrence in women taking estrogen after therapy after breast cancer. Although the consensus is that such a recommendation is contraindicated, **the data do not support this admonition**. (96-98)

In 2009, Drs. William T. Creasman and Philip J. DiSaia wrote a letter to the editor of Oncology, stating current data suggests no increased breast cancer recurrence with HRT in breast cancer survivors, writing:

> numerous published articles have noted that recurrence rates in breast cancer survivors who chose to take HRT (Hormone Replacement) for symptom relief were very low…. In view of the present data, we feel it is important for women to know there are choices, and **current data would suggest that there is no increased risk of recurrence with HRT.** (99)

Professor William T. Creasman served as the Professor of Gynecologic Oncology at Duke University Medical Center, and is currently Professor and Chairman at the Department of Obstetrics-Gynecology in the Medical University of South Carolina (MUSC). Professor Emeritus Philip J. DiSaia was Department Chair of Obstetrics and Gynecology at the University of California, Irvine. Sadly, Dr. Philip J. DiSaia passed away Sept. 27, 2018.

Dr Xydakis – Greece

In 2006, Dr. Antonios M. Xydakis from Greece writes in the Annals of the New York Academy of Science:

> No observational or retrospective study in breast cancer survivors (whether in pre- or postmenopausal women) has shown an increased risk of tumor recurrence or

increased mortality associated with HRT use. (100)

Dr. Eva Durna from Australia

In 2002 and 2004, Dr. Eva Durna from Australia reported two observational studies. One in 2002 in which hormone replacement was given to post-menopausal breast cancer survivors. A second in 2004 in which HRT was given to pre-menopausal (younger) breast cancer survivors. In both studies, Dr Eva Durna reports reduced mortality and reduced recurrence rates in HRT users. (101-102)

Dr. Pelin Batur of the Cleveland Clinic

In 2006, Dr. Pelin Batur from the Cleveland Clinic reviewed hormone replacement for breast cancer survivors. Dr. Pelin Batur identified seven studies that included a control group. The HRT users enjoyed a 50 percent reduction in recurrence compared to non-users. Among 1,416 HRT users, cancer recurrence was noted in 10.0%. However, for non-users, cancer recurrence was 20%. Mortality was also reduced in HRT users who enjoyed a 66 percent reduction in breast cancer mortality compared to non-users (2.6% vs. 7.8%). Dr. Batur writes:

> In our review, menopausal HT (hormone therapy) use in breast cancer survivors was not associated with increased cancer re-occurrence, cancer-related mortality, or total mortality. (103)

The Randomized Controlled Trial

When doing medical research, observational studies are usually the first to be done. As Dr. Creasman reports, observational studies agree that hormone replacement does not cause increased cancer recurrence in breast cancer survivors. Eventually, the observational studies are either confirmed or refuted by randomized controlled trials (RCTs) which are considered the Gold Standard for medical evidence. Two RCT studies were done, both from Sweden. The first is called HABITS, and the second is called the Stockholm study. Both studies gave HRT to breast cancer survivors in a randomized trial compared to placebo. The HABITS study (Dr. Lars Holmberg, 2004) showed three times greater breast cancer recurrence in HRT users, while the 2005 Stockholm study (Eva von Schoultz,2005) showed no increased recurrence. However, the HABITS study used a carcinogenic progestin, Norethisterone, discussed below. This is further discussed in Chapter 19, Hormone Replacement for Breast Cancer Survivors, Part Two. (104-106).

How to Explain Discrepant Findings?
Less Aggressive Cancers

In 2005, Dr. Rowan T. Chlebowski and Dr. Rose Culhane reported the HABITS trial (RCT) was discrepant from previous observational studies. Dr. Chlebowski explained this discrepancy by saying the observation studies tended to enroll women with less aggressive breast cancers that were axillary lymph node-negative. However, the real explanation for differing results of HRT studies of breast cancer survivors is related to the carcinogenic effects of synthetic progestins. This is discussed further in Chapters 19 and 20. (107-108)

Progestins Increase Breast Cancer!

It is clear after many years of studies that the horse estrogen, Premarin (CEE), when used alone, does not increase breast cancer risk. Perhaps the explanation for the reduction in breast cancer in the Premarin-treated women is the presence of unsaturated B-ring steroids in horse estrogen (CEE, Premarin) with preferentially binding to ER-beta, thus conferring a 23 percent reduction in breast cancer reported in the 2004 second arm of the WHI (Premarin-alone, post-hysterectomy). While the combined ovarian hormones estradiol and progesterone are not associated with increased breast cancer risk, the use of estradiol-alone was associated with a 1.3-fold increase in risk, as reported in 2008 by Dr. Agnès Fournier in the French Cohort study. This 1.3-fold increased

risk with estradiol was abrogated with the addition of natural micronized progesterone. As noted by Dr Agnes Fournier, adding a synthetic progestin is the major factor increasing breast cancer risk. For example, adding medroxyprogesterone resulted in a 1.48-fold increase, and adding norethisterone, a 2.11 increase in breast cancer. Similarly, increased breast cancer risk was reported by the first arm of the WHI (2004) using medroxyprogesterone. The ill-fated HABITS study from Sweden (Lars Holmberg, 2004) used norethisterone, a progestin that is known to be carcinogenic, associated with a 3-fold increase in breast cancer. On the other hand, the more favorable Stockholm Study had a larger number of women taking estrogen-alone, as well as the combination of estrogen and medroxyprogesterone, a synthetic progestin known to be carcinogenic, but less so than norethisterone. This is more completely discussed in Chapter 19, Part Two of this series. (5)(20)(33)(37) (104-106)

In both the above RCT studies, The HABITS and the Stockholm, bioidentical progesterone should have been used instead of carcinogenic progestin. Natural progesterone is not carcinogenic and has breast cancer preventive qualities. This was demonstrated in the 1999 French Cohort study by Dr. G. Plu-Bureau. In this study, topical progesterone conferred a significant reduction in breast cancer risk. Dr. G. Plu-Bureau writes:

> The added progestin plays a huge role in determining if the hormone replacement program will prevent or cause breast cancer. **The use of natural bioidential progesterone is preferable to synthetic progestins which are carcinogenic**. (109)

In 2005, Dr. Collins from Canada, independently agreed with Drs. Plu-Bureau. The added progestin determines breast cancer risk. Dr. Collins writes:

> valid evidence from randomized controlled trials (RCTs) indicates that **breast cancer risk is increased with estrogen-progestin**

use more than with estrogen alone. Note: estrogen-alone here means Premarin (CEE), the estrogen used in the WHI study, or estradiol alone. Progestin means the synthetic version of progesterone such as medroxyprogesterone or norethisterone. (110)

In 2008, Dr. Agnès Fournier published the French Cohort study in which she says the choice of progestogen (natural progesterone vs. synthetic progestin) is important, and the use of natural progesterone is preferable, writing:

> These findings suggest that the choice of the progestagen component in combined HRT is of importance regarding breast cancer risk; **it could be preferable to use [natural] progesterone.** (38)

Dr. Avrum Bluming, Author of "Estrogen Matters"

In 2022, Dr. Avrum Bluming reviewed 25 studies from 1980 to 2013 of breast cancer survivors on hormone replacement (and 20 reviews of these studies from 1994-2021) finding only one of the 25 studies shows increased breast cancer recurrence in hormone users, the HABITS study (Lars Holmberg, 2004) which used a synthetic progestin, norethisterone, known to be carcinogenic, as mentioned above. None of the 25 studies showed increased breast cancer mortality in the hormone-treated group. Thus, in 2022, Dr. Avrum Bluming agrees with Drs. William T. Creasman and Philip J. DiSaia. The data shows breast cancer survivors have no increased risk of breast cancer recurrence when using bioidentical hormone replacement. Dr. Avrum Bluming writes:

> Twenty-five studies of HRT after a breast cancer diagnosis, published between 1980 and 2013, are discussed, as are the 20 reviews of those studies published between 1994 and 2021...Only one of the 25 studies, the HABITS trial [Lars Holmberg, 2004] demonstrated an increased risk of recurrence, which was limited to local or contralateral, and not distant, recurrence.

None of the studies, including HABITS, reported increased breast cancer mortality associated with HRT. (44) (111-115)

BRCA Gene Positive Women and Hormone Replacement

Studies of women with BRCA1 gene mutations are most revealing. The BRCA gene is a breast cancer tumor suppressor, so mutation of the BRCA gene will void its tumor suppressor effect, and increase the risk for cancer. A BRCA1 mutation is associated with an 80% lifetime risk of breast cancer. The lifetime risk for ovarian cancer is 54% for BRCA1 and 23% for BRCA2 mutation carriers. BRCA gene carriers will frequently choose to have preventive oophorectomy, surgical removal of the ovaries, a form of surgically induced menopause, causing troublesome menopausal symptoms and they will then seek hormone replacement. Several studies of HRT use in BRCA gene carriers before or after oophorectomy show no increase in breast cancer risk. (116-122)

Dr. Andrea Eisen from Toronto

For example, in 2008, Dr. Andrea Eisen did a case-controlled study of HRT in 472 post-menopausal women with BRCA1 mutation. In these BRCA gene women who took estrogen-alone, Dr. Eisen found a 49% reduction in breast cancer recurrence. Dr. Eisen conducted two studies, one for estrogen-alone and a second study for the estrogen-progestin combination, finding estrogen alone gave superior results. Again, it is prudent to avoid carcinogenic synthetic progestins. (122)

The experience with hormone replacement in BRCA1 gene carriers shows that estrogen is preventive for breast cancer. As mentioned above in the discussion of the carcinogenic effect of 4-hydroxy estrogen quinones, the current theory of breast cancer is oxidative damage to the DNA of breast cells by estrogen metabolites. This mechanism applies to BRCA1 gene mutation carriers in which a malfunction in the anti-oxidant system leads to oxidative damage to the DNA of the breast cell. According to Dr. Haixia Chen (2018), breast cancers in BRCA gene carriers tend to be triple negative, ie. estrogen, progesterone, and HERS (human epidermal growth factor) receptor-negative. This triple negative cell type is the most aggressive and usually associated with poor prognosis. (123)

Selenium Supplementation for BRCA Gene Carriers

Three studies from Poland by Kowalska (2005), Huzarski (2006), and Dziaman (2009) show selenium supplementation is beneficial for the BRCA gene carrier patient. The BRCA gene is involved in DNA repair of oxidative damage using the selenoprotein repair system. Thus, BRCA gene carriers have more DNA oxidative damage, leading to an increased risk for cancer. In 2005, Dr. Elzbieta Kowalska studied lymphocytes in blood samples from 55 women BRCA1 carriers, finding excess rates of chromosomal breakage from DNA oxidative damage. BRCA gene carriers had twice the DNA damage compared to their normal siblings. The fifty-five women BRCA1 gene mutation carriers were then given selenium for 8 weeks (275 µg/d). The selenium supplement reduced DNA damage to normal levels, the same levels as their normal siblings. In 2006, a larger study by Drs. Huzarski and Kowalska verified that selenium does indeed reduce breast cancer incidence in BRCA1 gene carriers. After two years of selenium supplementation in 100 BRCA gene carriers, the expected BRCA1-associated cancers were reduced in half compared to matched control BRCA carriers not given selenium. This is a very impressive finding! Dr. Kowalska writes:

> The product of the BRCA1 gene is involved in the repair of double-stranded DNA breaks and it is believed that increased susceptibility to DNA breakage contributes to the cancer phenotype...the frequency of chromosome breaks was measured in cultured blood lymphocytes..BRCA1 serves multiple functions, including DNA damage

response, nucleotide excision repair, and protection against oxidative stress...**The frequency of chromosome breaks was greatly reduced following 1 to 3 months of oral selenium supplementation** (mean, 0.63 breaks per cell versus 0.40; P < 10–10). The mean level of chromosome breaks in carriers following supplementation was similar to that of the noncarrier controls (0.40 versus 0.39). **Oral selenium is a good candidate for chemoprevention in women who carry a mutation in the BRCA1 gene**...We performed two pilot studies involving 200 healthy BRCA1 mutation carriers (100 matched pairs – cases and controls). After two years of oral selenium administration the frequency of **BRCA1-associated tumors was two times lower in women who supplemented their diet with selenium,** as compared to women without supplementation. (124-125)

In 2009, Dr. Tomasz Dziaman from Poland examined DNA damage in BRCA1 gene carriers, measuring serum and urinary products of DNA oxidation (8-oxodG and 8-oxoGua) with and without selenium supplementation. Dr. Tomasz Dziaman found that damaged DNA products were higher in women with BRCA1 mutations, and were reduced by selenium supplementation. Their results suggest that BRCA1 mutation contributes to oxidative damage and breaks in cellular DNA, which may be responsible for cancer development. In addition, selenium supplementation is beneficial because it protects from oxidative DNA damage and restores the DNA repair mechanisms, reducing cancer risk in BRCA1 carriers. Dr. Dziaman writes:

> BRCA1 plays a role in repair of oxidative DNA damage...we determined 8-oxodG level in cellular DNA and urinary excretion of 8-oxodG and 8-oxoGua in the mutation carriers. We found that 8-oxodG level in leukocytes DNA is significantly higher in BRCA1 mutation carriers...In the distinct subpopulation of BRCA1 mutation carriers without symptoms of cancer who underwent adnexectomy and were supplemented with selenium, the level of 8-oxodG in DNA decreased significantly in comparison with

the subgroup without supplementation. Simultaneously in the same group, an increase of urinary 8-oxoGua, the product of base excision repair (hOGG1 glycosylase), was observed. Therefore, it is likely that the selenium supplementation of the patients is responsible for the increase of BER enzymes activities, which in turn may result in reduction of oxidative DNA damage. **Importantly, in a double-blinded placebo control prospective study, it was shown that in the same patient groups, reduction in cancer incidents was observed.** Altogether, these results suggest that BRCA1 deficiency contributes to 8-oxodG accumulation in cellular DNA, which in turn may be a factor responsible for cancer development in women with mutations, and that the risk to developed breast cancer in BRCA1 mutation carriers may be reduced in selenium-supplemented patients who underwent adnexectomy. Note: BER is base excision repair, a mechanism for DNA repair. Note: In view of the above information, all BRCA gene carriers should have serum selenium testing and should be given selenium supplements to maintain a good selenium level. For more on Selenium supplementation see my free 2011 book, *Bioidentical Hormones 101*, available at https://bioidenticalhormones101.com/. (126-130)

Contra-Indications to HRT in the Breast Cancer Patient

Breast cancer may occur in the younger pre-menopausal age group. Most commonly, this is an indolent form of cancer called DCIS (ductal carcinoma in situ) which has a very good prognosis. However, another form of cancer called infiltrating ductal can be very aggressive in the younger age group with poor prognosis regardless of treatment. This cancer cell type tends to be estrogen receptor-positive (ER+) and highly aggressive with a median survival of 26 months. In ER+ breast cancer in younger pre-menopausal women, the mainstream medical treatment is ovarian ablation, either surgical or drug-induced, to eliminate endogenous estrogen and prevent cancer cells from making their own estrogens. The drugs used are luteinizing hormone-releasing hormone agonists, tamoxifen or aromatase inhibitors letrazole, anastrazole, and exemestane. In 2021, Dr. Yen-Shen Lu reviewed the suppression of ovarian function suppression with luteinizing hormone-releasing hormone agonists (LHRHa) as treatment of hormone receptor-positive early breast cancer in the premenopausal female, writing:

As treatment options have rapidly expanded, management of adjuvant treatment of premenopausal women with early and advanced breast cancer has become more complicated. The most recent evidence suggests that addition of LHRHa to adjuvant endocrine therapy, with both tamoxifen and Ais [Aromatase Inhibitors], can provide significant benefits in some premenopausal patients who are at high risk of recurrence and have poor prognostic characteristics. Longer-acting depot and implant LHRHa formulations may help to overcome some of the barriers to adding OFS [ovarian function suppression] to endocrine therapy in the adjuvant setting in premenopausal women. (131)

in 2004, Dr. Tatiana Prowell did a metanalysis including 2,100 pre-menopausal estrogen receptor positive breast cancer patients in which **ovarian ablation** improved 15-year survival from 46% to 52%. This is a disappointing six percent absolute benefit. Dr. Prowell writes:

This overview of 12 randomized controlled trials enrolling a total of 2,102 patients reported that women under the age of 50 with early invasive breast cancer who underwent oophorectomy or ovarian irradiation, experienced approximately a 25% relative reduction in the risks of recurrence and mortality at 15 years of follow-up compared with those receiving no adjuvant therapy...In summary, virtually all premenopausal women with early-stage HR+ breast cancer should receive adjuvant endocrine therapy. Combined endocrine therapy appears to be at least as effective as adjuvant CMF chemotherapy [cyclophosphamide, methotrexate, and fluorouracil] in this population...Among 2102 women aged under 50 when randomised, most of whom would have been premenopausal at diagnosis, 1130 deaths and an additional 153 recurrences were reported. 15-year survival was highly significantly improved among those allocated **ovarian ablation (52.4 vs 46.1%,** 6.3 fewer deaths per 100 women, as was recurrence-free survival (**45.0 vs 39.0%,** 2p = 0.0007). (132)

Again, this poor prognosis underscores the aggressive nature of this cell type and the disappointing results of mainstream treatment. Obviously, in this scenario, hormone replacement therapy (HRT) would be contra-indicated. For post-menopausal women over the age of 50, ovarian ablation is of no benefit regardless of tumor receptor status, since these women are post-menopausal with very little ovarian function.

Which Breast Cancer Survivors to Treat with HRT?

My opinion differs from mainstream medicine's dogma that all HRT is contraindicated in the breast cancer survivor. My position is this: breast cancer survivors who are cancer-free after treatment are candidates for HRT. Those who request HRT because of severe menopausal symptoms should be allowed to use HRT, assuming such treatment is not contraindicated. The patient's age, tumor cell type and grade, hormone receptor status, and disease-free number of years since treatment, are important considerations when considering whether to offer HRT to the breast cancer survivor. The physician must exercise judgment, and decline HRT for patients with active disease, or any other interfering factors. The prescribing physician must be knowledgeable about various hormone formulations, avoiding the use of synthetic progestins, minimizing activation of

ER-alpha and maximizing activation of ER-beta. Breast cancer recurrence with such a program will be reduced compared to placebo. However, a small percentage of women on such a carefully devised bioidentical hormone program will have breast cancer recurrence. This is inevitable and must be discussed with each patient, to be included in the consent form.

Family History of Breast Cancer

The benefits of hormone replacement (HRT) extend to women with a family history of breast cancer. A 1997 study by Dr. Thomas A. Sellers from Iowa followed 41,800 women for 8 years. Those women using HRT with a family history of breast cancer enjoyed a 50% reduction in overall mortality compared to HRT non-users. Dr. Sellars writes:

> These data suggest that HRT [hormone replacement therapy] use in women with a family history of breast cancer is not associated with a significantly increased incidence of breast cancer but is associated with a significantly reduced total mortality rate. (133)

Opposing Opinions – Confusing Progestins with Progesterone

To be fair, a number mainstream authors oppose Dr. DiSaia and Creasman's opinion, such as Dr. Dan Labriola in a 2009 rebuttal letter. Again, this opposing view is usually based on the confusion of chemically altered progestins which are known to be carcinogenic, with the ovarian hormone, natural progesterone. Synthetic progestins are very different from natural progesterone. They have an entirely different chemical structure and very different biological activity. In this book, we define progestins as chemically altered versions of progesterone and define progesterone as not a progestin. (134-135)

As you read through the medical literature you will find a common mistake. Many reference articles on this topic use mistaken terminology, referring to a "progestin" hormone as "progesterone" which it is not. For example, in 2008, Dr. Andrea Eisen's BRCA gene article says the first arm of the WHI used a combination pill consisting of estrogen and **progesterone** which increased breast cancer risk. The WHI study used MPA, which is not natural progesterone. In 2005, Dr. William Creasman was guilty of this same mistake. In 2021, Dr. Nanette Santoro made this same mistake, writing:

> In the Women's Health Initiative (WHI), women with VMS [Vasomotor Symptoms] experienced an 85% reduction in symptoms after treatment with estrogen **plus progesterone**. Note: this is an error. The WHI study did not use natural bioidentical progesterone. (96) (122) (136)

The WHI study used medroxyprogesterone (MPA), a carcinogenic synthetic progestin. This mistake permeates the women's hormone literature explaining the many discrepancies in findings and opinions. When you see the statement that progesterone causes breast cancer in the medical literature, you will know this is an error. The author is making this same mistake, confusing progestins and progesterone. When you look closely, you will find that studies claiming progesterone causes breast cancer use a progestin, not natural progesterone, thus explaining the results of increased breast cancer. Progesterone does not cause breast cancer, progesterone prevents it, as discussed in 2017 by Dr. Allan Lieberman. (56)

Randomized Trials with Bioidentical Estradiol, Progesterone and Testosterone Urgently Needed

Unfortunately, after all these years, there are still no randomized controlled trials of postmenopausal hormone replacement with commonly used bioidentical hormone preparations. My dream would be to do such a menopausal hormone replacement RCT study using our standard office compounded hormone formula: 5 mg/gram of Bi-Est (80% estriol and 20%

estradiol) and 50 mg/ gram progesterone as a topical cream. Half gram of cream is applied twice a day. Next is the oral progesterone capsule 100 mg/qhs. Next is the topical testosterone 3 mg/d. In addition, I would combine this with the dietary modification and nutritional supplements for breast cancer prevention mentioned above. There is an urgent need for RCT studies treating breast cancer survivors with these bioidentical hormone formulas.

Conclusion: I would agree with Dr. Puthugramam K Natrajan (2002) who says that breast cancer survivors may be offered estrogen therapy. Dr. Natarajan gave estrogen to 69 breast cancer survivors finding no increase in cancer recurrence compared to non-users, writing:

> Most physicians believe that estrogen replacement therapy is contraindicated once a patient is diagnosed with breast cancer... The study group consisted of 123 women (mean age, 65.4 +/- 8.85 years) who were diagnosed with breast cancer in our practice, including 69 patients who received estrogen replacement therapy for < or = 32 years after diagnosis. The comparative groups were 22 women who used nonestrogenic hormones for < or = 18 years and 32 women who used no hormones for < or = 12 years... **Estrogen replacement therapy apparently does not increase either the risk of recurrence or of death in patients with early breast cancer. These patients may be offered estrogen replacement therapy after a full explanation of the benefits, risks, and controversies.** (137)

Back to 1977

Returning to 1977, making rounds on the breast cancer surgery floor at Rush Presbyterian Hospital with Dr. Steven Economou, and his pop question, does estrogen cause breast cancer? The answer is more nuanced than a simple, YES or NO. Today, the correct answer involves a discussion of estrogen metabolism, the carcinogenic effect of 4-hydroxy-estrogens metabolites, and the anti-cancer effect of 2 methoxy-estradiol, more fully discussed in later chapters. The answer also includes a discussion of the proliferative effects of various ligands and whether they activate estrogen receptor alpha which is proliferative, or estrogen receptor beta, which is a tumor suppressor. The answer also involves mentioning the carcinogenic effects of synthetic progestins such as MPA which block androgenic signaling. The MPA mouse model of breast cancer reveals MPA is carcinogenic. However, there is no estradiol mouse model of breast cancer. Of course, Dr. Economou had no way of knowing about estrogen receptors alpha and beta, nor the carcinogenic effects of 4-hydroxy estrogens and the complexities of estrogen metabolism, information uncovered in new research years later.

Long Term Estrogen Deprivation

After long-term estrogen deprivation (LTED), instead of acting as a growth signal, estrogen is transformed into a death signal, causing apoptosis (programmed cell death) in breast cancer cells. This is the estrogen-paradox of V. Craig Jordan. Perhaps LTED explains the partial success of estrogen in the form of diethylstilbestrol (DES) as a mainstream treatment for metastatic breast cancer with a 30-40 percent response rate. Sir Alexander Haddow used DES from 1944 until the 1970s, at which time mainstream medicine switched from estrogen (DES) to tamoxifen, an antiestrogen drug. Dr. Economou knew about estrogen as a treatment for breast cancer because he lived through the era when DES was routinely used to treat metastatic breast cancer. Yet, Dr. Economou remained silent, rushing in and out of the patient rooms with little time to waste on debating such questions. (138-142)

♦ References for Chapter 6 Bioidentical Hormones for Breast Cancer Survivors

1) Steven G. Economou: 1922 – 2007 Surgeon, Writer, Doodler. April 28, 2007, By Trevor Jensen, Tribune staff reporter. http://articles.chicagotribune.com/2007-04-28/news/0704271171_1_surgery-department-rush-university-medical-center-presbyterian-hospital

2) Pesheva Ekaterina, Estrogen A More Powerful Breast Cancer Culprit Than We Realized. HMS Communications May 17, 2023. https://news.harvard.edu/gazette/story/2023/05/estrogen-a-more-powerful-breast-cancer-culprit-than-we-realized/

3) Lee, Jake June-Koo, et al. "ER-alpha-Associated Translocations Underlie Oncogene Amplifications in Breast Cancer." Nature 618.7967 (2023): 1024-1032.

4) Ke, Bin, Chunyu Li, and Huifang Shang. "Sex Hormones in The Risk Of Breast Cancer: A Two-Sample Mendelian Randomization Study." American Journal of Cancer Research 13.3 (2023): 1128.

5) LaCroix, Andrea Z., et al. "Health Outcomes After Stopping Conjugated Equine Estrogens Among Postmenopausal Women with Prior Hysterectomy: A Randomized Controlled Trial." Jama 305.13 (2011): 1305-1314.

6) Rossouw, Jacques E., et al. "Risks and Benefits of Estrogen Plus Progestin in Healthy Postmenopausal Women: Principal Results from The Women's Health Initiative Randomized Controlled Trial." JAMA 288.3 (2002): 321-333.

7) Hodis, Howard N., and P. M. Sarrel. "Menopausal Hormone Therapy and Breast Cancer: What Is The Evidence From Randomized Trials?" Climacteric 21.6 (2018): 521-528.

8) Manson, JoAnn E., et al. "Menopausal Hormone Therapy and Long-Term All-Cause and Cause-Specific Mortality: The Women's Health Initiative Randomized Trials." JAMA 318.10 (2017): 927-938

9) Russo, Jose, et al. "17-Beta-Estradiol Induces Transformation and Tumorigenesis in Human Breast Epithelial Cells." The FASEB journal 20.10 (2006): 1622-1634.

10) Russo, Jose, et al. "17β-Estradiol is Carcinogenic in Human Breast Epithelial Cells." The Journal Of Steroid Biochemistry And Molecular Biology 80.2 (2002): 149-162.

11) Russo, Jose, et al. "Estrogen-induced breast cancer is the result of disruption of asymmetric cell division of the stem cell." Hormone molecular biology and clinical investigation 1.2 (2010): 53-65.

12) Zattarin, Emma, et al. "Hormone Receptor Loss in Breast Cancer: Molecular Mechanisms, Clinical Settings, And Therapeutic Implications." Cells 9.12 (2020): 2644.

13) Nicotra, Raquel, et al. "Rat models of hormone receptor-positive breast cancer." Journal of Mammary Gland Biology and Neoplasia 29.1 (2024): 12.

14) Liu, Chong, et al. "Advances in rodent models for breast cancer formation, progression, and therapeutic testing." Frontiers in Oncology 11 (2021): 593337.

15) Hanahan, Douglas, and Robert A. Weinberg. "Hallmarks of cancer: the next generation." cell 144.5 (2011): 646-674.

16) Cavalieri EL et al. Molecular origin of cancer: Catechol estrogen-3,4-quinones as endogenous tumor initiators Proceedings of the National Academy of Sciences, 1997. 94(20): pp. 10937–10942.

17) Rajkumar, Lakshmanaswamy, et al. "Hormone-Induced Protection of Mammary Tumorigenesis In Genetically Engineered Mouse Models." Breast Cancer Research 9 (2007): 1-11.

18) Arumugam, Arunkumar, Elaine A. Lissner, and Rajkumar Lakshmanaswamy. "The role of hormones and aromatase inhibitors on breast tumor growth and general health in a postmenopausal mouse model." Reproductive Biology and Endocrinology 12 (2014): 1-13.

19) Shull, James D., et al. "Rat Models Of 17β-Estradiol-Induced Mammary Cancer Reveal Novel Insights into Breast Cancer Etiology And Prevention." Physiological Genomics 50.3 (2018): 215-234.

20) Chlebowski, Rowan T., et al. "Estrogen Plus Progestin and Breast Cancer Incidence And Mortality In Postmenopausal Women." JAMA 304.15 (2010): 1684-1692.

21) Schairer, Catherine, et al. "Menopausal Estrogen and Estrogen-Progestin Replacement Therapy And Breast Cancer Risk." JAMA 283.4 (2000): 485-491.

22) Persson, Ingemar, et al. "Cancer Incidence and Mortality in Women Receiving Estrogen And Estrogen-Progestin Replacement Therapy—Long-Term Follow-Up Of A Swedish Cohort." International journal of cancer 67.3 (1996): 327-332.

23) Beral, Valerie, et al. "Breast Cancer and Hormone-Replacement Therapy: the Million Women Study." The Lancet 362.9392 (2003): 1330-1331.

24) Ross, Ronald K., et al. "Effect of Hormone Replacement Therapy on Breast Cancer Risk: Estrogen Versus Estrogen Plus Progestin." Journal of the National Cancer Institute 92.4 (2000): 328-332.

25) Lanari, C., et al. "The MPA Mouse Breast Cancer Model: Evidence for A Role of Progesterone Receptors In Breast Cancer." Endocrine-related cancer 16.2 (2009): 333.

26) Kordon, Edith, et al. "Hormone Dependence of a Mouse Mammary Tumor Line Induced In Vivo By Medroxyprogesterone Acetate." Breast cancer research and treatment 17 (1990): 33-43.

27) Molinolo, A. A., et al. "Mouse Mammary Tumors Induced by Medroxyprogesterone Acetate: Immunohistochemistry and Hormonal Receptors." Journal of the National Cancer Institute 79.6 (1987): 1341-1350.

28) Chlebowski, Rowan T., et al. "Breast Cancer After Use of Estrogen Plus Progestin And Estrogen Alone: Analyses Of Data From 2 Women's Health Initiative Randomized Clinical Trials." JAMA oncology 1.3 (2015): 296-305.

29) Cherry, Nicola, et al. "Long-term Safety of Unopposed Estrogen Used By Women Surviving Myocardial Infarction: 14-Year Follow-Up Of The ESPRIT Randomised Controlled Trial." BJOG: An International Journal of Obstetrics & Gynaecology 121.6 (2014): 700-705.

30) Schierbeck, Louise Lind, et al. "Effect of Hormone Replacement Therapy on Cardiovascular Events In Recently Postmenopausal Women: Randomised Trial." BMJ 345 (2012).

31) Birrell, Stephen N., et al. "Disruption of Androgen Receptor Signaling by Synthetic Progestins May Increase Risk Of Developing Breast Cancer." The FASEB Journal 21.10 (2007): 2285-2293.

32) De Lignieres, B., et al. "Combined Hormone Replacement Therapy and Risk of Breast Cancer In A French Cohort Study Of 3175 Women." Climacteric 5.4 (2002): 332-340.

33) Fournier, Agnes, et al. "Breast Cancer Risk In Relation To Different Types Of Hormone Replacement Therapy In The E3N-EPIC Cohort." International Journal of Cancer 114.3 (2005): 448-454.

34) Horwitz, Kathryn B., and Carol A. Sartorius. "Progestins in Hormone Replacement Therapies Reactivate Cancer Stem Cells In Women With Preexisting Breast Cancers: A Hypothesis." The Journal of Clinical Endocrinology & Metabolism 93.9 (2008): 3295-3298.

35) Mohibi, Shakur, et al. "Mouse Models of Estrogen Receptor-Positive Breast Cancer." Journal of Carcinogenesis 10 (2011).

36) Holmberg, L., and H. Anderson. "HABITS (hormonal replacement therapy after breast cancer—is it safe?): a randomized comparison trial stopped." Obstetrical & gynecological survey 59.6 (2004): 442-443.

37) Lyytinen, Heli, et al. "Do the dose or route of administration of norethisterone acetate as a part of hormone therapy play a role in risk of breast cancer: National-wide case-control study from Finland." International journal of cancer 127.1 (2010): 185-189.

38) Fournier, Agnès, Franco Berrino, and Françoise Clavel-Chapelon. "Unequal risks for breast cancer associated with different hormone replacement therapies: results from the E3N cohort study." Breast cancer research and treatment 107 (2008): 103-111.

39) Poggio, Francesca, et al. "Safety of Systemic Hormone Replacement Therapy In Breast Cancer Survivors: A Systematic Review And Meta-Analysis." Breast Cancer Research and Treatment 191 (2022): 269-275.

40) Kenemans P, et al. Safety and Efficacy of Tibolone In Breast-Cancer Patients With Vasomotor Symptoms: A Doubleblind, Randomised, Non-Inferiority Trial. Lancet Oncol. 2009;10:135–146.

41) WADA Tooltip World Anti-Doping Agency List of Prohibited Drugs. https://www.wada-ama.org/en/prohibited-list

42) Lee, Jae Kyung, et al. "Tibolone and Breast Cancer." Journal of Menopausal Medicine 29.3 (2023): 92.

43) Cummings, Steven R., et al. "The Effects of Tibolone in Older Postmenopausal Women." New England Journal of Medicine 359.7 (2008): 697-708.

44) Bluming, Avrum Zvi. "Hormone replacement therapy after breast cancer: it is time." The Cancer Journal 28.3 (2022): 183-190.

45) Kenemans P, Bundred NJ, Foidart JM, et al. Safety and efficacy of tibolone in breast-cancer patients with vasomotor symptoms: a doubleblind, randomised, non-inferiority trial. Lancet Oncol. 2009;10:135–146.

46) Escande, Aurélie, et al. "Regulation of Activities Of Steroid Hormone Receptors By Tibolone And Its Primary Metabolites." The Journal of Steroid Biochemistry And Molecular Biology 116.1-2 (2009): 8-14.

47) Ylikorkala Olavi , Metsä-Heikkilä M. Hormone Replacement Therapy In Women With A History Of Breast Cancer. Gynecol Endocrinol. 2002; 16:469–478.

48) Grant, William B. "Review of Recent Advances in Understanding the Role Of Vitamin D In Reducing Cancer Risk: Breast, Colorectal, Prostate, And Overall Cancer." Anticancer Research 40.1 (2020): 491-499.

49) Torres, Antía, et al. "The Impact of Vitamin D and Its Dietary Supplementation in Breast Cancer Prevention: An Integrative Review." Nutrients 16.5 (2024): 573.

50) Blasiak, Janusz, et al. "Vitamin D May Protect against Breast Cancer through the Regulation of Long Noncoding RNAs by VDR Signaling." International Journal of Molecular Sciences 23.6 (2022).

51) Demircan, Kamil, et al. "Serum Selenium, Selenoprotein P And Glutathione Peroxidase 3 As Predictors of Mortality And Recurrence Following Breast Cancer Diagnosis: A Multicentre Cohort Study." Redox biology 47 (2021): 102145.

52) Szwiec, Marek, et al. "Serum Selenium Level Predicts 10-Year Survival After Breast Cancer." Nutrients 13.3 (2021): 953.

53) Sandsveden, Malte, et al. "Prediagnostic Serum Selenium Levels In Relation To Breast Cancer Survival And Tumor Characteristics." International Journal of Cancer 147.9 (2020): 2424-2436.

54) Bleys, Joachim, et al. "Serum Selenium Levels and All-Cause, Cancer, And Cardiovascular Mortality Among US Adults." Archives of Internal Medicine 168.4 (2008): 404-410.

55) Ibrahim, Raihan Syah, and Aisyah Elliyanti. "The Potential of Iodine as A Treatment for Breast Cancer: A Narrative Review." Jurnal Kesehatan Manarang 9.3 (2023): 159-165.

56) Lieberman, Allan, and Luke Curtis. "In Defense of Progesterone: A Review of The Literature." Alternative Therapies in Health & Medicine 23.7 (2017).

57) Abenhaim, Haim A., et al. "Menopausal Hormone Therapy Formulation and Breast Cancer Risk." Obstetrics and gynecology 139.6 (2022): 1103-1110.

58) Graham, S., et al. "Review of Menopausal Hormone Therapy with Estradiol And Progesterone Versus Other Estrogens And Progestins." Gynecological Endocrinology (2022): 1-20.

59) Mal, Rahul, et al. "Estrogen Receptor Beta (ErβB): A Ligand-Activated Tumor Suppressor." Frontiers in Oncology 10 (2020): 587386.

60) FDA Drug Approval Package, Prometrium (Progesterone) Capsules, Company: Schering Corporation, Application No.: 020843, Approval Date: 12/26/1998

61) Moyer, Dean L., et al. "Prevention of Endometrial Hyperplasia by Progesterone During Long-Term Estradiol Replacement: Influence of Bleeding Pattern and Secretory Changes." Fertility And Sterility 59.5 (1993): 992-997.

62) Fitzpatrick, Lorraine A., and Andrew Good. "Micronized Progesterone: Clinical Indications and Comparison with Current Treatments." Fertility and sterility 72.3 (1999): 389-397.

63) Judd, Howard L., et al. "Effects of Hormone Replacement Therapy on Endometrial Histology in Postmenopausal Women: The Postmenopausal Estrogen/Progestin Interventions (PEPI) Trial." JAMA 275.5 (1996): 370-375.

64) Prior, J. C. "Progesterone for the Prevention and Treatment of Osteoporosis in Women." Climacteric 21.4 (2018): 366-374.

65) Prior, J. C. "Progesterone for the Prevention and Treatment of Osteoporosis in Women." Climacteric 21.4 (2018): 366-374.

66) Prior, J. C. "Progesterone for Symptomatic Perimenopause Treatment–Progesterone Politics, Physiology and Potential for Perimenopause." Facts, Views & Vision in ObGyn 3.2 (2011): 109.

67) Prior, J. C. "Progesterone for Treatment of Symptomatic Menopausal Women." Climacteric 21.4 (2018): 358-365.

68) Joffe, Hadine, et al. "Impact of Estradiol Variability and Progesterone on Mood in Perimenopausal Women with Depressive Symptoms." The Journal of Clinical Endocrinology & Metabolism 105.3 (2020): e642-e650.

69) Gordon, Jennifer L., et al. "Efficacy of Transdermal Estradiol and Micronized Progesterone in The Prevention of Depressive Symptoms in The Menopause Transition: A Randomized Clinical Trial." JAMA psychiatry 75.2 (2018): 149-157.

70) Fan, Yubo, et al. "Paradigm Shift in Pathophysiology of Vasomotor Symptoms: Effects of Estradiol Withdrawal and Progesterone Therapy." Drug Discovery Today: Disease Models 32 (2020): 59-69.

71) Spark, M. Joy, and Jon Willis. "Systematic Review of Progesterone Use by Midlife and Menopausal Women." Maturitas 72.3 (2012): 192-202.

72) Donovitz, Gary, and Mandy Cotten. "Breast Cancer Incidence Reduction in Women Treated With Subcutaneous Testosterone: Testosterone Therapy And Breast Cancer Incidence Study." European Journal Of Breast Health 17.2 (2021): 150.

73) Glaser, Rebecca L., et al. "Incidence of Invasive Breast Cancer in Women Treated with Testosterone Implants: A Prospective 10-Year Cohort Study." BMC Cancer 19 (2019): 1-10.

74) Thomson, Cynthia A., et al. "Chemopreventive Properties Of 3, 3'-Diindolylmethane In Breast Cancer: Evidence From Experimental And Human Studies." Nutrition Reviews 74.7 (2016): 432-443.

75) Reyes-Hernández, Octavio Daniel, et al. "3, 3'-Diindolylmethane and Indole-3-Carbinol: Potential Therapeutic Molecules for Cancer Chemoprevention and Treatment Via Regulating Cellular Signaling Pathways." Cancer Cell International 23.1 (2023): 180.

76) Williams, David E. "Indoles Derived from Glucobrassicin: Cancer Chemoprevention by Indole-3-Carbinol And 3, 3'-Diindolylmethane." Frontiers in Nutrition 8 (2021): 734334.

77) Koli, Papita, et al. "Anticancer Activity Of 3, 3'-Diindolylmethane and The Molecular Mechanism Involved in Various Cancer Cell Lines." ChemistrySelect 5.37 (2020): 11540-11548.

78) Cavalieri, Ercole, and Eleanor Rogan. "Catechol Quinones of Estrogens in The Initiation of Breast, Prostate, And Other Human Cancers: Keynote Lecture." Annals of the New York Academy of Sciences 1089.1 (2006): 286-301.

79) Cavalieri, Ercole L., and Eleanor G. Rogan. "Depurinating Estrogen–DNA Adducts in The Etiology and Prevention of Breast and Other Human Cancers." Future Oncology 6.1 (2010): 75-91.

80) Cavalieri, Ercole, and Eleanor Rogan. "The 3, 4-Quinones of Estrone and Estradiol Are the Initiators of Cancer Whereas Resveratrol And N-Acetylcysteine Are the Preventers." International Journal of Molecular Sciences 22.15 (2021): 8238.

81) Miao, Suyu, et al. "4-Hydroxy estrogen metabolite, causing genomic instability by attenuating the function of spindle-assembly checkpoint, can serve as a biomarker for breast cancer." American Journal of Translational Research 11.8 (2019): 4992.

82) Samavat, Hamed, and Mindy S. Kurzer. "Estrogen Metabolism and Breast Cancer." Cancer Letters 356.2 (2015): 231-243.

83) Zahid, Muhammad, et al. "Resveratrol and N-Acetylcysteine Block the Cancer-Initiating Step In MCF-10F Cells." Free Radical Biology and Medicine 50.1 (2011): 78-85.

84) Yager, James D. "Mechanisms of Estrogen Carcinogenesis: The Role of E2/E1–Quinone Metabolites Suggests New Approaches to Preventive Intervention–A Review." Steroids 99 (2015): 56-60.

85) Palliyaguru, Dushani L., et al. "Sulforaphane Diminishes the Formation of Mammary Tumors in Rats Exposed To 17β-Estradiol." Nutrients 12.8 (2020): 2282.

86) Sung, Nam-Ji, and Sin-Aye Park. "Effect of Natural Compounds on Catechol Estrogen-Induced Carcinogenesis." Biomedical Science Letters 25.1 (2019): 1-6.

87) Kaiser, Anna E., et al. "Sulforaphane: A Broccoli Bioactive Phytocompound with Cancer Preventive Potential." Cancers 13.19 (2021): 4796.

88) Almeida, Micaela, et al. "Influence of Estrogenic Metabolic Pathway Genes Polymorphisms on Postmenopausal Breast Cancer Risk." Pharmaceuticals 14.2 (2021): 94.

89) Chlebowski, Rowan T., et al. "Low-Fat Dietary Pattern and Breast Cancer Mortality in The Women's Health Initiative Randomized Controlled Trial." Journal of Clinical Oncology 35.25 (2017): 2919.

90) Chlebowski, R. T., and G. L. Blackburn. "Final Survival Analyses from The Women's Intervention Nutrition Study (WINS) Evaluation Dietary Fat Reduction As Adjuvant Breast Cancer Therapy. SABCS 2014; abstract S5–08."

91) Pan, Kathy, et al. "Low-Fat Dietary Pattern and Breast Cancer Mortality By Metabolic Syndrome Components: A Secondary Analysis Of The Women's Health Initiative (WHI) Randomized Trial." British Journal of Cancer 125.3 (2021): 372-379.

92) Arnone, Alana A., et al. "Diet Modulates the Gut Microbiome, Metabolism, and Mammary Gland Inflammation to Influence Breast Cancer Risk." Cancer Prevention Research 17.9 (2024): 415-428.

93) Pendyala, Swaroop, et al. "A High-Fat Diet Is Associated with Endotoxemia That Originates from The Gut." Gastroenterology 142.5 (2012): 1100-1101.

94) Candido, Thalita Lin Netto, et al. "Dysbiosis and Metabolic Endotoxemia Induced by High-Fat Diet." Nutricion Hospitalaria (2018).

95) Erridge, Clett, et al. "A High-Fat Meal Induces Low-Grade Endotoxemia: Evidence of A Novel Mechanism of Postprandial Inflammation." The American Journal of Clinical Nutrition 86.5 (2007): 1286-1292.

96) Creasman, William T. "Hormone replacement therapy after cancers." Current opinion in oncology 17.5 (2005): 493-499.

97) DiSaia, Philip J., et al. Clinical Gynecologic Oncology. Elsevier Health Sciences, 2017.

98) DiSaia, Philip J., et al. Women's Health Review E-book: A Clinical Update in Obstetrics-Gynecology. Elsevier Health Sciences, 2012.

99) Creasman, William T., and Philip J. DiSaia. "Hormone Replacement and Breast Cancer Risk: Reconsidering the Data." Breast Cancer 23.12 (2009).

100) Xydakis, Antonios M., et al. "Hormone Replacement Therapy in Breast Cancer Survivors." Annals of the New York Academy of Sciences 1092.1 (2006): 349-360.

101) Durna, Eva M., et al. "Hormone Replacement Therapy After a Diagnosis of Breast Cancer: Cancer Recurrence and Mortality." Medical Journal of Australia 177.7 (2002): 347-351.

102) Durna, E. M., et al. "Breast Cancer in Premenopausal Women: Recurrence and Survival Rates and Relationship to Hormone Replacement Therapy." Climacteric 7.3 (2004): 284-291.

103) Batur, Pelin, et al. "Menopausal Hormone Therapy (HT) In Patients with Breast Cancer." Maturitas 53.2 (2006): 123-132.

104) Holmberg, Lars, et al. "Increased Risk Of Recurrence After Hormone Replacement Therapy In Breast Cancer Survivors." Journal of the National Cancer Institute 100.7 (2008): 475-482.

105) Holmberg, L., and H. Anderson. "HABITS (Hormonal Replacement Therapy After Breast Cancer—Is It Safe?): A Randomized Comparison Trial Stopped." Obstetrical & gynecological survey 59.6 (2004): 442-443.

106) von Schoultz, Eva, and Lars E. Rutqvist. "Menopausal Hormone Therapy After Breast Cancer: The Stockholm Randomized Trial." Journal of the National Cancer Institute 97.7 (2005): 533-535.

107) Col, Nananda F., Jung A. Kim, and Rowan T. Chlebowski. "Menopausal Hormone Therapy After Breast Cancer: A Meta-Analysis and Critical Appraisal of The Evidence." Breast Cancer Research 7 (2005): 1-6.

108) Culhane, Rose, et al. "Menopausal Hormone Therapy in Breast Cancer Survivors." Cancers 16.19 (2024): 3267.

109) Plu-Bureau, G., et al. "Percutaneous Progesterone Use and Risk of Breast Cancer: Results From A French Cohort Study Of Premenopausal Women With Benign Breast Disease." Cancer Detection and Prevention 23.4 (1999): 290-296.

110) Collins, John A., Jennifer M. Blake, and Pier Giorgio Crosignani. "Breast Cancer Risk With Postmenopausal Hormonal Treatment." Human Reproduction Update 11.6 (2005): 545-560.

111) Bluming, Avrum Z. "Safety of Systemic Hormone Replacement Therapy In Breast Cancer Survivors." Breast Cancer Research and Treatment 191.3 (2022): 685-686.

112) Bluming, Avrum Z., and Carol Tavris. "Hormone Replacement Therapy: Real Concerns And False Alarms." The Cancer Journal 15.2 (2009): 93-104.

113) Bluming, Avrum Z., Howard N. Hodis, and Robert D. Langer. "'Tis But A Scratch: A Critical Review Of The Women's Health Initiative Evidence Associating Menopausal Hormone Therapy With The Risk Of Breast Cancer." Menopause (2023): 10-1097.

114) Bluming, Avrum. Estrogen Matters: Why Taking Hormones in Menopause Can Improve Women's Well-Being and Lengthen Their Lives-Without Raising the Risk of Breast Cancer. Hachette UK, 2018.

115) Bluming, Avrum Z., et al. "A Decline In Breast-Cancer Incidence." N Engl J Med 357.5 (2007): 509.

116) Chlebowski, Rowan T., and Ross L. Prentice. "Menopausal Hormone Therapy In BRCA1 Mutation Carriers: Uncertainty And Caution." Journal of the National Cancer Institute 100.19 (2008): 1341-1343.

117) Rebbeck, Timothy R., et al. "Effect of Short-Term Hormone Replacement Therapy On Breast Cancer Risk Reduction After Bilateral Prophylactic Oophorectomy In BRCA1 And BRCA2 Mutation Carriers: The PROSE Study Group." Journal of Clinical Oncology 23.31 (2005): 7804-7810.

118) Armstrong, Katrina, et al. "Hormone Replacement Therapy And Life Expectancy After Prophylactic Oophorectomy In Women With BRCA1/2 Mutations: A Decision Analysis." Journal of Clinical Oncology 22.6 (2004): 1045-1054.

119) Griggs, Jennifer J., et al. "American Society of Clinical Oncology Endorsement Of The Cancer Care Ontario Practice Guideline On Adjuvant Ovarian Ablation In The Treatment Of Premenopausal Women With Early-Stage Invasive Breast Cancer." Journal of clinical oncology 29.29 (2011): 3939-3942.

120) King, Mary-Claire, Joan H. Marks, and Jessica B. Mandell. "Breast and ovarian cancer risks due to inherited mutations in BRCA1 and BRCA2." Science 302.5645 (2003): 643-646.

121) Neibergs, Holly. "Breast and Ovarian Cancer Risks Due To Inherited Mutations In BRCA1 And BRCA2." The Women's Oncology Review 4.1 (2004): 59-60.

122) Eisen, Andrea, et al. "Hormone Therapy And The Risk Of Breast Cancer In BRCA1 Mutation Carriers." Journal of the National Cancer Institute 100.19 (2008): 1361-1367.

123) Chen, Haixia, et al. "Association between BRCA status and triple-negative breast cancer: a meta-analysis." Frontiers in pharmacology 9 (2018): 909.

124) Kowalska, Elzbieta, et al. "Increased Rates of Chromosome Breakage In BRCA1 Carriers Are Normalized By Oral Selenium Supplementation." Cancer Epidemiology Biomarkers & Prevention 14.5 (2005): 1302-1306.

125) Huzarski, Tomasz, et al. "A Lowering of Breast And Ovarian Cancer Risk In Women With A BRCA1 Mutation By Selenium Supplementation Of Diet." Hereditary Cancer in Clinical Practice 4 (2006): 1-1.

126) Dziaman, Tomasz, et al. "Selenium Supplementation Reduced Oxidative DNA Damage in Adnexectomized BRCA1 Mutations Carriers." Cancer Epidemiology, Biomarkers & Prevention 18.11 (2009): 2923-2928.

127) Chen, Haixia, et al. "Association Between BRCA Status and Triple-Negative Breast Cancer: A Meta-Analysis." Frontiers in pharmacology 9 (2018): 909.

128) Demircan, Kamil, et al. "Serum Selenium, Selenoprotein P And Glutathione Peroxidase 3 As Predictors of Mortality And Recurrence Following Breast Cancer Diagnosis: A Multicentre Cohort Study." Redox Biology 47 (2021): 102145.

129) Szwiec, Marek, et al. "Serum Selenium Level Predicts 10-Year Survival After Breast Cancer." Nutrients 13.3 (2021): 953.

130) Sandsveden, Malte, et al. "Prediagnostic Serum Selenium Levels In Relation To Breast Cancer Survival And Tumor Characteristics." International Journal of Cancer 147.9 (2020): 2424-2436.

131) Lu, Yen-Shen, Andrea Wong, and Hee-Jeong Kim. "Ovarian Function Suppression With Luteinizing Hormone-Releasing Hormone Agonists For The Treatment Of Hormone Receptor-Positive Early Breast Cancer In Premenopausal Women." Frontiers in Oncology 11 (2021): 700722.

132) Prowell, Tatiana M., and Nancy E. Davidson. "What is the Role of Ovarian Ablation In The Management Of Primary And Metastatic Breast Cancer Today?" The oncologist 9.5 (2004): 507-517.

133) Sellers, Thomas A., et al. "The Role of Hormone Replacement Therapy in The Risk For Breast Cancer And Total Mortality In Women With A Family History Of Breast Cancer." Annals of Internal Medicine 127.11 (1997): 973-980.

134) Letter to the Editor Hormone Replacement and Breast Cancer Risk: The Authors Reply Nov 13, 2009 Volume: 23 Issue: 12 Breast Cancer, Oncology Journal. https://archive.fo/9Z74r#selection-585.0-645.16

135) Labriola, Dan, Kathleen Pratt, and Patrick Bufi. "Natural Hormone Replacement and Breast Cancer Risk: Evidence for Safety and Efficacy." Oncology 23.7 (2009): 639-639.

136) Santoro, Nanette, et al. "The Menopause Transition: Signs, Symptoms, And Management Options." The Journal of Clinical Endocrinology & Metabolism 106.1 (2021): 1-15.

137) Natrajan, Puthugramam K., and R. Don Gambrell Jr. "Estrogen Replacement Therapy In Patients With Early Breast Cancer." American Journal Of Obstetrics And Gynecology 187.2 (2002): 289-295.

138) Bennink, Herjan JT Coelingh, et al. "The Use of High-Dose Estrogens for The Treatment of Breast Cancer." Maturitas 95 (2017): 11-23.

139) Chimento, Adele, et al. "Estrogen Receptors-Mediated Apoptosis in Hormone-Dependent Cancers." International Journal of Molecular Sciences 23.3 (2022): 1242.

140) Shete, Nivida, et al. "Revisiting Estrogen for the Treatment of Endocrine-Resistant Breast Cancer: Novel Therapeutic Approaches." Cancers 15.14 (2023): 3647.

141) Abderrahman, Balkees, and V. Craig Jordan. "Estrogen For the Treatment and Prevention of Breast Cancer: A Tale Of 2 Karnofsky Lectures." The Cancer Journal 28.3 (2022): 163-168.

142) Haddow, Alexander, et al. "Influence of Synthetic Estrogens on Advanced Malignant Disease." British Medical Journal 2.4368 (1944): 393.

Chapter 7

Errors in Modern Medicine, the Fear of Estrogen

REBECCA IS A 52-YEAR-OLD POST-MENO-PAUSAL school teacher sitting in my office crying because she knows she needs menopausal hormone replacement if she wants to remain healthy, and yet her primary care doctor and OB/Gyne doctor have both told her estrogen is dangerous and causes breast cancer. Both doctors refused to prescribe hormone replacement for her. In addition, Rebecca's two closest friends have advised her against it, citing a family member currently diagnosed with breast cancer and undergoing chemotherapy. Rebecca is torn between two opposing viewpoints, creating mental tension and despair. This is a recurring scenario across the country. As is my usual practice, I explained to Rebecca there is no need to decide right now, there is plenty of time to study the issue. I even pointed her to reading material on my blogs, my newsletters, and books discussing the safety and efficacy of menopausal hormone replacement. I then told Rebecca that if she gains a greater understanding and feels more comfortable in the future, I would be glad to get her started on the hormone replacement program. There is no pressure, and I respect whatever decision she feels comfortable with. Finally, Rebecca agreed with the plan and left the office in good spirits. This is a recurring scenario in my office, duplicated in countless doctor's offices across the country.

Errors in Modern Medicine

One of the most glaring errors in conventional medicine is the false idea that estrogen causes breast cancer and blood clotting. Modern medicine dogmatically says estrogen should never be prescribed to post-menopausal females, and especially never to breast cancer survivors. The resulting mass media propaganda campaign creates fear of estrogen throughout the population. Nothing could be farther from the truth. The reality is menopausal hormone therapy (MHT) with estrogen is safe and effective when prescribed properly. The route of delivery and the exact hormone formulation matter. In this chapter, we will go through the studies showing estrogen prevents rather than causes breast cancer. We will also discuss the safety of transdermal estrogen. When estrogen is applied as a topical cream via the transdermal route, estrogen is safe and does not cause blood clotting. That is why we avoid oral estrogen tablets which are associated with blood clots, increased coagulation, and deep venous thromboembolism (DVT). For more on this topic see Chapters 4 and 5 on The Safety of Transdermal vs Oral Estrogen.

Premarin-Alone (Estrogen) Users Have 23 percent less Breast Cancer Than Placebo

The second arm of the WHI study included 10,739 post-menopausal women after hysterectomy randomized to estrogen (Premarin, CEE) or to placebo. In 2004, the second arm of the Women's Health Initiative (WHI) was published in the Journal of the American Medical Association (JAMA). This data included 6.8 years of follow-up showing 23% less invasive breast cancer in the Premarin (CEE, conjugated equine estrogen) treated group compared to the placebo group. There were 94 breast cancer cases in the estrogen-treated group (Premarin, CEE) and 124 cases of breast cancer in the placebo group. (1-2)

Estrogen Group Second Arm of WHI: 45% Reduction in Mortality from Breast Cancer

In 2018, Dr. Howard Hodis reviewed the 18-year follow-up of the 2004, second arm of the WHI (Premarin-alone, CEE) showing a **45**

percent reduction in breast cancer mortality in the estrogen user group!! Remember, this group took estrogen for 7.2 years, and then stopped and followed for an additional 11 years. Dr. Howard Hodis writes:

> After 18 years of cumulative follow-up of the WHI-CEE cohort, **breast cancer mortality was statistically significantly reduced by 45%** (HR, 0.55; 95% CI, 0.33–0.92). This may well be the most significant and most overlooked finding of the WHI-CEE trial (3).

Breast Cancer Mortality in Both Arms of WHI – 18 Year Follow-Up

First Arm Breast Cancer Mortality CEE+MPA = 61 vs. Placebo = 40

Hazard Ratio (HR)=1.44 meaning a 44% increase in breast cancer mortality.

Second Arm Breast Cancer Mortality CEE-alone = 22 vs. Placebo = 41

Hazard Ratio (HR)= **0.55** = 45% reduction in breast cancer mortality

The First Arm of the Women's Health Initiative used Premarin (CEE) Plus Medroxyprogesterone yielding a 26 percent increased risk of breast cancer in the hormone-treated group. Remember there are two arms of the WHI. We discussed the second arm from 2004 above. Now, we will go back two years to 2002 and discuss the first Arm. (3)

First Arm of WHI, 2002

The 2002, first arm of the WHI randomized 16,608 post-menopausal women to either Prempro (Premarin plus Medroxyprogesterone, MPA) or placebo. After 5.2 years of follow-up, there were 290 cases of breast cancer in the Prempro group which is 26% greater than the placebo group. The authors write:

> The 26% increase (38 vs 30 per 10,000 person-years) observed in the estrogen plus progestin group almost reached nominal statistical significance and, as noted

herein, the weighted test statistic used for monitoring was highly significant...Estimated hazard ratios (HRs) (nominal 95% confidence intervals [CIs]) were as follows: CHD [coronary heart disease], **1.29** (1.02-1.63) with 286 cases; breast cancer, **1.26 (1.00-1.59)** with 290 cases; stroke, **1.41** (1.07-1.85) with 212 cases; PE [pulmonary embolus], **2.13** (1.39-3.25) with 101 cases; colorectal cancer, **0.63** (0.43-0.92) with 112 cases; endometrial cancer, **0.83** (0.47-1.47) with 47 cases; hip fracture, **0.66** (0.45-0.98) with 106 cases; and death due to other causes, **0.92** (0.74-1.14) with 331 cases. (4)

Note: CHD=Coronary Heart Disease. PE=Pulmonary embolus. Note the increased risk for CHD, breast cancer, stroke, PE (pulmonary embolus), and reduced risk for colorectal cancer, endometrial cancer, and hip fracture in the hormone-treated (CEE/MPA) group. Increased stroke and pulmonary embolus, venous thromboembolism (VTE) is due to the oral estrogen route of delivery and can be avoided with transdermal estrogen. For more on this see Chapters 5-6 on the Safety of Transdermal Estrogen Parts One and Two.

Hazard Ratio for Breast Cancer HR=1.26 Not Statistically Significant

In 2002, Dr. Jacques E. Rossouw, a cardiologist working for the NIH, published the first arm of the Women's Health Initiative (WHI) in JAMA. This study was halted early after 5.2 years because of increased breast cancer in the estrogen-treated group. Questions were raised about unusual aspects of how the study came to be published. The leadership made a press release before the publication date over the objections of statisticians in the research group, as discussed in 2017 by Dr. Robert Langer. In 2003, Dr. Ronald Strickler, Chair of Women's Heath, OB/GYN Senior Staff at Henry Ford Hospital, Detroit reviewed the data from the 2002 first arm WHI. Dr. Strickler points out the study was halted early after 5.2 years based on data that was not statistically significant. The hazard ratio (HR) was 1.26. However, the confidence interval included 1.0 meaning the HR was not statistically significant. However, the committee had set a predetermined safety threshold which had been exceeded. Referring to the 2002, first-arm WHI study, Dr. Ronald Strickler writes:

> The breast cancer hazard ratio **(HR) of 1.26**, with 95% CI of **1.00** –1.59, exceeded the committee's predetermined threshold, and hence the study was stopped. Notice that the 95% CI includes 1.00: the HR is not statistically significant. (126) (7)

In 2024, Drs. Avrum Z. Bluming, Howard N. Hodis, and Robert D. Langer responded to a letter to the journal, Menopause, written by Dr. Rowan T. Chlebowski. Dr. Bluming makes a few objections to the WHI study. The results of the 2002, first arm study using Premarin and medroxyprogesterone (MPA) were misrepresented and were not statistically significant, writing:

> We agree that the Women's Health Initiative (WHI)'s repeated assertion that conjugated equine estrogen (CEE) + medroxyprogesterone acetate (MPA) increases the risk of breast cancer is not funny. Neither is their failure to follow protocol, their abandonment of basic statistical norms, and their repeated misrepresentation of their own data. It is not funny that these claims have scared millions of women away from hormone therapy (HT), the safest and most effective treatment for menopausal symptoms, prevention of bone loss and fractures, improvement in quality of life, and reduction of all-cause mortality when started within 10 y of menopause…. Although the 2002 WHI article generated headlines reporting an increased risk of breast cancer among CEE + MPA recipients, the authors actually reported that breast cancer "almost reached nominal statistical significance" with a 95% confidence interval (CI) of 1.00 to 1.59. It was clearly

non-significant with the appropriate adjustment for multiple comparisons (95% CI, 0.83-1.92)… In the randomized trial phase of the WHI, the increase in the hazard rate of breast cancer for women on CEE + MPA, compared with women on placebo, is attributable to the unexplained extremely low rate of incident breast cancer in the women with prior HT assigned to placebo, not to an increased breast cancer risk among those randomized to CEE + MPA… Although the WHI has consistently ignored these critiques, at worst these investigators claim a rare increase of 1 additional case of non-fatal breast cancer for every 1,000 women on CEE + MPA for 1 year. **Note:** The carcinogenic effects of synthetic progestin drugs such as MPA is more completely discussed in later chapters. To cut to the chase, MPA is carcinogenic because of its ability to antagonize testosterone receptors. However, Premarin has anti-cancer activity, which offsets the carcinogenic properties of MPA, so the final result was neutral in this particular WHI study. The exact hormone formula is important. (127)

Dismantling HRT Medical Care

In 1996, the lead researcher for the WHI, Dr. Jacques E. Rossouw had already expressed an anti-hormone replacement therapy (HRT) bias. The title of Dr. Roussouw's article was: "Putting the Brakes on the [HRT] Bandwagon." Thus, a hidden agenda opposing HRT was at play. The press release of the 2002 WHI trial created havoc in medical practice and the general population. Fear of estrogen caused a precipitous decline in HRT use. Following the 2002 WHI publication in JAMA, the entire medical system of menopausal hormone replacement was dismantled. Training programs disappeared. New doctors entering practice lacked the expertise to prescribe hormone replacement. This is still the current situation in the U.S. HRT use has declined by 90 percent from 26.9 to 2.8 percent of menopausal women, according to Dr. Lin Yang writing in 2024. (4-8)

Synthetic Progestins Are Carcinogenic

The synthetic progestin, medroxyprogesterone (MPA) is a known breast cancer carcinogen and is commonly used to induce breast cancer in animal models. On the other hand, natural progesterone is non-carcinogenic and breast cancer-protective. So, the question here is: if MPA is carcinogenic, why didn't the 2002 study show a statistically significant increase in breast cancer in the Premarin/MPA group. The HR=1.26 was not statistically significant. The answer is the anti-cancer effects of Premarin. Premarin contains unsaturated B-ring steroids which preferentially target ER-beta, opposing the proliferative effects of MPA. This is more completely discussed in Chapter 3, Do Not Monkey with My Hormones. (9-14) (53)

Natural Progesterone Incapable of Causing Breast Cancer

In 2020, Dr. Kathryn B. Horwitz, Distinguished Professor at the University of Colorado, reviewed the past 90 years of progesterone receptor research and the role of the progesterone receptor (PR) in tumorigenesis. Firstly, Dr Horwitz debunks the idea that natural progesterone is carcinogenic, stating **progesterone is incapable of causing breast cancer**. On the other hand, synthetic progestins, such as MPA [medroxyprogesterone], norethindrone acetate, and levonorgestrel are confirmed breast carcinogens with 200% increased incidence of breast cancer in long-term users. Dr. Horwitz recognizes the two terms, natural progesterone, and synthetic progestins are commonly lumped together creating confusion in the medical literature. One must take care in differentiating natural progesterone from synthetic chemically altered progestins, writing:

> In the early 2000s, the somewhat surprising finding that prolonged use of synthetic progestin-containing menopausal hormone therapies was associated with increased breast cancer incidence raised new questions about the role of PR in

'tumorigenesis'…First, we need to debunk the notion that progesterone 'causes' breast cancers. There is considerable experimental and clinical evidence that, alone and at physiological levels, **progesterone is incapable of causing breast cancers** so that its reputation as a 'tumorigenic' or 'carcinogenic' hormone is undeserved…A 2019 meta-analysis by the Collaborative Group on Hormonal Factors in Breast Cancer confirmed the increased risk of breast cancer for MHT [Menopausal Hormone Therapy] containing MPA [medroxyprogesterone], norethindrone acetate, or levonorgestrol, compared to never users or estrogen-only users (Collaborative Group on Hormonal Factors in Breast 2019). This was especially pronounced for long-term (>10 year) progestin users, who had twice the risk of developing breast cancer. Notably, this meta-analysis did not include bioidentical progesterone formulations, which had either no additional risk or even decreased breast cancer risk (discussed in Piette 2018)…Furthermore, **it is important to distinguish between progestins and natural progesterone.** Currently, these tend to be lumped together leading to the view that progesterone is 'carcinogenic' (i.e. cancer causer). It is our opinion that **natural progesterone does not 'cause' breast cancer** but can expand it (see subsequent section). Hence, despite widespread linkage between the terms 'progestins' and 'carcinogenesis', we suggest that care must be taken with these ideas, as with the term 'bioidentical', until solid data are available, in women, differentiating between the natural hormone and any biosynthetic ones. (15)

The WHI Trial Did Not Study Women's Hormones at All!!

In 2011, Dr. Cynthia L. Bethea, Emeritus Scientist and Professor, Oregon National Primate Research Center, Oregon Health and Science University, reviewed the catastrophic findings of the first arm of the WHI study showing increased breast cancer risk from menopausal hormone replacement. Dr. Bethea remarks these WHI findings were considered an indictment of menopausal hormone replacement, **when in fact, the WHI trial did not study women's hormones at all, which would have required the use of natural hormones, estradiol, and progesterone**. Remember this WHI first arm study used Premarin (CEE) and Provera (MPA, medroxyprogesterone). Premarin is estrogen derived from a pregnant horse, and MPA is a synthetic progestin that does not occur naturally in the human body or anywhere else in nature. **To study women's hormones would have required the use of human bioidentical hormones** in a combination of human estrogen (usually estradiol and estriol), natural progesterone, and natural testosterone, all present naturally in the human body. Dr. Cynthia L. Bethea writes:

> While the WHI trial [2002, first arm] made a valuable contribution in revealing the risks associated with conjugated equine estrogens [Premarin, CEE] plus MPA [medroxyprogesterone] treatment in postmenopausal women, it unfortunately generated considerable controversy in the field because it was interpreted as an indictment of postmenopausal hormone replacement, when in fact, **it did not study hormone replacement at all: that would have required use of the natural hormones, estradiol and progesterone.** The actions of the natural hormones are significantly different from those of Premarin and MPA. (16)

Dr. Rowan Chlebowski, 18.3 Year Follow Up Estrogen Alone 44% reduction in Mortality from Breast Cancer

In 2019, Rowan Chlebowski MD, PhD, chief of the Division of Medical Oncology and Hematology at Harbor-UCLA Medical Center reviewed the **18.3-year follow-up** data from the second arm, 2004 WHI study showing Premarin (CEE)-only users were 23% less likely to be diagnosed with breast cancer, and 44% less likely to die from breast cancer. However, when MPA is added to Premarin as in the 2002 first arm (CEE/MPA), then hormone users had a **29 percent increase in breast cancer**, writing:

> **After 16.1 years** of cumulative follow-up... Compared with women who had received placebo, those who had received CEE [Premarin] were 23 percent less likely to have been diagnosed with breast cancer and **44 percent less likely to die** from the disease...**After 18.3 years** of cumulative follow-up, among those who received CEE plus MPA, ...Compared with women who had received placebo, those who had received CEE plus MPA were **29 percent more likely to have been diagnosed with breast cancer**...Estrogen alone decreased, while adding progestin increased, breast cancer incidence...CEE-alone [Premarin-Alone] and CEE plus MPA [medroxyprogesterone] use have opposite effects on breast cancer incidence. CEE alone significantly decreases breast cancer incidence which is long term and persists over a decade after discontinuing use. **CEE plus MPA use significantly increases breast cancer incidence which is long term and persists over a decade after discontinuing use.** (17)

Re-Analysis of First Arm WHI Data Shows Glaring Error

In 2018, Drs. Howard Hodis and Phillip Sarrel re-analyzed the data from the WHI 2002, first arm. They found an error in the study. Some of the women in the placebo group had a history of prior HRT use. These women should have been removed from the placebo group, yet they were not. The prior use of estrogen confers protection from breast cancer and falsely reduces the incidence of breast cancer in the placebo group. If the women with prior HRT use are removed from the placebo group, the data chart is entirely different. There is a null effect, meaning no difference between the Prempro group (CEE/MPA) and the placebo group in breast cancer incidence. Dr. Howard Hodis writes:

> In fact, the increased HR [Hazard Ratio] was not due to an increased breast cancer incidence rate in women randomized to

CEE + MPA [Prempro] therapy but rather due to a decreased and unexpectedly low breast cancer rate in the subgroup of women with prior HT [Hormone Therapy] use randomized to placebo. For women who were HT [Hormone Therapy] naïve when randomized to the WHI [Women's Health Initiative], the breast cancer incidence rate was not affected by CEE + MPA therapy relative to placebo for up to 11 years of follow-up…the data clearly show that CEE+MPA therapy had a null effect on breast cancer risk particularly in the subgroup of women representing the typical population of women treated with HT who are HT naïve before receiving menopausal HT…it is clear that breast outcome data from the WHI-CEE+MPA trial has been misinterpreted and overgeneralized. (3)

Dr. Lindsey Berkson: Estrogen is Breast Cancer Protective.

On October 24, 2023, Dr Lindsey Berkson discussed this issue on her substack blog, writing:

The WHI 1 [2002 first arm] authors forgot to ask, and thus did not remove, in the placebo group, any women who had already taken estrogen therapies…**Since estrogen was ultimately, on re-analysis found to be "breast protective",** and the synthetic progestins were huge adverse contributing issues, this made the incidence of breast cancer in the placebo group of ladies, lower, as women had already taken the protective estrogen…This made the experimental arm, women taking hormones, falsely appear as though they had more cases of breast cancer…When the data was re-analyzed, thus "righted" by taking out this "confounding issue", and longer term effects of estrogen looked at with a fine tooth comb, the same original authors, re-published their re-analysis… This is now what I call the WHI 2 [2004, second arm]…Women taking estrogen replacement therapy (ERT) had **23% less incidence or chance of getting breast cancer in the first place**. And if you had been on ERT, and got breast cancer,

you had a **44% decreased chance of dying from it**. Progestins were more the breast damaging issue. Not estrogen. So, stunning as it is, being on ERT [estrogen replacement therapy] put you in a better position, even if you went on to get breast cancer! This is something that is now replicated. Substantiated. But not taught in most med schools or appreciated by most docs and women. Or lawyers!…**Estrogen "protects" healthy breast tissue from getting breast cancer in the first place.** And, if a women has been on estrogen therapies for an average of 5 years and then gets breast cancer, **the estrogen therapies "reduce" her risk of dying from breast cancer by 44%. Nothing else like estrogen therapies has ever been shown to be so breast protective.** (18)

Natural Progesterone is also Breast Cancer Preventive

In 2017, Dr. Allan Lieberman reviewed the medical literature on progesterone as a breast cancer preventive agent, finding that a common error is confusing natural progesterone with synthetic progestins such as MPA, thus falsely attributing to progesterone the carcinogenic properties of progestins. Dr. Allan Lieberman writes:

The literature is extensive on the effects of estrogen and progesterone and their relationships to breast and other cancers and other health-related effects. Much of the medical literature on progesterone or progesterone-like compounds is contradictory, with progesterone sometimes implicated as a cause of breast cancer. These contradictory results are the result of researchers **confusing the effects of synthetic progestins with those of natural progesterone**…To avoid confusion surrounding the long-term health benefits and consequences of using progestogenic drugs, we recommend that the term progesterone be used only for the naturally occurring progestogen, P4, whereas the term progestin be used for any of the synthetic versions. The interchangeable use of these terms in scientific, medical, and lay articles confounds the interpretation of data from these different classes of progesterone receptor (PR) ligands and their implications for human health...**The evidence strongly suggests that natural progesterone is protective and preventive of breast cancer**... The authors wish to emphasize that natural progesterone is preventive of breast and endometrial cancer, and physicians should have no hesitation prescribing it. (19)

Two French Studies Using Estradiol and Micronized Progesterone

In 2007, Dr. Stephen Birrell reviewed the medical literature on synthetic progestins increasing risk for breast cancer via disruption of androgen receptor signaling. Dr. Birrell comments that when natural progesterone (micronized) is used instead of synthetic progestins, no such carcinogenic effects are observed. Dr. Birrell cites two French studies using micronized progesterone and estradiol for menopausal hormone replacement in which there was no observed increase in breast cancer, the E3N-EPIC cohort of 54,548 women by Dr. Fournier (2005) and a smaller study by Dr. de Lignieres (2002) of 3,175 women, writing:

Notably, in France the majority of women taking combined HRT [Hormone replacement Therapy] receive oral micronized progesterone rather than a synthetic progestin. In two French studies- the E3N-EPIC cohort of 54,548 women and a smaller study of 3175 women, no significant increase in breast cancer risk due to HRT use with micronized progesterone was observed compared with untreated women. (20-22)

The Estrogen Paradox

Notice the above WHI second arm study showed a reduced incidence of breast cancer with the use of estrogen. However, this estrogen was not human 17-beta-estradiol. The WHI 2nd arm used Premarin, pregnant horse estro-

gen called CEE (conjugated equine estrogen). In addition, the average age of the women was 63 years, meaning they were started on HRT after long term estrogen deprivation (LTED). After LTED, estrogen changes from a cancer growth factor to a cancer death factor. This is called the estrogen paradox described by V. Craig Jordan. (23-25)

Estradiol Alone without LTED

What if the 2004, second arm WHI study had used human estrogen (estradiol) rather than horse estrogen (Premarin, CEE)? And what if menopausal women started HRT earlier, immediately after the menopausal transition in the 50 to 55-year age group. In this scenario, HRT is started without LTED, and without the added natural progesterone. We have three large observational studies looking at this, the 2008 French E3N Cohort study by Dr. Agnes Fournier, the 2006 Finland study by Dr. Heli Lyytinen, and the 2011 EPIC study by Dr. Kjersti Bakke. All three observational studies are in close agreement that estradiol alone without LTED, and without natural progesterone increases breast cancer risk by about 25-30 percent. Increased breast proliferation with estradiol was also found by Dr. Charles Wood's primate studies comparing proliferative effects of estradiol (E2) to that of CEE (Premarin) which is discussed below. When natural progesterone is added to the estradiol, breast cancer risk is eliminated. For more on this see Chapters 16 and 17 on The Safety of Bioidentical Hormones. (26-28)

ER-alpha and ER-beta

Estrogen exerts its effects through two estrogen receptors, ER-alpha and ER-beta. Much of the early work on estrogen receptors was done by Elwood Jensen and Jan-Åke Gustafsson. ER-alpha was cloned in 1986 from human breast cancer (MCF-7) cells and ER-beta was cloned in 1996 from rat prostate cells. Estrogen receptors alpha (ER-alpha) and beta (ER-beta) have opposing roles. ER-alpha is proliferative and activates the oncogenes, Cyclin D1 and

C-Myc. ER-beta opposes the ER-alpha activation of Cyclin D1, thus serving as a tumor suppressor, downregulating the proliferative effects of ER-alpha. (29-36)

Premarin (CEE) is Less Proliferative than Estradiol

In 2008, Dr. Charles E. Wood in Winston Salem studied a post-menopausal primate model to compare the proliferative effects of CEE (Premarin, horse estrogen) to that of 17-beta estradiol (human estrogen). Proliferation was measured with the KI-67 marker, a commonly used proliferative marker in clinical pathology. Dr. Wood found a highly significant 259-330% increase in breast cell proliferation in estradiol-treated monkeys compared to controls. However, in the Premarin (CEE) treated monkeys there was far less breast cell proliferation, only 75% compared to controls. This difference in proliferative effect was explained in 2024 by Dr. Barbara Levy writing that horse estrogen, CEE, contains unsaturated B ring steroids which preferentially bind to and activate ER-beta, thus explaining less proliferative effects compared to estradiol. There are no unsaturated B-ring steroids in human estrogen. However, human hormones with ER-beta binding preference are estriol (E3) and testosterone metabolite 3Beta-diol, thus explaining their cancer-protective properties. These are discussed in later chapters. (37-40)

B-Ring Steroids in CEE are Breast Cancer Preventive

In 2015, Dr. Valerie A. Flores, Assistant Professor of Obstetrics, Gynecology at the Yale School of Medicine, discusses the mechanism of action to explain the less proliferative effect of CEE (horse estrogen) compared to estradiol (human estrogen). Horse estrogen (CEE) contains a mixture of eleven estrogens, many of which are unsaturated B-ring steroids binding preferentially to ER-beta, thus downregulating the proliferative ER-alpha receptor, thus explaining the less proliferative effect of CEE compared to estradiol. Dr. Valerie A. Flores suggests B-ring steroids may be beneficial in the treatment of breast cancer, writing:

> The choice of CEE [horse estrogen] in the ET arm [estrogen only second arm] of the WHI may explain the favorable effects seen on the breast. CEE contains a mixture of multiple estrogens, and each estrogen-type not only preferentially binds the two estrogen receptors, but may also exert differential actions depending on the target tissue. While E2 [human estradiol] is the well characterized estrogen, less is known about the many estrogenic components of CEE. Unlike E2, these other estrogens differ in their B-ring saturation and in their chemical moieties at the 17-position...Bhavnani et al analyzed the effects of 11 equine estrogens (in CEE preparations) on the transcriptional activity of ER alpha and beta, and found that **many of the equine estrogens preferentially bind ER beta. ER beta activation can inhibit ER alpha activity on cell proliferation. This inhibition induced by equine estrogens may in part explain the decreased risk of breast cancer observed in the WHI ET study [2004 second arm Womens Health Initiative Estrogen Therapy-only]**. (41)

With the above knowledge that CEE (horse estrogen) has less proliferative effects compared to estradiol (E2), one might ask the next logical question, why not use CEE (horse estrogen) for all menopausal hormone replacement? After all, horse estrogen (CEE) was studied in the WHI (Women's Health Initiative) randomized controlled trial and was found to reduce breast cancer risk by 23 percent compared to placebo, thus proving horse estrogen (CEE) is breast cancer preventive. My answer is CEE is certainly a good choice. The B-Ring steroid component of CEE targets preferentially the ER-beta, thus downregulating proliferative effects, serving as breast cancer prevention. However, oral Premarin (CEE) increases venous thromboembolism (VTE), while transdermal estrogen does not. A better approach is the use transdermal estradiol (E2) combined with estriol (E3). Although estradiol (E2) binds equally to ER-alpha and ER-beta, estriol (E3) preferentially binds to and activates ER-beta, a tumor suppressor.

Progesterone Acts as Proliferative Brake

Natural progesterone is an important addition to our hormone replacement formula. In 2015, Dr. Hisham Mohammed, Assistant Professor of Molecular and Medical Genetics CEDAR, OHSU Knight Cancer Institute, School of Medicine, states that natural progesterone functions as a proliferative brake on ER-alpha and downregulates estradiol-stimulated ER-alpha proliferative effects, writings:

> We conclude that activation of PR [progesterone receptor] results in a robust association between PR and the ER-alpha complex… PR is a critical determinant of ER-alpha function due to crosstalk between PR and ER-alpha. In this scenario, under estrogenic conditions, an **activated PR functions as a proliferative brake in ER-alpha+ breast tumors** by re-directing ER-alpha chromatin binding and altering the expression of target genes that induce a **switch from a proliferative to a more differentiated state**. (42)

Hormone Symphony

Testosterone is metabolized to 3-beta-diol which preferentially binds to and activates ER-beta. So, as you can see, we have a combination of hormones that create a hormone symphony, i.e. a combination of estriol, progesterone, and testosterone serving to downregulate the proliferative effects of estradiol. We know from the French Cohort study by Dr. Agnes Fournier that the addition of natural progesterone to estradiol (E2) reduces the HR 1.25 for breast cancer down to 1.00. In other words, the overall risk of breast cancer is reduced to that of the general population serving as controls. Testosterone is an excellent breast cancer preventive, reducing breast cancer risk by 40 percent based on studies by Drs. Rebecca Glaser

and Gary Donovitz. The net result is an excellent breast cancer preventive HRT program. For more on this, see Chapter 18 on Testosterone for Prevention and Treatment of Breast Cancer. (43-52)

ER-Beta as Tumor Suppressor

In 2020, Dr. Rahul Mal reviewed the medical literature on ER-Beta as a broad-spectrum ligand-activated tumor suppressor that could potentially treat breast cancer and many other cancers without the side effects of chemotherapy, writing:

> High ER-beta1 expression is associated with improved overall survival in women with breast cancer. The promise of ER-beta activation, as a potential targeted therapy, is based on concurrent activation of multiple tumor suppressor pathways with few side effects compared to chemotherapy. **Thus, ER-beta is a nuclear receptor with broad-spectrum tumor suppressor activity,** which could serve as a potential treatment target in a variety of human cancers including breast cancer. Further development of highly selective agonists that lack ER-beta agonist activity, will be necessary to fully harness the potential of ER-beta… As with ER-alpha, estrogenic compounds including **estradiol, estrone, and estriol activate ER-beta**. Relative to ER-alpha, **ER-beta binds estriol [E3] and ring B unsaturated estrogens [CEE] with higher affinity, while the reverse is true of 17beta-estradiol and estrone**. On the other hand, the **dihydrotestosterone metabolites** 5-androstenediol and 3-beta androstanediol **are relatively selective (3-fold) for ER-beta over ER-alpha**. (53)

Dr. Henry Lemon, Dr. Jonathan V. Wright, and Estriol

In 1978, Alvin H. Follingstad, MD called estriol (E3), the "forgotten hormone". No longer forgotten, Estriol (E3) now plays a prominent role in bioidentical menopausal hormone replacement as it represents 80 percent of the Bi-est formula (80% E3, 20%

E2) developed by pioneer Jonathan Wright MD, a graduate of Harvard University and the University of Michigan Medical School (1969). Dr. Wright established Tahoma Clinic in 1973 in Washington State, and Meridian Valley Laboratory in 1976. In 1982, Dr. Wright was the first to develop and introduce bio-identical menopause hormone therapy using the Bi-est formula. After many years of experience with urinary hormone testing, Dr. Wright discovered increased urinary estriol (E3) excretion after iodine supplementation, indicating a beneficial effect of iodine on estrogen metabolism. Dr. Wright's interest in iodine was stimulated by the earlier work on estriol (E3) by Dr. Henry Lemon's work showing estriol prevented breast cancer in animal models of chemically induced breast cancer. Another study by Dr. Lemon showed decreased urinary excretion of estriol (E3) in breast cancer patients compared to normal controls. As mentioned above, estriol (E3) preferentially binds to ER-beta thus serving as a breast cancer preventive. Similarly, Dr. Wright was one of the first to draw attention to the cancer-preventive properties of testosterone metabolite 3-beta-diol which preferentially binds to ER-beta. (54-71)

Our Breast Cancer Prevention Program

1) **Natural Progesterone:** Our office uses human bioidentical progesterone, avoiding carcinogenic synthetic progestins such as medroxyprogesterone (MPA) and norethindrone. In 2014, the mechanism of carcinogenicity of MPA was elucidated by Dr. Aleksandra M. Ochnik, finding it is the anti-androgenic effect of MPA causing the problem, writing:

> DHT [Di-Hydro-Testosterone] inhibited the proliferation of breast epithelial cells in an AR [Androgen Receptor]- dependent manner within tissues from postmenopausal women, and **MPA significantly antagonized this androgenic effect**...In a subset of postmenopausal women, MPA [medroxyprogesterone] exerts an antiandrogenic effect on breast

epithelial cells that is associated with increased proliferation and destabilization of AR protein [Androgen Receptor]. **This activity may contribute mechanistically to the increased risk of breast cancer in women taking MPA-containing EPT [estrogen/progestin therapy].** (72)

2) Our office uses Bi-Est which is 80% estriol (E3) and 20% estradiol (E2). Estriol (E3) is breast cancer preventive. Both estriol (E3) and metabolites of testosterone predominantly target the ER-beta receptor which is a tumor suppressor as discussed in 2020 by Dr. Rahul Mal. (53)

3) **Natural Progesterone:** My office uses oral micronized progesterone, FDA-approved (Solvay December 1998) for the prevention of endometrial hyperplasia. The usual dosage is micronized progesterone 100 mg oral capsule taken at bedtime. (19) (73-75)

4) **Iodine:** My office uses the spot urine test for iodine and we provide iodine supplements when low. Iodine decreases breast cancer risk and is useful as an adjunctive treatment for breast cancer. See Chapters 22-23, Iodine Prevents and Treats Breast Cancer. (76-77)

5) My office uses testosterone for routine menopausal HRT. Testosterone is an androgen, and as such is breast cancer preventive, decreasing breast cancer risk by 39-40 percent. The mechanism of action is preferential activation of ER-beta by a metabolites of testosterone as discussed by Dr. Glaser and Donovitz. See the Chapter 18 on Testosterone for Prevention and Treatment of Breast Cancer. (48-49)

6) **DIM:** My office routinely uses DIM (Di-Indole-Methane) which diverts estrogen metabolism toward favorable metabolic pathways. For more on DIM, see Chapter 21. (78-81)

Dr. Ercole Cavalieri Breast Cancer Caused by Catechol Estrogen 3,4, Quinones

Although estrogen itself is protective of breast cancer, some metabolites, estrogen-quinones are carcinogenic. Normal estrogen metabolism disposes of these carcinogenic quinone-estrogen metabolites. However, some women have genetic abnormalities causing bottlenecks in estrogen metabolism pathways which accumulate the carcinogenic 4-hydroxy-quinones. In 2021, Dr. Ercole Cavalieri studied breast cancer formation caused by catechol estrogen 3,4, quinones. These metabolites of estrogen are oxidative and will attach to estrogen receptors on DNA, DNA adducts causing oxidative damage to the DNA. Oxidative damage leads to DNA mutations and cancer. How do we avoid these harmful DNA adducts, the 4-hydroxy estrogen quinones? For more on this see Chapter 21 on Estrogen Metabolism Iodine, DIM and 2MeOE2. Dr. Ercole Cavalieri writes:

> Cancer can be initiated by increased formation of reactive estrogen metabolites called catechol estrogen-3,4-quinones. If estrogen metabolism becomes unbalanced and significant amounts of these quinones arise, depurinating estrogen-DNA adducts are primarily formed, leading to cancer-causing mutations. (82)

7) **Methyl-Folate:** My office routinely gives methyl-folate contained in a high quality multivitamin: Methylation defects (MTHFR polymorphisms) increase breast cancer risk. This may be treated with methyl-folate, provided in most high-quality multivitamins. Resveratrol, NAC (N-Acetyl cysteine), and sulforaphane have been found beneficial in patients with genetic mutations (SNPs) which impair estrogen metabolism and leading to accumulation of carcinogenic quinone-estrogen adducts. (83-88)

8) **Selenium:** My office tests for selenium levels and supplements when found low. Selenium deficiency is a risk factor for all cancers. Selenium is a cancer preventive agent. (89-91)

9) **What About Screening Mammography?**

My office does not recommend screening mammography for the general population because screening mammography results in harm related to overdiagnosis and overtreatment. In JAMA in 2009, Dr. Laura Esserman, professor of surgery and radiology at UCSF and director of the UCSF Breast Care Center since 1996, reviewed 20 years of national breast cancer mortality data finding the national program of screening mammography has not resulted in the anticipated reduction in cancer mortality. In 2021, Dr. Amanda E. Kowalski showed any mortality reduction associated with screening mammography is outweighed by increased mortality from over-treatment. In 2023 in JAMA, Dr. Michael Bretthauer, tenured professor of medicine, and gastroenterologist at Oslo University Hospital, Norway, Associate Editor of the Annals of Internal Medicine, conducted a meta-analysis of all screening tests showing screening mammography does not save lives by extending lifetime, writing:

> The findings of this meta-analysis suggest that current evidence does not substantiate the claim that common cancer screening tests save lives by extending lifetime, except possibly for colorectal cancer screening with sigmoidoscopy. (92)

The second reason we do not do routine screening mammography is that because of the superficial location of breast tissue, most breast cancers are detected fairly early, becoming self-evident to the patient. The physician will then order a diagnostic mammogram and ultrasound exam. The third reason is heavy mass media marketing of screening mammography has convinced the public that screening mammography prevents cancer. This is a false conception. Screening mammography may serve as early detection. However, it is not preventive, rather it gives a radiographic image of the breast, a snapshot in time. While indolent breast cancers such as DCIS (ductal carcinoma in situ) will be detected by the telltale punctate calcifications on screening mammograms, aggressive breast cancers will arise in between screenings. The patient will notice breast distortion, a mass or lump and bring it to their doctor's attention. A true breast cancer prevention program will involve the use of ER-beta ligands estriol, testosterone, and protective hormone progesterone, as mentioned above. Breast cancer prevention also requires supplements such as iodine, selenium, vitamin D3, DIM, methyl-folate, sulforaphane, resveratrol, etc. For more on this topic, see Chapter 29 on Screening Mammography. (93-97)

Synthetic Progestins Are Carcinogenic

Although natural progesterone is breast cancer preventive, all other synthetic progestins are carcinogenic. Examples are medroxyprogesterone (MPA), etonogestrel, levonorgestrel, drospirenone, norethindrone, and megestrol. Notice progesterone has been mistakenly included as a progestin, accounting for the frequent errors in many medical studies that conclude that progesterone is carcinogenic. Pharmaceutical progesterone is chemically identical to the progesterone made by the corpus luteum of the ovary. Progestins are synthetic compounds, chemically altered versions of progesterone that do not occur in nature, and are carcinogenic to breast tissue. (98)

Exposure to Exogenous Estrogen (ERT) Prevents Breast Cancer.

In 2022, Dr. Isaac Manyonda, Professor of Obstetrics and Gynecology at St. George's, University of London, U.K., reviewed the 20-year follow-up of the WHI (estrogen only, second arm 2004 JAMA) showing exposure to exogenous estrogen prevents breast cancer and is protective against osteoporosis and cardiovascular disease, writing:

> The WHI [Women's Health Initiative] study of ERT [CEE estrogen replacement, second arm, 2004] versus placebo in women with a prior hysterectomy is a most robust piece of research: prospective, randomized, placebo-controlled and with a 20-year follow-up, which now compels a direct interpretation of its finding, namely that **exposure to exogenous estrogen (ERT) prevents breast cancer.** This is of profound importance, not only in relation to the prevention of the most common cancer in women in the Western

world but also because estrogen, whilst being cost-effective and well-tolerated also has other **preventative properties against osteoporosis and cardiovascular disease, to name but two**. Note: ERT is estrogen replacement therapy. (99)

Estrogen Induces Apoptosis of Breast Cancer Cells

In 2011, Dr. Andrea Vasconsuelo Research Scientist Universidad Nacional del Sur, Argentina, and Ph.D. in Molecular Sciences, discussed the role of 17β-estradiol for the induction of apoptosis of breast cancer cells, finding estrogen will induce apoptosis after long-term estrogen deprivation or after long-term treatment with anti-estrogen drugs (tamoxifen or aromatase inhibitors) causing upregulation of ER-receptors on the breast cancer cells so that estrogen no longer functions as a survival signal, but as a death signal, writing:

> However, under some specific conditions E2 [estradiol] could trigger apoptosis in breast cancer cells, as opposed to its well-studied antiapoptotic role. This peculiar hormone behavior has been observed in cells from breast cancer which have been **long-term estrogen-deprived (LTED) or treated exhaustively with antiestrogens**. Curiously, the paradoxical induction of apoptosis by estrogen has been established under several unusual circumstances. For example, in this case, the pre-conditions of prolonged estrogen depletion or exhaustive treatment with anti-estrogens of the breast cancer cells are mandatory requisites to trigger apoptosis by E2 and could explain the dual action of the steroid to stimulate growth or apoptosis. Thus, the development of antihormone resistance over years of therapy reprograms the survival mechanism of the breast cancer cell so that **estrogen no longer functions as a survival factor but as a death signal.** (100)

Dr. V. Craig Jordan and the Estrogen Paradox

Dr. V. Craig Jordan was known as the father of tamoxifen, one of the first successful estrogen-blocking drugs used for breast cancer treatment. Dr. Jordan was a prolific breast cancer researcher and served as Professor of Breast Medical Oncology in Houston at the MD Anderson Cancer Center. In 2020, Dr. Jordan explained the counter-intuitive "estrogen paradox", accounting for the reduction in breast cancer in the estrogen-alone arm of the 2004 WHI second arm. These women had a 5-year gap of long-term estrogen deprivation which changes estrogen from a growth signal to a death signal, transforming high-dose estrogen into a successful treatment for breast cancer, writing:

> A sustained beneficial antibreast cancer action of estrogen alone noted in the WHI study [2004, second arm] is counter intuitive because the dogma is that estrogen, through the estrogen receptor (ER), is the primary signal for the initiation and growth of breast cancer...However, the paradox... is **estrogen causes a decrease in mortality and a decrease in the incidence of new breast cancers**. **This is counter intuitive to the scientific and medical community** unless one embraces and understands the known clinical evidence that governs safe estrogen use for the treatment of breast cancer after menopause. These were established 70 years ago....An estrogen deprivation gap of 5 years after menopause is required for high-dose estrogen to be an effective treatment for breast cancer...In addition, the same applies to 5 years of adjuvant tamoxifen therapy when recurrence and mortality continue to decrease after adjuvant tamoxifen treatment is stopped. (101)

In 2022, Dr. V. Craig Jordan provided insight into the use of estrogen for the prevention and treatment of breast cancer. Estrogen therapy for menopausal women after long-term estrogen deprivation (LTED) produces a sustained decrease in breast cancer, while the addition of MPA not only reverses this estrogen protective effect but also increases carcinogenesis. Dr. V. Craig Jordan writes:

The clinical description and discovery of estrogen-induced apoptosis [programmed cell death] with further clinical application in two Karnofsky lectures, separated by 38 years, has now provided a mechanistic insight into the adjuvant treatment of breast cancer, an insight into the "unexpected" results of the Women's Health Initiative investigation of estrogen and estrogen/progestin given to women as hormone replacement at the age of 60 vs the Million Women Study of hormone replacement therapies in the general population. The results of the two epidemiological interventional studies were not comparable but instructive about mechanisms of hormone action in the real world **if long-term estrogen deprivation occurs at menopause prior to HRT administration of estrogen alone produces a sustained decrease in breast cancer and the addition of medroxyprogesterone acetate [MPA] not only reverses but increases breast carcinogenesis.** (102)

Extensive research on estrogen as a treatment for breast cancer and inducer of apoptosis shows Dr. V. Craig Jordan is quite correct about all the above. **Note:** Sadly, Dr. V. Craig Jordan passed away on June 9, 2024, at the age of 76. (103-111)

How to Blow a Light Bulb

Estrogen deprivation upregulates estrogen receptor alpha (ER-alpha) in the breast cancer cells, which upon later reintroduction of estrogen, triggers apoptosis. One can think of this as analogous to blowing a 110 Volt light bulb by mistakenly plugging in the cord to 220 Volt outlet. In 2021, Dr. Nicole Traphagen, PhD in Cancer Biology from Dartmouth College, did preclinical work on two breast cancer cell lines finding estrogen deprivation leads to ER-alpha overexpression and high estrogen receptor activation, which converts estrogen from a growth promoter to a growth suppressor. Dr. Nicole Traphagen found breast cancer xenografts implanted into animals could be entirely controlled by oscillating between two alternating treatments, estrogen deprivation alternating with estrogen treatment, writing:

Herein, we demonstrate that ER [Estrogen Receptor] overexpression confers resistance to estrogen deprivation through ER activation in human ER+ breast cancer cells and xenografts grown in mice. However, **ER overexpression and the associated high levels of ER transcriptional activation converted 17b-estradiol from a growth-promoter to a growth-suppressor,** offering a targetable therapeutic vulnerability and a potential means of identifying patients likely to benefit from estrogen therapy. Since ER+ breast cancer cells and tumors ultimately developed resistance to continuous estrogen deprivation or continuous 17b-estradiol treatment, we tested schedules of alternating treatments. Oscillation of ER activity through cycling of 17b-estradiol and estrogen deprivation provided long-term control of patient-derived xenografts, offering a novel endocrine-only strategy to manage ER+ breast cancer...Although the anti-cancer effects of estrogens have been known since the 1940s, clinical use of estrogen therapy has been limited since the introduction of anti-estrogens that have improved adverse events profile. More recent clinical studies have demonstrated that treatment with **estrogens [i.e., 17β-estradiol (E2), diethylstilbestrol, or ethinylestradiol] elicits anti-tumor effects in approximately 30% of patients with advanced ER+ breast cancer previously treated with anti-estrogens and/or AIs [Aromatase Inhibitors].** Despite the proven efficacy of estrogen therapy, its clinical use has been hindered by a lack of mechanistic understanding and a predictive biomarker to identify patients likely to benefit from this treatment...Clinical evidence indicates that the patients most likely to benefit from estrogen therapy are post-menopausal, suggesting that a period of adaption to low levels of endogenous estrogens increases tumor sensitivity to estrogen therapy... Preclinical studies using MCF-7 cells and the WHIM16 patient-derived xenograft model,

as well as a clinical case report, suggest that amplification of the gene encoding ER (ESR1) is associated with therapeutic response to estrogen. Herein, we demonstrate that high ER levels confer resistance to estrogen deprivation and increase estrogen-independent ER activity. These high levels of ER elicit therapeutic responses to E2 [estradiol]...the effects of high ER expression are context-dependent, providing a growth advantage under estrogen deprivation but a disadvantage in E2-replete conditions. **High expression of ER is therefore targetable with estrogen therapy, and modulating ER activation by cycling estrogen and estrogen deprivation provides long-term control of tumor growth.** (112).

In 2022, Dr. Philipp Maximov, Research Assistant Professor, MD Anderson Cancer Center, working with V. Craig Jordan studied the mechanism of estrogen-induced apoptosis LTED (long-term estrogen deprived) breast cancer cells, finding upregulation of ER-alpha triggers the response, writing:

Only knockdown of ER-alpha prevents E2-induced apoptosis, confirming that E2 induces apoptosis via the ER-alpha... Induction of estrogen-induced apoptosis in LTED breast cancer may become a feasible and safe alternative therapy for many patients with lower side effects for many patients in the future. (113)

More on Animal Models of Breast Cancer – Dr. Lakshmanaswamy Rajkumar

Estrogen and Progesterone Induce a Long-lasting Protective Effect on Mammary Tumorigenesis in Two Genetically Engineered Mouse Models.

As mentioned above, breast cancer researchers have a well-known method of inducing breast cancer in animals called the MPA (medroxyprogesterone) mouse model of breast cancer. MPA is carcinogenic and will induce breast cancer in wild-type mice. Medroxyprogesterone is a synthetic progestin, which means natural progesterone has been chemically altered to create a carcinogenic "Frankenstein" molecule. These "Frankenstein" progestins do not occur anywhere in nature or the human body. What about natural estrogen and progesterone? Can they induce breast cancer in animal models?

Hormone Induced Breast Cancer in Animal Models

As mentioned in Chapter 6, we have two animal models of estradiol-induced breast cancer using inbred rats, the Noble (Nb) and August–Copenhagen–Irish (ACI) rats. However, wild-type mice and rats are not good models for hormone-induced breast cancer. What if we try to induce breast cancer in animals using human bioidentical hormones estrogen and progesterone? Will this be successful? The answer is no, natural estrogen and progesterone are breast cancer preventive. As mentioned above, they do not induce breast cancer in animal models. This was studied in 2021 by Dr. Chong Liu who reviewed the various breast cancer models in mice finding the simple administration of estrogen to wild-type mice was insufficient to create a viable model of breast cancer. The mouse models for breast cancer require inbreeding, genetic manipulation, or the use of carcinogenic chemicals. Estrogen and progesterone are breast cancer protective in animal models, as explained below. (114)

Hormones of Pregnancy Confer Long-Term Protection from Breast Cancer

In 2007, Dr. Rajkumar Lakshmanaswamy, Chair of the Department of Molecular and Translational Medicine and Professor of Biomedical Sciences Texas Tech University Health Sciences Center El Paso, recognized that high hormone levels of pregnancy confer long-term protection against breast cancer. Dr. Lakshmanaswamy studied natural estrogen and progesterone in two genetically engineered mouse models of breast cancer, finding short doses of estrogen and progesterone induce long-lasting protection from breast cancer

formation. Dr. Lakshmanaswamy's first experiment used genetically engineered mice with a deleted P53 tumor suppressor gene. This is called the p53-null mammary transplant model. These mice suffer spontaneous breast cancer. In this model, exposing the mouse for two weeks to estrogen and progesterone reduced mammary cancer by 70-80%. The second model is called the HuHER2 transgenic mouse model. These transgenic mice develop spontaneous mammary gland tumors within 6 to 12 months. Short-term estradiol or estradiol plus progesterone treatment decreased mammary tumor incidence by 66%. Dr. Lakshmanaswamy writes:

> These studies demonstrate that short doses of the hormones estrogen and progesterone induce a **long-lasting protective effect on mammary tumorigenesis** in two genetically engineered mouse models. (115-122)

Breast Cancer Treated with Natural Hormones - Dr. Rajkumar Lakshmanaswamy

What if we inject mice with human breast cancer cells (called a xenograft), and then treat half the mice with an aromatase inhibitor (AI), an antiestrogen drug. The other half will be treated with estrogen and progesterone. Would the aromatase inhibitor-treated mice fare better or worse than estrogen hormone treated mice? Medical dogma says the AI treated mice should fare better. But they do not. In 2014, Drs. Arumugam and Lakshmanaswamy performed a series of in-vivo experiments in ovariectomized mice transplanted with human breast cancer xenografts (MCF-7 cells). The cancer cells were "transfected" (i.e. genetically modified) with the human aromatase enzyme which converts testosterone to estrogen, making the cancer cells able to synthesize their own estrogen, a major growth factor. After human breast cancer cells were injected into the mice, they were then treated with various combinations of natural hormones, estrogen, progesterone, testosterone, and DHEA. Dr. Rajkumar Lakshmanaswamy found the natural hormone-treated mice had maximum reduction in tumor growth, and better outcomes than the AI (aromatase inhibitor) treated mice, writing:

> Because estrogen-blocking aromatase inhibitors are the current adjuvant treatment after hormone-sensitive breast cancer, common sense would lead to the assumption that any treatment containing estrogen itself would lead to opposite, highly negative impact on tumor growth. **However, this turned out not to be the case.** As was the case for general health markers, **maximal reduction in tumor growth was achieved by E [estrogen] plus P [progesterone] plus T [testosterone] treatment….current standard of practice considers hormones of any type absolutely contraindicated after hormone-receptor-positive breast cancer, with the assumption being that hormones "throw fuel on the fire" of cancer.** This assumption makes intuitive sense, since current treatment is to block remaining estrogens with aromatase inhibitors, the exact opposite…**Our results thus did not confirm the "throwing fuel on the fire" conception prevalent among clinicians**….In summary, our results indicate that the use of appropriate combinations of natural hormones along with, or instead of, classical breast cancer treatments [aromatase inhibitors] is beneficial against postmenopausal symptoms and improves cardiac and osteoporotic health in the mouse model. **The natural hormone combinations tested in this study provide evidence for a better alternative to standard aromatase inhibitor treatment following breast cancer in women**. (123)

Adding Aromatase Inhibitor Increases Recurrence

Dr. Rajkumar's above study suggests that bio-identical hormone therapy is a better alternative to aromatase inhibitors in the treatment of breast cancer. This is counter-intuitive since conventional treatment for breast cancer is aromatase inhibitor (AI) drugs that block estrogen. Adding an aromatase inhibitor

drug (AI) to post-menopausal HRT was examined in 2022 by Dr. Søren Cold MD, PhD, senior oncologist Department of Oncology, Odense University Hospital, Denmark, with an observational cohort study of 8461 women following a breast cancer diagnosis. Of the 8,461, about 1,957 women used vaginal estrogen therapy (VET) and 133 used menopausal hormone therapy (MHT). The cohort was followed 9.8 years for recurrence and 15.2 years for mortality. The overall mortality was decreased by 22 percent for VET and 6 percent for MHT. Dr. Cold found no increased recurrence of breast cancer for either of the VET or MRT groups. However, in women receiving VET or MRT and receiving adjuvant AI drug, breast cancer recurrence was **increased by 39%** compared to non-users. These findings suggest a better outcome is achieved **without adding adjuvant AI drugs to HRT for breast cancer survivors**, thus answering the above question by Dr. R. Lakshmanaswamy. Dr. Cold writes:

> In postmenopausal women treated for early-stage estrogen receptor–positive BC [breast cancer], neither VET nor MHT was associated with increased risk of recurrence or mortality. **A subgroup analysis revealed an increased risk of recurrence, but not mortality, in patients receiving VET with adjuvant aromatase inhibitors.** (124)

A Paradigm Shift in the Old Dogma

The above discussion highlights estrogen as a breast cancer preventive. This represents a paradigm shift in the mainstream dogmatic belief that estrogen causes breast cancer. If this is true, then what is the causal mechanism for breast cancer? In 2010, Dr. Wei Yue, Associate Professor of Research at the University of Virginia Medical School, Charlottesville, Virginia, reviewed the two widely accepted theories of breast cancer carcinogenesis. The first is estradiol (E2) acting through ER-alpha stimulates cell proliferation and mutations, leading to breast cancer. The second mechanism described Dr. Ercole Cavalieri is estrogen receptor-independent formation of estrogen quinone metabolites which bind to DNA and form adducts. Dr. Wei Yue writes:

> The most widely accepted theory, supported by extensive experimental evidence holds that E2, acting through ER-alpha, stimulates cell proliferation and initiates mutations that occur as a function of errors during DNA replication. The promotional effect of E2 then supports the growth of cells harboring mutations, which then accumulate until cancer ultimately results. Clinical and experimental data also suggest the possibility that **receptor independent effects of E2 may be mechanistically involved**. In a recent review, Yager and Davidson describe in detail how estrogen metabolites can exert genotoxic effects, which contribute to the development of breast cancer. **Estrogens are converted to quinone metabolites, which directly bind to DNA and form adducts...** Error prone DNA repair then results in the formation of mutations at the depurinated sites. **Accumulation of these mutations would then contribute to the development of breast cancer.** As predicted from the "estrogen genotoxic metabolite" hypothesis, a predisposition to breast cancer would be expected in women with **combinations of mutations of estrogen metabolizing enzymes**, a finding reported by Park et al. and Ritchie et al... Finally, a speculative consideration for the future is that blockade of estradiol metabolism with CYP1B1 and 1A1 inhibitors might be a means to reduce breast cancer incidence without blocking formation of E2 itself. (125)

Hormone Symphony: A Well-Designed Bioidentical Program

The optimal bioidentical hormone program is designed to decrease the risk of breast cancer. For post-menopausal women having natural or drug-induced LTED (long-term estrogen deprivation), estrogen induces apoptosis in latent breast cancer cells having upregulated estrogen receptor-alpha (ER-alpha). An example of this is the 18-year follow-up data from the WHI second arm (Premarin alone, CEE) shows a 45 percent reduction in mortality from breast cancer in the estrogen-treated group compared to the placebo. Of course, for women who have an intact uterus, hormone replacement with estrogen always includes progesterone for endometrial protection. Although breast cancer survivors will benefit from hormone replacement, estrogen is a growth factor for breast cancer. So, it is not advisable to give estrogen to women with active cancer, unless of course, they have been treated for a time with estrogen-blocking drugs (tamoxifen or aromatase inhibitors). This mimics a period of estrogen deprivation, and the cancer cells are now ripe for apoptosis induced by estrogen. However, this type of oscillating breast cancer treatment requires an experienced integrative oncologist familiar with this technique. For more on this, see Chapters 6 and 19 on Hormone Replacement for Breast Cancer Survivors.

Conclusion: Thanks, and credit goes to Dr. Lindsey Berkson for bringing to my attention the above re-analysis of the WHI by Dr. Howard Hodis, who points out the placebo group included women with prior hormone use which confers protection from breast cancer and gives the false impression of increased breast cancer risk in the WHI first arm using Premarin and MPA. When prior hormone use is removed from the placebo group, trend lines for breast cancer are superimposed for both groups. Overwhelming evidence was presented in this chapter showing bioidentical hormone replacement prevents breast cancer rather than causing it. It is important to avoid carcinogenic synthetic progestins. Unfortunately, most physicians are unaware of this because of the blurring of the distinction between natural progesterone and synthetic progestins throughout the medical literature. This is an error that favors increased pharmaceutical drug sales to treat menopausal symptoms of women who were refused or declined HRT. The needless fear of estrogen is the tragedy of our time. Faced with a medical system unresponsive to their needs, many women are left in a void to fend for themselves. Our mission is to offer the highest level of bioidentical HRT to all eligible menopausal women.

♦ References for Chapter 7 Errors in Modern Medicine, the Fear of Estrogen

1) Anderson, Garnet L., et al. "Effects of Conjugated Equine Estrogen in Postmenopausal Women With Hysterectomy: The Women's Health Initiative Randomized Controlled Trial." JAMA 291.14 (2004): 1701-1712.

2) Hulley, Stephen B., and Deborah Grady. "The WHI Estrogen-Alone Trial—Do Things Look Any Better?" JAMA 291.14 (2004): 1769-1771.

3) Hodis, Howard N., and P. M. Sarrel. "Menopausal Hormone Therapy And Breast Cancer: What Is The Evidence From Randomized Trials?" Climacteric 21.6 (2018): 521-528.

4) Rossouw, Jacques E., et al. "Risks and Benefits of Estrogen Plus Progestin In Healthy Postmenopausal Women: Principal Results From the Women's Health Initiative randomized controlled trial." Jama 288.3 (2002): 321-333.

5) Rossouw, Jacques E. "Estrogens for Prevention of Coronary Heart Disease: Putting The Brakes On The Bandwagon." Circulation 94.11 (1996): 2982-2985.

6) Rossouw, Jacques E. "Long-term Hormone Replacement Therapy: A Contrary Point of View." Hormonal Carcinogenesis III: Proceedings of the Third International Symposium. New York, NY: Springer New York, 2000.

7) Langer, R. D. "The Evidence Base For HRT: What Can We Believe?" Climacteric 20.2 (2017): 91-96.

8) Yang, Lin, and Adetunji T. Toriola. "Menopausal Hormone Therapy Use Among Postmenopausal Women." JAMA Health Forum. Vol. 5. No. 9. American Medical Association, 2024.

9) Jerry, D. Joseph. "Roles for Estrogen and Progesterone In Breast Cancer Prevention." Breast Cancer Research 9 (2007): 1-3.

10) Ferretti, Gianluigi, Alessandra Felici, and Francesco Cognetti. "The Protective Side Of Progesterone." Breast Cancer Research 9 (2007): 1-1.

11) Lanari, Claudia, et al. "The MPA Mouse Breast Cancer Model: Evidence for A Role Of Progesterone Receptors In Breast Cancer." Endocrine-Related Cancer 16.2 (2009): 333.

12) Aldaz, C. Marcelo, et al." Medroxyprogesterone Acetate Accelerates the Development and Increases The Incidence Of Mouse Mammary Tumors Induced By Dimethylbenzanthracene." Carcinogenesis 17.9 (1996): 2069-2072.

13) Pazos, P., et al. "Mammary Carcinogenesis Induced By N-Methyl-N-Nitrosourea (MNU) And Medroxyprogesterone Acetate (MPA) In BALB/C Mice." Breast cancer research and treatment 20.2 (1992): 133-138.

14) Nagasawa, H., et al. "Medroxyprogesterone Acetate Enhances Spontaneous Mammary Tumorigenesis and Uterine Adenomyosis In Mice." Breast cancer research and treatment 12.1 (1988): 59-66.

15) Horwitz, Kathryn B., and Carol A. Sartorius. "90 Years of Progesterone: Progesterone And Progesterone Receptors In Breast Cancer: Past, Present, Future." Journal of Molecular Endocrinology 65.1 (2020): T49-T63.

16) Bethea, Cynthia L. "MPA: Medroxy-Progesterone Acetate Contributes to Much Poor Advice for Women." Endocrinology. (2011): 343-345.

17) Chlebowski RT. Abstract GS5-00. Presented at: San Antonio Breast Cancer Symposium; Dec. 10-14, 2019; San Antonio.

18) Mercola and Guest's Analysis - WRONG on Estrogen by Devaki Berkson, Oct. 24, 2023. https://drlindseyberkson.substack.com/p/merco-la-and-guests-analysis-wrong

19) Lieberman, Allan, and Luke Curtis. "In Defense of Progesterone: A Review of The Literature." Alternative Therapies in Health & Medicine 23.7 (2017).

20) Birrell, Stephen N., et al. "Disruption of androgen receptor signaling by synthetic progestins may increase risk of developing breast cancer." The FASEB Journal 21.10 (2007): 2285-2293.

21) de Lignieres, B., de Vathaire, F., Fournier, S., Urbinelli, R., Allaert, F., Le, M. G., and Kuttenn, F. Combined hormone replacement therapy and risk of breast cancer in a French cohort study of 3175 women. Climacteric. 5, (2002): 332–340

22) Fournier, A., Berrino, F., Riboli, E., Avenel, V., and Clavel-Chapelon, F. (2005) Breast cancer risk in relation to different types of hormone replacement therapy in the E3N-EPIC cohort. Int. J. Cancer. 114.3, (2005): 448–454

23) Abderrahman, Balkees, and V. Craig Jordan. "Estrogen for the Treatment And Prevention Of Breast Cancer: A Tale Of 2 Karnofsky Lectures." The Cancer Journal 28.3 (2022): 163-168.

24) Jordan, V. Craig. "The New Biology Of Estrogen-Induced Apoptosis Applied To Treat And Prevent Breast Cancer." Endocrine-Related Cancer 22.1 (2015): R1-R31.

25) Jordan, V. Craig, and Leslie G. Ford. "Paradoxical Clinical Effect Of Estrogen On Breast Cancer Risk: A "New" Biology Of Estrogen-Induced Apoptosis." Cancer prevention research 4.5 (2011): 633-637.

26) Fournier, Agnès, et al "Unequal risks for breast cancer associated with different hormone replacement therapies: results from the E3N cohort study." Breast cancer research and treatment 107 (2008): 103-111.

27) Lyytinen, Heli, et al. "Breast cancer risk in post-menopausal women using estrogen-only therapy." Obstetrics & Gynecology 108.6 (2006): 1354-1360.

28) Bakken, Kjersti, et al. "Menopausal hormone therapy and breast cancer risk: impact of different treatments. The European Prospective Investigation into Cancer and Nutrition." International journal of cancer 128.1 (2011): 144-156.

29) Lee, Hye-Rim, Tae-Hee Kim, and Kyung-Chul Choi. "Functions and physiological roles of two types of estrogen receptors, ER-alpha and ER-beta, identified by estrogen receptor knockout mouse." Laboratory animal research 28.2 (2012): 71-76.

30) Zhao, Chunyan, Karin Dahlman-Wright, and Jan-Åke Gustafsson. "Estrogen receptor beta: an overview and update." Nuclear receptor signaling 6.1 (2008): nrs-06003.

31) Giulianelli, Sebastián, et al. "Estrogen receptor alpha mediates progestin-induced mammary tumor growth by interacting with progesterone receptors at the cyclin D1/MYC promoters." Cancer research 72.9 (2012): 2416-2427.

32) Liu, Meng-Min, et al. "Opposing action of estrogen receptors alpha and beta on cyclin D1 gene expression." Journal of Biological Chemistry 277.27 (2002): 24353-24360.

33) Vogiatzi, Paraskevi. "Cyclin D1 antagonizes BRCA1 repression of estrogen receptor [alpha] activity." Women's Oncology Review 6.1/2 (2006): 55.

34) Treeck, Oliver, et al. "Estrogen receptor beta exerts growth-inhibitory effects on human mammary epithelial cells." Breast cancer research and treatment 120 (2010): 557-565.

35) Ström, Anders, et al. "Estrogen receptor beta inhibits 17β-estradiol-stimulated proliferation of the breast cancer cell line T47D." Proceedings of the National Academy of Sciences 101.6 (2004): 1566-1571.

36) Bardin, Allison, et al. "Loss of ER-beta expression as a common step in estrogen-dependent tumor progression." Endocrine-related cancer 11.3 (2004): 537-551.

37) Wood, Charles E., et al. "Comparative effects of oral conjugated equine estrogens and micronized 17β-estradiol on breast proliferation: a retrospective analysis." Menopause 15.5 (2008): 978-983.

38) Levy, Barbara, and James A. Simon. "A Contemporary View of Menopausal Hormone Therapy." Obstetrics & Gynecology 144.1 (2024): 12-23.

39) Bhavnani, Bhagu R., and Frank Z. Stanczyk. "Pharmacology of conjugated equine estrogens: efficacy, safety and mechanism of action." The Journal of steroid biochemistry and molecular biology 142 (2014): 16-29.

40) Bhavnani, Bhagu R., Shui-Pang Tam, and XiaoFeng Lu. "Structure Activity Relationships And Differential Interactions And Functional Activity Of Various Equine Estrogens Mediated Via Estrogen Receptors (Ers) ER-Alpha And ER-Beta." Endocrinology 149.10 (2008): 4857-4870.

41) Flores, Valerie A., and Hugh S. Taylor. "The Effect Of Menopausal Hormone Therapies On Breast Cancer: Avoiding The Risk." Endocrinology and Metabolism Clinics 44.3 (2015): 587-602.

42) Mohammed, Hisham, et al. "Progesterone Receptor Modulates ER-Alpha Action In Breast Cancer." Nature 523.7560 (2015): 313-317.

43) Perkins MS, et al. A comparative characterization of estrogens used in hormone therapy via estrogen receptor (ER)-alpha and -beta. J Steroid Biochem Mol Biol. (2017) 174:27–39.

44) Bhavnani BR, Tam SP, Lu X. Structure activity relationships and differential interactions and functional activity of various equine estrogens mediated via estrogen receptors (ERs) ERα and ERβ. Endocrinology. (2008) 149:4857–70.

45) Barkhem T, et al. Differential response of estrogen receptor alpha and estrogen receptor beta to partial estrogen agonists/antagonists. Mol Pharmacol. (1998) 54:105–12.

46) Katzenellenbogen BS. Biology and Receptor Interactions Of Estriol And Estriol Derivatives In Vitro And In Vivo. J Steroid Biochem. (1984) 20:1033–7.

47) Paterni I, et al. Estrogen receptors alpha (ER-alpha) and beta (ER-beta): subtype-selective ligands and clinical potential. Steroids. (2014) 90:13–29.

48) Glaser, Rebecca L., Anne E. York, and Constantine Dimitrakakis. "Incidence of invasive breast cancer in women treated with testosterone implants: a prospective 10-year cohort study." BMC cancer 19 (2019): 1-10.

49) Donovitz, Gary, and Mandy Cotten. "Breast cancer incidence reduction in women treated with subcutaneous testosterone: testosterone therapy and breast cancer incidence study." European journal of breast health 17.2 (2021): 150.

50) L'Hermite M, et al. Could Transdermal Estradiol + Progesterone Be A Safer Postmenopausal HRT? A Review. Maturitas. 2008 Jul-Aug;60(3-4):185-201.

51) Fournier, A., et al. (2008). Unequal Risks For Breast Cancer Associated With Different Hormone Replacement Therapies: Results From The E3N Cohort Study. Breast Cancer Res Tr, 107(1), 103-111.

52) Fournier, A., et al. (2008). Use of Different Postmenopausal Hormone Therapies And Risk Of Histology- And Hormone Receptor-Defined Invasive Breast Cancer. J Clin Oncol, 26(8), 1260-1268.

53) Mal, Rahul, et al. "Estrogen Receptor Beta (ER-beta): A Ligand Activated Tumor Suppressor." Frontiers in Oncology 10 (2020): 587386.

54) Wright, Jonathan V. "Bio-Identical Steroid Hormone Replacement: Selected Observations from 23 Years of Clinical and Laboratory Practice." Annals of the New York Academy of Sciences 1057.1 (2005): 506-524.

55) Head, Kathleen A. "Estriol: safety and efficacy." Alternative medicine review: a journal of clinical therapeutic 3.2 (1998): 101-113.

56) Follingstad A. Estriol, the forgotten estrogen? JAMA. 1978;239(1):29-30.

57) The Anticancer Testosterone Metabolite 3β-Adiol. Townsend Letter by Dr. Jonathan V. Wright, MD https://www.meridianvalleylab.com/the-anticancer-testosterone-metabolite-3%CE%B2-adiol/

58) Guerini V, Poletti A et al. The Androgen Derivative 5A-Androstane-3B,17B-Diol Inhibits Prostate Cancer Cell Migration Through Activation of the Estrogen Receptor B Subtype Cancer Res 2005;65:(12), June 15, 2005

59) Oliveira AG et al. 5a-Androstane-3β,17β–diol, an estrogenic metabolite of 5a-dihydrotestosterone, is a potent modulator of estrogen receptor β (ERβ) in the ventral prostate of adult rats. Steroids 2007;72:914-922

60) Dondi D, Poletti A et al. Estrogen receptor b and the progression of prostate cancer role of 5a- androstane-3b,17b-diol Endocrine-Related Cancer (2010) 17 731–742

61) Lemon HM, Wotiz HH, Parsons L, Mozden PJ. Reduced estriol excretion in patients with breast cancer prior to endocrine therapy. JAMA. 1966;196(13):1128-1136.

62) Lemon, Henry M. "Estriol prevention of mammary carcinoma induced by 7, 12-dimethylbenzanthracene and procarbazine." Cancer Research 35.5 (1975): 1341-1353.

63) Lemon, Henry M. "Antimammary carcinogenic activity of 17-alpha-ethinyl estriol." Cancer 60.12 (1987): 2873-2881.

64) Lemon, Henry M., et al. "Inhibition of Radiogenic Mammary Carcinoma In Rats By Estriol Or Tamoxifen." Cancer 63.9 (1989): 1685-1692.

65) Marsden, Tracy. "Bioidentical Hormone Replacement: Guiding Principles for Practice." Nat Med J 2.3 (2010): 2010-03.

66) Holtorf K. The Bioidentical Hormone Debate: Are Bioidentical Hormones (Estradiol, Estriol, And Progesterone) Safer or More Efficacious Than Commonly Used Synthetic Versions In Hormone Replacement Therapy? PostGrad Med. 2009;121(1):1-13.

67) Taylor M. Unconventional estrogens: Estriol, Biest and Triest. Clin Obstet Gyn. 2001;4(4):864-869.

68) Sicotte NL, et al. Treatment of multiple sclerosis with the pregnancy hormone estriol. Ann Neurol. 2002;52(4):421-428.

69) Dessole S, et al. Efficacy of low-dose intravaginal estriol on urogenital aging in postmenopausal women. Menopause. 2004;11(1):49-56.

70) Granberg S, et al. The effects of oral estriol on the endometrium in postmenopausal women. Maturitas. 2002;42(2):149-156.

71) Takahashi K, et al. Efficacy and safety of oral estriol for managing postmenopausal symptoms. Maturitas. 2000;34(2):169-177.

72) Ochnik, Aleksandra M., et al. "Antiandrogenic Actions Of Medroxyprogesterone Acetate On Epithelial Cells Within Normal Human Breast Tissues Cultured Ex Vivo." Menopause 21.1 (2014): 79-88.

73) Moyer, Dean L., et al. "Prevention of endometrial hyperplasia by progesterone during long-term estradiol replacement: influence of bleeding pattern and secretory changes." Fertility and sterility 59.5 (1993): 992-997.

74) Fitzpatrick, Lorraine A., and Andrew Good. "Micronized progesterone: clinical indications and comparison with current treatments." Fertility and sterility 72.3 (1999): 389-397.

75) Judd, Howard L., et al. "Effects of hormone replacement therapy on endometrial histology in postmenopausal women: the Postmenopausal Estrogen/Progestin Interventions (PEPI) Trial." JAMA 275.5 (1996): 370-375.

76) Ibrahim, Raihan Syah, and Aisyah Elliyanti. "The Potential of Iodine as A Treatment for Breast Cancer: A Narrative Review." Jurnal Kesehatan Manarang 9.3 (2023): 159-165.

77) Manjer, Jonas, et al. "Serum Iodine and Breast Cancer Risk: A Prospective Nested Case–Control Study Stratified for Selenium Levels." Cancer Epidemiology, Biomarkers & Prevention 29.7 (2020): 1335-1340.

78) Thomson, Cynthia A., Emily Ho, and Meghan B. Strom. "Chemopreventive properties of 3, 3'-diindolylmethane in breast cancer: evidence from experimental and human studies." Nutrition reviews 74.7 (2016): 432-443.

79) Reyes-Hernández, Octavio Daniel, et al. "3, 3'-Diindolylmethane and indole-3-carbinol: potential therapeutic molecules for cancer chemoprevention and treatment via regulating cellular signaling pathways." Cancer Cell International 23.1 (2023): 180.

80) Williams, David E. "Indoles derived from gluco-brassicin: Cancer chemoprevention by indole-3-carbi-nol and 3, 3'-diindolylmethane." Frontiers in Nutrition 8 (2021): 734334.

81) Koli, Papita, et al. "Anticancer activity of 3, 3'-diindolylmethane and the molecular mechanism involved in various cancer cell lines." ChemistrySelect 5.37 (2020): 11540-11548.

82) Cavalieri, Ercole, and Eleanor Rogan. "The 3, 4-Quinones Of Estrone And Estradiol Are The Initiators Of Cancer Whereas Resveratrol And N-Acetylcysteine Are The Preventers." International Journal Of Molecular Sciences 22.15 (2021): 8238.

83) Ergul, Emel, et al. "Polymorphisms in the MTHFR gene are associated with breast cancer." Tumor biology 24.6 (2004): 286-290.

84) Almeida, Micaela, et al. "Influence of estrogenic metabolic pathway genes polymorphisms on post-menopausal breast cancer risk." Pharmaceuticals 14.2 (2021): 94.

85) Zahid, Muhammad, et al. "Resveratrol and N-acetylcysteine block the cancer-initiating step in MCF-10F cells." Free Radical Biology and Medicine 50.1 (2011): 78-85.

86) Yager, James D. "Mechanisms of estrogen car-cinogenesis: The role of E2/E1–quinone metabolites suggests new approaches to preventive interven-tion–A review." Steroids 99 (2015): 56-60.

87) Palliyaguru, Dushani L., et al. "Sulforaphane diminishes the formation of mammary tumors in rats exposed to 17β-estradiol." Nutrients 12.8 (2020): 2282.

88) Sung, Nam-Ji, and Sin-Aye Park. "Effect of Natural Compounds on Catechol Estrogen-Induced Carcinogenesis." Biomedical Science Letters 25.1 (2019): 1-6.

89) Zeng, Huawei, and Gerald F. Combs Jr. "Selenium as an anticancer nutrient: roles in cell proliferation and tumor cell invasion." The Journal of nutritional biochemistry 19.1 (2008): 1-7.

90) Jackson, Matthew I., and Gerald F. Combs Jr. "Selenium as a cancer preventive agent." Selenium: its molecular biology and role in human health. New York, NY: Springer New York, 2011. 313-323.

91) Reid, Mary E, et al. "The Nutritional Prevention of Cancer: 400 Mcg Per Day Selenium Treatment." Nutrition & Cancer 60.2 (2008).

92) Bretthauer, Michael, et al. "Estimated lifetime gained with cancer screening tests: a meta-analysis of randomized clinical trials." JAMA Internal Medicine 183.11 (2023): 1196-1203.

93) Esserman, Laura, et al. "Rethinking screening for breast cancer and prostate cancer." JAMA 302.15 (2009): 1685-1692.

94) Kowalski, Amanda E. "Mammograms and mortality: how has the evidence evolved?" Journal of Economic Perspectives 35.2 (2021): 119-140.

95) Peper, Erik, and Richard Harvey. Rethink the monies spent on cancer screening tests . November 24, 2023. https://peperperspective.com/2023/11/24/rethinking-the-monies-spent-on-cancer-screening-tests/

96) Bretthauer, Michael, et al. "Estimated lifetime gained with cancer screening tests: a meta-analysis of randomized clinical trials." JAMA Internal Medicine 183.11 (2023): 1196-1203.

97) Autier, Philippe, et al. "Breast Cancer Mortality In Neighboring European Countries With Different Levels Of Screening But Similar Access To Treatment: Trend Analysis Of WHO Mortality Database." Bmj 343 (2011).

98) Siddique, Y. H., and M. Afzal. "A Review on The Genotoxic Effects of Some Synthetic Progestins." Int J Pharmacol 4.6 (2008): 410-30.

99) Manyonda, Isaac, et al. "Could Perimenopausal Estrogen Prevent Breast Cancer? Exploring The Differential Effects of Estrogen-Only Versus Combined Hormone Replacement Therapy." Journal of Clinical Medicine Research 14.1 (2022): 1-7.

100) Vasconsuelo, Andrea, et al. "Role of 17β-estradiol and Testosterone In Apoptosis." Steroids 76.12 (2011): 1223-1231.

101) Jordan, V. Craig. "Molecular Mechanism For Breast Cancer Incidence In The Women's Health Initiative." Cancer Prevention Research 13.10 (2020): 807-816.

102) Abderrahman, Balkees, and V. Craig Jordan. "Estrogen for the Treatment And Prevention Of Breast Cancer: A Tale Of 2 Karnofsky Lectures." The Cancer Journal 28.3 (2022): 163-168.

103) Shete, Nivida, et al. "Revisiting Estrogen for the Treatment of Endocrine-Resistant Breast Cancer: Novel Therapeutic Approaches." Cancers 15.14 (2023): 3647.

104) Chimento, Adele, et al. "Estrogen Receptors-Mediated Apoptosis In Hormone-Dependent Cancers." International Journal Of Molecular Sciences 23.3 (2022): 1242.

105) Maximov, Philipp Y., et al. "Estrogen receptor complex to trigger or delay estrogen-induced apoptosis in long-term estrogen deprived breast cancer." Frontiers in Endocrinology 13 (2022): 869562.

106) Jordan, V. Craig. "Molecular mechanism for breast cancer incidence in the Women's Health Initiative." Cancer Prevention Research 13.10 (2020): 807-816.

107) Jordan, V. Craig. "The new biology of estrogen-induced apoptosis applied to treat and prevent breast cancer." Endocrine-related cancer 22.1 (2015): R1-R31.

108) Jordan, V. Craig, and Leslie G. Ford. "Paradoxical clinical effect of estrogen on breast cancer risk: a "new" biology of estrogen-induced apoptosis." Cancer prevention research 4.5 (2011): 633-637.

109) Sweeney, Elizabeth E., Ping Fan, and V. Craig Jordan. "Mechanisms underlying differential response to estrogen-induced apoptosis in long-term estrogen-deprived breast cancer cells." International journal of oncology 44.5 (2014): 1529-1538.

110) Obiorah, Ifeyinwa E., Ping Fan, and V. Craig Jordan. "Breast cancer cell apoptosis with phytoestrogens is dependent on an estrogen-deprived state." Cancer prevention research 7.9 (2014): 939-949.

111) Song, Robert X-D., et al. "Effect of long-term estrogen deprivation on apoptotic responses of breast cancer cells to 17β-estradiol." Journal of the National Cancer Institute 93.22 (2001): 1714-1723.

112) Traphagen, Nicole A., et al. "High Estrogen Receptor Alpha Activation Confers Resistance to Estrogen Deprivation and Is Required for Therapeutic Response to Estrogen in Breast Cancer." Oncogene 40.19 (2021): 3408-3421.

113) Maximov, Philipp Y., et al. "Estrogen receptor complex to trigger or delay estrogen-induced apoptosis in long-term estrogen deprived breast cancer." Frontiers in Endocrinology 13 (2022): 869562.

114) Liu, Chong, et al. "Advances in Rodent Models For Breast Cancer Formation, Progression, And Therapeutic Testing." Frontiers in Oncology 11 (2021): 593337.

115) Lakshmanaswamy, Rajkumar. Approaches to understanding breast cancer. Vol. 151. Academic Press, 2017.

116) Arumugam, Arunkumar, et al. "Short-term treatment with pregnancy levels of estradiol prevents breast cancer by delaying promotion and progression." Cancer Research 73.8_Supplement (2013): 199-199.

117) Lakshmanaswamy, Rajkumar, Raphael C. Guzman, and Satyabrata Nandi. "Hormonal prevention of breast cancer: significance of promotional environment." Hormonal Carcinogenesis V. New York, NY: Springer New York, 2008. 469-475.

118) Lakshmanaswamy, Rajkumar, et al. "Hormone-induced protection of mammary tumorigenesis in genetically engineered mouse models." Breast Cancer Research 9 (2007): 1-11.

119) Nandi, Satyabrata, et al. "Estrogen can prevent breast cancer by mimicking the protective effect of pregnancy." Hormonal Carcinogenesis IV (2005): 153-165.

120) Lakshmanaswamy, Rajkumar, et al. "Prevention of mammary carcinogenesis by short-term estrogen and progestin treatments." Breast Cancer Research 6.1 (2003): 1-7.

121) Lakshmanaswamy, Rajkumar, et al. "Short-term exposure to pregnancy levels of estrogen prevents mammary carcinogenesis." Proceedings of the National Academy of Sciences 98.20 (2001): 11755-11759.

122) Guzman, Raphael C., et al. "Hormonal prevention of breast cancer: mimicking the protective effect of pregnancy." Proceedings of the National Academy of Sciences 96.5 (1999): 2520-2525.

123) Arumugam, Arunkumar, Elaine A. Lissner, and Rajkumar Lakshmanaswamy. "The Role Of Hormones And Aromatase Inhibitors On Breast Tumor Growth And General Health In A Postmenopausal Mouse Model." Reproductive Biology and Endocrinology 12 (2014): 1-13.

124) Cold, Søren, et al. "Systemic or Vaginal Hormone Therapy After Early Breast Cancer: A Danish Observational Cohort Study." JNCI Journal of the National Cancer Institute 114.10 (2022): 1347.

125) Yue, Wei, et al. "Pro-Apoptotic Effects Of Estetrol On Long-Term Estrogen-Deprived Breast Cancer Cells And At Low Doses On Hormone-Sensitive Cells." Breast cancer: basic and clinical research 13 (2019): 1178223419844198.

126) Strickler, Ronald C. "Women's Health Initiative Results: A Glass More Empty Than Full." Fertility and Sterility 80.3 (2003): 488-490.

127) Bluming, Avrum Z., Howard N. Hodis, and Robert D. Langer. "RESPONSE TO LETTER TO EDITOR: Enough: the WHI's continued misrepresentation of its breast cancer claims." Menopause (New York, NY) 31.3 (2024): 243.

Chapter 8

Estrogen for Osteoporosis Prevention and Treatment

SALLY IS A 54-YEAR-OLD POST-MENOPAUSAL patient, one of the few who recovered completely from Graves' hyperthyroidism after medical treatment. However the bouts of hyperthyroidism took their toll, and Sally's latest DEXA scan showed a loss of bone density. The T-score at the spine and hip was -2.5 and -2.6. Sally is asking what is the best treatment for osteoporosis? Estrogen is now accepted as first-line treatment for the prevention and treatment of post-menopausal loss of bone density. (1-4)

We started Sally on our standard office protocol for bioidentical hormone replacement including topical estrogen and progesterone. After 6 months, Sally's follow up hormone lab panel showed a serum estradiol level of 32 pg/ml. Sally has a PhD in biological sciences and does her own research. She informed me that based on a 1992 study by Dr. Jean Reginster the minimal level of estradiol to prevent post-menopausal bone loss is 60 pg/ml, not the 32 pg/ml she has on her lab report. Sally says the estradiol level should be higher. (5)

I explained to Sally two things. Firstly, all patients in my office have the liberty of gradually increasing the topical estrogen cream dosage at home. We use the Topi-click dispenser which makes it easy to adjust dosage by increasing the number of clicks. For every rotation of the ring at the base of the dispenser, a quarter gram of cream is dispensed. As the number of clicks is increased, the dosage is increased. Eventually, one reaches the point of estrogen excess, recognizable with the obvious symptoms of breast enlargement, breast discomfort, nipple sensitivity, etc. Estrogen stimulates breast tissue. This will be obvious to the patient. A small amount of breast tissue stimulation is expected. Too much becomes annoying and uncomfortable. If this happens, the patient

is instructed to take a break from the hormone cream for 5-7 days, and return at half dosage once symptoms have resolved. The second symptom of estrogen excess is fluid retention. The patient's clothes fit more tightly. The analogy is an over-inflated beach ball. Again, this will be obvious to the patient. We have found that many women using Bi-Est will have symptoms of estrogen excess when estradiol levels exceed 60 pg/ml. Our treatment goal is not to target a serum estradiol level. Our treatment goal is a dosage of estradiol which provides complete relief from menopausal symptoms. In my office, women are at liberty to increase estrogen dosage based on their own judgement. However, higher dosages of estradiol may be associated with adverse side effects. Another factor to consider is the use of Bi-est rather than estradiol-alone means serum measurement of estradiol underestimates total estrogenic effect which is estradiol plus 1/8th the estriol dosage. A basic rule of thumb is estriol is 1/8 equivalent to estradiol in terms of estrogenic activity. (personal communication Dr. Lindsey Berkson)

Paucity of Evidence to Set Estradiol Target Goals

Secondly, Dr. Jean Reginster's old 1992 study is outdated. Since then, the science has changed and target goals for serum estradiol levels have been abandoned. In 2002, Dr. Armston reviewed this issue of target goal for estradiol to prevent bone loss, saying there is a paucity of evidence to support the setting of target levels at present, writing.

Evidence in the literature for using a particular cut-off value for plasma estradiol, which is sufficient to prevent osteoporosis, is compromised by:

- Lack of documentation of the estradiol method used.
- Differences in bias between laboratory estradiol assays.
- Lack of reference to the timing of sampling in relation to patch change or gel application.
- The small number of women included in many of the studies.
- In the case of patch studies, a failure to specify the type of patch studied.
- For gel preparations, lack of standardization of gel application area.

Despite these limitations, the trend to use HRT preparations that deliver lower doses of estradiol means that the clinical need for establishing a bone density response is greater than ever. Early measurement of estradiol levels within the first few months of treatment has been proposed to fulfill this need. However, there is a paucity of evidence to support the setting of target levels at present. Further studies are clearly required in this area. (6)

Trans-Dermal Estrogen Patch Vs. Topical Cream

In 2000, Dr. Tomas Andersson studied two trans-dermal estradiol delivery systems (Menorest® and Climara®), finding 25 pg/ml sufficient to relieve menopausal symptoms, 40 pcg/ml was the minimal concentration for preventing bone loss, and 60 pg/ml the upper limit because of tolerability. Dr. Tomas Andersson writes:

> The quantitative effects of estradiol are described by relating plasma levels to three proposed therapeutic or safety concentration limits for transdermal estradiol with 25 pg/ml as the minimum concentration for therapeutic effect on menopausal symptoms, 40 pg/ml as the estimated concentration required for osteoporosis prevention and 60 pg/ml as an upper limit because of tolerability. (7)

The above target of 40 mcg/ml does not apply to our type of HRT program because we use a topical cream applied twice a day. In my office, we do not routinely use the Menorest® or Climara® estradiol skin patches studied by Dr. Andersson. Instead, we use a compounded hormone cream (Biest, 80% estriol, 20% estradiol) applied topically to the skin twice a day. The skin patch stays in place delivering a steady dose of estrogen to the blood stream while the patch remains in place. When the patient presents herself for a blood draw at the lab, the estrogen patch remains in place. The steady release of estrogen from the patch makes measurement of serum estradiol level useful. Not so for application of topical estrogen twice a day which has peaks and troughs in serum estrogen levels. For topical creams, the patient is instructed to hold the morning application before the blood draw. This gives the trough for serum estradiol for the blood sample, and misses the estradiol peak. Therefore, at the time of the blood draw, the estradiol level with the topical cream may be lower than the comparable estradiol patch. Yet, the serum estradiol level one hour after application of topical compounded estrogen cream may be considerably higher than the comparable estradiol skin patch.

When it comes to estrogen dosage for the preservation of bone density, less is more. Excess estrogen can be uncomfortable and may lead to undesirable side effects. It is futile to set target levels for serum estradiol. Serum levels fluctuate and are higher immediately after the last dose. Estrogen serum targets are unreliable for topical creams for the reasons listed above by Dr. Armston. Rather, estrogen dosage is adjusted based on clinical symptoms with two goals in mind. The first goal of treatment is full relief of menopausal vasomotor symptoms (hot flashes, night sweats, and quality of sleep). The second goal of treatment is avoiding estrogen excess symptoms such as annoying breast symptoms (fullness, tenderness, pain, etc). If this happens, the patient is instructed to hold the estrogen cream for 4-5 days, and once estrogen excess symptoms resolve, return to it with a reduced dosage. Excess estrogen may cause uterine bleeding which promptly resolves when the dosage is reduced. If uterine bleeding persists, this triggers the usual work-up for post-menopausal bleeding which includes a pelvic sonogram and referral to a local gynecologist for pelvic exam. For patients with sonographic evidence of endometrial thickening, endometrial biopsy is indicated. Using our program of topical Bi-est (80% estriol, 20% estradiol), progesterone, and testosterone, we have found bone density increasing annually for all post-menopausal patients. Of course, we also use a bone-building nutritional supplement program, weight-bearing exercise program, and alkalinizing diet program.

Bone Building Nutritional Supplement Program

Vitamin C (ascorbate) fully buffered 500-1500 mg per day

Vitamin D3 5,000-10,000 iu/d optimized to blood levels above 50 ng/ml

Vitamin K1- 750 mcg/d

Vitamin K2- MK7 – 180 -250 mcg/day

Biotin 250 mcg/d

Calcium (chelated) 500 mg/d

Magnesium (chelated) 500 mg/d

Zinc Citrate 10 mg/d

Minerals (manganese, chromium, selenium, copper)

Molecular Iodine 3 mg/d

Potassium Iodide 150 mcg/d

Boron Citrate 6 mg/d

Vanadium citrate 250 mcg/d

Silica 10 mg/d

Strontium Citrate 50 mg/d

Adequate Dietary Protein Intake (10-11)

Medical Disorders Associated with Bone Fragility and Osteoporosis

The astute physician will be aware of and recognize various medical disorders associated with loss of bone density caused by malabsorption of calcium or protein. Such disorders include gluten sensitivity, atrophic gastritis, and previous bariatric gastric procedures. Other gastrointestinal disorders associated with increased bone fragility include inflammatory bowel disease, celiac disease, H. Pylori infection, chronic gastritis, gastric surgery, and intestinal microbiota dysbiosis as discussed in 2022 by Dr. Daniela Merlotti. (12-13)

Alkalinizing Diet Program

The basic concept of an alkalinizing diet is described by Susan Brown and Dr. Russell Jaffe in their books and articles. The American diet is acid-forming. To eliminate the acid, it must be buffered with calcium. Where does this buffering calcium come from? It comes from the skeleton where calcium is removed from the bones to provide buffering capacity. This leads to gradual loss of calcium in the skeleton over time, loss of bone density, and osteoporosis reversible with an alkaline diet. Dr Russell Jaffe's Alkaline Food Chart is a handy guide for those interested in the alkaline diet. (14-16)

Osteoporosis, the Basics

In 2017, Dr. Maria Almeida reviewed estrogen and testosterone in the basic physiology of the skeleton. There are about 2 million osteoporotic fractures annually in the United States. Post-menopausal women have a lifetime risk of osteoporotic fractures as high as 50% versus 25% in men. Post-menopausal women will lose on average 20-25 percent of their bone mass from age 50 to age 75. In both females and males, both estrogen deficiency and aging contribute to osteoporosis. What is the mechanism of post-menopausal osteoporosis? Acute estrogen deficiency causes a prolonged life span of "killer osteoclasts" which generate increased hydrogen peroxide used for dissolving, erod-ing, and removing cortical bone in both females and males. Post-menopausal estrogen replace-ment (ERT) shortens the lifespan of osteoclasts and prolongs the lifespan of osteoblasts, thus preventing "killer osteoclasts" from inducing accelerated osteoporosis. **Note:** osteoclasts remove bone, while osteoblasts produce new bone. The effects of estrogen and testoster-one on osteoclasts and osteoblasts are medi-ated through estrogen receptors (ER-alpha) and androgen receptors in these cells. For over 50 years, menopausal estrogen replacement therapy (ERT) was the treatment of choice for osteoporosis. However, since the 2002 release of the WHI study showing increased breast cancer, stroke, and venous thrombo-embo-lism in the hormone treated group using oral estrogen (Premarin, CEE) combined with syn-thetic progestin (medroxyprogesterone), ERT use has plummeted, leaving a void to be filled by anti-resorptive drugs such as the bisphos-phonates and more recently denosumab (anti-RANKL antibody). One of the tragic errors of modern medicine is the menopausal women concerned about bone density is more likely to be prescribed a bisphosphonate drug than hor-mone replacement. Dr. Maria Almeida writes:

> **The acceleration of cancellous bone loss that ensues with estrogen deficiency** results predominantly from a decrease in trabecular number caused by trabecular perforation and loss of connectivity. The precise mechanism of this pathological event is unclear, but histological evidence from human bone biopsies suggests that **acute estrogen deficiency gives rise to "killer osteoclasts"** capable of complete perforation of trabeculae in the cancellous bone compartment, such that it precludes subsequent refilling of the cavities by the bone-forming osteoblasts...The evidence from the study of mice with cell-specific deletion of the ERα, ... supports the notion that the **acceleration of trabecular bone loss following menopause is due to the prolongation of osteoclast lifespan,**

> **which results from the loss of the direct pro-apoptotic effects of estrogens on osteoclasts**...Replacement therapy with estrogens, commonly known as hormone replacement therapy (HRT), was the mainstay treatment for postmenopausal osteoporosis (and chronic disease prevention in older women in general) for almost 50 years. However, during the last decade, the use of HRT has diminished dramatically for two reasons: 1) the risk of adverse effects, such as breast cancer, venous thromboembolism, stroke, and coronary heart disease revealed by the Women's Health Initiative (WHI) trial; and 2) the availability of more potent and effective **antiresorptive drugs** that work regardless of whether increased resorption is the result of estrogen deficiency or other pathogenetic mechanisms. (17-18)

Mechanisms of Estrogen Prevention of Osteoporosis

In 2021, Dr. Sonja Gamsjaeger studied human iliac crest bone biopsies in post-meno-pausal women at baseline and after two years of estrogen treatment. Ten healthy women were studied using cyclic oral HRT (estradiol [2 mg]/norethisterone acetate [1 mg]) treatment. Dr. Sonja Gamsjaeger used Raman microspec-troscopic analysis of the paired bone biopsy samples, writing:

> Estrogen depletion due to menopause has been shown to increase osteoclast pre-cursor cells, decrease osteoclast differentiation...and delaying osteoclast apoptosis...**Hormone replacement therapy (HRT) prevents the effects on bone of estrogen depletion due to menopause,** by reducing the resorptive activity at the BMU [bone mineral unit] level and averting both increased osteoclast recruitment and delayed apoptosis [of osteoclasts]... In summary, the results of the present study offer indications that in addition to changes in bone turnover rates, **HRT affects the osteoid composition, mineralization regulation mechanisms, and possibly**

fibrillogenesis. Note: osteoid is premature bone secreted by osteoblasts composed of collagen fibrils, chondroitin and osteocalcin. The osteoid eventually becomes mineralized (calcified) and becomes bone. Note: The above study by Dr. Sonja Gamsjaeger used norethisterone, a carcinogenic synthetic progestin known to increase the risk of breast cancer. Natural progesterone is a superior choice. (19)

Adverse Effects of Anti-Resorptive Drugs

In 2022, Dr. Innocent U. Okagu discussed the adverse side effects of anti-resorptive drugs (bisphosphonates) pointing out these drugs are linked to several bone pathologies such as atypical femur fractures, osteonecrosis of the jaw, inflammatory eye disease, depression, anxiety, etc., writing:

> However, numerous findings are linking the usage of bisphosphonates to pathologic diseases including fractures of the femur, depression and anxiety, inflammatory eye disease, and medication-related osteonecrosis of the jaw (avascular necrosis of the jaw), the jawbone becomes exposed. Frequently, the tooth that was above it comes out, leaving a painful, non-healing sore behind. The lower jaw suffers from osteonecrosis more frequently than the upper jaw, and **this has increased scrutiny of the currently popular bisphosphonate treatment**. (20-22)

Estrogen Deficiency Triggers Inflammatory Disease

In 2006, Dr Neale Weitzmann discovered that estrogen deficiency accelerates bone loss through an inflammatory mechanism, resembling an inflammatory auto-immune disease. This is an astounding discovery! According to Dr. Henry Kronenberg from Harvard University, this accelerated post-menopausal bone loss resembles the post-partum state in which a sudden drop in estrogen level triggers bone resorption to meet the increased calcium demand of milk production for the newborn. All bone cells have functioning estrogen receptors, ER-alpha and ER-beta which control their activity, thus explaining estrogen's key role in maintaining bone mass, writing:

> postmenopausal osteoporosis should be regarded as the product of an **inflammatory disease** bearing many characteristics of an organ-limited **autoimmune disorder, triggered by estrogen deficiency,** and brought about by chronic mild decreases in T cell tolerance...One explanation is suggested by the need to stimulate bone resorption in the immediate postpartum period in order to meet the markedly increased maternal demand for calcium brought about by milk production. **The signal for this event is the drop in estrogen levels early postpartum.** Henry Kronenberg (Harvard University, Boston, Massachusetts, USA) has suggested that **postmenopausal bone loss should be regarded as an unintended recapitulation of this phenomenon** (personal communication)...Prior to 1987, bone cells were not generally considered direct targets of estrogen. However, it is now firmly established that osteoblasts (OBs), osteocytes, and osteoclasts (OCs) express functional estrogen receptors (ERs). These receptors are also expressed in bone marrow stromal cells (SCs), the precursors of OBs, which provide physical support for nascent OCs, T cells, B cells, and most other cells in human and mouse bone marrow. Estrogen signals through 2 receptors, ER-alpha and ER-beta. In humans, ER-alpha is the predominant isoform in cortical bone, while ER-beta is the predominant species in trabecular bone. In general, ER-alpha mediates most actions of estrogen on bone cells. (23)

Estrogen Therapy for Osteoporosis Prevention

In 1988, the FDA approved estrogen for the prevention of osteoporosis as an indicated use. Estrogen therapy prevents the annual loss of bone density caused by estrogen deficiency, thus reducing fracture risk. The Women's Health

Initiative (WHI) was the first randomized trial to show postmenopausal estrogen therapy provides a 34% reduction in osteoporotic fractures of the hip and vertebral bodies. These same results for estrogen use reducing fracture rates were confirmed independently by the Danish Osteoporosis Study (DOP) conducted by Dr. Leif Mosekilde (2000). Estrogen replacement therapy is now accepted by mainstream medicine as first-line prevention and treatment for osteoporosis. In my opinion, anti-resorptive drugs such as bisphosphonates and the newer osteoporosis drugs should be avoided. The new drugs, Denosumab (Prolia, FDA approved 2010) and Romosozumab (Evenity, FDA approved 2019) are even more dangerous. Rather than use anti-resorptive drugs, estrogen hormone replacement should be offered as the preferred treatment for all post-menopausal women concerned about bone density. One must keep in mind contraindications to estrogen therapy such as active breast cancer. In patients with thrombophilia with a history of deep venous thrombosis, oral estrogen is contra-indicated. The use of transdermal delivery of estrogen in selected cases is acceptable since there is no first pass through the liver and no increase in blood clotting factors as discussed in 2022 By Dr. Talia Sobel and in 2023 by Dr. Guy Morris. This topic is discussed further in Chapters 4-5. (24-33)

Menopausal Estrogen Therapy Has Anabolic Effect and Increases Bone Density in All Women

In 2001, Dr. Gautam Khastgir conducted a 6-year study of bone density in 22 older post-menopausal women (mean age 65.4 years) using 75 mg estradiol implant and oral MPA (medroxyprogesterone acetate) a synthetic progestin, 5 mg/day x 10 days each month. Iliac bone biopsies were done to measure bone response to hormone therapy. Dr Khastgir found supraphysiolgic serum levels of estradiol (450 pmol/L = 122 pg/ml) having an anabolic effect, and bone mineral density improved in all women, writing:

> The mean serum E2 level after 6 yr of treatment was 1,077 (range, 180-2568) pmol/L. Bone mineral density improved in every patient, with a median increase of 31.4% at the lumbar spine and 15.1% at the proximal femur. Bone histomorphometry showed an increase in cancellous bone volume from 10.75% to 17.31% (P < 0.001)... **This is the first report showing histological evidence for an increase in cancellous bone volume, together with an increase in wall thickness, in a longitudinal follow-up study of ERT in older postmenopausal women.** Our results show that E2 is capable of exerting an anabolic effect in women with osteoporosis, even when started well into the menopause. (34)

Author's Note: My main criticism of this study by Dr. Gautam Khastgiris that the 75 mg estradiol pellet resulted in supraphysiologic serum levels of estradiol equal to 122 pg/ml and above. In my opinion, this is an extremely high estradiol level that very few post-menopausal women can tolerate. My second criticism is this study used MPA, medroxyprogesterone, a carcinogenic synthetic progestin. MPA induces breast cancer in animals and is a well-known animal model called the MPA mouse model of breast cancer. The use of MPA for menopausal hormone replacement is no longer advised. Instead, the use of natural micronized progesterone is breast cancer preventive and a much safer alternative as discussed by Dr. Abenhaim (2022) and Dr. Asi (2016). (35-37)

Natural Estrogen and Progesterone Increases Bone Density

In 2004, Dr. Therese Nielsen conducted a 2-year randomized placebo-controlled trial of post-menopausal hormone therapy (MHT) using natural estrogen and progesterone in 386 women. Women were randomized to either placebo or intra-nasal 17beta-estradiol 150 or 300 mcg/day and micronized oral progesterone

200 mg/day. Dr. Therese Nielsen found bone mineral density (DEXA scan) increased at all measured sites in a dose-related manner. Bone density increased 5-7 percent at the spine and 3-5 percent at the hip over two years of hormone use. However, bone density decreased 3.2% for placebo users over two years. My clinical experience prescribing MHT over the last 20 years agrees with Dr. Nielsen. For post-menopausal women receiving estrogen hormone replacement, bone density increases annually. Quite the opposite for women on placebo, their bone density decreases annually. Dr. Therese Nielsen writes:

> BMD (Bone Mineral Density) increased at all measured sites in women receiving active treatment in a dose-related manner, the difference compared to placebo being 5.2% and 6.7% at the spine, and 3.2% and 4.7 % at the hip, respectively, with 150 µg and 300 µg (P<0.001). On the other hand, a decrease versus baseline of −3.2% and −3.3% at the spine and hip, respectively, was observed in women receiving a placebo. (P<0.001). (38)

Do We Even Need DEXA Scans?

The answer is no we do not. Primary care physicians are trained to routinely order a DEXA bone mineral density scan on all post-menopausal females, and then prescribe a bisphosphonate drug for osteoporosis defined as a T-score more negative than -2.5. As noted above by Dr. Therese Nielsen, bone density declines 1.6 percent annually in untreated menopausal women. However, for menopausal women on estrogen, bone density increases substantially from 1.5 to 3 percent annually. This is true for all healthy women, and is the reason why DEXA scans are unnecessary for women taking estrogen. My clinical experience confirms this observation. I no longer recommend serial DEXA scans on our patients taking estrogen, and I agree with the U.K. Royal Osteoporosis Society writing on their osteoporosis fact sheet:

You probably will not need regular bone density scans either, if you are taking HRT [hormone replacement therapy] to help strengthen your bones. (63-64)

Estrogen Therapy for Primary Prevention and Treatment of Osteoporosis

In 2018, Dr. V.A. Levin reviewed estrogen therapy for osteoporosis, proposing that estrogen-alone or estrogen combined with progesterone should be considered as primary prevention and treatment for osteoporosis, writing:

> Menopause predisposes women to osteoporosis due to declining estrogen levels. This results in a decrease in bone mineral density (BMD) and an increase in fractures. Osteoporotic fractures lead to substantial morbidity and mortality...The Women's Health Initiative (WHI) randomized controlled trial first proved hormonal therapy (HT) [with estrogen] reduces the incidence of all osteoporosis-related fractures in postmenopausal women... These studies support that HT [estrogen] improves BMD [bone mineral density] and reduces fracture risk in women with and without osteoporosis...**we propose that HT should be considered for the primary prevention and treatment of osteoporosis in appropriate candidates.** HT should be individualized and the once "lowest dose for shortest period of time" concept should no longer be used. ... low-dose and transdermal HT that have been shown to be safer than oral standard-dose HT...HT in the form of either **combined estrogen and progesterone or estrogen alone** has been shown to be effective in reducing the number of both vertebral and non-vertebral fractures in postmenopausal women, with efficacy equivalent to that of bisphosphonates [drugs such as Fosamax ®]...Randomized controlled trials and observational studies show that **standard-dose HT,** which was proposed by the manufacturer and approved by registration authorities to suit the average patient needs, reduces postmenopausal osteoporotic fractures of the hip, spine, and

all non-spine fractures in women with and without osteoporosis. (39)

Estrogen Superior to Anti-Resorptive Drugs

In 2019, Dr. Jan Stepan reviewed menopausal hormone therapy for fracture prevention stating that estrogen is superior to anti-resorptive drugs because estrogen can attenuate the inflammatory bone-microenvironment associated with estrogen deficiency, a feature lacking with anti-resorptive drugs, writing:

> **Estrogen alone or combined with progestin to protect the uterus from cancer significantly reduces the risk of osteoporosis-related fractures.** MHT [Menopausal Hormone Therapy] **increases type 1 collagen production and osteoblast survival** and maintains the equilibrium between bone resorption and bone formation by modulating osteoblast/osteocyte and T cell regulation of osteoclasts. Estrogens have positive effects on muscle and cartilage. **Estrogen, but not antiresorptive therapies, can attenuate the inflammatory bone-microenvironment associated with estrogen deficiency.** However, already on second year of administration, MHT is associated with excess breast cancer risk, increasing steadily with duration of use...MHT should be considered in women with premature estrogen deficiency and increased risk of bone loss and osteoporotic fractures. However, MHT use for the prevention of bone loss is hindered by increase in breast cancer risk even in women younger than 60 years old or who are within 10 years of menopause onset. (40)

Authors Note: Here, Dr. Jan Stepan makes the common error of attributing to progesterone the carcinogenic properties of synthetic progestins such as medroxyprogesterone, which increases breast cancer risk. Additionally, Dr. Jan Stepan makes the common error of attributing increased breast cancer risk to all MHT (Menopausal Hormone Therapy). This is incorrect as increased cancer risk is due to the synthetic progestin component, not to the natural estrogen, progesterone or testosterone components of MHT. The exact hormone formula matters. See the below discussion by Dr. Anna Gosset citing four studies supporting the observation that when natural progesterone is used instead of synthetic progestins, there is no increased risk of breast cancer. For more, see Chapter 7 on Errors in Modern Medicine: Fear of Estrogen.

Testosterone has Anabolic Bone Building Effects

Our routine menopausal hormone replacement program includes testosterone for all women. Testosterone has anabolic effects on bone formation, useful for prevention and treatment of osteoporosis. In 2011, Dr. Chevon M. Rariy studied testosterone blood levels and bone density in 232 elderly females (67-94 yrs), finding higher bone density associated with higher circulating testosterone levels. Dr. Rariy writes:

> In the setting of the low estradiol levels found in older women, circulating T [Testosterone] levels were associated with bone density. Women with higher free T levels had greater lean body mass, consistent with the anabolic effect of T. (41-47)

In 2023, Dr. Elsa Nunes studied hormone levels in elderly women over age 65 yrs. finding a significant correlation between testosterone levels and bone mineral density (BMD) at the hip, suggesting lower testosterone increases the risk for osteoporosis, writing:

> Using a highly sensitive hormone assay method, our study identified a significant association between testosterone and BMD [Bone Mineral Density] of the hip in women over 65 years of age, suggesting that **lower testosterone increases the risk of osteoporosis**. (48)

In 2006, Dr. Karen K. Miller conducted a randomized placebo-controlled trial giving testosterone replacement to 51 women of reproductive age with severe testosterone deficiency due to hypopituitarism, finding all women receiving testosterone showed an increase in bone density over 12 months, writing:

> This is the first randomized, double-blind, placebo-controlled study to show a positive effect of testosterone on bone density, body composition, and neurobehavioral function in women with severe androgen deficiency due to hypopituitarism. (49)

Conventional Medicine Now Accepts Estrogen Replacement as First Line Treatment for Prevention of Post Menopausal Bone Loss.

In 2021, Dr. Anna Gosset reviewed menopausal hormone therapy for the management of osteoporosis, stating bone is an estrogen-dependent tissue, and post-menopausal bone loss is due to estrogen deficiency. Estrogen hormone replacement to maintain bone mass and reduce fracture risk has been used since 1947 and is now endorsed as the first-line treatment for bone health by several international menopause societies. We have repeatedly made the point in this book that breast cancer risk from menopausal hormone replacement is due to synthetic progestins which are carcinogenic. When natural progesterone is used, studies show no increased risk of breast cancer. Dr. Anna Gosset agrees with this, citing four supportive observational studies, writing:

> In 1947, Fuller Albright established the principles that **exogenous estrogen** given to postmenopausal women was efficient to maintain bone mass and thus reduce fracture risk. Since then, menopausal hormone therapy (MHT) has long been used to prevent the risk of osteoporosis and related fractures in postmenopausal women. Accordingly, in the mid-1990s and early 2000s, it was widely prescribed in most European and North American countries for this specific purpose...Furthermore, the

> notion that MHT is an appropriate first-line therapy option for the maintenance of bone health in early postmenopausal women is now endorsed by several international menopause societies including among others the International Menopause Society (IMS), the North American Menopause Society, the Endocrine Society or the European Menopause and Andropause Society (EMAS)... we suggest that MHT should be considered and discussed with the patient as a **first-line treatment** even in the absence of major climacteric symptoms provided there are no contraindication... most studies have reported that a large majority of women who initiate low dose estrogen therapy will benefit protection against early postmenopausal bone loss... **Postmenopausal osteoporosis is for a large part the consequence of both quantitative and qualitative bone alterations induced by estrogen deficiency which occurs within the first years of the menopause transition.** Bone is a strong estrogen-dependent tissue and estrogens play a major role in the acquisition and maintenance of bone mineral content throughout life...**The risk of breast cancer remains the other concern and very often the main limitation to the use and acceptance of MHT by women**... the risk appears to depend on the type of progestogen used. In the large cohort study E3N, **there was no increase in the risk of breast cancer in women who were given a combination of E2 plus natural progesterone** or dydrogesterone for an average 5-year period of treatment whereas a significant increase was found when a synthetic progestogen was combined to E2 (Fournier, 2008) (Fournier, 2014). **Three other European cohort studies also reported a significantly lower risk of breast cancer when E2 was combined with progesterone** or dydrogesterone than with other synthetic progestogens (Lyytinen, 2009) (Cordina-Duverger, 2013) (Bakken, 2011). (50-61)

Conclusion: One of the tragic errors of modern medicine caused by the fear of estrogen, is the wide use of dangerous anti-resorptive drugs given to menopausal women concerned about bone density. Estrogen replacement is now first-line treatment for the prevention of post-menopausal bone loss even in the frail elderly over the age of 75 as discussed in 2001 by Dr. Dennis T. Villareal. In my office, our post-menopausal hormone replacement formula is: low-dose topical estrogen in the form of Bi-est (80% estriol/ 20% estradiol), progesterone, testosterone, bone-building nutrients, and exercise. This formula is very effective for maintaining bone density. In my opinion, the use of anti-resorptive drugs such as bisphosphonates and the newer antibody drugs (denosumab) should be discouraged, as these carry adverse side effects and modify the histology of the bone matrix to make the bones weaker, not stronger. Bone cells have estrogen receptors and estrogen signaling is required to maintain bone density. The prevention and treatment of osteoporosis is bioidentical estrogen, not chemical manipulation with anti-resorptive drugs which paradoxically cause osteonecrosis of the jaw and spontaneous femur fractures. (62)

♦ References for Chapter 8 Estrogen for Osteoporosis Prevention and Treatment

1) Tella, Sri Harsha, and J. Christopher Gallagher. "Prevention and Treatment of Post-Menopausal Osteoporosis." The Journal of Steroid Biochemistry and Molecular Biology 142 (2014): 155-170.

2) Nilas, L., and C. Christiansen. "The Pathophysiology of Peri- and Post-Menopausal Bone Loss." BJOG: An International Journal of Obstetrics & Gynaecology 96.5 (1989): 580-587.

3) Capozzi, Anna, et al. "Calcium, Vitamin D, Vitamin K2, and Magnesium Supplementation and Skeletal Health." Maturitas 140 (2020): 55-63.

4) Stuenkel, Cynthia A. "Menopausal Hormone Therapy and the Role of Estrogen." Clinical Obstetrics and Gynecology 64.4 (2021): 757-771.

5) Reginster, Jean Yves, et al. "Minimal Levels Of Serum Estradiol Prevent Postmenopausal Bone Loss." Calcified Tissue International 51 (1992): 340-343.

6) Armston, Annie, and Peter Wood. "Hormone Replacement Therapy (Estradiol-Only Preparations): Can The Laboratory Recommend A Concentration Of Plasma Estradiol To Protect Against Osteoporosis?" Annals Of Clinical Biochemistry 39.3 (2002): 184-193.

7) Andersson, Tomas LG, et al. "Drug Concentration Effect Relationship Of Estradiol From Two Matrix Transdermal Delivery Systems: Menorest® and Climara®." Maturitas 35.3 (2000): 245-252.

8) Smith-Bindman, R., E. Weiss, and V. Feldstein. "How Thick Is Too Thick? When Endometrial Thickness Should Prompt Biopsy in Postmenopausal Women Without Vaginal Bleeding." Ultrasound in Obstetrics and Gynecology. 24.5 (2004): 558-565.

9) Williams, Pamela M., and Heidi L. Gaddey. "Endometrial Biopsy: Tips and Pitfalls." American Family Physician 101.9 (2020): 551-556.

10) Skalny, Anatoly V., et al. "Role of Vitamins Beyond Vitamin D 3 In Bone Health And Osteoporosis." International Journal Of Molecular Medicine 53.1 (2024): 1-21.

11) Groenendijk, Inge, et al. "Discussion on Protein Recommendations For Supporting Muscle And Bone Health In Older Adults: A Mini Review." Frontiers in Nutrition 11 (2024): 1394916.

12) Walters, J. R., et al. "Detection of Low Bone Mineral Density by Dual Energy X Ray Absorptiometry In Unsuspected Suboptimally Treated Coeliac Disease." Gut 37.2 (1995): 220-224.

13) Merlotti, Daniela, et al. "Bone Fragility in Gastrointestinal Disorders." International Journal of Molecular Sciences 23.5 (2022): 2713.

14) Natural Bone Health: A Practitioner's Guide to Healthy Bone, Joints, and Muscles Paperback – September 15, 2022 by Russell M. Jaffe (Author) HSC Press (September 15, 2022)

15) Brown, Susan E., and R. Jaffe. "Acid-alkaline balance and its effect on bone health." International Journal of Integrative Medicine 2.6 (2000): 1-12.

16) Alkaline Food Chart by Russell Jaffe MD https://www.drrusselljaffe.com/alkaline-food-chart/

17) Almeida, Maria, et al. "Estrogens and Androgens In Skeletal Physiology And Pathophysiology." Physiological reviews 97.1 (2017): 135-187.

18) Albright F, Bloomberg E, Smith PH. Postmenopausal Osteoporosis. Trans Assoc Am Physicians 298–205, 1940.

19) Gamsjaeger, Sonja, et al. "Effect of hormone replacement therapy on bone formation quality and mineralization regulation mechanisms in early postmenopausal women." Bone reports 14 (2021): 101055.

20) Okagu, Innocent U., et al. "A Review On The Molecular Mechanisms Of Action Of Natural Products In Preventing Bone Diseases." International Journal of Molecular Sciences 23.15 (2022): 8468.

21) Zhang, Jie, Kenneth G. Saag, and Jeffrey R. Curtis. "Long-term Safety Concerns Of Antiresorptive Therapy." Rheumatic Disease Clinics 37.3 (2011): 387-400.

22) Whitaker, Marcea, et al. "Bisphosphonates For Osteoporosis—Where Do We Go From Here?." New England Journal of Medicine 366.22 (2012): 2048-2051.

23) Weitzmann, M. Neale, and Roberto Pacifici. "Estrogen Deficiency And Bone Loss: An Inflammatory Tale." Journal of Clinical Investigation 116.5 (2006): 1186.

24) Mosekilde, Leif, et al. "Hormonal Replacement Therapy Reduces Forearm Fracture Incidence In Recent Postmenopausal Women—Results of the Danish Osteoporosis Prevention Study." Maturitas 36.3 (2000): 181-193.

25) Schierbeck, Louise Lind, et al. "Effect of hormone replacement therapy on cardiovascular events in recently postmenopausal women: randomised trial." Bmj 345 (2012).

26) Sobel, Talia H., and Wen Shen. "Transdermal Estrogen Therapy In Menopausal Women At Increased Risk For Thrombotic Events: A Scoping Review." Menopause 29.4 (2022): 483-490.

27) Morris, Guy, and Vikram Talaulikar. "Hormone Replacement Therapy In Women With History Of Thrombosis Or A Thrombophilia." Post Reproductive Health 29.1 (2023): 33-41.

28) Na, Zhao, et al. "Role of Menopausal Hormone Therapy In The Prevention Of Postmenopausal Osteoporosis." Open Life Sciences 18.1 (2023): 20220759.

29) Anagnostis, Panagiotis, et al. "Estrogen and Bones After Menopause: A Reappraisal Of Data And Future Perspectives." Hormones 20 (2021): 13-21.

30) Rozenberg, Serge, et al. "Is There A Role For Menopausal Hormone Therapy In The Management Of Postmenopausal Osteoporosis?" Osteoporosis International 31 (2020): 2271-2286.

31) Chang, Cherry Yin-Yi, et al. "Timing and Dosage Of And Adherence To Hormone Replacement Therapy And Fracture Risk In Women With Menopausal Syndrome In Taiwan: A Nested Case-Control Study." Maturitas 146 (2021): 1-8.

32) Newson, Louise, Sarah Ball, and Rebecca Lewis. "Letter to Editor: Our Concerns About HRT Not Having A Priority As A Treatment For Osteoporosis In The NOGG Guidelines." Osteoporosis International 34.4 (2023): 815-816.

33) Bagger, Yu Z., et al. "Two To Three Years Of Hormone Replacement Treatment In Healthy Women Have Long-Term Preventive Effects On Bone Mass And Osteoporotic Fractures: The PERF Study." Bone 34.4 (2004): 728-735.

34) Khastgir, Gautam, et al. "Anabolic Effect Of Estrogen Replacement On Bone In Postmenopausal Women With Osteoporosis: Histomorphometric Evidence In A Longitudinal Study." The Journal of Clinical Endocrinology & Metabolism 86.1 (2001): 289-295.

35) Abenhaim, Haim A., et al. "Menopausal Hormone Therapy Formulation and Breast Cancer Risk." Obstetrics & Gynecology 139.6 (2022): 1103-1110.

36) Asi, Noor, et al. "Progesterone vs. Synthetic Progestins and The Risk Of Breast Cancer: A Systematic Review And Meta-Analysis." Systematic Reviews 5 (2016): 1-8.

37) Lanari, Claudia, et al. "The MPA Mouse Breast Cancer Model: Evidence For A Role Of Progesterone Receptors In Breast Cancer." Endocrine-Related Cancer 16.2 (2009): 333.

38) Nielsen, Therese F., et al. "Pulsed Estrogen Therapy In Prevention Of Postmenopausal Osteoporosis. A 2-Year Randomized, Double Blind, Placebo-Controlled Study." Osteoporosis International 15 (2004): 168-174.

39) Levin, V. A., X. Jiang, and R. Kagan. "Estrogen therapy for osteoporosis in the modern era." Osteoporosis International 29 (2018): 1049-1055.

40) Stepan, Jan J., et al. "Update on Menopausal Hormone Therapy For Fracture Prevention." Current Osteoporosis Reports 17 (2019): 465-473.

41) Rariy, Chevon M., et al. "Higher Serum Free Testosterone Concentration In Older Women Is Associated With Greater Bone Mineral Density, Lean Body Mass, And Total Fat Mass: The Cardiovascular Health Study." The Journal of Clinical Endocrinology & Metabolism 96.4 (2011): 989-996.

42) Zhang, Han, et al. "Association Between Testosterone Levels And Bone Mineral Density In Females Aged 40–60 Years From NHANES 2011–2016." Scientific Reports 12.1 (2022): 16426.

43) Yang, JinXiao, et al. "Association between serum total testosterone level and bone mineral density in middle-aged postmenopausal women." International Journal of Endocrinology 2022 (2022).

44) van Geel, Tineke ACM, et al. "Measures of Bioavailable Serum Testosterone And Estradiol And Their Relationships With Muscle Mass, Muscle Strength And Bone Mineral Density In Postmenopausal Women: A Cross-Sectional Study." European Journal of Endocrinology 160.4 (2009): 681-687

45) Tok, Ekrem C., et al. "The Effect Of Circulating Androgens On Bone Mineral Density In Postmenopausal Women." Maturitas 48.3 (2004): 235-242.

46) Jassal, Simerjot K., Elizabeth Barrett-Connor, and Sharon L. Edelstein. "Low Bioavailable Testosterone Levels Predict Future Height Loss In Postmenopausal Women." Journal of Bone and Mineral Research 10.4 (1995): 650-654.

47) Wang, Nan, et al. "Association of Total Testosterone Status With Bone Mineral Density In Adults Aged 40–60 Years." Journal of Orthopaedic Surgery and Research 16 (2021): 1-7.

48) Nunes, Elsa, et al. "Steroid Hormone Levels And Bone Mineral Density In Women Over 65 Years Of Age." Scientific Reports 13.1 (2023): 4925.

49) Miller KK, et al. Effects of Testosterone Replacement In Androgen-Deficient Women With Hypopituitarism: A Randomized, Double-Blind, Placebo-Controlled Study. J Clin Endocrinol Metab. 2006;91(5):1683–90.

50) Gosset, Anna, Jean-Michel Pouillès, and Florence Trémollieres. "Menopausal hormone therapy for the management of osteoporosis." Best Practice & Research Clinical Endocrinology & Metabolism 35.6 (2021): 101551.

51) Wronski, T. J., et al. "Estrogen Treatment Prevents Osteopenia And Depresses Bone Turnover In Ovariectomized Rats." Endocrinology 123.2 (1988): 681-686.

52) Tella, Sri Harsha, and J. Christopher Gallagher. "Prevention and Treatment of Post-Menopausal Osteoporosis." The Journal of Steroid Biochemistry and Molecular Biology 142 (2014): 155-170.

53) Nilas, L., and C. Christiansen. "The Pathophysiology of Peri- and Post-Menopausal Bone Loss." BJOG: An International Journal of Obstetrics & Gynaecology 96.5 (1989): 580-587.

54) Capozzi, Anna, et al. "Calcium, Vitamin D, Vitamin K2, and Magnesium Supplementation and Skeletal Health." Maturitas 140 (2020): 55-63.

55) Stuenkel, Cynthia A. "Menopausal Hormone Therapy and the Role of Estrogen." Clinical Obstetrics and Gynecology 64.4 (2021): 757-771.

56) Rozenberg, Serge, et al. "Is There a Role for Menopausal Hormone Therapy in the Management of Postmenopausal Osteoporosis?" Osteoporosis International 31 (2020): 2271-2286.

57) Fournier A, Berrino F, Clavel-Chapelon F. Unequal Risks for Breast Cancer Associated With Different Hormone Replacement Therapies: Results from the E3N Cohort Study. Breast Cancer Res Treat 2008; 107: 103–111.

58) Fournier A, Mesrine S, Dossus L, et al. Risk of Breast Cancer After Stopping Menopausal Hormone Therapy In The E3N Cohort. Breast Cancer Res Treat 2014; 145: 535–543.59

59) Lyytinen H, Pukkala E, Ylikorkala O. Breast Cancer Risk In Postmenopausal Women Using Estradiol–Progestogen Therapy. Obstet Gynecol 2009; 113: 65–73.

60) Cordina-Duverger E, Truong T, Anger A, et al. Risk of Breast Cancer By Type Of Menopausal Hormone Therapy: A Case-Control Study Among Post-Menopausal Women In France. PLoS One 2013; 8: e78016.

61) Bakken K, Fournier A, Lund E, et al. Menopausal Hormone Therapy And Breast Cancer Risk: Impact Of Different Treatments. The European Prospective Investigation into Cancer and Nutrition. Int J Cancer 2011; 128: 144-156.

62) Villareal, Dennis T., et al. "Bone Mineral Density Response to Estrogen Replacement In Frail Elderly Women: A Randomized Controlled Trial." JAMA 286.7 (2001): 815-820.

63) Royal Osteoporosis Society, Hormone Replacement Therapy (HRT) Fact Sheet. https://theros.org.uk/information-and-support/osteoporosis/treatment/hormone-replacement-therapy/

64) Finkelstein, Joel S., et al. "Bone Mineral Density Changes During the Menopause Transition In a Multiethnic Cohort Of Women." The Journal of Clinical Endocrinology & Metabolism 93.3 (2008): 861-868.

Chapter 9

Estrogen Prevents Heart Disease Part One

MARY IS A 52-YEAR-OLD STAY-AT-HOME Mom who is concerned about her risk for heart attack. Her father, uncle, and grandfather all died of heart attacks. Mary went to see her primary care doctor who said her cholesterol was high and gave her a statin drug as a preventive measure. Mary arrived in my office for a second opinion. I explained to Mary that estrogen is a far better preventive measure than statin drugs, and Mary was started on her menopausal hormone replacement program containing estrogen, progesterone, and testosterone. Mary should decline the statin drug, and I will explain why.

The "Flip Flop" on Estrogen and Heart Disease

Medical science has done a "Flip-Flop" on estrogen and coronary heart disease. Before 2003, forty observational studies convinced mainstream doctors that estrogen prevents heart disease in women. Doctors considered it unethical to withhold hormone replacement therapy (HRT) for post-menopausal women, and three medical organizations published guidelines said so, the American College of Physicians, the American College of Obstetrics and Gynecology, and the American Heart Association guidelines. Before 2003, physicians widely believed that estrogen bestowed protection from coronary heart disease (CHD) and that estrogen replacement in post-menopausal women prevented coronary artery disease (CAD). (1-3)

In 2017, Dr. Andrea Iorga reviewed the protective role of estrogen and its receptors in prevention of cardiovascular disease writing:

> While cardiovascular disease (CVD) is the leading cause of death among women, epidemiologic studies indicate that females prior to menopause are somewhat protected against the development of CVD when compared to men. Women have reduced incidence of CVD when compared to age-matched males and present with CVD 10 years later than men... Female protection against CVD is associated with sex hormone levels as the incidence and severity of CVD increases in women postmenopause, and the prevalence of coronary artery disease (CAD) is greater in young women who had an oophorectomy compared to those with intact ovaries ...The results of years of research into female sex hormones indicate that the main circulating female hormone, **estrogen (E2), is cardioprotective through a plethora of mechanisms** that are highlighted throughout this review. (4-5)

Premature Ovarian Failure and Coronary Artery Disease

Studies of women with premature ovarian failure (POF) provide supportive evidence that estrogen is cardioprotective. In 2010, Dr. Kate Maclaran studied the long-term sequelae of POF, finding an **80 percent increased mortality from cardiovascular disease** for POV under age 40 compared to menopause at age 49, writing:

> Life-expectancy: POF has been associated with a 50% higher mortality than women with menopause at age 52–55....An association between early menopause and increased mortality from cardiovascular disease has been established for many years, with **an estimated 80% increase risk of mortality from ischaemic heart disease in those with menopause under the age of 40 compared with those with menopause at 49–55**....This risk of ischaemic heart disease was more pronounced in those who

had never used estrogens. It has also been demonstrated that cardiovascular mortality in women oophorectomized before the age of 45 and not receiving HRT was significantly increased (**HR 1.84**; 95% CI 1.27–2.6) and that estrogen replacement may reduce this risk....Impaired ovarian function is thought to cause increased atherosclerosis progression based on non-human primate models and angiographic studies of estrogen deficiency of hypothalamic nature. (6-8)

2002 WHI Study with Premarin (CEE) and Medroxyprogesterone (MPA)

The tables were turned in 2002, when the catastrophic Women's Health Initiative Randomized Controlled Trial (WHI, first arm) showed increased heart disease in the hormone-treated group (CEE/MPA), for which the study was terminated early after 5.2 years. **Note:** the 2002 WHI first arm hormone-treated group was given Premarin (CEE) combined with medroxyprogesterone (MPA) also called Prempro, which many have argued is not real hormone replacement at all! True hormone replacement requires natural human bioidentical hormones, estradiol, and progesterone as discussed by Dr. Cynthia L. Bethea (2011) in Chapter 7 who writes:

It [the WHI trial] did not study hormone replacement at all: that would have required use of the natural hormones, estradiol and progesterone. **Author's Note:** The 2002 WHI first arm, the HERS study, and other randomized controlled trials (RCTs) in the 1990s used the wrong hormone, a synthetic progestin medroxyprogesterone MPA. They also used the wrong age group, older women in their 60's, fifteen years beyond the menopausal transition. (9)

We digress and will get back to this later. The failure of the 2002 first-arm WHI trial to show any cardiac benefit for the Premarin/MPA users was the final straw, arriving on the shirt-tails of the 2002 HERS II trial by Dr. Deborah Grady, expecting the same cardio-protection shown in earlier observational and animal studies, and failing to find it. HERS and WHI were randomized trials, yet both were an utter disappointment, failing to show the expected reduction in coronary artery disease in the hormone-treated group. (10-11)

Dr. Barrett-Conner 2003 - Estrogen is a Failure!

In her 2003 publication, Dr. Barrett-Conner laments the failure of estrogen to prevent coronary artery disease (CHD) in post-menopausal women: Dr. Barrett-Conner's dramatic quote:

The failure to find cardioprotective effects in any of the several clinical trials with CHD [coronary heart disease] outcomes **offers little hope that postmenopausal HT (Hormone Therapy) will prevent heart disease**. The results, unexpected and unwelcome, are nonetheless definitive and are likely to extend beyond the studied treatment regimens. (12)

Of course, the above comments by Dr. Barrett-Conner are **completely wrong,** as you will see below.

2011 - Massive Flip-Flop, Estrogen Is Indeed Cardio-Protective

Unlike the 2002 first-arm WHI study, the 2004 second-arm WHI study used estrogen-alone in women post hysterectomy yielding quite different results. The 11-year follow-up of the 2004, second arm, estrogen-alone, Women's Health Initiative (WHI) published in 2011 in JAMA, clearly showed that estrogen reduces mortality from heart disease, reduces heart attacks and reduces all-cause mortality (see data below). However, protection was conferred in the 50-60 year age group, and not for older women, after 15 years of menopausal estrogen deficiency. In addition, protection from cardiovascular disease (CVD) was lost or reduced when a synthetic progestin, medroxyprogesterone (MPA) was added to the hormone replacement cocktail. Here (below) is the data from the 11-year follow-up of the 2004 second-arm WHI study, in which women after hysterectomy (surgical menopause) were treated with estrogen-alone (Premarin, CEE).

Events by Group WHI Second Arm 2004 11-Year Follow Up (estrogen-alone)

AGE | **Coronary Artery Disease Hazard Ratio**

50-59 yrs............**0.59**
60-69 yrs............1.0
70-79 yrs............1.06

Total MI Hazard Ratio

50-59 yrs............**0.54**
60-69 yrs............1.05
70-79 yrs............1.23

All-Cause Mortality Hazard Ratio

50-59 yrs............**0.73**
60-69 yrs............1.04
70-79 yrs............1.12

Invasive Breast Cancer Hazard Ratio

50-59 yrs............**0.80**
60-69 yrs............**0.73**
70-79 yrs............**0.81**

Note: MI=Myocardial Infarction. Notice that for the "younger" age group (age 50-60 yrs), HRT with Premarin-alone provided an astounding **41% reduction in coronary artery disease, 46% reduction in myocardial infarctions, and 27 % reduction in all-cause mortality.** In addition, there is a 20-27 percent reduction in breast cancer. When one lumps all the age groups together, the benefit is NULL. However, when one looks at the 50-60 age group, starting hormone replacement immediately after the hormone decline of menopause, the benefits are striking. (13-14)

Degenerative Diseases Are Irreversible

The explanation for this "Flip-Flop" is quite simple. As Mark Houston MD so aptly describes in his 2018 book, the *Truth About Heart Disease*, coronary artery disease is a chronic degenerative disease caused by inflammation, preventable when measures are started early, before irreversible tissue damage. Once the degenerative disease has a chance to develop, irreversible tissue damage has occurred and the preventive measures are ineffective. This is true for other types of degenerative diseases such as osteoarthritis. In the younger age group, estrogen is effective in preventing atherosclerotic coronary artery disease when hormone replacement therapy (HRT) is started immediately after the menopausal transition. However, HRT is ineffective when introduced ten or twenty years later after enough time has elapsed to develop atherosclerotic coronary artery disease. **Note:** Although thought to be irreversible, coronary artery disease may be reversed, at least in part, by the Calcium Score Protocol as discussed in my 2018 book, *Heart Book*. (15-16)

23 RCT's Re-Examined - Behold a Benefit for Younger Post-Menopause

In 2006, Drs. Shelley R. Salpeter, and Judith Walsh, reviewed 23 clinical trials of hormone replacement from the 1990s and teased out the data for the younger women (50-59 years) who started early on HRT immediately after menopause transition, the authors found a **40 percent reduction in cardiovascular disease and 40 percent reduction in total mortality**, replicating the data from the 2004 second arm estrogen-alone WHI study. Dr. Shelley Salpeter writes:

> The literature search identified 23 trials that met inclusion criteria...We included randomized-controlled trials of at least 6 months duration that compared HT [Hormone Therapy] with placebo or no hormone therapy in postmenopausal women...The analysis included 39,049 participants, with a mean trial duration

of 4.9±1.7 years...**In younger women, the reduction in CHD events seen with HT is similar to the reduction in total mortality that has been seen in pooled trial data** (OR 0.6 [CI, 0.4 to 0.95]).4 This reduction in cardiac morbidity and mortality is similar to that found in the observational Nurses' Health Study, which followed a cohort of 120,000 women below the age of 55 years. After adjusting for potential confounding variables, such as age, cardiovascular risk factors, and socioeconomic status, **HT use was associated with a 40% reduction in CHD events and total mortality**. Note: OR=Odds Ratio. (17-19)

Two Coronary Calcium Score Studies Are Consistent with Cardio-Protection

The coronary artery calcium score has emerged over the last decade as the most sensitive and accurate test for determining future heart attack risk. One would be correct to say the cholesterol panel has been replaced by the calcium score as the best predictive tool. The Women's Health Initiative included two coronary calcium score studies showing striking calcium score benefits in the Premarin-alone users, but not in the Premarin-Progestin (MPA) combination users. In 2007, Dr. JoAnn Manson measured calcium score at the end of the 8.7-year observation period for "younger" post-menopausal women in the 50-59 age group. This group, using estrogen-alone (Premarin or CEE, 0.625 mg per day) **showed 60% reduction in calcium score compared to placebo users** (calcium score of 83.1 vs.123). Dr. JoAnn Manson writes:

> The mean coronary-artery calcium score after trial completion was lower among women receiving estrogen (83.1) than among those receiving placebo (123.1). (20)

Annual Progression of Calcium Score with Estrogen-Alone

Annual progression of calcium score has also emerged as a very accurate way to monitor treatment for coronary atherosclerosis. Dr. Paolo Raggi found that less than 15 percent annual progression conferred protection from future heart attack, indicating treatment success. However, annual progression of calcium score of greater than 15% denoted a poor prognosis with high risk for heart attack and treatment failure. In 2005, Dr. Matthew Budoff conducted a second calcium score study from women in the WHI clinical trial. This study performed two consecutive calcium scores one year apart examining progression of calcium score. Dr. Budoff found a 9 percent annual increase in calcium score for the estrogen-alone treated women, compared to 22% annual increase for placebo users. Prempro (Premarin plus MPA) users showed a 24% annual increase in the calcium score, about the same as placebo. This is highly significant because of the 2004 study by Paolo Raggi study which showed annual progression of calcium score over 15 percent is associated with high risk for future heart attack, while under 15 percent annual progression is protective with good prognosis. In 2018, Dr. Joshua D. Mitchell, Associate Professor of Medicine at Washington University in St. Louis showed that for women with a calcium score from 0 to 100, a statin drug has no clinical benefit regardless of cholesterol level. The calcium score determines who the cardiologist treats with a statin drug. This is the paradigm shift in mainstream cardiology. (21-23)

MPA Negates Cardiovascular Benefit of Estrogen

In 2007, Dr. Alexander Becker in Germany studied calcium score progression over 3 years in 277 women taking the Premarin-progestin (MPA) combination compared to matched non-users. This study showed no calcium score benefit when medroxyprogesterone is added to the estrogen. The 3-year increase in calcium score was the same for both groups, hormone users and non-users with no cardiovascular benefit in the hormone users, thus confirming the above 2005 calcium score study by

Dr. Mathew Budoff showing adding progestin (MPA) to the estrogen negates the cardiovascular benefits of estrogen. (24)

Animal Studies in Apo-E Deficient Mice

The Apo-E mouse is a favorite animal model for the study of atherosclerosis because it spontaneously develops atherosclerosis. In 1996, Dr. P.A. Bourassa, studied atherosclerosis in apolipoprotein E-deficient male and female mice, finding "the accelerated progression of [atherosclerotic] lesions resulting from ovariectomy was completely reversed with 17-Beta-estradiol treatment." Similarly, in **male** Apo-E deficient mice, estrogen reversed atherosclerotic lesions. Dr. P.A. Bourassa writes:

> Homologous recombination techniques targeting the apolipoprotein E (apoE) gene have recently generated mice that develop atherosclerosis, providing **a convenient source of large numbers of animals in which to study atherogenesis.** These apoE-deficient mice reproducibly develop hypercholesterolemia with progressively complex and widespread [atherosclerotic] lesions resembling inflammatory-fibrous plaques seen in humans... The elimination of circulating ovarian steroids by bilateral ovariectomy resulted in a significant increase in atherosclerotic lesions. ApoE-deficient mice spontaneously develop lesions in the aortic valve and throughout the arterial tree...**The accelerated progression of lesions resulting from ovariectomy was completely reversed with 17-Beta-estradiol treatment**... Consistent with studies in females, 17-Beta-estradiol treatment for 90 days significantly reduced lesion area and lesion progression in male mice. (25-26)

Statins Do The Opposite

Although estrogen reverses the progression of atherosclerotic plaque in the Apo-E deficient mouse model, statin drugs do the opposite, pravastatin and simvastatin increase the plaque size in the Apo-E mouse model. Perhaps this explains why estrogen is far more effective than statin drugs for cardioprotection in post-menopausal women. This is discussed further below. (27-29)

Cynomolgus Macaque Monkey Model

In 2002, Dr. Tomi Mikkola and Thomas Clarkson reviewed the role of estrogen in cynomolgus macaque monkeys who share 90% DNA homology with humans, and closely replicate the human hormone profile, with 28-day menstrual cycle, and having a menopause. After ovariectomy, the female monkeys will develop accelerated atherosclerosis. However, restoring estrogen levels using 30 months of estradiol (pellets) **reduced atherosclerotic plaque size by 50 percent**. Thirty months of oral estrogen (CEE, Premarin) **reduced plaque size by 72 percent**. However, when continuous medroxyprogesterone (MPA) was given either alone or with estrogen to the ovariectomized monkeys, plaque size was the same as the untreated controls. The benefits of estrogen were abrogated by continuous MPA. Dr. Mikkola writes:

> surgically menopausal female cynomolgus macaque monkey model is useful in evaluating the cardioprotective aspects of ERT [Estrogen Replacement Therapy]. This animal model is often used since these monkeys share with humans DNA greater than 90% homology, and their hormonal profile resemble those of women; distinct menarche, 28-day menstrual cycles, and menopause....Direct evidence for a beneficial effect of ERT on progression of coronary artery atherosclerosis was found in the results of a study from Adams et al. In this study, ovariectomized monkeys that were treated for 30 months with parenterally administered 17β-estradiol showed approximately **50% reduction in coronary artery atherosclerosis** compared to the control animals. These findings were confirmed in a subsequent 30-month study evaluating the effects of continuous oral conjugated equine estrogen (CEE) treatment on coronary artery atherosclerosis. **Oral CEE treatment caused a 72% reduction**

in coronary artery plaque size relative to untreated, estrogen-deficient controls. These studies strongly support the observational findings that estrogens are cardioprotective...There is strong evidence from both human and nonhuman primate studies supporting the conclusion that **estrogen deficiency increases the progression of atherosclerosis.** More controversial is the conclusion that postmenopausal estrogen replacement inhibits the progression of atherosclerosis. Estrogen treatment of older women (>65 years) with pre-existing coronary artery atherosclerosis had no beneficial effects. In contrast, **estrogen treatment of younger postmenopausal women or monkeys in the early stages of atherosclerosis progression has marked beneficial effects.** Whether progestogens attenuate the cardiovascular benefits of estrogen replacement therapy has been controversial for more than a decade. Current evidence from studies of both monkeys and women suggest little or no attenuation of estrogen benefits for coronary artery atherosclerosis. Lack of compliance with estrogen replacement therapy, usually because of fear of breast cancer, remains a major problem. (30)

Monkey Studies: Adding MPA to CEE

In 1997, Dr. Michael R. Adams studied the combination of estrogen (CEE)/MPA given to ovariectomized monkeys. His study showed MPA antagonizes the atheroprotective effects of estrogen (CEE). This identical finding was observed in the 2002 first arm WHI trial which used this same combination, Premarin/MPA, also showing no atheroprotective effect. Dr Michael R. Adams writes:

> Treatment with CEE [Premarin®] alone resulted in atherosclerosis extent that was reduced 72% relative to untreated (estrogen-deficient) controls (P<.004). Atherosclerosis extent in animals treated with CEE plus MPA [medroxyprogesterone] or MPA alone **did not differ from that of untreated controls**... Although the mechanism(s) remains

unclear, we conclude that oral CEE inhibits the initiation and progression of coronary artery atherosclerosis and that **continuously administered oral MPA antagonizes this atheroprotective effect.** (31-32)

Adding MPA Reverses Cardioprotection of Estradiol

In 2007, Dr. Erin A. Booth studied the cardio-protective effects of estradiol in New Zealand White rabbits, in a myocardial ischemia/infarct model. Dr. Booth found myocardial infarction size reduced by administration of estradiol before induction of myocardial ischemia (19.5 vs 55.7 percent). However, when MPA was added to the estradiol, there was no reduction in infarct size (52 percent). In addition, beneficial effects on other cardiac parameters were lost by the addition of MPA, writing:

> One possible explanation for the divergent results is the addition of progestin [MPA] to the hormone regimen in the Women's Health Initiative and the Heart and Estrogen/ Progestin Replacement Study trials. The aim of the present study was to examine the effects of a combination of 17beta-estradiol (E(2), 20 microg) and medroxyprogesterone acetate (MPA, 80 microg) on infarct size in New Zealand White rabbits. Infarct size as a percentage of the area at risk was significantly reduced by administration of E(2) 30 min before induction of myocardial ischemia compared with vehicle (19.5 +/- 3.1 vs. 55.7 +/- 2.6%, P < 0.001). However, E(2) + MPA failed to elicit a reduction in infarct size (52.5 +/- 4.6%), and MPA had no effect (50.8 +/- 2.6%). E(2) also reduced serum levels of cardiac troponin I, immune complex deposition in myocardial tissue, activation of the complement system, and lipid peroxidation. All these effects were reversed by MPA. The results suggest that MPA antagonizes the infarct-sparing effects of E(2), possibly through modulation of the immune response after ischemia and reperfusion. (33-35)

Surgical Menopause Associated with Greater Risk of Myocardial Infarction

In 1981, Dr. L. Rosenberg found that surgical menopause with bilateral oophorectomy, removal of ovaries, before age 35 carries a **7.2 times greater risk of MI** (myocardial infarction) compared to normal cycling premenopausal women, writing:

> Among women who became menopausal because of bilateral oophorectomy, the estimated relative risk of MI [myocardial infraction] increased with decreasing age at menopause, and women who underwent **bilateral oophorectomy before age 35 were estimated to have a risk of hospitalization for MI approximately 7.2 times** (95% confidence interval, 4.5 to 11.4) that of premenopausal women. Hysterectomy without the removal of both ovaries was only weakly associated with an increased risk. The data support the hypothesis that premature cessation of ovarian function [estrogen deficiency] increases the risk of nonfatal MI. (36-38)

Estrogen Protects Women After Myocardial Infarction

In 2001, Dr. Michael G. Shlipak studied 114,724 postmenopausal women **after myocardial infarction**, 7,353 (6.4%) of whom were using hormone replacement (HRT). The HRT users had a **59 percent reduction in unadjusted mortality** compared to non-users. Thus, indicating considerable protective effects of HRT after myocardial infarction, writing:

> The present study was performed with 114,724 women of age ≥55 years with confirmed myocardial infarction who presented between April 1998 and January 2000 to 1 of 1674 hospitals participating in the National Registry of Myocardial Infarction-3. Presenting characteristics, treatment, and clinical outcome data were obtained by chart review. At time of hospitalization, 7353 (6.4%) women reported current use of HRT, defined as use of estrogen, progestin, or estrogen/progestin

for reasons other than contraception. **Unadjusted mortality was 7.4% in users of HRT and 16.2% in nonusers (odds ratio 0.41, 95% confidence interval 0.36 to 0.43).** After adjustments were made for prior medical history, clinical characteristics, treatments received in-hospital, and likelihood of receiving HRT, HRT remained associated with an improved rate of survival (**odds ratio 0.65,** 95% confidence interval 0.59 to 0.72). **Significant association of HRT with decreased mortality after myocardial infarction was observed in all age strata.** (39)

Estrogen Up-Regulates EPCs, Conferring Myocardial Protection

The next logical question is, what is the mechanism for estrogen's protective effect post-myocardial infarction? Preclinical mouse models show endothelial progenitor cells (EPCs) confer protection post-myocardial infarction. Estrogen upregulates the EPCs, thus conferring protection. In 2007 Dr. Hiromichi Hamada studied wild-type and ER-alpha-knockout mice (ER-alphaKO), finding the administration of estrogen stimulates faster recovery after ischemic injury by increasing EPCs which neo-angiogenisis (new vessel formation) in wild-type mice, but not in ER-alphaKO mice. This supports the 2001 study by Dr. Michael G. Shlipak mentioned above in which post-menopausal women using HRT enjoyed improved prognosis with 59 percent reduction in mortality after myocardial infarction compared to non-users. Dr. Hiromichi Hamada writes:

> Both ER-alpha and ER-beta contribute to E2 [estradiol]-mediated EPC activation and tissue incorporation and to preservation of cardiac function after myocardial infarction. ER-alpha plays a more prominent role in this process. Moreover, ER-alpha contributes to upregulation of vascular endothelial growth factor, revealing possible mechanisms of an effect of E2 on EPC biology. Finally, **these data provide additional evidence of the**

importance of bone marrow–derived EPC phenotype in ischemic tissue repair. (40)

How Does Estrogen Prevent Coronary Artery Disease?

Estrogen Inhibits LDL (Lipoprotein) Oxidation

in 2022, Dr. Hui Jiang writes: "OxLDL [Oxidized Low Density Lipoproteins] represents the main culprit in current theories of atherosclerosis". The oxidized LDL is engulfed by macrophages which form foam cells, a characteristic feature of the atherosclerotic plaque. Indeed, the oxidation of LDL (low density lipoproteins) represents the first step in the atherosclerosis process, as described in 2004 by Dr. Mohamad Navab. Numerous natural anti-oxidants prevent LDL oxidation such as Vitamin E and estrogen shown in preclinical and human trials to inhibit lipoprotein oxidation, thus explaining cardioprotective benefits. (41-48)

Estrogen Prevents Endothelial Dysfunction

An excellent 2009 article by Dr. Bechlioulis reviews the benefits of estrogen in preventing endothelial dysfunction, the initiating step in atherosclerosis. Flow-mediated dilatation of the brachial artery (FMD) is a handy tool to evaluate endothelial dysfunction. Studies using FMD show that estrogen deficiency is linked to endothelial dysfunction in postmenopausal women. (49)

Estrogen is Anti-Inflammatory - Basic Science Studies

Several basic science studies in animals and humans show that estrogen has a profound anti-inflammatory effect, explaining cardio-protection. For more on the anti-inflammatory properties of estrogen, see Chapter 10, Estrogen Prevents Coronary Artery Disease, Part Two. (50-52)

Carotid Plaque Study from Italy - Less inflammation in the Plaque

In 2008, Dr. Raffaele Marfella from Italy conducted an elegant study of carotid artery plaques obtained from surgical specimens from 20 post-menopausal women on HRT (hormone replacement therapy) and 32 untreated women. Dr. Marfella studies the microscopic histology of the plaques looking for inflammatory markers and cells. These findings were then compared to carotid artery samples from non-hormone-user controls. In women using hormone replacement, the authors found considerable reduction of inflammatory cells and markers in the plaque material. Dr. Raffaele Marfella concludes that hormone replacement therapy inhibits the activation of nuclear factor kappa-B (NF-kB) dependent inflammation, responsible for plaque rupture. Dr. Raffaele Marfella writes:

this study examined the differences in inflammatory infiltration, as well as ubiquitin-proteasome activity, between asymptomatic carotid plaques of postmenopausal women with and without concomitant hormone replacement therapy. Plaques were obtained from 20 postmenopausal women treated with hormone replacement therapy (current users) and 32 nontreated women (never-users) enlisted to undergo carotid endarterectomy for extracranial high-grade (>70%) internal carotid artery stenosis. Plaques were analyzed for macrophages, T lymphocytes, human leukocyte antigen-DR cells, ubiquitinproteasome system, nuclear factor KappaB, inhibitor of nuclear factor KappaB, tumor necrosis factor-alpha, nitrotyrosine, matrix metalloproteinase-9, and collagen content (immunohistochemistry and ELISA). [Note: these are markers for inflammation] Compared with plaques from current users, **plaques from never-users had more macrophages , T lymphocytes, and human leukocyte antigen-DRcells [reactive immune cells], more ubiquitin-proteasome activity, tumor necrosis factor-alpha, and nuclear factor KappaB** (P<0.001); and more nitrotyrosine and matrix metalloproteinase-9 (P<0.001), along with a lesser collagen content and inhibitor of nuclear factor KappaB levels (P<0.001). This study supports the hypothesis that **hormone replacement therapy inhibits plaque ubiquitin-proteasome activity by decreasing oxidative stress generation in postmenopausal women**. This effect, in turn, might contribute to plaque stabilization by inhibiting the activation of nuclear factor KappaB–dependent inflammation, responsible for plaque rupture. (53)

Note: The ubiquitin-proteasome system is the principal degradation route of intracellular and oxidized proteins. In human carotid plaques increased oxidative stress is associated with inhibition of the proteasome activity and accumulation of ubiquitin conjugates, particularly in symptomatic patients. (54)

Drs. Howard Hodis and Wendy Mack

The cardio-protection of estrogen is best described by "The Timing Hypothesis" of Drs. Howard Hodis and Wendy Mack in a series of articles from 2008 to 2022. Dr. Howard Hodis is Professor of Cardiology, and Medicine and Director of Atherosclerosis Research at Keck School of Medicine USC. For younger postmenopausal women in the 50-60-year age group, all randomized controlled trials (RCTs) agree. Preclinical animal and human observational studies show estrogen HRT confers a **40-50 percent reduction** in cardiovascular disease in the 50-60 year old age group. However, for older women, there is a null effect. This 50-60 age group is called the "Window of Opportunity" for post-menopausal hormone replacement. The rapid estrogen decline of early menopause, or premature ovarian failure POV, is associated with increased mortality, and the onset of degenerative diseases such as coronary artery disease, osteoporosis, osteoarthritis, cognitive decline, and urogenital syndrome. This can be reversed with estrogen HRT. Lipid-lowering with statin drugs for primary prevention of cor-

onary artery disease in menopausal women is ill-advised. The accumulated data shows lipid lowering with statin drugs has no clinical benefit for women. Dr. Howard Hodis writes:

Lipid-lowering therapy, predominantly with HMG-CoA reductase inhibitors (statin drugs) is the mainstay for the primary prevention of CHD [coronary heart disease] in women. The cumulated data, however, do not provide convincing evidence for the significant reduction of CHD with lipid-lowering therapy relative to placebo when used for primary prevention of CHD in women and **there is no evidence that such therapy reduces overall mortality.** Note: for more on this see Chapter 27 on the Failure of Cholesterol Lowering Drugs. (55-57)

DOP Study Shows Cardiovascular Benefit in Younger Women

The 2012, Dr. Louise Schierbeck published the cardiovascular prevention data from the Danish Osteoporosis Prevention (DOP) study. The DOP was a randomized controlled trial (RCT) of hormone replacement with estradiol and norethisterone acetate (a progestin) in 1,006 healthy recently post-menopausal women in the 45–56-year age group, treated over 11 years. The primary endpoint of death, admission to hospital for heart failure, or myocardial infarction was **reduced 52 percent in the hormone treated group. All-cause mortality was reduced 43 percent in the hormone treated group.** Dr. Schierbeck writes:

After 10 years of randomized treatment, women receiving hormone replacement therapy early after menopause had a significantly reduced risk of mortality, heart failure, or myocardial infarction, without any apparent increase in risk of cancer, venous thromboembolism, or stroke. (58)

Estrogen Prevents Onset of Degenerative Disease

If one considers menopause as heralding the onset of various degenerative diseases, then this explains why hormone replacement is protective for early post-menopausal and is ineffective 10 to 15 years later. This also explains why Premarin (estrogen) is dramatically effective for reducing risk for heart disease in the 50-60 age group, yet ineffective for the over 60 age group as demonstrated in the 11-year follow data for the WHI (estrogen-alone, second arm, 2004). (13-14)

Avoid Synthetic Monster Hormones

In addition, it is now clear that women must avoid synthetic hormones, such as medroxy-progesterone (MPA), the "progestin" found in the PremPro (CEE/MPA). This synthetic monster hormone increases risk for heart attacks, coronary artery disease, and opposes the cardio-protective effects of estrogen. Around 20 observational studies show the progestin, medroxyprogesterone (MPA), increases risk for breast cancer. A commonly used animal model of breast cancer uses MPA to induce breast cancer in mice (Lanari, 2009). Avoid the synthetic progestins. Instead, use human bioidentical homones, estradiol, progesterone, and testosterone. These natural hormones cannot be patented. Without patent protection there is no profitability for the drug industry. As mentioned in Chapters 4-5, the transdermal delivery of estrogen avoids the blood clotting complications associated with oral estrogen pills. (59-64)

Statin Anti-Cholesterol Drugs

It is commonplace for postmenopausal women to be prescribed a statin drug by their primary care doctor intended to prevent heart disease. Prescribing statin drugs for post-menopausal women for primary prevention of heart disease is a form of mistreatment and abuse that should be halted. Women are terrorized by drug marketing advertising on television and worried about their cholesterol levels. Commonly, the first question during office lab reviews I hear is: "What is my cholesterol level?" Statin drugs come with horrendous adverse side effects, muscle pain, neuropathy, and cognitive dysfunction. In 2008, Dr. Beatrice Golomb summarized the adverse effects of statin drugs writing that statin drugs are mitochondrial toxins and deplete Co-Enzyme Q10 levels, causing neuropathy, myopathy, congestive heart failure, transient global amnesia, erectile dysfunction, dementia, and blood sugar elevation, diabetes, etc. For more on this, see Chapter 7 on the Failure of Cholesterol Lowering Drugs. (65-73)

No Benefit for Women with Statin Drugs

The reality is that statin drugs do not reduce mortality from heart disease in women, nor do they reduce all-cause mortality. This data is summarized nicely in articles by Dr. Judith Walsh in JAMA 2004, Dr. Mario Petretta in Int Journal of Cardiology (2020), and Dr. Kausik K. Ray in Archives of Int Med (2010). (74-76).

History of Hormone Replacement Therapy (HRT) and Heart Disease

1920–1939: Estradiol synthesized (1938).
1940–1949: Premarin introduced (1942).
1983–1987: The first observational studies of the benefits of HRT were published by Trudy Bush in 1983 and 1987 showing a **reduction in cardiovascular disease and total mortality** in hormone users. (77-78)
1991–Dr. Barrett-Connor reports eleven studies show a **50% reduction in mortality** from heart disease for post-menopausal women using estrogen-alone, without progestin (unopposed estrogen-alone). Dr. Conner writes:

> Most, but not all, studies of hormone replacement therapy in post-menopausal women show around a 50% reduction in risk of a coronary event in women using unopposed oral estrogen. (79)

1992–Dr. Deborah Grady reviewed the medical literature since 1970 showing that post-menopausal estrogen replacement reduces the risk of heart disease in women, and reduces osteoporotic fracture risk as well. (1)
1992–Three Medical Organizations Endorse

Estrogen Therapy to Prevent Heart Disease. The American College of Physicians, the American College of Obstetrics and Gynecology, and the American Heart Association guidelines: all post-menopausal women should be offered estrogen hormone replacement to prevent heart disease. (2-3)

1995–PremPro, (Premarin+MPA medroxy-progesterone), the first combination HRT pill is introduced.

1998–HERS study using Prempro reports early increased heart disease risk.

2002–WHI (First Arm-Prempro) reports increased heart disease, stroke, and breast cancer, FDA requires black box warnings for all postmenopausal estrogens with or without progestin.

2004–WHI Second Arm (Premarin-alone) shows 23 percent less breast cancer and less heart disease in estrogen users. (13-14)

2011–WHI Second Arm 11-year follow-up data shows 50 percent reduction in coronary artery disease "younger" 50-60 year post-meno-pausal women using estrogen alone (CEE, Premarin®) without MPA progestin. (13-14)

2018–WHI second arm (estro-gen-alone) 18-year follow-up shows a 45 per-cent reduction in breast cancer mortality in the estrogen-alone treated group. (80-81)

2016: More recent studies, KEEPS, and ELITE show vascular benefits of HRT with a good safety profile, supporting the Timing Hypothesis (Window of Opportunity) of Dr. Howard Hodis. In 2024, Dr. Felice Gersh writes:

the Kronos Early Estrogen Prevention Study (KEEPS) and Early Versus Late Intervention Trial with Estradiol (ELITE) studies, both showed no harm from HRT, with improvement in quality of life (QoL), and of vascular benefits for younger, recently PM [Post Menopausal] women in the ELITE data, which [is] similar to the WHI, was supportive of the "Timing Hypothesis." Note: ELITE= Early vs Late Treatment with Estradiol. (82-85)

In 2024, Dr. Jane Yang reviewed the entire medical literature on menopausal estrogen deficiency from 1972 to 2023, writing:

We posit that, because estrogen and ERs [estrogen receptors] are critical to the functioning of many systems of the body beyond the female reproductive system, there is great utility in replacing the declining supply of estrogen in postmenopausal women....An initial observation showed that **premenopausal women have lower rates of coronary artery disease and myocardial infarction than men of the same age**...Recent publications of the American Heart Association affirm that beneficial outcomes and **reductions in all-cause mortality may occur when HT is initiated in women <60 years of age or <10 years since menopause**, whereas null or harmful effects may occur when HT is initiated at older ages...Some studies have reported an increased risk of myocardial infarction in postmenopausal women, particularly after bilateral oophorectomy. Another study showed that women receiving postmenopausal HT had a decreased risk of mortality after MI... Overall, these findings suggest that estrogen directly affects the heart both in a cardioprotective manner and in response to injury. For this reason, women may benefit from cardiac tissue estrogen exposure, and this is particularly pronounced in the event of ischemic injury... The Early versus Late Intervention Trial with Estradiol (ELITE) demonstrated that HT led to **decreased progression of atherosclerosis in early menopausal women but did not have an effect on late menopausal women.** In that study, atherosclerotic change was quantified through carotid intima-media thickness (CIMT) measurements. Posttrial analyses show that increased E2 levels were associated with lower CIMT progression in early menopause but had no effect on CIMT progression if initiated in late menopause.... In mice, estrogen supplementation was found to prevent the formation of new atherosclerotic lesions but had little effect on the progression of established lesions or plaque stability in terms of intraplaque hemorrhage or medial erosion....The route of estrogen delivery appears to play an important role, as many studies report that **transdermal estrogen does not lead to thrombosis risk as compared to oral estrogen that does increase clotting risk**... **estrogen suppresses the production of proinflammatory cytokines such as IL-1, IL-6, and TNF-α, many of which are regulated by transcription factor NF-κB**. (86)

Conclusion - Don't Wait! The cardio-protective benefits of estrogen replacement are greatest at the initiation of menopause, and health benefits diminish after degenerative disease is allowed to progress. The calcium score data is clear. Estrogen reduces calcium score progression, while statins do the opposite. With the above information, we can propose mainstream primary doctors and cardiologists should be prescribing bioidentical hormone replacement for their post-menopausal patients, rather than statin drugs for primary prevention of coronary artery disease. In the next Chapter 10, we discuss how estrogen prevents atherosclerosis by maintaining the gut mucosal barrier, reversing leaky gut and reducing inflammation.

◆ **References for Chapter 9 Estrogen Prevents Heart Disease**

1) Grady, Deborah, et al. "Hormone Therapy to Prevent Disease and Prolong Life in Postmenopausal Women." Annals of Internal Medicine 117 (1992): 1016-1037.

2) Grady, Deborah, et al. "Guidelines for Counseling Postmenopausal Women About Preventive Hormone-Therapy." Annals of Internal Medicine 117.12 (1992): 1038-1041.

3) Andrews, W., et al. "Guidelines for Counseling Women on The Management Of Menopause." Washington, DC: Jacobs Institute of Women's Health Expert Panel on Menopause Counseling (2000).

4) Iorga, Andrea, et al. "The Protective Role of Estrogen and Estrogen Receptors In Cardiovascular Disease and The Controversial Use Of Estrogen Therapy." Biology of sex differences 8 (2017): 1-16.

5) Wake, Ryotaro, and Minoru Yoshiyama. "Gender Differences In Ischemic Heart Disease." Recent Patents on Cardiovascular Drug Discovery (Discontinued) 4.3 (2009): 234-240.

6) Maclaran, Kate, Etienne Horner, and Nick Panay. "Premature Ovarian Failure: Long-Term Sequelae." Menopause International 16.1 (2010): 38-41.

7) Blümel, Juan E., et al. "Is Premature Ovarian Insufficiency Associated with Mortality? A Three-Decade Follow-Up Cohort." Maturitas 163 (2022): 82-87.

8) Liu, Jiajun, et al. "The Risk Of Long-Term Cardiometabolic Disease In Women With Premature Or Early Menopause: A Systematic Review And Meta-Analysis." Frontiers in Cardiovascular Medicine 10 (2023): 1131251.

9) Bethea, Cynthia L. "MPA: Medroxy-Progesterone Acetate Contributes to Much Poor Advice for Women." Endocrinology. (2011): 343-345.

10) Rossouw, Jacques E., et al. "Risks and Benefits of Estrogen Plus Progestin in Healthy Postmenopausal Women: Principal Results From The Women's Health Initiative Randomized Controlled Trial." Jama 288.3 (2002): 321-333.

11) Grady, Deborah, et al. "Cardiovascular Disease Outcomes During 6.8 Years of Hormone Therapy: Heart And Estrogen/Progestin Replacement Study Follow-Up (HERS II)." Jama 288.1 (2002): 49-57.

12) Barrett-Connor, Elizabeth. "An Epidemiologist Looks at Hormones and Heart Disease In Women." The Journal of Clinical Endocrinology & Metabolism 88.9 (2003): 4031-4042.

13) LaCroix, Andrea Z., et al. "Health Outcomes After Stopping Conjugated Equine Estrogens Among Postmenopausal Women with Prior Hysterectomy: A Randomized Controlled Trial." JAMA 305.13 (2011): 1305-1314.

14) Anderson, Garnet L., et al. "Conjugated Equine Estrogen and Breast Cancer Incidence and Mortality In Postmenopausal Women With Hysterectomy: Extended Follow-Up Of The Women's Health Initiative Randomized Placebo-Controlled Trial." The Lancet Oncology 13.5 (2012): 476-486.

15) Houston, Mark. The Truth about Heart Disease: How to Prevent Coronary Heart Disease and Personalize Your Treatment with Nutrition, Nutritional Supplements, Exercise and Lifestyle Tailored to Your Genetics. CRC Press, 2022.

16) Dach, Jeffrey. Heart Book, How to Keep Your Heart Healthy. Medical Muse Press, 2018.

17) Salpeter, Shelley R., et al. "Brief Report: Coronary Heart Disease Events Associated With Hormone Therapy In Younger And Older Women: A Meta-Analysis." Journal Of General Internal Medicine 21 (2006): 363-366.

18) Grodstein, Francine, et al. "Postmenopausal estrogen and progestin use and the risk of cardio-vascular disease." New England Journal of Medicine 335.7 (1996): 453-461.

19) Falkeborn, Margareta, et al. "The risk of acute myocardial infarction after estrogen and estrogen-pro-gestogen replacement." BJOG: An International Journal of Obstetrics & Gynaecology 99.10 (1992): 821-828.

20) Manson, JoAnn E., et al. "Estrogen Therapy and Coronary-Artery Calcification." New England Journal of Medicine 356.25 (2007): 2591-2602.

21) Budoff, Matthew J., et al. "Effects of Hormone Replacement on Progression of Coronary Calcium As Measured By Electron Beam Tomography." Journal of women's health 14.5 (2005): 410-417.

22) Raggi, Paolo, Tracy Q. Callister, and Leslee J. Shaw. "Progression of Coronary Artery Calcium and Risk of First Myocardial Infarction in Patients Receiving Cholesterol-Lowering Therapy." Arteriosclerosis, Thrombosis, And Vascular Biology 24.7 (2004): 1272-1277.

23) Mitchell, Joshua D., et al. "Impact of Statins on Cardiovascular Outcomes Following Coronary Artery Calcium Scoring." Journal of the American College of Cardiology 72.25 (2018): 3233-3242.

24) Becker, Alexander, et al. "Comparison of Progression Of Coronary Calcium In Postmenopausal Women On Versus Not On Estrogen/Progestin Therapy." The American journal of cardiology 99.3 (2007): 374-378.

25) Bourassa, P. A., et al. "Estrogen Reduces Atherosclerotic Lesion Development In Apolipoprotein E-Deficient Mice." Proceedings of the National Academy of Sciences 93.19 (1996): 10022-10027.

26) Davezac, Morgane, et al. "Estrogen Receptor And Vascular Aging." Frontiers in aging 2 (2021): 727380.

27) Bea, Florian, et al. "Simvastatin Promotes Atherosclerotic Plaque Stability In Apoe-Deficient Mice Independently Of Lipid Lowering." Arteriosclerosis, thrombosis, and vascular biology 22.11 (2002): 1832-1837.

28) Zhang, Xiaoling, et L. "Pravastatin polarizes the phenotype of macrophages toward M2 and elevates serum cholesterol levels in apolipoprotein E knockout mice." Journal of International Medical Research 46.8 (2018): 3365-3373.

29) Tse, Jenny, et al. "Accelerated Atherosclerosis and Premature Calcified Cartilaginous Metaplasia In The Aorta Of Diabetic Male Apo E Knockout Mice Can Be Prevented By Chronic Treatment With 17β-Estradiol." Atherosclerosis 144.2 (1999): 303-313.

30) Mikkola, Tomi S., and Thomas B. Clarkson. "Estrogen Replacement Therapy, Atherosclerosis, And Vascular Function." Cardiovascular research 53.3 (2002): 605-619.

31) Adams, Michael R., et al. "Inhibition of Coronary Artery Atherosclerosis By 17-Beta Estradiol in Ovariectomized Monkeys. Lack Of an Effect of Added Progesterone." Arteriosclerosis: 10.6 (1990): 1051-1057.

32) Adams, Michael R., et al. "Medroxyprogesterone Acetate Antagonizes Inhibitory Effects of Conjugated Equine Estrogens on Coronary Artery Atherosclerosis." Arteriosclerosis, Thrombosis, And Vascular Biology 17.1 (1997): 217-221.

33) Booth, Erin A., and Benedict R. Lucchesi. "Medroxyprogesterone Acetate Prevents The Cardioprotective And Anti-Inflammatory Effects Of 17β-Estradiol In An In Vivo Model Of Myocardial Ischemia And Reperfusion." American Journal of Physiology-Heart and Circulatory Physiology 293.3 (2007): H1408-H1415.

34) Jeanes, Helen L., et al. "Medroxyprogesterone Acetate Inhibits the Cardioprotective Effect Of Estrogen In Experimental Ischemia-Reperfusion Injury." Menopause 13.1 (2006): 80-86.

35) Bernstein, Paula, and Gerald Pohost. "Progesterone, Progestins, And the Heart." Reviews In Cardiovascular Medicine 11.3 (2010): e141-9.

36) Rosenberg L, et al. Early Menopause and The Risk of Myocardial Infarction. Am J Obstet Gynecol 1981; 139:47-51.

37) Parker, William H., et al. "Ovarian Conservation at The Time of Hysterectomy and Long-Term Health Outcomes In The Nurses' Health Study." Obstetrics & Gynecology 113.5 (2009): 1027-1037.

38) Rivera, Cathleen M., et al. "Increased Cardiovascular Mortality After Early Bilateral Oophorectomy." Menopause 16.1 (2009): 15-23.

39) Shlipak Michael G., et al. Hormone Therapy and In-Hospital Survival After Myocardial Infarction in Postmenopausal Women. Circulation 2001; 104:2300-2304.

40) Hamada, Hiromichi, et al. "Estrogen Receptors A and B Mediate Contribution of Bone Marrow–Derived Endothelial Progenitor Cells to Functional Recovery After Myocardial Infarction." Circulation 114.21 (2006): 2261-2270.

41) Navab, Mohamad, et al. "Thematic Review Series: The Pathogenesis of Atherosclerosis The Oxidation Hypothesis Of Atherogenesis: The Role Of Oxidized Phospholipids And HDL." Journal Of Lipid Research 45.6 (2004): 993-1007.

42) Jiang, Hui, et al. "Mechanisms of Oxidized LDL-Mediated Endothelial Dysfunction and Its Consequences For The Development Of Atherosclerosis." Frontiers in Cardiovascular Medicine 9 (2022): 925923.

43) Escalante Gómez, et al. "HRT Decreases DNA And Lipid Oxidation in Postmenopausal Women." Climacteric 16.1 (2012): 104-110.

44) Escalante, Carlos Gómez, et al. "Hormone Replacement Therapy Reduces Lipid Oxidation Directly at The Arterial Wall: A Possible Link To Estrogens' Cardioprotective Effect Through Atherosclerosis Prevention." Journal Of Mid-Life Health 8.1 (2017): 11-16.

45) Violi, Francesco, et al. "Interventional Study with Vitamin E In Cardiovascular Disease And Meta-Analysis." Free Radical Biology and Medicine 178 (2022): 26-41.

46) Folahan, Joy Temiloluwa, et al. "Oxidized Dietary Lipids Induce Vascular Inflammation And Atherogenesis In Post-Menopausal Rats: Estradiol And Selected Antihyperlipidemic Drugs Restore Vascular Health In Vivo." Lipids in Health and Disease 22.1 (2023): 107.

47) Keaney Jr, John F., et al. "17 Beta-Estradiol Preserves Endothelial Vasodilator Function and Limits Low-Density Lipoprotein Oxidation in Hypercholesterolemic Swine." Circulation 89.5 (1994): 2251-2259.

48) Sugioka, Katsuaki, et al. "Estrogens as Natural Antioxidants of Membrane Phospholipid Peroxidation." FEBS letters 210.1 (1987): 37-39.

49) Bechlioulis, Aris, et al. "Menopause and Hormone Therapy: From Vascular Endothelial Function To Cardiovascular Disease." Hellenic J Cardiol 50.4 (2009): 303-15.

50) Meng, Qinghai, et al. "Activation of Estrogen Receptor Alpha Inhibits TLR4 Signaling In Macrophages And Alleviates The Instability Of Atherosclerotic Plaques In The Postmenopausal Stage." International Immunopharmacology 116 (2023): 109825.

51) Xie, Fei, et al. "Estrogen Mediates an Atherosclerotic-Protective Action Via Estrogen Receptor Alpha/SREBP-1 Signaling." Frontiers in Cardiovascular Medicine 9 (2022): 895916.

52) Nofer, Jerzy-Roch. "Estrogens and Atherosclerosis: Insights from Animal Models And Cell Systems." Journal Of Molecular Endocrinology 48.2 (2012): R13-R29.

53) Marfella, Raffaele, et al. "Proteasome Activity as a Target of Hormone Replacement Therapy–Dependent Plaque Stabilization in Postmenopausal Women." Hypertension 51.4 (2008): 1135-1141.

54) Versari, Daniele, et al. "Dysregulation of the Ubiquitin-Proteasome System In Human Carotid Atherosclerosis." Arteriosclerosis, Thrombosis, And Vascular Biology 26.9 (2006): 2132-2139.

55) Hodis, Howard N., and Wendy J. Mack. "Menopausal Hormone Replacement Therapy and Reduction of All-Cause Mortality and Cardiovascular Disease: It Is About Time and Timing." The Cancer Journal 28.3 (2022): 208-223.

56) Hodis, Howard N., et al. "Vascular Effects of Early Versus Late Postmenopausal Treatment with Estradiol." New England Journal of Medicine 374.13 (2016): 1221-1231.

57) Hodis, Howard N., and Wendy J. Mack. "In Perspective: Estrogen Therapy Proves to Safely and Effectively Reduce Total Mortality and Coronary Heart Disease In Recently Postmenopausal Women." Menopause management 17.2 (2008): 27.

58) Schierbeck, Louise Lind, et al. "Effect of Hormone Replacement Therapy On Cardiovascular Events In Recently Postmenopausal Women: Randomised Trial." BMJ 345 (2012). Denmark, 1990-93.

59) Ruan, Xiangyan, and Alfred O. Mueck. "The Choice of Progestogen for HRT In Menopausal Women: Breast Cancer Risk Is a Major Issue." Hormone Molecular Biology and Clinical Investigation 37.1 (2019): 20180019.

60) Booth, Erin A., and Benedict R. Lucchesi. "Medroxyprogesterone Acetate Prevents The Cardioprotective And Anti-Inflammatory Effects Of 17β-Estradiol In An In Vivo Model Of Myocardial Ischemia And Reperfusion." American Journal of Physiology-Heart and Circulatory Physiology 293.3 (2007): H1408-H1415.

61) Jeanes, Helen L., et al. "Medroxyprogesterone Acetate Inhibits the Cardioprotective Effect of Estrogen in Experimental Ischemia-Reperfusion Injury." Menopause 13.1 (2006): 80-86.

62) Bernstein, Paula, and Gerald Pohost. "Progesterone, Progestins, And the Heart." Reviews in cardiovascular medicine 11.3 (2010): e141-9.

63) Adams, Michael R., et al. "Medroxyprogesterone Acetate Antagonizes Inhibitory Effects of Conjugated Equine Estrogens on Coronary Artery Atherosclerosis." Arteriosclerosis, thrombosis, and vascular biology 17.1 (1997): 217-221.

64) Lanari, Claudia, et al. "The MPA Mouse Breast Cancer Model: Evidence for A Role Of Progesterone Receptors In Breast Cancer." Endocrine-related cancer 16.2 (2009): 333.

65) Golomb, Beatrice A., and Marcella A. Evans. "Statin Adverse Effects: A Review of The Literature and Evidence for A Mitochondrial Mechanism." American Journal of Cardiovascular Drugs 8 (2008): 373-418.

66) Rizvi, Kash, John P. Hampson, and John N. Harvey. "Do Lipid-Lowering Drugs Cause Erectile Dysfunction? A Systematic Review." Family Practice 19.1 (2002): 95-98.

67) Sattar, Naveed, et al. "Statins and Risk of Incident Diabetes: A Collaborative Meta-Analysis Of Randomized Statin Trials." The Lancet 375.9716 (2010): 735-742.

68) Laakso, Markku, and Lilian Fernandes Silva. "Statins and risk Of Type 2 Diabetes: Mechanism And Clinical Implications." Frontiers in Endocrinology 14 (2023): 1239335.

69) Tan, Brendan, et al. "Evidence and Mechanisms for Statin-Induced Cognitive Decline." Pharmacology 12.5 (2019): 397-406.

70) Attardo, Silvia, et al. "Statins Neuromuscular Adverse Effects." International Journal Of Molecular Sciences 23.15 (2022): 8364.

71) Langsjoen, Peter H., et al. "Statin-Associated Cardiomyopathy Responds to Statin Withdrawal And Administration Of Coenzyme Q10." The Permanente Journal 23 (2019).

72) Okuyama, Harumi, et al. "Statins Stimulate Atherosclerosis and Heart Failure: Pharmacological Mechanisms." Expert Review of Clinical Pharmacology 8.2 (2015): 189-199.

73) Langsjoen, Peter H., et al. "Treatment of Statin Adverse Effects with Supplemental Coenzyme Q10 And Statin Drug Discontinuation." Biofactors 25.1-4 (2005): 147-152.

74) Walsh, Judith ME, and Michael Pignone. "Drug Treatment of Hyperlipidemia in Women." Jama 291.18 (2004): 2243-2252.

75) Petretta, Mario, et al. "Impact of Gender in Primary Prevention of Coronary Heart Disease with Statin Therapy: A Meta-Analysis." International Journal of Cardiology 138.1 (2010): 25-31.

76) Ray, Kausik K., et al. "Statins and All-Cause Mortality in High-Risk Primary Prevention: A Meta-Analysis Of 11 Randomized Controlled Trials Involving 65 229 Participants." Archives of Internal Medicine 170.12 (2010): 1024-1031.

77) Bush, Trudy L., et al. "Estrogen use And All-Cause Mortality: Preliminary Results from The Lipid Research Clinics Program Follow-Up Study." Jama 249.7 (1983): 903-906.

78) Bush, Trudy L., et al. "Cardiovascular Mortality and Noncontraceptive Use of Estrogen In Women: results from the Lipid Research Clinics Program Follow-up Study." Circulation 75.6 (1987): 1102-1109.

79) Barrett-Connor, Elizabeth, and Trudy L. Bush. "Estrogen and Coronary Heart Disease in Women." JAMA 265.14 (1991): 1861-1867.

80) Manson, JoAnn E., et al. "Menopausal Hormone Therapy and Long-Term All-Cause and Cause-Specific Mortality: The Women's Health Initiative Randomized Trials." JAMA 318.10 (2017): 927-938.

81) Hodis, Howard N., and P. M. Sarrel. "Menopausal Hormone Therapy and Breast Cancer: What Is the Evidence From Randomized Trials?" Climacteric 21.6 (2018): 521-528.

82) Gersh Felice, et al. Estrogen and Cardiovascular Disease. Progress in Cardiovascular Diseases. January 2024.

83) Miller, Virginia. M., et al. "Lessons from KEEPS: the Kronos Early Estrogen Prevention Study." Climacteric 24.2 (2021): 139-145.

84) Miller, Virginia M., et al. "The Kronos early estrogen prevention study (KEEPS): what have we learned?" Menopause 26.9 (2019): 1071-1084.

85) Hodis, Howard N., et al. Vascular Effects of Early Versus Late Postmenopausal Treatment with Estradiol. New England Journal of Medicine 374.13 (2016): 1221-1231.

86) Yang, Jane L., et al. "Estrogen Deficiency in Menopause and The Role of Hormone Therapy: Integrating the Findings of Basic Science Research with Clinical Trials." Menopause (2024): 10-1097.

Chapter 10

Estrogen Prevents Heart Disease Part Two

IN CHAPTER 9, PART ONE, we discussed the history of hormone replacement therapy (HRT) related to the prevention of coronary artery disease. We found that **HRT confers a 50 percent reduction in mortality from heart disease** depending on the study and the age of initiating estrogen replacement in post-menopausal women. The greatest cardiovascular benefit from HRT was found in post-hysterectomy women with surgically induced menopause. Excellent cardiovascular mortality reductions are obtained with HRT in the early postmenopausal age group, starting HRT within 5 years of the menopausal transition. This is called the "Timing Hypothesis" of Dr. Howard Hodis. Cardiovascular benefits of HRT are lost in women starting HRT more than 5 years after menopausal transition. (1)

How to Explain the Timing Hypothesis

The next logical question is how can we explain this "Timing Hypothesis?" What is the underlying pathophysiology that explains why estrogen replacement prevents coronary artery disease when started early, but not when started later? One obvious explanation is the abrupt decline in estrogen levels in post-menopausal women triggers accelerated coronary artery disease, a degenerative change in the walls of our arteries, with calcified plaque formation. Once coronary artery plaque has been established, this is difficult to reverse later. This explanation is not completely satisfying. Perhaps something is missing. The subject of Chapter 10 is to expand on the etiology of coronary artery disease and provide a better explanation of the "Timing Hypothesis" by Dr. Howard Hodis.

Gut Permeability, Leaky Gut and Atherosclerosis

My 2018 book on coronary artery disease, entitled *Heart Book*, discusses the role of intestinal permeability, "Leaky Gut", and low-level endotoxemia in the pathogenesis of coronary artery disease. The low-level endotoxemia arising from a "leaky gut" leads to colonization of the arterial wall by polymicrobial biofilm. This results in an intense inflammatory reaction that incites calcification in the artery's wall, demonstrated by the coronary calcium score test performed with a CAT scan machine. In Chapter 9 (part one), we learned of two coronary calcium score studies derived from the WHI patient population showing the estrogen-treated group had lower calcium scores and reduced progression of calcium scores compared to the placebo-treated group. Again, vascular calcification is demonstrated on a CAT scan test known as the coronary calcium score. Coronary artery calcification is a response to inflammation caused by polymicrobial infection in the arterial wall seeded from periodontal and gut micro-organisms. (2-3)

Heart Book, 2018

In my 2018 *Heart Book*, a mechanism was proposed to explain the train of events leading to coronary artery calcification. The originating event is "Leaky Gut," a disruption of the mucosal barrier of the gut leading to leakage of LPS (lipopolysaccharide) and gram-negative microorganisms into the bloodstream, defined as low-level endotoxemia. Both LDL (low-density lipoprotein) and macrophages take up the LPS and whole micro-organisms, which then migrate into the arterial plaque, causing infection with polymicrobial biofilm. This incites an inflammatory reaction which leads to vascular calcification. **Note:** LPS is the outer coat of gram-negative bacteria, one of the most toxic substances known, and a highly lethal cause of gram-negative septicemia. (4)

Causes of Leaky Gut

There are many causes of "Leaky Gut." The list includes fatty meals, high blood sugar (diabetes), dysbiosis with pathogenic bacteria, wheat gluten sensitivity which prolongs the opening of tight junctions, stress-altered permeability, etc. We can now add abrupt menopausal decline in estrogen levels to this list of causes of "Leaky Gut." (5-6)

Menopausal Estrogen Decline Triggers Leaky Gut

The next obvious question is: can the sudden menopausal decline in estrogen trigger leaky gut and subsequent low-level endotoxemia? This is not intuitively obvious nor well-known. However, the medical literature is overwhelming. Menopausal decline in estrogen does indeed trigger disruption of the mucosal barrier of the gut and causes increased gut permeability (i.e. Leaky Gut). We can now propose the menopausal decline in estrogen as the trigger for accelerated coronary atherosclerosis. Estrogen has two mechanisms for prevention of coronary artery disease. Firstly, estrogen is anti-inflammatory. Secondly, estrogen controls gut mucosal integrity. The anti-inflammatory effects of estrogen are through the inhibition of nuclear factor kappa-B (NF-KB), the inflammatory master controller. Estrogen downregulates proinflammatory cytokines IL-1, IL-6, and TNF-alpha. Control of gut barrier integrity is mediated through estrogen receptor beta (ER-beta) which regulates tight junctions between epithelial cells of the gut lining. Thus, menopausal estrogen deficiency leads to disruption of the gut barrier, low-level endotoxemia, increased pro-inflammatory cytokines, and acceleration of atherosclerotic vascular disease, all reversed by estrogen replacement. In 2020, Dr. Albert

Shieh studied animal models showing menopausal estrogen deficiency leads to increased gut permeability and inflammation, two major predicates for coronary artery disease and loss of bone mineral density (BMD), writing:

> Inflammation is implicated in many aging-related disorders. In animal models, **menopause [estrogen deficiency] leads to increased gut permeability and inflammation**...Gut permeability increases during the MT [Menopausal Transition]. Greater gut permeability is associated with more inflammation and lower BMD [Bone Mineral Density]. Future studies should examine the longitudinal associations of gut permeability, inflammation, and BMD. Note: lower bone mineral density (BMD) is associated with increased gut permeability and inflammation. (7-12)

In 2024, Dr. Xiuting Xiang studied the role of estrogen in the digestive system, noting ER-beta signaling regulates the permeability of the intestinal barrier, writing:

> Overall, estrogen influences the composition and function of the gastrointestinal barrier and also impacts inflammatory processes within the digestive system. ER-beta signaling has been shown to regulate the permeability of the intestinal barrier by increasing the integrity of tight junctions through the expression of occludin and junctional adhesion molecule A. (13)

In 2024, Dr. Qinghai Meng from Nanjing, China found menopausal women have disordered gut microbiota which amplifies intestinal tight junction damage, accelerating atherosclerosis. Dr. Meng studied both women and mouse models before and after menopause finding estrogen deficiency promotes microbiome disturbance, intestinal barrier damage, inflammation, and accelerates atherosclerosis, writing:

> This study examined aortic estrogen receptor expression, histological changes, and gut microbiota in women before and after menopause, and tested serum estrogen

levels, systemic inflammation, intestinal estrogen receptor expression, histological changes, atherosclerosis, and gut microbiota in low-density lipoprotein receptor knockout (LDLR-/-) female mice before and after ovariectomy. We demonstrated that the **downregulation of estrogen and estrogen receptors after menopause promotes gut microbiota disturbance in both women and female mice.** We found that gut microbiota disturbance amplifies the intestinal barrier damage and aggravates systemic inflammation, **thereby promoting atherosclerosis in female mice**. Note: LDL receptor knockout mice are a good animal model for the study of atherosclerosis since they spontaneously develop atherosclerotic lesions. (14-16)

In 2023, Dr. Yuanuan Li used mice to show chronic psychological stress leads to decreased estrogen levels, disruption of the intestinal barrier with increased intestinal permeability, and increased pro-inflammatory cytokines. Estrogen replacement is protective and prevents these changes, writing:

> Chronic psychological stress resulted in colonic mucosal injury, pro-inflammatory reaction, and decreased the diversity and richness of the colonic microbiota in pregnant mice. It was interesting that 25 pg/mL E2 provides better protective effect on intestinal epithelial cells...In summary, chronic stress significantly induces stress responses in maternal mice, leading to reduced estrogen secretion, disruption of the intestinal barrier, increased secretion of pro-inflammatory cytokines, decreased secretion of anti-inflammatory cytokines, and dysbiosis of the gut microbiota. **Moreover, 25 pg/mL E2 protected the intestine**. Our research demonstrates that maintaining stable maternal estrogen levels during pregnancy can safeguard intestinal health... (17)

Dr. Jane Yang on Estrogen and the GI Tract

In 2024, Dr. Jane Yang reviewed the entire medical literature on menopausal estrogen deficiency from 1972 to 2023. Within this larger topic, Dr. Jane Wang discusses estrogen's importance for the gastrointestinal (GI) tract, pointing out the extensive presence of estrogen receptors throughout the GI tract. If you thought estrogen receptors play a controlling role, then you would be quite correct. ER-beta signaling controls the permeability of the gut barrier by maintaining the integrity of the tight junctions. When estrogen levels decline after menopause, this gut barrier integrity is lost, triggering leaky gut, low level endotoxemia, and accelerated atherosclerosis. In my practice, we have seen the beneficial effects of estrogen in colitis patients. Dr. Jane Wang writes:

> ER-alpha, ER-beta, and GPER1 [G-protein coupled receptor 1] are found widely throughout the gastrointestinal (GI) tract... Overall, estrogen influences the composition and function of the gastrointestinal barrier and also impacts inflammatory processes within the digestive system. **ER-beta signaling has been shown to regulate the permeability of the intestinal barrier by increasing the integrity of tight junctions** through the expression of occludin and junctional adhesion molecule A... **In multiple experimental models of colitis, estrogen has been shown to decrease intestinal inflammation and tissue damage**. Note: GPER1 is the estrogen receptor on the cell membrane involved in immediate signaling. The other two receptors must enter the nucleus which takes more time. (18-26)

Menopause as an Inflammatory Event

There is considerable evidence for viewing the menopausal transition as an inflammatory event. In 2007, Dr. Toshiyuki Yasui found that inflammatory cytokine IL-6 is inversely correlated with serum estradiol levels. As estradiol decreased, the pro-inflammatory cytokine IL-6 increased. In 2021, Dr. Haidong Wang found estradiol has an inhibitory effect on NF-KappaB, the inflammatory master controller. (27-29)

In 2005, Dr Serena Ghisletti studied the mechanism by which estradiol exerts its anti-inflammatory effects, finding estrogen blocks the p65 transcription factor, a member of the NF-KappaB family, inhibiting the p65 intracellular transport to the nucleus. This anti-inflammatory effect of estrogen is mediated through ER-alpha, writing:

> Estrogen is an immunoregulatory agent, in that hormone deprivation increases while 17β-estradiol (E2) administration blocks the inflammatory response; however, the underlying mechanism is still unknown. The transcription factor **p65**/relA, a member of the **nuclear factor κB (NF-κB) family**, plays a major role in inflammation and drives the expression of proinflammatory mediators. Here we report a novel mechanism of action of E2 in inflammation. We observe that in macrophages **E2 blocks lipopolysaccharide-induced DNA binding and transcriptional activity of p65 by preventing its nuclear translocation**. This effect is selectively activated in macrophages to prevent p65 activation by inflammatory agents and extends to other members of the NF-κB family, including c-Rel and p50. We observe that E2 activates a rapid and persistent response that involves the activation of phosphatidylinositol 3-kinase, without requiring de novo protein synthesis or modifying Iκ-Bα degradation and mitogen-activated protein kinase activation. Using a time course experiment and the microtubule-disrupting agent nocodazole, we observe that the **hormone inhibits p65 intracellular transport to the nucleus. This activity is selectively mediated by estrogen receptor alpha (ER-alpha)** and not ERβ and is not shared by conventional anti-inflammatory drugs. These results unravel a novel and unique mechanism for E2 anti-inflammatory activity, which may be useful for identifying more selective ligands for the prevention of the inflammatory response. (30-35)

In 2020, Dr. Micheline McCarthy reviewed the menopausal transition as an inflammatory event, suggesting the menopausal decline in estrogen level drives a systemic pro-inflammatory state, writing:

> It is now known that one of **the key functions of estrogen is to work as a potent anti-inflammatory factor** ...The presence of the inflammasome complex in the cerebrospinal fluid of post-menopausal women suggests that **the decline in estrogens induces a pro-inflammatory state**...There is increasing and compelling evidence showing that **estrogen decline during the menopausal transition drives a systemic inflammatory state.** This state is characterized by **systemic pro-inflammatory cytokines** derived from reproductive tissues, alteration in the cellular immune profile, increased availability of inflammasome proteins in the CNS, and a pro-inflammatory microenvironment which makes the brain more susceptible to ischemic and other stressors.... The use of ER-beta-selective agonists may constitute a safer and more effective target for future therapeutic research than an ER-alpha agonist or E2 [estradiol]. ER-beta activation in the brain confers ischemic protection, stimulates mitochondrial functions, and inhibits inflammasome activation. ER-beta agonists may be safer in that ER-beta lacks the ability to stimulate the proliferation of breast or endometrial tissue. The ER-beta agonist may be able to act both on the cerebro- and cardiovascular system to reduce the ischemic burden. Thus, ER-beta signaling is a guide for future translational research to reduce cognitive decline and cerebral ischemia incidents and impact in post-menopausal women, while avoiding the side effects produced by chronic E2 [estradiol] treatment...Emerging evidence is showing that peri-menopause is pro-inflammatory and disrupts estrogen-regulated neurological systems. Estrogen is a master regulator that functions through a network of estrogen receptors subtypes alpha (ER-α) and beta (ER-β). Estrogen receptor-beta has been shown to regulate a key component of

the innate immune response known as the inflammasome, and it also is involved in regulation of neuronal mitochondrial function. (36-38)

Low Grade Endotoxemia, LPS and Infection in Atherosclerotic Plaque

In 2023, Dr. Francesco Violi reviewed the link between gut-derived low grade endotoxemia, atherosclerosis and cardiovascular disease. Gut dysbiosis (Leaky Gut) is implicated in atherosclerosis via leakage of live bacteria and LPS into the bloodstream causing low-level endotoxemia. Studies show the presence of LPS adjacent to macrophages within atherosclerotic plaques. Dr. Francesco Violi writes:

> A growing body of evidence indicates that gut dysbiosis is implicated in the atherothrombotic process via **increased translocation of viable bacteria or bacterial products such as lipopolysaccharides (LPS) and trimethylamine-N-oxide (TMAO) into the systemic circulation** ....Support for the putative role of LPS in atherosclerosis has been provided by immunohistochemistry analysis of carotid atherosclerotic plaques from patients undergoing endarterectomy, which revealed the **presence of LPS adjacent to plaque macrophages** with high TLR4 levels. **By contrast, LPS was not detected in atherosclerosis-free thyroid arteries from the same patient**s....Experimental data support the association between low-grade endotoxemia, atherosclerosis, and thrombosis, and indicate that **gut dysbiosis-induced changes in intestinal permeability are a key step for LPS translocation into the systemic circulation.** (39)

16s Ribosome Sequencing Finds Bacteria Within Plaques

A new technology for identifying microbial infection is the 16s ribosome technique. In 2011, Dr. Omry Koren performed 16s ribosomal RNA sequencing to find bacterial DNA present within atherosclerotic plaques, writing:

Using qPCR [Quantitative polymerase chain reaction], we show that **bacterial DNA was present in the atherosclerotic plaque and that the amount of DNA correlated with the amount of leukocytes in the atherosclerotic plaque.** To investigate the microbial composition of atherosclerotic plaques and test the hypothesis that the oral or gut microbiota may contribute to atherosclerosis in humans, we used 454 pyrosequencing of **16S rRNA genes** to survey the bacterial diversity of atherosclerotic plaque, oral, and gut samples of 15 patients with atherosclerosis, and oral and gut samples of healthy controls...Our analysis also revealed several OTUs [operational taxonomic units] shared between the atherosclerotic plaque and the gut, **suggesting that bacteria present in the atherosclerotic plaque could also be derived from the distal gut as well as the oral cavity.** One mechanism by which bacteria could reach the atherosclerotic plaque is phagocytosis by macrophages at epithelial linings (e.g., the oral cavity, gut, and the lung). **Upon phagocytosis, the macrophages become activated, and when they reach the activated endothelium of the atheroma, they leave the blood stream to enter the atheroma and transform into cholesterol-laden foam cells**. In support of this mechanism, patients with cardiovascular disease have a twofold increase of C. pneumonia-infected peripheral blood mononuclear cells compared with controls. Furthermore, bacteria are only present in atheromas and not in healthy aortic tissues in mice and have been identified in human atherosclerotic plaques. **Thus, infected macrophages may specifically target bacteria to atheromas**... In summary, we detected key bacterial members of dental plaque in atherosclerotic plaques in humans, as well as a novel common member, Chryseomonas, in all atherosclerotic plaques. In addition, the atherosclerotic plaques contained numerous bacteria from different phyla. **Our findings strongly support the hypothesis that the oral cavity and gut can be sources for atherosclerotic plaque-associated bacteria. (40)**

In 2018, Dr. Roberto Canevale found LPS from E. Coli (gram-negative bacteria) within atherosclerotic plaque material. (41-42)

In 2022, Dr. Iman Razeghian-Jahromi reviewed the medical literature on the prevalence of micro-organisms within atherosclerotic plaques of the coronary arteries finding 44 supportive studies, writing:

> In this systematic review and meta-analysis, the existence of pathogens in atherosclerotic plaques of coronary arteries was investigated in CAD patients...**44 studies were selected**... Infection of vascular cells, detection of the microbes within the atherosclerotic plaque, and the development of atherosclerotic lesions after microbial infection in animal models reinforce the direct association of infection with atherosclerosis....**Our studies show the presence of different pathogens (bacteria and virus) in the atherosclerotic plaques of coronary arteries**...This may show the implication of a microbial center for the initiation and development of atherosclerotic plaque...**Several studies reported the presence of more than one pathogen in atherosclerotic plaques**. The simultaneous existence of some bacteria synergistically enhances their virulence. Virulence factors of bacteria including fimbriae, degradative enzymes, exopolysaccharide capsules, toxins, and atypical lipopolysaccharides trigger the process of inflammation and affect vital organs like the cardiovascular system. These adverse effects will be more detrimental in mixed infections...Atherosclerosis may be originated from bacterial infection in terms of microbial symbiosis and inflammatory stimulus]. **Microbial agents contribute to the atherosclerosis process directly by infecting the vascular cells or indirectly by activation of inflammatory cytokines**.
> ... It seems that the entry of bacteria and other microbiome populations into the bloodstream is a continuous flow which inevitably leads to a surge in the expression of inflammatory cytokines and chemokines. These factors could be drivers of CAD [71]. For example, it was shown that bacterial lipopolysaccharide (LPS) upregulates LDL

levels, increasing the risk of CAD. Indeed, the remnants of bacteria like DNA or membrane phospholipids provoke CAD via modulating adipose or vascular tissues. **However, it was reported that the development of atherosclerotic lesions is largely accelerated by live organisms rather than heat-killed ones or their LPS. Reports on successful culturing of pathogens after their isolation from atheroma only existed about C. pneumoniae and E. hormaechei.** (43-57)

Menopausal Hypertension

As mentioned above, menopausal estrogen deficiency heralds the onset of leaky gut, low level endotoxemia, and release of inflammatory cytokines inciting pathological events in the coronary arteries. The next logical question is what is the effect of all this on the systemic circulation and blood pressure? The short answer is that hypertension is another estrogen deficiency disease of menopause. In 2022, Dr. Lama Ghazi explored the connection between menopause and hypertension, writing:

> Menopause per se appears to activate a cluster of genes that lead to hypertension... Other studies have shown that estrogen replacement therapy is associated with a decrease in ambulatory BP [blood pressure]. (58)

In 2023, Dr. Jan Pitha reviewed the pre-clinical and clinical studies on hypertension related to estrogen deficiency, finding a lower incidence of hypertension in pre-menopausal women compared to males, However, this ratio is reversed after menopause with greater incidence in females than males. Most animal models of menopausal hypertension show protective effects of estrogen against hypertension, which is abolished by ovariectomy, and reversed by estrogen replacement. Dr. Jan Pitha writes:

> premenopausal women have a lower incidence of hypertension and other cardiovascular events than men of the same

age, but the reduced sex differences after menopause suggest that 17β-estradiol (E2) is a protective agent... To summarize, in contrast to human studies, **most animal studies show a favorable effect of estrogens on the protection against hypertension, coronary heart disease, stroke, and heart failure.** Female rodents are protected against increased BP [blood pressure], vascular injury, and heart failure compared to males, but this cardioprotection is abolished after OVX [ovariectomy] and reversed back by estrogen substitution. (59)

In 2011, Dr. D.Y. Lee studied the effect of estrogen therapy with CEE (Premarin) combined with micronized progesterone on blood pressure in post-menopausal Korean women finding beneficial effects, writing:

CEE [Premarin®] increased the daytime blood pressure in women with normal blood pressure, but reduced it in women with high blood pressure. Micronized progesterone may provide beneficial effects on blood pressure when combined with CEE. (60)

Conclusion: When reviewing estrogen's role in leaky gut, endotoxemia, and microbial colonization of atherosclerotic plaque, one begins to realize the role of estrogen in the prevention of coronary artery disease. Firstly, estrogen has strong anti-inflammatory effects, and secondly, estrogen maintains the integrity of the gut barrier. Menopausal loss of estrogen means disruption of the gut barrier, increased low-level endotoxemia, increased systemic inflammation, and increased atherosclerosis. It becomes clear estrogen is a major factor in the prevention of coronary artery disease and hypertension in the immediate post-menopausal time frame. It also becomes clear why lowering serum cholesterol with statin drugs fails in primary prevention. Although statin drugs have pleiotropic effects, namely anti-inflammatory and antimicrobial effects, statin drugs do not address the integrity of the gut membrane barrier. Statin drugs are associated with horrendous adverse side effects, muscle pain, neuropathy, and

loss of cognitive function. Estrogen hormone replacement is more effective and spares the adverse effects of statin drugs. We can now explain the timing hypothesis of Dr. Howard Hodis. Promptly starting estrogen replacement at the onset of menopause prevents the disruption of the gut barrier, low level endotoxemia, and microbial infiltration of the atherosclerotic plaque, the underlying pathology of coronary artery disease. Menopausal estrogen replacement is truly preventive for coronary artery disease. The most casual observer can see the overwhelming evidence for the infection theory of coronary artery disease, yet the cholesterol theory is too entrenched and the financial stakes are too high for mainstream medicine to make the paradigm shift. Mainstream medicine clings to the old dogmas, and ignores and refuses to accept all the studies listed here that prove microbial infection is the over-riding pathology in the coronary artery plaque which causes calcification seen on the Calcium Score, a CAT scan test. For more, see my previous book entitled *Heart Book* (2018). (4)

♦ **References for Chapter 10 Estrogen Prevents Heart Disease Part Two**

1) Hodis, Howard N., and Wendy J. Mack. "Menopausal hormone replacement therapy and reduction of all-cause mortality and cardiovascular disease: it is about time and timing." The Cancer Journal 28.3 (2022): 208-223.

2) Manson, JoAnn E., et al. "Estrogen therapy and coronary-artery calcification." New England Journal of Medicine 356.25 (2007): 2591-2602.

3) Budoff, Matthew J., et al. "Effects of hormone replacement on progression of coronary calcium as measured by electron beam tomography." Journal of women's health 14.5 (2005): 410-417.

4) Dach, Jeffrey, "Heart Book", Medical Muse Press, August 2, 2018.

5) Camilleri, Michael. "Leaky gut: mechanisms, measurement and clinical implications in humans." Gut 68.8 (2019): 1516-1526.

6) Stewart, Amy Stieler, Shannon Pratt-Phillips, and Liara M. Gonzalez. "Alterations in intestinal permeability: the role of the "leaky gut" in health and disease." Journal of equine veterinary science 52 (2017): 10-22.

7) Shieh, Albert, et al. "Gut permeability, inflammation, and bone density across the menopause transition." JCI insight 5.2 (2020).

8) Ricardo-da-Silva, Fernanda Yamamoto, et al. "Estradiol prevented intestinal ischemia and reperfusion-induced changes in intestinal permeability and motility in male rats." Clinics 76 (2021): e2683.

9) Li, Yuanyuan, et al. "Chronic stress that changed intestinal permeability and induced inflammation was restored by estrogen." International Journal of Molecular Sciences 24.16 (2023): 12822.

10) Acharya, Kalpana D., et al. "Distinct changes in gut microbiota are associated with estradiol-mediated protection from diet-induced obesity in female mice." Metabolites 11.8 (2021): 499.

11) Fidya, et al. "Protective role of estrogen through G-protein coupled receptor 30 in a colitis mouse model." Histochemistry and Cell Biology 161.1 (2024): 81-93.

12) Jin, Ye, et al. "Estrogen deficiency aggravates fluoride-induced small intestinal mucosa damage and junctional complexes proteins expression disorder in rats." Ecotoxicology and Environmental Safety 246 (2022): 114181.

13) Xiang, Xiuting, et al. "The Role of Estrogen across Multiple Disease Mechanisms." Current Issues in Molecular Biology 46.8 (2024): 8170-8196.

14) Meng, Qinghai, et al. "Disordered gut microbiota in postmenopausal stage amplifies intestinal tight junction damage to accelerate atherosclerosis." Beneficial Microbes 1.aop (2024): 1-23.

15) Meng, Qinghai, et al. "The gut microbiota during the progression of atherosclerosis in the perimenopausal period shows specific compositional changes and significant correlations with circulating lipid metabolites." Gut Microbes 13.1 (2021): 1880220.

16) Brandsma, Eelke, et al. "A proinflammatory gut microbiota increases systemic inflammation and accelerates atherosclerosis." Circulation research 124.1 (2019): 94-100.

17) Li, Yuanyuan, et al. "Chronic stress that changed intestinal permeability and induced inflammation was restored by estrogen." International Journal of Molecular Sciences 24.16 (2023): 12822.

18) Yang, Jane L., et al. "Estrogen deficiency in the menopause and the role of hormone therapy: integrating the findings of basic science research with clinical trials." Menopause (2024): 10-1097.

19) Braniste, Viorica, et al. "Estradiol decreases colonic permeability through estrogen receptor beta-mediated up-regulation of occludin and junctional adhesion molecule-A in epithelial cells." The Journal of physiology 587.13 (2009): 3317-3328.

20) Song, Chin-Hee, et al. "Effects of 17β-estradiol on colonic permeability and inflammation in an azoxymethane/dextran sulfate sodium-induced colitis mouse model." Gut and liver 12.6 (2018): 682.

21) Looijer-van Langen, Mirjam, et al. "Estrogen receptor-beta signaling modulates epithelial barrier function." American Journal of Physiology-Gastrointestinal and Liver Physiology 300.4 (2011): G621-G626.

22) Collins, Fraser L., et al. "Temporal and regional intestinal changes in permeability, tight junction, and cytokine gene expression following ovariectomy-induced estrogen deficiency." Physiological reports 5.9 (2017): e13263.

23) Zhou, Zejun, et al. "Progesterone decreases gut permeability through upregulating occludin expression in primary human gut tissues and Caco-2 cells." Scientific reports 9.1 (2019): 8367.

24) Pigrau, M., et al. "The joint power of sex and stress to modulate brain–gut–microbiota axis and intestinal barrier homeostasis: implications for irritable bowel syndrome." Neurogastroenterology & Motility 28.4 (2016): 463-486.

25) van der Giessen, Janine, et al. "A direct effect of sex hormones on epithelial barrier function in inflammatory bowel disease models." Cells 8.3 (2019): 261.

26) Roomruangwong, Chutima, et al. "The menstrual cycle may not be limited to the endometrium but also may impact gut permeability." Acta Neuropsychiatrica 31.6 (2019): 294-304.

27) Yasui, Toshiyuki, et al. "Changes in serum cytokine concentrations during the menopausal transition." Maturitas 56.4 (2007): 396-403.

28) Wang, Haidong, et al. "17β-Estradiol alleviates intervertebral disc degeneration by inhibiting NF-κB signal pathway." Life Sciences 284 (2021): 119874.

29) Wen, Yi, et al. "Estrogen attenuates nuclear factor-kappa B activation induced by transient cerebral ischemia." Brain research 1008.2 (2004): 147-154.

30) Ghisletti, Serena, et al. "17β-estradiol inhibits inflammatory gene expression by controlling NF-κB intracellular localization." Molecular and cellular biology 25.8 (2005): 2957-2968.

31) Xing, Dongqi, et al. "Estrogen modulates NFκB signaling by enhancing IκBα levels and blocking p65 binding at the promoters of inflammatory genes via estrogen receptor-β." PloS one 7.6 (2012): e36890.

32) Liu, Hui, Kenian Liu, and Donald L. Bodenner. "Estrogen receptor inhibits interleukin-6 gene expression by disruption of nuclear factor κB transactivation." Cytokine 31.4 (2005): 251-257.

33) Okamoto, Mariko, et al. "The membrane-type estrogen receptor G-protein-coupled estrogen receptor suppresses lipopolysaccharide-induced interleukin 6 via inhibition of nuclear factor-kappa B pathway in murine macrophage cells." Animal Science Journal 88.11 (2017): 1870-1879.

33) Deshpande, Rohini, et al. "Estradiol down-regulates LPS-induced cytokine production and NFkB activation in murine macrophages." American journal of reproductive immunology 38.1 (1997): 46-54.

34) Zang, Ying CQ, et al. "Regulatory effects of estriol on T cell migration and cytokine profile: inhibition of transcription factor NF-κB." Journal of neuroimmunology 124.1-2 (2002): 106-114.

35) Vegeto, Elisabetta, et al. "Estrogen receptor-α mediates the brain antiinflammatory activity of estradiol." Proceedings of the National Academy of Sciences 100.16 (2003): 9614-9619.

36) McCarthy, Micheline, and Ami P. Raval. "The peri-menopause in a woman's life: a systemic inflammatory phase that enables later neurodegenerative disease." Journal of neuroinflammation 17 (2020): 1-14.

37) Zhang, Wen-yuan, et al. "Neuroprotective effects of vitamin D and 17ß-estradiol against ovariectomy-induced neuroinflammation and depressive-like state: Role of the AMPK/NF-κB pathway." International Immunopharmacology 86 (2020): 106734.

38) Son, Hee Jin, et al. "17β-Estradiol reduces inflammation and modulates antioxidant enzymes in colonic epithelial cells." The Korean Journal of Internal Medicine 35.2 (2020): 310.

39) Violi, Francesco, et al. "Gut-derived low-grade endotoxaemia, atherothrombosis and cardiovascular disease." Nature reviews cardiology 20.1 (2023): 24-37.

40) Koren, Omry, et al. "Human oral, gut, and plaque microbiota in patients with atherosclerosis." Proceedings of the National Academy of Science. Vol. 108. 2011.

41) Carnevale, Roberto, et al. "Localization of lipopolysaccharide from Escherichia Coli into human atherosclerotic plaque." Scientific Reports 8 (2018).

42) Wiedermann CJ, et al. Association of endotoxemia with carotid atherosclerosis and cardiovascular disease: prospective results from the Bruneck study. J. Am. Coll. Cardiol. 1999;34:1975–1981.

43) Razeghian-Jahromi, Iman, et al. "Prevalence of Microorganisms in Atherosclerotic Plaques of Coronary Arteries: A Systematic Review and Meta-Analysis." Evidence-Based Complementary and Alternative Medicine 2022.1 (2022): 8678967.

44) Khan, Ikram, et al. "Analysis of the blood bacterial composition of patients with acute coronary syndrome and chronic coronary syndrome." (2022).

45) Cretoiu, Dragos, et al. "Gut microbiome, functional food, atherosclerosis, and vascular calcifications—Is there a missing link?." Microorganisms 9.9 (2021): 1913.

46) Pisano, Eugenia, et al. "Microbial signature of plaque and gut in acute coronary syndrome." Scientific Reports 13 (2023).

47) Joshi, Chaitanya, et al. "Detection of periodontal microorganisms in coronary atheromatous plaque specimens of myocardial infarction patients: A systematic review and meta-analysis." Trends in cardiovascular medicine 31.1 (2021): 69-82.

48) Rao, Amita, et al. "Molecular analysis shows the presence of periodontal bacterial DNA in atherosclerotic plaques from patients with coronary artery disease." Indian Heart Journal 73.2 (2021): 218-220.

49) Jonsson, Annika Lindskog, et al. "Bacterial profile in human atherosclerotic plaques." Atherosclerosis 263 (2017): 177-183.

50) Pavlic, Verica, et al. "Identification of periopathogens in atheromatous plaques obtained from carotid and coronary arteries." BioMed Research International 2021.1 (2021): 9986375.

51) Masoumi, Omid, et al. "Detection of fungal elements in atherosclerotic plaques using mycological, pathological and molecular methods." Iranian Journal of Public Health 44.8 (2015): 1121.

52) Sato, Ayako, et al. "Metagenomic Analysis of Bacterial Microflora in Dental and Atherosclerotic Plaques of Patients With Internal Carotid Artery Stenosis." Clinical Medicine Insights: Cardiology 18 (2024): 11795468231225852.

53) Ravnskov, Uffe, Abdullah Alabdulgader, and Kilmer S. McCully. "Infections may cause arterial inflammation, atherosclerosis, myocarditis and cardiovascular disease." Medical Research Archives 11.5 (2023).

54) Lv, Hang, et al. "Characteristics of the gut microbiota of patients with symptomatic carotid atherosclerotic plaques positive for bacterial genetic material." Frontiers in Cellular and Infection Microbiology 13 (2023).

55) Kissinger, Dohn. "Are bacterial infections a major cause of cardiovascular disease?." Frontiers in Cardiovascular Medicine 11 (2024): 1389109.

56) Huang, Xiaofei, et al. "The Roles of Periodontal Bacteria in Atherosclerosis." International Journal of Molecular Sciences 24.16 (2023): 12861.

57) Jayasinghe, Thilini N., et al. "Identification of oral bacteria in the gut, atherosclerotic plaque, and cultured blood samples of patients with cardiovascular diseases–A secondary analysis of metagenomic microbiome data." (2023).

58) Ghazi, Lama, et al. "Hypertension across a woman's life cycle." Current hypertension reports 24.12 (2022): 723-733.

59) Pitha, Jan, Ivana Vaneckova, and Josef Zicha. "Hypertension after the menopause: what can we learn from experimental studies?" Physiological Research 72.Suppl 2 (2023): S91.

60) Lee, D. Y., et al. "Effects of hormone therapy on ambulatory blood pressure in postmenopausal Korean women." Climacteric 14.1 (2011): 92-99.

Chapter 11

Bioidentical Hormones Found Beneficial After Hysterectomy

Should My Ovaries Be Removed?

SHIRLEY IS A 43-YEAR-OLD EXECUTIVE for an internet tech company who came to the office because of irregular bleeding from massive uterine fibroids. The continuous bleeding interfered with her lifestyle and caused severe fatigue from the iron deficiency anemia. I proposed a procedure called hysterectomy, the surgical removal of the uterus because of the enlarging fibroids. Shirley went home to think it over, A few days later Shirley finally decided to schedule the operation with her OB/Gyne surgeon.

What are Uterine Fibroids?

Uterine fibroids are benign growths in the uterus that may enlarge, causing pelvic pressure symptoms and irregular bleeding. The heavy bleeding can lead to iron deficiency anemia and chronic fatigue. The laboratory panel will show low hemoglobin and hematocrit, and the MCV, or mean corpuscular volume, will be decreased. Serum iron and ferritin will be decreased. Uterine fibroids are quite common and can be detected by an enlarged uterus on pelvic examination, and confirmed with a pelvic sonogram and/or MRI scan. Massive uterine fibroids are usually treated with hysterectomy which removes the uterus and fibroid tumors, usually with a good outcome. Having a surgical operation with general anesthesia is not to be taken lightly. However, in some cases where medical treatments have failed and the patient's quality of life is negatively impacted, hysterectomy is justified.

Visiting the OB Gyne Surgeon

Next, Shirley visited her OB-GYN surgeon to discuss the operation. Her surgeon recommended a complete hysterectomy with removal of both ovaries. Sitting in the surgeon's office, Shirley timidly protested. "Why remove my ovaries?" asked Shirley. The surgeon replied, "Removing the ovaries eliminates the chance of ovarian cancer, and you don't need them anyway."

To Remove or Not to Remove the Ovaries? That is the Question

So, now Shirley was again in my office asking for my opinion. Should she go against the surgeon's advice and insist on preserving her ovaries, or should she follow the surgeon's advice to have a complete hysterectomy with removal of the ovaries? The ovaries are the hormone factories that pump out women's hormones, estrogen, and testosterone daily. They also make progesterone during the luteal phase of the menstrual cycle, after ovulation on day 11 with rupture of the follicle and formation of the corpus luteum. Removing the ovaries removes the source of these three hormones inducing surgical menopause.

Studies Show Removing Ovaries Increases Mortality

Luckily, the answer to Shirley's question about whether or not the ovaries should be removed can be found in the medical literature. In 2009, Drs. William Parker reported that removing the ovaries is detrimental to overall health and results in increased mortality. Dr. Parker followed 30,000 women for 24 years after their hysterectomy. Half the patients had the ovaries removed and half had ovaries preserved. The group with ovaries removed did have a lower rate of ovarian and breast cancer. However, this was overshadowed by a marked increase in deaths from heart disease and other cancers. The group with the ovaries removed had a higher all-cause mortality rate, and therefore Dr. Parker recommended women in the pre-menopausal age group should preserve the ovaries. Dr. William Parker also found that post-operative hormone replacement is very beneficial in reducing the risk of heart disease. (1)

In 2009, Dr. Cathleen Rivera followed 1,000 pre-menopausal women, under age 45 after a hysterectomy, finding removal of the ovaries resulted in a disturbing **84% increase in death** from heart disease. However, if these women were given estrogen replacement after ovarian removal, they were protected with a **35% decrease in mortality** from heart disease. I thought this was rather impressive. (2-4)

Bioidentical Hormones After Hysterectomy

These two studies provide convincing evidence of the health benefits of hormone replacement after hysterectomy. Although the patients in these two studies were given Premarin (CEE Conjugated Equine Estrogen) which is a natural hormone from a pregnant horse, we find that a cocktail of bioidentical hormones including estradiol, estriol, progesterone, DHEA, and testosterone is equally effective. What about preventing ovarian cancer and breast cancer in high-risk groups? For women at high risk with familial breast and ovarian cancer, and positive BRCA genetic markers, Dr. Parker says it makes sense to go ahead with removing the ovaries for these people in high-risk groups.

Dr. William Parker and the Nurses' Health Study

In 2013, Dr. William Parker reviewed the Nurses' Health study, a prospective cohort study of 30,117 women having a hysterectomy for benign disease. Dr. Parker studied the long-term mortality associated with oophorectomy, removal of ovaries inducing surgical menopause, compared to ovarian preservation. Dr. William Parker found that surgical menopause before age 50 is associated with **41 percent** increased all-cause mortality. However, no increased mortality was found if the patient used estrogen hormone replacement. Mortality from coronary artery disease was more than doubled in the oophorectomy patients who were not using estrogen replacement. Dr. Parker writes:

> For women younger than 50 at the time of hysterectomy, bilateral oophorectomy was associated with significantly increased mortality in women who had never-used estrogen therapy, but not in past and current users: all-cause mortality (**HR=1.41**); lung cancer mortality (**HR=1.44);** and CHD mortality (HR=2.35)...Conclusions: For women younger than 50 at the time of hysterectomy, bilateral oophorectomy was associated with significantly increased mortality in women who had never-used estrogen therapy. At no age was oophorectomy associated with increased overall survival. Note: CHD=coronary heart disease. (5)

Dr. Phillip Sarrel Calculates 50,000 Women Died Needlessly

In 2013, Dr. Phillip Sarrell used the data from the 2004, second arm of the WHI study, women post hysterectomy given Premarin-alone or placebo. This study showed the 50-60 year age group taking Premarin-alone had reduced all-cause mortality by **27 percent**, and reduced cardiovascular mortality by **41 percent** compared to placebo users. Because of bad press in 2002 after the publication of the first arm WHI study using combination of Premarin and MPA, HRT use with declined precipitously. Before 2002, of all women with surgical menopause after hysterectomy, 90 percent were using HRT (Premarin-alone). However, after 2002, **only 33 percent** were using ET. Dr. Sarrell blames the media for distorting the data, causing an **estimated 50,000 needless deaths because of estrogen avoidance**, writing:

> We therefore undertook an analysis, based on published data, to determine the likely

toll of excess, premature death among hysterectomized women aged 50 to 59 years in the United States following the WHI publication in 2002... ET [estrogen therapy] reduces total mortality primarily through reducing CHD-related deaths. Since 1959 there have been many reports showing increased risk of CHD [coronary heart disease] mortality after early surgical menopause and especially after oophorectomy. Essentially, estradiol inhibits the development of atherosclerosis and helps maintain normal arterial blood flow...A 1998 meta-analysis of 25 observational studies reported a **30% lower risk of CHD in ET users.** The California Teachers Study, in which 97% of the oophorectomized women used HT (almost entirely ET), shows an all-cause mortality reduction similar to the WHI-ET results [about 30 percent]... The WHI-ET and the California Teachers Study results indicate that the decrease in CHD [coronary heart disease] events and all-cause mortality are limited to hysterectomized women younger than 60 years or within 10 years of menopause. A meta-analysis of 23 trials found that HT significantly reduced CHD events in these women. In fact, for CHD, the WHI-ET [estrogen-only, post hysterectomy] findings among women aged 50 to 59 years are in line with the reduced risks reported in the observational studies and support a "timing hypothesis" [of Dr. Howard Hodis] for ET cardioprotection when ET is started close to the time of menopause. The current thinking is that by age 60 years, pathological changes in vascular endothelial cells compromise the ability of estrogen to inhibit atherosclerosis and promote blood flow...Our analysis suggests that between 2002 and 2011 a minimum of 18,601 and as many as 91,610 excess deaths occurred among hysterectomized women aged 50 to 59 years following the publication of the original WHI findings because of the resulting aversion to hormone replacement therapy of all kinds that ensued among doctors and patients alike. The actual toll of excess mortality is likely to be between 40,292 and 48,835...**Estrogen Therapy (ET) in younger postmenopausal women is** associated with a decisive reduction in all-cause mortality. (6)

The Fear of Breast Cancer from Hormone Therapy

In 2022, Dr. Delfin A. Tan, OB/Gyne at St. Luke's Medical Center, Quezon City, Philippines states the fear of breast cancer associated with menopausal hormone replacement is no longer valid. Increased breast cancer risk applies only to the combination of estrogen with progestin (medroxyprogesterone, MPA) and not to estrogen alone, nor the combination of estrogen and natural progesterone. In addition, Dr. Tan says avoidance of menopausal hormone replacement because of fear of breast cancer places women in harm's way, writing:

> The threat that women may develop breast cancer is the major reason why both physicians and women are afraid to use menopausal hormone therapy (MHT). The fear pertains to estrogen-progestin [medroxyprogesterone] replacement therapy (EPRT) as estrogen-alone replacement therapy has no, or even a reduced, breast cancer risk... and **avoiding MHT use when indicated puts a woman in harm's way. (7)**

2004 Second Arm Women's Health Initiative Study 11.8 years Follow Up

23% Reduction in Breast Cancer in Estrogen Users

In 2012, Drs. Garnet L Anderson PhD, and Rowan T Chlebowski reported the data on the 11.8-year follow-up of the 2004 second arm WHI study, women after hysterectomy treated with Premarin (CEE) alone. After 11.8 years of follow-up, the Premarin (CEE, horse estrogen) had **151** cases of invasive breast cancer and the placebo group had **199** cases. Estrogen-alone users (Premarin, CEE) had a **23 percent reduction in breast cancer, a 62 percent reduction in breast cancer mortality, and a 40 percent reduction in all-cause mortality** in the estrogen-treated group. (8)

In 2012, Drs. Anthony Howell and Jack Cuzick commented on the above study, writing:

> In The Lancet Oncology, the Women's Health Initiative (WHI) investigators report that receipt of conjugated equine estrogen for a median of 5·9 years reduced the risk of invasive breast cancer by 23% compared with placebo (**151 cases in 5310 women who received estrogen vs 199 cases in 5429 controls**; p=0·02). Women who did develop breast cancer after receipt of estrogen had significantly reduced breast cancer-specific mortality (**six deaths in the estrogen group vs 16 deaths in controls; p=0·03**) and all-cause mortality (**30 deaths vs 50 deaths**; p=0·04). This preventive effect occurred at all ages and continued beyond the period of estrogen use, a carryover effect also noted in prevention trials of tamoxifen. (9)

Avoid Carcinogenic Synthetic Progestins

Synthetic versions of progesterone called progestins such as medroxyprogesterone (MPA) are known to be carcinogenic based on data from the 2002 first arm WHI study with a planned duration of 8.5 years. This study was halted early after 5.2 years when they found a non-significant increase risk of breast cancer, **HR= 1.26 (1.00-1.59)** with 290 cases. Since the confidence interval included 1.0, the 1.26 hazard ratio (HR) is not statistically significant. The exact number of breast cancer cases in the placebo arm was not published. These women were taking MPA (medroxyprogesterone), a synthetic progestin known to increase the risk of breast cancer. Cancer researchers commonly use MPA to induce breast cancer in mice. This is called the MPA mouse model of breast cancer in BALB-c mice by Claudia Lanari (2009). Avoid carcinogenic progestins. Instead, use natural progesterone. (10-22)

The Derailed and Fragmented Care of Mid-Life Women

In 2016, Dr. JoAnn E. Manson wrote an editorial in the New England Journal of Medicine about the sad situation in our medical system regarding the reluctance to treat menopausal women with hormone replacement (HRT). Dr. JoAnn Manson writes:

> Despite the availability of effective hormonal …treatments for menopausal symptoms, few women with these symptoms are evaluated or treated…**Leading medical societies agree that hormone therapy is the most effective treatment currently available for (menopausal) symptoms and should be recommended**…Yet, 20% of women in early menopause… remain untreated despite having symptoms that adversely affect their daily activities, sleep, and quality of life… the new generation of primary care providers often lacks training and core competencies in management of menopausal symptoms and prescribing of hormonal treatments… **Reluctance to treat menopausal symptoms has derailed and fragmented the clinical care of midlife women, creating a large and unnecessary burden of suffering**. (23-25)

Post-Menopausal Calcium Score Study 2017

In 2017, Dr. Yoav Arnson did a prospective cohort study of 4286 consecutive asymptomatic post-menopausal females (average age 62.4) who underwent CAC (calcium score) scanning in their institution between 1998 and 2012. The women were followed for mortality over 8.4 years. Of the 4286 patients, 41% reported taking hormone replacement therapy (HRT) at the time of the scan. Women using HRT were:

1) **Thirty percent less likely to die** than those not on hormone therapy (i.e. **thirty percent reduction in mortalit**y).

2) **Twenty percent** more likely to have a coronary calcium score of zero (the lowest possible score, indicating a low likelihood of heart attack)

3) **Thirty-six percent** less likely to have a coronary calcium score that indicates extensive atherosclerosis and a 10-fold increase in heart attack risk. (26)

Surgical Menopause, Premature Ovarian Failure, Early Menopause

About 5 percent of women will undergo early menopause before age 45. All three entities, surgical menopause, premature ovarian failure (POI), and early menopause are remarkably similar and refer to the same pattern of estrogen deficiency, menopausal symptoms, increased all-cause and cardiovascular mortality. All three have the same pathophysiology, ovarian failure with estrogen deficiency. These patients will have a cessation of menstrual periods. Laboratory studies will show elevated FSH levels, low estradiol and low progesterone levels. The standard of care is hormone replacement therapy (HRT), yet only half of these women are treated with HRT. In 2016, Drs. Shannon Sullivan and Philip M. Sarrel discuss HRT for young women with primary ovarian insufficiency (POI) and early menopause, writing:

> Primary ovarian insufficiency (POI) is characterized by menopausal levels of follicle stimulating hormone (FSH) and absent or irregular menstrual cycles prior to age 40. Because the average age of natural menopause is 50–51 years, women exhibiting these findings after age 40 but prior to age 45 are said to have early menopause...An estimated 5% of women undergo early menopause prior to age 45...**in contrast to women with normal menopause, the situation in young women with POI and early menopause is in fact a pathologic state of estrogen deficiency compared to their peers with normal ovarian function**...Health complications of POI include menopausal symptoms (hot flashes, night sweats, insomnia, dyspareunia, decreased sexual desire, and vaginal dryness); decreased bone mineral density (BMD) and increased risk of fracture; infertility; increased risk of mood disorders, namely depression and anxiety; cognitive decline; sexual dysfunction; increased rates of auto-immune disease; increased risk of cardiovascular disease; increased risk of Type 2 diabetes mellitus (T2DM) or pre-DM;

> and dry eye syndrome...Physiologic EPT [estrogen/progestin therapy] ameliorates many of these health risks and is considered standard of care for women with POI or early menopause...Women with POI, regardless of the etiology, have an increased risk of cardiovascular disease and ischemic stroke... POI is a pathologic condition in which young women have low serum estradiol levels as compared with their peers. For young women with estradiol deficiency, hormone therapy is indeed "replacement," whereas in women with normal menopause, hormone therapy is in fact hormone "extension." It is important to make this distinction clear to patients. Unfortunately, a recent study showed that **more than half (52%) of young women with POI either never take HRT**, start HRT many years after their diagnosis, and/or discontinue HRT use prior to age 45..The weight of evidence now favors **transdermal or transvaginal estradiol therapy** as the first line of HRT for young women with POI or early menopause. **Young women who develop POI require long-term ovarian sex steroid replacement. Some will require this therapy for decades.** Current therapies are prescribed to control symptoms and help prevent disease related to estradiol deficiency...The transdermal patch and the vaginal ring that deliver 0.100 mg of estradiol per day are a first rudimentary step in this direction. These formulations mimic the daily ovarian production rate of estradiol and achieve average serum estradiol levels of 100 pg/ml; this is the average level women with normal ovarian function experience across the menstrual cycle. An equivalent dose of oral estradiol is also effective replacement, however, the transdermal and transvaginal routes of administration deliver hormone directly into the circulation, which avoids complications associated with the first pass effect on the liver when estrogen is given orally. The risk of venous thromboembolism is increased by oral estrogen compared to transdermal estrogen use. Note: Although Drs. Sullivan and Sarell (above) suggest using HRT formula with an estrogen/progestin combination, I would suggest using

natural progesterone rather than synthetic progestin. (27)

Conclusion: Surgical menopause (hysterectomy), premature ovarian insufficiency (POI), and early menopause are associated with increased mortality from coronary artery disease. The standard of care is hormone replacement therapy. Yet, because of the fear of estrogen, only half of these women are treated leading to increased mortality. The 2004 second-arm WHI study using Premarin-alone revealed that estrogen therapy does not cause breast cancer. Therefore, the fear of estrogen is unfounded. Yet, the media deceived the population by emphasizing the 2002, first arm WHI study using Premarin (CEE)/MPA showing a non-significant increase in breast cancer (HR=1.26). This created so much fear of estrogen therapy that an estimated 50,000 post-hysterectomy women died from estrogen avoidance. This is only one example of "Fake News" resulting in widespread death in the population. There have been many others. This is a tragedy that falls squarely on the mass media who have never been held accountable and probably never will be. Even though it has been known for decades that estrogen deficiency after hysterectomy is a health risk, and associated with increased mortality, our misguided medical system has been denying estrogen replacement to millions of women, causing needless suffering and increased mortality. Over the past two decades, we have quietly and diligently worked to counter this trend by prescribing bioidentical hormone replacement for every menopausal woman requesting it. That has been our mission. (28-29)

♦ References for Chapter 11. Bioidentical Hormones Found Beneficial After Hysterectomy

1) Parker, William H., et al. "Ovarian Conservation at The Time of Hysterectomy and Long-Term Health Outcomes in The Nurses' Health Study." Obstetrics & Gynecology 113.5 (2009): 1027-1037.

2) Rivera, Cathleen M., et al. "Increased Cardiovascular Mortality After Early Bilateral Oophorectomy." Menopause 16.1 (2009): 15-23.

3) Cusimano, Maria C., et al. "Association of Bilateral Salpingo-Oophorectomy With All Cause And Cause Specific Mortality: Population Based Cohort Study." BMJ 375 (2021).

4) Mann, Shivani N., et al. "17α-Estradiol Prevents Ovariectomy-Mediated Obesity And Bone Loss." Experimental Gerontology 142 (2020): 111113.

5) Parker, William H., et al. "Long-Term Mortality Associated with Oophorectomy Compared With Ovarian Conservation In The Nurses' Health Study." Obstetrics & Gynecology 121.4 (2013): 709-716.

6) Sarrel, Philip M., et al. "The Mortality Toll of Estrogen Avoidance: An Analysis of Excess Deaths Among Hysterectomized Women Aged 50 To 59 Years." American journal of public health 103.9 (2013): 1583-1588.

7) Tan, Delphin A., and A. Dayu. "Menopausal Hormone Therapy: Why We Should No Longer Be Afraid of The Breast Cancer Risk." Climacteric 25.4 (2022): 362-368.

8) Anderson, Garnet L., et al. "Conjugated Equine Estrogen and Breast Cancer Incidence And Mortality In Postmenopausal Women With Hysterectomy: Extended Follow-Up Of The Women's Health Initiative Randomised Placebo-Controlled Trial." The Lancet Oncology 13.5 (2012): 476-486.

9) Howell, Anthony, and Jack Cuzick. "Estrogen and Breast Cancer: Results from The WHI Trial." The Lancet Oncology 13.5 (2012): 437-438.

10) Lanari, Claudia, et al. "The MPA Mouse Breast Cancer Model: Evidence for A Role Of Progesterone Receptors In Breast Cancer." (2009). The MPA mouse breast cancer model Lanari Claudia 2009

11) Aldaz, C. Marcelo, et al. "Medroxyprogesterone Acetate Accelerates the Development and Increases the Incidence Of Mouse Mammary Tumors Induced By Dimethylbenzanthracene." Carcinogenesis 17.9 (1996): 2069-2072.

12) Lanari, Claudia, et al. "Induction of Mammary Adenocarcinomas By Medroxyprogesterone Acetate In Balbc Female Mice." Cancer letters 33.2 (1986): 215-223.

13) Lanari, Claudia, et al. "Mammary Adenocarcinomas Induced by Medroxyprogesterone Acetate: Hormone Dependence and EGF Receptors Of BALB/C In Vivo Sublines." International Journal of Cancer 43.5 (1989): 845-850.

14) Molinolo, A. A., et al. "Mouse Mammary Tumors Induced by Medroxyprogesterone Acetate: Immunohistochemistry and Hormonal Receptors." Journal of the National Cancer Institute 79.6 (1987): 1341-1350.

15) Nagasawa, Hiroshi, et al. "Medroxyprogesterone Acetate Enhances Spontaneous Mammary Tumorigenesis and Uterine Adenomyosis in Mice." Breast cancer research and treatment 12 (1988): 59-66.

16) Liang, Yayun, et al. "Synthetic Progestins Induce Growth And Metastasis Of BT-474 Human Breast Cancer Xenografts In Nude Mice." Menopause (New York, NY) 17.5 (2010): 1040.

17) Kordon, Edith, et al. "Hormone Dependence Of A Mouse Mammary Tumor Line Induced In Vivo By Medroxyprogesterone Acetate." Breast cancer research and treatment 17 (1990): 33-43.

18) Sartorius, Carol A., et al. "Progestins initiate A Luminal To Myoepithelial Switch In Estrogen-Dependent Human Breast Tumors Without Altering Growth." Cancer research 65.21 (2005): 9779-9788.

19) Pazos, Patricia, et al. "Mammary Carcinogenesis Induced By N-Methyl-N-Nitrosourea (MNU) And Medroxyprogesterone Acetate (MPA) In BALB/C Mice." Breast cancer research and treatment 20 (1991): 133-138.

20) Molinolo, Alfredo, et al. "Involvement of EGF in medroxyprogesterone Acetate (MPA)-Induced Mammary Gland Hyperplasia And Its Role In MPA-Induced Mammary Tumors In BALB/C Mice." Cancer letters 126.1 (1998): 49-57.

21) Kordon, Edith, et al. "Estrogen Inhibition of MPA-Induced Mouse Mammary Tumor Transplants." International Journal of Cancer 49.6 (1991): 900-905.

22) Buqué, Aitziber, et al. "MPA/DMBA-Driven Mammary Carcinomas." Methods in Cell Biology. Vol. 163. Academic Press, 2021. 1-19.

23) Manson, JoAnn E., and Andrew M. Kaunitz. "Menopause Management—Getting Clinical Care Back on Track." New England Journal of Medicine 374.9 (2016): 803-806.

24) Anderson, G. L., et al. "Women's Health Initiative Steering Committee. Effects of conjugated equine estrogen in postmenopausal women with hysterectomy: the Women's Health Initiative randomized controlled trial." JAMA 291.14 (2004): 1701-1712.

25) Writing Group for the Women's Health Initiative Investigators. "Risks and Benefits of Estrogen Plus Progestin in Healthy Postmenopausal Women: Principal Results from The Women's Health Initiative Randomized Controlled Trial." JAMA 288.3 (2002): 321-333.

26) Arnson, Yoav, et al. "Hormone Replacement Therapy Is Associated With Less Coronary Atherosclerosis And Lower Mortality." Journal of the American College of Cardiology 69.11S (2017): 1408-1408.

27) Sullivan, Shannon D., Philip M. Sarrel, and Lawrence M. Nelson. "Hormone Replacement Therapy In Young Women With Primary Ovarian Insufficiency And Early Menopause." Fertility and sterility 106.7 (2016): 1588-1599.

28) Wilson, Louise F., et al. "Hysterectomy Status and All-Cause Mortality In A 21-Year Australian Population-Based Cohort Study." American Journal of Obstetrics & Gynecology 220.1 (2019): 83-e1.

29) Ferris, Jennifer S., et al. "Excess Morbidity and Mortality Associated with Underuse Of Estrogen Replacement Therapy In Premenopausal Women Who Undergo Surgical Menopause." American Journal of Obstetrics and Gynecology 230.6 (2024): 653-e1.

Chapter 12

Bioidentical Hormones Prevent Arthritis

MARY, A 53-YEAR-OLD POST-MENOPAUSAL MOM, has a chief complaint of arthritis in her fingers. She has trouble opening jars because her finger joints are swollen and tender. On physical examination of her hands, the osteo-arthritis is obvious with swollen enlarged joints and Heberden's nodes. Mary was sent for Xrays of the hands which confirmed the diagnosis. Laboratory testing for rheumatoid arthritis was negative. Mary was then started on a meno-pausal hormone replacement program includ-ing estrogen, progesterone, testosterone, and DHEA, all prepared in an oil vehicle applied topically. Mary was instructed to rub the topi-cal hormone oil directly into the painful finger joints. Six weeks later during a follow up call by telephone, Mary reports her arthritis pain has greatly improved.

Estrogen Deficiency Causes Degenerative Osteoarthritis

Degenerative osteoarthritis is common in post-menopausal women resulting in defor-mity, swelling, and pain in finger joints and knee joints. Recent medical research shows that menopausal estrogen deficiency is the direct cause of osteoarthritis. Menopausal hormone replacement with estrogen prevents osteoarthritis, and may partially reverse some of the symptoms. In 1998, Dr. David T. Felson reviewed menopausal arthritis finding osteo-arthritis increases dramatically in women after menopausal age of 50 years. Post-menopausal women using hormone replacement have less osteoarthritis when compared to non-users, thus suggesting a role for menopausal estro-gen replacement for the prevention of osteo-arthritis in post-menopausal women. In 2009, Dr. Herrero- Beaumont from Spain reviewed

the medical literature from 1952 to 2008 and found three causes for osteoarthritis, estrogen deficiency, genetic causes, and aging. In 2009, Dr Herrero Beaumont writes:

> There is now increasing evidence that estrogens influence the activity of joint tissues through complex molecular pathways that act at multiple levels. (1-3)

Estrogen Replacement Reduces Osteoarthritis of the HIP by 43%

In 1996, Drs. Michael C Nevitt and Harry Genant at the University of California, San Francisco examined 4,366 post-menopausal women over the age of 65 years. Hip X-rays were used to assess osteoarthritis of the hip joint. The authors found that women who took oral estrogen had a **38% reduced risk of osteoarthritis** (OA) of the hip. Women who used estrogen for 10 years or longer had a **43% reduction in OA of the hip.** The authors con-cluded that:

> Postmenopausal estrogen replacement therapy may protect against Osteoarthritis (OA) of the hip. Note: Dr. Harry Genant was Professor of Radiology, Medicine, Epidemiology and Orthopaedic Surgery, and Chief of Musculoskeletal Radiology at UCSF for 30 years from 1974 to 2004. He was known as the "Godfather of Skeletal Imaging in Osteoporosis" Sadly Dr. Genant passed away in 2021. (4-5)

Framingham Study - Arthritis of the Knee Reduced by 60%

In 1998, Dr. Yuqing Zhang of the Boston University School of Medicine studied knee arthritis in the famous Framingham study, ask-ing the question: Does estrogen replacement therapy (ERT) prevent radiographic worsening of osteoarthritis (OA) of the knee in elderly women? Dr. Zhang followed 551 post-menopausal women over the age of 63 for 8 years with serial knee X-Rays looking for worsening of osteoarthritis over time. Dr. Zhang found a **60% decrease in osteoarthritis in estrogen users compared to non-users**. (6)

Women's Health Initiative
WHI- Lower Incidence of Joint
Replacements and Fractures

In 2006, Dr. Dominic Cirillo reviewed data from the Women's Health Initiative study showing that women receiving oral Premarin-alone (CEE) had 12% lower rates for joint replacement for osteoarthritis. Bone mineral density (BMD) increased 3.7 percent over 3 years in HRT users (Premarin/MPA). For placebo users, their BMD increased a negligible 0.14 percent. Hip and clinical vertebral fractures were significantly reduced in HRT users by **34%.** (7-8)

Genetic Causes of Arthritis -
Disturbed Estrogen Metabolism

A 2010 study by Dr. Jose Riancho from Spain in the Journal of Osteoarthritis and Cartilage explored the association of genetic abnormalities with severe osteoarthritis (OA) in 3,147 patients compared with 2,381 controls. Dr. Jose Riancho examined two gene mutations that reduce estrogen activity and their association with severe osteoarthritis (OA). These are the genes for the aromatase enzyme and for the estrogen receptor (ER-alpha). Women with one of these gene mutations had a **60 percent increased risk for knee arthritis**. Thus, genetic studies indicate mutations that reduce estrogen are associated with severe osteoarthritis. Dr. Jose Riancho writes:

> Common genetic variations of the aromatase and ER [estrogen receptor] genes are associated with the risk of severe OA [osteoarthritis] of the large joints of the lower limb in a sex-specific manner. These

results are consistent with the hypothesis that estrogen activity may influence the development of large-joint OA. (9-11)

Cellular Mechanisms of Estrogen on Cartilage

In 2008, Dr. Laszlo B. Tanko from Denmark summarized three decades of medical research on the cellular mechanism of osteoarthritis, finding estrogen receptors present in the cartilage of animals and humans, and estrogen is necessary to maintain bone density and articular cartilage in postmenopausal women, writing:

> Estrogen receptors have been identified in articular chondrocytes from various animals and humans...the effects of estrogen on articular cartilage further corroborate the due consideration of estrogen therapy for maintaining not only bone but also cartilage health in postmenopausal women. (12)

Monkey Study - Estrogen Prevents Arthritis

In 2002, Dr. Kimberly Ham from the University of Minnesota examined the effect of estrogen replacement therapy on the severity of osteoarthritis of the knee joint in postmenopausal female monkeys, after surgical menopause with removal of the ovaries. After three years of estrogen treatment using Premarin (CEE), the monkeys were sacrificed and knee joints were examined under the microscope. The authors found that cartilage lesions of osteoarthritis were significantly less severe in the animals given estrogen replacement compared with those in the control group. Dr. Kimberly Ham writes:

> These results demonstrate that long-term estrogen replacement significantly reduces the severity of osteoarthritis. (13)

Mice - Estrogen Protects Cartilage

A 2006 study by Dr. Svetlana Oestergaard and Laszlo B. Tanko in Arthritis and Rheumatism found that estrogen treatment prevents joint and cartilage degradation in rats. Cartilage turnover was estimated by measuring the serum levels of C-telopeptide of type II collagen. Dr. Oestergaard found that estrogen treatment in mice prevents collagen deterioration. The greatest benefit was seen when the estrogen is given early, immediately after menopause. Estrogen protects the cartilage cells and chondrocytes from deterioration. (14)

Pigs - Estrogen Preserves Cartilage

In 2002, Dr. Horst Claassen from Germany studied pigs, finding estrogen treatment prevents cartilage degradation. On the other hand, estrogen deficiency degrades the articular cartilage. (15)

Guinea Pig - Electron Microscope Study of Estrogen Deficiency

In 2005, Dr. Dai Guofeng from China studied guinea pig joints using scanning electron microscopy (SEM) and transmission electron microscope (TEM) to analyze cartilage degeneration after ovariectomy, surgical removal of the ovaries, with induction of estrogen deficiency. Dr. Dai Guofeng found estrogen receptors (ER) within the articular cartilage of the guinea pigs. Estrogen deficiency-induced joint cartilage degeneration was detected by electron microscopy at 6 weeks, and more severe degeneration at 12 weeks after ovariectomy, compared to controls. Dr. Dai Guofeng writes: "Bilateral ovariectomy in the guinea pig leads to severe osteoarthritis." (16)

Sheep – Estrogen Deficiency Causes Cartilage Degeneration

In 1997, Dr. Simon Turner from Colorado State University studied estrogen replacement with estradiol implants in ovariectomized sheep. Twelve months after surgical menopause, the articular cartilage from the knee joint was carefully evaluated. Ovariectomy-induced estrogen deficiency had a significant deleterious effect on articular cartilage. Treatment with estradiol reversed these deleterious effects and maintained the structural integrity of the joints. (17)

In 2005, Dr. Martin A. Cake from Australia studied arthritis in sheep. After 26 weeks of estrogen depletion cartilage thickness was reduced and arthritic changes were present. Dr. Martin A. Cake writes:

> estrogen depletion caused regional thinning of femoro-tibial cartilage, with biomechanical and histological changes suggestive of a disturbance in proteoglycan and collagen. (18)

Animal Model Summaries

In 2008, Dr. Yvonne H. Sniekers from the Netherlands summarized all preceding animal studies stating that animal models are useful to evaluate the dramatic increase in post-menopausal osteoarthritis after age 50. The author found 11 of 14 animal studies showing ovariectomy-induced estrogen deficiency resulted in cartilage damage, indicating estrogen deficiency causes cartilage degeneration. (19)

In 2023, Dr. Gabrielle Gilmer from the University of Pittsburg found 38 animal studies in menopausal animal models showing estrogen treatment ameliorates osteoarthritis. Overall, cartilage outcomes were worse in post-menopausal animals compared to controls. This was evidenced by measurements of cartilage histological scoring, cartilage thickness, type II collagen, and c-terminal cross-linked telopeptide. Another finding is that earlier initiation and higher doses of estrogen produced greater improvement in cartilage health. In 2023, Dr. G. Gilmer writes:

> Thirty-eight manuscripts were eligible for inclusion. The most common menopause model used was ovariectomy (92%), and most animals were young at the time of menopause induction (86%)...Cartilage outcomes were worse in post-menopausal animals compared to age-matched, non-menopausal animals, as evidenced by cartilage histological scoring [0.75, 1.72],

cartilage thickness [−4.96, −0.96], type II collagen [−4.87, −0.56], and c-terminal cross-linked telopeptide of type II collagen (CTX-II) [2.43, 5.79] (95% CI of Effect Size (+greater in menopause, −greater in non-menopause)). Moreover, modeling suggests that cartilage health may be improved with **early initiation and higher doses of estrogen treatment**. (20)

Inducing Estrogen Deficiency with Aromatase Inhibitors

The adjuvant hormone treatment for breast cancer involves anti-estrogen drugs such as aromatase inhibitors such as exemestane, letrozole, or anastrozole to induce estrogen deficiency. Joint pain and osteoarthritis are known adverse side effects of antiestrogen drug treatment. Numerous studies reveal women on anti-estrogen drugs report joint pain, suggesting estrogen plays an important role in maintaining cartilage in the joints. (21-28)

Still Have Doubts?

Despite the overwhelming evidence presented above, there is still some confusion in the medical literature about the role of estrogen in post-menopausal osteoarthritis. For example, in 2023, Dr. Uyen-sa Nguyen writes in the European Journal of Rheumatology that the evidence of a protective effect of estrogen in osteoarthritis is still inconclusive, and obesity and aging may be more relevant causative factors:

> Although several studies show that hormone replacement therapy has the potential to be protective of OA [osteoarthritis] for some joints, there are studies that showed no protective effect or even adverse effect. Taken together, the evidence for the protective effect of estrogen therapy depends on OA joint, OA outcome, and study design. Although this area has been studied for decades, more exclusively since the 1990s, there is a lack of high-quality experimental research in this topic. **The lack of definitive conclusion on whether estrogen can play a role in the development in OA of either the knee, hip, spine, or hand** is often in part due to the noncomparability of studies existing within the literature. Differences in diagnostic criteria, imaging modalities, populations studied, study designs, and outcome measures, as well as random error, have all contributed to inconclusive evidence. Future research on the role of estrogen in OA is needed, particularly as global demographic shifts in increasing overweight/obesity prevalence and ageing populations may contribute to widening OA-related health inequalities. (29)

I agree with Dr. Uyen-sa Nguyen (above) that obesity and aging are additional causative factors in the osteoarthritis story. Both obesity and aging play a role in the etiology of osteoarthritis. I do not argue with this. Indeed, for obese patients, weight loss may be the best treatment for preventing joint erosion and destruction, thereby delaying joint replacement. However, regarding the role of estrogen deficiency in causing osteoarthritis, the evidence is overwhelming. It would be an error to hold the contrary view or suggest the evidence is inconclusive. The massive evidence in the medical literature supports the conclusion that estrogen deficiency is a major cause of postmenopausal osteoarthritis, and estrogen therapy relieves joint pain and allows cartilage regeneration. The benefits of estrogen for post-menopausal joint pain are blatantly obvious to clinicians treating these patients.

NSAIDS and Steroid Injections Accelerate Joint Destruction

Steroid injections into the joint and non-steroidal anti-inflammatory drugs (NSAIDs) are commonly prescribed and may provide short-term pain relief. However, long-term use of NSAID drugs and steroid injections may accelerate joint destruction and hasten joint replacement. Gastro-intestinal bleeding complications of NSAIDs cause about 16,000 deaths annually in the U.S as discussed by Dr. Michael M. Wolfe (1999). Switching from oral to topical NSAIDs may reduce gastrointestinal complications. In 2019, Dr. Cyrus Cooper studied the adverse effects of NSAIDs, writing:

> NSAIDs have been associated with wide-ranging adverse events affecting the gastrointestinal, cardiovascular, and renal systems. Gastrointestinal toxicity is found with all NSAIDs, which may be of particular concern when treating older patients with osteoarthritis, and gastric adverse events may be reduced by taking a concomitant gastroprotective agent, although intestinal adverse events are not ameliorated. Cardiovascular toxicity is associated with all NSAIDs to some extent and the degree of risk appears to be pharmacotherapy specific. An increased risk of acute myocardial infarction and heart failure is observed with all NSAIDs, while an elevated risk of hemorrhagic stroke appears to be restricted to the use of diclofenac and meloxicam. All NSAIDs have the potential to induce acute kidney injury, and patients with osteoarthritis with co-morbid conditions including hypertension, heart failure, and diabetes mellitus are at increased risk. Osteoarthritis is associated with **excess mortality,** which may be explained by reduced levels of physical activity owing to lower limb pain, presence of comorbid conditions, **and the adverse effects of anti-osteoarthritis medications especially NSAIDs.** (30-37)

Other Treatments for Osteoarthritis

Aging-associated nutritional deficiencies may be addressed with dietary and herbal supplements such as glucosamine, chondroitin, MSM (methylsulfonylmethane), hydrolyzed collagen, and hyaluronic acid. In 2021, Dr. Alessandro Colletti writes:

> Among the most used nutraceuticals in OA [osteoarthritis], chondroitin sulphate, glucosamine sulphate, collagen, hyaluronic acid, and methylsulfonylmethane were shown to be impressive in the improvement of clinical symptoms and in decreasing

inflammatory indices in subjects with OA. (38)

Botanicals with anti-inflammatory effects are safer than NSAIDs without the adverse effects on the gastrointestinal tract. These include Curcumin, Boswellia, Chinese Skullcap, Ginger, Artemisinin, Ginseng, etc. Other treatments include PEMF (pulsed electromagnetic field) therapy and Low-Level Laser therapy, both showing considerable benefits for osteoarthritis. A recent mainstream orthopedic treatment is Platelet-Rich Plasma (PRP) injections which stimulate the healing. (39-51)

My Clinical Experience with Topical Estrogen Oil for Osteoarthritis (OA)

A 56-year-old female came to my office with a chief complaint of knee arthritis made worse by standing long hours at her job. She was instructed to apply a topical estrogen/progesterone oil to both knees twice a day. 6 weeks later she reports considerable improvement, and 12 weeks later her knee pain has resolved. As of this writing, we have used a topical HRT formula for post-menopausal patients with OA, finding all patients have dramatically improved with the resolution of joint pain. The formula is Biest 5 mg and Progesterone 50 mg in one gram of olive oil. The patient is instructed to apply half gram of oil twice a day to the affected joint, taking a few minutes to thoroughly rub the oil in. The topical oil should have good penetration into the joint, inducing cartilage regeneration. In 2023, Dr. Huiwen Pang writes:

> Estrogen supplementation **has been shown to be effective** at ameliorating IVD [intervertebral disc] degeneration and OA progression, indicating its potential use as a therapeutic agent for people with LBP [low back pain] and OA pain. (52)

Conclusion: There is now overwhelming evidence from preclinical animal studies and human clinical trials that estrogen deficiency induces cartilage damage and osteoarthritis.

Postmenopausal HRT is the prevention and treatment of menopausal arthritis. Estrogen relieves post-menopausal joint pain and regenerates cartilage. Because of the fear of estrogen, women with menopausal arthritis are offered NSAIDs, steroid injections, and finally, joint replacement. Although lucrative for the drug and medical industries, the failure to offer menopausal HRT is yet another tragic error of modern medicine. (53-54)

♦ **References for Chapter 12 Bioidentical Hormones Prevent Arthritis**

1) Herrero-Beaumont, Gabriel, et al. "Primary Osteoarthritis No Longer Primary: Three Subsets with Distinct Etiological, Clinical, And Therapeutic Characteristics." Seminars in arthritis and rheumatism. Vol. 39. No. 2. WB Saunders, 2009.

2) Roman-Blas, Jorge A., et al. "Osteoarthritis Associated with Estrogen Deficiency." Arthritis research & therapy 11 (2009): 1-14.

3) Felson, David T., and Michael C. Nevitt. "The Effects of Estrogen on Osteoarthritis." Current Opinion In Rheumatology 10.3 (1998): 269-272.

4) Nevitt, Michael C., et al. "Association of Estrogen Replacement Therapy with The Risk Of Osteoarthritis Of The Hip In Elderly White Women." Archives Of Internal Medicine 156.18 (1996): 2073-2080.

5) Yu, Wei, et al. "In Memory of Prof. Harry Genant, MD/PhD." Journal of Orthopaedic Translation 27 (2021): A1.

6) Zhang, Yuqing, et al. "Estrogen Replacement Therapy and Worsening of Radiographic Knee Osteoarthritis: The Framingham Study." Arthritis & Rheumatism: Official Journal of The American College of Rheumatology 41.10 (1998): 1867-1873.

7) Cirillo, Dominic J., et al. "Effect of Hormone Therapy on Risk of Hip and Knee Joint Replacement in the Women's Health Initiative." Arthritis & Rheumatism: Official Journal of the American College of Rheumatology 54.10 (2006): 3194-3204.

8) Cauley, Jane A., et al. "Effects of Estrogen Plus Progestin on Risk of Fracture and Bone Mineral Density: The Women's Health Initiative Randomized Trial." JAMA 290.13 (2003): 1729-1738.

9) Riancho, Jose A., et al. "Common Variations in Estrogen-Related Genes Are Associated with Severe Large-Joint Osteoarthritis: A Multicenter Genetic And Functional Study." Osteoarthritis and Cartilage 18.7 (2010): 927-933.

10) Dai, Xiaoyu, et al. "Association of Single Nucleotide Polymorphisms in Estrogen Receptor Alpha Gene with Susceptibility to Knee Osteoarthritis: A Case-Control Study in a Chinese Han Population." BioMed Research International 2014.1 (2014): 151457.

11) Martín-Millán, Marta, and Santos Castañeda. "Estrogens, Osteoarthritis and Inflammation." Joint Bone Spine 80.4 (2013): 368-373.

12) Tanko, Laszlo B., et al. "An Update Review of Cellular Mechanisms Conferring the Indirect and Direct Effects Of Estrogen On Articular Cartilage." Climacteric 11.1 (2008): 4-16.

13) Ham, Kimberley D., et al. "Effects of Long-Term Estrogen Replacement Therapy on Osteoarthritis Severity in Cynomolgus Monkeys." Arthritis & Rheumatism: Official Journal of the American College of Rheumatology 46.7 (2002): 1956-1964.

14) Oestergaard, Svetlana, et al. "Effects of Ovariectomy and Estrogen Therapy on Type II Collagen Degradation and Structural Integrity Of Articular Cartilage In Rats: Implications Of The Time Of Initiation." Arthritis and Rheumatism. 54.8 (2006): 2441-2451.

15) Claassen, Horst, et al. "The effect of Estrogens and Dietary Calcium Deficiency On The Extracellular Matrix Of Articular Cartilage In Göttingen Miniature Pigs." Annals of Anatomy-Anatomischer Anzeiger 184.2 (2002): 141-148.

16) Guofeng, Dai, et al. "The Relationship of The Expression of Estrogen Receptor In Cartilage Cell And Osteoarthritis Induced By Bilateral Ovariectomy In Guinea Pig." Journal of Huazhong University of Science and Technology [Medical Sciences] 25 (2005): 683-686.

17) Turner, A. Simon, et al. "Biochemical Effects of Estrogen on Articular Cartilage In Ovariectomized Sheep." Osteoarthritis and Cartilage 5.1 (1997): 63-69.

18) Cake, Martin A., et al. "Ovariectomy Alters the Structural and Biomechanical Properties of Ovine Femoro-Tibial Articular Cartilage and Increases Cartilage Loss." Osteoarthritis and Cartilage 13.12 (2005): 1066-1075.

19) Sniekers, Yvonne H., et al. "Animal Models for Osteoarthritis: The Effect Of Ovariectomy And Estrogen Treatment—A Systematic Approach." Osteoarthritis and Cartilage 16.5 (2008): 533-541

20) Gilmer, Gabrielle, et al. "Uncovering The "Riddle of Femininity" In Osteoarthritis: A Systematic Review and Meta-Analysis of Menopausal Animal Models and Mathematical Modeling of Estrogen Treatment." Osteoarthritis And Cartilage 31.4 (2023): 447-457.

21) Camejo, Natalia, et al. "Arthralgia and Myalgia Associated with Aromatase Inhibitors: Frequency And Characterization In Real-Life Patients." Ecancer Medical Science 18 (2024).

22) Tenti, Sara, et al. "Aromatase Inhibitors—induced musculoskeletal disorders: current knowledge on clinical and molecular aspects." International Journal of Molecular Sciences 21.16 (2020): 5625.

23) Kim, Sara, Nan Chen, and Pankti Reid. "Current and future advances in practice: aromatase inhibitor—induced arthralgia." Rheumatology Advances in Practice 8.2 (2024): rkae024.

24) Chien, Hsu-Chih, et al. "Aromatase inhibitors and risk of arthritis and carpal tunnel syndrome among Taiwanese women with breast cancer: a nationwide claims data analysis." Journal of Clinical Medicine 9.2 (2020): 566.

25) Grigorian, Nelly, and Steven J. Baumrucker. "Aromatase inhibitor—associated musculoskeletal pain: An overview of pathophysiology and treatment modalities." SAGE Open Medicine 10 (2022): 20503121221078722.

26) Baba, Ozge, Hakan Kisaoglu, and Mukaddes Kalyoncu. "Letrozole-induced inflammatory arthritis and tendinopathy in pediatric rheumatology setting." International Journal of Rheumatic Diseases 26.11 (2023): 2314-2316.

27) Gaudio, Agostino, et al. "Therapeutic options for the management of aromatase inhibitor-associated bone loss." Endocrine, Metabolic & Immune Disorders-Drug Targets (Formerly Current Drug Targets-Immune, Endocrine & Metabolic Disorders) 22.3 (2022): 259-273.

28) Wang, Tao, et al. "Prevalence and correlates of joint pain among Chinese breast cancer survivors receiving aromatase inhibitor treatment." Supportive Care in Cancer 30.11 (2022): 9279-9288.

29) Nguyen, Uyen-sa, et al. "Sex Difference In OA: Should We Blame Estrogen?" European Journal of Rheumatology (2023).

30) Cooper, Cyrus, et al. "Safety of Oral Non-Selective Non-Steroidal Anti-Inflammatory Drugs In Osteoarthritis: What Does The Literature Say?" Drugs & Aging 36.Suppl 1 (2019): 15-24.

31) Wolfe, M. Michael, David R. Lichtenstein, and Gurkirpal Singh. "Gastrointestinal Toxicity of Nonsteroidal Antiinflammatory Drugs." New England Journal of Medicine 340.24 (1999): 1888-1899.

32) Rosen, Zach. 'Generally safe' NSAIDs? American Family Physician 63.4 (2001): 637.

33) Cryer, Byron. "Management of NSAID Associated Upper Gastrointestinal Problems." Journal of Managed Care Pharmacy 11.2 Supp A (2005): S2-S9.

34) Kompel, Andrew J., et al. "Intra-articular Corticosteroid Injections In The Hip And Knee: Perhaps Not As Safe As We Thought?" Radiology 293.3 (2019): 656-663.

35) Wijn, Stan RW, et al. "Intra-Articular Corticosteroid Injections Increase the Risk of Requiring Knee Arthroplasty: A Multicentre Longitudinal Observational Study Using Data from the Osteoarthritis Initiative." The Bone & Joint Journal 102.5 (2020): 586-592.

36) Hauser, Ross A. "The Acceleration of Articular Cartilage Degeneration In Osteoarthritis By Nonsteroidal Anti-Inflammatory Drugs." Journal of Prolotherapy 2.1 (2010): 305-322.

37) Wang, Yuhui, et al. "Relative Safety and Efficacy of Topical And Oral NSAIDs In The Treatment Of Osteoarthritis: A Systematic Review And Meta-Analysis." Medicine 101.36 (2022): e30354.

38) Colletti, Alessandro, and Arrigo FG Cicero. "Nutraceutical approach to chronic osteoarthritis: from molecular research to clinical evidence." International Journal of Molecular Sciences 22.23 (2021): 12920.

39) Fini, M., et al. "Pulsed Electromagnetic Fields Reduce Knee Osteoarthritic Lesion Progression In The Aged Dunkin Hartley Guinea Pig." Journal of Orthopaedic Research 23.4 (2005): 899-908.

40) Fini, Milena, et al. "Effect of pulsed Electromagnetic Field Stimulation on Knee Cartilage, Subchondral and Epyphiseal Trabecular Bone Of Aged Dunkin Hartley Guinea Pigs." Biomedicine & pharmacotherapy 62.10 (2008): 709-715.

41) Cadossi, Ruggero, et al. "Pulsed Electromagnetic Field Stimulation of Bone Healing and Joint Preservation: Cellular Mechanisms of Skeletal Response." Journal of the American Academy of Orthopaedic Surgeons. Global Research & Reviews 4.5 (2020): e1900155.

42) Hegedűs, Béla, et al. "The Effect of Low-Level Laser In Knee Osteoarthritis: A Double-Blind, Randomized, Placebo-Controlled Trial." Photomedicine And Laser Surgery 27.4 (2009): 577-584.

43) Kahn, F., R. Liboro, and F. Saraga. "Laser Therapy for The Treatment Of Arthritic Knees: A Clinical Study." Mechanisms for Low-Light Therapy V. Vol. 7552. SPIE, 2010.

44) Cho, Hyung-Jin, et al. "Effect of Low-Level Laser Therapy on Osteoarthropathy In Rabbit." in vivo 18.5 (2004): 585-592.

45) Kim, Hye In, et al. "Clinical Effects of Korean Red Ginseng In Postmenopausal Women With Hand Osteoarthritis: A Double-Blind, Randomized Controlled Trial." Frontiers in Pharmacology 12 (2021): 745568.

46) Chen, Jincai, et al. "Protective Effects of Ginseng And Ginsenosides In The Development Of Osteoarthritis." Experimental and Therapeutic Medicine 26.4 (2023): 1-12.

47) Vaishya, Raju, et al. "Current Status of Top 10 Nutraceuticals Used for Knee Osteoarthritis In India." Journal Of Clinical Orthopaedics and Trauma 9.4 (2018): 338-348.

48) Sethi, Vidhu, et al. "Potential Complementary And/Or Synergistic Effects of Curcumin and Boswellic Acids for Management Of Osteoarthritis." Therapeutic Advances in Musculoskeletal Disease 14 (2022): 1759720X221124545.

49) Marana, Rosana Rodrigues, et al. "Omega 3 Polyunsaturated Fatty Acids: Potential Anti-Inflammatory Effect in A Model Of Ovariectomy And Temporomandibular Joint Arthritis Induction In Rats." Archives of Oral Biology 134 (2022): 105340.

50) Zhong, Gang, et al. "Artemisinin Ameliorates Osteoarthritis By Inhibiting The Wnt/B-Catenin Signaling Pathway." Cellular Physiology and Biochemistry 51.6 (2019): 2575-2590.

51) Mende, Emily, et al. "A Comprehensive Summary of the Meta-Analyses and Systematic Reviews on Platelet-Rich Plasma Therapies for Knee Osteoarthritis." Military Medicine (2024): usae022.

52) Pang, Huiwen, et al. "Low Back Pain and Osteoarthritis Pain: A Perspective of Estrogen." Bone Research 11.1 (2023): 42.

53) Karsdal, M. A., et al. "The Pathogenesis of Osteoarthritis Involves Bone, Cartilage and Synovial Inflammation: May Estrogen Be A Magic Bullet?" Menopause International 18.4 (2012): 139-146.

54) Mei, Yixue, et al. "Roles of Hormone Replacement Therapy and Menopause on Osteoarthritis and Cardiovascular Disease Outcomes: A Narrative Review." Frontiers in rehabilitation sciences 3 (2022): 825147.

Chapter 13

Safety and Adverse Effects of Natural Progesterone Part One

NANCY, A 27-YEAR-OLD FEMALE NEWLY married housewife, arrived in my office with a chief complaint of PMS (premenstrual syndrome). Her PMS symptoms included anxiety, depression, anger, irritability, bloating, breast tenderness, headache, and water retention, all reported during the last week of the menstrual cycle. For her PMS symptoms, Nancy was started on natural progesterone taken on a schedule, days 12-26 of the menstrual cycle. About a week after the office visit, I received this email from Nancy.

> Hello Dr Dach,
> I mentioned to my mom this evening that I would start to take progesterone pills next month for a portion of my cycle. She said that a few years ago, she used a progesterone cream. She said she could not remember the reason why her doctor put her on a cream instead of a pill so she wanted me to ask about the health risks that go along with taking a progesterone pill. I did not think to ask Dr. Dach when I spoke with him today about the side effects of using progesterone (pill or cream). Are there any side effects that I should be concerned about or aware of? Is there any difference in health risks between the progesterone pill or cream? Thank you all for the time that you spend with me answering questions and planning my treatment! I really appreciate it!
> Sincerely, Nancy

My Reply to Nancy:

Hi Nancy,
Progesterone is the natural hormone made by the ovary after ovulation, so it is very safe with no adverse effects. Excess dosage however, can cause drowsiness, which is helpful for treating insomnia if taken before bedtime to get a good night's sleep. For the cycling female, the usual dosage is 100 mg capsule twice a day with food for days 12-26 of the cycle. This may be increased to 200 mg of micronized oral progesterone at bedtime and 100 mg in the AM with breakfast or lunch as a treatment for PMS (Premenstrual Syndrome). If the morning progesterone dosage causes drowsiness, then this is omitted and instead both capsules are taken at night around bed time. Regards from Jeffrey Dach MD

Progesterone is NOT a Progestin

Be careful not to confuse natural progesterone with the synthetic progestins made in the laboratory by modifying the progesterone chemical structure. The commonly used progestin, MPA (medroxyprogesterone) has an added methyl group at position 6 and an added acetoxy group at position 17. This new chemical structure can now be patented, a prerequisite for protecting drug company profits. Occasionally, a patient or even their doctor mistakenly confuses progesterone with a progestin, thinking progesterone is a progestin. Natural progesterone, also called bioidentical progesterone is not a synthetic progestin and the two should not be confused. In 2018, Dr. Paul Piette from Brussels, Belgium discusses the difference between natural progesterone and synthetic progestins, writing:

> There is a **lot of confusion** in the literature about natural progesterone, progestagens, progestogens and progestins. **The term progesterone should only be used for the natural hormone produced by the ovaries or included in a registered drug, qualified as 'body identical' or 'bioidentical'** ...The term 'progestin' is used for synthetic compounds, designed to target the progesterone

receptors, and belonging to different classes of molecules with sometimes very different pharmacological properties and modes of action. (1-3)

Confusion in the Medical Literature

There are many examples of this confusion in the medical literature which confuses progesterone with progestins. One illustrative example in 2021 by Dr. David C. Slawson in American Family Physician. His article is entitled, "Twenty-Year Follow-Up of the Women's Health Initiative Trials: Lower Breast Cancer Mortality with Estrogen Alone, No Difference with Estrogen Plus **Progesterone**". This is an error. As you know, the 2002, first arm WHI study **did not use progesterone**. Rather, medroxyprogesterone (MPA) a carcinogenic synthetic version of progesterone was used. This confusion was created by drug industry propaganda as part of the drug industry's war on natural substances. Progesterone is a natural substance and therefore competes with synthetic progestins marketed by the drug industry. Remember, the drug industry business model is to obtain a patent for chemically altering a natural substance. This patent gives the drug company a monopoly which protects profits. A natural substance without this chemical alteration cannot be patented, making the natural substance (or repurposed drug) a competitor and economic enemy of the drug industry. As mentioned above, the drug industry alters the chemical structure of progesterone. This new chemical structure is called a progestin, a synthetic hormone known to cause cancer and heart disease, and other adverse effects not shared by natural progesterone. I do not recommend progestins in my office except for dysfunctional uterine bleeding where synthetic progestins have a justifiable use, as discussed below. For all other uses, I recommend safe natural progesterone. For clarity, this book uses the word progesterone only for natural human bioidentical progesterone and reserves the word progestin for all other synthetic, chemical alterations of progesterone. (4)

The Criminal Drug Industry Has Captured Conventional Medicine

The drug industry has captured the medical literature, the medical societies, the medical journals the mass media, the House and Senate of Congress, and all the regulatory agencies. The drug industry has paid out more than 26 billion dollars in civil and criminal penalties since 2001, thus proving the drug industry is a criminal racketeering organization. Nothing happens in conventional medicine that is not controlled by the drug industry. The medical literature and the mass media is completely captured by the drug industry. When we find the two terms, natural progesterone and synthetic progestins blurred together and confused throughout the medical literature, this is no accident. It is intentional malfeasance by the drug industry. How do we know this? Follow the money. Natural substances are the economic competitor and enemy of the patented drug model. Patented synthetic progestins are competing with natural progesterone. By convincing all the doctors these two drugs are the same thing, then doctors will unknowingly prescribe their synthetic patented progestins, rather than the superior natural progesterone, and the drug industry wins a great economic windfall. Likewise, the 2002 first WHI study created another windfall by creating fear of estrogen. Menopausal women and doctors stopped using menopausal hormone replacement, resulting in a massive windfall for the drug and medical industries. Instead of hormone replacement, women are offered a list of drugs to treat menopausal symptoms and estrogen deficiency diseases. What are all these drugs? To name a few: statin drugs for menopausal coronary artery disease, SSRI antidepressants for menopausal depression, bisphosphonates for menopausal osteoporosis, NSAID drugs for menopausal arthritis, dementia drugs, etc. (5-13)

The MPA Mouse Model of Breast Cancer

The progestin, medroxyprogesterone (MPA) is still being prescribed by mainstream OB/

GYNE doctors and primary care doctors for menopausal hormone replacement. MPA and other progestins can be found in birth control pills. MPA is the progestin used by 74 million women globally for contraception, the same synthetic hormone used in the 2002 first arm Women's Health Initiative Study halted early because of increased breast cancer (HR=1.26). This is not surprising because MPA is routinely used in research laboratories to induce breast cancer in mice. This is called the MPA Mouse Model of Breast Cancer by Claudia Lanari (2009). Would you use a carcinogenic synthetic hormone that is routinely used to induce breast cancer in mice? I would not recommend it. Birth control pills contain carcinogenic progestins which cause breast cancer, a fact conveniently ignored by the medical system. In 2024, Dr. Aline Zurcher from the University of Bern, Switzerland reviewed the medical literature asking the question: does MPA increase the risk for breast cancer? Dr. Zurcher writes:

> Regarding the progestin [MPA]-only pill, there was evidence to suggest that they may slightly increase the risk of breast cancer, at least with current or recent use. The overall risk, however, was still considered relatively low…The same applies to hormone replacement therapy (HRT) with medroxyprogesterone acetate (MPA). The association between HRT and breast cancer has been a topic of extensive research and debate. Recent research has provided a more nuanced understanding of this relationship. Several studies indicated that the **long-term use of combined estrogen-progestin therapy (including MPA) seemed to be associated with a slight increase in the risk of breast cancer,** while the use of estrogen alone significantly reduced the breast cancer incidence. Nonetheless, the absolute breast cancer risk remained small. (14-18)

French Cohort Study

One of the many studies showing progestins increase breast cancer is the 2008 E3N French Cohort study by Dr. Agnes Fournier finding that adding a synthetic progestin to estrogen (estradiol) is the most significant factor that increases breast cancer risk. For example, adding medroxyprogesterone resulted in a 1.48-fold increase. However, Dr. Fournier found no increased risk for breast cancer with the combination of estradiol plus natural progesterone. If the added progestin is norethisterone, then this gives a 2.11-fold increase in breast cancer. A different study in 2008 by Dr. Heli Lyytinen from Finland confirmed Dr. Fournier's findings of a doubling (RR=1.94) of breast cancer risk for women using the synthetic progestin, norethisterone, writing:

> In Finland, the most common progestogen as a part of EPT [estrogen progestin therapy] is norethisterone acetate (NETA), which can be given both orally and transdermally. The use of a "low" dose NETA-regimen was associated with an increased risk for breast cancer already in 3 years of use (**RR=1.94**; 1.39–2.70). (19-20)

Progesterone for PMS

Let us now turn to the discussion of natural progesterone for premenstrual syndrome (PMS). In 1953, Katharina Dalton coined the term, PMS, premenstrual syndrome, and she was the first to popularize progesterone as an effective treatment for PMS. In 1953, Dr. Dalton established a PMS clinic for over 40 years using injectable progesterone at first, and later micronized oral progesterone as it became available. (21-29)

In 1985, Dr. Lorraine Dennerstein, Professor in the Department of Psychiatry at the University of Melbourne, Australia studied the use of (300mg/d) oral micronized progesterone for treatment of PMS. The progesterone was given on a schedule for 10 days starting three days after ovulation, which was determined by urine estrogen and pregnanediol concentrations (pregnanediol is a metabolite of progesterone). This is roughly days 12-26 of the menstrual cycle, with day one defined as the first day of menstrual bleeding. Dr. Lorraine Dennerstein found beneficial effects of the oral progesterone which alleviates many of the PMS symptoms of anxiety, stress, depression, hot flashes, swelling, and water retention. Dr. Lorraine Dennerstein writes:

> Women were instructed to take one 100 mg capsule in the morning and two 100 mg capsules at night as there had been reports of drowsiness of short duration. Treatment was prescribed for 10 days of each menstrual cycle starting roughly three days after ovulation. In each cycle ovulation was confirmed by determinations of urinary 24 hour pregnanediol and total oestrogen concentrations...**Our findings confirm descriptive reports of beneficial effects of progesterone on the symptoms of premenstrual tension.** Improvements were attained both in mood symptoms such as anxiety, depression, and stress and in the physical complaints of swelling and hot flushes. Although not all variables reached a significant level of improvement, the direction of change for premenstrual complaints, with the exception of arousal, was always in favour of progesterone treatment...Taken together the analyses also show the general positive effects of treatment. **There was a trend to general improvement in almost all the physical and psychological variables over the four months of treatment,** an improvement even more appreciable for the months of progesterone treatment alone...**This study showed that an oral formulation of micronised progesterone was effective in alleviating many premenstrual complaints including those of anxiety, stress, depression, hot flushes, swelling, and water retention**. (30)

Vitex Consumption by Wild Chimpanzees in Gombe Park

Vitex, also called ChasteBerry, is a commonly used over-the-counter (OTC) herbal remedy for PMS and cyclic mastalgia, breast pain and tenderness caused by high prolactin levels. Dr. Tori Hudson considers Chaste Berry the single most important plant extract for the treatment of premenstrual syndrome. Vitex works on the hypothalamus-hypophysis axis (HPA) to increase secretion of luteinizing hormone which induces ovulation and production of progesterone. In 2008, Dr. Melissa Emery Thompson observed wild chimpanzees in Africa consuming Vitex fruit, an herbal remedy for PMS commonly used by women. Dr. Thompson did further studies measuring urinary hormone concentrations in the chimpanzees finding "dramatic and abrupt elevation" of progesterone during the time of intense Vitex fruit consumption, writing:

> Chimpanzees in Gombe National Park consume fruits of Vitex fischeri during a short annual fruiting season. This fruit species is a member of a genus widely studied for phytoestrogen composition and varied physiological effects. One particularly well-studied species, V. agnus-castus, is noted for its documented effects on female reproductive function, evidenced in increased progesterone levels and consequent regulation of luteal function. We examined reproductive hormone levels in both male and female chimpanzees during a 6-week period of intense V. fischeri consumption. V. fischeri consumption was associated with an **abrupt and dramatic increase in urinary progesterone levels of female chimpanzees to levels far exceeding the normal range of variation**. Female estrogen levels were not significantly impacted, nor were male testosterone levels. These are some of the first data indicating that phytochemicals in the natural diet of a primate can have significant impacts on the endocrine system...(31-34)

Vitex's mechanism of action is on the hypothalamic pituitary axis (HPA), where the herb stimulates production of LH and FSH by the pituitary which induces ovulation and subsequent progesterone production by the corpus luteum. Vitex has been found to reduce prolactin levels and is useful for premenstrual mastalgia symptoms (breast pain) thought to be caused by elevated prolactin levels during the premenstrual phase of the menstrual cycle. Vitex has anti-inflammatory activity and is a COX (Cyclo-Oxygenase) Inhibitor. Vitex also has anti-cancer activity in numerous preclinical studies. **Note:** LH and FSH are pituitary hormones that control ovulation and progesterone production. (35-43)

Progesterone for Peri-Menopause

In 2011, Dr. Jerilynn Prior suggested progesterone would be useful in the treatment of the peri-menopausal transition. This is the time of fluctuating hormone levels, skipped menstrual periods, and mood disorders. This transition period may last 3-6 months, finally ending with full menopause with declining estrogen levels and vasomotor symptoms of hot flashes and night sweats. Notice Dr. Prior gives higher doses of progesterone, 300 mg, at bedtime, writing:

> Because P4 [progesterone] and E2 [estradiol] complement/counterbalance each other's tissue effects, oral micronized P4 (OMP4 300 mg at bedtime) is a physiological therapy for treatment-seeking, symptomatic perimenopausal women. Given cyclically (cycle day 14-27, or 14 on/off) in menstruating midlife women, OMP4 [oral micronized progesterone] decreases cyclic VMS [vasomotor symptoms], improves sleep and premenstrual mastalgia [breast pain]. (44)

Progestins for DUB

Although the use of progestins such as MPA is not recommended for menopausal hormone replacement, progestins such as MPA, levo-norgestrel, and norethisterone are commonly prescribed for the short-term control of dysfunctional uterine bleeding (DUB). In my opinion, the use of synthetic progestins is justified for this purpose because the patient may avoid hysterectomy (surgical removal of the uterus), estimated to occur in about 30 percent of patients with DUB. In Europe, other progestins such as norethisterone may be used instead of MPA whose carcinogenicity is overlooked when dealing with uterine bleeding. (45-51)

Progesterone as Breast Cancer Preventive Agent

A commonly used animal model of breast cancer is DMBA, a carcinogenic chemical used to induce breast cancer in mice. What if we gave progesterone to the mice before trying to induce breast cancer with DMBA. Would the progesterone be protective? In 1982, Dr. Anne G Jabara found pretreatment of the mice with progesterone inhibited DMBA-induced breast carcinogenesis. This study nicely illustrates the protective effect of progesterone. Dr. Anne Jabara concluded, "Progesterone acts directly on the mammary gland to inhibit carcinogenesis." (52)

In 1998, Dr. Bent Formby and Teresa S. Wiley studied two breast cancer cell lines in-vitro showing progesterone "exhibited a strong anti-proliferative effect" and induced apoptosis in the cancer cell line expressing the progesterone receptor. (53)

In 2007, Drs. Gianluigi Ferretti and Joseph Jerry studied the protective effects of progesterone, citing the work of Dr. Rajkumar Lakshmanaswamy, Professor of Biomedical Sciences at Texas Tech in El Paso, who showed a protective effect of combining estrogen and progesterone in animal models of breast cancer, writing:

> Rajkumar [Lakshmanaswamy] and coworkers have now demonstrated that these hormones [estrogen and progesterone] protect mice from mammary tumors initiated by a spectrum of oncogenic alterations that are common in breast cancers. Although differences between rodent models and humans remain, the **results reveal that exogenous estrogen and progesterone potently inhibit tumorigenesis through multiple pathways and establish a foundation for strategies to prevent breast cancer.** (54-56)

Human Breast Biopsy Study
Progesterone vs. MPA

As mentioned above, in both human and animal studies, progesterone counteracts the proliferative effect of estrogen on breast tissue. However, the synthetic progestin MPA (medroxyprogesterone) does the opposite, markedly increasing estrogen-induced breast proliferation. This is graphically demonstrated in 1999 by Dr. Lorne Hofseth from Michigan State University, Associate Dean for Research and Professor, College of Pharmacy at the University of South Carolina who studied histology slides from breast biopsies in three groups of postmenopausal women. Group one was the control group receiving placebo. Group two was treated with estrogen alone, and group three was treated with estrogen combined with MPA (medroxyprogesterone). In group three, treated with a combination of estrogen and MPA, Dr. Lorne Hofseth found the TDLUs (terminal ductal lobular units) demonstrated worsening proliferation compared to estrogen alone (group two). The addition of the synthetic progestin, MPA, made the estrogen more proliferative, meaning a greater risk for carcinogenesis. **Note:** the type of estrogen was not specified in the abstract, but I assume it was Premarin (CEE) since it was done in the U.S. in 1999. (57)

Human Breast Biopsy Study of Estrogen + Progesterone vs. Estrogen + MPA

In 2011, Dr. Daniel Murkes from Sweden ran a prospective randomized controlled trial of the proliferation effect of progesterone compared to MPA. Dr. Murkes replicated the hormones used in the WHI study, using CEE (0.625mg oral equine estrogen) and MPA (5mg oral medroxyprogesterone) in 77 post-menopausal women for 2 months. This first group of women was then compared to a second group of women given bioidentical estradiol (1.5 mg topical) combined with natural progesterone (200mg oral) for 14 days each month, for two months. The proliferative effect of the two different hormone treatments was studied by obtaining breast cells by core needle biopsy before, and two months after hormone treatment. The core breast biopsy cells were then immune-stained for KI-67 and the antiapoptotic protein Bcl-2. **Note:** Ki-67 is a commonly used marker for cell proliferation. The CEE/MPA group had greater proliferation than the estradiol/natural progesterone group. Dr. Daniel Murkes writes:

> Seventy-seven women were assigned randomly to receive sequential HT [hormone therapy] with two 28-day cycles of either oral 0.625 mg conjugated equine estrogens or 2.5 g 0.06% percutaneous E2 [estradiol] gel (1.5 mg E2), daily, with the addition of respectively 5 mg of oral medroxyprogesterone acetate (MPA) or 200 mg of oral P, daily, for the last 14 days of each cycle. Core biopsy breast tissue samples for the CEE+MPA treated group revealed increased numbers of cells with nuclei staining for KI-67 indicating greater proliferation when compared to the combination of estrogen [estradiol] with natural progesterone. Note: greater numbers of nuclei staining for KI-67 means greater proliferation. Again, natural progesterone is found to be safer than the synthetic progestin, MPA. (58)

Dr. Daniel Murkes compares his study (above) with the French E3N cohort study. Remember, the French Cohort study found estrogen (topical estradiol) combined with progesterone preferable because there was no increased risk of breast cancer for the estradiol/progesterone combination. On the other hand, in the French Cohort study, the estradiol/MPA combination **did** increase the risk for breast cancer. Dr. Daniel Murkes suggested this could be explained by his finding of higher breast cell proliferative activity for the Premarin (CEE)/MPA combination compared to the less proliferative activity seen with the E2 (estradiol) /natural progesterone combination, writing:

> In the French E3N cohort there was an absence of breast cancer risk increase for women taking estrogen [topical estradiol, E2]

in combination with natural P [progesterone] for at least 5 years of treatment. This is in line with the indication in the current study of a **higher proliferative activity in the breast imposed by oral conjugated equine estrogens [CEE]–MPA versus percutaneous E2 [estradiol]–micronized P [progesterone] orally,** maintaining an E2 dose of 1.5 mg daily, which is needed by many women at least in an initial phase of postmenopausal symptoms. (59)

In Defense of Progesterone

In 2017, Dr. Alan Lieberman reviewed the medical literature to examine the benefits and safety of natural progesterone as compared to synthetic progestins such as MPA (medroxy-progesterone). Dr. Lieberman found 3 studies involving 87,000 postmenopausal women in which the use of natural progesterone carried **significantly less risk of breast cancer than synthetic progestins**. In addition, progesterone is protective for many other cancers such as endometrial, colon, ovarian, melanoma, mesothelioma, and prostate cancer. Progesterone has benefits in preventing cardiovascular disease, and neuroprotective effects in stroke and traumatic brain injury patients. Dr. Lieberman writes:

> Use of progesterone has been linked to lower rates of uterine and colon cancers and may also be useful in treating other cancers such as ovarian, melanoma, mesothelioma, and prostate. Progesterone may also be helpful in preventing cardiovascular disease and preventing and treating neurodegenerative conditions such a stroke and traumatic brain injury...Physicians should have no hesitation prescribing natural progesterone, as the evidence is clear that progesterone does not cause breast cancer. **Indeed, progesterone is protective and preventative of breast cancer.** (60)

Progesterone Acts as Proliferative Brake

Given Dr. Lieberman's claim that progesterone is anti-proliferative and protective of breast cancer, one might then ask what is the mechanism of action? In 2015, Dr. Hisham Mohammed was thinking the same thing. Dr Mohammed did a series of studies using in vitro and in vivo mouse xenografts, finding activation of the progesterone receptor (PR), with exogenous progesterone, results in a "robust association between PR and the ER-alpha complex" which then acts as a "proliferative brake" and blocks breast tumor growth in mouse xenografts. Remember, ER-alpha is the proliferative receptor, while ER-beta is the protective, tumor suppressor receptor. In this scenario, progesterone receptor protein binds to ER-alpha and acts as a brake, halting ER-alpha-induced proliferation. This results in a good clinical outcome. Dr. Hisham Mohammed writes:

> PR [progesterone receptor] is a critical determinant of ER-alpha [estrogen receptor alpha] function due to crosstalk between PR and ER-alpha. In this scenario, under estrogenic conditions, **an activated PR functions as a proliferative brake in ER-alpha positive breast tumours** by re-directing ER-alpha chromatin binding and altering the expression of target genes that **induce a switch from a proliferative to a more differentiated state**...In the presence of agonist ligands, PR associates with ER-alpha to direct ER-alpha chromatin binding events within breast cancer cells, **resulting in a unique gene expression program that is associated with good clinical outcomes.** Progesterone inhibited estrogen-mediated growth of ER-alpha positive cell line xenografts and primary ER-alpha positive breast tumor explants and had increased anti-proliferative effects when coupled with an ER-alpha antagonist. Copy number loss of PgR [progesterone receptor] is a common feature in ER-alpha positive breast cancers, explaining lower PR levels in a subset of cases. Our findings indicate that **PR functions as a molecular rheostat to control ER-alpha chromatin binding and transcriptional activity**, which has important implications for prognosis and therapeutic interventions...**the increased**

risk of breast cancer associated with progestogen-containing HRT is mainly attributed to specific synthetic progestins, in particular medroxyprogesterone acetate (MPA), which is known to also have androgenic properties. The relative risk is not significant when native progesterone is used. In ER-alpha positive breast cancers, PR is often used as a positive prognostic marker of disease outcome… progesterone treatment has been shown to be **antiproliferative** in ER-alpha positive PR positive breast cancer cell lines and progestogens [natural progesterone, not progestins] have been shown to oppose estrogen-stimulated growth of an ER-alpha positive PR postive patient-derived xenograft. In addition, **exogenous expression of PR in ER-alpha positive breast cancer cells blocks estrogen-mediated proliferation and ER-alpha transcriptional activity.**…These observations imply that **PR activation in the context of estrogen-driven, ER-alpha positive breast cancer, can have an anti-tumourigenic effect.** In support of this, PR agonists [progesterone] can exert clinical benefit in ER-alpha positive breast cancer patients that have relapsed on ER-alpha antagonists [estrogen blocking drugs]. (61)

Use of Progesterone Pre-Operatively

Interest in using progesterone preoperatively arose because of the discovery that pre-operative menstrual cycle timing can benefit the outcome after breast surgery. In other words, scheduling the breast cancer surgery during the luteal phase of the menstrual cycle yields better patient outcomes, in terms of reduced cancer recurrence and reduced mortality. **Note:** The luteal phase is when progesterone levels are the highest. For post-menopausal patients, several studies show giving the patient progesterone before breast cancer surgery reduces the recurrence rate and improves mortality post-operatively. (62-70)

Conclusion: Progesterone is safe, while the synthetic progestin, MPA is not. This is demonstrated by the studies showing the addition of natural progesterone to estrogen reduces rather than increases breast proliferation. Progesterone acts as a proliferative brake. MPA does the exact opposite, making estrogen more proliferative, meaning increasing the risk for carcinogenesis. This is exactly the point I made in 1977, in my reply to Dr. Steven Economou who asked me during hospital rounds: "Does estrogen cause breast cancer? My answer was not entirely correct. Rather than estrogen, the breast cancer culprit is the MPA. For more on this see Chapter 6, Bioidentical Hormone for Breast Cancer Survivors. There are other uses for progesterone, such as alleviating PMS symptoms in young women. Progesterone is also useful for the perimenopausal transition, characterized by anovulation with low progesterone levels. For the treatment of dysfunctional uterine bleeding, the use of MPA and other progestins is justified despite inherent carcinogenicity.

♦ **References for Chapter 13 Safety and Adverse Effects of Natural Progesterone Part One**

1) Piette, Paul. "The History of Natural Progesterone, The Never-Ending Story." Climacteric 21.4 (2018): 308-314.

2) Prior, Jerilynn C. "Progesterone or Progestin as Menopausal Ovarian Hormone Therapy: Recent Physiology-Based Clinical Evidence." Current Opinion in Endocrinology, Diabetes and Obesity 22.6 (2015): 495-501.

3) Prior, Jerilynn C. "Progesterone is NOT a Progestogen/Progestin—It's Estrogen's Unique Biological Partner." https://cemcor.ca/resources/progesterone-not-progestogen-progestin%E2%80%94-it%E2%80%99s-estrogen%E2%80%99s-unique-biological-partner

4) Slawson, David C. "Twenty-Year Follow-Up of the Women's Health Initiative Trials: Lower Breast Cancer Mortality with Estrogen Alone, No Difference with Estrogen Plus Progesterone." American Family Physician 103.2 (2021)

5) List of Largest Pharmaceutical Settlements on Wikipedia. https://en.wikipedia.org/wiki/List_of_largest_pharmaceutical_settlements

6) Gotzsche, Peter. Deadly Medicines and Organised Crime: How Big Pharma Has Corrupted Healthcare. CRC Press, 2019.

7) Gotzsche, Peter C. "Big Pharma Often Commits Corporate Crime, And This Must Be Stopped." BMJ 345 (2012).

8) Gotzsche, Peter C., and DrMedSci MD. "Corporate Crime in The Pharmaceutical Industry Is Common, Serious and Repetitive." BMJ 345 (2012): e8462.

9) Davis, Courtney, and John Abraham. "Is There a Cure for Corporate Crime in The Drug Industry?" Bmj 346 (2013).

10) Braillon, Alain. "Drug Industry Is Now Biggest Defrauder of US Government." BMJ 344 (2012).

11) Roehr, B. "GlaxoSmithKline is fined record $3 bn in US." BMJ (Clinical Research ed.) 345 (2012): e4568-e4568.

12) Kmietowicz, Zosia. "Eli Lilly pays record $1.4 bn for promoting off-label use of olanzapine." (2009).

13) Dyer, Owen. "Opioid lawsuits: Sackler family agree final $6 bn civil settlement with US states." (2022).

14) Zurcher, Aline, et al. "Depot Medroxyprogesterone Acetate And Breast Cancer: A Systematic Review." Archives of gynecology and obstetrics 309.4 (2024): 1175-1181.

15) Lanari, Claudia, et al. "The MPA mouse breast cancer model: evidence for a role of progesterone receptors in breast cancer." Endocrine-related cancer 16.2 (2009): 333-350.

16) Kordon, Edith, et al. "Hormone dependence of a mouse mammary tumor line induced in vivo by medroxyprogesterone acetate." Breast cancer research and treatment 17 (1990): 33-43.

17) Molinolo, A. A., et al. "Mouse mammary tumors induced by medroxyprogesterone acetate: immuno-histochemistry and hormonal receptors." Journal of the National Cancer Institute 79.6 (1987): 1341-1350.

18) Montecchia, María Fernanda, et al. "Progesterone receptor involvement in independent tumor growth in MPA-induced murine mammary adenocarcinomas." The Journal of Steroid Biochemistry and Molecular Biology 68.1-2 (1999): 11-21.

19) Lyytinen, Heli, et al. "Do the Dose or Route of Administration of Norethisterone Acetate as A Part of Hormone Therapy Play a Role in Risk of Breast Cancer: National-Wide Case-Control Study from Finland." International Journal of Cancer 127.1 (2010): 185-189.

20) Fournier, Agnès, et al. "Unequal Risks for Breast Cancer Associated with Different Hormone Replacement Therapies: Results from The E3N Cohort Study." Breast Cancer Research and Treatment 107 (2008): 103-111.

21) Interview: Katharina Dalton, MD: Progesterone and Related Topics, Int Journal of Pharmaceutical Compounding 1999

22) Greene, Raymond, and Katharina Dalton. "The Premenstrual Syndrome." British Medical Journal 1.4818 (1953): 1007.

23) O'Brien, P. M. "Helping Women With Premenstrual Syndrome." BMJ: British Medical Journal 307.6917 (1993): 1471.

24) The Prophet of PMS: New York Times. Dec 26, 2004 https://www.nytimes.com/2004/12/26/maga-zine/the-prophet-of-pms.html

25) Katharina Dorothea Dalton (1916–2004) By: Bianca Zietal Published: 2017-05-24 by Arizona State University. https://keep-dev.lib.asu.edu/items/173405

26) Tiranini, Lara, and Rossella E. Nappi. "Recent Advances in Understanding/Management Of Premenstrual Dysphoric Disorder/Premenstrual Syndrome." Faculty reviews 11 (2022).

27) Roomruangwong, Chutima, et al. "Lowered Plasma Steady-State Levels of Progesterone Combined with Declining Progesterone Levels During The Luteal Phase Predict Peri-Menstrual Syndrome And Its Major Subdomains." Frontiers in psychology 10 (2019): 488965.

28) Stefaniak, Małgorzata, et al. "Progesterone and Its Metabolites Play A Beneficial Role In Affect Regulation In The Female Brain." Pharmaceuticals 16.4 (2023): 520.

29) Stiernman, Louise, et al. "Emotion-Induced Brain Activation Across the Menstrual Cycle In Individuals With Premenstrual Dysphoric Disorder And Associations To Serum Levels Of Progesterone-Derived Neurosteroids." Translational psychiatry 13.1 (2023): 124.

30) Dennerstein, Lorraine, et al. "Progesterone and the Premenstrual Syndrome: A Double Blind Crossover Trial." Br Med J (Clin Res Ed) 290.6482 (1985): 1617-1621.

31) Emery Thompson, Melissa, et al. "Hyperprogesteronemia in Response To Vitex Fischeri Consumption In Wild Chimpanzees (Pan Troglodytes Schweinfurthii)." American Journal of Primatology: Official Journal of the American Society of Primatologists 70.11 (2008): 1064-1071.

32) Watt-Boolsen, S., A. N. Andersen, and M. Blichert-Toft. "Serum prolactin and oestradiol levels in women with cyclical mastalgia." Hormone and Metabolic Research 13.12 (1981): 700-702.

33) Nazli, K., et al. "Controlled trial of the prolactin inhibitor bromocriptine (Parlodel) in the treatment of severe cyclical mastalgia." The British Journal of Clinical Practice 43.9 (1989): 322-327.

34) Ooi, Soo Liang, et al. "Vitex agnus-castus for the treatment of cyclic mastalgia: A systematic review and meta-analysis." Journal of Women's Health 29.2 (2020): 262-278.

35) Premenstrual Syndrome; A Natural Approach by Tori Hudson, N.D. Chaste Tree (Vitex agnus castus) by Tori Hudson https://drtorihudson.com/articles/premenstrual-syndrome-a-natural-approach/

36) Farhoodi, Moghadam M., and S. M. A. Khalafi. "The Effect Vitex Agnus-Castus On Serum Concentration Of Cortisol, Progesterone And Luteinizing Hormone In Dairy Cows." (2018): 2597-2606.

37) Haerifar, Naiyereh, et al. "The Effect Of Vitex Agnus Castus Extract On The Blood Level Of Prolactin, Sex Hormones Levels, And The Histological Effects On The Endometrial Tissue In Hyperprolactinemic Women." Crescent Journal of Medical & Biological Sciences 7.4 (2020).

38) Zahid, Hina, et al. "Phytopharmacological Review On Vitex Agnus-Castus: A Potential Medicinal Plant." Chinese Herbal Medicines 8.1 (2016): 24-29.

39) Niroumand, Mina Cheraghi, et al. "Pharmacological and Therapeutic Effects of Vitex agnus-castus L.: A Review." Pharmacognosy Reviews 12.23 (2018).

40) Seidlova-Wuttke, Dana, and Wolfgang Wuttke. "The Premenstrual Syndrome, Premenstrual Mastodynia, Fibrocystic Mastopathy And Infertility Have Often Common Roots: Effects Of Extracts Of Chasteberry (Vitex Agnus Castus) As A Solution." Clinical Phytoscience 3 (2017): 1-11.

41) Carmichael, A. R. "Can Vitex Agnus Castus Be Used for The Treatment Of Mastalgia? What Is the Current Evidence?" Evidence-Based Complementary and Alternative Medicine 5.3 (2008): 247-250.

42) Verkaik, Saskia, et al. "The Treatment of Premenstrual Syndrome with Preparations Of Vitex Agnus Castus: A Systematic Review And Meta-Analysis." American Journal of Obstetrics and Gynecology 217.2 (2017): 150-166.

43) Schellenberg, Ruediger. "Treatment for the Premenstrual Syndrome with Agnus Castus Fruit Extract: Prospective, Randomised, Placebo Controlled Study." Bmj 322.7279 (2001): 134-137

44) Prior, J. C. "Progesterone for Symptomatic Perimenopause Treatment–Progesterone Politics, Physiology and Potential For Perimenopause." Facts, Views & Vision in ObGyn 3.2 (2011): 109.

45) Maness, David L., et al. "How Best to Manage Dysfunctional Uterine Bleeding." Journal of Family Practice 59.8 (2010).

46) Leal, Caio RV, et al. "Abnormal Uterine Bleeding: The Well-Known and The Hidden Face." Journal Of Endometriosis and Uterine Disorders (2024): 100071.

47) Ely, John W., et al. "Abnormal Uterine Bleeding: A Management Algorithm." The Journal of the American Board of Family Medicine 19.6 (2006): 590-602.

48) Jain, Varsha, et al. "Uterine Bleeding: How Understanding Endometrial Physiology Underpins Menstrual Health." Nature Reviews Endocrinology 18.5 (2022): 290-308.

49) Shoupe, Donna. "The Progestin Revolution: Progestins Are Arising as The Dominant Players in The Tight Interlink Between Contraceptives And Bleeding Control." Contraception and Reproductive Medicine 6.1 (2021): 3.

50) Dikke, Galina B., et al. "Experience of Treating Patients with Abnormal Uterine Bleeding Associated With Ovulatory Dysfunction." Obstetrics and Gynecology 3 (2024): 142-152.

51) Kader, Mohammad Irfan Abdul, V. Karthikeyan, and J. Sabitha. "A Comparative Study on Efficacy of Norethisterone and Medroxyprogestrone in The Management of Dysfunctional Uterine Bleeding: A Prospective Observational Study."

52) Jabara, Anne G., and P. S. Anderson. "Effects of Progesterone On Mammary Carcinogenesis When Various Doses Of DMBA Were Applied Directly To Rat Mammae." Pathology 14.3 (1982): 313-316.

53) Formby, Bent, and Teresa S. Wiley. "Progesterone Inhibits Growth And Induces Apoptosis In Breast Cancer Cells: Inverse Effects on Bcl-2 and p53." Annals of Clinical & Laboratory Science 28.6 (1998): 360-369.

54) Ferretti, Gianluigi, Alessandra Felici, and Francesco Cognetti. "The protective side of progesterone." Breast Cancer Research 9.6 (2007): 402.

55) Jerry, D. Joseph. "Roles for estrogen and progesterone in breast cancer prevention." Breast Cancer Research 9 (2007): 102.

56) Lakshmanaswamy, Rajkumar, et al. "Hormone-induced protection of mammary tumorigenesis in genetically engineered mouse models." Breast Cancer Research 9.1 (2007): R12.

57) Hofseth, Lorne J., et al. "Hormone Replacement Therapy with Estrogen or Estrogen Plus Medroxyprogesterone Acetate Is Associated with Increased Epithelial Proliferation in The Normal Postmenopausal Breast." The Journal of Clinical Endocrinology & Metabolism 84.12 (1999): 4559-4565.

58) Murkes D, Conner P, Leifland et al. Effects of Percutaneous Estradiol-Oral Progesterone Versus Oral Conjugated Equine Estrogens-Medroxyprogesterone Acetate on Breast Cell Proliferation And Bcl-2 Protein In Healthy Women. Fertil Steril. 2011 Mar 1;95(3): 1188-91.

59) Murkes D, Lalitkumar PG, Leifland et al. Percutaneous Estradiol/Oral Micronized Progesterone Has Less-Adverse Effects and Different Gene Regulations Than Oral Conjugated Equine Estrogens/ Medroxyprogesterone Acetate In The Breasts Of Healthy Women In Vivo. Gynecol Endocrinol. 2012 Oct;28 Suppl 2:12-5.

60) Lieberman, Allan, and Luke Curtis. "In Defense Of Progesterone: A Review Of The Literature." Alternative Therapies in Health & Medicine 23.7 (2017).

61) Mohammed, Hisham, et al. "Progesterone Receptor Modulates ER-alpha Action In Breast Cancer." Nature 523.7560 (2015): 313-317.

62) Hrushesky, William J. M., Avrum Z Bluming, and ScottA Gruber. "Menstrual Influence on Surgical Cure of Breast Cancer." The Lancet 335.8695 (1990): 984.

63) Ratajczak, H. V., et al. "Estrous Influence on Surgical Cure Of A Mouse Breast Cancer." The Journal of Experimental Medicine 168.1 (1988): 73-83.

64) Badwe, R. A., et al. "Timing of surgery During Menstrual Cycle and Survival Of Premenopausal Women With Operable Breast Cancer." The Lancet 337.8752 (1991): 1261-1264.

65) Badwe, R. A., et al. "Serum Progesterone at The Time Of Surgery And Survival In Women With Premenopausal Operable Breast Cancer." European Journal of Cancer 30.4 (1994): 445-448.

66) Badwe, Rajendra, et al. "Single-Injection Depot Progesterone Before Surgery and Survival in Women With Operable Breast Cancer: A Randomized Controlled Trial." (2011).

67) Pujol, Pascal, et al. "A Prospective Prognostic Study of The Hormonal Milieu At The Time Of Surgery In Premenopausal Breast Carcinoma." Cancer: Interdisciplinary International Journal of the American Cancer Society 91.10 (2001): 1854-1861.

68) Zhang, Baoning. "Prognosis of Patients with Breast Cancer Related to The Timing Of Operation During Menstrual Cycle." Chinese Journal of Cancer Research 10 (1998): 138-142.

69) Atif, Fahim, et al. "Progesterone Treatment Attenuates Glycolytic Metabolism and Induces Senescence In Glioblastoma." Scientific Reports 9.1 (2019): 988.

70) Mohamed, Omyma Shehata, et al. "Cytoprotective Effect and Clinical Outcome Of Perioperative Progesterone In Brain Tumors, A Randomized Microscopically Evidence Study." Egyptian Journal of Anaesthesia 38.1 (2022): 466-475.

Chapter 14

Progesterone for Perimenopause Part Two

AS WE HAVE SEEN FROM the previous Chapter 13, progesterone is major hormone of the female menstrual cycle, ovulation, and fertility. The name progesterone means "for gestation", meaning progesterone is required for pregnancy, i.e. the implantation of the oocyte (egg) and growth of the developing fetus. If progesterone receptors are blocked in early pregnancy with a drug such as mifepristone (RU-486) this leads to immediate abortion of the fetus. As it turns out, mifepristone is not only an abortion drug, it is also an excellent repurposed anti-cancer drug. However, for political reasons, access to mifepristone is restricted in the U.S., and in some states only prescribed by abortion clinics or approved clinical trials. The status of mifepristone depends on individual state legislation. About half the states allow online prescribing and postal delivery of mifepristone. (1-4)

Understanding the Normal Menstrual Cycle with Hormone Charts

Imagine a monthly menstrual hormone chart for estrogen and progesterone from college-level biology class showing the idealized normal menstrual cycle. These are widely available on the internet by typing in key words, menstrual hormone chart, into your search engine. To understand progesterone and estrogen, one may review such a chart showing the major events of the menstrual cycle. Estrogen (estradiol) and progesterone are natural human hormones produced by the ovary. Estradiol is produced throughout most of the month, from days 5-26 of the menstrual cycle with a peak just before ovulation on day 11, and another peak in estrogen production around day 19. Sudden decline in estrogen level on days 27-28 leads to sloughing off the endometrium which is no longer supported by hormonal stimulation.

Idealized Normal Menstrual Cycle

For the idealized menstrual cycle in the normal female, the corpus luteum in the ovary makes progesterone on days 12-26 of the monthly cycle. The body temperature will spike up about one degree Fahrenheit at the time of ovulation around day 11 at the same time as the LH and FSH spike. FSH is follicle stimulating hormone, and LH is luteinizing hormone. Both are pituitary hormones which stimulate the ovary. Estrogen (estradiol) spikes just before ovulation, and then has a second peak around day 19 and then falls off around day 27. Progesterone starts rising after ovulation, peaks around day 19 and then falls off around day 27. The endometrial lining thickens with hormonal stimulation and then sloughs off as hormone levels drop off at day 27, thus giving rise to bleeding. (5)

The Corpus Luteum Makes Progesterone

Progesterone is produced the last two weeks of the monthly cycle by the corpus luteum, a specialized ovarian structure, representing the remnant of the ovulatory follicle after ovulation. During ovulation, the egg (i.e. ovum, oocyte) is expelled from the ovarian follicle and starts its long journey down the fallopian tube to the endometrial cavity where fertilization and implantation takes place. Ovulation usually occurs on day 11, after which progesterone levels start rising and peak on days 19-20. Again, progesterone and estrogen levels decline abruptly in the final days of the menstrual month. This sudden drop in hormone levels initiates menstrual bleeding. The endometrial lining has lost hormonal support and sloughs off, thus the menstrual bleeding. Home urinary ovulation tests are widely available for determining ovulation and fertility. Other methods for ovulation detection are transvaginal ultrasonography, home basal body temperature charting, urinary luteinizing hormone (LH) levels, serum progesterone and urinary pregnanediol 3-glucuronide levels, urinary follicular stimulating hormone levels, cervical mucus and salivary fern pattern as discussed by Dr. Su (2017). For our purposes in the office, we try to time the patient's blood draw for their laboratory panel on day 19 (plus or minus one day) of the menstrual cycle to catch the hormonal peak in progesterone level after ovulation. (6-7)

Day 19 for Progesterone Level

We have two different laboratory techniques for measuring serum progesterone. To determine if ovulation is occurring normally, serum progesterone levels are checked on day 19 of the monthly cycle with RAI (radio-immunoassay) or HPLC (High-Pressure Liquid Chromatography). If the patient is taking progesterone capsules on a schedule for days 12-26, progesterone serum levels for day 19 indicate ovulation when progesterone rises above 10 ng/ml. However, progesterone metabolites from the oral progesterone capsules give a false elevated reading for RIA, but not for HPLC. The RIA reading may be falsely elevated eight times higher than the HPLC. In this case, the HPLC test is more accurate and is preferred over RIA. (8)

Anovulation, Estrogen Dominance, and Menstrual Irregularities

What happens if ovulation does not occur? This is called anovulation, a common problem in young women in which the ovum (egg) is not expelled from the follicle, no corpus luteum is formed, and **no progesterone is produced.** When ovulation occurs, laboratory testing on day 18-21 shows progesterone levels above 10 ng/ml. If the patient is not ovulating, the menstrual cycle length becomes variable and menstrual bleeding becomes irregular.

This leads to a condition known as "estrogen dominance" in which elevated estrogen stimulatory effects have no opposing progesterone inhibitory effects. Once ovulation occurs, usually around day 11, progesterone production by the corpus luteum inhibits further ovulation by remaining follicles. The anti-ovulatory effect of progesterone was first demonstrated in 1937 by Dr. A. W. Makepeace by injecting progesterone into mated female rabbits. Dr. Makepeace found progesterone inhibits ovulation and serves as a birth control drug. Thus, the first birth control drug was natural bioidentical progesterone. However, early forms of oral progesterone were poorly absorbed, necessitating injections which were inconvenient and not practical for mass birth control. Eventually, the drug industry invented synthetic progestins which could be taken as an oral pill. The improvement in oral absorption made synthetic progestins more practical for mass birth control. A more compelling reason to market synthetic progestins is patent protection making the product more profitable than natural progesterone which cannot be patented. Patent law states natural substances are not eligible for patent protection. Thus, progestins dominate the birth control market. The anti-ovulatory effect of natural progesterone and synthetic progestins is important in younger cycling females. To avoid any inhibition of ovulation in the younger age group, progesterone (or progestin) is given on a schedule beginning just after ovulation, for days 12-26 of the menstrual cycle. If progesterone is taken before ovulation on days 1-10, this will create an anovulatory cycle. Irregular bleeding will alarm the patient unless pre-warned about this effect. In older post-menopausal women who are not ovulating, progesterone may be given straight through the month with no concern, since there is no menstrual cycle. (9-13)

PMS - Premenstrual Syndrome - Estrogen Dominance

Progesterone deficiency in the anovulatory female is associated with bloating, breast tenderness, and fluid retention. Estrogen causes breast tissue stimulation and fluid retention. Both estrogen and progesterone are neurosteroids. Estrogen is a stimulatory neurosteroid, and progesterone an inhibitory neurosteroid. Without the calming effect of progesterone, the patient may complain of insomnia, anxiety and mood disorders and breast tenderness (mastalgia). As estrogen level drops at the end of the luteal phase, the serum binding protein level also drops, thus liberating testosterone to cause acne, oily skin, and aggressive, "snappy" behavior. In some cases, ovulation occurs, yet progesterone production is insufficient to balance the massive amounts of estrogen, in which case these patients will also suffer from "estrogen dominance" and a similar symptom complex called "PMS" premenstrual syndrome. (14)

Treatment for PMS and Irregular Menstrual Cycles

Since progesterone is the missing element in the PMS syndrome, it would be logical to assume providing progesterone in the form of creams or capsules would benefit the patient. In fact, many doctors and patients report dramatic resolution of PMS symptoms with oral micronized progesterone capsules. The oral route generates a progesterone metabolite, allopregnanolone, more than the transdermal route. Allopregnanolone has a calming and anti-inflammatory effect. Allopregnanolone acts on the GABA receptors to relieve anxiety and provide sedation, in a mechanism very similar to the benzodiazepine drugs (benzos) without the adverse effects of benzo drugs, namely addiction and withdrawal effects. Although laboratory testing of progesterone level on day 19 of the menstrual cycle can be used to detect ovulation, this type of lab testing is not required by mainstream medicine, and in fact, rarely done. For the patients with irregular uterine bleeding, a pelvic exam by the local gynecologist, and trans-vaginal pelvic sonogram is routine. For more on Progesterone for PMS, see Chapter 15. (15-17)

Progesterone for Perimenopause

The transition from normal ovulation with regular cycles to full menopause with cessation of menstrual cycles usually takes 6-12 months. This is called perimenopause characterized by irregular menstrual cycles, high or fluctuating estrogen levels, and low or non-existent progesterone levels. Symptoms during perimenopause include irregular or heavy menstrual bleeding, infertility, breast tenderness, insomnia, and night sweats. During the menopausal transition, hormone fluctuations cause mood disturbance, which may be relieved using topical or oral micronized progesterone, 300-400 mg at bedtime. In 2005, Jerilynn C. Prior, MD, Professor of Endocrinology and Metabolism at the University of British Columbia in Vancouver, BC. clears up the confusion about perimenopause, writing:

> At present, **no perimenopause therapies** have been adequately validated in randomized controlled trials. However, based on the endocrine changes of perimenopause, **cyclic or daily oral micronized progesterone in doses of 300 to 400 mg at bedtime appears to help with heavy flow, hot flushes, breast tenderness, and sleep.** (18)

In 2011, Dr. Jerilynn Prior discussed laboratory findings in the peri-menopausal transition, finding wild fluctuations in estrogen levels, and absent progesterone levels. For the treatment of perimenopausal transition, Dr. Jerilynn Prior recommends 300 mg of oral micronized progesterone at bedtime. If the patient is still having regular menstrual cycles, then the progesterone is taken on a schedule for days 12-26 of the monthly cycle. Dr. Prior writes:

> Evidence shows that with disturbed brain-ovary feedbacks, **E2 [estradiol] levels average 26% higher and soar erratically – some women describe feeling pregnant!** Also, ovulation and progesterone (P4) levels become insufficient or absent. **The most symptomatic women have higher E2 and**

lower P4 levels...Because P4 [progesterone] and E2 [estradiol] complement/counterbalance each other's tissue effects, oral micronized P4 (OMP4 300 mg at bedtime) [progesterone] is a physiological therapy for treatment-seeking, symptomatic perimenopausal women. [Progesterone is] given cyclically (cycle days 14-27, or 14 on/off) in menstruating midlife women. (19)

In 2020 Dr. Hadine Joffe reviews the utility of progesterone for relief of peri-menopausal depressive symptoms, finding an inverse relationship between depression symptoms and progesterone levels. The progesterone metabolite, allopregnanolone, has an antidepressant effect through its direct inhibition of GABA receptors in the brain. Thus, in 2019, allopregnanolone was FDA approved for intravenous use in treatment of post-partum depression. This drug version of allopregnanolone is called brexanolone, brand name Zulresso, effective for prompt relief of post-partum depression. Dr. Hadine Joffe writes:

Consistent with our observation of an inverse relationship between depressive symptoms and concurrent exposure to progesterone is the recent finding of an antidepressive effect of the progesterone-derived neurosteroid allopregnanolone for treatment of postpartum depression, leading to the approval of allopregnanolone by the FDA for postpartum depression. As a treatment for another reproductive hormone-associated mood disturbance, allopregnanolone's efficacy for postpartum depression challenges earlier presumptions that progestins adversely affect mood. Allopregnanolone may act as a neurosteroid to mediate the protective effect of peripheral progesterone on mood through **direct inhibition of gamma-aminobutyric acid (GABA) receptors.** (20-21)

Still Not Accepted by Mainstream Medicine

Although progesterone is an FDA-approved drug, mainstream medicine has yet to accept bioidentical, natural progesterone as a treatment for perimenopause and PMS. Instead, mainstream medicine relies on synthetic forms of progesterone called "progestins" such as medroxyprogesterone (MPA) which increased breast cancer by 26 percent in the ill-fated 2002 first arm Women's Health Initiative Study. In 2000, Dr. Catherine Schairer found that women taking the Premarin(CEE)/ MPA combination had increased breast cancer risk compared to women taking Premarin-alone. Increased breast cancer risk with the estrogen/MPA combination was confirmed in 2000 by Dr. Ronald K Ross, and in 2003 by Dr. Hakan L Olsson. Why has not mainstream medicine accepted natural progesterone and rejected synthetic versions progesterone such as MPA? This is explained by the "War Against Natural Medicine". Unpatented natural substances such as progesterone are the enemy of the drug industry because they compete with profits from their patented products. (22-25)

The War Against Natural Medicine

In my first book, *Natural Medicine 101*, I delved into the reasons for the information war between natural medicine and mainstream medicine. In short, mainstream medicine is dominated by the drug industry whose profits are dependent on the patented drug system, and using these exorbitant profits to capture the medical industry, the mass media, and all government agencies involved in health care regulation and research (FDA, NIH, CDC, HHS). A patented drug is a natural molecule whose chemical structure has been altered to obtain a patent. Molecular structures naturally occurring in the human body or nature cannot be patented, and are of no interest to the drug industry. Rather, any natural substance such as progesterone represents economic competition to the patented drug counterpart. This is why natural progesterone has been ignored by mainstream medicine. (26)

Dysfunctional Uterine Bleeding

Dysfunctional bleeding (DUB) related to anovulatory cycles may be caused by low thyroid function, usually responding to administration of thyroid hormone. Assuming normal thyroid function, mainstream OB/Gyne practice standard of care for DUB is a short course of synthetic progestin such as medroxyprogestrone, MPA. In my opinion, the use of synthetic progestins for DUB is justifiable despite carcinogenicity which is generally overlooked. Uterine bleeding in the post-menopausal female will trigger further evaluation using trans vaginal pelvic sonogram. If there is endometrial stripe thickening, endometrial biopsy is performed by the local OG/Gyne doctor. For the post-menopausal female on bioidentical hormone replacement, excessive estrogen dosage may cause bleeding. Temporary withdrawal of the estrogen hormone replacement usually resolves the bleeding and allows time for a pelvic sonogram which may demonstrate uterine fibroids or any other underlying pathology. In the event of a normal pelvic sonogram without thickening of the endometrial stripe, the hormone replacement may be resumed at a lower dosage, usually without the resumption of bleeding. (27-37)

Vitamin A for Uterine Bleeding - More than Just Fibroids!

After anovulatory cycles, perhaps the second most common cause for uterine bleeding is fibroids, benign tumors of the uterus that tend to bleed. A third cause is vitamin A deficiency. In 1952, Dr D.M. Lithgow found vitamin A deficiency decreases estrogen production and represents a common cause of menorrhagia (irregular uterine bleeding) in 68 percent of patients. Providing vitamin A supplementation alleviates menorrhagia in more than 92 percent of his patients. Dr. Lithgow writes:

> Hypovitaminosis A was found to be an important cause of menorrhagia, and a statistically significant difference between the fasting serum vitamin A values of healthy controls and patients with menorrhagia was noted. Vitamin A is a co-factor of 3,beta-dehydrogenase in steroidogenesis and deficiencies of this vitamin may result in impaired enzyme activity. The level of endogenous 17-beta-estradiol appears to be elevated with vitamin A therapy, and **menorrhagia was alleviated in more than 92% of patients**...Table 11 details the causes of menorrhagia in 174 patients. **Vitamin A deficiency appeared to be the major aetiological factor in 43, 68% of these women**. The aetiology was unknown in 17, 24%. A number of patients (11, 49%) had previously been subjected to sterilization. **Pyridoxine (vitamin B6)** deficiency was found in 9, 77% and was diagnosed clinically in 11 and subsequently biochemically."
> Eight patients with both vitamin B6 and vitamin A deficiency were classified as vitamin A-deficient. The causative factors most frequently observed were deficient diet, malabsorption, recent infections, overexposure to sunlight" and excessive intake of alcohol." ...The rise in 17,Beta-oestradiol following vitamin A therapy (Table III) is significant, since graphs of the menstrual cycle indicate an association in the peaks of 17-beta estradiol and vitamin A. Here is an example of a vitamin program used for uterine bleeding: Vitamin A (retinol) 25,000 i.u. BID, or 100,000 i.u. vitamin A per day for 15 days, Vitamin C, and Vitamin K (MK-7). (38-40)

Myo-Inositol for Restoring Ovulation

Myo-inositol has shown efficacy in restoring ovulation and stopping DUB. Since irregular uterine bleeding is usually caused by anovulation, restoring regular ovulatory cycles is a rational treatment goal. In 2024, Dr. Galina B. Dikke reported her experience treating 2,000 patients with myoinositol with the goal of restoring ovulation. Dr. Galina B. Dikke's combined treatment involved oral micronized progesterone, iron supplements and a complex containing myoinositol, D-chiroinositol (5:1), folic acid and manganese. Dr. Galina B. Dikke concludes inositols are highly effective in restoring ovulation, regular menstrual cycles, normal amount of menstrual blood loss, eliminating anemia and decreasing body weight, writing:

> Abnormal uterine bleeding (AUB) associated with ovulatory disfunction (OD) is the most common finding among women with chronic AUB, accounting for 57.7% of cases. Oral progestogens are often prescribed for irregular and copious menstruation. However, a course of hormonal rehabilitation after AUB-OD may not be enough. **Inositols have been shown to be highly effective in restoring ovulation, normalizing the menstrual cycle**, correcting carbohydrate and lipid metabolism, and reducing body weight...The multicentre study in real clinical practice included 2,042 women with OD [ovulatory dysfunction]. The patients received **dydrogesterone** or **micronized progesterone for 3 cycles** (from 14 to 25 days), a medication containing **myoinositol** 1000 mg, D-chiroinositol 200 mg, folic acid 200 mg, manganese 5 mg (Dikirogen) for 6 cycles, iron sulfate/ascorbic acid for 3–4 months (according to indications). The parameters of the menstrual cycle (MC), hemoglobin, serum ferritin, and body weight were assessed at 3, 6 and 12 months from the start of treatment. The age of the patients ranged from 18 to 45 years, the average age was 30 (25; 35) years. The number of patients with a normal MC [menstrual cycle] rhythm after 3 and 6 months was observed in 76.5 and 90.9% of patients versus 46.9% before treatment, p<0.001, and with a moderate volume of menstruation in 77.9 and 89.9% versus 45.4%, respectively, p<0.001; iron deficiency anemia decreased from 39.9% to 18.2% of patients after 3 months, p<0.001, and **there were no patients with anemia by 6 months.** Menstrual cyclicity remained at the achieved level, and the volume of blood loss decreased statistically significantly by 12 months. BMI decreased from 26.8 (21.3; 27.3) to 23.4 (21.3; 24.3) kg/m2 by 6 months of treatment, p=0.001, and stabilized at this level until 12 months...Therapy ... **is effective in achieving regular menstrual cycle and volume of menstrual blood loss, eliminating anemia and normalizing body weight**. Note: Dydrogesterone is an isomer of progesterone with similar pharmacology and better oral absorption. (41)

Progesterone Hypersensitivity (PH)

Progesterone is a mainstream treatment for women with threatened abortion or repeated miscarriage in the first trimester associated with low progesterone levels. Fertility doctors, i.e. reproductive endocrinologists, will commonly prescribe large doses of progesterone, usually transvaginal or intramuscular (I.M.) with generally good results. The assisted reproduction medical community reports that rarely their patients exhibit allergic reactions to the progesterone. The syndrome is called exogenous progesterone hypersensitivity (PH), a rare event that I have never seen in my clinical practice. This syndrome has been described in cycling women who report cyclical urticaria and angioedema during the luteal phase which resolves upon onset of menses. The syndrome seems related to previous exogenous progesterone or progestin exposure from birth control pills or other medications which sensitizes the patient to progesterone. The mechanism involves production of IgE antibodies that cross react with progesterone receptors in the skin. In 2019, Dr. Eun-Jung Jo reported this syndrome in 9 patients with symptoms such as hives, erythema and itching treated with antihistamines and corticosteroids. One patient with anaphylaxis and hypotension required an epinephrine injection, writing:

> Nine patients had exogenous PH... The mean latency to symptom onset was 5.8 days (range 1 h to 11 days). The patients complained of hives, erythema and itching, and one developed anaphylaxis. All patients were treated with antihistamines, and six patients were treated with systemic corticosteroids. Epinephrine was administered to one patient with

hypotension. The symptom duration was 1-14 days. Skin tests were performed in four patients; all were positive. Two patients were treated successfully by progesterone desensitization... The clinical features of exogenous PH were similar to those of type I hypersensitivity reactions, but tended to develop later and did not respond to antihistamines or steroids. As use of progesterone increases, an understanding of the clinical features of exogenous PH becomes ever-more important. ... Few cases of exogenous PH have been reported; they include erythema multiforme due to progesterone in a low-dose oral contraceptive pill; generalized, pruritic, intensely erythematous, morbilliform, scaling dermatitis in reaction to oral megestrol acetate; and urticaria due to synthetic intramuscular progesterone administered in association with in vitro fertilization (IVF). Foer et al. reported 24 cases of endogenous and exogenous PH, in which the most common symptoms were dermatitis, urticaria, and angioedema, followed by asthma and anaphylaxis... However, how patients become sensitive to progesterones is not clear. Endogenous PH is frequently associated with prior exposure to exogenous progesterone.5 Exogenous progesterone exposure leads to sensitization through generation of progesterone-specific IgE antibodies, which cross-react with the increasing endogenous progesterone level in the luteal phase of the menstrual cycle... the binding of [anti-progesterone IgE antibodies] to progesterone receptors in the skin and oral mucosa results in cutaneous inflammation...Previous studies showed that most women with endogenous PH had a previous exposure to exogenous progesterone. (42-50)

Although assisted reproductive techniques such as in-vitro fertilization (IVF) use intra-vaginal or IM progesterone, prior exogenous exposure could include oral micronized progesterone or progestins in birth control pills. The progesterone hypersensitivity syndrome is not fully understood. The description brings to mind a similarity to food hypersensitivity in patients with underlying achlorhydria and gut permeability (leaky gut). In these patients, undigested food proteins may leak into the bloodstream invoking an immune response that manifests as urticaria and eczema-type symptoms. Perhaps a similar mechanism is at play in which particles of clumped progesterone (or progestin) leak across the gut barrier into the bloodstream, thus eliciting an IgE allergic immune response. This is speculative and future studies may confirm or refute this hypothesis. (51-57)

Progesterone Beneficial for ALS

Amyotrophic lateral sclerosis (ALS, Lou Gehrig's disease) is an incurable progressive motor neuron disease of unknown etiology. Progesterone has neuroprotective properties and is beneficial in mouse models of ALS. In 2022, Dr. Lucie Kolatorova found progesterone increases BDNF (brain derived neurotrophic factor):

Amyotrophic lateral sclerosis is a motor neuron disease and, after Alzheimer's disease and Parkinson's disease, is the third most common neurodegenerative disorder. PROG [progesterone] has been implicated in **various neuroprotective properties**, of which longevity, muscle strength, cell health, lowered oxidative stress in the spinal cord, and nitric oxide are the most relevant. It has been shown to **increase brain-derived neurotrophic factor** and normalizes mRNA levels in components of the sodium-potassium pump, which is important for cell nutrition and neurotransmission and is also crucial for mitochondrial health. Moreover, PROG was observed to inhibit the activity of astrocytes, which have predominately deleterious effects in the context of amyotrophic lateral sclerosis because they correspond to increased inflammation. PROG also protects against glutamate excitotoxicity in vitro, one of the major sources of pathology in amyotrophic lateral sclerosis. (58-60)

Progesterone Benefits for the Brain and Nervous System

In 2021, Dr. Bernadett Nagy reviewed progesterone's neuroprotective effects for traumatic brain injury, and other benefits for the brain and nervous system. writing:

> Progesterone is synthesized by neurons and glial cells, and this neurosteroid is involved in the regulation of various molecular and cellular processes underlying neurogenesis, myelination, neuroprotection, neuromodulation, learning, memory, and mood. Clinical trials showed that progesterone infusion after traumatic brain injury results in reduced neuronal loss, enhanced remyelination, improved functional recovery, and an overall decrease in cerebral edema. (78) (90-91)

Progesterone as Anti-Inflammatory and Immunomodulatory Drug

Progesterone (P4) has strong anti-inflammatory and immunomodulatory effects described in 2022 by Dr. Tatiana A. Fedotcheva who suggests progesterone is a good candidate to replace glucocorticoids in the treatment of chronic inflammatory disease. I have seen a few patients with colitis resolving after starting progesterone for PMS or perimenopauuse. This observation attests to the anti-inflammatory activity of progesterone. Anti-inflammatory effects of progesterone are due to strong inhibition of Nuclear Factor Kappa-B (NF-kB), the inflammatory master controller, and inhibition of COX (Cyclo-Oxygenase). **Note:** An entire class of anti-inflammatory drugs has been based on COX-2 inhibition, the NSAID drug Celecoxib (brand name Celebrex®) as an example. Unfortunately, NSAID drugs are associated with adverse side effects such as GI bleeding and increased mortality from heart attacks, such as the example of Vioxx. Dr. Tatiana A Fedotcheva suggests progesterone can serve as a COX inhibitor without any adverse side effects. Progesterone is neuroprotective as demonstrated in animal and human studies showing benefits in traumatic brain injury. Progesterone invokes a state of immune tolerance as shown in pregnancy when progesterone levels reach 30-100 fold higher than the non-pregnant state. Exogenous progesterone given clinically supports high progesterone levels required to prevent miscarriage induced by the maternal immune system's rejection of the "non-self" genetic makeup of the fetus. Progesterone's immune tolerance is achieved through inhibition of proinflammatory cytokines, inhibition of Th1 (T-helper) cell activity, and secretion of progesterone induced blocking factor (PIBF). Dr. Tatiana A Fedotcheva writes:

> The anti-inflammatory effects of P4 can be defined as nonspecific, associated with the **inhibition of NF-κB and COX,** as well as the inhibition of prostaglandin synthesis, or as specific, associated with the regulation of T-cell activation, the regulation of the production of pro- and anti-inflammatory cytokines, and **the phenomenon of immune tolerance**. The specific anti-inflammatory effects of P4 and its derivatives (progestins) can also include the inhibition of proliferative signaling pathways...The goal of this paper is to highlight the possibility of using P4 and its derivatives as alternative steroid hormones to glucocorticoids in the treatment of inflammatory diseases, especially chronic inflammatory diseases accompanied by resistance to hormone therapy [glucocorticosteroids]...In addition to the gradually emerging resistance to GCs [glucocorticoids], serious side-effects of their usage are hyperglycemia with the further possibility of insulin resistance, sodium retention, and hypertension, muscle atrophy, osteoporosis, features of Cushing's syndrome, etc. [On the other hand,] The possible side-effects with progesterone usage as an anti-inflammatory drug should be extremely minimal or absent... The neuroprotective role of P4 has been confirmed in both men and women; there are results of a number of clinical studies confirming the high therapeutic efficacy of P4 against necrotic damage and behavioral abnormalities caused by traumatic brain

injury (TBI)...The immune tolerance during pregnancy is sufficient to assess its role. P4 causes immune tolerance by inhibiting the production of proinflammatory cytokines and reducing the activity of T-helper type 1 cells (Th1). In this context, it is important to note that P4 and its synthetic analogues are successfully used in clinical practice to prevent miscarriage. The progesterone-specific regulation of T-cell activation became evident through several observed facts: **a more active cell-mediated immune response to viral infections in women than in men**; the correlation of peak incidence of autoimmune pathology in women with periods of hormonal changes, i.e., puberty, early postpartum, or menopause; the phenomenon of immune tolerance during pregnancy, when the **concentration of P4 increases hundredfold**. It is believed that T-cell activation can be specifically inhibited by P4, but not by estrogens. Such a conclusion can be made as, in 1990, it was shown for the first time that the in vitro cytotoxic T lymphocyte (CTL) response was strongly inhibited by P4 but not by estrone, estradiol, or testosterone...This specific P4-dependent regulation may be due to the P4-dependent production of at least two mediators of the immune response: PIBF (progesterone-induced blocking factor) and LIF (leukemia inhibitory factor)...In general, PIBF modulates cytokine synthesis and T-cell cytotoxicity toward a more **immunotolerant state**....The analysis of the current data revealed the key fundamental mechanisms through which **P4 can be used in the treatment of hormone [corticosteroid]-resistant chronic inflammatory diseases.** (61-62) (92)

Progesterone as Anti-Inflammatory Neurosteroid

In 2024, Dr. Irina Balan highlighted the role progesterone as a neurosteroid produced by neurons in the brain, particularly pregnane-type progesterone metabolites, allopregnanolone, and pregnenolone. These have anti-inflammatory effects and have potential for use in neuropsychiatric disorders such as depression, post-traumatic stress disorder (PTSD), alcohol use disorder (AUD) and post-partum depression. In 2019, the FDA approved allopregnanolone (brexanolone) for treatment of post-partum depression. Allopregnanolone acts on the GABA receptor system leading to reduced anxiety, sedation, anti-convulsant activity, and the enhancement of inhibitory circuits in the brain. **Note:** the benzodiazepine class of drugs (Valium, Xanax, etc) act on the GABA receptors to relieve anxiety and insomnia. However, benzodiazepine drugs are highly addictive, whereas progesterone is not. In animal models, allopregnanolone has shown benefits for alcoholism, chronic stress-induced depression, traumatic brain injury (TBI), multiple sclerosis (MS), and Alzheimer's disease (AD)...Progesterone has shown efficacy in traumatic brain injury (TBI) and allopregnanolone in postpartum depression (PPD). Dr. Irina Balan writes:

> Pregnane neuroactive steroids, notably allopregnanolone and pregnenolone, exhibit efficacy in mitigating inflammatory signals triggered by toll-like receptor (TLR) activation, thus attenuating the production of inflammatory factors. Clinical studies highlight their therapeutic potential, particularly in conditions like postpartum depression (PPD), where the FDA-approved compound brexanolone, an intravenous formulation of allopregnanolone, effectively suppresses TLR-mediated inflammatory pathways, predicting symptom improvement. Additionally, pregnane neurosteroids exhibit trophic and anti-inflammatory properties, stimulating the production of vital trophic proteins and anti-inflammatory factors... Neuroactive steroids are synthesized within both endocrine glands and the brain. In the brain, neurons are the primary producers of neurosteroids. Neuroactive steroids, synthesized from cholesterol, can be classified into three categories: pregnane, androstane, and sulfated neuroactive steroids...Pregnanes, including allopregnanolone, pregnanolone...act as positive modulators of GABAA receptor

subtypes. **These compounds enhance inhibitory neurotransmission mediated by GABAA receptors, leading to anxiolysis, sedation, anti-convulsant activity, and the enhancement of inhibitory circuits in the brain.** Importantly, their anti-inflammatory actions are distinct from their GABAergic mechanisms...Preclinical and clinical studies have highlighted **reduced levels of pregnane neuroactive steroids, such as pregnenolone and allopregnanolone, in conditions like stress, depression, post-traumatic stress disorder (PTSD), and alcohol use disorder (AUD)...**Allopregnanolone and its precursors, pregnenolone and progesterone, have shown promise in animal models of AUD, chronic stress-induced depression, traumatic brain injury (TBI), multiple sclerosis (MS), and Alzheimer's disease (AD)...In clinical studies, progesterone has demonstrated efficacy in TBI and cocaine craving, while pregnenolone has shown benefits in alcohol and cannabis use disorders, and allopregnanolone has been effective in treating postpartum depression (PPD)... Clinical observations further support the therapeutic potential of compounds like brexanolone, a Food and Drug Administration (FDA)-approved intravenous formulation of allopregnanolone, in conditions such as PPD, attributed to their inhibition of TLR inflammatory pathways... pregnane neuroactive steroids, possess anti-inflammatory properties that operate independently of their effects on GABAA receptors. (63)

Post-Finasteride Syndrome

The above discussion highlighted the benefits of the progesterone metabolite allopregnanolone. What if the patient takes a drug that inhibits allopregnanolone. What does this do? Drugs which decrease brain allopregnanolone cause severe neuro-psychiatric adverse side effects. One example is the post-finasteride syndrome in which finasteride reduces brain allopregnanolone. Finasteride is a 5-alpha-reductase inhibitor used by urologists for treatment of BPH (benign prostatic hypertrophy).

Finasteride (Propecia) is also used by dermatologists for hair regrowth. After FDA approval of finasteride, unanticipated adverse side effects were discovered. 5-alpha-reductase inhibitors such as finasteride and Propecia decrease allopregnanolone in the brain by inhibiting the conversion of progesterone to allopregnanolone. This is called the post-finasteride syndrome, a constellation of symptoms such as depression, anxiety, cognitive dysfunction, loss of libido, erectile dysfunction, decreased sexual arousal, and difficulty achieving orgasm. The post-finasteride syndrome may resemble the loss of libido reported as adverse effects of SSRI drugs. I have seen several young men in my office reporting these symptoms after taking Propecia for hair regrowth. These symptoms may persist after stopping the drug. There is currently a class action lawsuit in progress. In 2020, Dr. Graziano writes:

> **Pharmacological treatments, including finasteride and oral contraceptives, that inhibit 5a-RI [5-alpha reductase Type 1], which results in a blood and brain allopregnanolone decrease** also affect subunit expression of GABAA receptor and are associated with mood symptoms and suicide and are part of post finasteride syndrome. Post finasteride syndrome, in addition to depression, anxiety, and cognitive deficits also induces sexually-related side effects, such as loss of libido, erectile dysfunction, decreased arousal, and difficulty in achieving an orgasm that persists after drug withdrawal. (64-68)

Better Outcomes with Pre-Operative Progesterone

In pre-menopausal women undergoing breast cancer surgery, the menstrual timing of surgery determines the long-term outcome. As mentioned in previous chapters, the luteal phase, days 12-26 of the menstrual cycle is the time when progesterone is produced by the corpus luteum after ovulation. In 1990, Dr. William Hrushesky discovered that operation during this luteal phase when progesterone is highest, days 7-20, of the menstrual cycle reduced postoperative cancer recurrence of cancer to one-quarter that of performing surgery in the follicular phase, days 1-7, when the progesterone levels are lowest. In 1990, Dr. William Hrushesky in Lancet performed a retrospective study of 44 premenopausal women undergoing resection for breast cancer, and followed for 5-12 years, writing:

> Patients who underwent resection during the perimenstrual period [follicular phase] had a more than quadrupled risk of recurrence and death compared with women operated upon during days 7 to 20 [luteal phase] of the menstrual cycle. (69-72)

The next logical question is what would happen if breast cancer patients are given pre-operative progesterone? Would this improve the outcome? In 2011, Dr. Rajendra Badwe was thinking the same thing. He performed a randomized controlled trial (RCT) of 471 lymph node-positive breast cancer patients given pre-operative progesterone compared to a control group. The 5-year overall and disease-free survival was considerably improved in the progesterone-treated group. However, patients who were lymph node-negative, with no cancer found in the lymph nodes, did not benefit, thus explaining why only node-positive patients were studied. **Note:** lymph node-positive means the breast cancer is aggressive and has spread to the draining lymph nodes in the axilla. (72-74)

In 2023, Dr. Gaurav Chakravorty was inspired by the above 2011 progesterone pre-operative study by Dr. Rajendra Badwe and did further work to elucidate the molecular basis and mechanism of progesterone's benefit when given preoperatively to lymph node-positive breast cancer patients. Dr. Gaurav Chakravorty performed whole transcriptome sequencing of breast cancer samples before and after progesterone. His study revealed that **pre-operative progesterone downregulates the inflamma-**

tory response and production of TNF (tumor necrosis factor). The inflammatory response is "known to induce proliferative, invasive, and malignant behavior of breast cancer cells." So, downregulating inflammation is very beneficial. This effect was independent of the progesterone receptor status of the breast cancer cells. Dr. Gaurav Chakravorty writes:

> We investigated the molecular basis of action of hydroxyprogesterone, a biosimilar form of natural progesterone, that patients with operable breast cancer received as a single injection prior to surgery. In the study, we performed whole transcriptome sequencing (RNA-Seq) of primary breast tumor samples collected before and after hydroxyprogesterone exposure, as well as a control group of patients who underwent only surgery. The results suggested 207 genes to be significantly altered between the progesterone-exposed and -unexposed groups; of which, **142 genes were upregulated post-surgery in progesterone-exposed patients... Specifically, the study revealed that preoperative hydroxyprogesterone manifests its effect in patients with breast cancer by downregulating genes involved in inflammatory response and production of TNF that are known to induce proliferative, invasive, and malignant behavior of breast cancer cells...**The study also identified **upregulation of a tumor metastasis suppressor gene, N-Myc** downstream regulated gene 1, NDRG1, along with increased expression of the AP-1 network genes, suggesting that preoperative progesterone intervention may modulate the effect of surgical stress on breast cancer by altering the expression of several protein-coding genes, thereby improving patient survival...Overall, these findings indicate that **pre-operative hydroxyprogesterone may act to curb cellular stress responses and potentially mediate the pro-survival effects through a negative regulation of inflammation**...As this effect was consistently observed in both PR-positive and -negative breast cancer cells, it suggests that the

benefits of preoperative progesterone exposure may not be limited to patients with PR-positive cancers. This finding is of clinical importance as a significant proportion of patients have PR-negative cancers and may fail to respond to adjuvant or neoadjuvant endocrine therapies that target the PR pathway...The preoperative hydroxyprogesterone administration improves disease-free and overall survival in patients with node-positive breast cancer. The mechanism behind this improvement is thought to be related to the modulation of cellular stress response and the negative regulation of inflammation through the upregulation of the kinase gene SGK1...Progesterone-induced modification of the PR and ER genomic binding pattern orchestrates estrogen signaling in breast cancer, preventing cell migration and invasion, and improving the prognosis of patients with breast cancer. (75)

Progesterone Acts as a Proliferative Brake on ER-alpha

In 2015, Dr. Hisham Mohammed deciphered progesterone's beneficial effects in breast cancer, finding progesterone blocks ER-alpha-induced tumor growth via a robust association between the progesterone receptor (PR) and ER-alpha, a form of crosstalk between PR and ER-alpha, acting as a proliferative brake. Note: ER-alpha is estrogen receptor alpha, the proliferative receptor. Dr. Hisham Mohammed writes:

We conclude that activation of PR results in a robust association between PR and the ER-alpha complex. Progesterone blocks ER-alpha+ tumour growth. PR is a critical determinant of ER-alpha function due to crosstalk between PR and ER-alpha. In this scenario, under estrogenic conditions, an activated PR functions as a proliferative brake in ERα+ breast tumours by re-directing ERα chromatin binding and altering the expression of target genes that induce a switch from a proliferative to a more differentiated state... We now show that PR is not merely an ER-alpha-induced gene

target, but is also an ER-alpha-associated protein that modulates its behaviour. In the presence of agonist ligands, PR associates with ER-alpha to direct ER-alpha chromatin binding events within breast cancer cells, resulting in a unique gene expression programme that is associated with good clinical outcome. Progesterone inhibited estrogen-mediated growth of ER-alpha+ cell line xenografts and primary ER-alpha+ breast tumour explants and had increased anti-proliferative effects when coupled with an ER-alpha antagonist. ...Our findings indicate that PR functions as a molecular rheostat to control ER-alpha chromatin binding and transcriptional activity, which has important implications for prognosis and therapeutic interventions...There is compelling evidence that inclusion of a progestogen as part of hormone replacement therapy (HRT) increases risk of breast cancer, implying that PR signalling can contribute towards tumour formation. However, the increased risk of breast cancer associated with progestogen-containing HRT is mainly attributed to specific synthetic progestins, in particular medroxyprogesterone acetate (MPA), which is known to also have androgenic properties. The relative risk is not significant when native progesterone is used. In ER-alpha+ breast cancers, PR is often used as a positive prognostic marker of disease outcome4, but the functional role of PR signalling remains unclear. While activation of PR may promote breast cancer in some women and in some model systems, progesterone treatment has been shown to be antiproliferative in ER-alpha+ PR+ breast cancer cell lines and progestogens [both MPA and natural progesterone] have been shown to oppose estrogen-stimulated growth of an ER-alpha+ PR+ patient-derived xenograft (Kabos, Peter, 2012). In addition, exogenous expression of PR in ER-alpha+ breast cancer cells blocks estrogen-mediated proliferation and ERα transcriptional activity. Furthermore, in ERalpha+ breast cancer patients, PR is an independent predictor of response to adjuvant tamoxifen, high levels of PR correlate with decreased metastatic events in early stage disease and administration of a progesterone injection prior to surgery can provide improved clinical benefit. **These observations imply that PR activation in the context of estrogen-driven, ER-alpha+ breast cancer, can have an anti-tumourigenic effect.** In support of this, PR agonists can exert clinical benefit in ER-alpha+ breast cancer patients that have relapsed on ER-alpha antagonists. (76-77)

Progesterone Anti-Cancer Effects

We have made the case that progesterone downregulates the proliferative effects of estradiol, and is breast cancer preventive. Progesterone also prevents endometrial cancer. Progesterone is FDA-approved for the prevention of endometrial hyperplasia and endometrial cancer and is accepted by mainstream medicine for this role. What about other cancers? Is progesterone protective for other cancers? Yes. Progesterone reduces the incidence of colon cancers and useful in treating endometrial carcinoma, ovarian carcinoma, melanoma, mesothelioma, and prostate tumors. In 2021, Dr. Bernadett Nagy reviewed the clinical uses of progesterone for cancer prevention and treatment, writing:

> **The use of progesterone has been linked to lower rates of uterine and colon cancers and may also be useful in treating endometrial carcinoma, ovarian carcinoma, melanoma, mesothelioma, and prostate tumors**... a recent meta-analysis involving 86,881 postmenopausal women reported that the use of natural progesterone was associated with a significantly lower risk of breast cancer compared with that of synthetic progestins [Noor Asi, 2016]. These results support the view that **natural progesterone does not cause breast cancer**...Progesterone is necessary for successful embryo implantation and pregnancy maintenance. Vaginal progesterone treatment minimizes the risk of recurrent miscarriage and decreases the risk of preterm birth, saving many fetal lives. However, progesterone is far more than a gestational agent. Progesterone is an essential steroidogenetic **precursor** of other gonadal and non-gonadal hormones such as **aldosterone, cortisol, estradiol and testosterone**. These hormones are responsible for innumerable functions such as sodium conservation in the kidney, regulation of blood pressure, response to stress and low blood-glucose concentration, development of female and male secondary sexual characteristics. Progesterone also plays an important role in the nervous system. Its neurogenic effect is essential for normal brain development in fetuses, while the neuroprotective effect of progesterone improves the patient's survival after traumatic brain injury. Progesterone and novel progestins have many important functions, including contraception, luteal phase support, treatment of dysfunctional uterine bleeding, and endometriosis. Progesterone has an important role in immune response and also in the prevention and treatment of various cancers. (78-89)

Conclusion: The more one studies progesterone, one realizes this is a miraculous hormone with many health benefits, and a very safe profile even at very high doses. Not only is progesterone preventive of breast cancer, it is also beneficial for many other cancers. Progesterone is useful for PMS and perimenopausal transition and has neuroprotective properties for traumatic brain injury, depression, ALS, and many other neuropsychiatric conditions. The progesterone metabolite, allopregnanolone has been developed into an IV and oral drug FDA-approved for treatment of depression and is poised to replace SSRI antidepressants. One might make the case that many patients on psychiatric drugs would do better taking natural progesterone instead.

♦ **References for Chapter 14 Progesterone for Perimenopause Part Two**

1) Maria, Bernard, et al. "Termination of Early Pregnancy by A Single Dose of Mifepristone (RU 486), A Progesterone Antagonist." European Journal of Obstetrics & Gynecology and Reproductive Biology 28.3 (1988): 249-255.

2) Check, Jerome H., and Diane L. Check. "The Role of Progesterone and The Progesterone Receptor in Cancer: Progress in The Last 5 Years." Expert Review of Endocrinology & Metabolism 18.1 (2023): 5-18.

3) Check, Jerome H., and Diane Check. "New Insights as To Why Progesterone Receptor Modulators, Such as Mifepristone, Seem to Be More Effective in Treating Cancers That Are Devoid of The Classical Nuclear Progesterone Receptor." Anticancer Research 41.12 (2021): 5873-5880.

4) Check, Jerome H., and Diane L. Check. "Progesterone And Glucocorticoid Receptor Modulator Mifepristone (RU-486) As Treatment for Advanced Cancers." Drug Repurposing-Molecular Aspects and Therapeutic Applications (2020).

5) Reed, Beverly G., and Bruce R. Carr. "The Normal Menstrual Cycle and the Control of Ovulation." Endotext [Internet]. MDText. com, Inc., 2018.

6) Owen, Martin. "Physiological Signs of Ovulation and Fertility Readily Observable by Women." The Linacre Quarterly 80.1 (2013): 17-23.

7) Su, Hsiu-Wei, et al. "Detection of Ovulation, A Review of Currently Available Methods." Bioengineering & Translational Medicine 2.3 (2017): 238-246.

8) Levine, Howard, and N. Watson. "Comparison of the Pharmacokinetics of Crinone 8% Administered Vaginally Versus Prometrium Administered Orally in Postmenopausal Women." Fertility and sterility 73.3 (2000): 516-521.

9) Makepeace, A. W., et al. "The Effect of Progestin and Progesterone on Ovulation in The Rabbit." American Journal of Physiology-Legacy Content 119.3 (1937): 512-516.

10) Campagnoli, Carlo, et al. "Progestins and Progesterone in Hormone Replacement Therapy and The Risk of Breast Cancer." The Journal of Steroid Biochemistry and Molecular Biology 96.2 (2005): 95-108.

11) Dhont, Marc. "History of Oral Contraception." The European Journal of Contraception & Reproductive Health Care 15.sup2 (2010): S12-S18.

12) Foran, Terri. "A Tale of Two Hormones." Fertility & Reproduction 1.01 (2019): 39-42.

13) A Garside, Deborah, et al. "An Update on Developments in Female Hormonal Contraception." Current Women's Health Reviews 8.4 (2012): 276-288.

14) Gudipally, Pratyusha R., and Gyanendra K. Sharma. "Premenstrual Syndrome." StatPearls [Internet]. StatPearls Publishing, 2023.

15) Gao, Q., et al. "Role of Allopregnanolone-Mediated Gamma-Aminobutyric Acid A Receptor Sensitivity In The Pathogenesis Of Premenstrual Dysphoric Disorder: Toward Precise Targets For Translational Medicine And Drug Development." Frontiers in Psychiatry 14 (2023).

16) Drexler, Berthold, et al. "Allopregnanolone Enhances Gabaergic Inhibition in Spinal Motor Networks." International Journal of Molecular Sciences 21.19 (2020): 7399.

17) Stiernman, Louise, et al. "Emotion-Induced Brain Activation Across the Menstrual Cycle In Individuals With Premenstrual Dysphoric Disorder And Associations To Serum Levels Of Progesterone-Derived Neurosteroids." Translational Psychiatry 13 (2023).

18) Prior, Jerilynn C. "Clearing Confusion About Perimenopause." BC Med J 47.10 (2005): 538-42.

19) Prior, J. C. "Progesterone for Symptomatic Perimenopause Treatment–Progesterone Politics, Physiology And Potential For Perimenopause." Facts, Views & Vision in ObGyn 3.2 (2011): 109.

20) Joffe, Hadine, et al. "Impact of Estradiol Variability and Progesterone on Mood In Perimenopausal Women With Depressive Symptoms." The Journal of Clinical Endocrinology & Metabolism 105.3 (2020): e642-e650.

21) Gordon, Jennifer L., et al. "Efficacy of Transdermal Estradiol and Micronized Progesterone In The Prevention Of Depressive Symptoms In The Menopause Transition: A Randomized Clinical Trial." JAMA psychiatry 75.2 (2018): 149-157.

22) Rossouw, Jacques E., et al. "Risks and Benefits Of Estrogen Plus Progestin In Healthy Postmenopausal Women: Principal Results From The Women's Health Initiative Randomized Controlled Trial." JAMA 288.3 (2002): 321-333.

23) Schairer, Catherine, et al. "Menopausal Estrogen and Estrogen-Progestin Replacement Therapy And Breast Cancer Risk." JAMA 283.4 (2000): 485-491.

24) Ross, Ronald K., et al. "Effect of Hormone Replacement Therapy on Breast Cancer Risk: Estrogen Versus Estrogen Plus Progestin." Journal of the National Cancer Institute 92.4 (2000): 328-332.

25) Olsson, Hakan L., et al. "Hormone Replacement Therapy Containing Progestins and Given Continuously Increases Breast Carcinoma Risk in Sweden." Cancer: Interdisciplinary International Journal of the American Cancer Society 97.6 (2003): 1387-1392.

26) Dach, Jeffrey. Natural Medicine 101: How to Win the Medical Information War and Take Control of Your Health. Booksurge Publishing; First Edition (9 February 2009)

27) Nekkanti, Dhanusha, and Vasantha S. Kumar. "The Study of Thyroid Dysfunction in Patients with Abnormal Uterine Bleeding." International Journal of Reproduction, Contraception, Obstetrics and Gynecology 12.12 (2023): 3498-3503.

28) Mounika, Koduru. "Study on Thyroid Dysfunction in Patients of Dysfunctional Uterine Bleeding." Journal of Pharmaceutical Research International 33.40A (2021): 44-49.

29) Singh, Shipra, et al. "Evaluation of Thyroid Dysfunction in Reproductive-age Women with Menstrual Disorders-A Case Control Study." European Journal of Cardiovascular Medicine 14.1 (2024).

30) Leal, Caio RV, et al. "Abnormal Uterine Bleeding: The Well-Known and The Hidden Face." Journal Of Endometriosis and Uterine Disorders (2024): 100071.

31) Black, D. "Diagnosis and Medical Management of Abnormal Premenopausal and Postmenopausal Bleeding." Climacteric 26.3 (2023): 222-228.

32) Papakonstantinou, Efthymia, and Georgios Adonakis. "Management of Pre-, Peri-, And Post-Menopausal Abnormal Uterine Bleeding: When to Perform Endometrial Sampling?" International Journal of Gynecology & Obstetrics 158.2 (2022): 252-259.

33) Jain, Varsha, et al. "Uterine Bleeding: How Understanding Endometrial Physiology Underpins Menstrual Health." Nature Reviews Endocrinology 18.5 (2022): 290-308.

34) Shoupe, Donna. "The Progestin Revolution: Progestins Are Arising as The Dominant Players In The Tight Interlink Between Contraceptives And Bleeding Control." Contraception and Reproductive Medicine 6.1 (2021): 3.

35) Bitzer, Johannes, et al. "Medical Management of Heavy Menstrual Bleeding: A Comprehensive Review of the Literature." Obstetrical & Gynecological Survey 70.2 (2015): 115-130.

36) Maness, David L., et al. "How Best to Manage Dysfunctional Uterine Bleeding." Journal of Family Practice 59.8 (2010).

37) Aksu, M. Feridun, et al. "High-Dose Medroxyprogesterone Acetate for the Treatment of Dysfunctional Uterine Bleeding in 24 Adolescents." Australian and New Zealand Journal of Obstetrics and Gynaecology 37.2 (1997): 228-231.

38) Lithgow, DM and Politzer, WM. "Vitamin A in the Treatment of Menorrhagia." South African Medical Journal 51.7 (1977): 191-193.

39) Zekavat, Omid Reza, et al. "Acquired vitamin K deficiency as unusual cause of bleeding tendency in adults: a case report of a nonhospitalized student presenting with severe menorrhagia." Case Reports in Obstetrics and Gynecology 2017.1 (2017): 4239148.

40) Abdou, Merna M., Kevin P. Forey, and Leslie Padrnos. "24-Year-Old Woman With Menorrhagia, Mucosal Bleeding, and Easy Bruising." Mayo Clinic Proceedings. Vol. 97. No. 7. Elsevier, 2022.

41) Dikke, Galina B., et al. "Experience of Treating Patients with Abnormal Uterine Bleeding Associated with Ovulatory Dysfunction." Obstetrics and Gynecology 3 (2024): 142-152.

42) Jo, Eun-Jung, et al. "Clinical Characteristics of Exogenous Progestogen Hypersensitivity." Asian Pacific Journal of Allergy and Immunology 37.3 (2019): 183-187.

43) Azadi, Negar, et al. "Effect of Progesterone Administration on Tissue Mast Cell Population and Histamine Content in Mice Uterus After Ovulation Induction." JBRA Assisted Reproduction 27.3 (2023): 436.

44) Uchida, Hitoshi, et al. "Neurosteroid-Induced Hyperalgesia Through a Histamine Release Is Inhibited by Progesterone and P, P'-DDE, an Endocrine Disrupting Chemical." Neurochemistry international 42.5 (2003): 401-407.

45) Mittman, Robert J., et al. "Progesterone-Responsive Urticaria and Eosinophilia." Journal Of Allergy and Clinical Immunology 84.3 (1989): 304-310.

46) Vliagoftis, Harissis, et al. "Progesterone Triggers Selective Mast Cell Secretion Of 5-Hydroxytryptamine." International Archives of Allergy and Immunology 93.2-3 (1990): 113-119.

47) Sandru, Florica, et al. "Progesterone Hypersensitivity in Assisted Reproductive Technologies: Implications for Safety and Efficacy." Journal of Personalized Medicine 14.1 (2024): 79.

48) Sashidhar, Nivedita, et al. "Exogenous Progestogen Hypersensitivity and its Increasing Association with Assisted Reproductive Techniques (ART)/in vitro Fertilization (IVF)." Indian Dermatology Online Journal 15.1 (2024): 24-32.

49) Dhaliwal, Gurnoor, et al. "Progesterone Hypersensitivity Induced by Exogenous Progesterone Exposure." Cureus 15.9 (2023).

50) Bernstein, I. Leonard, et al. "A Case of Progesterone-Induced Anaphylaxis, Cyclic Urticaria/Angioedema, And Autoimmune Dermatitis." Journal of Women's Health 20.4 (2011): 643-648.

51) Pike, Michael G., et Al. "Increased Intestinal Permeability in Atopic Eczema." Journal of Investigative Dermatology 86.2 (1986): 101-104.

52) Charlesworth, E. N., et al. "Cutaneous Late-Phase Response In Food-Allergic Children And Adolescents With Atopic Dermatitis." Clinical & Experimental Allergy 23.5 (1993): 391-397.

53) Wright, Jonathan. "Treatment of Childhood Asthma with Parenteral Vitamin B12, Gastric Re-Acidification, and Attention to Food Allergy, Magnesium and Pyridoxine: Three Case Reports with Background and An Integrated Hypothesis." Journal of Nutritional Medicine 1.4 (1990): 277-282.

54) Wright, Jonathan V., and Lane Lenard. Why Stomach Acid Is Good for You: Natural Relief from Heartburn, Indigestion, Reflux and GERD. Rowman & Littlefield, 2001.

55) Guilliams, Thomas G., and Lindsey E. Drake. "Meal-Time Supplementation with Betaine HCl for Functional Hypochlorhydria: What is the Evidence?" Integrative Medicine: A Clinician's Journal 19.1 (2020): 32.

56) Untersmayr, Eva, and Erika Jensen-Jarolim. "The Role of Protein Digestibility and Antacids on Food Allergy Outcomes." Journal of Allergy and Clinical Immunology 121.6 (2008): 1301-1308.

57) Bulletti, Carlo, et al. "Progesterone: The Key Factor of The Beginning of Life." International Journal of Molecular Sciences 23.22 (2022): 14138.

58) Kolatorova, Lucie, et al. "Progesterone: A Steroid with Wide Range of Effects in Physiology as Well As Human Medicine." International Journal of Molecular Sciences 23.14 (2022): 7989.

59) Deniselle, Maria Claudia Gonzalez, et al. "Basis of Progesterone Protection in Spinal Cord Neurodegeneration." The Journal of Steroid Biochemistry and Molecular Biology 83.1-5 (2002): 199-209.

60) De Nicola, Alejandro F., et al. "Progesterone and Allopregnanolone Neuroprotective Effects in The Wobbler Mouse Model of Amyotrophic Lateral Sclerosis." Cellular and Molecular Neurobiology 42.1 (2022): 23-40.

61) Fedotcheva, Tatiana A., et al. "Progesterone as an Anti-Inflammatory Drug and Immunomodulator: New Aspects In Hormonal Regulation Of The Inflammation." Biomolecules 12.9 (2022): 1299.

62) Al-Kuraishy, Hayder M., et al. "New Insights on The Potential Effect Of Progesterone In Covid-19: Anti-Inflammatory And Immunosuppressive Effects." Immunity, Inflammation and Disease 11.11 (2023): e1100.

63) Balan, Irina, et al. "Neuroactive Steroids, Toll-like Receptors, and Neuroimmune Regulation: Insights into Their Impact on Neuropsychiatric Disorders." Life 14.5 (2024): 582.

64) Pinna, Graziano. "Allopregnanolone, the Neuromodulator Turned Therapeutic Agent: Thank You, Next?" Frontiers In Endocrinology 11 (2020): 236.

65) Pinna, Graziano. "Role of PPAR-Allopregnanolone Signaling In Behavioral And Inflammatory Gut-Brain Axis Communications." Biological Psychiatry 94.8 (2023): 609-618.

66) Traish, Abdulmaged M. "Post-finasteride syndrome: a surmountable challenge for clinicians." Fertility and sterility 113.1 (2020): 21-50.

67) Giatti, Silvia, et al. "Post-Finasteride Syndrome and Post-SSRI Sexual Dysfunction: Two Clinical Conditions Apparently Distant, But Very Close." Frontiers in Neuroendocrinology 72 (2024): 101114.

68) Lonergan, Eibhlin Marie, et al. "8679 Case: The Sexual, Physical and Psychological Effects of Post-Finasteride Syndrome." Journal of the Endocrine Society 8. Supplement_1 (2024): bvae163-1570.

69) Hrushesky, WilliamJ M., Avrum Z Bluming, and ScottA Gruber. "Menstrual Influence on Surgical Cure Of Breast Cancer." The Lancet 335.8695 (1990): 984.

70) Ratajczak, H. V., R. B. Sothern, and W. J. Hrushesky. "Estrous Influence on Surgical Cure of a Mouse Breast Cancer." The Journal of Experimental Medicine 168.1 (1988): 73-83.

71) Zhang, Baoning. "Prognosis of Patients with Breast Cancer Related to The Timing of Operation During Menstrual Cycle." Chinese Journal of Cancer Research 10 (1998): 138-142.

72) Pujol, Pascal, et al. "A Prospective Prognostic Study of The Hormonal Milieu at The Time of Surgery in Premenopausal Breast Carcinoma." Cancer: Interdisciplinary International Journal of the American Cancer Society 91.10 (2001): 1854-1861.

72) Badwe, Rajendra, et al. "Single-Injection Depot Progesterone Before Surgery and Survival in Women with Operable Breast Cancer: A Randomized Controlled Trial." (2011).

73) Badwe, R. A., et al. "Timing of Surgery During Menstrual Cycle and Survival of Premenopausal Women with Operable Breast Cancer." The Lancet 337.8752 (1991): 1261-1264.

74) Badwe, R. A., et al. "Serum Progesterone at The Time of Surgery and Survival in Women With Premenopausal Operable Breast Cancer." European Journal of Cancer 30.4 (1994): 445-448.

75) Chakravorty, Gaurav, et al. "Deciphering the Mechanisms of Action of Progesterone In Breast Cancer." Oncotarget 14 (2023): 660.

76) Mohammed, Hisham, et al. "Progesterone Receptor Modulates ER-Alpha Action in Breast Cancer." Nature 523.7560 (2015): 313-317.

77) Leo, Joyce CL, et al. "Gene Regulation Profile Reveals Consistent Anticancer Properties of Progesterone in Hormone-Independent Breast Cancer Cells Transfected with Progesterone Receptor." International Journal of Cancer 117.4 (2005): 561-568.

78) Nagy, Bernadett, et al. "Key to Life: Physiological Role and Clinical Implications Of Progesterone." International Journal of Molecular Sciences 22.20 (2021): 11039.

79) Lieberman, A.; Curtis, L. In Defense of Progesterone: A Review of the Literature. Altern. Ther. Health Med. 2017, 23, 24–32.

80) Asi, Noor, et al. "Progesterone vs. synthetic Progestins and The Risk of Breast Cancer: A Systematic Review and Meta-Analysis." Systematic Reviews 5 (2016): 1-8.

81) Atif, Fahim, et al. "Progesterone Treatment Attenuates Glycolytic Metabolism and Induces Senescence in Glioblastoma." Scientific Reports 9.1 (2019): 988.

82) Mohamed, Omyma Shehata, et al. "Cytoprotective Effect and Clinical Outcome Of Perioperative Progesterone In Brain Tumors, A Randomized Microscopically Evidence Study." Egyptian Journal of Anaesthesia 38.1 (2022): 466-475.

83) Luo, Hui, et al. "Prognostic Value of Progesterone Receptor Expression In Ovarian Cancer: A Meta-Analysis." Oncotarget 8.22 (2017): 36845.

84) Kim, Olga, et al. "Targeting Progesterone Signaling Prevents Metastatic Ovarian Cancer." Proceedings of the National Academy of Sciences 117.50 (2020): 31993-32004.

85) Tamburello, Mariangela, et al. "Preclinical Evidence of Progesterone as A New Pharmacological Strategy in Human Adrenocortical Carcinoma Cell Lines." International Journal of Molecular Sciences 24.7 (2023): 6829.

86) Mahbub, Amani A., et al. "Enhanced Anti-Cancer Effects of Estrogen and Progesterone Co-Therapy Against Colorectal Cancer in Males." Frontiers in Endocrinology 13 (2022): 941834.

87) Jabeen, Muafia, et al. "Synthesis, Pharmacological Evaluation and Docking Studies of Progesterone and Testosterone Derivatives as Anticancer Agents." Steroids 136 (2018): 22-31.

88) Chakraborty, Ajanta, et al. "Progesterone Receptor Agonists and Antagonists as Anticancer Agents." Mini Reviews in Medicinal Chemistry 10.6 (2010): 506-517.

89) Zhou, Yasi, et al. "Progesterone Induces Glioblastoma Cell Apoptosis by Coactivating Extrinsic and Intrinsic Apoptotic Pathways." Molecular & Cellular Toxicology 20.1 (2024): 107-117.

90) Gonzalez-Orozco, J.C.; Camacho-Arroyo, I. Progesterone Actions During Central Nervous System Development. Front. Neurosci.2019, 13, 503.

91) Espinoza, T.R.; Wright, D.W. The Role of Progesterone in Traumatic Brain Injury. J. Head Trauma Rehabil. 2011, 26, 497–499.

92) Karha, Juhana, and Eric J. Topol. "The Sad Story of Vioxx, And What We Should Learn from It." Cleve Clin J Med 71.12 (2004): 933-934.

Chapter 15

Diagnosis and Treatment of Premenstrual Syndrome Part Three

GILDA IS A 42-YEAR-OLD ZUMBA instructor and mother of three children. She has been doing well on 120 mg (2 grains) of natural desiccated thyroid for Hashimoto's thyroid disease. Lately, Gilda's premenstrual syndrome (PMS) symptoms have been worsening. Gilda's symptoms occur during the last two weeks of the menstrual cycle, during the luteal phase, when she reports anxiety and irritability, breast tenderness, and headaches. In the past, Gilda's 28-day menstrual cycles have always been regular, however recent laboratory testing on day 19 of her monthly cycle showed a normal estradiol level of 170 pg/ml, and a low progesterone level of 0.1 ng/ml. The low progesterone indicates anovulation and "estrogen dominance". Gilda was started on progesterone topical cream 50 mg for days 12-26 of the menstrual cycle. A few months later, I received this email from Gilda:

Dear Dr. Dach,
Hello and good morning, I hope everything is well with you all! My gynecologist asked me to have the composition of the progesterone cream I am using, the reasons I am using it, and for how long I will be using it. Thank you!
From Gilda

My Reply to Gilda:

Dear Gilda,
The natural progesterone dosage is 50 mg per day as a topical cream. Progesterone is a treatment for PMS associated with anovulatory cycles. The length of treatment is yet to be determined. If there are any other questions, I am happy to discuss with your doctor. Give them my office number to call me: 954-792-4663.
Regards, from Dr. D.

Symptoms of PMS

It is not unusual for mainstream OB/GYN doctors to have no knowledge of natural progesterone for PMS. It may even seem foreign to them. OB/GYN medical training and educational journals tell them natural progesterone does not work for PMS, so do not even think about it. For the mainstream OB/GYN doctor, the first-line treatments for PMS are SSRI antidepressants and oral contraceptive pills (OCPs). Although both are effective, their adverse side effects preclude their use except under the most extreme circumstances. Effective safer measures are available for most women as described in this chapter. The treatment strategy is important since PMS affects more than half of reproductive-age women worldwide. The common symptoms during the luteal phase of the menstrual cycle are breast tenderness and pain, weight gain, bloating, headache, mood swings, depression, anxiety, anger, and irritability. **Note:** The follicular phase is the first 11 days before ovulation. The luteal phase is the last 14-17 days after ovulation. (1-5)

Katharina Dalton Progesterone for PMS

In 1953, Katharina Dalton (1916–2004), a British MD coined premenstrual syndrome (PMS) and opened a women's clinic in London. For the next 40 years, Dr. Dalton specialized in diagnosis and treatment of PMS with injectable progesterone (25 mg) initially and later, oral micronized progesterone. One might wonder, what were Dr. Dalton's thoughts in 1994 when the American Psychiatric Association (APA) renamed PMS as a mental disorder called premenstrual dysphoric disorder (PMDD)? The answer comes from Dr. Sally King in 2020, who says Dr. Dalton openly criticized the APA, saying they were hijacking our understanding of PMS. Dr. Dalton felt that to understand PMS, the emphasis should be placed on physical symptoms and hormonal fluctuations rather than mood disturbances and psychological aspects. Sally King PhD, doctorate in medical sociology at King's College London, writes:

> Greene and Dalton (1953) argued that premenstrual symptoms were far more extensive than just 'nervous tension.' The most prominent 'PMS expert' at this time, Dr. Katharina Dalton, attempted to counter the undue emphasis on mood-based menstrual symptoms, and openly criticized what she called "the hijacking of PMS by psychologists" (Dalton and Holton 2000). (6-8)

Initially, Dr. Dalton treated PMS with injectable progesterone, and later vaginal progesterone suppositories with starting dosage of 400 mg twice a day (800 mg/day). Oral micronized progesterone had been widely available in Europe since 1980. Dr. Dalton's classic 1953 paper on PMS in the British Medical Journal describes how the PMS was closely aligned with the menstrual cycle and was not coincidental. The cause of PMS is high estrogen to progesterone ratio. Dr. Dalton found a 93 percent success rate (cured or improved) in treating PMS with 25 mg of injectable progesterone in 88 patients. Dr. Dalton writes:

> All the cases had attacks of various symptoms, which occurred during the premenstrual phase, during menstruation or ovulation, or at the time of a missed period, and all cases were symptom-free at other times. All cases included in the series had experienced attacks during each of the last three menstrual cycles; thus any chance coincidence between the attack and menstruation was eliminated...patients were asked to keep a calendar...Symptoms: **Headache, nausea, irritability, depression, joint pain, pitting edema, asthma, epilepsy, mastalgia, acne, eczema, glossitis**...Is Water Retention the Cause of the Syndrome ? That the symptoms are due to **water retention**

is strongly suggested by the oliguria and gain in weight which announces their arrival and the diuresis and loss of weight which accompanies their relief at, or soon after, the onset of menstruation...the trouble was not so much a high level of estradiol as a lack of antagonism by progesterone... the cause of the syndrome is an **abnormally high estradiol/progesterone ratio**... painful breasts, so common a feature of the syndrome, are due to a high estrogenic level...**The results with intramuscular progesterone in 61 cases were more satisfactory- (83.5%) became free from symptoms, 4 (6.6%) improved**, and 4 (6.6%) obtained no relief...**Treatment with a progestogen is almost invariably successful.** (9-11)

PMS, Crime and Suicide

Dr. Dalton found a connection between PMS, crime, and suicide, finding half of all female crimes and suicides in England between 1950 and 1970 occurred in the four days before menstruation, typically the worst PMS days of the menstrual cycle. However, Dr. Dalton's success with progesterone for PMS was met with opposition from the medical community which failed to replicate her success, citing numerous randomized controlled trials (RCTs) showing progesterone ineffective for PMS. For example, in 1986 Dr. Sarah Maddocks did an RCT of 20 women with PMS finding "response to vaginal progesterone in these dosages [200 mg] is at best, marginal and not significantly different from response with placebo use." Mainstream opposition to progesterone as a treatment for PMS continues to the current day. (12-18)

Insufficient Progesterone Dosage – The Progesterone Paradox

Dr. Dalton's reply to her critics was that the progesterone dosage was insufficient to achieve success. Because of receptor cross-talk, small doses of progesterone may aggravate rather than calm estrogen dominance symptoms, making all the PMS symptoms worse. This can be quite frustrating to the patient and doctor representing a pitfall when initiating progesterone. After a few days, this resolves and progesterone returns to its calming effects. Here is an example of this paradoxical progesterone effect when using small doses. This newsletter from Women's International Compounding Pharmacy (now renamed Belmar Compounding Pharmacy) explains progesterone stimulates estrogen receptors which potentiated estrogen effect requiring higher doses of progesterone for a good outcome, writing:

> Ironically, estrogen side-effects may occur when progesterone therapy is initiated. Estrogen symptoms such as headaches, nausea, and depression sometimes get worse with progesterone replacement, **particularly when the dose is small. Progesterone stimulates estrogen receptors for estrogen. The initial stimulation occurs and potentiates the estrogen effect.** When an activity is potentiated, the amount of non-converted progesterone may not be enough to counter or balance these symptoms. Higher doses of progesterone may be needed. (19)

Although mainstream medicine acknowledges that small doses make PMS symptoms worse (the progesterone paradox), there is no acknowledgment that higher doses are needed. In 2020, Dr. Inger Sundström-Poromaa claims ignorance of the mechanism of progesterone's benefits and harms, writing:

> As far as progesterone is concerned, we might be dealing with a two-edged sword: while its metabolite allopregnanolone has been proven useful for treatment of PPD [post-partum depression], **it may trigger negative symptoms in women with PMS and PMDD**. Overall, our current knowledge on the beneficial and harmful effects of progesterone is limited and further research is imperative. Note: Premenstrual Dysphoric Disorder (PMDD) is another name for PMS. (20)

In clinical practice, many PMS patients require higher doses of progesterone to achieve adequate blood levels as discussed below by Dr. Phyllis Bronson. Both Drs. Dalton and Bronson's views on PMS were validated in 2019 by Dr. Chutima Roomruangwong's study of 21 women with PMS. Dr. Roomruangwong found that PMS symptoms could be predicted by lower steady-state serum levels of progesterone, as well as the declining levels of progesterone during the luteal phase. In 1998 Dr. Lena Seippel and in 2021, Dr. Nur Indah Noviyanti both found PMS associated with higher estrogen levels in the luteal phase. These findings agree with Dr. Bronson's observations that the combination of high estrogen and low progesterone levels during the luteal phase predicts PMS symptoms (21-23)

Studies Showing Efficacy of Progesterone for PMS

Several positive studies showing the efficacy of progesterone for PMS have been largely ignored. For example, in 1985, Dr. Professor Lorraine Dennerstein Professor Emeritus in the Department of Psychiatry, at the University of Melbourne, Australia, ran a four-month double-blind crossover trial of progesterone for PMS using 300 mg of micronized oral progesterone for the luteal phase (days 14-24) of the menstrual cycle (100 mg PO with breakfast, and 200 mg PO at bedtime). Dr. Dennerstein found oral micronized progesterone effective for PMS, writing:

> This study showed that an oral formulation of micronized progesterone **was effective** in alleviating many premenstrual complaints including those of anxiety, stress, depression, hot flushes, swelling, and water retention. (24)

In 2017, Dr. Olha Horbatiuk from Ukraine studied 37 premenopausal women with regular menstrual cycles suffering from severe PMS, compared to 32 controls without PMS. The women were treated with sublingual micronized progesterone for days 11-25 of the menstrual cycle. Full resolution of PMS symptoms was noted in 86 percent of patients, writing:

> 86.5% of all women of the main group were observed to have had full regression of severe PMS clinical effects, while 13.5% of all women suffering from severe PMS and who were within the decompensated stage were observed to have a decrease in PMS symptoms. (25)

Phyllis Bronson PhD, Progesterone for PMS

Phyllis Bronson, Ph.D., is a biochemist and progesterone expert from Aspen, Colorado. Her book is entitled, *Moods, Emotions, and Aging: Hormones and the Mind-body Connection (2013).* Dr. Bronson was mentored by the late Abram Hoffer MD, the father of orthomolecular medicine. I first saw Phyllis Bronson PhD in 2007 at the podium of the Orthomolecular Medicine meeting in Toronto. She was a keynote speaker on Women's Bio-Identical Hormone Replacement discussing the biochemistry of estradiol and progesterone, and their effects on mood, depression, and brain function. One of the things she explained was how hormonal imbalance causes Pre-Menstrual Syndrome (PMS), and natural progesterone is the preferred treatment rather than SSRI anti-depressant drugs. Over many decades, Dr. Phyllis Bronson worked as a team member managing patient care in a busy bioidentical hormone practice dealing with PMS, women's hormonal imbalance and associated emotional disturbances. Replicating Dr. Katharina Dalton's work, Dr. Bronson found success using topical progesterone 100 mg. applied twice a day. Again, Dr. Phyllis Bronson found higher doses of progesterone were needed in some patients who required up to 600 mg per day of topical progesterone. Dr. Bronson's treatment protocol included frequent lab testing with serum estradiol and progesterone levels, finding patients felt best when progesterone levels reached at least 4 ng/ml and at the same time, estradiol below 100 pg/ml. In some cases, relief of mood symptoms required progesterone levels of 8-15 ng/ml. Dr. Bronson found that estrogen is neuro-excitatory, and high levels of estrogen may predispose to anxiety and panic attacks. Natural progesterone, on the other hand, is neuro-inhibitory, with calming effects. In 2001, Dr. Phyllis Bronson did a clinical study of patients with PMS at the Aspen Clinic for Preventive and Environmental Medicine, writing:

> Interestingly, for subject 2, all her adult life she had thought she was prone to significant depression. She had been treated with psychotropic medications, specifically Paxil and Prozac with little relief. After treatment with natural progesterone for fifteen months, from October of 1996 until December of 1997, she finally felt really well premenstrually for the first times in years. This has continued but with some variations...She reported feeling best at times when progesterone levels came up at least to 4 ng/ml, and estrogen levels decreased below 100 pg/ml...**After the introduction of micronized progesterone, in the form of transdermal progesterone cream, premenstrual tension has essentially been obliterated**....we are seeing this trend show up in many more subjects not in the original study but being seen at our clinic...The evidence points toward the premise that in anxiety prone women, **when progesterone levels are low relative to estrogen**, these subjects feel tense and irritable, and exhibit other criteria of general anxiety. Relief of these symptoms is generally seen when progesterone is measured at **8-15 ng/ml.** The mood changes were qualified as follows. **As the progesterone levels rose gradually, most symptoms of acute anxiety disappeared**. Chronic symptoms took longer to dissipate, although they too diminished over time. The subjects also reported that if they were not diligent about using progesterone, symptoms recurred...(26-28)

The Safety of High-Dose Progesterone

The safety of high doses of progesterone is evident from fact that in pregnancy, progesterone levels are greater than that of exogenous progesterone by any route. In 2012, Dr. Pratap Kumar says progesterone levels in the third trimester of pregnancy may reach 100-200 ng/ml. Compare this to the 8-15 ng/ml described above by Dr. Bronson for progesterone levels obtained when treating PMS. Dr. Pratap Kumar writes:

> When the pregnancy reaches term gestation, progesterone levels range from 100-200 ng/ml and the placenta produces about 250 mg/day. (29)

Chaste Tree: Vitex Agnus Castus for PMS

Perhaps the most successful botanical extract for PMS is called Chaste Tree or Vitex Agnus used for centuries in folk medicine for menstrual disorders. In 2006, Dr. Tori Hudson, Adjunct Clinical Professor of Naturopathic Medicine at Bastyr University, described Vitex Agnus as "the single most important plant for the treatment of premenstrual syndrome". (30-33)

Vitex Consumption by Wild Female Chimpanzees

In 2008, Dr. Melissa Emery Thompson, Professor and Director of the Comparative Human and Primate Physiology Center at the University of New Mexico studied urinary hormone levels in wild female chimpanzees consuming Vitex fischeri, finding consumption of Vitex fruit caused an abrupt and dramatic increase in progesterone levels, writing:

> V. [Vitex] fischeri consumption was associated with an abrupt and dramatic increase in urinary progesterone levels of female chimpanzees to levels far exceeding the normal range of variation. Female estrogen levels were not significantly impacted, nor were male testosterone levels. (34)

Dopamine like Compounds in Vitex for PMS

In 2003, Dr. Wolfgang Wuttke reviewed the pharmacology and clinical indications for Chaste Tree (Vitex) , finding widespread use of Vitex for PMS symptoms. Dr. Wuttke found the fruit extract of Vitex Agnes Castus (AC) contains dopaminergic compounds, which are perhaps the most clinically useful for treatment of premenstrual mastodynia (breast pain, tenderness) and PMS (premenstrual syndrome) symptoms, writing:

> Extracts of the fruits of chaste tree, Vitex agnus castus [VAC] are widely used to treat premenstrual symptoms. Double-blind placebo-controlled studies indicate that one of the most common premenstrual symptoms, i.e. premenstrual mastodynia (mastalgia) is beneficially influenced by VAC extract. In addition, numerous less rigidly controlled studies indicate that VAC extracts have also beneficial effects on other psychic and somatic symptoms of the PMS. Premenstrual mastodynia is most likely due to a latent hyperprolactinemia, i.e. patients release more than physiologic amounts of prolactin in response to stressful situations and during deep sleep phases which appear to stimulate the mammary gland. Premenstrually this unphysiological prolactin release is so high that the serum prolactin levels often approach heights which are misinterpreted as prolactinomas. Since VAC extracts were shown to have beneficial effects on premenstrual mastodynia serum prolactin levels in such patients were also studied in one double-blind, placebo-controlled clinical study. Serum prolactin levels were indeed reduced in the patients treated with the extract. The search for the prolactin-suppressive principle(s) yielded a number of compounds with dopaminergic properties...almost identical in their prolactin-suppressive properties than dopamine itself. Hence, it is concluded that dopaminergic compounds present in Vitex agnus castus are clinically the important compounds which improve premenstrual mastodynia and possibly also other

symptoms of the premenstrual syndrome. (35-37)

In 2017, Dr. Wolfgang Wuttke again studied Vitex Agnes as a treatment for PMS. Dr. Wuttke found that elevated prolactin levels were causing anovulatory cycles, breast pain, and tenderness. The pituitary release of elevated prolactin levels can be inhibited by the dopamine-like compounds in Vitex Agnes castus (VAC), thus restoring ovulation and eliminating breast pain and other symptoms of PMS. Dr. Wolfgang Wuttke writes:

> The dried fruits of the chaste tree Vitex agnus castus (VAC) were traditionally used by monks as a substitute for pepper and was therefore also called Monk's pepper. For the last 50 years it is commercially provided for the treatment of premenstrual symptoms, particularly to prevent premenstrual mastodynia (mastalgia)...A number of placebo controlled studies gave proof that extracts of VAC had beneficial effects on premenstrual breast pain. This breast sensation is induced by latent hyperprolactinemia which is characterized by secretory episodes of prolactin release by the pituitary in response to stress and deep sleep phases. This latent hyperprolactinemia induces...a corpus luteum insufficiency which is a common reason for infertility [anovulation]...it is well accepted that prolactin release can be reduced by dopamine and dopaminergic drugs. The efficacy of VAC extracts to ameliorate prolactin induced premenstrual mastodynia was therefore suggestive that VAC may contain dopaminergic compounds... Consequently, prolactin release in vitro from dispersed pituitary cells and in vivo in rats and postmenopausal women was inhibited by VAC 1095. Placebo controlled studies proved also the efficacy of VAC extracts to ameliorate premenstrual symptoms. In several placebo-controlled studies a clear relation between reduction of breast pain and reduction of serum prolactin levels could be established. In addition, VAC extracts

were also highly effective in women suffering from fibrocystic mastopathy. In many of these women serum prolactin levels were also elevated and reduced by VAC extracts... The results from all trials suggested that VAC extracts ameliorated premenstrual symptoms including mastodynia, premenstrual dysphoric disorder and latent hyperprolactinemia. Cystic mastopathy and sterility due to corpus luteum insufficiencies were also beneficially influenced. Adverse events with VAC were mild and generally infrequent. (38-41)

Vitex Treatment for Hyperprolactinemia

In 2020, Dr. Naiyereh Haerifar from Iran studied 105 women of reproductive age with elevated prolactin levels. The women were divided into three groups and treated with bromocriptine, Dostinex [cabergoline], or Vitagnus (Vitex) over 4 months. **Note:** bromocriptine and cabergoline are dopaminergic drugs (dopamine D2 receptor agonists) commonly used to treat hyperprolactinemia caused by pituitary adenoma. The women treated with Vitex had a significant increase in estradiol and progesterone levels, and a decrease in prolactin levels. Dr. Naiyereh Haerifar writes:

> The results further revealed that the amounts of estradiol in the Vitagnus group had a significant increase compared to other groups (P < 0.05)...Eventually, the effect of the Vitagnus tablet on progesterone was remarkable compared to the other two medications in the 2nd and 3rd cycles. Conclusions: Similar to other drugs, Vitagnus has a significant effect on the amount of prolactin and sex hormones and thus can be successfully used in treating hyperprolactinemia. Finally, reductions in endometrial thickness were significant in the Vitagnus group [this is a progesterone effect which thins the endometrial stripe visible on pelvic sonogram] compared to the other two groups. (42)

Vitex Meta-Analysis of 17 RCT's by Dr. Verkaik

In 2017, Dr. Saskia Verkaik from Erasmus University Medical Center, Rotterdam published a meta-analysis of 17 RCTs of Vitex agnes castus for the treatment of PMS, finding efficacy equal to oral contraceptives and SSRI antidepressants. In addition, Vitex efficacy was better than all the other herbal extracts and supplements. The obvious question to ask is: why take oral contraceptives or SSRIs when an over-the-counter botanical extract is safer and just as effective for PMS? (43-49)

The Serotonin Connection - 5HTP for PMS and Mood Disorders

In the above discussion of Dr. Phylis Bronson, I failed to mention Dr. Bronsin's attention to neurotransmitter biochemistry and the use of supplements such as tryptophan, an amino acid in our diet and a precursor to the synthesis of 5-hydroxy-tryptophan (5-HTP). In the brain, tryptophan is converted to 5-HTP which is then converted into serotonin, an important neurotransmitter that regulates mood and behavior. The idea here is interventions which increase serotonin levels are beneficial for the PMS patient. In 2009, Dr. Miles Berger discusses the role of serotonin in mood and behavior, writing:

> The behavioral and neuropsychological processes modulated by serotonin include mood, perception, reward, anger, aggression, appetite, memory, sexuality, and attention, among others. Indeed, it is difficult to find a human behavior that is not regulated by serotonin. (50)

5-HTP instead of SSRI Antidepressants for PMS

Multiple placebo-controlled studies have found SSRI antidepressants effective for PMS, thus explaining their wide acceptance by mainstream medicine. **Note:** SSRI drugs are designed to inhibit serotonin re-uptake by the terminal synapse, thus increasing serotonin availability at the synaptic cleft between the pre-and post-synaptic neurons. The next logical thought is: instead of an SSRI drug, why not try oral tryptophan or 5-hydroxy-tryptophan (5-HTP) capsules as a safer way to increase brain serotonin levels? Tryptophan and 5-HTP are both widely available at health food stores, with no prescription required. They are safer because they avoid the adverse effects of SSRI drugs, as discussed in Chapter 26, The Depressing State of Antidepressants. In 2021, Dr. İpek Ayhan studied the mechanism of PMS within the brain, pointing out many of the symptoms of PMS are associated with the serotonergic system, and are made worse by tryptophan depletion, writing:

> Aggression, impulse control, anxiety, sexual behavior, pain, sleep, and eating difficulties are some of the symptoms of PMS, and these symptoms are associated with the serotonergic system. Some studies demonstrated that tryptophan, a serotonin precursor, has antidepressant properties and that acute tryptophan deficiency exacerbates premenstrual symptoms. (51-52)

Tryptophan For PMS

In 1999, Dr. Susanne Steinberg studied the effect of tryptophan on PMS with a 3-month RCT, finding a beneficial effect on PMS symptoms. 37 PMS patients were given oral L-tryptophan, 6 grams /day, and 34 given a placebo. The patients were treated during the luteal phase of the menstrual cycle for four cycles. Dr. Steinberg writes:

> The Visual Analogue Scales (VAS) revealed a significant (p < .004) therapeutic effect of L-tryptophan relative to placebo for the cluster of mood symptoms comprising the items of **dysphoria, mood swings, tension, and irritability**. The magnitude of the reduction from baseline in maximum luteal phase VAS-mood scores was **34.5% with L-tryptophan compared to 10.4% with placebo.** Conclusions: These results suggest that **increasing serotonin synthesis during**

the late luteal phase of the menstrual cycle has a beneficial effect in patients with premenstrual dysphoric disorder. Note: Tryptophan is converted to 5-HTP, the direct precursor to serotonin, and is more effective at lower doses than tryptophan. (53-54)

Progesterone, Allopreganolone and Depression

The major metabolite of progesterone, allopregnanolone, is a neuroactive steroid FDA-approved for the treatment of postpartum depression. Allopregnanolone has far-reaching benefits for mood and emotions. In 2014, Dr. Cornelius Schüle studied the role of allopregnanolone in depression and anxiety, suggesting allopregnanolone acts not only through the GABA neurotransmitter system, but also through enhancement of neurogenesis, myelination, neuroprotection, and HPA axis function, writing:

Reduced levels of allopregnanolone in the peripheral blood or cerebrospinal fluid were found to be associated with **major depression, anxiety disorders, premenstrual dysphoric disorder, negative symptoms in schizophrenia, or impulsive aggression.** The importance of allopregnanolone for the regulation of emotion and its therapeutic use in depression and anxiety may not only involve GABAergic mechanisms, but probably also includes enhancement of neurogenesis, myelination, neuroprotection, and regulatory effects on HPA axis function. Note: Gamma-aminobutyric acid (GABA) is the main inhibitory neurotransmitter in the brain responsible for slowing or reducing brain activity. Oral GABA supplements are available at health food stores and have anti-anxiety and calming effects on stress and sleep. (55-58)

Neurosteroids Poised to Replace SSRI Antidepressants

The progesterone metabolite, allopregnanolone, is an attractive candidate for drug development. Sage Pharmaceuticals has converted allopregnanolone into two drugs. The first drug is Zulresso®, brexanolone, FDA approved on March 19, 2019, for intravenous use in the treatment of post-partum depression. The second drug is Zuranolone, FDA approved on August 4, 2023 (Biogen Inc. and Sage Therapeutics) for the treatment of major depressive disorder and postpartum depression. In 2024, Dr. Jamie Maguire says that because these two allopregnanolone drugs derived from progesterone are safer and more effective than SSRI drugs, they are poised to replace them, writing:

neuroactive steroids have been studied for decades but have risen as a new class of **rapid-acting, durable antidepressants** with a distinct mechanism of action from previous antidepressant treatments and from other compounds covered in this issue. **Neuroactive steroids are natural derivatives of progesterone but are proving effective as exogenous treatments.** The best understood mechanism is that of **positive allosteric modulation of GABAA receptors,** where subunit selectivity may promote their profile of action. Mechanistically, there is some reason to think that neuroactive steroids may separate themselves from the liabilities of other GABA modulators [benzodiazepines], although research is ongoing. (59)

Oral Micronized Progesterone Metabolizes to Allopregnanolone

Oral micronized progesterone, the economic competitor and enemy of the drug industry is an excellent source of allopregnanolone. Because of the first pass through the liver, oral progesterone is promptly metabolized to allopregnanolone. The oral route is more effective than the transdermal route in which progesterone directly enters the circulation without the first pass effect. Thus, oral micronized progesterone is an excellent way to raise allopregnanolone levels and provide benefits for depression, anxiety, and other neuropsychiatric symptoms. In clinical practice, I have found progesterone useful for SSRI withdrawal syndrome helping to alleviate symptoms of anxiety, depression, and insomnia in women tapering off SSRI antidepressants. This has not been studied and probably should be. For more on this topic see Chapter 26, The Depressing State of Antidepressants. (60-63)

Myo-Inositol to Restore Regular Menstrual Cycles and Stop Abnormal Bleeding

Another supplement at the health food store useful for the anovulatory young female is myo-inositol, sometimes used in combination with Vitex agnes castus. For young females with abnormal uterine bleeding associated with anovulatory menstrual cycles, anemia, and PMS symptoms, myoinositol is a safe nutritional supplement shown to be highly effective in restoring ovulation and normalizing the menstrual cycle. In 2024, Dr. Galina Dikke from St. Petersburg, Russia, studied the use of micronized progesterone combined with myoinositol, D-chiroinositol (5:1 ratio) in 2,042 women, average age 30 years, having anovulatory heavy menstrual bleeding and PMS. The women were treated for 6 months. For the first three months, the women were treated with myoinositol combined with progesterone, and for the second three months with myoinositol alone. **After 6 months, 91 percent of the women achieved normal menstrual cycles with resolution of anemia.** Dr. Galina Dikke writes:

> Abnormal uterine bleeding (AUB) associated with ovulatory dysfunction (OD) [anovulation] is the most common finding among women with chronic AUB, accounting for 57.7% of cases. Oral progestogens [synthetic progestins] are often prescribed for irregular and copious menstruation. However, a course of hormonal rehabilitation after AUB-OD may not be enough. **Inositols** have been shown to be highly effective in restoring ovulation, normalizing the menstrual cycle. (64-68)

Mainstream Use of Progestins for Abnormal, Dysfunctional Uterine Bleeding

Synthetic progestins such as medroxyprogesterone (MPA) are mainstream medical treatment for dysfunctional uterine bleeding (DUB), and their carcinogenicity is overlooked. Failing medical therapy leads to hysterectomy in an estimated 30 percent of women with DUB as discussed in 2015 by Dr. Simar Kauer. (69-73)

Vitamin B6

Vitamin B6 is a cofactor in the synthesis of serotonin from tryptophan and 5-HTP and for the synthesis of dopamine from tyrosine. Consuming vitamin B6, or its biologically active form, P5P, pyridoxal 5-phosphate, improves brain production of both serotonin and dopamine. This suggests the utility of B6 in PMS and other neuropsychiatric syndromes. Several studies show taking B6 or P5P provides a benefit over a placebo for PMS. The P5P version of this vitamin is preferable since P5P avoids toxicity associated with higher doses of B6 as discussed in 2022 by Dr. Pramod Reddy. In 2016, Dr. S. Samieipoor reviewed Vitamin B6 in the treatment of PMS, finding vitamin B6 is effective treatment for PMS symptoms. Dr. S. Samieipoor writes:

> Conclusion: The results of our meta-analysis confirmed **vitamin B6 as a beneficial, inexpensive, and effective treatment for PMS symptoms.** Therefore, the administration of this treatment option will enable midwives to achieve the important goal of reducing PMS symptoms. (74-78)

Cannabidiol (CBD) for PMS

Cannabidiol (CBD is the non-psychoactive component of cannabis. Several studies suggest CBD is useful in ameliorating PMS symptoms. In 2017, Dr Melissa Slavin reviews the historical use of cannabis for menstrual-related symptoms, writing:

Little formal research has been done to address the use of cannabis for women suffering from PMS/PMDD, but it has been suggested as an efficacious and safe alternative treatment for a wide range of women's conditions including dysmenorrhea (Russo 2002), as well as PMS itself, in doses low enough not to cause cognitive impairment (Grinspoon & Bakalar 1997). Cannabis has a long medicinal history dating back to cultivation in China in 4000 B.C., where Chinese emperor Chen Nung recommended its use for 'female disorders' amongst many other illnesses (Grinspoon & Bakalar 1997). Anecdotally, cannabis was prescribed by Queen Victoria's personal physician, Sir John Russell Reynolds, for her menstrual discomfort throughout her adult life (Russo 2002). After 30 years of experience with cannabis, Reynolds reported that cannabis helped alleviate symptoms associated with menstruation (Reynolds 1890). Current research has found cannabis to ameliorate a variety of symptoms that overlap with those of PMS and PMDD, including sleep problems, irritability, depression, and joint pain (Earleywine 2005; Russo et al. 2007; Russo & Hohmann 2013). (79-81)

Conclusion: Although mainstream medicine considers SSRI antidepressants and oral contraceptives (OCPs) effective standard of care for PMS, their adverse side effects make them a poor choice. As seen in the above meta-analysis by Dr. Saskia Verkaik, a naturally occurring botanical called Vitex is equally effective without the side effects associated with OCPs and SSRIs. Natural substances such as Vitex, inositol, 5-HTP, and vitamin B6 have efficacy as well. Although natural progesterone treatment of PMS has been rejected by mainstream medicine in favor of pharmaceuticals, this is one of the tragic errors of modern medicine. High-dose natural progesterone is safe and effective for PMS, and is the preferred treatment. Progesterone is available over the counter (OTC), or from compounding pharmacies as topical creams, micronized oral capsules, and vaginal suppositories. Various natural remedies may be added as needed: Vitex agnus castus, myoinositol, 5-HTP, vitamin B6 (P5P), and CBD.

♦ **References for Chapter 15. Diagnosis and Treatment of Premenstrual Syndrome Part Three**

1) Takeda, Takashi. "Premenstrual Disorders: Premenstrual Syndrome and Premenstrual Dysphoric Disorder." Journal of Obstetrics and Gynaecology Research 49.2 (2023): 510-518.

2) Yoshimi, Kana, et al. "Current Status and Problems in The Diagnosis and Treatment of Premenstrual Syndrome and Premenstrual Dysphoric Disorder from The Perspective of Obstetricians and Gynecologists in Japan." Journal of Obstetrics and Gynaecology Research 49.5 (2023): 1375-1382.

3) Tiranini, Lara, and Rossella E. Nappi. "Recent Advances in Understanding/Management Of Premenstrual Dysphoric Disorder/Premenstrual Syndrome." Faculty reviews 11 (2022).

4) Hofmeister, Sabrina, and Seth Bodden. "Premenstrual syndrome and premenstrual dysphoric disorder." American Family Physician 94.3 (2016): 236-240.

5) Dickerson, Lori M., et al. "Premenstrual syndrome." American Family Physician 67.8 (2003): 1743-1752.

6) King, Sally. "Premenstrual Syndrome (PMS) And the Myth of The Irrational Female." The Palgrave Handbook of Critical Menstruation Studies (2020): 287-302.

7) Dalton, Katharina, and Wendy M. Holton. 2000. The PMS Bible: The Guide to Understanding and Treating PMS. 6th ed. London: Vermilion.

8) Zachar, Peter, and Kenneth S. Kendler. "A Diagnostic and Statistical Manual of Mental Disorders History of Premenstrual Dysphoric Disorder." The Journal of Nervous and Mental Disease 202.4 (2014): 346-352.

9) Greene, Raymond, and Katharina Dalton. "The Premenstrual Syndrome." British Medical Journal 1.4818 (1953): 1007.

10) Dalton, Katharina. "What is this PMS?" The Journal of the Royal College of General Practitioners 32.245 (1982): 717.

11) de Lignières, Bruno. "Oral Micronized Progesterone." Clinical Therapeutics 21.1 (1999): 41-60.

12) Slater, Lauren. "The Prophet of PMS." The New York Times Magazine, 26 Dec. 2004, p. 34. https://www.nytimes.com/2004/12/26/magazine/the-prophet-of-pms.html

13) Zietal, Bianca, "Katharina Dorothea Dalton (1916–2004)". Embryo Project Encyclopedia (2017-05-24). Arizona State University. School of Life Sciences. (2017)

14) Maddocks, Sarah., et al. "A Double-Blind Placebo-Controlled Trial of Progesterone Vaginal Suppositories in The Treatment of Premenstrual Syndrome." American Journal of Obstetrics and Gynecology 154.3 (1986): 573-581.

15) Freeman, E., et al. "Ineffectiveness of Progesterone Suppository Treatment for Premenstrual Syndrome." JAMA 264.3 (1990): 349-353.

16) Wyatt, Katrina, et al. "Efficacy of Progesterone and Progestogens In Management Of Premenstrual Syndrome: Systematic Review." BMJ 323.7316 (2001): 776.

17) O'Brien, P. M. "Helping Women with Premenstrual Syndrome." BMJ: British Medical Journal 307.6917 (1993): 1471.

18) Sampson, G. A. "Premenstrual syndrome: a Double-Blind Controlled Trial of Progesterone and Placebo." The British Journal of Psychiatry: The Journal of Mental Science 135 (1979): 209-215.

19) A Connection with Yeast a Generalized Imbalance That May Leave You Feeling "Bad All Over" by Women's International Compounding Pharmacy Newsletter Online. https://womensinternational.com/blog/portfolio-items/yeast/

(20) Sundström-Poromaa, Inger, et al. "Progesterone–Friend or Foe?" Frontiers in Neuroendocrinology 59 (2020): 100856.

21) Roomruangwong, Chutima, et al. "Lowered Plasma Steady-State Levels of Progesterone Combined with Declining Progesterone Levels During the Luteal Phase Predict Peri-Menstrual Syndrome And Its Major Subdomains." Frontiers in Psychology 10 (2019): 488965.

22) Noviyanti, Nur Indah, et al. "The Effect of Estrogen Hormone on Premenstrual Syndrome (PMS) Occurrences in Teenage Girls at Pesantren Darul Arqam Makassar." Gaceta Sanitaria 35 (2021): S571-S575.

23) Seippel, Lena, and Torbjörn Bäckström. "Luteal-Phase Estradiol Relates To Symptom Severity In Patients With Premenstrual Syndrome." The Journal of Clinical Endocrinology & Metabolism 83.6 (1998): 1988-1992.

24) Dennerstein, Lorraine, et al. "Progesterone and the Premenstrual Syndrome: A Double Blind Crossover Trial." Br Med J (Clin Res Ed) 290.6482 (1985): 1617-1621.

25) Horbatiuk, Olha, et al. "Using Micronized Progesterone for Treatment Of Premenopausal Age Women Suffering From Severe Premenstrual Syndrome." Current Issues in Pharmacy and Medical Sciences 30.3 (2017): 138-141.

26) Bronson, Phyllis J. "Mood Biochemistry of Women at Mid-life." Journal of Orthomolecular Medicine 16.3 (2001): 141-154.

27) Bronson, Phyllis J. Moods, Emotions, and Aging: Hormones and the Mind-body Connection. Rowman & Littlefield Publishers, 2013.

28) Bronson, Phyllis J. "In Defence of Estrogen." Journal of Orthomolecular Medicine 22.3 (2007): 147-152.

29) Kumar, Pratap, and Navneet Magon. "Hormones in pregnancy." Nigerian Medical Journal 53.4 (2012): 179-183.

30) Davies, Sasha, and Tori Hudson. The Menopause Companion: A Beginner's Guide to Owning Your Transition, from Peri to Post. Shambhala Publications, 2023.

31) Hudson, Tori. "Premenstrual Syndrome--A Review of Herbal and Nutritional Interventions." Townsend Letter for Doctors and Patients 270 (2006): 126-132.

32) Hudson, Tori. Premenstrual Syndrome; A Natural Approach by Tori Hudson, N.D. https://drtorihudson.com/articles/premenstrual-syndrome-a-natural-approach/ Accessed 2/5/25.

33) Hudon, Tori. Vitex Agnus Castus (Chaste tree). Jun 19th, 2010 by Tori Hudson, N.D. https://drtorihudson.com/botanicals/vitex-agnus-castus-chaste-tree/

34) Emery Thompson, Melissa, et al. "Hyperprogesteronemia in Response to Vitex Fischeri Consumption in Wild Chimpanzees (Pan Troglodytes Schweinfurthii)." American Journal of Primatology: Official Journal of the American Society of Primatologists 70.11 (2008): 1064-1071.

35) Wuttke, Wolfgang, et al. "Chaste Tree (Vitex Agnus-Castus)–Pharmacology and Clinical Indications." Phytomedicine 10.4 (2003): 348-357.

36) Zahid, Hina, et al. "Phytopharmacological Review on Vitex Agnus-Castus: A Potential Medicinal Plant." Chinese Herbal Medicines 8.1 (2016): 24-29.

37) Niroumand, Mina Cheraghi, et al. "Pharmacological and Therapeutic Effects of Vitex Agnus-Castus L.: A Review." Pharmacognosy Reviews 12.23 (2018).

38) Seidlova-Wuttke, Dana, and Wolfgang Wuttke. "The Premenstrual Syndrome, Premenstrual Mastodynia, Fibrocystic Mastopathy, and Infertility Have Often Common Roots: Effects of Extracts of Chasteberry (Vitex Agnus Castus) As A Solution." Clinical Phytoscience 3 (2017): 1-11.

39) Carmichael, A. R. "Can Vitex Agnus Castus Be Used for The Treatment of Mastalgia? What Is the Current Evidence?" Evidence-Based Complementary and Alternative Medicine 5.3 (2008): 247-250.

40) Schellenberg, Ruediger. "Treatment For the Premenstrual Syndrome with Agnus Castus Fruit Extract: Prospective, Randomised, Placebo-Controlled Study." BMJ 322.7279 (2001): 134-137.

41) Zamani, Mehrangiz, et al. "Therapeutic Effect of Vitex Agnus Castus in Patients with Premenstrual Syndrome." Acta Medica Iranica 50.2 (2012): 101-106.

42) Haerifar, Naiyereh, et al. "The Effect of Vitex Agnus Castus Extract on The Blood Level of Prolactin, Sex Hormones Levels, and the Histological Effects on The Endometrial Tissue in Hyperprolactinemic Women." Crescent Journal of Medical & Biological Sciences 7.4 (2020).

43) Verkaik, Saskia, et al. "The Treatment of Premenstrual Syndrome With Preparations Of Vitex Agnus Castus: A Systematic Review And Meta-Analysis." American Journal of Obstetrics and Gynecology 217.2 (2017): 150-166.

44) Pauli GF, Farnsworth NR, Wang ZJ. Opioidergic Mechanisms Underlying The Actions Of Vitex Agnus-Castus L. Biochem Pharmacol. 2011;81(1):170–177.

45) Ibrahim NA, et al. Gynecological Efficacy And Chemical Investigation Of Vitex Agnus-Castus L. Fruits Growing in Egypt. Nat Prod Res. 2008;22(6):537–546.

46) Momoeda M, et al. Efficacy and Safety Of Vitex Agnus-Castus Extract For Treatment Of Premenstrual Syndrome In Japanese Patients: A Prospective, Open-Label Study. Adv Ther. 2014;31(3):362–373. doi:10.1007/s12325-014-0106-z124.

47) Ambrosini A, Di Lorenzo C, Coppola G, Pierelli F. Use of Vitex Agnus-Castus In Migrainous Women With Premenstrual Syndrome: An Open-Label Clinical Observation. Acta Neurol Belg. 2013;113(1):25–29.

48) He Z, Chen R, Zhou Y, et al. Treatment for Premenstrual Syndrome With Vitex Agnus Castus: A Prospective, Randomized, Multi-Center Placebo Controlled Study In China. Maturitas. 2009;63(1):99–103.

49) Ma L, Lin S, Chen R, Wang X. Treatment of Moderate to Severe Premenstrual Syndrome With Vitex Agnus Castus (BNO 1095) In Chinese Women. Gynecol Endocrinol. 2010;26(8):612–616.

50) Berger, Miles, et al. "The Expanded Biology of Serotonin." Annual Review of Medicine 60.1 (2009): 355-366.

51) Ayhan, İpek, et al. "Premenstrual Syndrome Mechanism In The Brain." Demiroglu Science University Florence Nightingale Journal of Medicine 7.2 (2021): 213-224.

52) Menkes DB, et al. Acute Tryptophan Depletion Aggravates Premenstrual Syndrome. J Affect Disord. 1994; 32:37–44.

53) Steinberg, Susanne, et al. "A Placebo-Controlled Clinical Trial Of L-Tryptophan in Premenstrual Dysphoria." Biological Psychiatry 45.3 (1999): 313-320.

54) Weinberg-Wolf, Hannah, et al. "The Effects Of 5-Hydroxytryptophan on Attention And Central Serotonin Neurochemistry In The Rhesus Macaque." Neuropsychopharmacology 43.7 (2018): 1589-1598.

55) Schüle, Cornelius, et al. "The Role Of Allopregnanolone In Depression And Anxiety." Progress in neurobiology 113 (2014): 79-87.

56) Hepsomali, Piril, et al. "Effects of Oral Gamma-Aminobutyric Acid (GABA) Administration on Stress And Sleep In Humans: A Systematic Review." Frontiers in Neuroscience 14 (2020): 559962.

57) Bruni, Oliviero, et al. "Herbal Remedies and Their Possible Effect on The GABAergic System And Sleep." Nutrients 13.2 (2021): 530.

58) Zhu, Wenwen, et al. "GABA and its Receptors' Mechanisms In The Treatment Of Insomnia." Heliyon (2024).

59) Maguire, Jamie L., and Steven Mennerick. "Neurosteroids: Mechanistic Considerations And Clinical Prospects." Neuropsychopharmacology 49.1 (2024): 73-82.

60) Fornaro, Michele, et al. "Antidepressant discontinuation syndrome: A state-of-the-art clinical review." European Neuropsychopharmacology 66 (2023): 1-10.

61) Henssler, Jonathan, et al. "Incidence of Antidepressant Discontinuation Symptoms: A Systematic Review and Meta-Analysis." The Lancet Psychiatry (2024).

62) Andréen, Lotta, et al. "Pharmacokinetics of Progesterone and Its Metabolites Allopregnanolone And Pregnanolone After Oral Administration Of Low-Dose Progesterone." Maturitas 54.3 (2006): 238-244.

63) Arafat, ElSayed S., et al. "Sedative and Hypnotic Effects of Oral Administration of Micronized Progesterone May Be Mediated Through Its Metabolites." American Journal of Obstetrics And Gynecology 159.5 (1988): 1203-1209.

64) Dikke, Galina B., et al. "Experience of Treating Patients With Abnormal Uterine Bleeding Associated With Ovulatory Dysfunction." Obstetrics and Gynecology 3 (2024): 142-152.

65) Carlini, Sara V., et al. "Management of Premenstrual Dysphoric Disorder: A Scoping Review." International Journal of Women's Health (2022): 1783-1801.

66) Mukai T, Kishi T, Matsuda Y, Iwata N. A Meta-Analysis of Inositol for Depression And Anxiety Disorders. Hum Psychopharmacol. 2014;29 (1):55–63. doi:10.1002/hup.2369

67) Carlomagno, Gianfranco, et al. "Myo-Inositol in The Treatment of Premenstrual Dysphoric Disorder." Human Psychopharmacology: Clinical and Experimental 26.7 (2011): 526-530.

68) Sharma, Neha, et al. "Myo-Inositol: A Potential Prophylaxis Against Premature Onset of Labour and Preterm Birth." Nutrition research reviews 36.1 (2023): 60-68.

69) Kaur, Simar, et al. "Hysterectomy for Dysfunctional Uterine Bleeding In The Era Of Uterine Conservation." International Journal of Reproduction, Contraception, Obstetrics and Gynecology 4.4 (2015): 1133-1137.

70) Dikke, Galina B., et al. "Experience of Treating Patients with Abnormal Uterine Bleeding Associated with Ovulatory Dysfunction." Obstetrics and Gynecology 3 (2024): 142-152.

71) Leal, Caio RV, et al. "Abnormal Uterine Bleeding: The Well-Known and The Hidden Face." Journal of endometriosis and uterine disorders (2024): 100071.

72) Jain, Varsha, et al. "Uterine bleeding: How Understanding Endometrial Physiology Underpins Menstrual Health." Nature Reviews Endocrinology 18.5 (2022): 290-308.

73) Shoupe, Donna. "The Progestin Revolution: Progestins Are Arising as The Dominant Players in The Tight Interlink Between Contraceptives and Bleeding Control." Contraception and Reproductive Medicine 6.1 (2021): 3.

74) Samieipoor S, et al. Effects of Vitamin B6 On Premenstrual Syndrome: A Systematic Review And Meta-Analysis. J Chem Pharm Sci. 2016;9(3):1346–1353

75) Calderón-Guzmán, David, et al. "Pyridoxine, Regardless Of Serotonin Levels, Increases Production Of 5-Hydroxytryptophan In Rat Brain." Archives Of Medical Research 35.4 (2004): 271-274.

76) Hartvig, P., et al. "Pyridoxine Effect on Synthesis Rate of Serotonin in The Monkey Brain Measured with Positron Emission Tomography." Journal of Neural Transmission/General Section JNT 102 (1995): 91-97.

77) Jung, Hyo Young, et al. "Pyridoxine Improves Hippocampal Cognitive Function Via Increases Of Serotonin Turnover And Tyrosine Hydroxylase, And Its Association With CB1 Cannabinoid Receptor-Interacting Protein And The CB1 Cannabinoid Receptor Pathway." Biochimica et Biophysica Acta (BBA)-General Subjects 1861.12 (2017): 3142-3153.

78) Reddy, Pramod. "Preventing Vitamin B6–Related Neurotoxicity." American Journal of Therapeutics 29.6 (2022): e637-e643.

79) Slavin, Melissa, et al. "Cannabis and Symptoms of PMS and PMDD." Addiction Research & Theory 25.5 (2017): 383-389.

80) Ferretti, Morgan L. "The Effects of Cannabidiol Isolate on Menstrual-Related Symptoms." (2022).

81) Ferretti, Morgan L., et al. "Examination of the Effects of Cannabidiol on Menstrual-Related Symptoms." Experimental and Clinical Psychopharmacology (2024).

Chapter 16

The Safety and Importance of Bioidentical Hormones Part One

HORMONES ARE THE MESSENGERS THAT orchestrate how our bodies function on a cellular level. We must have good hormone levels to maintain our health. Our hormones are considered safe because we could not maintain our health without them. However, we are interested in a slightly different question, what is the safety of the medical administration of hormones? What is the safety of bio-identical hormones as routinely used in medical practice for menopausal hormone replacement? Let us try to answer that question.

A 50 Million Year Medical Experiment

Endogenous bio-identical hormones are produced by our bodies and circulate in our bloodstream throughout our lifespan. How do we know that endogenous hormones are safe? Just think about this. Any chemical produced by the human body that impairs survival would lead to the extinction of the human species. For example, chemicals in our environment that interfere with our hormone system are called endocrine-disrupting chemicals (EDCs). These EDCs lead to infertility and eventual extinction of the species, as described in 2022 by Shanna H. Swan. What if our bodies are manufacturing EDCs? If our bodies made unsafe hormones, over time, natural selection would eliminate the harmful mutations making them. This is the basic idea of Darwinian evolution accepted by mainstream medical science. For example, if my genetic profile manufactured EDCs in my own body, I would be infertile and unable to pass these harmful genes to the next generation, thus removing this unfavorable genetic profile from the population. Consider the following medical experiment, bioidentical hormones, estrogen, progesterone, and testosterone have been present in mammals for at least 120 mil-lion years, and in us primates for 60 million years. Are we still here on the planet? If you said yes, then we have just completed a successful medical experiment on the safety of endogenous bioidentical hormones. **Note:** we humans are primates. (1-6)

The Concept of Hormone Replacement for Menopausal Hormone Deficiency

As we age, our hormone levels decline leading to poor health outcomes. By restoring our hormone levels, we can regain health. The idea of hormone replacement for menopausal hormone deficiency is an old idea going back to ancient China where the aging nobility ingested hormones from crystalline urine obtained from young women. Every period in history had its version of HRT. During medieval times the nobility had preparations of urinary steroids. Modern hormone replacement started in 1966 when Robert A. Wilson, a New York gynecologist, published his best-selling book, *Feminine Forever,* thus promoting the idea that estrogen deficiency is a disease that can be treated with hormone replacement. Dr. Wilson's book promoted Premarin (CEE), an estrogen obtained from pregnant horses, writing:

> Many physicians simply refuse to recognize menopause for what it is — a serious, painful, and often crippling disease. Every woman alive today has the option of remaining feminine forever.

Thus, began a most bitter and contentious debate in medicine between those who agree with Dr. Robert Wilson and those who disagree and claim that for safety reasons, hormone replacement should be rejected. One of the goals of this book is to lay out for the reader this debate, the arguments for each side on the health benefits and safety of hormone replacement. Thus, the reader will be empowered with the knowledge to come to their own conclusions. We will do this by clearing up the confusion created by the drug industry between synthetic hormones and bioidentical hormones. We will examine the studies in the medical literature showing the exact hormone formula used is important for determining health outcomes and long-term safety. It is hoped the reader will come to an understanding of which hormone formulas are safe and which are not. We will show you the formulas we have been using for 20 years that are both safe and effective, containing estradiol, estriol, progesterone, and testosterone in the correct dosage, and route of delivery. (7-10)

Either Excess or Deficiency of Anything Can be Harmful

One of our routine laboratory tests is called the chemistry panel which measures electrolytes, sodium, potassium, bicarbonate, chloride, glucose, calcium, albumin, total protein, ALP (alkaline phosphatase), ALT (alanine transaminase), and AST (aspartate aminotransferase), bilirubin, BUN (blood urea nitrogen) and creatinine. Our bodies automatically maintain electrolyte levels within narrow ranges to maintain health. If levels deviate above or below normal ranges, this may cause serious health disturbances. For example, elevated potassium levels cause cardiac arrest. Magnesium deficiency causes muscle spasms and arrhythmia. Excessive amounts of Vitamins A or D can lead to toxicity. Compared to the strict narrow limits for electrolytes, our hormone levels enjoy a considerably wide range of acceptable limits. (11-13)

Safety of Progesterone

How do we know that progesterone is safe, even at very large doses? Progesterone levels during the third trimester of pregnancy may reach 150-300 ng/ml, about 15-30 times higher than luteal phase progesterone levels in young females with a plateau at 10–20 ng/ml. And remember, the fetus is bathed in high progesterone serum levels of pregnancy. What this means is that extremely high serum levels of progesterone are perfectly safe. Administration of even very large doses of progesterone cannot come close to the high levels found in pregnancy. Having said that, large doses of oral progesterone may induce a sedative effect which may be unwanted or annoying if one needs to stay alert during the day. In addition, as mentioned previously, the rare patient may have an allergic response to progesterone if sensitized by previous exogenous progesterone exposure. Also, because of crosstalk with estrogen receptors, administration of insufficient doses of progesterone may induce a paradoxical worsening of PMS symptoms or peri-menopausal symptoms, usually relieved by increasing the progesterone dosage. For more on this, see Chapters 13-15 on Progesterone Safety and Adverse Effects, parts 1-3. However, a deficiency or excess of estrogen, or a deficiency of progesterone may produce symptoms. This is called estrogen deficiency/ excess, and progesterone deficiency, and they each have typical signs and symptoms easily recognized. (14-20)

Common Signs of Estrogen Deficiency

Mental Fogginess, Forgetfulness, Depression, Minor Anxiety, Mood Change, Difficulty Falling Asleep (Insomnia), Hot Flashes, Night Sweats, Temperature Swings, Day-Long Fatigue, Reduced Stamina, Decreased Sense Of Sexuality, Lessened Self-Image and Attention To Appearance, Dry Eyes, Skin, And Vagina, Loss of Skin Radiance, Feel Balanced 2nd Part of The Cycle, Sagging Breasts and Loss of Fullness, Pain With Sexual Activity, Weight Gain, Increased Back and Joint Pain, Episodes of Rapid Heartbeat, Headaches and Migraines, Gastrointestinal Discomfort, Constipation.

Common Signs of Excess Estrogen

Breast Tenderness or Pain, Increased Breast Size, Bloating, Water Retention, Impatient,

Snappy Behavior with Clear Mind, Pelvic Cramps, Nausea, Fibrocystic Breast Disease, Uterine Fibroids, Heavy Menstrual Periods, Worsening PMS.

Common Signs of Progesterone Deficiency

No period at all (no ovulation), The period comes infrequently (every few months), Heavy and frequent periods (large clots, due to buildup in the uterus), Spotting a few days before the period. (Progesterone level is dropping), PMS (premenstrual syndrome), Cystic breasts, Painful breasts, Breasts with lumps (fibrocystic breast disease), endometriosis, adenomyosis, fibroids, anxiety, irritability, nervousness and water retention. Credit and thanks go to Uzzi Reiss, MD for the above list found in his book, *Natural Hormone Balance for Women* by Uzzi Reiss and Martin Zucker (2002). (21)

Testosterone for Women: Signs and Symptoms of Testosterone Deficiency/Excess

Testosterone is breast cancer preventive and reduces breast cancer by about 50 percent. We know this is true because of the work of Rebecca Glaser MD, Assistant Clinical Professor at Wright State University, and a breast surgeon in Dayton, Ohio who treats menopausal women with testosterone pellets. Dr. Glaser treated 1,267 with testosterone pellets. The 10-year follow-up showed a 40 percent reduction in breast cancer compared to national averages. The 15-year follow-up as of March 1, 2023, revealed 16 cases of invasive breast cancer versus 30 cases expected, indicating a 47% reduction.

Ignoring for the moment testosterone benefits for breast cancer prevention, what about safety for the rest of the body? How do we know testosterone is safe for women? Compare testosterone levels in men and women. Post-menopausal women have testosterone levels in the 20-30 ng/dl range. However, in the younger male, testosterone may reach 1200 ng/dl, roughly 50-60 times higher than the post-menopausal female. If higher testoster-

one levels are safe in males, then the same holds for females. Having said that, excess testosterone in females causes annoying virilizing symptoms, acne and increased facial hair. These can be avoided by taking a break from testosterone and cutting back on the dosage. For more on this see Chapter 18, Testosterone for Prevention and Treatment of Breast Cancer.

Signs and Symptoms of Testosterone Deficiency:

Loss Of Libido, Low Motivation, Fatigue, Loss of Muscle Mass (Sarcopenia), Loss of Bone Density, Osteopenia, Osteoporosis), Cognitive Dysfunction, Mood Disturbance, Dry Skin, Dry Eye Syndrome, Increased Risk for Breast Cancer. (23-25)

Signs and Symptoms of Testosterone Excess

Excess Oily Skin, Acne (Pimples), Excess or Aggravation of Facial Hair, Excess Libido, Aggression, Hostility, Mood Disturbance, Fluid Retention, Joint Pain (From Fluid Accumulation). (26-28)

Relieving Symptoms of Hormone Excess

When patients use oral, topical, or vaginal hormone preparations of estrogen, progesterone, or testosterone, they may experience hormonal excess symptoms. If this happens, the patient is instructed to simply take a break from the hormone preparation for 3-7 days. Once symptoms resolve, do not throw the tube into the garbage can, because you will want to resume the hormone preparation at half dosage. The advantage of topical or vaginal hormone delivery is the ease of adjusting dosage. Compare topical hormones to hormone pellet insertion. The disadvantage of the hormone pellet is: insertion requires a surgical procedure with local anesthesia. Once the pellet is in, it is difficult to remove. The dosage cannot be adjusted without digging out the pellet, something most people would rather avoid.

How Safe Are NSAID Pills?

Over-the-counter pain pills such as non-steroidal anti-inflammatory drugs (NSAIDs) such as aspirin, naproxen and ibuprofen are considered safe. After all, they are over-the-counter (OTC), and you do not need a prescription to buy them. In 1999, Dr. Michael Wolfe found NSAID use is responsible for an estimated 16,500 annual deaths in the US, mostly from gastric bleeding. Compare this to the adverse events of hormone excess as listed above, easily treated by adjusting the dosage. The subject of safety is more complicated. For example, what are the benefits and adverse effects of HRT on the cardiovascular system, on the brain, on the risk for breast cancer, on the risk for all other cancers, and on the risk for blood clotting? These questions will be discussed throughout the book. (29)

The Elephant in the Room

Of course, the elephant in the room is the question, does menopausal hormone treatment increase the risk for breast cancer? The answer to this question is more complicated than a simple yes or no, and requires us to evaluate the hormone formulation. What are the individual hormones used in the formulation? What are the types and routes of delivery for estrogen, progesterone, and testosterone separately, and in combination? When thinking about the role of hormones in the mammalian body, one must look at the big picture. The way hormones work together is analogous to a symphony with many gifted musicians playing different instruments to make beautiful music. A good menopausal hormone replacement program uses combinations of hormones in the proper ratios to achieve the highest safety and efficacy. In addition, do not forget the importance of diet, supplements and avoiding endocrine disrupting chemicals (EDCs) in plastics and food. (30-32)

Do Bio-Identical Hormones Cause Breast Cancer? The French Cohort Study

According to the 2008 French Cohort study by Dr. Agnes Fournier, there is no increased risk of breast cancer in women using the bio-identical hormone combination of estradiol and natural progesterone. However, breast cancer risk is increased with the estradiol/ synthetic progestin combination. (33-35)

Synthetic Progestins Increase Proliferation, Act as Carcinogens

Synthetic progestins are carcinogenic via two mechanisms. Firstly, as suggested by Dr. Sebastián Giulianelli in 2012, progestins activate ER-alpha which then activates Cyclin D1 and MYC oncogenes. Secondly, as Dr. Steven Birrell suggests in 2007, synthetic progestins are anti-androgens that interfere with androgenic activation of ER-beta, the tumor suppressor receptor, thus increasing the risk for breast cancer. (36-40)

Testosterone Inhibits Breast Cell Proliferation

Dr. Birrell's above hypothesis was confirmed in 2014 by Dr. Aleksandra M. Ochnik who obtained normal breast tissue from surgical specimens and cultured the breast cells in the presence of androgens (DHT) and MPA (medroxyprogesterone), finding the androgen DHT inhibits breast cell proliferation. MPA exerts an antiandrogen effect which increases proliferation by destabilizing the androgen receptor protein. Dr. Ochnik writes:

> **DHT [Di-Hydro-Testosterone] inhibited the proliferation of breast epithelial cells in an AR-[Androgen Receptor] dependent manner within tissues from postmenopausal women, and MPA significantly antagonized this androgenic effect**....In a subset of postmenopausal women, **MPA exerts an antiandrogenic effect on breast epithelial cells that is associated with increased proliferation and destabilization of AR [Androgen Receptor] protein**. This activity may contribute mechanistically to the

increased risk of breast cancer in women taking MPA-containing EPT [estrogen-progestin therapy]. (41)

Progesterone Acts as Proliferative Brake

In 2015, Dr. Hisham Mohammed did a series of studies using in vitro and in vivo mouse xenografts, finding activation of the progesterone receptor (with exogenous pro-gesterone) results in a "robust association between PR and the ER-alpha complex which then acts as a "proliferative brake" and blocks breast tumor growth in mouse xenografts. Dr. Hisham Mohammed also states the progester-one receptor "functions as a molecular rheo-stat to control ER-alpha chromatin binding and transcriptional activity. Remember, ER-alpha is the proliferative receptor, while ER-beta is the protective one. In this scenario, progesterone receptor protein binds to ER-alpha and halts ER-alpha induced proliferation, resulting in a good clinical outcome. Dr. Hisham Mohammed writes: "The increased risk of breast cancer is mainly attributed to specific synthetic proges-tins, in particular medroxyprogesterone acetate (MPA), which is known to also have androgenic properties. The relative risk is not significant when native progesterone is used." :

> under estrogenic conditions, an activated PR [progesterone receptor] functions as a proliferative brake in ER alpha+ breast tumors by re-directing ER alpha chromatin binding and altering the expression of target genes that induce a switch from a proliferative to a more differentiated state … Our findings indicate that PR functions as a molecular rheostat to control ER alpha chromatin binding and transcriptional activity, which has important implications for prognosis and therapeutic interventions…**the increased risk of breast cancer associated with progestogen-containing HRT is mainly attributed to specific synthetic progestins, in particular medroxyprogesterone acetate (MPA),** which is known to also have androgenic properties. **The relative risk is not significant when native progesterone**

is used. In ER alpha+ breast cancers, PR is often used as a positive prognostic marker of disease outcome4… progesterone treatment has been shown to be antiproliferative in ER alpha+ PR+ breast cancer cell lines and progestogens [natural progesterone, not progestins] have been shown to oppose estrogen-stimulated growth of an ER alpha+ PR+ patient-derived xenograft. In addition, exogenous expression of PR in ER alpha+ breast cancer cells blocks estrogen-mediated proliferation and ER -alpha transcriptional activity…**These observations imply that PR activation in the context of estrogen-driven, ER-alpha+ breast cancer, can have an anti-tumorigenic effect.** Note: ER-alpha is the proliferative, pro-carcinogenic receptor, while ER-beta is the tumor suppressor receptor. (42-48)

French Cohort Study

In 2008, Dr. Agnes Fournier, Doctorate in Epidemiology from Paris, France studied 80,377 postmenopausal women using various hormone replacement regimens for a mean duration of 8.1 years. This is called the E3N/EPIC study which used self-administered ques-tionnaires. Pathology reports for 2,354 cases of invasive breast cancer were reviewed in 96 percent of cases. Almost all (98 percent) of the women used estradiol, and only 2 percent used CEE (Premarin). Estradiol-alone users had a **29 percent increase in breast cancer** due to proliferative effects of estradiol. However, the estradiol/progesterone combination group had **no increase in risk for breast cancer** compared to the general population, thus sug-gesting an anti-proliferative effect of natural progesterone. However, when a synthetic pro-gestin is added to the estradiol, these women had a **69 percent increased risk for breast cancer**, thus suggesting a procarcinogenic effect of progestins. These results are similar to those of Dr. Charles Wood using a primate model of post-menopausal hormone replace-ment. For more on primate hormone stud-ies, see: Chapter 3 on Don't Monkey with my

Hormones. Dr. Agnès Fournier writes:

> Recently, Charles Wood et al. [22] compared the effects of estradiol given with either medroxyprogesterone acetate or micronized progesterone on risk biomarkers for breast cancer in a postmenopausal primate model. In this randomized crossover trial, they found that, compared to placebo, estradiol + medroxyprogesterone acetate resulted in significantly greater proliferation (as measured by Ki67 expression) in lobular and ductal breast epithelium, while estradiol + micronized progesterone did not. This result supports our findings suggesting that, when combined with an estrogen, progesterone may have a safer risk profile in the breast compared with some other progestagens. Note: the word "progestagens" used here refers to synthetic progestins. (33-35)

Error in the Medical Literature from the Netherlands

In 2023, Dr. Bennink Coelingh from the Netherlands makes a false statement. Dr. Bennink Coelingh says that natural progesterone causes breast cancer. Of course, this is completely wrong. Synthetic progestins cause breast cancer, not natural progesterone. In 2023, Drs. Gompel and Jerilynn Prior argue the E3N French Cohort data is a rebuttal to Dr. Bennink Coelingh. The French E3N Cohort study shows 29 percent increased risk with estradiol alone (RR=1.29). However, when natural progesterone is combined with estradiol, this increased risk for breast cancer is abolished (RR=1.00). However, when the synthetic progestin, MPA is combined with estradiol, the risk is the greatest (RR=1.69). In 2023, Drs. Gompel and Jerilynn Prior write:

> the breast cancer results from an 8-year prospective observational study with more than 80,000 participants (in France called E3N cohort) which allowed a comparison of breast cancer risk on various single and combined menopausal hormonal therapies (MHT). This study documented breast cancer risk in untreated postmenopausal women as controls versus those taking MHT [Menopausal Hormone Therapy] with estradiol alone (**risk ratio 1.29**, 95% confidence interval 1.02, 1.65) versus estradiol plus progesterone (**risk ratio 1.00**, 95% confidence interval 0.83, 1.22). Although there was no increased breast cancer risk with estradiol–progesterone, estradiol with synthetic progestin MHT was related to significantly increased risk (**risk ratio 1.69**, 95% confidence interval 1.50, 1.91). A recent review [Hipolito, 2021] summarizes publications on a lower breast cancer risk in MHT of estradiol with micronized progesterone and dydrogesterone versus other progestins. (49-51)

In 2021, Dr. Rodrigues and Ann Gompel compare hormone replacement formulas containing either micronized progesterone (P) or synthetic progestins (PGs), citing observational studies showing formulas containing natural micronized progesterone have significantly less breast cancer than those containing synthetic progestins, writing:

> Existing evidence from clinical studies on the use of micronized P [progesterone] in MHT [menopausal hormone therapy], for the most part, shows favorable outcomes, without deleterious effects. Micronized P is able to prevent endometrial hyperplasia in combination with estrogens, does not increase the risk of VTE [venous thromboembolism] and stroke when used with transdermal estrogens. Micronized P does not seem to attenuate the cardiovascular benefits of estrogens and is likely safer than PGs [synthetic progestins]. **The breast cancer issue is of great concern, and according to observational studies, MHT regimens containing micronized P are associated with a significantly lower risk of breast cancer than those containing PGs.** (52)

As mentioned above, the French Cohort is an observational study, not a randomized, placebo-controlled trial (RCT). Whether or not estradiol alone increases or decreases risk of breast cancer has been debated in the medical literature over decades. Part of this debate relates to whether the menopausal patients studied have or have not been long-term estrogen-deprived (LTED). Dr. V. Craig Jordan cites LTED as his explanation for the 23 percent reduction in breast cancer in 2004 second arm of the WHI study (Premarin-alone). The women were older when starting hormone replacement, thus a gap of 5 years or greater is the criteria for long-term estrogen deprivation (LTED). As it turns out, LTED transforms estrogen from a growth factor to a cancer death factor (apoptosis inducer). So, it would be nice if we could find other supportive studies to confirm the French Cohort findings of increased risk of breast cancer for estradiol alone. HR=1.29 for estradiol alone, and HR=1.0 for combined estradiol/progesterone. Below here are two supportive studies showing exactly that.

In 2006, Dr. Heli Lyytinen from Finland studied breast cancer risk in 84,729 post-menopausal women using estrogen-alone therapy. Dr. Heli Lyytinen assumed the women started HRT at age 52, so they were not estrogen deprived. The breast cancer risk was increased 34 percent after 5 years of estradiol-alone use (HR=1.34), 57 percent for 10-20 years use, and 75 percent for greater than 20 years of use. Dr. Heli Lyytinen, writes:

> we estimated the duration of estradiol use in women using estradiol since 1994 by assuming that they had started use at the age of 52 years, which is the case generally in Finland. The standardized incidence ratio related to estimated estradiol use of 5–10 years was 1.34 (95% CI 1.16–1.54; 193 observed cases of breast cancer), for use of 10–20 years 1.57 (1.31–1.86; 125 cases), and for use of more than 20 years, 1.75 (1.16 –2.55; 27 cases)...Estradiol for 5 years or more, either orally or transdermally, means 2–3 extra cases of breast cancer per 1,000 women who are followed for 10 years. (53)

In 2011, Dr. Kjersti Bakken studied different types of menopausal hormone replacement and breast cancer risk in 133,744 women over 8.6 years of follow-up. The most frequently used estrogen was estradiol, except for Germany and the United Kingdom, where CEE dominated. Predominant synthetic progestins were testosterone derivatives in Denmark, Germany, Norway, the Netherlands and the United Kingdom, whereas progesterone derivatives were more used in France, Italy and Spain. The most frequently used progestin was norethisterone acetate (NETA), followed by progesterone and norgestrel. Among women who used estrogen-only therapy, the breast cancer risk was similar for estradiol vs. CEE. Women using estrogen-alone experienced 4,312 primary breast cancers, a 42 percent increased relative risk, writing:

> The EPIC-cohort is a multicentre prospective cohort with 23 contributing centres in 10 European countries...We investigated the association between the use of different types of MHT and breast cancer risk... Approximately 133,744 postmenopausal women contributed to this analysis. Information on MHT was derived from country-specific self-administered questionnaires with a single baseline assessment. Incident breast cancers were identified through population cancer registries or by active follow-up (mean: 8.6 yr). A total of 4312 primary breast cancers were diagnosed during 1,153,747 person-years of follow-up...Compared with MHT never users, breast cancer risk was higher among current users of **estrogen only** (RR: 1.42, 95% CI 1.23–1.64). (54)

Of course, randomized controlled trials (RCT's) are considered higher level of evidence than observational studies. In 2024 Dr. Rowan Chlebowski did a meta-analysis of 10 RCT's of estrogen-alone and breast cancer incidence, finding a 23 percent reduction in breast cancer with estrogen alone. This 23 percent reduction is the same as the second arm of the WHI study using Premarin-alone (CEE, conjugated equine estrogen). I suspect CEE was used in

Dr. Chlebowski's meta-analysis, and I would ask if the patients were started after a period of estrogen deprivation, all of which could influence the results. Dr. Rowan Chlebowski writes:

> Findings from 10 randomized trials included 14,282 participants and 591 incident breast cancers.... Combining the 10 trials, 3.6% (262 of 7339) vs 4.7% (329 of 6943) randomized to estrogen-alone vs placebo (overall RR 0.77 95% CI 0.65–0.91, P = 0.002). Conclusion: The totality of randomized clinical trial evidence supports a **conclusion that estrogen-alone use significantly reduces breast cancer incidence.** (55)

To be fair, one must mention the 2016 UK Generations Study by Dr. Michael Jones which also found no increase in breast cancer for estrogen-alone use (HR=1.0). However, this estrogen-alone study included mostly Premarin (CEE) use, which excludes this as an estradiol-alone study. (56)

B-Ring Steroids

Another explanation for the 23 percent reduced risk of breast cancer in the second arm of the WHI is the presence of B-ring unsaturated steroids in Premarin (CEE from pregnant horses). B-ring unsaturated steroids are not made from cholesterol and are not present in humans. These B-ring steroids act preferentially on ER-beta, the tumor suppressor receptor, and thus are cancer-protective, much like estriol and testosterone metabolites 3-beta-diol which also preferentially target ER-beta. In 2008, Dr. Bhagu Bhavnani studied relative binding affinities of various estrogens to ER-alpha and ER-beta, writing:

> some of these unique estrogens [ring B unsaturated estrogens] had two to four times greater affinity for ER beta than ER alpha... the effects of ring B unsaturated estrogens are mainly mediated via ER-beta and that the presence of both ER subtypes further enhances their activity. It is now possible to develop hormone replacement therapy using selective ring B unsaturated

estrogens for target tissues where ER beta is the predominant ER. Note ER=estrogen receptor. (57)

The Tragedy of DES

We should mention here breast cancer prevention means avoiding endocrine-disrupting chemicals (EDCs) in the environment, pesticides, Phthalates, Bisphenol-A and C, plastics etc. You might be surprised to learn that many of the EDC's are synthetic hormones made by the drug industry. In 1938, Diethylstilbestrol (DES) was the first widely used synthetic estrogen and handed our freely to an estimated 10 million people, pregnant mothers and breast cancer patients. Tragically, 37 years later, DES was banned when doctors learned DES is an endocrine-disrupting chemical (EDC) causing cancer in the offspring of users. In 2005, Dr. Marieke Veurink from the Netherlands writes:

> In 1971, it became clear that this apparently innocent treatment proved to be a time bomb for the infants exposed to DES during the first trimester of pregnancy. DES is now associated with an increased risk of breast cancer, clear cell adenocarcinoma (CCAC) of the vagina and cervix, and reproductive anomalies...[There are] potential long-term health implications of DES on the mother, DES daughters and DES sons, and the possible side effects on the third generation. (58-59)

In 2021, Dr. Pilar Zamora-León from Chile writes:

> Diethylstilbestrol (DES), a transplacental endocrine-disrupting chemical, was prescribed to pregnant women for several decades. The number of women who took DES is hard to know precisely, but it has been estimated that over 10 million people have been exposed around the world. DES was classified in the year 2000 as carcinogenic to humans. The deleterious effects induced by DES are very extensive, such as abnormalities or cancers of the genital tract and breast, neurodevelopmental alterations, problems

associated with socio-sexual behavior, and immune, pancreatic and cardiovascular disorders. Not only pregnant women but also their children and grandchildren have been affected. Epigenetic alterations have been detected, and intergenerational effects have been observed. More cohort follow-up studies are needed to establish if DES effects are transgenerational. Even though DES is not currently in use, its effects are still present, and families previously exposed and their later generations deserve the continuity of the research studies. (60)

The story of DES is an important lesson. What about all the other endocrine-disrupting chemicals and synthetic hormones in wide use today? The drug industry also makes synthetic progestins which are known carcinogens and EDCs. Will we find out 40 years later these are transmitting carcinogenic effects to generations of offspring just like DES? Will we find out 40 years later EDCs are causing the extinction of the human race? I will leave these questions to future historians. Two recommended books for learning about EDC's are: *Hormone Deception* by Dr. Lindsey Berkson, and *Our Stolen Future: Are We Threatening our Fertility, Intelligence, and Survival?* by Theo Colburn. (61-70)

What Causes Breast Cancer?

If estrogen/progesterone bioidentical hormone replacement does not cause cancer, then the next obvious question is what does cause breast cancer? Carcinogenic chemicals in the environment, water, and food supply cause cancer. Here is a partial list of carcinogens in our food supply- Bisphenol A (BPA), Phthalates, Pesticides, Styrene, Vinyl Chloride, etc. When researchers wish to study breast cancer in mice, they induce breast cancer with the carcinogenic chemical DMBA, the synthetic progestin medroxyprogesterone MPA, or they use genetically modified mice. There is no reliable animal model of breast cancer using estrogen to induce breast cancer in wild-type mice. (71-75)

Etiology of Breast Cancer in Humans

In humans, breast cancer is thought related to carcinogenic properties of estrogen metabolites called 4-hydroxy-quinones acting as DNA adducts causing oxidative damage to DNA. Normally, estrogen is preferentially metabolized towards the favorable pathways leading to 2MeOE2 (2 methoxy-estradiol) which is cancer preventive. However, in some cases of nutritional deficiency, or genetic disorders involving methylation pathways, unfavorable 4-hydroxy quinones may accumulate in a bottleneck, thus increasing the risk for breast cancer as suggested in 2021 by Dr. Ercole Cavalieri. That is where a knowledgeable doctor is helpful for laboratory testing, and providing supplements such as iodine, DIM, methylfolate, selenium, resveratrol, N-acetylcysteine and vitamin D3 to reduce the risk for breast cancer. (76-78)

Do Bio-Identical Hormones Cause Heart Disease?

Again, the answer is NO. Coronary calcium score has emerged as the most sensitive and accurate technique for determining risk for cardiovascular disease. The higher the calcium score, the greater the risk. In 2007, Dr. JoAnn E. Manson studied calcium scores in women from the 2004 second arm of the WHI taking Premarin(CEE) alone. The study showed lower coronary artery calcium in women taking Premarin-alone. These women had previous hysterectomy and synthetic progestins were not needed for endometrial protection. Similar results were obtained in a 2005 calcium score study by Dr. Mathew Budoff. A third study of calcium scores in 2012 by Dr. Nicole Weinberg enrolled 544 women, almost half were using hormone therapy (HT). Mean coronary calcium scores were about 40 Agatson units lower in hormone therapy (HT) users. Mean score for HT users was 57 Agatson units vs. non-users with 96 units. Dr. Nicole Weinberg writes:

The mean age at which HT use was initiated was 54 (±6) years. The mean number of years of HT use among HT users was 2.2 years (±4.8), ranging from zero to 33 years…Of the 544 enrolled women aged 50–80 years, 252 (46.3%) were hormone therapy users. Hormone therapy users had a significantly lower prevalence of any coronary artery calcium (defined as coronary artery calcium score > 0; 37%), than non-users (50%, p = 0.04), as well as significantly lower mean calcium scores (p = 0.02)…Among HT users, the mean CAC score was significantly lower (56.8 [±173.9]) than among HT non-users (96.4 [±257.6]) (p < 0.001). On average, the HT users had a 39.6-unit lower CAC score than HT non-users (p < 0.001). (79-81)

These three coronary calcium studies show estrogen hormone replacement reduces the risk for coronary artery disease. However as mentioned previously, the cardiovascular protection of hormone replacement therapy (HRT) from coronary artery disease depends on timing. For the greatest benefit, women should start HRT soon after menopausal transition. Women older than 60 years before starting HRT have no cardiovascular benefit. This is the timing hypothesis of Dr. Howard Hodis. (82)

A Closer Look at the Women's Health Initiative WHI Study

Understanding the 2002 Women's Health Initiative (WHI) study is not difficult, and is very important to answer the question of hormone safety. The WHI study was a large NIH-sponsored medical study that compared synthetic hormones to placebo in two large groups of women. The WHI study consisted of two arms. The first arm used the synthetic hormone Premarin (CEE) combined with medroxyprogesterone (MPA), and the second arm used Premarin (CEE) alone in women after hysterectomy. **Note:** CEE=conjugated equine estrogen.

What is Premarin and MPA?

Premarin and MPA are not bio-identical hormones. Premarin is obtained from the urine of pregnant horses and contains a combination of many hormones, including the unsaturated B-ring steroids which preferentially bind to ER-beta, the tumor suppressor receptor. Thus, Premarin is breast cancer protective. Medroxyprogesterone (MPA) is a synthetic hormone not found anywhere in the natural world. Its chemical structure is obtained by adding methyl groups to the progesterone backbone. When the two hormones Premarin and MPA are combined into one pill, we have the oral hormone pill commonly prescribed by mainstream medicine for relief of menopausal symptoms. This is the same hormone preparation used in the first arm of the 2002 WHI study.

WHI Study First Arm:

The 2002 WHI study, first arm study published in JAMA was terminated early, after 5.2 years, because the combination of Premarin and Provera (Prempro) caused increased breast cancer (HR=1.26), stroke (HR=1.41) and coronary heart disease (CHD HR =1.29). Immediately after this study was published, the mass media spread sensationalized the fear, causing doctors and menopausal women to steer clear of any form of HRT. Training programs in the use of HRT quickly dried up, and doctors lost the knowledge and expertise required to prescribe HRT. Another thing happened. Knowledgeable doctors and menopausal women made the switch from the Prempro pill (Premarin and MPA) to safer hormone formulas. After 2002, drug maker Wyeth reported a 50 percent decline in sales of Premarin and Prempro hormone products from 2 billion in sales in 2002 to about one billion in 2006. (70) (83-84)

In 2008, Dr. Steven Hotze describes the findings of the 2002 WHI study (first arm), writing:

The [2002 first arm] WHI study demonstrated that the combination of Premarin [CEE] with Provera [MPA] produces the following adverse effects: a 26% increase in risk of invasive breast cancer; a 29% increase in risk of myocardial infarction or death from CHD [coronary heart disease]; a 41% increase in risk of stroke; and a 200% increase in risk of thromboembolism. (85)

WHI Study (Second Arm):

All women in the 2004, second arm WHI study had prior hysterectomy, surgical removal of the uterus, so they did not need the synthetic progestin, MPA, to prevent endometrial cancer. Rather, these women were given Premarin-alone (CEE). Unlike the 2002, first arm WHI study, women taking Premarin-alone had no increase in breast cancer risk. Rather than a 26 percent increase in breast cancer seen in the 2002 first arm WHI study, the 2004 second arm study showed a 23 percent decrease in breast cancer in the Premarin-alone treated group!

In 2022, Dr. Isaac Manyonda from the U.K. reviewed the 2004 second-arm WHI using Premarin (CEE) alone, suggesting that estrogen is a breast cancer preventive, compelling a paradigm shift in thinking, writing:

> The evidence now compels a paradigm shift from the traditional thinking that estrogen could cause breast cancer **to a recognition that it actually prevents the disease**, and that when the disease does occur (no preventative intervention achieves a 100% preventative effect), it is often picked up early and **mortality is reduced by up to 44%.** (86)

Dr. Zsuzsanna Suba Discusses the Catastrophic 2002 WHI Study

Dr. Zsuzsanna Suba MD, PhD is professor emeritus at the Department of Molecular Pathology, National Institute of Oncology, Budapest, Hungary. She has board certification in pathology and her area of expertise is breast cancer genetics. Dr. Zsuzsanna Suba has published extensively for decades on most of the topics covered in this book relating to menopausal hormone replacement and breast cancer prevention and treatment. In 2023, Dr. Zsuzsanna Suba reviewed the results of the 2002 WHI clinical trial using Prempro

(Premarin, CEE combined with MPA). Dr. Suba concluded that the catastrophic results of this 2002 WHI study was entirely due to the toxic effect of medroxyprogesterone (MPA), a synthetic progestin, and not the horse estrogen (CEE, Premarin) component, writing:

> In 2002, the results of a great, prospective, placebo-controlled Women's Health Initiative (WHI) study strengthened that combined CEE plus medroxyprogesterone-acetate (MPA) treatment (PremPro, Pfizer) increased the risk of breast cancer, thromboembolism, and cardiovascular diseases. Following these serious experiences, there was a consequential precipitous decrease in MHT [Menopausal Hormone Therapy] use among postmenopausal women and a thorough re-evaluation of MHT practice became necessary. Later, in a prospective MHT study, the highly toxic effects of MPA [medroxyprogesterone] were published as compared with other synthetic progestins. This finding illuminated that in the WHI study published in 2002, **the MPA component of PremPro [combined Premarin and MPA] may be blamed for the catastrophic results of MHT instead of the horse urine deriving Premarin**. (87-88)

Dr. Marty Makary on the Women's Health Initiative (WHI) Study

On November 5, 2024, in her weekly podcast, Dr. Louise Newsome was joined by Dr. Marty Makary, a pancreatic cancer surgeon at Johns Hopkins Medical Center and newly appointed commissioner for the U.S. Food and Drug Administration (FDA). Dr. Makary is the author of the best-selling book, *Blind Spots in Medicine*. In this podcast, Dr. Marty Makary discusses the back story of what happened with the WHI study. The principal authors led by Jacques E. Rossouw, a cardiologist and NIH researcher decided to publish data that was not statistically significant, over objections from statisticians in the group, and then held a press conference to announce the findings before the release of data. Dr. Makary says,"It turns out they had deceived their co-investigators. They had bamboozled the general public. They had played the media by not releasing their data until long after the announcement. And they had even crushed dissenters, ruining, trying to ruin their careers. So it's a very incredible back story of how basically a small group of people decided to call hormone replacement therapy a carcinogen, when in fact, for the vast majority of women going through menopause, it is a miracle":

Dr Marty Makary: [00:05:43] ...They initially tried to claim when the Women's Health Initiative study was announced in 2002 that breast cancer rates had gone down after the announcement as if they had rescued these women from the perils of HRT. But a deeper analysis revealed those were decreases in breast cancer deaths started just before the announcement, and you wouldn't see an effect within months if people stopped taking their hormone replacement therapy. So that was debunked...And there's been many false claims about hormone replacement therapy... **the dogma that taking hormone replacement therapy at the time of menopause causes breast cancer is probably the biggest screw-up in modern medicine**. There's probably no medication that improves the health outcomes of a population more than hormone replacement therapy for women who start it within ten years of the onset of menopause, arguably with the exception of antibiotics. Women live longer, feel better. **The benefits are overwhelming.** This is something I'm sure you've covered many times...But when I took a deep dive into the data on the incredible health benefits, reducing cardiovascular disease, preventing Alzheimer's, avoiding cognitive decline, making bones healthier, maybe even reducing the risk of diabetes and cancer, in some studies. **The benefits are overwhelming.** If hormone therapy did increase the risk of breast cancer as they had claimed it would, the risk would be far eclipsed by the overwhelming health benefits. **Now, I don't I don't believe that claim that it causes breast cancer.** And in

the book *Blind Spots* that I just am putting out now, I did a deep investigative journalism sort of review of what happened when they made that announcement. **It turns out they had deceived their co-investigators. They had bamboozled the general public. They had played the media by not releasing their data until long after the announcement. And they had even crushed dissenters, ruining, trying to ruin their careers. So it's a very incredible back story of how basically a small group of people decided to call hormone replacement therapy a carcinogen, when in fact, for the vast majority of women going through menopause, it is a miracle...**And I think there's been a narrative. There's been a group think on hormone replacement therapy. I discovered that the few doctors who declared that it causes breast cancer without supporting data had already really kind of made up their mind. The lead investigator who made the announcement had said on record prior years that, "we have to stop the HRT bandwagon." [Note: this is from Jacques E. Rossouw's article in 1996, he is a retired cardiologist and academic researcher from National Institutes of Health (NIH).] Yeah. Well, you're leading the largest study ever done in the history of medicine, a clinical trial. You're supposed to wait for the results, not declare we got to stop this HRT bandwagon. So we have not had good leaders with this study. They've deceived the public. Deceived. I interviewed the lead guy who made this announcement in my book *Blind Spots*, and it was unbelievable what I had discovered. But yeah, if you don't like data, doesn't mean you dismiss it. It means you've got to discuss it and we have to have a civil discourse...And so I think it's pretty amazing how the medical establishment got this wrong. The group think, the sort of intellectual laziness where people will to this day cite the 2002 Women's Health Initiative, saying that's the reason why they believe hormone therapy causes breast cancer. And now I'm telling some of those doctors, Well, it's amazing that you think that's the reason why it causes breast cancer, because I interviewed the lead author of that study, showed him his data. And **he acknowledged to me that hormone therapy did not increase the risk of breast cancer mortality**. So, people believe it. And he didn't you know, he didn't even acknowledge. Note: In the above discussion between Dr. Louise Newsome and Dr. Marty Makary, there is no mention of how the exact HRT formula determines breast cancer risk. Synthetic progestins such as medroxyprogesterone (MPA), the progestin used in the first arm 2002 WHI study, is an endocrine-disrupting chemical and a known carcinogen. This should be avoided. (89-90)

The Media Says Hormones Cause Cancer and Heart Disease

If bio-identical hormones are so safe, then why do the newspapers say that women's hormones cause breast cancer and heart disease? The answer is the media and the medical profession routinely confuse synthetic chemically altered monster hormones with bio-identical hormones. The drug companies intentionally create this confusion because they want to hide the fact that synthetic hormones are Frankenstein monsters that cause cancer. The drug industry must chemically alter a hormone to obtain a patent to protect drug profits. By law, naturally occurring bio-identical hormones cannot be patented. However, natural hormones compete for market share with synthetic hormones, and that is why bioidentical hormones are the enemy of the drug industry. In 2008, Dr. Steven Hotze reviews the case for bioidentical hormones and says the synthetic progestin, medroxyprogesterone (MPA), is not really a hormone, it is a "hormone-mimicking compound", and FDA approval does not magically turn it into a hormone. I would call MPA an endocrine disrupting chemical (EDC), Dr. Steven Hotze writes:

a lot of money is spent to manipulate physicians—through sponsoring speakers, organizing symposia, and even conducting studies published as scholarly articles in prestigious journals. All these efforts are designed to give the impression that "evidence-based medicine" means the use of patented exogenous compounds [synthetic progestins]…Much of the confusion about bioidentical hormone replacement flows from the failure to distinguish hormones from non-hormones. Obtaining FDA approval for a hormone-mimicking compound, such as medroxyprogesterone (Provera) or conjugated equine estrogens (Premarin), does not turn it into a hormone. Unfortunately, many scholarly articles have even referred to Provera as "progesterone," and to conjugated equine estrogens as "estrogen." (85)

A Listing of a Few Monster Hormones:

Chemically Altered forms of progesterone:

Dienogest, Desogestrel, Drospirenone, Dydrogesterone, Ethisterone, Etonogestrel, Ethynodiol diacetate, Gestodene, Gestonorone, Levonorgestrel, Lynestrenol, Medroxyprogesterone, Megestrol, Norelgestromin, Norethisterone, Norethynodrel, Norgestimate, Norgestrel, Norgestrienone, Tibolone

Chemically Altered Forms of Estrogen:

Dienestrol, Diethylstilbestrol, Ethinylestradiol, Fosfestrol, Mestranol

Chemically Altered Hormones in Oral Contraceptive Pills (OCPs):

levonorgestrel and ethinyl estradiol [oral contraceptive] (ALESSE 28, AVIANE, NORDETTE, SEASONALE, TRIPHASIL, TRIVORA-28); norethindrone and ethinyl estradiol (COMBI PATCH, LOESTRIN FE 1/20, NEOCON 1/35, ORTHO-NOVUM 7/7/7, OVCON 35); norgestimate and ethinyl estradiol (ORTHO-CYCLEN, ORTHOTRI-CYCLEN, TRINESSA); norgestrel and ethinyl estradiol (LO/OVRAL 28, LOW-OGESTREL), desogestrel and ethinyl estradiol (DESOGEN, MIRCETTE, ORTHO-CEPT), drospirenone and ethinyl estradiol (YASMIN)

Chemically Altered Forms of Testosterone:

Androstanolone, Fluoxymesterone, Mesterolone, Methyltestosterone

Pregnancy is Breast Cancer Protective

In Italy in 1713, breast cancer was quite rare. An Italian doctor, Bernardino Ramazzini (1633–1714) observed a relatively higher incidence of breast cancer in nuns at the local convent compared to married women, and wondered if this was related to celibate lifestyle. In 2005, Dr. Jose Russo was intrigued with Bernardino Ramazzini's idea that pregnancy confers protection from breast cancer. Dr. Russo did a series of studies finding single and multiple pregnancies confer life-time protection from breast cancer, while no pregnancies, being nulliparous, leads to increased risk of breast cancer. Dr. Russo writes:

> Women who gave birth to a child when they were younger than 24 years of age **exhibit a decrease in their lifetime risk of developing breast cancer**, and additional pregnancies increase the protection. The protective effect of full-term pregnancy is a well-established concept not only in humans, but also in experimental rodent models. (91-92)

What is the Molecular Mechanism of Pregnancy Induced Protection from Breast Cancer? Pregnancy Reprograms the Epigenome of the Breast

If high hormone levels of pregnancy confer protection from breast cancer, perhaps this is the secret to breast cancer prevention. In 2019 and 2020, Dr. Mary Feigman studied this question in a transgenic mouse model called CAGMYC. These are mice inbred to have overexpression of the cMYC oncogene, thus these genetically modified mice have spontaneous breast cancer. Dr. Mary Feigman found that in this CAGMYC mouse model, pregnancy conferred protection from cancer by increasing the P53 protein content in the MECs (mammary epithelial cells), blocking the development of malignancy. Pregnancy modified the epigenome of the mice, and made the MEC's less responsive to cMYC oncogene overexpression, thus blocking the development of premalignant lesions.

Dr. Mary Feigman concludes **pregnancy reprograms the epigenome of mammary epithelial cells** and blocks the development of premalignant lesions, writing:

> But how can pregnancy decrease mammary oncogenesis?...To characterize the influence of a pregnancy-induced epigenome on the response to oncogene expression, we used a transgenic mouse strain (CAGMYC), in which **overexpression of the oncogene cMYC, an inducer of mammary tumor development**, is driven in a doxycycline (DOX)-dependent manner...Using this transgenic mouse strain, we found that **the postpregnancy epigenome was incompatible with cMYC overexpression, blocking the activation of MYC-downstream signals and their progression to oncogenesis**... Our study revealed a **substantial increase of p53 protein** in post-pregnancy CAGMYC [transgenic mice with cMYC overexpression] organoids, possibly promoting a senescent state that could block the development of malignant phenotypes...Using inducible cMYC overexpression, we demonstrate that post-pregnancy MECs are resistant to the downstream molecular programs induced by cMYC, a response that blunts carcinoma initiation, but does not perturb the normal pregnancy-induced epigenomic landscape... **Using this system, we confirmed post-pregnancy MECs are less responsive to cMYC overexpression**. Note: P53 is called "the Guardian of the Genome" because the P53 protein is involved in repairing DNA damage, or else inducing apoptosis if the DNA damage cannot be repaired. The P53 performs cancer surveillance...Pregnancy reprograms the epigenome of mammary epithelial cells and blocks the development of premalignant lesions. (93-94)

A quick question for you. What does our friend Mr. Charles Darwin say about the above? Nulliparous women with no children will have a higher mortality from breast cancer, while women with multiple children will be relatively protected from breast cancer. Thus, mothers having children receive preferential treatment by Mr. Darwin's theories, while women without children receive the blunt end of a stick called natural selection. Nature can be cruel. I will leave these philosophical questions for future historians to ponder.

Bioidentical Hormones Prevent Breast Cancer in Mouse Models

Protection from breast cancer is conferred by high hormone levels of estrogen and progesterone in pregnancy. This hypothesis was confirmed in 2007 by Dr. Rajkumar Lakshmanaswamy who studied two different models of mice genetically engineered to spontaneously develop breast cancer. When Dr. Rajkumar recreated pregnancy hormone levels in mice by treating them with estrogen, progesterone, and testosterone, the incidence of breast cancer in the mouse models was drastically reduced, demonstrating that hormone treatment protected genetically engineered mice from developing breast cancer. In a third model of breast cancer using human breast cancer cells xenografted in mice, natural, bioidentical hormone treatment again markedly inhibited cancer progression. (95-98)

In 2013, Dr. Arunkumar Arumugam studied a mouse model in which breast cancer is induced by the injection of a carcinogenic chemical, N-methyl-N-nitrosourea. Before injection of the carcinogen chemical, Dr Arumugam gave the mice short treatment with pregnancy levels of estradiol, finding this very effective in preventing carcinogenesis, writing:

> **We have earlier demonstrated that short-term treatment with pregnancy levels of estradiol (STET) is very effective in preventing mammary carcinogenesis.** Rats were injected with N-methyl-N-nitrosourea at 7 weeks of age and divided into 2 groups. **Short-term treatment with pregnancy levels of estradiol drastically reduced the incidence of overt mammary cancers**... These findings demonstrate that STET confers protection against mammary cancer

development by blocking promotion and progression of transformed cells. (99)

Adding Progesterone Inhibits Estrogen Stimulated Breast Cancer in Mouse Model

In 1985, Dr. Akira Inoh from Japan showed progesterone prevented breast cancer in inbred W/Fu strain mice treated with estrogens. These inbred strains of mice were prone to develop spontaneous cancers of various types, including breast cancer. When the mice were ovariectomized and given prolonged treatment with diethylstilbestrol [synthetic estrogen] or 17 beta-estradiol [natural estrogen], the mice developed multiple breast tumors. However, when the mice were simultaneously given progesterone along with the estrogen, the "multiplicity and size of estrogen-induced MTs [multiple tumors] were reduced by the simultaneous administration of either progesterone or tamoxifen". This same anti-proliferative, protective effect of progesterone when added to estradiol was demonstrated in the 2008 French Cohort study by Dr. Agnes Fournier. The Relative Risk (RR) of breast cancer in the transdermal estradiol-treated group was increased (1.28). However, when progesterone was combined with the transdermal estradiol, the risk is decreased to RR=1.08. Further discussion of progesterone as breast cancer prevention can be found in chapters 13-15, on the Safety and Adverse Effects of Progesterone. (100) (33-35)

Radiation Induced Breast Cancer Prevented with Progesterone

One of the adverse side effects of radiation treatment is the carcinogenic effect of radiation. Another model of breast cancer is radiation-induced breast cancer in mice, studied in 1973 by Dr. Albert Segaloff from New Orleans. Mice promptly develop breast cancer after radiation treatment. When the mice were pre-treated with progesterone pellets, none of them developed radiation-induced breast cancer, writing:

In this preliminary experiment none of the animals that bore just progesterone pellets and were radiated developed mammary carcinoma. (101)

Progesterone Deficiency Increases Risk for Breast Cancer 5.4-Fold

In 1981, Dr. Linda Cowan studied the incidence of breast cancer in women with progesterone deficiency. Dr. Cowan studied 1,083 women evaluated and treated for infertility between 1945 and 1965, then followed through April 1978 to determine the incidence of breast cancer. Women with progesterone deficiency had **5.4 times greater incidence of breast cancer** compared to women with infertility due to non-hormonal causes. **The women with progesterone deficiency experienced a 10-fold increased mortality from all cancers**. Thus, strongly suggesting progesterone serves as an all-purpose anti-cancer agent. Dr. Cowan writes:

These [1,083 progesterone deficient] women were categorized as to the cause of infertility into two groups, those with endogenous progesterone deficiency (PD) and those with nonhormonal causes (NH). Women in the PD group had 5.4 times the risk of premenopausal breast cancer compared to women in the NH group...Women in the PD group also experienced a 10-fold increase in deaths from all malignant neoplasms compared to the NH group. (102)

Progesterone Induces Apoptosis in Breast Cancer, In Vitro Study

In 1998, Drs. Bent Formby, and T. S. Wiley studied in-vitro the effect of progesterone on T47-D breast cancer cells exhibiting the progesterone receptor, finding a strong anti-proliferative effect. Progesterone at the same concentration as the third trimester of pregnancy-induced apoptosis in the breast cancer cells with downregulation of the anti-apoptosis protein, bcl-2, and upregulation of the p53 protein, known as the "guardian of the genome". Dr.

Bent Formby writes:

[after progesterone] the expression by T47-D cancer cells of bcl-2 was down-regulated, and that of p53 was up-regulated as detected by semiquantitative RT-PCR analysis. These results demonstrate that progesterone at a concentration **similar to that seen during the third trimester of pregnancy exhibited a strong antiproliferative effect** on at least two breast cancer cell lines. **Apoptosis was induced** in the progesterone receptor expressing T47-D breast cancer cells. (103)

Natural Progesterone, the First Birth Control Drug

Initially, back in the 1940s, natural progesterone was an injectable drug used to inhibit ovulation and thus explored as a birth control drug (contraceptive). This was before the availability of oral micronized progesterone. To make oral contraceptives more convenient to use, synthetic progestins were developed. Oral tablets are more convenient than injections. Secondly, the drug maker may obtain patent protection for the altered chemical formula. Natural progesterone cannot be patented, since it is a natural substance exempt from patent protection. Birth control pills are effective for preventing pregnancy by suppressing ovulation and shutting down ovarian function, thus acting as a form of chemical castration. Unfortunately, birth control pills contain synthetic progestins which are carcinogenic and cause breast cancer. The pills have many other adverse side effects such as blood clots and strokes. It is advisable to avoid synthetic progestins, as these are monster Frankenstein hormones. Instead, it is advisable to use non-hormonal means of contraception, such as barrier methods with the diaphragm and cervical cap. The non-hormonal copper T-IUD is no longer recommended as it is in litigation for product defect, with breakage and retention of the plastic arm upon attempted removal. For more on this see Chapter 25, The Adverse Side Effects of Birth Control Pills. (104-105)

More on Breast Cancer and Hormone Levels

If higher estrogen levels are the primary cause of breast cancer, we would expect to find more breast cancer mortality in women with higher hormone levels at age 30, and less breast cancer in women with lower post-menopausal hormone levels after age 50. What we find is the exact opposite. 77 percent of the breast cancer mortality is in post-menopausal women with low hormone levels, and only 23 percent in younger premenopausal women. In 2001, Dr. Rosemary Yancik writes:

> Postmenopausal women aged 55 years and older have 66% of incident breast tumors and **experience 77% of breast cancer mortality**. (106)

In 2024, Dr. Zsuzsanna Suba from Budapest says 80 percent of breast cancers are above age 50, and 20 percent are younger premenopausal women, writing:

> Among breast cancer cases, about 80% are above 50, being predominantly postmenopausal, while only 20% of them are younger, premenopausal women. (107-108)

Conclusion: Bio-identical hormones used at appropriate dosages are safe, and effective for the relief of menopausal symptoms. Health benefits extend to all organ systems. Bioidentical hormone treatments are safest when used in the correct combination, much like a symphony with multiple musical instruments making beautiful music. Transdermal forms of estrogen are not associated with abnormal coagulation, blood clots and venous thromboembolism as is the oral (pill) form of estrogen. Thus, transdermal estrogen is preferable. Estradiol is the most proliferative estrogen, and when used alone there is a small increase in breast cancer risk which is abolished when combined with natural progesterone. The safest estrogens reduce the risk of breast cancer and are those that target ER beta, the tumor suppressor receptor, such as unsaturated B-ring steroids found in Premarin (CEE), estriol (E3), and testosterone metabolite 3-beta-diol. Exposure to pregnancy levels of endogenous hormones reprograms the epigenome, conferring protection from breast cancer. This is thought to be the case for exogenous exposure as well. Long-term estrogen deprivation (LTED) transforms estrogen from a growth factor to a death factor in breast cancer cells. When treating breast cancer survivors with hormone replacement, the length of LTED is an important factor in determining recurrence rate, with higher recurrence for patients with less than 2 years of LTED. Any chemical alteration of a human hormone creates a Frankenstein monster. One example is medroxyprogesterone (MPA), a chemically altered version of progesterone used to induce cancer in the MPA mouse model of breast cancer. Chemically altered monster hormones should never have been FDA-approved for use as hormone replacement or for oral birth control (contraceptives). The one exception is dysfunctional uterine bleeding, a justifiable use for synthetic hormones, and their carcinogenicity is overlooked.

♦ References for Chapter 16 The Safety and Importance of Bioidentical Hormones Part One

1) Swan, Shanna H., and Stacey Colino. Count Down: How Our Modern World Is Threatening Sperm Counts, Altering Male and Female Reproductive Development, And Imperiling the Future Of The Human Race. Simon and Schuster, 2022.

2) Stone, Irwin. "The Natural History of Ascorbic Acid in The Evolution of The Mammals and Primates and Its Significance for Present-Day Man." Orthomolecular Psychiatry 1.2-3 (1972): 82-89.

3) Stack, David. "Charles Darwin: Theory Of Natural Selection." Encyclopedia of Evolutionary Psychological Science. Cham: Springer International Publishing, 2021. 1000-1011.

4) Levine, Hagai, et al. "Temporal Trends in Sperm Count: A Systematic Review and Meta-Regression Analysis of Samples Collected Globally in the 20th And 21st Centuries." Human Reproduction Update 29.2 (2023): 157-176.

5) Trasande, Leonardo, and Robert M. Sargis. "Endocrine-Disrupting Chemicals: Mainstream Recognition of Health Effects and Implications for The Practicing Internist." Journal of Internal Medicine 295.2 (2024): 259-274.

6) Coburn, T., D. Dumanoski, and J. P. Myers. "Our Stolen Future, 1996." Plume: New York.

7) Kohn, Grace E., et al. "The History of Estrogen Therapy." Sexual Medicine Reviews 7.3 (2019): 416-421.

8) Gwei-Djen, Lu, and Joseph Needham. "Medieval Preparations of Urinary Steroid Hormones." Medical History 8.2 (1964): 101-121.

9) Dhont, Marc. "Treatment Of the Menopause: The Swinging Pendulum." Facts, Views & Vision in Obgyn 2.3 (2010): 173.

10) Wilson, Robert A. "Feminine Forever. New York: M. Evans and Company." (1966).

11) Fhadil, Sadeer, and Paul Wright. "Electrolytes in Cardiology." The Pharmaceutical Journal 294.7849 (2015): 181-4.

12) Penniston, Kristina L., and Sherry A. Tanumihardjo. "The Acute and Chronic Toxic Effects of Vitamin A." The American Journal of Clinical Nutrition 83.2 (2006): 191-201.

13) Marcinowska-Suchowierska, Ewa, et al. "Vitamin D toxicity–a Clinical Perspective." Frontiers in Endocrinology 9 (2018): 550.

14) Lindberg, Bo S., Bo A. Nilsson, and Elof DB Johansson. "Plasma Progesterone Levels in Normal and Abnormal Pregnancies." Acta Obstetricia Et Gynecologica Scandinavica 53.4 (1974): 329-335.

15) Johansson, Elof DB. "Progesterone Levels in Peripheral Plasma During the Luteal Phase of The Normal Human Menstrual Cycle Measured by A Rapid Competitive Protein Binding Technique." European Journal of Endocrinology 61.4 (1969): 592-606.

16) Holmdahl, Tore, et al. "Peripheral Plasma Levels Of 17α-Hydroxyprogesterone, Progesterone and Estradiol During Normal Menstrual Cycles in Women." European Journal of Endocrinology 71.4 (1972): 743-754.

17) Sandru, Florica, et al. "Progesterone Hypersensitivity in Assisted Reproductive Technologies: Implications for Safety and Efficacy." Journal of Personalized Medicine 14.1 (2024): 79.

18) Sashidhar, Nivedita, et al. "Exogenous Progestogen Hypersensitivity and its Increasing Association with Assisted Reproductive Techniques (ART)/In Vitro Fertilization (IVF)." Indian Dermatology Online Journal 15.1 (2024): 24-32.

19) Foer, Dinah, et al. "Progestogen Hypersensitivity In 24 Cases: Diagnosis, Management, And Proposed Renaming and Classification." The Journal of Allergy and Clinical Immunology: In Practice 4.4 (2016): 723-729.

20) Lipman, Zoe M., et al. "Autoimmune Progesterone Dermatitis: A Systematic Review." Dermatitis 33.4 (2022): 249-256.

21) Reiss, Uzzi, and Martin Zucker. Natural Hormone Balance for Women: Look Younger, Feel Stronger, And Live Life With Exuberance. Simon and Schuster, 2002.

22) Glaser, Rebecca L., et al. "Incidence of Invasive Breast Cancer in Women Treated With Testosterone Implants: Dayton Prospective Cohort Study, 15-Year Update." (2024). Preprint.

23) Dimitrakakis, Constantine, et al. "Low salivary testosterone levels in patients with breast cancer." BMC cancer 10 (2010): 1-8.

24) Islam, Rakibul M., et al. "Safety and efficacy of testosterone for women: a systematic review and meta-analysis of randomised controlled trial data." The lancet Diabetes & endocrinology 7.10 (2019): 754-766.

25) Glaser, Rebecca, and Constantine Dimitrakakis. "Testosterone therapy in women: Myths and misconceptions." Maturitas 74.3 (2013): 230-234.

26) Redmond, Geoffrey P., and Wilma F. Bergfeld. "Diagnostic approach to androgen disorders in women: acne, hirsutism, and alopecia." Cleve Clin J Med 57.5 (1990): 423-7.

27) Carmina, Enrico, et al. "Female adult acne and androgen excess: a report from the multidisciplinary androgen excess and PCOS committee." Journal of the Endocrine Society 6.3 (2022): bvac003.

28) Glaser, Rebecca, Anne E. York, and Constantine Dimitrakakis. "Beneficial effects of testosterone therapy

29) Wolfe, M. Michael, et al. "Gastrointestinal Toxicity of Nonsteroidal Antiinflammatory Drugs." New England Journal of Medicine 340.24 (1999): 1888-1899.

30) Interdonato, Livia, et al. "Endocrine Disruptor Compounds in Environment: Focus On Women's Reproductive Health And Endometriosis." International Journal of Molecular Sciences 24.6 (2023): 5682.

31) Flaws, Jodi, et al. "Plastics, EDCs and health." Endocrine Society: Washington, DC, USA (2020).

32) Bertram, Michael G., et al. "Endocrine-Disrupting Chemicals." Current Biology 32.13 (2022): R727-R730.

33) Fournier A, et al. Unequal Risks for Breast Cancer Associated with Different Hormone Replacement Therapies: Results from The E3N Cohort Study. Breast Cancer Res Treat 2008; 107:103–11

34) Fournier A, Berrino F, Riboli E, et al. Breast Cancer Risk in Relation to Different Types Of Hormone Replacement Therapy in the E3N-EPIC cohort. Int J Cancer 2005; 114:448–54.

35) De Lignieres, B., et al. "Combined Hormone Replacement Therapy and Risk of Breast Cancer In A French Cohort Study Of 3175 Women." Climacteric 5.4 (2002): 332-340.

36) Birrell, Stephen N., et al. "Disruption of Androgen Receptor Signaling by Synthetic Progestins May Increase Risk Of Developing Breast Cancer." The FASEB Journal 21.10 (2007): 2285-2293.

37) Giulianelli, Sebastián, et al. "Estrogen Receptor Alpha Mediates Progestin-Induced Mammary Tumor Growth by Interacting with Progesterone Receptors at The Cyclin D1/MYC Promoters." Cancer research 72.9 (2012): 2416-2427.

38) Dhanasekaran, Renumathy, et al. "The MYC oncogene—The Grand Orchestrator of Cancer Growth And Immune Evasion." Nature reviews Clinical oncology 19.1 (2022): 23-36.

39) Kim, Jong Kyong, and J. Alan Diehl. "Nuclear Cyclin D1: An Oncogenic Driver In Human Cancer." Journal of cellular physiology 220.2 (2009): 292-296.

40) Perkins, Meghan S., et al. "Upregulation of an Estrogen Receptor-Regulated Gene By First Generation Progestins Requires Both The Progesterone Receptor And Estrogen Receptor Alpha." Frontiers in Endocrinology 13 (2022): 959396.

41) Ochnik, Aleksandra M., et al. "Antiandrogenic Actions Of Medroxyprogesterone Acetate On Epithelial Cells Within Normal Human Breast Tissues Cultured Ex Vivo." Menopause (New York, NY) 21.1 (2014): 79-88.

42) Mohammed, Hisham, et al. "Progesterone Receptor Modulates ER-alpha Action In Breast Cancer." Nature 523.7560 (2015): 313-317.

43) Campagnoli C, et al. Progestins and Progesterone In Hormone Replacement Therapy And The Risk Of Breast Cancer. J Steroid Biochem Mol Biol 2005;96:95–108.

44) Ory K, Lebeau J, Levalois C, et al. Apoptosis Inhibition Mediated by Medroxyprogesterone Acetate Treatment Of Breast Cancer Cell Lines. Breast Cancer Res Treat 2001;68:187–98. 554

45) Hofseth LJ, Raafat AM, Osuch JR, et al. Hormone Replacement Therapy with Estrogen Or Estrogen Plus Medroxyprogesterone Acetate Is Associated With Increased Epithelial Proliferation In The Normal Postmenopausal Breast. J Clin Endocrinol Metab 1999; 84:4559–65.

46) Jeng MH, Parker CJ, Jordan VC. Estrogenic Potential of Progestins In Oral Contraceptives Stimulate Human Breast Cancer Cell Proliferation. Cancer Res 1992;52:6539–46.

47) Kalkhoven E, Kwakkenbos-Isbrücker L, de Laat SW, et al. Synthetic progestins induce proliferation of breast tumor cell lines via the progesterone or estrogen receptor. Mol Cell Endocrinol 1994;102:45–52.

48) Cline JM, Soderqvist G, von Schoultz E, et al. Effects of Conjugated Estrogens, Medroxyprogesterone Acetate, And Tamoxifen on The Mammary Glands Of Macaques. Breast Cancer Res Treat 1998;48:221–9

49) Coelingh Bennink, Herjan JT, et al. "Progesterone From Ovulatory Menstrual Cycles Is An Important Cause Of Breast Cancer." Breast Cancer Research 25.1 (2023): 60.

50) Coelingh Bennink, H. J. T., and F. Z. Stanczyk. "Progesterone and Not Estrogens Or Androgens Causes Breast Cancer." Climacteric 27.2 (2024): 217-222.

51) Gompel, A., Prior, J.C. et al. "Lack of Evidence That Progesterone In Ovulatory Cycles Causes Breast Cancer." Climacteric 26.6 (2023): 634-637.

52) Hipolito Rodrigues, Marcio Alexandre, and Anne Gompel. "Micronized Progesterone, Progestins, And Menopause Hormone Therapy." Women & Health 61.1 (2021): 3-14.

53) Lyytinen, Heli, et al. "Breast Cancer Risk In Postmenopausal Women Using Estrogen-Only Therapy." Obstetrics & Gynecology 108.6 (2006): 1354-1360.

54) Bakken, Kjersti, et al. "Menopausal Hormone Therapy and Breast Cancer Risk: Impact of Different Treatments. The European Prospective Investigation into Cancer and Nutrition." International Journal of Cancer 128.1 (2011): 144-156.

55) Chlebowski, Rowan T., et al. "Randomized Trials Of Estrogen-Alone And Breast Cancer Incidence: A Meta-Analysis." Breast Cancer Research and Treatment (2024): 1-8.

56) Jones, Michael E., et al. "Menopausal Hormone Therapy and Breast Cancer: What Is The True Size Of The Increased Risk?" British Journal of Cancer 115.5 (2016): 607-615.

57) Bhavnani, Bhagu R., Shui-Pang Tam, and XiaoFeng Lu. "Structure Activity Relationships And Differential Interactions And Functional Activity Of Various Equine Estrogens Mediated Via Estrogen Receptors (ERs) ER-Alpha And ER-Beta." Endocrinology 149.10 (2008): 4857-4870.

58) Veurink, Marieke, et al. "The History Of DES, Lessons To Be Learned." Pharmacy World and Science 27 (2005): 139-143.

59) Herbst, Arthur L., et al. "Adenocarcinoma of the Vagina: Association Of Maternal Stilbestrol Therapy With Tumor Appearance In Young Women." New England journal of medicine 284.16 (1971): 878-881.

60) Zamora-León, Pilar. "Are The Effects of DES Over? A Tragic Lesson from The Past." International Journal of Environmental Research and Public Health 18.19 (2021): 10309.

61) Berkson, D. Lindsey. Hormone Deception. Contemporary Books., 2000.

62) Colborn, Theo, et al. Our Stolen Future: Are We Threatening Our Fertility, Intelligence, And Survival? --A Scientific Detective Story. Penguin, 1997.

63) Gore, Andrea C., et al. "Introduction to Endocrine Disrupting Chemicals (EDCS)." A Guide For Public Interest Organizations And Policy-Makers (2014): 21-22.

64) Kahn, Linda G., et al. "Endocrine-Disrupting Chemicals: Implications for Human Health." The Lancet Diabetes & endocrinology 8.8 (2020): 703-718.

65) Hansel, Megan C., et al. "Exposure to Synthetic Endocrine-Disrupting Chemicals in Relation to Maternal and Fetal Sex Steroid Hormones: A Scoping Review." Current Environmental Health Reports (2024): 1-24.

66) Kek, Tina, et al. "Exposure to Endocrine Disrupting Chemicals (Bisphenols, Parabens, And Triclosan) And Their Associations with Preterm Birth In Humans." Reproductive Toxicology (2024): 108580.

67) Pan, Jing, et al. "The Adverse Role Of Endocrine Disrupting Chemicals In The Reproductive System." Frontiers in Endocrinology 14 (2024): 1324993.

68) Liu, Xiaohui, et al. "Bisphenol-C is the Strongest Bifunctional ER-Alpha-Agonist And ER-Beta-Antagonist Due To Magnified Halogen Bonding." PloS one 16.2 (2021): e0246583.

69) Gaudillière, Jean-Paul. "DES, Cancer, And Endocrine Disruptors." Powerless Science (2013): 65-94.

70) Laws, Mary Jo, et al. "Endocrine Disrupting Chemicals and Reproductive Disorders In Women, Men, And Animal Models." Advances in pharmacology. Vol. 92. Academic Press, 2021. 151-190.

71) Cavalier, Haleigh, et al "Exposures to Pesticides And Risk Of Cancer: Evaluation Of Recent Epidemiological Evidence In Humans And Paths Forward." International journal of cancer 152.5 (2023): 879-912.

72) Baj, Jacek, et al. "Derivatives of Plastics As Potential Carcinogenic Factors: The Current State Of Knowledge." Cancers 14.19 (2022): 4637.

73) Escrich, Edward. "Validity Of The DMBA-Induced Mammary Cancer Model For The Study Of Human Breast Cancer." The International Journal Of Biological Markers 2.3 (1987): 197-206.

74) Buqué, Aitziber, et al. "MPA/DMBA-driven mammary carcinomas." Methods in cell biology. Vol. 163. Academic Press, 2021. 1-19.

75) Lanari, Claudia Lee Malvina, et al. "The MPA Mouse Breast Cancer Model: Evidence For A Role Of Progesterone Receptors In Breast Cancer." (2009).

76) Cavalieri, Ercole, and Eleanor Rogan. "The 3, 4-Quinones of Estrone and Estradiol Are the Initiators of Cancer Whereas Resveratrol And N-Acetylcysteine Are The Preventers." International journal of molecular sciences 22.15 (2021): 8238.

77) Cavalieri, Ercole L., and Eleanor G. Rogan. "Inhibition of Depurinating Estrogen-DNA Adduct Formation In The Prevention Of Breast And Other Cancers." Trends in breast cancer prevention (2016): 113-145.

78) Cavalieri, Ercole L., and Eleanor G. Rogan. "The Etiology and Prevention of Breast Cancer." Drug Discovery Today: Disease Mechanisms 9.1-2 (2012): e55-e69.

79) Weinberg, Nicole, et al. "Physical activity, hormone replacement therapy, and the presence of coronary calcium in midlife women." Women & health 52.5 (2012): 423-436.11)

80) Manson, JoAnn E., et al. "Estrogen therapy and coronary-artery calcification." New England Journal of Medicine 356.25 (2007): 2591-2602.

81) Budoff, Matthew J., et al. "Effects of hormone replacement on progression of coronary calcium as measured by electron beam tomography." Journal of women's health 14.5 (2005): 410-417.

82) Hodis, Howard N., and Wendy J. Mack. "Menopausal Hormone Replacement Therapy And Reduction Of All-Cause Mortality And Cardiovascular Disease: It Is About Time And Timing." The Cancer Journal 28.3 (2022): 208-223.

83) Majumdar, Sumit R., Elizabeth A. Almasi, and Randall S. Stafford. "Promotion and prescribing of hormone therapy after report of harm by the Women's Health Initiative." Jama 292.16 (2004): 1983-1988.

84) Wyeth Reports Earnings Results for the 2006 Fourth Quarter and Full Year. News Release. http://www.wyeth.com/irj/servlet/prt/portal/prtroot/com.sap.km.cm.docs/wyeth_xml/home/news/announcements/1170158273391.pdf

85) Hotze, Steven F., and Donald P. Ellsworth. the Case for Bioidentical. Journal of American Physicians and Surgeons Volume 13.2 (2008): 43.

86) Manyonda, Isaac, et al. "Could Perimenopausal Estrogen Prevent Breast Cancer? Exploring the Differential Effects of Estrogen-Only Versus Combined Hormone Replacement Therapy." Journal of Clinical Medicine Research 14.1 (2022): 1.

87) Suba, Zsuzsanna. "Rosetta Stone For Cancer Cure: Comparison of the Anticancer Capacity of Endogenous Estrogens, Synthetic Estrogens and Antiestrogens." Oncology Reviews 17 (2023): 10708.

88) Clark, James H. "A Critique of Women's Health Initiative Studies (2002-2006)." Nuclear Receptor Signaling 4 (2006).

89) The Dr. Louise Newson Podcast 281 - Blind spots in modern medicine, with Dr. Marty Makary. Nov. 5, 2024. https://www.balance-menopause.com/menopause-library/blind-spots-in-modern-medicine-with-dr-marty-makary/

90) Rossouw, Jacques E. "Estrogens for Prevention of Coronary Heart Disease: Putting the Brakes On The Bandwagon." Circulation 94.11 (1996): 2982-2985.

91) Lukong, Kiven Erique. "Understanding Breast Cancer—The Long and Winding Road." BBA clinical 7 (2017): 64-77.

92) Russo, Jose, et al. "The Protective Role of Pregnancy In Breast Cancer." Breast cancer research 7 (2005): 1-12.

93) Feigman, Mary J., et al. "Pregnancy Reprograms The Epigenome Of Mammary Epithelial Cells And Blocks The Development Of Premalignant Lesions." Nature Communications 11.1 (2020): 2649.

94) Feigman, Mary J., et al. "Pregnancy Reprograms The Enhancer Landscape Of Mammary Epithelial Cells And Alters The Response To Cmyc-Driven Oncogenesis." bioRxiv (2019): 642330.

95) Lakshmanaswamy, Rajkumar, et al. "Hormone-Induced Protection of Mammary Tumorigenesis In Genetically Engineered Mouse Models." Breast Cancer Research 9 (2007): 1-11.

96) Lakshmanaswamy, Rajkumar, et al. "Short-Term Exposure to Pregnancy Levels of Estrogen Prevents Mammary Carcinogenesis." Proceedings of the National Academy of Sciences 98.20 (2001): 11755-11759.

97) Lakshmanaswamy, Rajkumar, et al. "Prevention of Mammary Carcinogenesis by Short-Term Estrogen and Progestin Treatments." Breast Cancer Research 6.1 (2003): 1-7.

98) Lakshmanaswamy, Rajkumar, et al. "Hormonal Prevention of Breast Cancer: Significance of Promotional Environment." Hormonal Carcinogenesis V. New York, NY: Springer New York, 2008. 469-475.

99) Arumugam, Arunkumar, et al. "Short-term Treatment With Pregnancy Levels Of Estradiol Prevents Breast Cancer By Delaying Promotion And Progression." Cancer Research 73.8_Supplement (2013): 199-199.

100) Inoh, Akira, et al. "Protective Effects of Progesterone and Tamoxifen In Estrogen-Induced Mammary Carcinogenesis In Ovariectomized W/Fu Rats." Japanese Journal of Cancer Research GANN 76.8 (1985): 699-704.

101) Segaloff, Albert. "Inhibition by Progesterone Of Radiation-Estrogen-Induced Mammary Cancer In The Rat." Cancer Research 33.5 (1973): 1136-1137.

102) Cowan, Linda D., et al. "Breast Cancer Incidence in Women with A History of Progesterone Deficiency." American Journal of Epidemiology 114.2 (1981): 209-217.

103) Formby, Bent, and T. S. Wiley. "Progesterone Inhibits Growth And Induces Apoptosis In Breast Cancer Cells: Inverse Effects On Bcl-2 And P53." Annals of Clinical & Laboratory Science 28.6 (1998): 360-369.

104) Liao, Pamela Verma, and Janet Dollin. "Half a Century Of The Oral Contraceptive Pill: Historical Review And View To The Future." Canadian Family Physician 58.12 (2012): e757-e760.

105) Dhont, Marc. "History of Oral Contraception." The European Journal of Contraception & Reproductive Health Care 15.sup2 (2010): S12-S18.

106) Yancik, Rosemary, et al. "Effect of Age And Comorbidity In Postmenopausal Breast Cancer Patients Aged 55 Years And Older." JAMA 285.7 (2001): 885-892.

107) Suba, Zsuzsanna. "Estrogen Regulated Genes Compel Apoptosis in Breast Cancer Cells, Whilst Stimulate Antitumor Activity in Peritumoral Immune Cells in a Janus-Faced Manner." Current Oncology 31.9 (2024): 4885-4907.

108) McGuire, Andrew, et al. "Effects of Age On The Detection And Management Of Breast Cancer." Cancers 7.2 (2015): 908-929.

109) Wile AG, Opfell RW, Margileth DA. Hormone Replacement Therapy In Previously Treated Breast Cancer Patients. Am J Surg 1993;165:372-5.

Chapter 17

The Safety and Importance of Bioidentical Hormones Part Two

WHY ARE BIOIDENTICAL HORMONES IMPORT-ANT? What do they do and why do we need them? First, let us take a look at the definition of a bioidentical hormone and their way they differ from synthetic hormones. Bioidentical hormones are the hormones that exist naturally in the human body. Since these are natural, they are not eligible for patent protection. On the other hand, synthetic hormones are chemically altered versions of natural hormones. These do not occur in nature anywhere and can be patented. It is that simple. Our Western medical system has rejected natural substances. Bioidentical hormones are natural substances and therefore the enemy and competitor of the patented drug industry. Off-patent and repurposed drugs are also placed into this same category. They are also the enemy and competitor of the patented drug industry.

Why Chemically Alter a Human Hormone?

The drug companies hire chemists to alter the structure of human hormones in the laboratory so the drug company can obtain a patent on the new chemical structure, which is a new drug. This alteration is required for the drug company to obtain a patent which gives exclusive marketing rights to the drug company. The patent is necessary to protect profits. Because of a quirk in our patent laws, only chemically altered substances can be patented. Natural substances like human hormones cannot be patented, and are therefore not as profitable to manufacture and market. When I first learned about this, I realized our entire medical system is based on this paradigm of patented, chemically altered substances. Why is this problematic? The original natural substance used in traditional medicine for centu-ries is almost always more effective with fewer adverse side effects. This is especially true when dealing with human hormones.

Chemically Altered Hormones are Frankenstein Hormones

Hormones fit onto their receptors just like a "lock and key", so any slight alteration of their chemical structure creates an endocrine disrupting chemical (EDC), a "Frankenstein hormone", never found in the human body or anywhere else in nature. The reality is that synthetically altered monster hormones should never have been approved for marketing and sale to the American people. Yet, that is exactly what your mainstream medical doctor will offer you if you ask for menopausal hormone replacement. Worse than that, they may withhold hormone replacement and instead offer one of many patented drugs such as SSRI antidepressants, bisphosphonate bone drugs, NSAIDs, and statins.

Women's Health Initiative Dogma and the Greatest Error in Medicine

The great error and tragedy of modern medicine is the fear of hormone replacement. This tragedy has resulted in increased mortality and lives lost. In 2013, Dr. Philip Sarrel, professor of Obstetrics and Gynecology at Yale Medical School, estimated the mortality toll of estrogen avoidance could be as high as 91,000 women over 10 years from 2002 to 2012. Dr. Sarrel writes:

> We examined the effect of estrogen avoidance on mortality rates among hysterectomized women aged 50 to 59 years...We derived a formula to relate the excess mortality among hysterectomized

women aged 50 to 59 years assigned to placebo in the Women's Health Initiative randomized controlled trial to the entire population of comparable women in the United States, incorporating the decline in estrogen use observed between 2002 and 2011...Over a 10-year span, starting in 2002, a minimum of 18 601 and as many as 91 610 postmenopausal women died prematurely because of the avoidance of estrogen therapy (ET)...**ET in younger postmenopausal women is associated with a decisive reduction in all-cause mortality,** but estrogen use in this population is low and continuing to fall. Our data indicate an associated annual mortality toll in the thousands of women aged 50 to 59 years. Informed discussion between these women and their health care providers about the effects of ET is a matter of considerable urgency. Note: this mortality estimate is only for women who have had hysterectomy, a small subset of the 50 million post-menopausal women not using HRT. The increased mortality in this larger group has not been included in Dr. Sarrel's calculations. (1)

It gets even worse than that, according to Dr. Martin Makary, a pancreatic cancer surgeon, Professor of Health Policy at Johns Hopkins Medicine, and bestselling author of *Blind Spots in Medicine*, a woman suffering from menopausal symptoms is more likely to be offered SSRI antidepressants than hormone replacement. In 2024, Dr. Martin Makary says:

> perhaps the dogma [arising from the 2002 Women's Health Initiative study] that taking hormone replacement therapy at the time of menopause causes breast cancer is probably the **biggest screw up in modern medicine**...basically a small group of people decided to call hormone replacement therapy a carcinogen, when in fact, **for the vast majority of women going through menopause, it is a miracle**...You know, tragically because of this Women's Health initiative dogma 22 years ago, medical schools just kind of concluded, well, there's nothing you can do for menopause, so

why teach it? And so menopause itself got ignored in medical school curricula and residencies...So today, tragically, at least in the United States, **a woman is more likely to be prescribed an antidepressant for menopause than they are hormone replacement.** (2)

Hormone Replacement Should Be Integral to Medical Practice

As Dr. Marty Makary says in the previous quote, all doctors should be offering menopausal hormone replacement, but they do not because of fears created after the 2002 Women Health Initiative (WHI) study. In 2008, Drs. Erika Schwartz and Kent Holtorf write that menopausal hormone replacement is so important, it should be an integral part of medical practice. Dr. Erika Schwartz writes:

> Confusion and controversy surround the use of estrogen, progesterone, testosterone, growth hormone, and thyroid hormones. [We] discuss age-related hormone loss and supplementation therapies for age-related hormonal deficiencies as possible first-line therapeutic modalities to be considered in our search to improve quality of life, prevent chronic illnesses, and maintain wellness. The well-informed use of hormones in wellness and disease prevention will result in symptomatic improvement and **should be considered an integral part in the armamentarium of options we offer our patients.** (3)

Bioidentical Hormones are the Preferred Form of HRT

The 2002 first arm WHI study was marketed as studying human hormone replacement (HRT), yet has been criticized for not studying human hormone replacement at all. The WHI trial studied Premarin, a horse hormone, and medroxyprogesterone (MPA), a synthetic hormone, not found anywhere in nature or the human body. The preferred form of HRT is bioidentical hormones. In 2009, Dr. Kent

Holtorf concluded in Postgraduate Medicine that bioidentical hormones are more efficacious with less adverse effects than synthetic hormones, writing:

> Physiological data and clinical outcomes demonstrate that bioidentical hormones are associated with lower risks, including the risk of breast cancer and cardiovascular disease, and are more efficacious than their synthetic [medroxyprogesterone, MPA] and animal derived [Premarin, CEE] counterparts. Until evidence is found to the contrary, **bioidentical hormones remain the preferred method of HRT [Hormone Replacement Therapy]**. Further randomized controlled trials are needed to delineate these differences more clearly. (4)

The Information War and Terminology

Many years ago, after the invention of synthetic hormones, an information war was launched by the drug industry to create confusion among the public and medical professionals about the difference between natural human hormones and synthetic hormones. The medical journals, academic medicine, key opinion leaders, mass media, government regulatory agencies, and Congress are all captured by the drug industry. Many of these agencies and institutions serve as a mouthpiece of the drug industry. Thus, the drug industry has a firm control over the narrative. Information warfare obscures the carcinogenicity of synthetic progestins by sowing confusion in the medical literature, making people think natural progesterone is the same as synthetic progestins. They are not the same. The second information war creates confusion about the carcinogenicity of estrogen, stoking unfounded fears that estrogen causes breast cancer. These fears were magnified by the 2002, first arm WHI study, causing a sudden decline in numbers of women using hormone replacement. Before 2002, 35 million women used HRT. This number declined to 10 million women a few years later. Who benefits from an additional 25 million women living in a state of hormone deficiency? Who benefits from our chronic disease epidemic? Follow the money. The pharmaceutical industry and the health care industry have the most to gain financially from this state of affairs, as you will read below. Because of this information war, we must now use the terminology, "bioidentical" hormones which means natural, human hormones. I consider this an embarrassment to medical science that we are forced to use the word "bio-identical" for natural hormones found in the human body. We should not feel that we are forced to do this. It should be sufficient to use the same old names in the biochemistry textbooks. The simple word "hormone" should suffice. Yet here we are again finding ourselves using the word "bio-identical hormone" thanks to the "Information War" going on between natural medicine and the drug industry. (5-12)

How Do Hormones Work?

Hormones are signaling messengers that attach to receptors in the DNA in trillions of our cells which signal our DNA to influence gene expression. We also have receptors on the cell membrane for immediate cell signaling, called GPER (G protein-coupled estrogen receptor). There are two places in the cell where we find DNA. One is in the cell nucleus, and the other is in the mitochondria, the little organelles that make energy. Mitochondrial DNA is only one strand inherited from the mother. Nuclear DNA consists of a double helix made of two strands of DNA. One strand is inherited from the mother and the other from the father. The double helix structure was originally discovered by Rosalind Franklin in 1951, more than a year before Watson and Crick's famous publication in 1953. However, Rosalind Franklin died in 1958, so in 1962, Watson and Crick received the Nobel Prize for Rosalind Franklin's work. Note the Nobel prize is not awarded posthumously (after someone dies). (13-16)

Hormones Control Transcription of Proteins

DNA resides in the nucleus of all living cells which contains the source code for protein synthesis and controls the growth, metabolism, and reproduction of the organism from embryo to adulthood. In essence, DNA is the master controller of the cell, like the source code that runs your computer. Mutations in our DNA cause genetic diseases, obviously not a good thing. Bio-identical hormones are messengers that attach to cellular receptors such as ER-alpha and ER-beta. We also have membrane receptors for immediate cell signaling called G-Protein coupled receptors. Once entering the cell, and attaching to a hormone receptor, a hormone-receptor dimer is formed which makes its way into the nucleus and then attaches response elements in our DNA, turning on DNA protein synthesis. In a process called induction of transcription, the hormone messenger instructs the source code in our DNA to produce specific proteins. Some hormone messages have the opposite effect suppressing the transcription of specific genes. This is called repression of transcription. After age 50, hormones decline leading to reduced protein synthesis and the onset of degenerative diseases. Bio-identical hormone replacement restores DNA protein synthesis, thereby delaying or reversing the onset of degenerative disease. In 1996, Dr. Miguel Beato writes:

> Regulation of gene expression by hormones involves an interaction of the DNA-bound receptors with other sequence-specific transcription factors …partly mediated by co-activators and co-repressors…Depending on the nature of these interactions, the final outcome can be induction or repression of transcription [of proteins]. (17- 18)

Why Are Proteins Important? Regenerative and Reparative Proteins

Proteins are the major building blocks for the human body and for all life forms, for that matter. Proteins serve a variety of functions. For example, "structural" proteins such as collagen make up the structural elements of the body such as bones, skin, arteries, hair, connective tissue, ligaments, tendons, and muscles. Other proteins are enzymes such as ATP-synthase, involved in energy production. There are proteins involved in communication, neurological function, and cognition called neurotransmitters. There are proteins involved in the immune system called antibodies and cytokines, and the list goes on. We need a constant supply of proteins to repair the body's wear and tear. For example, during the exertion of a marathon, the runner may endure micro tears to the tendons, ligaments, and muscles. Recovery time after a marathon depends on the speed of repair of these injuries based on the ability of the body to make new proteins and new cells. This requires good hormone levels of estrogen and testosterone. (19-22)

New Cell Layers Needed for Life

To live our daily lives, we need to make new cells. As our older cells and cell layers age and eventually die, we must have the ability to manufacture new cells. Examples are the blood cells that must be replaced by the bone marrow every 90 days, and the skin cells that slough off as the outer layer is replaced by a new layer of cells every 28 days. The epithelium of the gastrointestinal lining is generated at the basal cell layer which matures and replaces the surface layer every 5-7 days. The adult brain regenerates by neurogenesis in the Hippocampus with a full cycle every 7 weeks. All parts of our bodies require new cells to replace old ones. The turnover rate for our bodies is 80 grams of new cells every day. These new cells are made of proteins, so regeneration of new cell layers requires hormones to signal our DNA to make protein components of our cells. Hormone deficiency means a lack of regenerative proteins and the onset of degenerative diseases. (23-29)

Hormone Levels Decline with Age

We know from observational studies that hormone levels decline with age. Starting around age 50, women experience a sudden decline in estrogen levels called menopause. In men, hormonal decline after age 50 is called andropause, with a gradual decline in testosterone levels. Clinically, hormone deficiency causes a reduction in protein synthesis, muscle mass, and bone density. Body fat, fasting glucose, insulin levels are all increased. Hormone deficiency may cause fatigue, depression, loss of libido, and erectile dysfunction. (30-31)

The Linear Increase in Life Span

In 2019 Jim Oeppen and James W. Vaupel provided a useful chart of human life span over the centuries. Starting with the Industrial Revolution in 1820, we see a linear increase in human life span. I suspect this is due to improvements in living standards, sanitation, nutrition, and mass production of goods and services. Before 1900, most people did not live past age 50, so very few people were alive to experience the hormonal decline of aging. However, after 1900, an increasing population lived after the age of 50 with menopausal hormone deficiency. There is an even greater trend now, with the largest over-50 population in the history of Western Civilization. As of 2023, 125 million people in the U.S. (55% female), were over the age of 50, living with hormone deficiency and the associated degenerative diseases. Of the 62 million post-menopausal women in the U.S., only 10 million are currently using hormone replacement. That means 52 million women are suffering needlessly! (32-34)

Lack of Reparative Proteins Leads to Degenerative Diseases of Aging

Without the hormone signaling to our DNA turning on protein synthesis, we will not have the reparative and regenerative proteins to maintain a good quality of life and we will start

to experience degenerative diseases of aging. Bioidentical hormone replacement therapy (HRT) prevents many of the degenerative diseases of aging. However, rather than offering hormone replacement, mainstream medicine offers various drug pharmaceutical drugs or medical procedures, such as joint replacement and coronary artery stenting and bypass. Here is that list of pharmaceutical drugs:

List of the Degenerative Diseases	Drugs Offered
Osteoarthritis	Vioxx (rofecoxib), ibuprofen (Motrin, Advil) naproxen (Aleve)
Osteoporosis	Alendronate (Fosamax), Denosumab (Prolia)
Atherosclerotic Vascular Disease	Atorvastatin (Lipitor), Evolocumab (Repatha)
Cognitive Dysfunction	Donepezil (Aricept®) Rivastigmine (Exelon)
Immune System Dysfunction	Antibiotics (Z-pack, Azithromycin, Amoxicillin)
Loss of Libido	Sildenafil (Viagra) Tadalafil (Cialis)
Depression/anxiety	Fluoxetine (Prozac), Paroxetine (Paxil), Sertraline (Zoloft)
Insomnia	Zolpidem (Ambien), Zaleplon (Sonata), Benzodiazepines

More Degenerative Disease Means Greater Profits for Drug Companies

The major drug companies make most of their profits on blockbuster drugs such as Lipitor, Vioxx, and Zoloft treating the chronic disease epidemic. Notice these three drugs are listed in the above chart of chronic diseases caused by estrogen deficiency, all of which can be prevented by using bio-identical HRT. Thus, bioidentical HRT competes with the drug industry for market share and represents the financial enemy. If 50 million women not using HRT were treated with HRT, this would remove 50 million customers from the drug industry, and make a large dent in the chronic disease epidemic.

Why is the Chronic Disease Epidemic and Emergency?

The chronic disease epidemic is a national emergency because roughly half the US population has a chronic disease, and 86% of $4.1 Trillion in health care costs pay for treatment of chronic disease. Despite spending the most per capita on healthcare, we rank last among Western industrial nations in overall health. In 2022, Rita Numerof of Forbes Magazine tells us 40 percent of the adult US population has two or more chronic diseases, and 90 percent of our $4.1 trillion dollar health care expenditure goes to the treatment of chronic disease, writing:

> Today, almost half of the U.S. population, or 133 million Americans, are living with at least one chronic condition and 40% of adults suffer from two or more. Heart disease and diabetes are two leading causes of death, with diabetes affecting about 1 in 10 Americans...Managing chronic disease is and remains a costly business: according to the Centers for Disease Control (CDC), **90% of the roughly $4.1 trillion spent on healthcare goes to chronic and mental health conditions**. (35)

In 2024, Michael Bryant, Professor of History and Legal Studies in Rhode Island, calls the silent epidemic of chronic disease a national emergency, yet ignored by the corporate media and mainstream medicine, writing:

> There is no greater **national emergency** facing the people of the United States than the chronic disease epidemic currently affecting millions of children and adults....The corporate media has been **virtually silent**

about what could be called a crisis of chronic diseases in our country. Chronic Disease: What Is It? ...chronic disease...can be defined broadly as a condition that lasts one year or more, that requires ongoing medical attention or limits activities of daily living (or both), and that is one of the leading causes of illness, disability, and death. Most of these ongoing conditions, which range from heart disease to hypertension, from arthritis to asthma, from COPD to cancer, and from diabetes to depression, are **preventable** and some are reversible. They are for the most part manageable through early detection, improved diet, an exercise regimen, and some **sort of treatment**. Recent estimates from the CDC assert that 129 million adults in the United States have been diagnosed with at least one chronic condition. ...That means **more than half of American adults** are currently suffering from a chronic illness!....**By contrast, in the 1960s, the US Department of Health and Human Services records showed that only 6% of American children had chronic diseases**... the percentage of children with a chronic disease has been steadily rising. By 1986, the number had grown to **11.8%.** By 2006, according to some accounts, it reached a staggering **54%...** what is not up for debate is that over the past 50 years, the health of Americans has deteriorated to the point where the US now faces a health crisis of catastrophic proportions. **What is also not up for dispute is that the US has poorer health outcomes for both children and adults than other wealthy nations despite spending more per capita on health care than those other nations.** (36)

In 2020, Dr. Halstead Homan says mainstream medicine fails to recognize or respond to the chronic disease epidemic, creating a health care crisis bankrupting our nation, writing:

> Currently, some 50% of the US population has a chronic disease, creating an epidemic, and 86% of health care costs are attributable to chronic disease. The medical profession and its leadership did not recognize or respond appropriately to

the rising prevalence of chronic disease. As a consequence, a **health care crisis emerged**, with inadequate access to care and quality of care together with **excessive costs**...When will we awaken? (37)

The chronic disease epidemic is now a national emergency. How did this sad state of affairs come about? Follow the money. Profits for the drug industry are maximized by the sale of blockbuster drugs for chronic disease. The failure to provide menopausal HRT is part of it. Before 2002, 40 percent of menopausal women used HRT. By 2020 only 5 percent of menopausal women used HRT. 95% of menopausal women are untreated because of an unfounded fear of estrogen. The resulting list of estrogen deficiency diseases has contributed greatly to our current day epidemic of chronic disease, and maximized profits for the drug industry. (38)

Conclusion: The drug industry is not a benevolent grandfather whose goal is to make the nation healthier. They make the greatest profits when people are sick, not when they are healthy. In reality, the drug industry is a criminal racketeering enterprise that has paid out 62.3 billion in criminal and civil penalties over the last 30 years. Like any other corporation listed on the US stock exchange, the goal of the drug industry is to maximize profits for shareholders, and allowing the chronic disease epidemic to run rampant is good for the bottom line. What if another 50 million post-menopausal women started using bio-identical hormones? Would this mean massive declining sales and declining profits for the drug industry? Of course, it would. Are you starting to understand why there is intense animosity and competition between the drug industry and natural bioidentical hormones, and why we have a long-standing information war between the drug industry and bioidentical hormones? Part of this information war is to use the dirty trick playbook to firstly, create confusion and blur the distinction between progesterone and carcinogenic progestins, and, secondly, create fear of hormone replacement, thus creating a large population of loyal consumers of pharmaceutical drugs. Call me a conspiracy theorist, but it looks like the system is rigged. (39-41)

♦ References for Chapter 17 The Safety and Importance of Bioidentical Hormones Part Two

1) Sarrel, Philip M., et al. "The Mortality Toll of Estrogen Avoidance: An Analysis of Excess Deaths Among Hysterectomized Women Aged 50 To 59 Years." American journal of public health 103.9 (2013): 1583-1588.

2) The Dr. Louise Newson Podcast, Blind Spots in Modern Medicine, with Dr Marty Makary. Nov. 5, 2024. https://www.balance-menopause.com/menopause-library/blind-spots-in-modern-medicine-with-dr-marty-makary/

3) Schwartz, Erika T., and Kent Holtorf. "Hormones in Wellness and Disease Prevention: Common Practices, Current State Of The Evidence, And Questions For The Future." Primary Care: Clinics in Office Practice 35.4 (2008): 669-705.

4) Holtorf, Kent. "The Bioidentical Hormone Debate: Are Bioidentical Hormones (Estradiol, Estriol, And Progesterone) Safer or More Efficacious Than Commonly Used Synthetic Versions In Hormone Replacement Therapy?" Postgraduate Medicine 121.1 (2009): 73-85.

5) Smith, Richard. "Medical Journals Are an Extension of The Marketing Arm Of Pharmaceutical Companies." Plos medicine 2.5 (2005): e138.

6) Patsner, Bruce. "Capture of Academic Medicine by Big Pharma." Health Law Perspectives (University of Houston Health Law & Policy Institute) (2009).

7) Moynihan, Ray. "Key Opinion Leaders: Independent Experts or Drug Representatives In Disguise?" Bmj 336.7658 (2008): 1402-1403.

8) Reed, Genna, et al. "The Disinformation Playbook: How Industry Manipulates the Science-Policy Process—And How to Restore Scientific Integrity." Journal of Public Health Policy 42.4 (2021): 622.

9) Thomas, John P. "Mainstream News Media: Freedom of The Press or Controlled Propaganda." (2018).

10) Wouters, Olivier J. "Lobbying Expenditures and Campaign Contributions by The Pharmaceutical And Health Product Industry in the United States, 1999-2018." JAMA Internal Medicine 180.5 (2020): 688-697.

11) Schpero, William L., et al. "Lobbying Expenditures In The US Health Care Sector, 2000-2020." JAMA Health Forum. Vol. 3. No. 10. American Medical Association, 2022.

12) More Than Two-Thirds Of Congress Cashed A Pharma Campaign Check In 2020, New STAT Analysis Shows by Lev Facher, June 9, 2021. https://www.statnews.com/feature/prescription-politics/feder-al-full-data-set/

13) Braun, Gregory, et al. "How Rosalind Franklin Discovered the Helical Structure Of DNA: Experiments In Diffraction." The Physics Teacher 49.3 (2011): 140-143.

14) Danylova, T. V., and S. V. Komisarenko. "Standing on the Shoulders of Giants: James Watson, Francis Crick, Maurice Wilkins, Rosalind Franklin and The Birth Of Molecular Biology." Ukr Biochem J 92.4 (2020): 154-165.

15) Watson, James D., and F. H. Crick. "A Structure for Deoxyribose Nucleic Acid. 1953." Nature 421.6921 (2003): 397-8.

16) Watson, James D., and Francis HC Crick. "The Structure Of DNA." Cold Spring Harbor Symposia On Quantitative Biology. Vol. 18. Cold Spring Harbor Laboratory Press, 1953.

17) Beato, Miguel, et al. "Transcriptional Regulation by Steroid Hormones." Steroids 61.4 (1996): 240-251.

18) Gronemeyer, Hinrich. "Control of Transcription Activation by Steroid Hormone Receptors." The FASEB Journal 6.8 (1992): 2524-2529.

19) Collado-Boira, Eladio, et al. "Influence of Female Sex Hormones on Ultra-Running Performance And Post-Race Recovery: Role Of Testosterone." International Journal of Environmental Research And Public Health 18.19 (2021): 10403.

20) Chidi-Ogbolu, Nkechinyere, and Keith Baar. "Effect of Estrogen on Musculoskeletal Performance And Injury Risk." Frontiers in Physiology 9 (2019): 1834.

21) Whitford, David. Proteins: Structure And Function. John Wiley & Sons, 2013.

22) Liu, Zhenqi, et al. "The Regulation of Body and Skeletal Muscle Protein Metabolism by Hormones and Amino Acids." The Journal of Nutrition 136.1 (2006): 212S-217S.

23) Alberts, Bruce, et al. "Epidermis and Its Renewal by Stem Cells." Molecular Biology of the Cell. 4th edition. Garland Science, 2002.

24) Klein, Robert M., and James C. McKenzie. "The Role of Cell Renewal In The Ontogeny Of The Intestine. I. Cell Proliferation Patterns in Adult, Fetal, And Neonatal Intestine." Journal of Pediatric Gastroenterology and Nutrition 2.1 (1983): 10-43.

25) Fliedner, Theodor M. "The Role of Blood Stem Cells in Hematopoietic Cell Renewal." Stem cells 16.6 (1998): 361-374.

26) Sender, Ron, and Ron Milo. "The Distribution of Cellular Turnover in The Human Body." Nature medicine 27.1 (2021): 45-48.

27) Berlin, N. I., T. A. Waldmann, and S. M. Weissman. "Life Span of Red Blood Cell." Physiological Reviews 39.3 (1959): 577-616.

28) Barker, Nick. "Adult Intestinal Stem Cells: Critical Drivers of Epithelial Homeostasis and Regeneration." Nature Reviews Molecular Cell Biology 15.1 (2014): 19-33.

29) Peng, Lei, and Michael A. Bonaguidi. "Function and Dysfunction of Adult Hippocampal Neurogenesis In Regeneration And Disease." The American Journal of Pathology 188.1 (2018): 23-28.

30) Biagetti, Betina, and Manel Puig-Domingo. "Age-Related Hormones Changes and Its Impact on Health Status and Lifespan." Aging and Disease 14.3 (2023): 605.

31) Pataky, Mark W., et al. "Hormonal and Metabolic Changes of Aging and The Influence of Lifestyle Modifications." Mayo Clinic Proceedings. Vol. 96. No. 3. Elsevier, 2021.

32) Oeppen, Jim, and James W. Vaupel. "The Linear Rise in The Number of Our Days." Old and New Perspectives on Mortality Forecasting (2019): 159-166.

33) Oeppen, Jim, and James W. Vaupel. "Broken Limits to Life Expectancy." Science 296.5570 (2002): 1029-1031.

34) Resident Population Of The United States By Sex And Age As Of July 1, 2023 (in millions). Statistica. https://www.statista.com/statistics/241488/popula-tion-of-the-us-by-sex-and-age/

35) Our Nation's Chronic Disease Epidemic Is Getting Worse So, Who's Responsible? Rita Numerof Forbes Magazine. Nov 22, 2022. https://www.forbes.com/sites/ritanumerof/2022/11/22/our-nations-chronic-disease-epidemic-is-getting-worse-so-whos-responsible/

36) A National Emergency: The Silent Epidemic of Chronic Disease in the United States Part 1 of 3 By Michael Bryant September 17, 2024 https://healthfreedomdefense.org/a-national-emergency-the-silent-epidemic-of-chronic-disease-in-the-united-states/

37) Holman, Halsted R. "The Relation of The Chronic Disease Epidemic To The Health Care Crisis." ACR Open Rheumatology 2.3 (2020): 167-173.

38) US Health System Ranks Last Among Eleven Countries on Measures of Access, Equity, Quality, Efficiency, and Healthy Lives. June 16, 2014. The Commonwealth Fund. https://www.commonwealth-fund.org/press-release/2014/us-health-system-ranks-last-among-eleven-countries-measures-access-equity

39) Persistent Misconduct Forces Pharmaceutical Manufacturers to Pay $62.3 Billion in Penalties. Public Citizen, May 21, 2024. https://www.citizen.org/news/persistent-misconduct-forces-pharmaceutical-manufacturers-to-pay-62-3-billion-in-penalties/

40) Biddle, Justin B. "Deadly Medicines and Organised Crime: How Big Pharma has Corrupted Healthcare by Peter Gøtzsche." Kennedy Institute of Ethics Journal 26.2 (2016): E-40.

41) Gotzsche, Peter. Deadly Medicines and Organised Crime: How Big Pharma Has Corrupted Healthcare. CRC Press, 2019.

Chapter 18

Testosterone Prevents Breast Cancer

PEGGY IS A 44-YEAR-OLD INTERIOR designer from Idaho who arrived in my office to tell her story. Five years ago, Peggy felt a nodule in her right breast and presented to medical attention. An image-guided needle biopsy confirmed breast cancer, and Peggy underwent lumpectomy. The pathology report showed infiltrating ductal carcinoma, an invasive cell type. Surgical margins were good and no additional treatment offered. Peggy did well for two years at which time the cancer returned with a new mass in the right breast and palpable nodes in the right axilla. In addition, CAT scan showed multiple metastatic nodules in the lungs. Her doctors offered her chemotherapy which Peggy declined. Instead, she did her research and found a doctor by the name of Rebecca Glaser MD in Dayton, Ohio, who treats breast cancer with testosterone. Peggy traveled to Dr. Glaser's clinic and had a testosterone pellet implanted, along with letrozole, an aromatase inhibitor drug acting as an anti-estrogen by preventing the conversion of testosterone to estrogen, resulting in undetectable serum estradiol levels. Over time, the breast mass and axillary nodes gradually decreased in size and disappeared. A follow-up CAT scan one year later showed the lungs were clear. Peggy was in complete remission from breast cancer using Dr. Rebecca Glaser's testosterone/ letrozole protocol. I was impressed with this, as this information about using testosterone for the treatment of breast cancer was new to me. In this chapter, we will examine the medical literature on using testosterone as prevention and treatment for breast cancer, and secondly, discuss the benefits and adverse effects of testosterone for menopausal women. (1)

Women Using Testosterone Enjoy 39% Reduction in Breast Cancer

In 2013 and 2019, Drs. Rebecca Glaser studied the incidence of breast cancer in a 10-year prospective cohort study of 1,267 pre and post-menopausal women receiving at least two testosterone pellets. The 10-year follow-up data showed a **39 percent reduction** in invasive breast cancer in women treated with testosterone compared with the "age-matched" population. In 2019, Dr. Glaser's 10-year follow-up report revealed 11 cases of invasive cancer. Three of the eleven breast cancer patients (21 percent) were on combined Testosterone/AI (aromatase inhibitor), which is about the same percentage of women using the T/AI combination (about 21 percent). Based on these numbers, the addition of an AI drug to testosterone HRT did not add any additional protection from breast cancer. This raises the question: does adding an estrogen-blocking drug improve or worsen the outcome of menopausal women using HRT? This is further discussed by Dr. Søren Cold from Denmark (below) and in the next Chapter 19, Adjuvant Endocrine Therapy, in a discussion of the HABITS trial in Sweden in which women on HRT combined with an anti-estrogen drug, tamoxifen, had increased risk of breast cancer. (2-3)

Second Study: Testosterone Prevents Breast Cancer

In 2021, Dr. Gary Donovitz conducted a retrospective observational study over 10 years of 2,377 pre-and post-menopausal women treated with testosterone pellets, alone or combined with estradiol. This study showed 35-50% reduction in the incidence of breast cancer for testosterone-treated women, thus confirming the earlier findings of Dr. Rebecca Glaser. (4)

Case Reports of Treating Breast Cancer with Testosterone /AI pellets

In 2015, Dr. Rebecca Glaser published a case report of a patient with palpable breast mass and axillary nodes. This was an infiltrating lobular, hormone receptor-positive breast cancer (ER+, PR+, AR+). A baseline mammogram demonstrated the breast mass and axillary nodes. A follow-up mammogram 19 weeks after treatment a pellet contained testosterone and aromatase inhibitor (letrozole) revealed a **significant reduction in tumor size and absence of previously palpable axillary lymph nodes.** **Note:** ER= estrogen receptor, PR= progesterone receptor, AR=androgen receptor. (5)

Second Case Report

In 2021, Dr. Rebecca Glaser wrote a case report of a 67-year-old female with a large breast mass presenting with acute respiratory failure. A CAT scan of the chest showed multiple metastatic nodules. A core biopsy revealed the mass was invasive ductal carcinoma (ER+, estrogen receptor-positive). The patient declined conventional treatment, surgery, and chemotherapy. Instead, she agreed to treatment with testosterone (T)/ letrozole (AI) pellet which, fortunately, induced a favorable response with regression of disease. Dr. Rebecca Glaser writes:

> The patient refused conventional therapy and was treated with testosterone and letrozole pellet implants (320 mg T + 24 mg letrozole every 9 weeks). She also began a "whole food" low glycemic diet. One year later, CT scan ...showed considerable improvement in the size and number of nodules throughout the lungs. The patient lost 13.6 kg... remains asymptomatic, and "feels amazing." **The large 8-cm breast mass has markedly decreased in size and axillary nodes are no longer palpable.** (6)

Third Case Report

In 2021, Dr. Rebecca Glaser provided another case report of a 58-year-old female with a large breast cancer mass fixed to the sternum. The patient declined conventional therapy with surgery and chemotherapy and instead was treated with testosterone (T)/ letrozole (AI) pellets. After 14 weeks, the mass was no longer palpable, with a complete response. Dr. Glaser writes:

> [A] fifty-eight-year-old patient [was] referred with a large immobile breast cancer fixed to the sternum. [The patient] refused conventional therapy. She was treated with testosterone 180–240 mg +12 mg letrozole combination pellet implants at baseline, weeks 6, 14, and 26. She also implemented dietary changes...baseline [exam and the clinical photograph showed a], 6-cm tumor fixed to chest wall (sternum) UIQ R [Upper Inner Quadrant Right] breast, and skin discoloration... Baseline ultrasound, tumor invading periosteum (sternum) and skin— [tumor is] too large to be measured (extends off-screen)...Week 14 [Follow-Up Exam and Clinical Photograph showed] complete clinical response, mass no longer palpable... Week 26 [Follow-Up Ultrasound showed] complete response confirmed on ultrasound. The patient continues on T + AI pellets and remains healthy and disease-free at 2 years. (6)

Testosterone/AI Pellet with Chemotherapy for Breast Cancer

In 2017, Dr. Rebecca Glaser reported a 51-year-old female who developed breast cancer (ER +) 5 years after starting testosterone pellet therapy for menopausal symptoms. The patient had also been treated with an aromatase inhibitor (AI), anastrazole. The breast cancer was treated with high dose testosterone/letrozol pellets implanted around the cancer mass. Ultrasound exam 41 days later, before chemotherapy showed a 43 percent reduction in size of the mass. After 5 cycles of chemotherapy, the mass was gone. Dr. Glaser writes:

JR is a 51-year-old postmenopausal patient, with a 30-year smoking history (1978-2008) followed on a prospective IRB trial **examining the incidence of breast cancer in women treated with subcutaneous T implants**. She began treatment with T implants in October of 2008 for symptoms of hormone deficiency. Since February 2013, she occasionally received T in combination with a low dose (ie, **4 mg**) of anastrozole [AI] combined in the implant for **symptoms of excess estrogen including irritability, fluid retention, and weight gain**....Six weeks before starting neoadjuvant chemotherapy, the patient was treated with **subcutaneous testosterone-letrozole implants** and instructed to follow a low-glycemic diet...Ultrasound measurements demonstrated a **43% reduction in tumor volume** (12.28 vs 6.96 cc) 41 days after the patient's initial **180 mg testosterone + 12 mg letrozole** subcutaneous implant therapy and dietary changes... This significant reduction in tumor volume occurred before the initiation of chemotherapy...after five cycles of chemotherapy, the tumor was no longer palpable on clinical examination and unable to be identified on ultrasound, that is, complete clinical response. Most significantly, there was no residual invasive cancer at the time of definitive surgery, that is, complete pathologic response...Note: This is a woman on testosterone HRT who developed breast cancer after adding an AI drug, anastrozole to the HRT. This raises the question: does adding an estrogen-blocking drug improve or worsen the outcome of menopausal women using HRT? This is further discussed by Dr. Søren Cold from Denmark (below) and in the next Chapter 19, Adjuvant Endocrine Therapy regarding the HABITS trial in Sweden in which women on HRT and tamoxifen had worse outcomes than non-users. (7-8)

Aromatase Inhibitors as Mainstream Treatment for Breast Cancer

Note: AI=aromatase inhibitor drugs such as anastrozole, letrozole, exemestane, etc. Since the late 1990's, aromatase inhibition has been the mainstream oncology treatment for breast cancer. This family of drugs inhibits the aromatase enzyme responsible for the conversion of testosterone to estradiol. ER-positive breast cancer cells contain aromatase enzyme used to produce intracellular estrogen serving as a growth factor. In addition, fibroblast cells and adipose cells in microenvironment in the vicinity of a breast cancer mass are recruited to express high levels of aromatase enzyme which feed estradiol into the cancer cells. Blocking the conversion of testosterone to estradiol with an AI (aromatase inhibitor) essentially turns off estrogen as a cancer growth factor and causes regression of the cancer mass. (9-14)

Innovative Treatment for Breast Cancer: Placement of Multiple Testosterone Implants at Tumor Site.

In 2014, Dr. Glaser published a case report of a 90-year-old female with a left breast mass discovered incidentally on CAT scan. Subsequent mammogram and ultrasound showed a deep 2.3-centimeter mass. Ultrasound-guided core biopsy revealed infiltrating ductal carcinoma (estrogen receptor [ER] positive, progesterone receptor [PR] positive, AR positive, and **HER2 negative**). The patient refused lumpectomy, and instead took 20 mg tamoxifen daily. After 4 months of tamoxifen treatment, a follow up ultrasound showed no change in the mass, still measuring 2.3 cm. The patient was then treated with three testosterone (T)/anastrazole (AI, aromatase inhibitor) implants around the tumor site. The pellet treatment resulted in a a 12-fold reduction in tumor volume. Dr. Glaser writes:

> Through a 5-mm lateral incision, **three compounded 60 mg T [testosterone] + 4 mg A [anastrozole] pellets** were implanted into the breast tissue surrounding the tumor approximately 1 cm superior to, 1 cm inferior to, and anterior to the subareolar tumor through a disposable trocar. **Tamoxifen was discontinued**...Follow-up examination of the left breast 2 weeks after intramammary

T + AI pellet implantation revealed a marked decrease in tumor size on physical examination and office US [ultrasound]. The periareolar "thickening" was no longer palpable. By week 4, the patient's (previously unreported) left breast pain had subsided... 46 days after intramammary T + A therapy, follow-up left breast mammogram and US...the tumor measured 1.6 × 1.1 × 0.8 cm... indicating a sevenfold reduction in tumor volume compared with 5.12 mL [at baseline]...Three additional implants (ie, total dose of 180 mg T + 12 mg A) were again placed peritumorally in the left breast on [day 48]...Follow-up mammogram ...on week 13, ...revealed that the size of the carcinoma had continued to decrease, measuring 1.5 × 0.8 × 0.6 cm on US... a 12-fold reduction in tumor volume from the original measurement... **In the future, this combination [Testosterone/AI] may have the potential for both systemic and local therapies for breast cancer in subgroups of patients, possibly eliminating surgical operation, radiation therapy, and adverse effects of oral medication.** Note AR=androgen receptor, HER2 is Human epidermal growth factor receptor-2. (15-16)

Health Benefits of Testosterone

For the past 20 years, my office has used topical testosterone cream rather than implanted pellets to achieve a serum testosterone level of 70-120 ng/dl. The health benefits of testosterone for women include improvement in muscle mass, tone and strength, improved bone density, improved confidence, improved libido. Testosterone is the ultimate antidepressant. Testosterone is anti- inflammatory, improves immune function, and has benefits for the cardiovascular, neurological, gastrointestinal, pulmonary, endocrine, breast, and genitourinary health. Caution is advised when using serum levels to diagnose and monitor estrogen or testosterone deficiency, as these may be inaccurate. Instead, Dr. Glaser uses clinical symptoms. In 2021, Dr. Glaser writes:

T [Testosterone] has a **profound effect on lean muscle mass, bone density, and confidence as well as sex drive and performance in both sexes**... Adequate amounts of (local) bioavailable T [testosterone] at the AR [Androgen Receptor] are critical for overall health, immune function, and preventing inflammation, as well as cardiovascular, neurological, gastrointestinal, pulmonary, endocrine, breast, and genitourinary health...Thus, clinical indications for T therapy include many signs and symptoms caused by T deficiency...Unlike adipose tissue, which can contribute to the circulating pool of estrogens, **E2 [estradiol] from local aromatization would not be measurable in serum**. Therefore, similar to serum T levels, serum levels of E2 should be interpreted with caution and taken into context with clinical evaluation. (6)

Testosterone Health Benefits for Bone Density

In 1988 and 1992, Michael Savvas, Consultant Gynaecologist at King's College Hospital London, found the combination of estrogen and testosterone yielded a much better increase in bone density than estrogen alone. Bone density increased 5 percent after one year of combined estrogen and testosterone, writing in 1992:

Twenty women who were receiving long-term oral estrogen replacement. Ten changed to estradiol and testosterone implants; the remaining ten continued with oral estrogens. Bone density was measured using dual photon absorptiometry at the lumbar spine and neck of femur at the start of the study and **after one year**. The bone density increased significantly by **5.7% at the spine and by 5.2%** at the neck of femur in those women who changed to implant therapy **but remained unchanged in those women who continued with oral [estrogen] therapy**. Conclusion: Subcutaneous estradiol and testosterone implants will result in an increase in bone mass even after many years of oral estrogen replacement therapy. (17-18)

Testosterone Benefits in Women with CHF

In 2010, Dr. Ferdinando Iellamo MD from Rome Italy studied 36 elderly females with congestive heart failure (CHF) and an average ejection of 33 percent (normal is 55 percent). 24 of the 36 received a testosterone patch and 12 received a placebo. Although echocardiograms remained unchanged, there were other cardiac benefits in the testosterone group, writing:

Results Distance walked at 6MWT [6-min walking test] as well as peak oxygen consumption significantly improved in the T [testosterone] group, whereas they were unchanged in the P [placebo] group (p < 0.05 for all comparisons). The homeostasis model was significantly reduced in the T group in comparison with the P group (−16.5% vs. +5%, respectively; p < 0.05). Maximal voluntary contraction and peak torque increased significantly in the T group but did not change in the P group. Increase in distance walked at 6MWT was related to the increase in free testosterone levels (r = 0.593, p = 0.01). No significant changes in echocardiographic parameters were observed in either group. No side effects requiring discontinuation of T were detected. Conclusions: Testosterone supplementation improves functional capacity, insulin resistance, and muscle strength in women with advanced CHF. Testosterone seems to be an effective and safe therapy for elderly women with CHF. (19)

Clinical Signs and Symptoms of Androgen Deficiency

We routinely measure serum testosterone levels in our office, However, rather than relying on serum testosterone levels to initiate and monitor testosterone therapy, Dr. Glaser relies on clinical signs and symptoms of androgen deficiency, which are listed below. Dr. Glaser writes:

[Symptoms of androgen deficiency include] a diminished sense of well-being, dysphoric mood, anxiety, irritability, fatigue, decreased libido, sexual activity, and pleasure,

vasomotor instability, bone loss, decreased muscle strength, insomnia, changes in cognition and memory loss, urinary symptoms and incontinence, vaginal dryness and atrophy, joint and muscular pain. (6)

One may recognize the above list as identical to menopausal symptoms commonly associated with estrogen deficiency, all relieved by testosterone therapy, as discussed below. The ready conversion of testosterone to estrogen by the aromatase enzyme explains why testosterone relieves menopausal symptoms. Another explanation is the ability of testosterone to bind to estrogen response elements in DNA as discussed in 1976 by Dr. Henri Rochefort and in 2009 by Dr. Amelia Peters (below). (20)

Testosterone Pellet Implants Relieve Menopausal Symptoms

In 2011, Dr. Glaser conducted a prospective study of 300 pre-and post-menopausal women treated with testosterone pellet therapy for menopausal symptoms. No aromatase inhibitor was given to this cohort. Using a self-reported questionnaire, called the Menopause Rating Scale, Dr. Glaser showed testosterone pellet therapy alone was sufficient for complete relief of menopausal symptoms. Supplemental estrogen was not needed nor administered. This is explained by Dr. Glaser writing:

Adequate levels of continuous testosterone, provided by the subcutaneous implant, most likely protect against estrogen deficiency thus explaining why testosterone alone is effective therapy in post-menopausal patients. In our clinical practice (not included in this cohort of 300 patients), an aromatase inhibitor is used in combination with testosterone when estrogen is contraindicated (i.e. breast cancer survivors). (21)

The above 2011 study by Dr. Glaser suggests testosterone by itself can replace estrogen for the treatment of estrogen deficiency symptoms. In 2009, Dr. Amelia Peters studied 215 invasive ductal breast carcinomas, using an in vitro breast cancer model, showing the androgen receptor (AR) binds to a subset of estrogen response elements (ERE), potently inhibiting ER-alpha transactivation activity and estradiol-stimulated growth of breast cancer cells. Testosterone and tamoxifen have equal efficacy in treating breast cancer, writing:

> Androgens have been used as hormonal therapy for breast cancer, with efficacy comparable with that seen with the widely used estrogen receptor alpha (ER-alpha) antagonist, tamoxifen. Indeed, it is possible that **the greater therapeutic response of breast cancers to aromatase inhibitors compared with tamoxifen is due to a concomitant reduction in E2 and an increase in androgen signaling** (Macedo, 2006)...We further show that AR [androgen receptor] is a direct repressor of ER-alpha signaling in breast cancer cells and show for the first time that this is due to an association of the AR with response elements of estrogen target genes... Consistent with a role for AR in breast cancer outcome, **AR [androgen receptor] potently inhibited ER-alpha transactivation activity** and 17-beta-estradiol–stimulated growth of breast cancer cells. Transfection of MDA-MB-231 breast cancer cells with either functionally impaired AR variants or the DNA-binding domain of the AR indicated that the latter is both necessary and sufficient for inhibition of ER-alpha signaling. Consistent with molecular modeling, electrophoretic mobility shift assays showed **binding of the AR to an estrogen-responsive element (ERE)**. We conclude that, **by binding to a subset of EREs, the AR can prevent the activation of target genes that mediate the stimulatory effects of 17B-estradiol on breast cancer cells [ER-alpha stimulatory effects]. ..The effect of androgens on breast cancer cells derives primarily from inhibition of ER-alpha signaling rather than via activation of AR-regulated target genes** ...We hypothesize that binding of the AR to an ERE will interfere with the cyclic recruitment of ER-alpha and its coregulators

at these sites, **thereby preventing the estrogen-dependent progression of those loci to an active transcriptional regulatory sequence**...In summary, we have identified a previously unrecognized mechanism for the **specific and direct inhibition of ER-alpha activity by the AR** in breast cancer cells. (22-23)

As mentioned above for breast cancer survivors, an aromatase inhibitor (letrozole) is added to the compounded testosterone pellet formulation. This prevents conversion of testosterone to estrogen, rendering estradiol levels undetectable. A few of the oral aromatase inhibitors currently in clinical use include: Exemestane (Aromasin®), Anastrozole (Arimidex®), and Letrozole (Femara®). (24-26)

Testosterone/AI Treatment for Breast Cancer Survivors- No Breast Cancer Recurrence after 8 years of Follow Up

By 2014, Dr. Glaser had treated 1000 breast cancer survivors with Testosterone/Aromatase Inhibitor (AI) pellet therapy. Of this group, 72 breast cancer survivors were included in a formal prospective study. This study showed no breast cancer recurrence after 8 years of follow-up. (27)

Procarcinogenic Effect of MPA Due to Blocking Testosterone Receptors

Once it is understood that testosterone is breast cancer preventive, one then understands that interfering with the AR (androgen receptor) and blocking the protective effect of testosterone is a bad thing because this will increase breast cancer risk. This explains the increased breast cancer with the use of MPA (medroxyprogesterone) demonstrated in the 2002 Women's Health Initiative Study. Medroxyprogesterone (MPA) has anti-androgenic effects at the dose used in the Women's Health Initiative Study (WHI) and interferes with the AR (androgen receptor). **This explains why mortality from breast cancer was increased in the MPA group compared to placebo. (HR=1.44** p=0.07 nonsignificant) However, at high doses, MPA has stronger androgenic effects and serves as treatment for breast cancer as if it was testosterone. In the 1980s, MPA was used for palliative treatment of metastatic breast cancer with objective response in about 30% of cases. Overall results for high-dose MPA in metastatic breast cancer are roughly similar to the use of testosterone itself as discussed in 2014 by Dr. Corrado Boni. (28-31)

In 2007, Dr. Stephen Birrell suggested disruption of the androgen signal by synthetic progestins such as MPA, at low doses typical for HRT **may increase the risk of developing breast cancer.** This was also recognized by the WHO, citing progestin-containing contraceptives as potentially carcinogenic. Dr. Stephen Birrell cites the 2005 French Cohort study by Dr. Agnes Fournier which shows that removing the MPA and substituting with natural micronized progesterone should be done, writing:

> There is now [by 2007] considerable evidence that using a combination of synthetic progestins [MPA] and estrogens in hormone replacement therapy (HRT) increases the risk of breast cancer compared with estrogen alone. Furthermore, the World Health Organization has recently cited combination contraceptives, which contain synthetic progestins, as potentially carcinogenic to humans, particularly for increased breast cancer risk. Given the above observations and the current trend toward progestin-only contraception, it is important that we have a comprehensive understanding of how progestins act in the millions of women worldwide who regularly take these medications. While synthetic progestins, such as medroxyprogesterone acetate (MPA), which are currently used in both HRT and oral contraceptives were designed to act exclusively through the progesterone receptor, it is clear from both clinical and experimental settings that their effects may be mediated, in part, by binding to the androgen receptor (AR). **Disruption of androgen action by synthetic progestins may have serious deleterious side effects**

in the breast, where the balance between estrogen signaling and androgen signaling plays a critical role in breast homeostasis. Here, we review the role of androgen signaling in the normal breast and in breast cancer and present new data demonstrating that **androgen receptor function can be perturbed by low doses of MPA, similar to doses achieved in serum of women taking HRT...We propose that the observed excess of breast malignancies associated with combined HRT may be explained, in part, by synthetic progestins such as MPA acting as endocrine disruptors to negate the protective effects of androgen signaling in the breast...**MPA binds to the AR with an affinity comparable to the native androgenic ligand, 5 -dihydrotestosterone (DHT)... Notably, in France the majority of women taking combined HRT receive oral micronized progesterone rather than a synthetic progestin. In two French studies- the E3N-EPIC cohort of 54,548 women and a smaller study of 3175 women [Fournier, 2005 and De Lignieres, 2002], **no significant increase in breast cancer risk due to HRT use with micronized progesterone was observed compared with untreated women**. (32-34)

In 2015, Dr. Aleksandra M. Ochnik studied the anti-androgenic actions of medroxyprogesterone acetate (MPA) using breast cells obtained from postmenopausal women undergoing incidental breast surgery for benign disease. These histologically normal breast cells were cultured ex-vivo, and treated with 5-alpha-dihydrotestosterone (DHT), MPA, or the androgen receptor (AR) antagonist bicalutamide. Dr. Ochnik found that the potent androgen DHT inhibited breast cell proliferation. On the other hand, MPA significantly antagonized the beneficial androgenic effect of DHT. MPA disrupted the normal functioning of the androgen receptor. **"MPA exerts an antiandrogenic effect on breast epithelial cells that is associated with increased proliferation and destabilization of AR protein. This activity may contribute mechanistically to the increased risk of breast cancer in women taking MPA-**

containing EPT [estrogen progestin therapy]". Dr. Aleksandra M. Ochnik writes:

> Medroxyprogesterone acetate (MPA), a component of combined estrogen-progestin therapy (EPT), has been associated with **increased breast cancer risk in EPT [estrogen progestin therapy] users**...the aim of this study was to investigate the potential of MPA to disrupt AR [androgen receptor] signaling in an ex vivo culture model of normal human breast tissue...DHT inhibited the proliferation of breast epithelial cells in an AR-dependent manner within tissues from postmenopausal women, and MPA significantly antagonized this androgenic effect...In a subset of postmenopausal women, **MPA exerts an antiandrogenic effect on breast epithelial cells that is associated with increased proliferation and destabilization of AR protein. This activity may contribute mechanistically to the increased risk of breast cancer in women taking MPA-containing EPT.** (35)

In 2015, Dr. Valerie Flores confirmed the above finding of Dr. Ochnik, finding the androgen receptor as a mediator for the carcinogenic effects of medroxyprogesterone (MPA) by blocking the normal signaling of the androgen receptor, writing:

> A role for the AR [androgen receptor] as a mediator of progestogens' [MPA] carcinogenic effect was assessed in a study using an ex vivo culture system. Breast explant tissue from postmenopausal women was cultured and exposed to MPA at concentrations similar to seen in women taking an MPA containing formulation of EPT [Estrogen-Progestin-Therapy]. **While the normal physiologic role of AR signaling results in inhibition of breast cell proliferation, this study found that in postmenopausal women MPA blocked the normal signaling effect of AR, preventing AR from inhibiting epithelial cell growth.** (36)

Medroxyprogesterone as Treatment for Breast Cancer

As mentioned above, at low doses MPA is an androgen blocker. However, at high doses, MPA acts as an androgen and is transformed from a carcinogen to a cancer treatment. In the 1980s MPA was used as a treatment for metastatic breast cancer, with objective response rates of 30-40 percent, comparable to response rates of high-dose estrogen (DES) commonly used in the 1950s. In 2004, Dr. R. Zaucha reported a case of a 66-year-old woman with progressive metastatic breast cancer resistant to tamoxifen and chemotherapy treatments, who was then switched to high-dose oral MPA resulting in complete regression of the metastatic liver lesions, with a spectacular and long-lasting response. The patient was still alive and ambulatory 11 years later, writing:

> The prognosis of breast cancer patients with liver metastases is **extremely poor.** Here we present the case of a 66-year-old female breast cancer patient with multiple liver metastases diagnosed 2 years after a radical modified mastectomy followed by adjuvant tamoxifen. At progression, anthracycline-based chemotherapy [DNA-damaging drugs] was administered, but a CT scan following two cycles of FEC (5-fluorouracil, epirubicin, cyclophosphamide) showed **progression of the liver metastases.** Chemotherapy was therefore switched to medroxyprogesterone acetate (MPA). After 3 months the patient's general status improved, and disease stabilization was observed at the next CT scan. A further 4 months of MPA treatment resulted in a complete response of all liver lesions. Treatment with oral MPA was continued for 4 years. **At present, 11 years after the diagnosis of metastatic liver involvement, the patient is alive, free of cancer, and fully ambulatory. Despite bulky visceral disease and chemoresistance, hormonal treatment with MPA resulted in a spectacular and long-lasting response.** (37-40)

In 1989, Dr. R. Poulin studied the effect of

high-dose MPA on the human mammary carcinoma cell line ZR-75-1, in vitro, showing MPA inhibition of cancer cell proliferation in a dose-dependent manner. Dr. R. Poulin used 3H-estradiol in the study. This is estradiol radiolabeled with tritium (hydrogen-3) so estradiol uptake could be measured using the tritium radiotracer. When incubating the breast cancer cells with high-dose MPA over 12 days, **the MPA inhibited estradiol uptake to the same extent as the most biologically active androgen, DHT (dihydrotestosterone).** As you know, estradiol is a growth factor for breast cancer cells. This growth stimulation by estradiol was inhibited by high dose MPA, acting as an androgen. Dr. R. Poulin concluded the main effect of high-dose MPA on breast cancer cells was due to **androgen receptor-mediated inhibitory action.** At low doses, MPA acts as an androgen blocker. However, at high doses, MPA acts as an androgen which inhibits estrogen uptake by the breast cancer cells, writing:

> Medroxyprogesterone acetate (MPA) is a synthetic progestin, currently used in the adjuvant treatment of advanced breast cancer, **which induces remission rates (30-40%)** comparable to **those obtained with other types of endocrine therapies** [testosterone, DES, tamoxifen and AIs] ...A 12-day preincubation of ZR-75-1 [breast cancer] cells with increasing concentrations of MPA (10 (-12) to 3 x 10 (-6)M) decreased the specific uptake of [3H] estradiol (E2) by intact cell monolayers to the same extent as 10 nM DHT [dihydrotesosterone], an effect which was competitively blocked by the addition of OHF (3 microM) [OHF= hydroxyflutamide, an anti-androgen]... The present data indicate that **the main action of MPA on ZR-75-1 human breast cancer cell growth is due to its androgen receptor-mediated inhibitory action**...(41)

In 2005, Dr. Grant Buchanan found that patients with breast cancer who fail to respond to high-dose MPA have decreased function or decreased numbers of androgen receptors. (42)

Testosterone for Women: My Office Protocol

In 2013, Dr. Glaser studied pharmacological dosing of 100 mg testosterone pellets. About four weeks after Dr. Glaser inserts a 100 mg testosterone pellet, she will see a total testosterone serum levels of about 300 ng/dl plus or minus 100 ng/dl. Using a higher dose 200 mg pellet, serum testosterone levels one week after insertion are about 490 + 210 ng/dl. After about 4 months, testosterone levels decline and the patient returns for a new pellet. In my office, instead of pellets, we use topical testosterone cream (3 mg per day), providing a serum testosterone level of 70-120 ng/dl. In general, this testosterone dosage is well tolerated. A small amount of testosterone cream may be applied directly to the outer eyelids, and the remainder topically to the face around the eyes, rendering a cosmetic result. Testosterone application to the eyelid improves lubrication of the eye and relieves dry eye syndrome. Lubrication of the eye is due to oil secretion by the meibomian glands in the eyelid. This oil secretion is controlled by testosterone, as discussed in 2003 by Dr. Charles Connor from the Southern College of Optometry, in Memphis Tennessee, and later patented in 2007. The main advantage of topical testosterone cream is the ease of adjusting dosage. Once the pellet has been inserted subcutaneously, the dosage cannot be adjusted, without removing the pellet. (43-46)

Our Nutritional Supplement Protocol for Breast Cancer Prevention

A useful addition to the HRT program is our breast cancer prevention protocol using nutritional supplements:

1) Iodine testing and iodine supplementation to optimize urinary iodine excretion (customary dosage is a 12.5 mg Iodoral tablet daily for all female patients on hormone replacement). (47-49)

2) DIM (Di-Indole Methane) as a breast cancer preventive agent. (50-51)

3) Optimize vitamin D3 levels. (52)

4) Optimize selenium levels. (53-54)

5) Methyl-folate containing multivitamins are used to cover those patients with MTHFR mutation, and methylation defects which are risk factors for breast cancer. (55-56)

6) Low-fat diet. (57-58)

Our Complete HRT Program

In addition to the testosterone topical cream, my office also provides a second cream containing the Bi-est (20% estradiol E2, 80% estriol E3) and progesterone combination cream. The starting formula is 50 mg/gram progesterone and 5 mg per gram Bi-Est. The usual starting dosage is half a gram topically twice a day. This may be titrated up for complete relief of menopausal symptoms or titrated down for estrogen excess symptoms of breast enlargement and tenderness. A third component of the program is a progesterone capsule (100 mg) taken at bedtime. For women with vaginal atrophy symptoms, a similar formula is transferred to a vaginal capsule inserted nightly, or reduced to every other night based on symptoms. Because vaginal absorption is superior to topical skin absorption, the dosage for vaginal use is reduced by one-half to one-third that of topical use. Immediately after starting the vaginal HRT, the dosage is commonly adjusted down depending on estrogen excess symptoms of breast stimulation.

Testosterone Excess Symptoms

Testosterone excess symptoms will be obvious to the patient and easy to recognize. These include symptoms of acne and facial hair. If these symptoms occur, the topical cream is stopped for 4-7 days. After about a week or so, symptoms resolve, and the testosterone topical cream may be restarted at half dosage. Although androgenic alopecia is a widely recognized syndrome at the dermatology clinic, Dr. Glazer says this is not the case for testosterone implant patients. Most hair loss in women starting HRT is called telogen effluvium which is mistakenly diagnosed as androgenic hair loss. Thanks to Dr. Rebecca Glaser and Dimitrakakis

for their 2013 article on testosterone therapy in women, reporting two-thirds of women report hair regrowth on testosterone therapy, an anabolic effect of testosterone. They write:

> **There is no evidence that T [Testosterone] or T therapy is a cause of hair loss in either men or women**...Hair loss is a complicated, multi-factorial, genetically determined process that is poorly understood... Hair loss is common in both women and men with insulin resistance....Approximately one third of women experience hair loss and thinning with aging, coinciding with Testosterone decline... We have previously reported that two thirds of women treated with subcutaneous T implants have scalp hair re-growth on therapy. Women who did not re-grow hair on T were more likely to be **hypo or hyperthyroid, iron deficient or have elevated body mass index.** In addition, **none of 285 patients treated for up to 56 months with subcutaneous T therapy complained of hair loss**, despite pharmacologic serum T levels on therapy...**Fact: Testosterone therapy increases scalp hair growth in women.** (59)

Testosterone is Breast Cancer Preventive: Upregulation of ER-beta

The two estrogen receptors are called ER-alpha and ER-beta. The ER-alpha receptor is proliferative and pro-carcinogenic. The ER-beta receptor is a tumor suppressor. Estradiol attaches to both receptors equally in a ratio of 50:50. Estrone (E1) attaches predominantly to ER-alpha, while estriol (E3) attaches predominantly to ER-beta. In the above discussion by Dr. Amelia A Peters (2009), we saw how testosterone inhibits ER-alpha accounting for its breast cancer preventive effects. Another mechanism for testosterone's protective effect in breast cancer prevention is the testosterone metabolite 3-beta-adiol which binds to and activates estrogen receptor beta (ER-beta), the tumor suppressor that inhibits breast cell proliferation. (60-66)

In 2014 Dr. Pietro Rizza studied breast cancer cells in-vitro (MCF-7, ZR-75 and MDA-MD 231), finding activation of the androgen receptor upregulates the ER-beta gene expression, thus inhibiting breast cancer cell growth, writing:

> The two isoforms of estrogen receptor (ER) alpha and beta play opposite roles in regulating proliferation and differentiation of breast cancers, with **ER-alpha mediating mitogenic effects and ER-beta acting as a tumor suppressor**...Collectively, these data provide evidence for a novel mechanism by which **activated AR [Androgen Receptor], through an up-regulation of ER-beta gene expression, inhibits breast cancer cell growth**. Note: although Dr. Rizza used a synthetic androgen, mibolerone, her findings also apply to natural testosterone as well. (67)

Both Estradiol (E2) and Testosterone (T) Receptors Are Found in Mitochondria

Under conditions of long-term estrogen depletion, estrogen transforms from a growth signal to a death signal for the breast cancer cell. This process is called apoptosis or programmed cell death, controlled by mitochondria. In 2008, Dr. V. Craig Jordan described the paradoxical actions of estrogen in breast cancer, acting as a growth signal or as a death signal. Like estrogen, testosterone may act as a pro-apoptotic death signal to breast cancer cells as discussed below by Dr. Yu Wang (2013). In 2011, Dr. Andrea Vasconsuelo studied the effects of both estradiol and testosterone activating receptors in mitochondria of breast cancer cells to trigger apoptosis (programmed cell death). Dr. Vasconsuelo writes:

> 17beta-Estradiol (E2) and Testosterone (T) exert actions in most animal tissues, in addition to the reproductive system. Thus, both sex steroid hormones affect growth and different cell functions in several organs. Accordingly, the nuclear estrogen (ER) and androgen (AR) receptors are ubiquitously expressed...**In mitochondria, the presence of ER and AR and actions of estrogen and androgen have been shown, in keeping with the organelle being a control point of apoptosis.** The most recurrent action for each steroid hormone is the protection of mitochondria against different insults, resulting in antiapoptosis. **This review summarizes the molecular basis of the modulation of programmed cell death by E2 and T in several tissues**...However, **under some specific conditions** E2 could trigger apoptosis in breast cancer cells, opposed to its well studied antiapoptotic role. This peculiar hormone behavior has been observed in cells from breast cancer which have **been long-term estrogen-deprived (LTED) or treated exhaustively with antiestrogens**. Curiously, the paradoxical induction of apoptosis by estrogen has been established under several unusual circumstances...For example, in this case, the pre-conditions of prolonged estrogen depletion or exhaustive treatment with anti-estrogens [tamoxifen or aromatase inhibitors] of the breast cancer cells are mandatory requisites to trigger apoptosis by E2 and could explain the **dual action of the steroid [estrogen] to stimulate growth or apoptosis.** Thus, the **development of antihormone resistance** over years of [anti-estrogen] therapy, reprograms the survival mechanism of the breast cancer cell so that **estrogen no longer functions as a survival factor but as a death signal**...(68-69)

Testosterone Induces Apoptosis in Breast Cancer Cells

In 2013, Dr. Yu Wang from the Cleveland Clinic in Ohio was interested in elucidating the molecular pathway by which testosterone induces apoptosis in breast cancer cells. Dr. Yu Wang analyzed 188 normal and 1247 malignant breast cancer tissue samples using vitro and in vivo murine xenograft models. Dr. Yu Wang found androgen receptor expression induces apoptosis in breast cancer cells via upregulation of tumor suppressor gene PTEN and KLLN, directly promoting P53-induced apoptosis and cell cycle arrest. Dr. Yu Wang writes:

Androgen receptor (AR) expression by immunohistochemistry correlates with better prognosis and survival among breast cancer patients. We and others have shown that AR [androgen receptor] inhibits proliferation and **induces apoptosis** in breast cancer cells. However, the mechanism of AR's anti-tumor effect in breast cancer is still not fully understood. Our recent study indicates that **AR upregulates expression of tumor suppressor gene PTEN** by promoter activation in breast cancer. KLLN, encoding KLLN protein, is a newly identified gene, which shares a bidirectional promoter with PTEN and is transcribed in the opposite direction. So far, the function of KLLN has never been studied in tumorigenesis. Here, we define KLLN as a tumor suppressor in breast carcinomas, which inhibits tumor growth and invasiveness. After analyzing 188 normal breast and 1247 malignant breast cancer tissues, we observed the loss of KLLN in multiple breast cancer subtypes and **this decreased KLLN expression associates with tumor progression and increasing histological grade in invasive carcinomas**. We characterize KLLN, for the first time, **as a transcription factor, directly promoting the expression of TP53 and TP73, with consequent elevated apoptosis and cell cycle arrest in breast cancer cells**. We demonstrate, in vitro and in murine xenograft models, that **both KLLN and PTEN are AR-target genes, mediating androgen-induced growth inhibition and apoptosis in breast cancer cells**. Our observations suggest that KLLN might be used as a potential prognostic marker and novel therapy target for breast carcinomas. ..Androgens have been shown to inhibit ER-mediated tumor proliferation, induce apoptosis and cell cycle arrest in AR-breast cancer cell lines, such as CAMA-1, ZR75 −1 and T47D...We found that **androgen stimulates KLLN expression in breast cancer cells, which induces S-phase arrest and apoptosis**. (70)

In 2021 in Nature, Dr. Theresa E. Hickey from the University of Adelaide, South Australia studied testosterone as a treatment for breast cancer using an ex vivo patient-derived explant model in mice. Dr. Hickey found the anti-cancer effects of testosterone were superior to tamoxifen, and suggested testosterone, i.e. AR receptor agonism, could become a highly effective alternative treatment for estrogen receptor (ER) positive breast cancer, for all stages of the disease. In short, testosterone is an attractive, implementable therapeutic opportunity to treat ER-positive breast cancers. Dr. Theresa E. Hickey writes:

> Using a diverse, clinically relevant panel of [breast cancer] cell-line and patient-derived models, we demonstrate that **AR [androgen receptor] activation, not suppression, exerts potent antitumor activity in multiple disease context**s, including resistance to standard-of-care ER [estrogen receptor] and CDK4/6 inhibitors [Palbiciclib]. **Notably, AR agonists combined with standard-of-care agents enhanced therapeutic responses... These findings provide unambiguous evidence that AR has a tumor suppressor role in ER-positive breast cancer and support AR agonism as the optimal AR-directed treatment strategy, revealing a rational therapeutic opportunity...** This work has immediate implications for women with metastatic ER-positive breast cancer. It provides compelling evidence that exploitation of an endogenous, **ligand-activated AR pathway to inhibit ER [estrogen receptor]-mediated transcriptional activity constitutes an attractive, implementable therapeutic opportunity to treat ER-positive breast cancers**, even those resistant to current forms of endocrine therapy and those with genomic aberrations of ESR1 or CCND1. Moreover, **we provide new evidence that AR agonists can be more effective than existing (for example tamoxifen) or new (for example palbociclib) standard-of-care treatments** and, in the case of the latter, can be combined to enhance growth inhibition.... **Given the efficacy of AR agonism at multiple stages of ER positive disease, this treatment strategy has potential to become an alternative endocrine therapy for breast cancer...** (71)

In 2021 in BMJ, Dr. Theresa Hicky used advanced genome sequencing laboratory techniques to study the inhibitory effects of androgens on ER-positive breast cancer cells. Dr. Theresa Hicky found that androgens oppose estrogen receptor binding to key regulatory elements in the genome (DNA) which controls cell cycle genes, thus inhibiting cell proliferation, writings:

> The first evidence that sex hormone antagonism could be therapeutically harnessed to treat breast cancer arose in the 1950s, when androgenic drugs were used to successfully treat patients prior to advent of the selective ER modulator (SERM) tamoxifen. **Although androgen therapy equaled tamoxifen in therapeutic efficacy, masculinizing side effects underpinned its clinical demise**...Exploiting chromatin immunoprecipitation followed by massively parallel sequencing (ChIP-seq) to interrogate genome-wide interaction of transcription factors and associated co-factors with chromatin, **we clearly show that activation of AR opposes ER binding at key regulatory elements that control transcription of cell cycle genes as a means of inhibiting proliferation in vitro and in vivo.** (72)

In 2024, Dr. Amy H. Tien from Vancouver reviewed breast cancer treatments targeting the androgen receptor, commenting that the 2021 study by Dr. Theresa E. Hickey (above) showed **the androgen receptor (AR) behaves as a tumor suppressor in ER alpha-positive breast cancer.** Activation of the androgen receptor with an androgen drug suppresses ER signaling, and represses ER-regulated cell cycle genes, thus inhibiting tumor cell proliferation, writing:

> On the other hand, a study by Hickey and colleagues demonstrated that AR behaves as a tumor suppressor in ER alpha-positive breast cancer. They showed that AR transactivation with androgen inhibited ER-driven cell proliferation in an ex vivo patient-derived explant model. Using breast cancer cell lines, they also showed that **AR**

> **was detected at 42% of estrogen-stimulated ER-binding sites on chromatin when both AR and ER were activated.** This suggests that AR could directly affect ER transcriptional activity by redistributing ER. Interestingly, **the binding or recruitment of coactivators p300 and SRC-3, which are required for ER signaling, were both reduced and replaced by AR upon AR activation. This led to the repression of ER-regulated cell cycle genes and the inhibition of tumor cell proliferation. Thus, AR activation suppressed ER signaling in ER alpha-positive breast cancer cells**. (73)

Testosterone Works Best for ER-positive Cancers, not Triple Negative or HERS Positive

In 2016, Dr. Elisabetta Pietri from Italy wrote an exhaustive review of androgen receptor signaling as a target for breast cancer treatment, finding that testosterone works best when both ER and AR receptors are present, found in 60-80 percent of all breast cancers. In this most common type, androgen signaling antagonizes the cancer cell growth stimulation from estrogen. However, in ER-negative cell types, i.e. triple-negative, and ER-negative HERS, androgens act quite the opposite as a growth stimulator. Dr. Pietra suggests anti-androgen drugs would be more useful in these two ER-negative subtypes, i.e. Triple negative, and ER-negative HERS. Dr. Elisabetta Pietri writes:

> Evidence of a benefit from androgens in patients with advanced breast carcinoma was first reported in 1939 by Ulrich and Loeser using testosterone propionate and confirmed a few years later by Fels (1944) and Adair Frank (1946)...**The inhibitory effect of AR on ER activity seen in several in vitro BC [breast cancer] models indicates that we can consider androgens and AR modulators as a cancer treatment for ER/AR-positive subtype**...We offer an overview of AR signaling pathways in different breast cancer subtypes, providing evidence that its oncogenic role is likely to be different in distinct biological and clinical scenarios. In particular, in ER-positive breast

cancer, AR signaling often antagonizes the growth stimulatory effect of ER signaling; **in triple-negative breast cancer (TNBC), AR seems to drive tumor progression (at least in luminal AR subtype of TNBC with a gene expression profile mimicking luminal subtypes despite being negative to ER and enriched in AR expression); in HER2-positive breast cancer, in the absence of ER expression, AR signaling has a proliferative role.** These data represent the rationale for AR-targeting treatment as a potentially new target therapy in breast cancer subset using **androgen agonists in some AR-positive/ER-positive tumors, AR antagonists in triple-negative/AR-positive tumors and in combination with anti-HER2 agents** or with other signaling pathways inhibitors (including PI3K/MYC/ERK) in HER2-positive/AR-positive tumors. Note ER=estrogen receptor, AR=androgen receptor. HERS2=human epidermal growth factor receptor 2. (74-76)

Signs and Symptoms of Post-Menopausal Testosterone Deficiency in Women

Typical signs and symptoms of menopausal testosterone deficiency are: a diminished sense of well-being, mood disturbance, anxiety, irritability, fatigue, decreased libido, vasomotor instability, loss of bone density, decreased muscle strength, insomnia, cognition dysfunction and memory loss, recurrent urinary tract infection, stress incontinence (genito-urinary syndrome), vaginal dryness and atrophy, joint and muscular pain. (6)

Benefits and Adverse Effects of Testosterone in Women

Notice that the above list bears a striking resemblance to estrogen deficiency. When post-menopausal women are treated with topical testosterone, the above list of symptoms promptly resolve. Dr. Glaser finds that testosterone alone may provide complete relief from menopausal symptoms. In the event of testosterone excess, the adverse effects of acne and facial hair growth (hirsutism) become obvious to the patient. In this event, the patient is instructed to hold the testosterone topical cream for 3-5 days until complete resolution of the acne, and then resume at half dosage. When comparing testosterone pellets to topical creams, the advantage is ease of dosage adjustment. In the event of testosterone excess, the patient is instructed to take a few days off the topical cream. The testosterone levels will decline promptly, and the topical cream restarted after the acne symptoms have resolved. Dosage for the pellet can not be adjusted without removing the pellet. Dr. Glaser found testosterone alone provides complete relief from menopausal symptoms. This is useful in the post-menopausal breast cancer survivor treated with a testosterone pellet/AI protocol devised by Dr. Glaser. Note AI = aromatase inhibitor which blocks the conversion of testosterone to estradiol, serving as an anti-estrogen drug. Dr. Glaser found **no breast cancer recurrence** after 8 years of follow-up in 72 breast cancer survivors treated with her testosterone pellet/AI protocol. (27)

More on Synergy with Testosterone/ AI Combination

Regarding the excellent results for HRT in breast cancer survivors with the above protocol, perhaps the best discussion of intravaginal estrogen and testosterone is in 2022 by Drs. Abbie Laing, Louise Newson, and James A. Simon. In their 2022 article in the Cancer Journal, they cite an in vitro study on MCF-7 cancer cells by Dr. R. Chen showing considerable synergy with the combination of testosterone with the aromatase inhibitor (AI), anastrozole which **enhanced cytotoxicity and apoptosis more than two-fold in the cancer cells compared to either agent alone**. However, in-vitro does efficacy not always translate into efficacy for in-vivo or human trials. Because of studies by Dr. Soren Cold, and Dr. Rajkumar below, I am not entirely convinced adding an AI drug to testosterone is necessary in routine HRT. This would be a good subject for further study. Dr. Abbie Laing writes:

> An in vitro study reports that the antiproliferative effects of anastrozole on human breast cancer cells are significantly enhanced by combined treatment with testosterone, and another case study has concluded that **a higher letrozole dose enables a greater inhibitory effect of testosterone at the breast**. (77-78)

Aromatase Inhibitor Increases Recurrence in Breast Cancer Survivors on HRT

In 2022, Dr. Søren Cold from Denmark published an observational cohort study using a national prescription registry of 8,461 women not using HRT diagnosed with breast cancer. After the breast cancer diagnosis was confirmed, 1,957 of the original 8,467 women used VET [vaginal estrogen therapy] and 133 used MHT [systemic menopausal hormone therapy]. The women were followed 9.8 years for recurrence and 15.2 years for mortality. If estrogen causes breast cancer, then one would expect to find the HRT users would have increased cancer recurrence. Dr. Søren Cold found this idea to be false. In his study, **there was no increased risk of breast cancer recurrence for VET or MHT users, compared to non-users.** However, for women using VET in the subgroup also receiving adjuvant aromatase inhibitors, the **recurrence rate was increased by 39 percent. Note:** this study suggests adding an aromatase inhibitor may not be beneficial in this setting. Dr. Søren Cold writes:

> Overall, there was no increased risk of recurrence but notably **the risk of recurrence (but not mortality) was increased in women using VET [vaginal estrogen therapy] whilst also taking an aromatase inhibitor (HR 1.39).** (79-80)

Aromatase Inhibitor Reduces Outcome in HRT Mice with Breast Cancer

In 2014, Drs. A. Arumugam and Rajkumar Lakshmanaswamy studied ovarectomized mice xenografted with human breast cancer cells, after which the mice were treated with various combinations of natural hormones, estrogen, progesterone, testosterone, and DHEA. Dr. Rajkumar Lakshmanaswamy found the natural hormone-treated mice had a maximum reduction in tumor growth and better outcomes. However, **outcomes were reduced by adding an AI (aromatase inhibitor) drug to the treated mice. Note:** Aromatase inhibitors (AI) are estrogen-blocking drugs commonly used by oncologists to treat breast cancer patients. These two studies by Dr. Soren Cold (above) and Dr. Lakshmanaswamy are the reason I am not entirely convinced adding an AI to the testosterone is necessary. Perhaps future studies will clarify this. This topic will be discussed more completely in the next chapters, 19-20. (81-82)

Testosterone Improves Bone Density

In 1995, Dr. Susan Davis did a 2-year single-blind randomized trial of 34 post-menopausal women who received either estradiol alone or estradiol plus testosterone. Bone density was measured with DEXA scans. The women treated with testosterone added to estrogen enjoyed a more effective increase in bone density, and improved sexuality. Dr. Susan Davis concluded:

> in postmenopausal women, treatment with **combined estradiol and testosterone implants was more effective in increasing bone mineral density in the hip and lumbar spine than estradiol implants alone.** Significantly greater improvement in sexuality was observed with combined therapy, verifying the therapeutic value of testosterone implants for diminished libido in postmenopausal women. (83)

Testosterone for Postpartum Depression

Since the administration of exogenous testosterone does not appreciably increase levels in milk from the breastfeeding mother, postpartum depression may be treated with testosterone, the ultimate antidepressant. Testosterone is a much safer choice compared to psychotropic drugs. In 2009, Dr. Rebecca Glaser presented a case report and studied testosterone milk levels after testosterone administration to a breastfeeding mother with post-partum depression, writing:

> Maternal testosterone therapy is **safe for the breast fed infant**...Testosterone delivered by sublingual drops, vaginal cream and pellet implant is absorbed (measurable in maternal blood) but not measurably excreted into breast milk. Testosterone, delivered by pellet implant is effective in relieving symptoms of testosterone deficiency and **was not measurably increased in breast milk or measurable in infant serum. Testosterone, by pellet implant may be a safer and more physiologic alternative to psychotropic medications**. (84)

Safety of Testosterone for Women

In 2023, Dr. Guilherme Renke reviewed the cardiovascular safety and benefits of testosterone for women, writing:

> Subcutaneous testosterone therapy (STT) administered by silastic or bioabsorbable implants has been used successfully in women since the 1930s. Published studies demonstrate promising results and **safety in doses ranging from 75 to 225 mg**. Thus, higher doses of T [testosterone] (500–1800 mg) in subcutaneous implants have been used effectively to treat patients with breast cancer. Because it is not excreted in breast milk, T has been used to treat symptoms of postpartum depression and fatigue during lactation and the puerperium...T therapy alone, at physiologic doses, has been reported to be more effective than estradiol (E2)-T therapy or E2 alone for relieving

postmenopausal symptoms with the same cardiovascular (CV) safety. (85)

The Failure of Modern Medicine

In 2022, Drs. Abdulmaged M. Traish and Abraham Morgentaler reviewed the benefits of testosterone therapy for women such as improved sexual function, improved bone mineral density, improved metabolic function, improved muscle mass, and improved mood regarding anxiety, depression, and anorexia nervosa, all without serious safety concerns. Despite these benefits for women, the medical community lacks education and ignores testosterone for women. Primary care and OB/GYN doctors have no interest in prescribing testosterone for women. For more than 60 years now, England and Australia have licensed testosterone products for women. Yet in the U.S. there is no FDA-approved testosterone product for women. For women to receive a testosterone prescription, it must be off-label through a compounding pharmacy. Drs. Traish and Morgentaler write:

> The impact of this [testosterone insufficiency] is enormous, affecting millions of women in the United States alone, depriving them (and their partners) of satisfying sexual relations, strength, and vigor...the lack of education regarding women's sexual dysfunction, and T [testosterone] insufficiency, needs to be addressed during medical education and training...it will be helpful if professional organizations lend their support to the importance of diagnosing and treating women with T deficiency...it is to be hoped that one or more T products for women obtain regulatory approval in the United States and other countries...(86)

False Objections to the Use of Bioidentical Hormones

A common objection to the use of compounded bioidentical hormones is the claim that they are unregulated. This is untrue and is "fake news" in the mass media and the medical literature. In 2022, Dr. Gary Donovitz replies to this objection, writing:

> The statement that bioidentical hormones are not regulated is inaccurate. There currently exist long-standing regulations at the Federal and State levels. The Drug Quality and Security Act in 2014 gave the FDA additional powers to regulate compounding pharmacies, including establishing the requirement of compliance with Current Good Manufacturing Practice (cGMP) , as is required by the pharmaceutical industry. This allowed for improved testing for manufacturing processes, assuring purity, potency, and quality testing, in addition to improved sterility protocols. (87)

> The statement from the ACOG [American College of Obstetrics and Gynecology] that "evidence lacking to support superiority claims of compounded bioidentical hormones over conventional menopausal hormone therapy" **is unfounded and outdated based on multiple studies**... A more contemporary position statement is needed from both the ACOG and NAMS [North American Menopause Society]. Such statements should provide a more accurate depiction of the current state of compounded Bio-identical hormones. These statements should also note that the **stringent conditions imposed by the FDA on 503b outsourced pharmacies have improved the purity, potency, and sterility of these products, comparable to that of commercially available hormone products.** The question concerning the long-term safety of bio-identical hormones has been answered through years of experience utilizing **estradiol patches** (e.g.,Vivelle® ... Climara®)...and sub-cutaneous hormone pellet therapy with testosterone and estradiol. (88-89)

Conclusion: Thanks, and credit goes to Dr. Rebecca Glaser and Dr. Gary Donovitz for their work on testosterone for breast cancer pre-vention and treatment. We now have the two studies by Drs. Glaser and Donovitz with testosterone pellets for menopausal women showing 40-50% reduction in breast cancer compared to controls in the general population. Would we see similar results for other routes of administration such as topical testosterone or vaginal testosterone? I would think so. What if the first arm 2002 Women's Health Initiative had used vaginal estradiol, natural progesterone, and testosterone? My guess is the 2002 first-arm WHI results would have been much improved, with a 40 percent reduction in breast cancer in the hormone-treated group similar to Dr. Glaser's study. Also likely, adding in estriol (E3) by switching from estradiol (E2) to Bi-est (80% E3, 20% E2) would yield even better results. If everyone contacted the director of the NIH director (Dr. Jay Bhattacharya) or the secretary of HHS (John F Kennedy Jr.) and requested such an RCT study receive NIH funding, we would finally have good science on human bioidentical hormones for menopausal women. For more on hormone replacement therapy (HRT) for breast cancer survivors see Chapter 6 and Chapter 19 on this topic.

♦ References Chapter 18 Testosterone Prevents Breast Cancer

1) Scott, Alice, and Louise Newson. "Should we be Prescribing Testosterone to Perimenopausal and Menopausal Women? A Guide to Prescribing Testosterone For Women In Primary Care." British Journal of General Practice 70.693 (2020): 203-204.

2) Glaser RL, Dimitrakakis C. Reduced Breast Cancer Incidence in Women Treated With Subcutaneous Testosterone, Or Testosterone With Anastrozole: A Prospective, Observational Study. Maturitas. 2013;76(4):342–349.

3) Glaser, Rebecca L., et al. "Incidence of Invasive Breast Cancer in Women Treated with Testosterone Implants: A Prospective 10-Year Cohort Study." BMC cancer 19.1 (2019): 1271.

4) Donovitz, Gary, and Mandy Cotten. "Breast Cancer Incidence Reduction in Women Treated with Subcutaneous Testosterone: Testosterone Therapy and Breast Cancer Incidence Study." European Journal of Breast Health 17.2 (2021): 150.

5) Glaser, R., and C. Dimitrakakis. "Testosterone and Breast Cancer Prevention." Maturitas 82.3 (2015): 291-295.

6) Glaser, Rebecca, and Constantine Dimitrakakis. "Testosterone Implant Therapy in Women with and Without Breast Cancer: Rationale, Experience, Evidence." Clinical Research and Therapeutics 2.1 (2021): 94-110.

7) Glaser, Rebecca L., et al. "Subcutaneous Testosterone-Letrozole Therapy Before and Concurrent with Neoadjuvant Breast Chemotherapy: Clinical Response and Therapeutic Implications." Menopause (New York, NY) 24.7 (2017): 859.

8) Glaser, Rebecca L., et al. "Incidence of Invasive Breast Cancer in Women Treated with Testosterone Implants: A Prospective 10-Year Cohort Study." BMC Cancer 19.1 (2019): 1271.

9) Bulun, S. E., et al. "Regulation of Aromatase Expression in Breast Cancer Tissue." Annals of the New York Academy of Sciences 1155.1 (2009): 121-131.

10) Kijima, Ikuko, et al. "Growth Inhibition of Estrogen Receptor-Positive and Aromatase-Positive Human Breast Cancer Cells in Monolayer and Spheroid Cultures by Letrozole, Anastrozole, and Tamoxifen." The Journal of Steroid Biochemistry and Molecular Biology 97.4 (2005): 360-368.

11) Glaser RL, Dimitrakakis C. Reduced Breast Cancer Incidence in Women Treated with Subcutaneous Testosterone, or Testosterone with Anastrozole: A Prospective, Observational Study. Maturitas. 2013;76(4):342–349.

12) Chumsri, Saranya, et al. "Aromatase, Aromatase Inhibitors, And Breast Cancer." The Journal of Steroid Biochemistry and Molecular Biology 125.1-2 (2011): 13-22.

13) Fabian, Carol J. "The What, Why and How of Aromatase Inhibitors: Hormonal Agents for Treatment and Prevention of Breast Cancer." International Journal of Clinical Practice 61.12 (2007): 2051-2063.

14) Narashimamurthy, J., et al. "Aromatase Inhibitors: A New Paradigm in Breast Cancer Treatment." Current Medicinal Chemistry-Anti-Cancer Agents 4.6 (2004): 523-534.

15) Glaser, Rebecca L., and Constantine Dimitrakakis. "Rapid Response of Breast Cancer to Neoadjuvant Intramammary Testosterone-Anastrozole Therapy: Neoadjuvant Hormone Therapy in Breast Cancer." Menopause (New York, NY) 21.6 (2014): 673.

16) Gutierrez, Carolina, and Rachel Schiff. "HER2: Biology, Detection, And Clinical Implications." Archives Of Pathology & Laboratory Medicine 135.1 (2011): 55-62.

17) Savvas, Michael., et al. "Increase in Bone Mass After One Year of Percutaneous Estradiol And Testosterone Implants In Post-Menopausal Women Who Have Previously Received Long-Term Oral Estrogens." BJOG: An International Journal of Obstetrics & Gynaecology 99.9 (1992): 757-760.

18) Savvas, Michael., et al. "Skeletal Effects Of Oral Estrogen Compared With Subcutaneous Estrogen and Testosterone In Postmenopausal Women." British Medical Journal 297.6644 (1988): 331-333.

19) Iellamo F, et al. Testosterone Therapy in Women With Chronic Heart Failure: A Pilot Double-Blind, Randomized, Placebo-Controlled Study. J Am Coll Cardiol. 2010; 56:1310–1316.

20) Rochefort, Henri, and Marcel Garcia. "Androgen on the Estrogen Receptor I—Binding And In Vivo Nuclear Translocation." Steroids 28.4 (1976): 549-560.

21) Glaser, Rebecca, Anne E. York, and Constantine Dimitrakakis. "Beneficial Effects Of Testosterone Therapy In Women Measured By The Validated Menopause Rating Scale (MRS)." Maturitas 68.4 (2011): 355-361.

22) Peters, Amelia A., et al. "Androgen Receptor Inhibits Estrogen Receptor-A Activity and Is Prognostic in Breast Cancer." Cancer Research 69.15 (2009): 6131-40.

23) Macedo, Luciana F., et al. "Role of Androgens On MCF-7 Breast Cancer Cell Growth and On The Inhibitory Effect Of Letrozole." Cancer Research 66.15 (2006): 7775-7782.

24) Chumsri, Saranya, et al. "Aromatase, Aromatase Inhibitors, And Breast Cancer." The Journal of Steroid Biochemistry and Molecular Biology 125.1-2 (2011): 13-22.

25) Fabian, Carol J. "The What, Why and How of Aromatase Inhibitors: Hormonal Agents for Treatment And Prevention Of Breast Cancer." International Journal of Clinical Practice 61.12 (2007): 2051-2063.

26) Narashimamurthy, J., et al. "Aromatase Inhibitors: A New Paradigm in Breast Cancer Treatment." Current Medicinal Chemistry-Anti-Cancer Agents 4.6 (2004): 523-534.

27) Glaser, Rebecca L., et al. "Efficacy of Subcutaneous Testosterone on Menopausal Symptoms in Breast Cancer Survivors." J Clin Oncol 32. Suppl 2 (2014): 109.

28) Boni, Corrado, et al. "Therapeutic Activity of Testosterone in Metastatic Breast Cancer." Anticancer Research 34.3 (2014): 1287-1290.

29) Guarnieri, A., et al. "Oral Route Administration of Medroxyprogesterone Acetate (MAP) At High Doses in The Treatment of Advanced Breast Cancer: Clinical Results." Chemioterapia: International Journal of the Mediterranean Society of Chemotherapy 3.5 (1984): 320-323.

30) Davila, Enrique, et al. "Clinical Trial of High-Dose Oral Medroxyprogesterone Acetate in The Treatment of Metastatic Breast Cancer and Review of The Literature." Cancer 61.11 (1988): 2161-2167.

31) Lanari, Claudia Lee Malvina, et al. "The MPA Mouse Breast Cancer Model: Evidence for A Role Of Progesterone Receptors In Breast Cancer." (2009).

32) Birrell, Stephen N., et al. "Disruption of Androgen Receptor Signaling by Synthetic Progestins May Increase Risk of Developing Breast Cancer." The FASEB Journal 21.10 (2007): 2285-2293.

33) Fournier, Agnes, et al. "Breast Cancer Risk In Relation To Different Types Of Hormone Replacement Therapy In The E3N-EPIC Cohort." International Journal Of Cancer 114.3 (2005): 448-454.

34) De Lignieres, B., et al. "Combined Hormone Replacement Therapy and Risk of Breast Cancer in A French Cohort Study Of 3175 Women." Climacteric 5.4 (2002): 332-340.

35) Ochnik, Aleksandra M., et al. "Antiandrogenic Actions of Medroxyprogesterone Acetate On Epithelial Cells Within Normal Human Breast Tissues Cultured Ex Vivo." Menopause 21.1 (2014): 79-88.

36) Flores, Valerie A., and Hugh S. Taylor. "The Effect of Menopausal Hormone Therapies on Breast Cancer: Avoiding the Risk." Endocrinology and Metabolism Clinics 44.3 (2015): 587-602.

37) Zaucha, R., et al. "Long-term survival of a Patient With Primarily Chemo-Resistant Metastatic Breast Cancer Treated With Medroxyprogesterone Acetate." The Breast 13.4 (2004): 321-324.

38) Hackenberg, Reinhard, et al. "Medroxyprogesterone Acetate Inhibits The Proliferation Of Estrogen-And Progesterone-Receptor Negative MFM-223 Human Mammary Cancer Cells Via The Androgen Receptor." Breast Cancer Research and Treatment 25 (1993): 217-224.

39) Bentel, Jacqueline M., et al. "Androgen Receptor Agonist Activity of The Synthetic Progestin, Medroxyprogesterone Acetate, In Human Breast Cancer Cells." Molecular and cellular endocrinology 154.1-2 (1999): 11-20.

40) Dran, G., et al. "Effect of Medroxyprogesterone Acetate (MPA) And Serum Factors On Cell Proliferation In Primary Cultures Of An MPA-Induced Mammary Adenocarcinoma." Breast cancer research and treatment 35.2 (1995): 173-186.

41) Poulin, R., et al. "Androgen and Glucocorticoid Receptor-Mediated Inhibition of Cell Proliferation by Medroxyprogesterone Acetate In ZR-75-1 Human Breast Cancer Cells." Breast Cancer Research and Treatment 13.2 (1989): 161-172.

42) Buchanan, Grant, et al. "Decreased Androgen Receptor Levels and Receptor Function in Breast Cancer Contribute to The Failure of Response to Medroxyprogesterone Acetate." Cancer Research 65.18 (2005): 8487-8496.

43) Glaser, Rebecca L., at al. "Incidence of Invasive Breast Cancer in Women Treated With Testosterone Implants: A Prospective 10-Year Cohort Study." BMC cancer 19.1 (2019): 1271.

44) Glaser, Rebecca, et al. "Testosterone Implants in Women: Pharmacological Dosing for A Physiologic Effect." Maturitas 74.2 (2013): 179-184.

45) Connor, C. G. "Treatment of Dry Eye with A Transdermal 3% Testosterone Cream." Investigative Ophthalmology & Visual Science 44.13 (2003): 2450-2450.

46) Connor, Charles Gerald, and Charles L. Haine. "Treatment for Dry Eye." U.S. Patent Application No. 11/634,347.

47) Mendieta, Irasema, et al. "Molecular Iodine Exerts Antineoplastic Effects By Diminishing Proliferation And Invasive Potential And Activating The Immune Response In Mammary Cancer Xenografts." BMC cancer 19.1 (2019): 1-12.

48) Moreno-Vega, Aura, et al. "Adjuvant Effect Of Molecular Iodine In Conventional Chemotherapy For Breast Cancer. Randomized Pilot Study." Nutrients 11.7 (2019): 1623.

49) Manjer, Jonas, et al. "Serum Iodine and Breast Cancer Risk: A Prospective Nested Case-Control Study Stratified for Selenium Levels." Cancer Epidemiology, Biomarkers & Prevention 29.7 (2020): 1335-1340.

50) Williams, David E. "Indoles Derived From Glucobrassicin: Cancer Chemoprevention By Indole-3-Carbinol And 3, 3'-Diindolylmethane." Frontiers in Nutrition 8 (2021): 734334.

51) Amarakoon, Darshika, et al. "Indole-3-Carbinol: Occurrence, Health-Beneficial Properties, and Cellular/Molecular Mechanisms." Annual Review of Food Science and Technology 14 (2023): 347-366.

52) Benarba, Bachir, and Adel Gouri. "Role of Vitamin D In Breast Cancer Prevention and Therapy: Recent Findings." Journal of Medicine 21.1 (2020): 46.

53) Szwiec, Marek, et al. "Serum Selenium Level Predicts 10-Year Survival After Breast Cancer." Nutrients 13.3 (2021): 953.

54) Kim, Seung Jo, et al. "Antitumor Effects of Selenium." International Journal of Molecular Sciences 22.21 (2021): 11844.

55) Li, Zhen, et al. "The methylenetetrahydrofolate reductase (MTHFR) C677T gene polymorphism is associated with breast cancer subtype susceptibility in southwestern China." Plos one 16.7 (2021): e0254267.

56) Omran, Moataza H., et al. "Strong Correlation of MTHFR Gene Polymorphisms with Breast Cancer and its Prognostic Clinical Factors among Egyptian Females." Asian Pacific Journal of Cancer Prevention: APJCP 22.2 (2021): 617.

57) Chlebowski, Rowan T., et al. "Dietary Modification and Breast Cancer Mortality: Long-Term Follow-Up of The Women's Health Initiative Randomized Trial." Journal of Clinical Oncology 38.13 (2020): 1419-1428.

58) Montégut, Léa, et al. "Science-Driven Nutritional Interventions for The Prevention And Treatment Of Cancer." Cancer Discovery 12.10 (2022): 2258-2279.

59) Glaser, Rebecca, and Constantine Dimitrakakis. "Testosterone Therapy in Women: Myths And Misconceptions." Maturitas 74.3 (2013): 230-234.

60) Wright, Jonathan V. "The Anticancer Testosterone Metabolite 3Beta-Adiol." Townsend Letter July 1, 2012, Retrieved Nov 25 2024 from https://www.thefreelibrary.com/The+anticancer+testosterone+metabolite+3%5bbeta%5d+-+Adiol.-a0297056560

61) Oliveira AG et al. 5a-Androstane-3b,17b–diol, an Estrogenic Metabolite Of 5a-Dihydrotestoserone, Is A Potent Modulator of Estrogen Receptor Beta In The Ventral Prostate Of Adult Rats. Steroids. 2007; 72:914–922.

62) Warner, Margaret, et al. "25 years of ER Beta: A Personal Journey." Journal of Molecular Endocrinology 68.1 (2022): R1-R9.

63) Rymbai, Emdormi, et al. "Role of Estrogen Receptors in Cancer: A Special Emphasis On The Therapeutic Potential Of Estrogen Receptor Beta." Pharmaceutical Sciences Asia 49.5 (2022).

64) Wu, Wan-fu, et al. "Estrogen Receptor Beta and Treatment with A Phytoestrogen Are Associated with Inhibition of Nuclear Translocation of EGFR In the Prostate." Proceedings of the National Academy of Sciences 118.13 (2021): e2011269118.

65) Chimento, Adele, et al. "Estrogen Receptors-Mediated Apoptosis in Hormone-Dependent Cancers." International Journal of Molecular Sciences 23.3 (2022): 1242.

66) You, Chan-Ping, et al. "Modulating the Activity of Androgen Receptor for Treating Breast Cancer." International Journal of Molecular Sciences 23.23 (2022): 15342.

67) Rizza, Pietro, et al. "Estrogen Receptor Beta as A Novel Target of Androgen Receptor Action in Breast Cancer Cell Lines." Breast Cancer Research 16 (2014): 1-13.

68) Vasconsuelo, Andrea, et al. "Role of 17beta-Estra-diol and Testosterone in Apoptosis." Steroids 76.12 (2011): 1223-1231.

69) Jordan VC. The 38th David A. Karnofsky Lecture: The Paradoxical Actions of Estrogen in Breast Cancer—Survival or Death? J Clin Oncol 2008; 26:3073–82.

70) Wang, Yu, et al. "Androgen Receptor-Induced Tumor Suppressor, KLLN, Inhibits Breast Cancer Growth and Transcriptionally Activates P53/P73-Mediated Apoptosis in Breast Carcinomas." Human Molecular Genetics 22.11 (2013): 2263-2272.

71) Hickey, Theresa E., et al. "The Androgen Receptor Is a Tumor Suppressor in Estrogen Receptor–Positive Breast Cancer." Nature Medicine 27.2 (2021): 310-320.

72) Hickey, Theresa E., Amy R. Dwyer, and Wayne D. Tilley. "Arming Androgen Receptors To Oppose Oncogenic Estrogen Receptor Activity In Breast Cancer." British Journal of Cancer 125.12 (2021): 1599-1601.

73) Tien, Amy H., and Marianne D. Sadar. "Treatments Targeting the Androgen Receptor and Its Splice Variants in Breast Cancer." International Journal of Molecular Sciences 25.3 (2024): 1817.

74) Pietri, Elisabetta, et al. "Androgen Receptor Signaling Pathways as A Target for Breast Cancer Treatment." Endocrine-Related Cancer 23.10 (2016): R485-R498.

75) Fels, Eric. "Treatment of Breast Cancer with Testosterone Propionate." The Journal of Clinical Endocrinology 4.3 (1944): 121-125.

76) Adair, Frank E., and Julian B. Herrmann. "The use of Testosterone Propionate in The Treatment of Advanced Carcinoma of The Breast." Annals of Surgery 123.6 (1946): 1023-1035.

77) Laing, Abbie J., Louise Newson, and James A. Simon. "Individual Benefits and Risks of Intravaginal Estrogen and Systemic Testosterone In The Management Of Women In The Menopause, With A Discussion Of Any Associated Risks For Cancer Development." The Cancer Journal 28.3 (2022): 196-203.

78) Chen R, Cui J, Wang Q, et al. Antiproliferative Effects of Anastrozole On MCF-7 Human Breast Cancer Cells In Vitro Are Significantly Enhanced By Combined Treatment With Testosterone Undecanoate. Mol Med Rep. 2015; 12:769–775.

79) Cold, Søren, et al. "Systemic or Vaginal Hormone Therapy After Early Breast Cancer: A Danish Observational Cohort Study." Journal of the National Cancer Institute 114.10 (2022): 1347-1354.

80) Culhane, Rose, et al. "Menopausal Hormone Therapy in Breast Cancer Survivors." Cancers 16.19 (2024): 3267.

81) Lakshmanaswamy, Rajkumar, et al. "Hormone-Induced Protection of Mammary Tumorigenesis In Genetically Engineered Mouse Models." Breast Cancer Research 9 (2007): 1-11.

82) Arumugam, Arunkumar, Elaine A. Lissner, and Rajkumar Lakshmanaswamy. "The Role of Hormones and Aromatase Inhibitors On Breast Tumor Growth And General Health In A Postmenopausal Mouse Model." Reproductive Biology and Endocrinology 12 (2014): 1-13.

83) Davis, Susan R., et al. "Testosterone Enhances Estradiol's Effects On Postmenopausal Bone Density And Sexuality." Maturitas 21.3 (1995): 227-236.

84) Glaser, Rebecca L., et al. "Safety of Maternal Testosterone Therapy During Breast Feeding." Int J Pharm Compd 13.4 (2009): 314-7.

85) Renke, Guilherme, and Francisco Tostes. "Cardiovascular Safety and Benefits of Testosterone Implant Therapy in Postmenopausal Women: Where Are We?" Pharmaceuticals 16.4 (2023): 619.

86) Traish, Abdulmaged M., and Abraham Morgentaler. "Androgen Therapy In Women With Testosterone Insufficiency: Looking Back And Looking Ahead." Clinical Research and Therapeutics 3.1 (2022): 2-13.

87) Donovitz, Gary S. "A Personal Perspective on Testosterone Therapy in Women—What We Know in 2022." Journal of Personalized Medicine 12.8 (2022): 1194.

88) Donovitz, Gary S. "Society Position Statements on Bio-Identical Hormones-Misinformation Leads to a Dilemma in Women's Health." Healthcare. Vol. 9. No. 7. MDPI, 2021.

89) Committee on Gynecologic Practice, and American Society for Reproductive Medicine Practice Committee. "Committee opinion No. 532: Compounded Bioidentical Menopausal Hormone Therapy." Obstetrics And Gynecology 120.2 Pt 1 (2012): 411-415.

Chapter 19

Hormone Replacement for Breast Cancer Survivors Part Two

MICHELLE IS A 48-YEAR-OLD CEO of a high-tech company with severe menopausal symptoms of hot flashes, night sweats, insomnia, osteoarthritis in her finger joints, brain fog, chronic fatigue, and recurrent urinary tract infections. Three years ago, Michelle found a lump in her breast and had a lumpectomy for breast cancer (infiltrating ductal). Since then, for the past three years, her oncologist has been treating her with estrogen-blocking drugs, anastrozole, and Zoladex (goserelin). Michelle is now in my office seeking relief from her menopausal symptoms asking for bioidentical hormone replacement therapy (BHRT). The mainstream medical dogma is that hormone replacement is contraindicated for the breast cancer survivor (BCS) for fear of inducing recurrent cancer or a new cancer. In part one of this series, the opposite argument was presented, that this dogma should be overturned, and HRT made freely available to breast cancer survivors. We now have 25 studies showing hormone replacement is beneficial for the BCS's quality of life, without increased breast cancer recurrence compared to controls. Perhaps the strongest advocate for treating the BCS with BHRT is Avrum Bluming, MD, now retired after 50 years practicing oncology in Encino, California. Dr. Bluming was a clinical professor of medical oncology at the University of Southern California and spent four years as a senior investigator for the National Cancer Institute. He is author of the book, *Estrogen Matters*. Note: in 2021, Dr Bluming wrote an article on "Progesterone and Breast Cancer Pathogenesis." The title is misleading. The actual discussion is not about progesterone, it is about whether synthetic progestins cause breast cancer. (1-3)

The year 2002 saw the peak in menopausal hormone replacement therapy (HRT) with 35 million HRT users in the U.S. However shortly after, HRT use rapidly declined following the 2002 first arm Women's Health Initiative (WHI) study (Premarin/MPA). By 2011, HRT use had declined from 35 million to 10 million HRT users, a 70 percent drop. Another major shift after 2002, not widely known, is the change in the predominant formula for HRT from oral Premarin (CEE) combined with medroxyprogesterone (MPA) to topical estradiol combined with natural progesterone as discussed in 2011 by Dr. Agnes Fournier from France, citing the E3N and GAZEL, two French prospective cohort studies. The predominant estrogen in the U.S. was Premarin (CEE) while in France, the predominant estrogen was estradiol. (4-5)

How Many Breast Cancer Survivors Use HRT in the U.S.?

In 1993, Dr. Douglas J. Marchant from Tufts University estimated there were 2.5 million breast cancer survivors in the U.S., and about 10 percent or 250,000 used HRT. By 2024, the number of breast cancer survivors increased from 250,000 to 4 million, of which 400,000 use hormone replacement, writing:

> Two centuries ago, less than 30% of women lived long enough to reach menopause, but today 90% of women reach the climacteric [menopause]. In the United States, more than 30 million women have an average postmenopausal life expectancy of 28 years...With an average age at diagnosis of 60 years, coupled with a 25-year expected survival, it is estimated that the number of breast cancer survivors in the United States is **nearly 2.5 million women**. The short-term effects of estrogen deficiency (e.g., vasomotor instability and urogenital atrophy) and long-term consequences (e.g.,

osteoporosis and cardiovascular disease) are important quality-of-life and health issues for breast cancer survivors... However, prior studies have documented that up to **10% of breast cancer survivors use ERT [estrogen replacement therapy] for relief of menopausal symptoms**. (6)

Dr. Bluming's 26 Studies

Credit and thanks go to Dr. Avrum Bluming for his work on hormone therapy for breast cancer survivors, and his meticulous collection of 26 relevant medical studies. Of the 26 studies, 25 show no increased cancer recurrence in the hormone-treated group. Of the 26, the only study to show increased recurrence with HRT is the HABITS study by Dr. Lars Holberg from Sweden. This study used estradiol combined with norethisterone, a highly carcinogenic progestin derived from a testosterone backbone. Another tidbit of useful information: all patients in the HABITS trial on hormone replacement who suffered a recurrence of breast cancer were also taking **tamoxifen**, an estrogen receptor-blocking drug, a commonly used adjunctive endocrine therapy. Tamoxifen is discussed more completely in the next chapter. In 2020, Dr. Lindsey Berkson writes in the Townsend Letter:

> Of the 20 studies between 1980 and 2008 that showed estrogen was not only safe for breast cancer patients but was also protective, only the HABITS study found an increased risk of recurrence in breast cancer patients on HRT. As previously stated, this risk only occurred if the women were on tamoxifen, which "blocked" the action of estrogen. (7-8)

In a recent podcast September 5, 2024, Dr. Avrum Bluming says:

> there are now 26 studies in the medical literature that I could find in languages that I can understand, and I reviewed all of those studies. Of the 26 studies, 25 show no increased risk of recurrence of breast cancer among women who take hormone

replacement therapy (HRT) of any kind. And that is independent of whether they had estrogen receptor positive or estrogen receptor negative breast cancer. The one study that suggested an increased risk of recurrence is a study that was done in Sweden called the HABITS Study. HABITS stands for Hormones After Breast Cancer: Is it Safe? And that study reported an increased risk of contra, or local breast cancer, not an increased risk of distant recurrence. And there was no increased risk of death from breast cancer. (9)

In 2022, Drs. Abbie Laing and Louise Newson reviewed HRT for breast cancer survivors, noting only one study, the HABITS from Sweden showed increased recurrence, writing:

> Of all the data, there is currently only 1 study [HABITS] that describes an increased risk of recurrence in women taking hormonal replacement therapy after a history of breast cancer, whereas 3 studies demonstrate a decreased risk, and most studies demonstrate no difference whatsoever. (10)

Many of the 26 studies showed a protective effect of HRT with less recurrence in the hormone-treated group. In 2022, Dr Avrum Bluming, reviewed hormone replacement therapy (HRT) after breast cancer. Writing in the Cancer Journal, this is Dr. Bluming's list of 7 studies showing less recurrence in the hormone-treated group (1):

> Palshof et al. 1980 Prospective randomized trial: **Reduced recurrence.** (11)
>
> Palshof et al. 1985 Updates of the original study: **Reduced recurrence**, Reduced mortality. (12)
>
> Eden et al. 1995 Retrospective case-control: **Reduced recurrence**. (13)
>
> Durna et al. 2002 Retrospective observational: **Reduced recurrence** Reduced mortality.
>
> Guidozzi 1999 Prospective single-arm: **No recurrence**. (14)

O'Meara et al. 2001 Retrospective case-control: **Reduced recurrence**, Reduced mortality. (15)

Bluming 2008 Prospective cohort: **Reduced recurrence**. (16)

In Chapter 19, we will take a deep dive into these 26 studies compiled by Dr. Bluming, with attention to the protective effects of pregnancy and prior estrogen exposure, the carcinogenic effects of progestins, the beneficial effects of long-term estrogen deprivation (LTED), and finally, rank the various hormone formulas on a scale from most carcinogenic to least carcinogenic, and from most preventive of cancer recurrence to least preventive. Since many breast cancer survivors will be given adjuvant hormone therapy with tamoxifen, we also want to compare outcomes for HRT with the outcomes for tamoxifen and aromatase inhibitors.

Four Studies Comparing HRT to Tamoxifen

In the following 4 studies, breast cancer survivors using HRT had a similar reduction in recurrence and mortality compared to outcomes found in clinical trials of tamoxifen:

Eden, John, 1995: Most patients used combined continuous estrogen-progestin therapy, usually an oral estrogen with a moderate dosage of progestin. Among the 90 estrogen users, there were **no deaths. Breast cancer recurrence rate in HRT users was 60% less than nonuser controls,** (7% for HRT users compared to 17% for nonuser controls). (13)(110)

Guidozzi, 1999: 20 patients using Premarin and medroxyprogesterone and four tibolone. There were **no recurrences** to date after 24-44 months. (14)

O'Meara et al. 2001: Estrogens were unopposed by progestogens for 79% of the users; the rest combined a progestogen with estrogen for at least one monthly cycle. **Note:** "Unopposed estrogen" means no carcinogenic synthetic progestin was used in this study, thus explaining the excellent outcome. The breast cancer recurrence rate for HRT users was

reduced by **50%** compared to non-user controls. The breast cancer mortality rate for HRT users was reduced by **66% compared to** non-user controls. And, total mortality rate for HRT users was reduced by **50%** compared to non-user controls. (15)

Durna et al. 2002: Types of HRT used: Half the patients used oral Premarin plus Progestin. 18% used vaginal estrogen with or without oral progestin. 6% used transdermal estrogen alone. (total for all these groups = 74%) Results showed breast cancer recurrence was reduced by **38%** in HRT users compared to non-users. All-cause mortality in HRT users was reduced by **66%** in HRT users compared to non-users. Death from the primary tumor was reduced by **60%** in HRT users compared to non-users. (13)

I would venture to say the above four studies showed outcomes are vastly better than results obtained with estrogen-blocking drugs, tamoxifen and aromatase inhibitors, which are the standard of care. The added advantage of HRT is improvement in the quality of life (QOL). The obvious question to ask here is: what is preventing mainstream medicine from obtaining these same excellent outcomes? This answer is estrogen-blocking drugs (anti-estrogens) are the standard of care for mainstream medicine, which is difficult to change. The wheels of progress in medicine turn slowly, so I would not hold my breath expecting change anytime soon.

Estrogen is a Growth Factor which Stimulates Cancer Cells

Laboratory studies on breast cancer cells show that estrogen is a growth factor that stimulates growth and breast cancer cell division. In the 1970's, many in-vitro studies of breast cancer cells such as a 1976 study by Dr. Marc Lippman working in-vitro with MCF-7 breast cancer cells. Dr. Marc Lippman found that estrogen at physiologic levels is a growth signal and stimulates cell division. Anti-estrogen drugs that block estrogen receptors serve as growth inhibitors. However, the growth and cell division stimulation by estrogen is biphasic, with a cancer cell-killing effect at high estrogen concentrations. This cancer killing effect of estrogen can be amplified by submitting the cells to long-term estrogen deprivation (LTED). Because of the above 1976 study by Dr. Lippman and many others, the use of estrogen for breast cancer survivors is dogmatically contra-indicated thinking it would "throw gasoline on the fire" and increase breast cancer recurrence. (17)

The argument that estrogen causes breast cancer is supported by two facts: 1) There is a greater incidence of breast cancer in women than males by a factor of 100. This is thought to be due to higher estrogen levels in women than in males. Perhaps the overriding factor is not the high estrogen in females. Rather, it is the high testosterone in males that is breast cancer preventive. On the other hand, perhaps the most significant factor is the minuscule amount of breast tissue in males. 2) Estrogen-blocking drugs (anti-estrogens) are the standard of care for women at increased risk for breast cancer, as well as breast cancer survivors. This is often quoted as an argument supporting estrogen in the causation of breast cancer. However, what if the use of anti-estrogens is misguided ? We will discuss this in the next Chapter 20. The above two considerations have created a "fear of estrogen" among the general population, and also among the medical profession. This fear of estrogen was amplified in 2002 by the mass media after the publication of the 2002 first-arm WHI study showing 26 percent increased breast cancer in the HRT group (Premarin/MPA). However, the data from published studies and actual clinical practice do not support dogmatic denial of HRT for breast cancer survivors as we will discuss below. (18-19)

Cancer Treatments Cause Long Term Estrogen Deficiency

As mentioned above, breast cancer survivors are treated with anti-estrogens, tamoxifen, aromatase inhibitors, and LH-blocking drugs which block estrogen production by the ovaries (Zoladex). These drugs induce severe menopausal symptoms motivating women to seek relief with HRT. Especially distressing are symptoms related to genito-urinary syndrome affecting 70% of menopausal women, yet only 7% receive treatment, as discussed in 2022 by Drs. Abbie Laing and Louise Newson, writing:

> Symptoms [of genito-urinary syndrome] are very common, affecting approximately **70%** of all menopausal women. Symptoms include genital changes (dryness, itching, burning, pain, and irritation), sexual concerns (dyspareunia, decreased arousal, reduced lubrication, postcoital bleeding, reduced or absent orgasm), and urinary problems (urgency, nocturia, dysuria, recurrent UTIs, and urinary incontinence)... Collectively, these symptoms can cause significant distress, reduced quality of life, depression, anxiety, and other mood disorders and unlike vasomotor symptoms do not usually improve with time; instead, they are chronic and progressive. A response to hormonal treatments is usually rapid and sustained with most women obtaining substantial relief even after 3 weeks...Thus, it remains very concerning that **only 7% of women with GSM [genito-urinary-syndrome of menopause] receive treatment.** (10)

The use of intra-vaginal estrogen is very effective for GSM. In our office, we use vaginal estriol combined with testosterone for less severe cases, and vaginal Bi-est (estriol/estradiol 4:1 with progesterone) and topical testosterone for more severe cases. In Europe, vaginal estriol (E3) and estradiol (E2) are widely available at conventional pharmacies. However, in the U.S. vaginal estriol (E3) is available only from compounding pharmacies. Big Pharma has captured the medical journals, the mass media and the endocrine societies, and has vilified compounding pharmacies. The result is symptomatic women in the US rarely receive vaginal estriol, while this is more commonly prescribed in Europe. Withholding vaginal HRT from symptomatic menopausal women denies them health benefits for the urogenital system (genito-urinary syndrome), bones, joints, heart, and cognitive function. One of the catastrophic errors of modern medicine is the withholding of HRT from not only breast cancer survivors, it is withheld from all menopausal women in general. We will review the published data and make a few observations.

High Dose Estrogen Equal to Tamoxifen for Metastatic Breast Cancer

If estrogen is a growth factor for breast cancer cells, then how is it possible that in the 1940s, high-dose synthetic estrogen (DES) was the mainstream treatment for metastatic breast cancer? DES is 50 times more potent than estradiol (E2), with a response rate of about 30 percent for metastatic breast cancer. In 1944, Dr. Alexander Haddow was the first to use high-dose estrogen therapy for the treatment of metastatic breast cancer, and DES was quickly adopted as mainstream treatment until the 1970s when DES was replaced by tamoxifen (TAM). In 2009, Dr. Matthew J Ellis showed that a daily dose of 6 mg of oral estradiol may serve instead of DES, similarly providing clinical benefit in 30 percent of metastatic breast cancer patients who had failed aromatase inhibitors after 3-5 years of long-term estrogen deprivation (LTED). (20-24)

LTED Long Term Estrogen Deprivation

Studies comparing efficacy of estrogen (DES) with that of tamoxifen (TAM) in treating progressive metastatic breast cancer showed patients on DES had better survival than TAM treated patients and the duration of response was about the same for both. However, because TAM was easier to tolerate and had less toxicity, TAM replaced DES, and became standard of care in 1974. The rise of tamoxifen was due to the efforts of one man, Dr. V. Craig Jordan, the "father of tamoxifen," who spent his career in the laboratory trying to decipher the molecular biology of breast cancer. Using in-vitro breast cancer cells, Dr. V. Craig Jordan discovered that long-term estrogen deprivation (LTED)

transforms estrogen from a growth signal to a death signal. For breast cancer cells deprived of estrogen long term, the reintroduction of estrogen induces mitochondrial apoptosis, programmed cell death for the cancer cells. LTED up-regulates ER-alpha, so when estrogen is reintroduced, ER-alpha sends excess signals that stimulate mitochondrial apoptosis, programmed cell death. Think of it this way. If you plug your 110-volt lamp into a 220-volt socket for the dryer or washing machine, what will happen? The light bulb "pops" and burns out. This is the same concept for reintroduction of estrogen for the LTED cancer cell. The voltage is too high for the fragile light bulb, and the estrogen signal is too much for the LTED treated cancer cells. **Note:** Sadly, Dr. V. Craig Jordan passed away on June 9, 2024, at age 76 of renal cell carcinoma. (25-29)

Estrogen Receptors Alpha and Beta

In 1958, Elwood Jensen discovered the estrogen receptor, and ER-alpha was cloned in 1985. Likewise, in 1995, Dr. Jan-Åke Gustafsson cloned the ER-beta receptor. Knowledge of these two estrogen receptors provides a deeper understanding of why various hormone preparations increase or decrease breast cancer risk, depending on which receptor has preferential binding. ER-alpha is the pro-carcinogenic, proliferative receptor while ER-beta is the tumor suppressor receptor. The best results are obtained with hormones or natural agents that bind preferentially to ER-beta, while avoiding or downregulating ER-alpha receptors. Hormones that preferentially target ER-beta or down-regulate ER-alpha are breast cancer preventive. Specifically, E3 (estriol), Premarin (B-ring unsaturated steroids), testosterone (3-beta-diol), and progesterone are all breast cancer protective. Hormones or hormone-disrupting chemicals (HDCs) which preferentially target ER-alpha are proliferative and pro-carcinogenic. Bisphenol A (BPA) found in plastics is an endocrine-disrupting chemical (EDC). The carcinogenicity of Bisphenol A (BPA) arises from its ability to bind to and activate ER-alpha, while inhibiting ER-beta. (30-32)

Progestins Activate ER-Alpha

With knowledge of the two estrogen receptors, we can explain why estradiol (E2) is more proliferative than Premarin (CEE). Estradiol targets both ER-alpha and ER-beta equally, while Premarin contains B-ring unsaturated steroids which preferentially bind to ER-beta, thus reducing proliferative potential. Progesterone is breast cancer protective acting as a proliferative brake on ER-alpha. However, progestins do the exact opposite, activating ER-alpha and downstream oncogenes, cyclin D1 and cMYC. In 2012, Sebastián Giulianelli studied a mouse breast cancer model showing that progestin, medroxyprogesterone activates ER-alpha leading to oncogene activation of cMYC and Cyclin D1. This behavior is the exact opposite of natural progesterone which acts as a proliferative brake on ER-alpha based on the 2015 study by Hisham Mohammed. In 2012, Dr. Sebastián Giulianelli writes:

> In this study, we used a murine progestin-dependent tumor to investigate the role of ER-alpha in progestin-induced tumor cell proliferation. We found that treatment with the progestin medroxyprogesterone acetate (MPA) induced the expression and activation of ER-alpha, as well as rapid nuclear colocalization of activated ER-alpha with PR [progesterone receptor]. Treatment with the pure antiestrogen fulvestrant to block ER-alpha disrupted the interaction of ER-alpha and PR in vitro and induced the regression of MPA-dependent tumor growth in vivo. ER-alpha blockade also prevented an MPA-induced increase in CYCLIN D1 (CCND1) and MYC expression [both are oncogenes]. Chromatin immunoprecipitation studies showed that MPA triggered binding of ER-alpha and PR to the CCND1 and MYC promoters [oncogenes]. Interestingly, blockade or RNAi-mediated silencing of ER-alpha inhibited ER-alpha, but not PR binding to both regulatory sequences, indicating that an interaction between

ER-alpha and PR at these sites is necessary for MPA-induced gene expression and cell proliferation. We confirmed that nuclear colocalization of both receptors also occurred in human breast cancer samples. Together, our findings argued that ER-alpha–PR association on target gene promoters is essential for progestin-induced [MPA] cell proliferation. (33-36)

Natural Progesterone Acts as Proliferative Brake on ER-Alpha

In 2015, Dr. Hisham Mohammed found progesterone acts as a proliferative brake to the ER-alpha receptor, writing:

> In this scenario, under estrogenic conditions, an activated PR [progesterone receptor] functions as a **proliferative brake in ER-alpha+ breast tumors** by re-directing ER-alpha chromatin binding and altering the expression of target genes that induce a switch from a proliferative to a more differentiated state. (37)

Progestins Interfere with Androgenic Signaling

In the previous Chapter 18, we read how androgen signaling downregulates ER-alpha while upregulating ER-beta. Thus, androgen (testosterone) is breast cancer protective. In 2007, Dr. Stephen Birrell suggests carcinogenicity of synthetic progestins, MPA, and norethisterone, arises from interference with androgen signaling. The carcinogenicity of the many synthetic progestins will vary based on androgenic activity. The more androgenic, the greater the carcinogenicity. Norethisterone made from a testosterone backbone is more carcinogenic than medroxyprogesterone (MPA), made from a progesterone backbone. The carcinogenicity of progestins is dose-dependent. At lower doses used clinically, they compete with androgen receptors and are carcinogenic. At higher doses, they become actual androgens and are used for breast cancer treatment with efficacy like testosterone itself. (38-41)

Exogenous Estrogen vs Tumor Produced Estrogen

Even though estrogen is a growth factor for breast cancer, exogenous estrogen may not be a significant factor because breast tumors independently make their estrogen via upregulation of intracellular aromatase. Thus, the administration of exogenous estrogen to the patient may not add much to the already stimulated tumor growth. This point was made in 2001 by Dr. Ellen O'Meara, writing:

> Breast tumors can regulate and maintain internal levels of estradiol independent of levels outside the tumor, so exogenous estrogens may have relatively little effect on tumor growth. (15) (42)

LTED and Estrogen Induced Apoptosis

Things are more nuanced than this because one must also consider long-term estrogen deprivation (LTED) on the differing effects of the two receptors, the proliferative ER-alpha, and tumor suppressor ER-beta. Long-term anti-estrogen drug treatment with tamoxifen or aromatase inhibitors mimics LTED which **upregulates estrogen receptor-alpha (ER-alpha) four to ten-fold** so that estrogen re-introduction causes mitochondrial apoptosis of the breast cancer cells via ER-alpha hyperstimulation. One must consider the beneficial effect of LTED upon starting HRT which will induce apoptosis within pre-existing microscopic nests of breast cancer cells in the long-term estrogen-deprived breast cancer survivor. (43-44)

Prior Exposure to HRT or High Hormone Levels of Pregnancy Are Protective

Another factor to consider is the protective effect of high hormone levels of pregnancy, estradiol (E2), estriol (E3) and progesterone, conferring a 70 percent decrease in lifetime breast cancer risk. First documented by Bernardino Ramazzini in 1700, early first full-term birth is the most effective form of breast cancer prevention, providing a 50-70 percent lifetime reduction in the risk for breast cancer. High endogenous hormone levels of pregnancy reprogram the epigenome of breast cells making them less responsive to oncogene activation. Prior exposure to exogenous estrogen has a similar protective effect. In 2020, Dr. Lindsey Berkson discusses the breast cancer effects of pregnancy writing:

> There is a 70% decrease in breast cancer risk associated with a full-term pregnancy before the age of 18. It has also been shown that pregnancy is safe after treatment of breast cancer, even among estrogen receptor–positive women patients. Note: ER+ means pathologists identify estrogen receptors in the tumor. (7-8) (45-48)

When doing HRT studies in breast cancer survivors, prior HRT use and the number of pregnancies should be included in the data. In 2002, Dr. Durna studied HRT in breast cancer survivors and recorded the total number of pregnancies (parity) and prior use of HRT, two factors affecting outcome. The median parity was 2.2 children for both HRT users and non-users, so this was not a factor in the study. However, prior use of HRT (before breast cancer diagnosis) was greater in the HRT user group of breast cancer survivors (67% vs. 19%). Prior use of HRT mimics elevated hormone levels of pregnancy. This factor, by itself, could skew the data in favor of lower recurrence for the HRT-user group. Remember, high hormone levels of pregnancy confer protection from breast cancer by reprogramming the epigenome to be more resistant to oncogene activation, as discussed by Dr. Mary Feigman (2020). This same mechanism is thought to be in play for women with prior HRT exposure. Prior HRT exposure protection from breast cancer was demonstrated by Howard Hodis in his 2018 re-analysis of the WHI study. Upon removal of women with prior HRT exposure from the placebo group, there was no longer increased cancer risk in the HRT group using Premarin and MPA, 2004, first arm WHI. (13) (49-51)

How to Get Excellent Results with HRT

In 2002, Dr. Eva Durna from Australia studied HRT in breast cancer survivors. Dr. Eva Durna's results showed the HRT group had a **38 percent** reduction in cancer recurrence, a **66 percent** reduction in all-cause mortality and a **60 percent** reduction in mortality from breast cancer compared to non-users. I would venture to say these results are equal to or better than any anti-estrogen therapy with tamoxifen or AI drug. Dr. Durna made two pertinent observations explaining the excellent results of her study. Her first comment is a recognition of the beneficial effect of LTED, and the second comment is increased risk of breast cancer with synthetic progestins, writing:

> HRT users had reduced risk of cancer recurrence (adjusted relative risk [RR], **0.62**), all-cause mortality (RR, **0.34**) and death from primary tumour (RR, **0.40**)…A possible explanation of the [good] results is that women with estrogen deficiency tend to have better outcomes after breast cancer. [This is the beneficial effect of LTED] …Other recent observational studies also suggest that the use of sequential or cyclic progestins in HRT may increase the risk of breast cancer. [This is the recognition of progestins as carcinogenic]. (13)

Estrogen Metabolism Studies?

What causes breast cancer? Single Nucleotide Polymorphisms (SNPs), mutations in genes that code for enzymes involved in estrogen metabolism play a major role in the genesis of breast cancer as discussed in 2021 by Dr. Ercole Cavalier Professor at the University of Nebraska Medical Center. His career was devoted to understanding the initiation of cancer, and the role of estrogen metabolites. The 4-hydroxy-estrogen metabolites lead to carcinogenic DNA adducts called the 4-hydroxy-quinones. However, the 2-hydroxy-estrogen metabolites lead to 2-methoxy-estradiol, an endogenous anti-cancer drug. However, none of the above 26 studies of breast cancer survivors on HRT evaluated estrogen metabolites, or metabolic testing for activity of key enzymes: COMT, CYP1A1, and CYP1B1 Future breast cancer studies should include metabolic or genetic testing data. In this way, patients at higher risk can be identified and treated appropriately. For more on estrogen metabolism and breast cancer see Chapter 21 on Estrogen Metabolism, Iodine, 2MEO, Part Three. Sadly, Dr. Cavalier passed away at age 84 in 2021. (52)

Testosterone is Cancer Preventive

In 2019, a study by Dr. Rebecca Glaser and in 2021 by Dr. Gary Donowitz both show subcutaneous testosterone pellets inserted into menopausal women reduce the risk of breast cancer by about **40%**. In 2007, Dr. Marie Hofling from Stockholm, Sweden asked a slightly different question, could testosterone oppose the proliferative effects of norethisterone, the carcinogenic progestin used in the HABITS trial? Dr. Hofling studied 88 postmenopausal women over 6 months in a prospective, randomized, double-blind, placebo-controlled study looking at estrogen/progestin HRT-induced breast proliferation with breast biopsy specimens. All patients were given HRT, continuous combined oral estradiol 2 mg, and oral norethisterone acetate 1 mg. All 88 patients on HRT were randomized to a testosterone skin patch or placebo. After 6 months of treatment, breast cells were obtained by fine needle aspiration and the KI-67 proliferation index was obtained. Dr. Marie Hofling found 500 percent increase in KI-67 breast proliferation in the placebo group compared to the testosterone group. The addition of testosterone opposes the proliferative effects of the estradiol/synthetic progestin, norethisterone combination, writing:

> During the past few years, serious concern has been raised about the safety of combined estrogen/progestogen [norethisterone] hormone therapy, in particular about its effects on the breast. Several observations suggest that androgens may counteract the proliferative effects of

estrogen and progestogen in the mammary gland....In the placebo group there was a more than **fivefold increase** (P G 0.001) in total breast cell proliferation from baseline (median 1.1%) to 6 months (median 6.2%). During testosterone addition, no significant increase was recorded (1.6% vs 2.0%). The different effects of the two treatments were apparent in both epithelial and stromal cells. The addition of testosterone may counteract breast cell proliferation as induced by estrogen/progestogen therapy in postmenopausal women. (53-56)

Androgen Receptor Potently Inhibits ER-alpha Stimulation

As discussed in the previous Chapter 18, in 2009, Dr. Amelia Peters studied androgen receptors in 215 invasive ductal breast carcinomas, followed by in-vitro studies on these same breast cancer cells showing the androgen receptor (AR) binds to estrogen response elements (ERE), thus potently inhibiting ER-alpha transactional activity and estradiol-induced proliferation, writing:

There is emerging evidence that the balance between estrogen receptor-alpha (ER-alpha) and androgen receptor (AR) signaling is a critical determinant of growth in the normal and malignant breast. In this study, we assessed AR status in a cohort of 215 invasive ductal breast carcinomas. AR and (ER-alpha) were coexpressed in the majority (80-90%) of breast tumor cells... **AR potently inhibited (ER-alpha) transactivation activity and 17beta-estradiol-stimulated growth of breast cancer cells**. Transfection of MDA-MB-231 breast cancer cells with either functionally impaired AR variants or the DNA-binding domain of the AR indicated that the latter is both necessary and sufficient for inhibition of ER-alpha signaling. Consistent with molecular modeling, electrophoretic mobility shift assays showed **binding of the AR to an estrogen-responsive element (ERE)**. Evidence for a functional interaction of the AR with an ERE in vivo was provided by chromatin immunoprecipitation data,

revealing recruitment of the AR to the progesterone receptor promoter in T-47D breast cancer cells. We conclude that, **by binding to a subset of EREs, the AR can prevent activation of target genes that mediate the stimulatory effects of 17beta-estradiol [E2] on breast cancer cells**. (57)

DHT Inhibits E2-induced Expression of Cyclin D1 mRNA

in 2012, Dr. Natalija Eigeliene from Finland studied the effect of testosterone on human breast tissue obtained from reduction mammoplasty operations of postmenopausal women (explants). These tissue samples were then cultured and studied in-vitro with or without testosterone (T) and its more potent metabolite 5-alpha-dihydrotestosterone (DHT), or in combination with 17 beta-estradiol (E2) for 7 days and 14 days. Dr. Natalija Eigeliene found testosterone and its more active metabolite, DHT, inhibited the proliferation of the human breast tissue cells grown in culture. When estradiol (E2) was added to the culture media, testosterone opposed the estradiol (E2) stimulated proliferation, and DHT inhibited E2-induced expression of cyclin D1 mRNA. **Note:** cyclin D1 is an oncogene, implicated in breast cancer, so inhibiting mRNA expression of cyclin D1 is cancer-preventive. Dr. Eigeliene writes:

T [testosterone] and DHT [di-hydro-testosterone] reduced proliferation and increased apoptosis in breast epithelium, the effects of which were reversed by bicalutamide [androgen blocker]. In combination with E2 [estradiol], they [T and DHT] suppressed E2-stimulated proliferation and cell survival. DHT also inhibited basal (P < 0.05) and E2-induced expression of cyclin-D1 mRNA (P < 0.05). (58)

Testosterone Treats Progressive Metastatic Breast Cancer

In 2014, Dr. Corrado Boni from Emilia, Italy did a retrospective study of 53 breast cancer patients treated with intramuscular injections of testosterone propionate for progressive metastatic disease non-responsive to multiple chemotherapy and endocrine treatments. About 60 percent of the patients had a favorable outcome with regression in 10 patients and stabilization in 22 patients. About 40% of patients failed to respond, and despite testosterone treatment had progressive disease. Median survival after starting testosterone was about a year. I suspect the non-responders had ER negative cell types or lacked androgen receptors. (59)

Does the HRT Contain Testosterone?

Let us now return to Dr. Bluming's list of 26 studies of HRT in breast cancer survivors. The next logical question is: Does the HRT formula include testosterone? One such study in 1999 by Dr. Puthgraman Natrajan from Augusta, Georgia included testosterone pellets in 38 of 50 patients on HRT after a diagnosis of breast cancer. His study showed an amazing **81% reduction in mortality in HRT users compared to non-user controls**. As discussed above, testosterone is strongly cancer-preventive. I would expect similar excellent outcomes for any study including testosterone. On a practical basis, all HRT formulas must include testosterone for its superb ability to prevent breast cancer. (60)

Vaginal Estrogen Associated with Exceptionally Good Outcomes

In 2002, Dr. Eva Durna from Australia noticed the excellent outcomes with vaginal estrogen-alone in 32 breast cancer survivors in a retrospective observational study. These results were much better than all other hormone formulas and all other types of hormone delivery (oral and transdermal). This group of **32 patients** showed the lowest relative risk (RR for recurrence, **0.18** compared to non-users. The RR of 0.18 means HRT users had **82 percent less recurrence than HRT non-users, 70 percent reduction in all-cause mortality and 65 percent reduction in death from breast cancer.** This is astounding and is superior to anti-estrogen therapy! Dr. Durna writes:

Vaginal estrogens included estriol [E3] cream (0.5 g) and estradiol [E2] vaginal tablets (25 microg) used twice weekly...vaginal estrogen (n=32)...those who used vaginal estrogen alone [32 patients] had significantly lower risk of recurrence or new breast cancer (adjusted RR, **0.18**)... we believe that the current practice of withholding HRT from women with breast cancer who suffer menopausal symptoms may need review. (13)

In 2001, Dr. George N. Peters, a breast surgeon in Dallas Texas studied HRT use in breast cancer survivors, finding eight vaginal estrogen patients in his study group that had results so outstanding, they were reported separately so as not to skew the results of the main study. The exact type of vaginal estrogen cream, whether E2 or E3 was not specified. In these eight patients followed for 11.4 years, there were **no cancer recurrences and no deaths**. Dr. Peters writes:

Of the eight patients who have used only vaginal cream ERT [estrogen replacement therapy], median follow-up from diagnosis was 11.4 years, and median time on ERT since diagnosis was 4.0 years. There have been no contralateral breast cancers; no local, regional, or distant recurrences; and no cancer deaths in this group. it is reasonable to conclude that ERT does not appear to have an adverse effect on cancer outcome. (61)

More on Vaginal Estrogen

A 2019 study by the Collaborative Group in Lancet was a meta-analysis of 58 epidemiological studies from 21 countries including 143,887 postmenopausal women with invasive breast cancer and 424,972 without breast cancer serving as controls. All women in this 2019 Lancet study had been using post-menopausal HRT at the time of breast cancer diagnosis. This 2019, Lancet study reports the vaginal delivery of estrogen is the only form of menopausal hormone replacement **NOT associated** with increased breast cancer risk. **Note:** this 2019 Lancet study is **not a breast cancer survivor study**. It includes only **every-day postmenopausal HRT users.** This study also found greater breast cancer risk when synthetic progestins are used. The authors write:

> Every MHT [menopausal hormone therapy] type, **except vaginal estrogens**, was associated with excess breast cancer risks, which increased steadily with duration of use and were **greater for estrogen-progestagen** than estrogen-only preparations. (4)

In 2006, Dr. H. Lyytinen in Finland studied 110,000 post menopausal women using HRT. Dr. H. Lyytinen found vaginal use of any type of estrogen-alone (without progestin) **was not associated with increased breast cancer risk**, writing:

> The use of estradiol [without progestin] was associated with an increased risk of breast cancer after 5 years of use (incidence ratio IR=1.3-1.4). **Neither an oral estriol [E3] regimen nor vaginal use of any estrogen formulations were accompanied by a significantly increased risk of breast cancer.** (62)

Tamoxifen Users in the HABITS Trial

Going back to the HABITS trial in which HRT users had a 3-fold increase in recurrence compared to non-users, remember it was stated that all the patients with recurrence were also taking tamoxifen, thus raising suspicion about tamoxifen as a factor for increased recurrence. What about the use of vaginal estrogen (LHT local hormone therapy) in patients also taking tamoxifen or aromatase inhibitors? Has this been studied? Yes, it has. In 2012, Dr. Isabelle Le Ray from Dijon, France used a large UK database to do a cohort study (a nested case-control analysis) of adjuvant hormone treatment in breast cancer survivors. The breast cancer recurrence rate in controls was **2.6 percent**. Vaginal estrogen users had reduced recurrence (RR: **0.78**) compared to controls. **Tamoxifen-**treated vaginal estrogen users also had 17 percent less recurrence (RR: **0.83**). However, for **vaginal estrogen users on aromatase inhibitors, no patient had a recurrence.** It would be nice to know the vaginal estrogen formula, whether Premarin, estradiol (E2) or estriol (E3) was used, and at what dosage. We do not have this information. Dr. Isabelle Le Ray writes:

> A total of 13,479 women were included in the study, of which 2,673 received AIs [aromatase inhibitor], 10,806 received tamoxifen, and 271 received LHT [local hormone therapy, vaginal estrogen]. Mean age at cohort entry was 63.7 years, and mean follow-up was 3.5 years. The crude recurrence rate 25.9 per 1,000 per year. Overall, the use of LHT was not associated with an increased risk of recurrence (**RR: 0.78**, 95 % CI 0.48-1.25) compared with non-use. In stratified analyses, LHT did not increase the risk of recurrence among tamoxifen-treated patients (**RR: 0.83**, 95 % CI 0.51-1.34), while the risk was not estimable among AI-treated patients since **no patients receiving LHT experienced a recurrence.** The use of LHT is not associated with an increase in breast cancer recurrence among women receiving a hormone therapy. (63).

In 2024, Dr. Lauren McVicker studied breast cancer survivors using vaginal estrogen in two large cohorts of 49,237 females in Scotland and Wales. About 5% used vaginal estrogen, and these women had a **23% reduction in breast cancer mortality** compared to nonusers. Dr. Lauren McVicker writes:

> A recent observational Danish study showed no increase in recurrence in patients with breast cancer who received vaginal estrogen therapy aside from a sub group who received both vaginal estrogen therapy and aromatase inhibitors... the cohort comprised 49,237 females with breast cancer (between 40 and 79 years of age) and 5,795 breast cancer-specific deaths. **Five percent of patients with breast cancer used vaginal estrogen therapy after breast cancer**

diagnosis. In vaginal estrogen therapy users compared with HRT nonusers, **there was no evidence of a higher risk of breast cancer-specific mortality in the pooled fully adjusted model (HR, 0.77)**...Conclusions and relevance: Results of this study showed **no evidence of increased early breast cancer-specific mortality in patients who used vaginal estrogen therapy compared with patients who did not use HRT.** This finding may provide some reassurance to prescribing clinicians and support the guidelines suggesting that vaginal estrogen therapy can be considered in patients with breast cancer and genitourinary symptoms. (64)

What is the explanation for these outstanding results for vaginal estrogen use? We still do not know. Here are my speculative thoughts: Firstly, in the studies using estrogen alone via transvaginal delivery, there was no added carcinogenic progestin. This could explain improved outcomes. The second explanation is that many of the vaginal preparations contain estriol (E3) which preferentially binds to ER-beta, the tumor suppressor. Thirdly, vaginal route has a much better absorption rate compared to transdermal route, reaching higher blood levels without first pass through the liver. Thus, vaginal estrogen (Bi-est is 20% estradiol E2 and 80% estriol E3), used at systemic doses, **closely mimics the high hormone levels of pregnancy** which is known for centuries to confer protection from breast cancer. I speculate these factors may account for the excellent outcomes with vaginal estrogen alone. (65-69)

Jonathan V. Wright, Bi-Est and Pregnancy

Jonathan V. Wright MD, from Tahoma, Washington is a legendary bioidentical hormone pioneer who devised the Bi-est formula and followed his patients with urinary hormone metabolite testing in his Meridian Valley Laboraory. In 2005, Dr. Wright commented that oral estradiol [E2] of 1-2 mg per day may mimic pregnancy hormone levels, and the addition of estriol (E3) more closely mimics pregnancy

than estradiol (E2) alone. Thus, the vaginal Bi-est (80% E3/20%E2) formula may resemble pregnancy better than all other available hormone preparations and perhaps explains the superlative breast cancer prevention seen in the above studies. Dr. Wright writes:

> Estradiol replacement therapy at oral doses of 1-2 mg/day results in **serum and urine estrone concentrations resembling those seen in pregnancy**. However, there is an important difference between the estrogens present with estradiol replacement and those during pregnancy – **estriol concentrations are much higher during pregnancy than during estradiol replacement.** Experimental and epidemiological studies **suggest a high concentration of estriol may be necessary** to modulate the impact of the high concentrations of estradiol and estrone seen in pregnancy. (70)

Stopping Anti-Estrogens for a Pregnancy

We have been talking about breast cancer protection by mimicking the hormone levels of pregnancy. How about pregnancy itself? Here we are talking about premenopausal breast cancer survivors on adjuvant antiestrogen drug therapy (tamoxifen or AIs). What if these women stopped the estrogen-blocking drugs and then carried a pregnancy to term? Would the high estrogen levels of pregnancy "throw gasoline on the fire," so to speak, and cause breast cancer to go rampant? This is exactly what Dr. Ann Partridge wanted to know with her POSITIVE trial. Contrary to the expectations of worsening cancer recurrence, this study showed no increased recurrence in the interruption/pregnancy arm of the study compared to controls. The 3-year incidence of breast cancer events was 8.9% in the treatment-interruption group and 9.2% in the control cohort. So, the answer is no, high hormone levels of pregnancy do not increase cancer recurrence. Dr. Ann Partridge writes:

We conducted a single-group trial in which we evaluated the temporary interruption of adjuvant endocrine therapy to attempt pregnancy in young women with previous breast cancer. Eligible women were 42 years of age or younger; had had stage I, II, or III disease; had received adjuvant endocrine therapy for 18 to 30 months; and desired pregnancy...Among 516 women, the median age was 37 years, the median time from breast cancer diagnosis to enrollment was 29 months, and 93.4% had stage I or II disease. Among 497 women who were followed for pregnancy status, 368 (74.0%) had at least one pregnancy and 317 (63.8%) had at least one live birth. In total, 365 babies were born. At 1638 patient-years of follow-up (median follow-up, 41 months), 44 patients had a breast cancer event, a result that did not exceed the safety threshold. **The 3-year incidence of breast cancer events was 8.9%** in the treatment-interruption group and **9.2%** in the control cohort. Conclusions. Among select women with previous hormone receptor–positive early breast cancer, temporary interruption of endocrine therapy to attempt pregnancy did not confer a greater short-term risk of breast cancer events, including distant recurrence, than that in the external control cohort. (71)

The Value of LTED, Long Term Estrogen Deprivation

As mentioned above, long-term estrogen deprivation (LTED) transforms estrogen from a growth signal to a death signal, causing mitochondrial apoptosis in the breast cancer cell. Post-menopausal breast cancer survivors are usually offered tamoxifen. On the other hand, pre-menopausal breast cancer survivors are treated for 3-5 years with estrogen-blocking drugs such as aromatase inhibitors (letrozole, anastrozole) and LH-blocking drugs (Zoladex). These pre-menopausal women have been living through 3-5 years of estrogen deficiency and are generally miserable with climacteric symptoms. When started on estrogen hormone replacement after LTED, these women enjoy the benefits of estrogen-induced apoptosis rendering an excellent prognosis. Of the 26 studies compiled by Dr. Bluming, the one study best illuminating the value of LTED is a small study of 25 breast cancer survivors by Dr. Alan G. Wile from Orange County California. In this study, patients were divided into two groups based on HRT-free interval after diagnosis of breast cancer. For Group one, the LTED was less than 2 years (average 8 months), For Group II, the LTED was greater than 2 years (average 64.5 months). All three patients with breast cancer recurrence were in Group I with the shorter length of estrogen deprivation, less than 2 years. Dr. Wiles writes:

> The HRT-free interval for group I patients averaged 7.9 months and for group II patients averaged 64.5 months...Three of 25 patients have had a recurrence, all in group I. (72)

Increased Cancer Risk with Progestins, Medroxyprogesterone, Levonorgestrol and Northisterone

In 2021, Dr. Irene Lambrinoudaki reviewed the 2020 study by Dr. Vinogradova, a case-controlled study using two large UK general practice databases. Dr. Vinogradova's study revealed over 9 years of use, **estrogen alone** (84% CEE, equine estrogen, 16% estradiol) therapy showed only a marginal increase in breast cancer (OR=**1.14**), while the **combination of estrogen with a synthetic progestin** showed a more pronounced increase in breast cancer (OR=**1.87**), writing:

> Estrogen-only therapy for up to 9 years increased only marginally the risk of breast cancer (OR 1.14, CI 1.08–1.21), whereas estrogen-progestin combination therapy for the same duration was associated with a more pronounced increase in breast cancer risk (OR 1.70, CI 1.64 to 1.76). The risk differed according to the progestin used, being higher with medroxyprogesterone acetate [OR=**1.87**], levonorgestrel [OR=**1.79**] and norethisterone [**1.88**] ...and lower with

dydrogesterone [OR=1.24] for more than 5 years of therapy. The excess risk dissipated in past users...At the time of our study, two types of estrogen (conjugated equine estrogen and estradiol) and four types of progestogen (norethisterone acetate, levonorgestrel, medroxyprogesterone, and dydrogesterone) were commonly prescribed in the UK and were included in our analyses. (73-74)

The HABITS Study by Dr. Lars Holmberg of Sweden

This brings us to the next obvious question, which is why on earth would anyone give a carcinogenic progestin, norethisterone, for menopausal hormone replacement when cancer-preventive natural progesterone is available? This is the HABITS study by Dr. Lars Holmberg of Sweden which used estradiol combined with norethisterone, a highly carcinogenic progestin, resulting in **three times greater recurrence in the HRT group after 2.1 years of use.** One should take note that there was no increased recurrence in HABITS trial in these subgroups: **women using estrogen alone, or in women using Premarin (CEE)**. We have previously discussed how Premarin (CEE) has breast cancer protective properties conferred by the unsaturated B-ring steroids which preferentially bind to ER-beta, the tumor suppressor. The binding affinity of unsaturated B-ring steroids in Premarin (CEE) to ER-beta is 5-6 times greater than the binding affinity to ER-alpha. The use of Premarin-alone could explain the favorable results in the WHI second arm 2004 in which HRT users had a 23 percent reduction in breast cancer compared to non-users. The 20-year follow-up of the 2004 WHI second arm showed the Premarin-alone users had a 40 percent reduction in mortality from breast cancer. Mainstream medicine mindlessly lumps together estradiol and Premarin (CEE) calling them both estrogens, saying they are equal. They are not equal. Whenever the medical literature reports a study in which estrogen was used, we need to know what type

of estrogen. Did the study use Premarin (CEE) or estradiol (E2)? The cancer preventive properties of Premarin (CEE) vs. the proliferative properties of estradiol were recognized in 2019 by Richard J. Santen, writing:

Notably, pre-clinical studies demonstrated that conjugated equine estrogen [Premarin, CEE} , as used in the WHI [Women's Health Initiative, second arm, 2004], has unique, pro-apoptotic properties compared to the anti-apoptotic effects of estradiol, **a finding providing an explanation for the reduction in breast cancer with conjugated equine estrogen [CEE, Premarin].** (75)

The dismal outcome of the HRT group of the HABITS trial obscures the improved outcomes within subgroups. **Notice there was no increased recurrence in the estrogen-alone sub-group compared with estrogen combined with norethisterone**. This raises an accusing finger at the carcinogenicity of norethisterone, the synthetic progestin. All patients with breast cancer recurrence were also taking tamoxifen, an estrogen-blocking drug. In 2022, Dr. Avrum Bluming discusses the HABITS trial, writing:

The HABITS trial was prematurely terminated on December 17, 2003, after only 2 years of median follow-up and after only 434 women of the proposed 1300 had been enrolled. The reason for the sudden termination, according to the initial paper, was the disproportionate number of women randomized to HRT who developed another breast cancer (26 of 174 = **15%**), compared with only 7 of the 171 (**5%**) randomized to no HRT. The increase was seen only as local recurrences or contralateral tumors. There was no increase in the development of distant metastases, nor was there an increase in the risk of death. Further, there was **no increase among women randomized to estrogen alone; there was no increase when Premarin [CEE] (conjugated estrogens) was used as the source of estrogen**; there was no increase among women who had been initially diagnosed with lymph node involvement, and **the increase was noted only among women who were taking tamoxifen in conjunction with HRT**. (1) (76-77)

In the HABITS trial the increased recurrence was noted only in women on HRT (estradiol plus norethisterone) **taking tamoxifen**. There was no increased recurrence in women taking estrogen alone (without a progestin) and no increase when the estrogen was Premarin (with or without a progestin). This tells us several things. Firstly, never give estradiol and norethisterone to women on tamoxifen. This is a bad combination discussed in more detail in the next Chapter 20. But if you must use a progestin, use medroxyprogesterone, use it combined with Premarin. The unsaturated B ring steroids bind to ER-beta, a tumor suppressor, so Premarin tends to counteract the carcinogenicity of the progestin. Estrogen alone, without the progestin is OK with no recurrence reported in the HABITS trial for estrogen-alone. I would re-iterate a general rule that natural progesterone should be substituted for synthetic progestin. (78-80)

The Stockholm Study

The 2005, Stockholm Study by Dr. Eva von Schoultz, was a parallel study to the HABITS trial by Lars Holberg from Sweden. The main difference between these two studies was that Dr. Eva von Schoultz of the Stockholm study recognized the carcinogenic effect of progestins, and modified the Stockholm study design to minimize the use of the progestin, medroxyprogesterone (MPA), thought to be less carcinogenic than the norethisterone used in the HABITS study. This modification was a reduction in exposure time to the progestin. Instead of continuous progestin use, this was changed to sequential use for 10 days per month, every third month in 73 percent of the women. As a result, the Stockholm study showed much better results with an **RR value of 0.82** for recurrence in the hormone-treated group, meaning

there was **18 percent less recurrence** for HRT users compared to non-users in the Stockholm Study. I am rather perplexed why the Stockholm study designers could not use natural progesterone rather than medroxyprogesterone. The study results would have been so much better. In any event, by replacing norethisterone with medroxyprogesterone (MPA), a less carcinogenic progestin, and changing continuous use (30 days a month) to sequential use (10 days a month) and a "spacing out regimen" to reduce its use even further, the Stockholm study resulted in less recurrence for the HRT users (RR=0.82), and **both total mortality and breast cancer mortality was reduced in half compared to non-users.** The author, Dr. Eva von Schoultz stated in her discussion section that changing the progestin, and reducing its use was the main difference between the HABITS and the Stockholm studies, and was responsible for yielding much better results. In his 2022 review article, Dr. Avrum Bluming quite thoroughly discussed both studies, mentioning the larger percentage of women in the HRT group of the Stockholm study also taking tamoxifen (52% vs 34%), and criticizing the HABITS study for not doing baseline mammograms. Yet, Dr. Bluming was silent about the much larger issue, the carcinogenicity of synthetic progestins. I find this surprising because the stated goal of the Stockholm study was to minimize the use of synthetic progestins, a goal not shared by the HABITS trial. The rationale for minimizing the use of progestins was a 1996 monkey study (cynomolgus macaques) by Dr. J. Mark Cline at Wake Forest, North Carolina in which menopausal monkeys were treated with either conjugated equine estrogens (Premarin, CEE) and medroxyprogesterone acetate (MPA), or Premarin-alone (CEE), in doses equivalent to human HRT. D.J. Mark Cline found greater breast cell proliferation in the combined treatment (CEE plus MPA) compared to estrogen alone (Premarin, CEE). Dr. J. Mark Cline concludes the clinical implications of his monkey study is that **women receiving combined** therapy (CEE plus MPA) may have a greater risk for breast cancer than women receiving CEE alone. Dr. J. Mark Cline writes:

> These results [of Dr. Cline's monkey study] indicate a proliferative response of mammary gland epithelium to therapy with conjugated equine estrogens plus medroxyprogesterone acetate in postmenopausal macaques. The clinical implication of this finding may be a **greater risk for development of breast neoplasms in women receiving combined hormone replacement therapy**. (81-86)

Apparently, Dr. Eva von Schoultz took this to heart and made it the rationale for minimizing exposure to MPA in the Stockholm study, and then stated this reduced exposure to progestins is the reason for improved results in the Stockholm Study (less recurrence RR=**0.82**) compared to the HABITS study (Relative Hazard=**3.3**). In 2005, the first author of the Stockholm study, Dr. Eva von Schoultz, writes:

> In 1997 two independent randomized clinical trials, Hormonal Replacement Therapy After Breast Cancer--Is It Safe? (**HABITS; 434 patients) and the Stockholm trial (378 patients),** were initiated in Sweden to compare menopausal hormone therapy with no menopausal hormone therapy after diagnosis of early-stage breast cancer. Much of the design of both studies was similar; however, a goal of the Stockholm protocol, not shared with the HABITS trial, was to **minimize the use of progestogen** combined with estrogen. The HABITS trial was prematurely stopped in December 2003, because, at a median follow-up of 2.1 years, the risk for recurrence of breast cancer among patients receiving menopausal hormone therapy was statistically significantly higher (**relative hazard [RH] = 3.3**, 95% confidence interval [CI] = 1.5 to 7.4) than among those receiving no treatment. In the Stockholm trial, however, at a median follow-up of 4.1 years, the risk of breast cancer recurrence was not associated with menopausal hormone therapy (RH = **0.82**, 95% CI = 0.35 to 1.9)... Patients who had had a hysterectomy in the menopausal hormone therapy group were given continuous treatment with 2 mg of estradiol valerate daily... In addition, the Stockholm protocol attempted to minimize the use of progestogen in combination with estrogen. In contrast to the HABITS trial, the Stockholm trial recommended that patients avoid continuous combined treatment with estrogen and progestogen and use regimens that incorporated 1 week of no treatment every 1 (cyclic regimen) or 3 (spacing out regimen) months...These treatment recommendations were based on the following results, available when the trial was initiated, that indicated differential effects on the breast when treatment with estrogen alone was compared with combined treatment with estrogen and progestogen. First, breast cell proliferation is increased only during the luteal phase of the menstrual cycle when levels of both estrogen and progesterone are high. Second, using a relevant prospective monkey model for menopausal hormone therapy, we reported statistically significantly higher proliferation in the breast during continuous combined estrogen and progestogen treatment than during estrogen-only treatment. Finally, cyclic discontinuation of hormonal treatment was hypothesized to decrease the expression of local growth factors in breast tissue and to initiate and stimulate apoptosis...In the Stockholm trial, 73% of the women were first assigned to menopausal hormone therapy containing either estrogen alone or the spacing out regimen, in which progestogen was given for only 14 days at 3-month intervals. **This protocol could provide one explanation for the lack of difference in breast cancer recurrence between the menopausal hormone therapy group and the no treatment group in the Stockholm trial**. (1) (87-88)

Given the above, we should note that in his 2022 review, Dr. Bluming was silent on the carcinogenicity of progestins in the HABITS and Stockholm studies. Remember, Dr. Bluming's area of specialty was oncology, and the majority of oncologists have no knowledge or interest in prescribing hormone replacement to menopausal women. Most likely, if Dr. Bluming had a breast cancer patient using HRT, she would have been referred out to one of his colleagues in the surrounding community, most of whom were prescribing oral Premarin with medroxyprogesterone (MPA), a synthetic progestin, as this was the standard of care in southern California and still is. So, in retrospect, it is unrealistic to expect an oncologist, such as Dr. Bluming, to delve into the nuances of HRT formulation. (1)

Big Pharma Has Captured the Narrative

For many decades now, Big Pharma has "captured" the regulatory agencies, the media, the medical societies, and the medical journals and controls "the narrative", thus maximizing and protecting profits. Nothing happens in medicine without the blessing and sponsorship of Big Pharma. Thus, blurring the distinction between natural progesterone and synthetic progestins works to their favor, as this obscures the fact that progestins are carcinogenic while natural progesterone is not. If the medical literature singles out progestins as carcinogenic monsters while holding natural progesterone on a pedestal as non-carcinogenic and therefore preferable, this will decrease market share for synthetic progestins. Meanwhile, market share will increase to their main competitor, natural progesterone, resulting in economic loss to Big Pharma. (89-91)

More on the HABITS Study

Dr. Lars Holmberg, the author the 2008 HABITS study says medroxyprogesterone "resembles natural progesterone". The synthesis of medroxyprogesterone uses a progesterone backbone, and the resulting progestin is less androgenic, and therefore less carcinogenic than the other two, norethisterone and levonorgestrel which are synthesized from a testosterone backbone. Dr. Lars Holmberg writes:

The progestin preferred in the USA is medroxyprogesterone-acetate (MPA), which resembles natural progesterone and is administered either cyclically or continuously. In Europe the predominant regimen prescribed is 17-b-estradiol opposed by testosterone-derived progestins, mainly norethistosterone acetate (NETA) or levonorgestrel (LNG); while the less androgenic progestin MPA is used to a lesser extent. **Note:** there is a slight resemblance of MPA to natural progesterone, however, they are chemically distinct. (77)

How Not to Do a Hormone Replacement Study

Let us do a thought experiment on a hypothetical hormone replacement study done by Dr. Evil who wants the results to show HRT increases breast cancer recurrence. This is easy, just use a highly carcinogenic progestin such as norethisterone with a proliferative estrogen, estradiol. This was used in the HABITS trial. What else can Dr. Evil do to increase breast cancer recurrence? Make sure the hormone-treated group includes more advanced, more aggressive cancers, as these are associated with a greater recurrence rate. Make sure the HRT group has no long-term estrogen deprivation (LTED). Make sure most of the HRT group are women who have never been exposed to estrogen and have never been pregnant (nulliparous). Make sure you do not use any agents that target ER-beta or otherwise prevent breast cancer such as Premarin (CEE), estriol (E3) Testosterone (3-beta-Adiol), natural progesterone, DIM, and Iodoral. I think you are getting the idea.

Ranking Hormone Formulas by Carcinogenicity

Most Carcinogenic: oral estradiol combined with oral norethisterone **while on tamoxifen**. This is the HABITS trial finding 3-fold greater breast cancer recurrence in the HRT group, and all were using tamoxifen, an estrogen blocker.

Note: If only the HABITs trial had added testosterone to the hormone cocktail! The addition of testosterone reverses the proliferation induced by this hormone combination, as shown by Marie Hofling, 2007. (53)

Less Carcinogenic: Oral estradiol plus oral medroxyprogesterone (Dr. Agnes Fournier French Cohort Study showed an increased risk of breast cancer (RR=**1.69).** (92)

Neutral: Oral Premarin plus oral medroxy-progesterone (The corrected 2002 first arm WHI study shows no increased breast cancer, once prior estrogen exposure is removed from the placebo group as discussed by Drs. Avrum Bluming and Howard Hodis, 2023.) (49-51)

Preventive: Oral Premarin Alone Greater Efficacy than Tamoxifen in ER-negative Breast Cancer: In 1999, Dr. Rena Vassilopoulou-Sellin followed 319 breast cancer survivors prospectively followed for four years. 39 of these were treated with **oral Premarin-alone** (CEE) 0.625 mg on days 1–25 of each month which reduced breast cancer recurrence to one-half the recurrence rate of non-users (1/39=**2.6%** vs. 14/280=**5.0%**). There were no deaths in the Premarin (CEE) treated group. Among the 39 patients in this study, 75% were ER-negative and the other 25% had unknown ER status. **Note:** ER-negative breast cancers are usually not treated with Tamoxifen, as this drug is ineffective in ER-negative breast cancer. Hence, a 50% reduction in recurrence with Premarin is a marked improvement compared to the expected disease progression in all patients on Tamoxifen for ER-negative breast cancer. This study is remarkable because it shows a good outcome using estrogen (CEE) therapy for ER-negative cell types that do not respond to estrogen-blocking drugs. In 2016, Dr. Subrata Manna writes:

> many clinical studies involving **tamoxifen treatment in women with ER-negative breast cancers had not produced significant benefits in terms of mortality reduction or reduction of recurrence of breast cancer** in these patients. (93-94)

Total of 6644 node-positive breast cancer patients receiving adjuvant treatment.... For ER-negative tumors, chemotherapy improvements reduced the relative risk of recurrence by 21, 25 and 23 percent in the three studies, respectively, and 55 percent comparing the lowest dose in the first study with biweekly cycles in the third study. Corresponding relative risk reductions for ER-positive tumors treated with tamoxifen were 9, 12, and 8 percent in the three studies, and 26 percent overall. The overall mortality rate reductions associated with chemotherapy improvements were 55 and 23 percent among ER-negative and ER-positive patients, respectively. All individual ER-negative comparisons and no ER-positive comparisons were statistically significant. Absolute benefits due to chemotherapy were greater for patients with ER-negative compared with ER-positive tumors: 22.8 percent more ER-negative patients survived to 5 years disease-free if receiving chemotherapy versus 7.0 percent for ER-positive patients; corresponding improvements for overall survival were 16.7 versus 4.0 percent.(94)

In 2003, Dr. Janet Daling studied HRT in two groups of breast cancer survivors, the first group used **oral Premarin-alone** (CEE). The second group used **Premarin combined with medroxyprogesterone (MPA)**. The Premarin-alone group had a **30%** reduction in risk of being diagnosed with later-stage disease (regional or distant). However, the Premarin and medroxyprogesterone group (MPA) had **70%** increased recurrence (an odds ratio of **1.7**) compared to non-users. This 2002 study by Dr. Daling again shows Premarin reduces breast cancer recurrence. However, adding a synthetic progestin, MPA reverses this effect and increases recurrence by 70 percent. (95)

In 2018, Dr. Zexian Zeng from the Northwestern University Feinberg School of Medicine searched the Northwestern Medicine Enterprise Data Warehouse for menopausal HRT use. Dr. Zexian Zeng studied breast cancer risk associated with various HRT preparations, finding the lowest reduced risk with Premarin (CEE) alone (**HR =0.31**), followed by the next lowest reduced risk with bioidentical estrogen (estradiol E2) (**HR=0.65)**. Synthetic estrogen showed the highest risk of **HR= 1.49**. Dr. Zexian Zeng's study shows Premarin-alone use is associated with an impressive **69 percent reduction** in risk for breast cancer. This result far exceeds the results obtained with estrogen-blocking drugs, tamoxifen and AI. Dr. Zexian Zeng writes:

> Significant results include CEE [Premarin] Alone is associated with decreased breast cancer risk **(HR =0.31)**, other Synthetic Estrogen Alone is associated with increased breast cancer risk (**HR =1.49**), Bioidentical Estrogen Alone is associated with decreased breast cancer risk (**HR =0.65)**. (96)

Most Preventive: Oral estrogen alone. Dr. Rowan T. Chlebowski MD PhD is Chief of Medical Oncology and Hematology at the Harbor-UCLA Medical Center and was one of the Women's Health Initiative (WHI) investigators. In 2024, Dr. Chlebowski published a meta-analysis of 10 randomized trials of breast cancer incidence in 14,282 participants using estrogen alone. Dr. Rowan T. Chlebowski concludes: "**The totality of randomized clinical trial evidence supports a conclusion that estrogen-alone use significantly reduces breast cancer incidence**," writing:

> Findings from 10 randomized trials included 14,282 participants and 591 incident breast cancers. In 9 smaller trials, with 1.2% (24 of 2029) vs 2.2% (33 of 1514) randomized to estrogen-alone vs placebo (open label, one trial) (RR 0.65 95% CI 0.38-1.11, P = 0.12). For 5 trials evaluating **estradiol formulations**, RR = 0.**63** 95% CI 0.34-1.16, P = 0.15. Combining the 10 trials, 3.6% (262 of 7339) vs 4.7% (329 of 6943) randomized to estrogen-alone vs placebo (overall RR 0.77 95% CI 0.65-0.91, P = 0.002). Conclusion: **The totality of randomized clinical trial evidence supports a conclusion that estrogen-alone use significantly reduces breast cancer incidence. (97)**

Most Preventive: Adding Testosterone to the HRT:

In 1999, Puthgraman K Natrajan from Augusta, Georgia studied 50 breast cancer survivors on HRT which included testosterone pellets in 40 of the 50 patients. The HRT/testosterone users had an **81% reduction in mortality compared to non-users.** (60)

Most Preventive: Vaginal Estradiol-alone or Vaginal Premarin-alone are most breast cancer preventive. The unsaturated B ring steroids bind to ER-beta making Premarin (CEE) breast cancer preventive, and less proliferative than estradiol.

In 2001, Dr. George Peters from Texas studied 8 patients using vaginal estrogen alone after breast cancer followed for 11.4 years. There were no cancer recurrences and no cancer deaths in this group. (61)

In 2002, Dr. Eva Durna from Australia reported a group of 32 vaginal estrogen users with an **82 percent reduction** in breast cancer recurrence compared to non-users. (13)

Most Preventive: Vaginal Biest/Progesterone with Testosterone.

Bi-est contains 20 percent estradiol E2, and 80 percent estriol (E3). The E3 binds preferentially to ER-Beta and is breast cancer preventive. Progesterone and Testosterone both act as a proliferative break on ER-alpha activity. Testosterone metabolite 3-beta-Adiol preferentially binds to ER-beta. The vaginal route of delivery may more closely mimic the protective hormone levels of pregnancy. This is the preferred formula used in my office.

The Big Lump Together

So, now we can see breast cancer outcomes will vary depending on exact hormone formulations. It will now be obvious that lumping together all hormone formulations into one category and calling it HRT, is not a good idea, as this blurs the important distinctions between formulations. When reviewing the 26 studies of HRT compiled by Dr. Bluming, and the totality of medical literature on hormone replacement, one is struck by the failure to understand the differences between hormone formulations. Instead, what we find is a tendency to lump together natural progesterone with synthetic progestins, and to lump together different estrogens, Premarin (CEE, horse estrogen), estradiol (E2), and Estriol (E3). Many of the authors of the above 26 studies of HRT in breast cancer survivors describe the HRT formula as estrogen and progestogen, words which lump together different estrogens, and lump together natural progesterone with synthetic progestins. As you have seen with the above discussion, there is considerable variations in breast cancer outcomes depending on the exact hormone formulation. When looking for a health care provider, it is advisable to seek a physician knowledgeable about the various HRT formulations, which formulas are safest, and which formulas to avoid. (98-102)

Future Studies of Vaginal Estrogen in Breast Cancer Survivors

I would suggest a future prospective controlled study of HRT in breast cancer survivors with the following modifications: Use a large enough number of patients to reach statistical significance. Use vaginal estriol (E3)/ estradiol (E2) in a 4:1 ratio. This is called Bi-est, first pioneered by Dr. Jonathan Wright. The estriol (E3) binds preferentially to ER-beta, the tumor suppressor receptor. The Bi-est should be combined with natural progesterone, rather than synthetic progestins MPA or norethisterone. Testosterone should be included in the formula. Give nutritional supplements, such as iodine and DIM, as described below. (103-105)

The final hormone to add is testosterone, which downregulates ER-alpha activity and its metabolite 3beta-diol selectively binds to ER-beta and is highly breast cancer preventive. What about providing supplements to HRT users? Both Iodine and DIM divert estrogen metabolism towards beneficial pathways, and these should be included. Low vitamin D and low selenium levels are associated with increased cancer risk. Why not check these levels and make sure all patients are given supplements if deficient? This is our office protocol. It is imperative that natural progesterone rather than synthetic progestins are used for all patients regardless of previous hysterectomy. Synthetic progestins are known to be carcinogenic and should not be used, especially when natural progesterone is widely available. A prospective controlled trial is suggested because a randomized controlled trial (RCT) may not be feasible because "one cannot offer a placebo for a symptomatic climacteric patient wishing to receive ERT (estrogen replacement therapy)" as stated in 2001 by Dr. Merja Marttunen. The two Swedish studies, HABITS and Stockholm) were randomized prospective trials, not randomized placebo-controlled (106)

Conclusion: Credit and thanks go to Dr. Avrum Bluming for compiling the 26 studies of HRT in breast cancer survivors. By reviewing these studies, we have gained a greater understanding of how various HRT components affect breast cancer risk. The first lesson from HABITS and Stockholm studies is to avoid synthetic progestins which are carcinogenic. Use natural progesterone for endometrial protection, instead. Progesterone acts as a proliferative brake on ER-alpha, and is breast cancer preventive. Progestins do the exact opposite. They activate ER-alpha and its downstream oncogenes. The second lesson is testosterone selectively binds to ER-beta and is strongly breast cancer preventive. Testosterone is capable of reversing the breast proliferation induced by the combination of oral estradiol E2 and norethisterone, the ill-fated combination causing 3-fold increase in breast recurrence in the HABITS trial. The third lesson is the unique safety profile of vaginal estrogen. The fourth lesson is the value of LTED, long term estrogen deprivation. The fifth lesson is the breast cancer protection conferred by high hormone levels of pregnancy or previous estrogen exposure. The sixth lesson is the importance of using agents that bind preferentially to ER-beta, the tumor suppressor, such as Premarin, CEE, with its unsaturated B-ring steroids, estriol (E3), and testosterone. The seventh lesson is to add DIM and Iodine to divert estrogen metabolism towards beneficial pathways, away from the 4-hydroxy-quinones, and towards 2-methoxy-estradiol. Credit and thanks go to Dr. Lindsey Berkson and her teaching course on "Everything Hormones" for many of the HRT insights used in writing this chapter. (47-48) (55-56) (107-109)

♦ References for Chapter 19 Hormone Replacement for Breast Cancer Survivors Part Two

1) Bluming, Avrum Zvi. "Hormone Replacement Therapy After Breast Cancer: It Is Time." The Cancer Journal 28.3 (2022): 183-190.

2) Bluming, Avrum Z. "Progesterone and Breast Cancer Pathogenesis." Journal Of Molecular Endocrinology 66.1 (2021): C1-C2.

3) Bluming, Avrum Z, Estrogen Matters, Little, Brown Spark; 1st edition (September 4, 2018)

4) Collaborative Group on Hormonal Factors in Breast Cancer. "Type And Timing of Menopausal Hormone Therapy and Breast Cancer Risk: Individual Participant Meta-Analysis of The Worldwide Epidemiological Evidence." The Lancet 394.10204 (2019): 1159-1168.

5) Fournier, Agnes, et al. "Postmenopausal Hormone Therapy Initiation Before and After the Women's Health Initiative in Two French Cohorts." Menopause (New York, NY) 18.2 (2011): 219-223.

6) Marchant, Douglas J. Estrogen-Replacement Therapy After Breast Cancer. Risks Versus Benefits. Cancer 1993;71(6 Suppl):2169–76.

7) Berkson, Devaki Lindsey. "Estrogen Vindication, Part 1: Estrogen and the WHI." Townsend Letter 445-446 (2020): 36-40.

8) Berkson, Devaki Lindsey. "Estrogen Vindication, Part 2: Estrogen, Cancer Stem Cells, and Studies." Townsend Letter 447 (2020): 60-66. Berkson Devaki Lindsey Estrogen Vindication Part 1 and 2 Townsend Letter 445-446 2020 36-40

9) Menopause, Breast Cancer, and What Comes Next, Conversation with Dr. Avrum Bluming. September 5, 2024 Dr. Corrine Menn, Monica Molinarr. https://youtu.be/2QpaYmTWSvs

10) Laing, Abbie J., Louise Newson, and James A. Simon. "Individual Benefits and Risks of Intravaginal Estrogen and Systemic Testosterone in the Management of Women in The Menopause, with a Discussion of Any Associated Risks for Cancer Development." The Cancer Journal 28.3 (2022): 196-203.

11) Palshof T, Mouridsen HT, Daehnfeldt JL. Adjuvant Endocrine Therapy of Breast Cancer—A Controlled Clinical Trial of Estrogen and Anti-estrogen: Preliminary Results of the Copenhagen Breast Cancer Trials. Recent Results Cancer Res. 1980; 71:185–189.

12) Palshof T, Carstensen B, Mouridsen HT, et al. Adjuvant Endocrine Therapy in Pre- and Postmenopausal Women with Operable Breast Cancer. Reviews on. Endocr Relat Cancer. 1985;(suppl 17):43–50. No abstract available

13) Durna, Eva M., et al. "Hormone Replacement Therapy After a Diagnosis of Breast Cancer: Cancer Recurrence and Mortality." Medical Journal of Australia 177.7 (2002): 347-351.

14) Guidozzi F. Estrogen replacement therapy in breast cancer survivors. Int J Gynecol Obstet. 1999; 64:59–63.

15) O'Meara, Ellen S., et al. "Hormone Replacement Therapy After a Diagnosis of Breast Cancer in Relation To Recurrence And Mortality." Journal of the National Cancer Institute 93.10 (2001): 754-761.

16) Bluming AZ. Hormone Replacement Therapy (HRT) In Women with Previously Treated Primary Breast Cancer: Update XIV. Proc ASCO. J Clin Oncol.2008;15 s:20693.

17) Lippman, Marc, et al. "The Effects of Estrogens and Antiestrogens on Hormone-Responsive Human Breast Cancer in Long-Term Tissue Culture." Cancer research 36.12 (1976): 4595-4601.

18) Ly, Diana, et al. "An International Comparison of Male and Female Breast Cancer Incidence Rates." International Journal of Cancer 132.8 (2013): 1918-1926.

19) Owens, Douglas K., et al. "Medication Use to Reduce Risk of Breast Cancer: US Preventive Services Task Force Recommendation Statement." JAMA 322.9 (2019): 857-867.

20) Ellis, Matthew J., et al. "Lower-Dose Vs High-Dose Oral Estradiol Therapy of Hormone Receptor–Positive, Aromatase Inhibitor–Resistant Advanced Breast Cancer: A Phase 2 Randomized Study." Jama 302.7 (2009): 774-780.

21) Bennink, Herjan JT Coelingh, et al. "The Use of High-Dose Estrogens for The Treatment of Breast Cancer." Maturitas 95 (2017): 11-23.

22) Lønning, Per E., et al. "High-Dose Estrogen Treatment in Postmenopausal Breast Cancer Patients Heavily Exposed to Endocrine Therapy." Breast Cancer Research and Treatment 67 (2001): 111-116.

23) Haddow, Alexander, et al. "Influence of Synthetic Estrogens on Advanced Malignant Disease." British Medical Journal 2.4368 (1944): 393.

24) Shete, Nivida, Jordan Calabrese, and Debra A. Tonetti. "Revisiting Estrogen for the Treatment of Endocrine-Resistant Breast Cancer: Novel Therapeutic Approaches." Cancers 15.14 (2023): 3647.

25) Jordan, V. Craig. "The New Biology of Estrogen-Induced Apoptosis Applied to Treat and Prevent Breast Cancer." Endocrine-Related Cancer 22.1 (2015): R1.

26) Ingle, James N., et al. "Randomized Clinical Trial of Diethylstilbestrol Versus Tamoxifen In Postmenopausal Women With Advanced Breast Cancer." New England Journal of Medicine 304.1 (1981): 16-21.

27) Peethambaram, Prema P., et al. "Randomized Trial of Diethylstilbestrol Vs. Tamoxifen in Postmenopausal Women with Metastatic Breast Cancer. An Updated Analysis." Breast Cancer Research and Treatment 54 (1999): 117-122.

28) Traphagen, Nicole A., et al. "High Estrogen Receptor Alpha Activation Confers Resistance to Estrogen Deprivation and Is Required For Therapeutic Response To Estrogen In Breast Cancer." Oncogene 40.19 (2021): 3408-3421.

29) V. Craig Jordan, PhD, a Founding Father of Targeted Therapy in Cancer, Dies at Age 76 By Ronald Piana July 10, 2024. The ASCO Post. https://ascopost.com/issues/july-10-2024/v-craig-jordan-a-founding-father-of-targeted-therapy-in-cancer-dies-at-age-76/

30) Sengupta, S., et al. "Molecular Mechanism of Action of Bisphenol and Bisphenol a Mediated By Estrogen Receptor Alpha In Growth And Apoptosis Of Breast Cancer Cells." British Journal of Pharmacology 169.1 (2013): 167-178.

31) Iwamoto, Masaki, et al. "Bisphenol A Derivatives Act as Novel Coactivator-Binding Inhibitors for Estrogen Receptor Beta." Journal of Biological Chemistry 297.5 (2021).

32) Khan, Nadeem Ghani, et al. "A Comprehensive Review on The Carcinogenic Potential of Bisphenol A: Clues and Evidence." Environmental Science and Pollution Research 28 (2021): 19643-19663.

33) Giulianelli, Sebastián, et al. "Estrogen Receptor Alpha Mediates Progestin-Induced Mammary Tumor Growth by Interacting with Progesterone Receptors at The Cyclin D1/MYC Promoters." Cancer Research 72.9 (2012): 2416-2427.

34) O'Malley, Bert W., and Sohaib Khan. "Elwood V. Jensen (1920–2012): Father of The Nuclear Receptors." Proceedings of the National Academy of Sciences 110.10 (2013): 3707-3708.

35) Gustafsson, Jan-Åke. "Estrogen Receptor beta—A Multifaceted Player." Women's Health and Menopause: New Strategies—Improved Quality of Life. Boston, MA: Springer US, 2002. 1-5.

36) Koehler, Konrad F., et al. "Reflections on the Discovery and Significance of Estrogen Receptor Beta." Endocrine Reviews 26.3 (2005): 465-478.

37) Mohammed, Hisham, et al. "Progesterone Receptor Modulates ER-alpha Action in Breast Cancer." Nature 523.7560 (2015): 313-317.

38) Ganzina, Fabrizio. "High-Dose Medroxyprogesterone Acetate (MPA) Treatment In Advanced Breast Cancer. A Review." Tumori Journal 65.5 (1979): 563-585.

39) Bentel, Jacqueline M., et al. "Androgen Receptor Agonist Activity of The Synthetic Progestin, Medroxyprogesterone Acetate, In Human Breast Cancer Cells." Molecular and cellular endocrinology 154.1-2 (1999): 11-20.

40) Wood, Charles E., et al. "Effects of Estradiol with Micronized Progesterone or Medroxyprogesterone Acetate on Risk Markers for Breast Cancer in Postmenopausal Monkeys." Breast cancer research and treatment 101 (2007): 125-134.

41) Birrell, Stephen N., et al. "Disruption of Androgen Receptor Signaling by Synthetic Progestins May Increase Risk of Developing Breast Cancer." The FASEB Journal 21.10 (2007): 2285-2293.

42) Mukhopadhyay, Keya De, et al. "Aromatase Expression Increases the Survival and Malignancy Of Estrogen Receptor Positive Breast Cancer Cells." Plos One 10.4 (2015): E0121136.

43) Maximov, Philipp Y., et al. "Estrogen Receptor Complex to Trigger or Delay Estrogen-Induced Apoptosis in Long-Term Estrogen Deprived Breast Cancer." Frontiers in Endocrinology 13 (2022): 869562.

44) Santen, Richard J., et al. "Adaptation to Estradiol Deprivation Causes Up-Regulation of Growth Factor Pathways and Hypersensitivity to Estradiol in Breast Cancer Cells." Innovative endocrinology of cancer (2008): 19-34.

45) Feigman, Mary J., et al. "Pregnancy reprograms the epigenome of mammary epithelial cells and blocks the development of premalignant lesions." Nature Communications 11.1 (2020): 2649.

46) Subramani, Ramadevi, et al. "Pregnancy Inhibits Mammary Carcinogenesis by Persistently Altering the Hypothalamic–Pituitary Axis." Cancers 13.13 (2021): 3207.

47) Barton, Maria, Julia Santucci-Pereira, and Jose Russo. "Molecular Pathways Involved In Pregnancy-Induced Prevention Against Breast Cancer." Frontiers in Endocrinology 5 (2014): 213.

48) Katz, Tiffany A. "Potential Mechanisms Underlying the Protective Effect of Pregnancy Against Breast Cancer: A Focus on the IGF Pathway." Frontiers in Oncology 6 (2016): 228.

49) Bluming, Avrum Z., Howard N. Hodis, and Robert D. Langer. "'Tis but a Scratch: A Critical Review of The Women's Health Initiative Evidence Associating Menopausal Hormone Therapy with The Risk of Breast Cancer." Menopause (2023): 10-1097.

50) Feigman, Mary J., et al. "Pregnancy Reprograms the Epigenome of Mammary Epithelial Cells and Blocks the Development of Premalignant Lesions." Nature Communications 11.1 (2020): 2649.

51) Hodis, Howard N., and P. M. Sarrel. "Menopausal Hormone Therapy and Breast Cancer: What Is the Evidence from Randomized Trials?" Climacteric 21.6 (2018): 521-528.

52) Cavalieri, Ercole, and Eleanor Rogan. "The 3, 4-Quinones of Estrone and Estradiol Are the Initiators of Cancer Whereas Resveratrol And N-Acetylcysteine Are the Preventers." International Journal of Molecular Sciences 22.15 (2021): 8238.

53) Hofling, Marie, et al. "Testosterone inhibits estrogen/progestogen-induced breast cell proliferation in postmenopausal women." Menopause: The Journal of The North American Menopause Society 14.2 (2007): 183-190.

54) Glaser, Rebecca L., Anne E. York, and Constantine Dimitrakakis. "Incidence of invasive breast cancer in women treated with testosterone implants: a prospective 10-year cohort study." BMC cancer 19.1 (2019): 1271.

55) Glaser, Rebecca L., A. E. York, and C. Dimitrakakis. "Abstract P6-13-02: Reduced incidence of breast cancer with testosterone implant therapy: A 10-year cohort study." Cancer Research 79.4_Supplement (2019): P6-13.

56) Donovitz, Gary, and Mandy Cotten. "Breast cancer incidence reduction in women treated with subcutaneous testosterone: testosterone therapy and breast cancer incidence study." European journal of breast health 17.2 (2021): 150.

57) Peters, Amelia A., et al. "Androgen Receptor Inhibits Estrogen Receptor-Alpha Activity and Is Prognostic in Breast Cancer." Cancer Research 69.15 (2009): 6131-6140.

58) Eigeliene N, Elo T, LinhalaM, et al. Androgens Inhibit the Stimulatory Action Of 17-beta-Estradiol On Normal Human Breast Tissue In Explant Cultures. J Clin Endocrinol Metab. 2012;97: E1116–E1127.

59) Boni, Corrado, et al. "Therapeutic Activity of Testosterone In Metastatic Breast Cancer." Anticancer Research 34.3 (2014): 1287-1290.

60) Natrajan, Puthgraman K., et al. "Estrogen Replacement Therapy in Women with Previous Breast Cancer." American Journal of Obstetrics and Gynecology 181.2 (1999): 288-295.

61) Peters, George N., et al. "Estrogen Replacement Therapy After Breast Cancer: A 12-Year Follow-Up." Annals Of Surgical Oncology 8 (2001): 828-832.

62) Lyytinen H, et al. Breast Cancer Risk in Postmenopausal Women Using Estrogen-Only Therapy. Obstet Gynecol. 2006;108(6):1354-1360.

63) Le Ray, Isabelle, et al. "Local Estrogen Therapy and Risk of Breast Cancer Recurrence Among Hormone-Treated Patients: A Nested Case-Control Study." Breast cancer research and treatment 135.2 (2012): 603-609.

64) McVicker, Lauren, et al. "Vaginal Estrogen Therapy Use and Survival In Females With Breast Cancer." JAMA Oncology 10.1 (2024): 103-108.

65) Mal, Rahul, et al. "Estrogen Receptor Beta (ER-Beta): A Ligand Activated Tumor Suppressor." Frontiers in Oncology 10 (2020): 587386.

66) Beste, Mary E., et al. "Vaginal Estrogen Use in Breast Cancer Survivors: A Systematic Review and Meta-Analysis of Recurrence and Mortality Risks." American Journal of Obstetrics and Gynecology (2024).

67) Agrawal, Pranjal, et al. "Safety of Vaginal Estrogen Therapy for Genitourinary Syndrome of Menopause in Women with A History Of Breast Cancer." Obstetrics & Gynecology (2022): 10-1097.

68) Moegele, M., et al. "Vaginal Estrogen Therapy for Patients with Breast Cancer." Geburtshilfe und Frauenheilkunde 73.10 (2013): 1017-1022.

69) Dew, Jennifer. E., Barry G. Wren, and John A. Eden. "A Cohort Study of Topical Vaginal Estrogen Therapy in Women Previously Treated For Breast Cancer." Climacteric 6.1 (2003): 45-52.

70) Friel, Patrick N., Christa Hinchcliffe, and Jonathan V. Wright. "Hormone Replacement with Estradiol: Conventional Oral Doses Result in Excessive Exposure to Estrone." Altern Med Rev 10.1 (2005): 36-41.

71) Partridge, Ann H., et al. "Interrupting Endocrine Therapy to Attempt Pregnancy After Breast Cancer." New England Journal of Medicine 388.18 (2023): 1645-1656.

72) Wile AG, Opfell RW, Margileth DA. Hormone Replacement Therapy in Previously Treated Breast Cancer Patients. Am J Surg 1993; 165:372-5.

73) Lambrinoudaki, Irene. "Menopausal Hormone Therapy and Breast Cancer Risk: All Progestogens Are Not The Same." Case Reports in Women's Health 29 (2021).

74) Vinogradova Y. Use of Hormone Replacement Therapy and Risk of Breast Cancer: Nested Case-Control Studies Using the Qresearch And CPRD Databases. BMJ. 2020 Oct 28;371

75) Santen, R. J., and W. Yue. "Cause or Prevention of Breast Cancer with Estrogens: Analysis from Tumor Biologic Data, Growth Kinetic Model and Women's Health Initiative study." Climacteric 22.1 (2019): 3-12.

76) Holmberg L, Anderson H. HABITS (Hormonal Replacement Therapy After Breast Cancer—Is It Safe?), A Randomised Comparison: Trial Stopped. Lancet. 2004; 363:453–455.

77) Holmberg, Lars, et al. "Increased Risk of Recurrence After Hormone Replacement Therapy in Breast Cancer Survivors." Journal of the National Cancer Institute 100.7 (2008): 475-482.

78) Hsieh, Robert W., et al. "Molecular Characterization of a B-Ring Unsaturated Estrogen: Implications for Conjugated Equine Estrogen Components of Premarin." Steroids 73.1 (2008): 59-68.

79) Bhavnani, Bhagu R., et al. "Structure-Activity Relationships and Differential Interactions and Functional Activity Of Various Equine Estrogens Mediated Via Estrogen Receptors (ERs) ER-alpha and ER-beta." Endocrinology 149.10 (2008): 4857-4870.

80) Flores, Valerie A., and Hugh S. Taylor. "The Effect of Menopausal Hormone Therapies on Breast Cancer: Avoiding The Risk." Endocrinology and Metabolism Clinics 44.3 (2015): 587-602.

81) Cline, J. M., et al. "Effects of Hormone Replacement Therapy on The Mammary Gland of Surgically Postmenopausal Cynomolgus Macaques." American Journal of Obstetrics and Gynecology 174.1 Pt 1 (1996): 93-100.

82) Reid SE, et al. Endocrine and Paracrine Hormones in The Promotion, Progression And Recurrence Of Breast Cancer. Br J Surg 1996; 83: 1037 – 46.

83) Santen RJ, et al. Risk of Breast Cancer with Progestins in Combination with Estrogen As Hormone Replacement Therapy. J Clin Endocrinol Metab. 2001; 86: 16 – 23.

84) Weiss LK, et al. Hormone Replacement Therapy Regimens and Breast Cancer Risk. Obstet Gynecol 2002; 100: 1148 – 58.

85) Olsson HL, Ingvar C, Bladström A. Hormone Replacement Therapy Containing Progestins and Given Continuously Increases Breast Carcinoma Risk in Sweden. Cancer 2003; 97: 1387 – 92.

86) Chlebowski RT, et al. Influence of Estrogen Plus Progestin on Breast Cancer And Mammography In Healthy Postmenopausal Women: the Women's Health Initiative Randomized Trial. JAMA 2003; 289: 3243 – 53.

87) von Schoultz Eva, Rutqvist LE, Stockholm Breast Cancer Study Group. Menopausal Hormone Therapy After Breast Cancer: The Stockholm Randomized Trial. J Natl Cancer Inst. 2005; 97:533–535.

88) Fahlén M, Fornander T, Johansson H, et al. Hormone Replacement Therapy After Breast Cancer: 10 Year Follow-Up of The Stockholm Randomized Trial. Eur J Cancer. 2013; 49:52–59.

89) Saltelli, Andrea, et al. "Science, the Endless Frontier of Regulatory Capture." Futures 135 (2022): 102860.

90) Nelson, Mark M. "What is to be Done? Options For Combating the Menace of Media Capture." In the Service of Power: Media Capture and the Threat to Democracy (2017): 143.

91) Abraham, John, and Rachel Ballinger. "Science, Politics, And Health in The Brave New World of Pharmaceutical Carcinogenic Risk Assessment: Technical Progress Or Cycle Of Regulatory Capture?" Social Science & Medicine 75.8 (2012): 1433-1440.

92) Fournier, Agnès, et al. "Unequal Risks For Breast Cancer Associated With Different Hormone Replacement Therapies: Results From The E3N Cohort Study." Breast Cancer Research And Treatment 107 (2008): 103-111.

93) Manna, Subrata, and Marina K. Holz. "Tamoxifen Action in ER-Negative Breast Cancer." Signal transduction insights 5 (2016): 1.

94) Berry, Donald A., et al. "Estrogen-receptor status and outcomes of modern chemotherapy for patients with node-positive breast cancer." Jama 295.14 (2006): 1658-1667.

95) Daling, Janet R., et al. "Association of Regimens of Hormone Replacement Therapy to Prognostic Factors Among Women Diagnosed with Breast Cancer Aged 50–64 Years." Cancer Epidemiology Biomarkers & Prevention 12.11 (2003): 1175-1181.

96) Zeng, Zexian, et al. "Conjugated Equine Estrogen and Medroxyprogesterone Acetate Are Associated with Decreased Risk Of Breast Cancer Relative To Bioidentical Hormone Therapy And Controls." PLoS One 13.5 (2018): e0197064.

97) Chlebowski, Rowan T., et al. "Randomized Trials of Estrogen-Alone and Breast Cancer Incidence: A Meta-Analysis." Breast Cancer Research and Treatment (2024): 1-8.

98) Fournier, Agnes, et al. "Breast Cancer Risk in Relation To Different Types Of Hormone Replacement Therapy In The E3N-EPIC Cohort." International Journal of Cancer 114.3 (2005): 448-454.

99) Fournier, Agnès, et al. "Unequal Risks for Breast Cancer Associated with Different Hormone Replacement Therapies: Results from The E3N Cohort Study." Breast Cancer Research and Treatment 107 (2008): 103-111.

100) Fournier, Agnès, et al. "Use of Different Postmenopausal Hormone Therapies and Risk of Histology-And Hormone Receptor–Defined Invasive Breast Cancer." Journal of Clinical Oncology 26.8 (2008): 1260-1268.

101) Fournier, Agnès, et al. "Estrogen-Progestagen Menopausal Hormone Therapy and Breast Cancer: Does Delay from Menopause Onset to Treatment Initiation Influence Risks?" Journal of Clinical Oncology 27.31 (2009): 5138-5143.

102) Fournier, Agnès, et al. "Risk of Breast Cancer After Stopping Menopausal Hormone Therapy in The E3N Cohort." Breast Cancer Research and Treatment 145 (2014): 535-543.

103) Marsden, Tracy. "Bioidentical Hormone Replacement: Guiding Principles for Practice." Nat Med J 2.3 (2010): 2010-03.

104) Plu-Bureau G, Le M, Thalabard J, et al. Percutaneous Progesterone Use and Risk Of Breast Cancer: Results From A French Cohort Study Of Premenopausal Women With Benign Breast Disease. Cancer Detect Prev. 1999; 23:290-296.

105) L'Hermite M, et al. Could Transdermal Estradiol + Progesterone Be a Safer Postmenopausal HRT? A Review. Maturitas. 2008; 60:185-201.

106) Marttunen, Merja B., et al. "A Prospective Study on Women with a History of Breast Cancer and With Or Without Estrogen Replacement Therapy." Maturitas 39.3 (2001): 217-225.

107) Lieberman, Allan, and Luke Curtis. "In Defense of Progesterone: A Review of The Literature." Alternative Therapies in Health & Medicine 23.7 (2017).

108) Ferretti, Gianluigi, Alessandra Felici, and Francesco Cognetti. "The Protective Side of Progesterone." Breast Cancer Research 9 (2007): 1-1.

109) Stevenson, John C., et al. "Progestogens as a Component of Menopausal Hormone Therapy: The Right Molecule Makes the Difference." Drugs in Context 9 (2020).

110) Eden JA, Bush T, Nand S, Wren BG. A case–control study of combined continuous estrogen–progestin replacement therapy among women with personal history of breast cancer. J North Am Menopause Soc 1995; 2: 67-72

Chapter 20

Adjuvant Endocrine Therapy, Dr. Jekyll and Mr. Hyde

MARJORIE IS A 54-YEAR-OLD STAY-AT-HOME mom who had a lumpectomy and radiation therapy for breast cancer about four years ago. Ann had been taking an estrogen-blocking drug called letrozole, an aromatase inhibitor (AI) that inhibits the production of estrogen, leading to undetectable serum estrogen levels. The letrozole estrogen-blocking drug is causing severe menopausal symptoms, vaginal dryness, recurrent urinary tract infections, hot flashes, night sweats, and insomnia. Marjorie is so miserable, she is thinking of stopping the estrogen-blocking drug and starting hormone replacement, a plan vigorously opposed by her oncologist. In this Chapter, we review the mainstream oncology treatment of breast cancer with the two most common estrogen-blocking drugs, tamoxifen (TAM) and aromatase inhibitors (AIs). Even mainstream medicine recognizes the Dr. Jekyll and Mr. Hyde qualities of this class of medications, in some cases, they help and in other cases, they harm. The 5-year survival rate for breast cancer was approximately 90% in the U.S. from 2012 to 2018. However, this survival rate is only 28% if there is metastatic spread, suggesting a need for better treatments in patients with metastatic disease. Note: tamoxifen is an estrogen receptor blocker, while AIs prevent the conversion of testosterone to estrogen. In 2022, Dr. Evelina Arzanova writes:

> The highest 5-year survival rate (99.1%) in the USA is seen in BC cases with tumors localized to the initial place of origin (31). In comparison, BC tumors with a regional lymph node spread have a 29% survival rate while women with metastatic BC have only a 6% rate of survival (1-3).

The History of Breast Cancer and Its Treatment

In 1896, Dr. George Beatson was the first to observe oophorectomy (surgical removal of ovaries) induced a favorable tumor response in a premenopausal patient with metastatic breast cancer. The oophorectomy procedure removed the source of estrogen production, creating an estrogen depletion state, thus oophorectomy and estrogen depletion became mainstream treatment of breast cancer. By 1944, the oophorectomy cases were compiled and reviewed by Dr. Alexander Haddow from the U.K. finding a **tumor response rate of about 30 percent**. Despite this low efficacy rate, a 30 percent response rate became the "magic number", and the future outcome for all endocrine therapies to follow. Sadly, tumor responses after oophorectomy were only transient. This state of affairs left much to be desired, leading to the search for better treatment. (4-5)

High Dose Synthetic Estrogen as Treatment for Breast Cancer 1944-1972

In 1944, medicine reversed course, switching from estrogen depletion to the opposite treatment, giving high-dose estrogen to women with metastatic breast cancer. Dr. Alexander Haddow reported treatment of metastatic breast cancer with high-dose synthetic estrogen (DES) provided a **30 percent tumor response rate**. Note this 30 percent response rate, "magic number" is the same for oophorectomy. DES treatment was quickly adopted, replacing oophorectomy, since it gave the same 30 percent response rate without submitting the patient to a surgical procedure. DES seemed to work best in women more than 5 years post-menopause, women who had been deprived of estrogen for the long term. **Note:** first synthesized in 1938, DES is diethylstilbestrol, an endocrine-disrupting chemical, later banned in 1971 in the U.S. when it was discovered to cause clear cell vaginal and cervical cancer in daughters of mothers who used DES during pregnancy. Sons were also affected with reproductive problems. (6-8)

Invention of Tamoxifen and Aromatase Inhibitors in the early 1970s.

Since 1896, more than 128 years ago, estrogen hormones have been mistakenly regarded as causing breast cancer, thus leading to the idea that estrogen-blocking drugs should be the ideal prevention and treatment for breast cancer. Thus, in the early 1970s estrogen blocking drugs (anti-estrogens) were invented and adopted for prevention and treatment of breast cancer. Invented as a "morning after" contraceptive pill in 1962, tamoxifen languished as a failed drug. In 1972, Dr. V. Craig Jordan, the "father of tamoxifen", proposed the drug for the prevention and treatment of breast cancer. Tamoxifen is an estrogen receptor blocker, thus preventing the proliferative effects of estrogen signaling. Later, aromatase inhibitors (AIs) such as anastrozole, letrozole, and exemestane were invented. AIs prevent the conversion of testosterone to estrogen, the main pathway for estrogen production. Thus, creating a chemical form of oophorectomy. Dr. V. Craig Jordan also made the paradoxical discovery that once exhaustive anti-tamoxifen resistance develops in the breast cancer patient, estrogen treatment can now be used with success. The cancer cells are now sensitized to low-dose estrogen which induces apoptosis, programmed cell death. (9-10)

Opposite Hormonal Treatments Have the Same Efficacy

In 1981, clinical trials were done to answer the question, which drug is better for treating breast cancer, the estrogen receptor blocking drug tamoxifen, or the high dose synthetic estrogen drug DES? These two drugs have opposite mechanisms of action, yet both achieve the same result of a 30 percent tumor response rate. Despite equivalent efficacy, tamoxifen quickly replaced DES and became the standard of care because it was easier to tolerate with less patient drop-out. The estrogen treatment of breast cancer was sent to the medical museum and forgotten. **Note:** this magic number of 30 percent response rate for tamoxifen applies to all breast cancer cases grouped, both ER-positive and negative. When only estrogen receptor (ER) positive cases are considered, this number increases to 40-50 percent, as ER-negative cases rarely respond to tamoxifen. (11-12)

What is the Mechanism of Opposite Hormonal Manipulations?

The next obvious question is how can opposite endocrine treatments give the same results? And if anti-estrogen is the correct treatment for breast cancer, how do we explain the development of anti-estrogen resistance with the transformation of the cancer into a more aggressive cell type? What is the biological mechanism here? In 2015, Zsuzsanna Suba MD PhD, professor emeritus at Semmelweis University in Budapest, provided the answer. Medical science has misunderstood the connection between estrogen signaling and the development of breast cancer. Mainstream medicine ignores the role of estrogen in DNA stabilization and maintenance of genome stability. Estrogen works closely in association with the BRCA gene, a tumor suppressor gene involved in DNA repair, cell cycle control, apoptosis, and genome stability. (13-16)

Crosstalk Between Estrogen and DNA Repair

In 2022, Dr. Lia Yedidia-Aryeh studied the tight crosstalk between estrogen and DNA repair gene activation, finding estrogen is required for repairing DNA damage. In postmenopausal long-term estrogen-deprived (LTED) patients or patients treated with anti-estrogen drugs, the estrogen-deficient state will impair DNA repair mechanisms and lead to chronic DNA damage and mutation, a pro-carcinogenic mechanism, writing:

> We and others have shown that estrogen is required for intact DSB [double stranded break] [DNA] repair, since we found that DSB repair is impaired in ER-positive cells depleted for estrogen...Studies indicate that there is a tight crosstalk between estrogen and [DNA] repair. Estrogen regulates the expression of [DNA repair] factors and the activity of [DNA repair]. (17)

Cell survival depends on maintaining DNA repair mechanisms and genome integrity. This in turn, is dependent upon maintaining estrogenic signaling. Thus, all cells including tumor cells will desperately try to preserve estrogen signaling whenever threatened with estrogen deprivation. If genome integrity cannot be maintained in the damaged cell which is beyond repair, then estrogen signaling triggers apoptosis to eliminate the defective cell. This paradoxical estrogen-induced apoptosis after exhaustive tamoxifen treatment, or a period of long-term estrogen deprivation, was recognized early on by Dr. V. Craig Jordan. (18-20)

Explaining the Mechanism of How Opposing Endocrine Treatments Yield Similar Results

In 2015, Dr. Zsuzsanna Suba explained how opposing endocrine treatments yield similar results. Both low and high estrogen levels upregulate estrogen receptor signaling to maintain genome stability. ER-alpha signaling is the major player in maintaining genome stability, working closely with the BRCA genes. ER-alpha signaling triggers cancer cell apoptosis, i.e. programmed cell death. In 2015, Dr. Zsuzsanna Suba writes:

> Comprehensive analysis of experimental and clinical–epidemiologic results suggests that ER-alpha signaling is the chief safeguard of genome stability in strong interplay with DNA controlling and repairing systems, such as BRCA genes and their protein products...estrogen signaling recognizes and

destroys malignant tumor cells by means of apoptotic mechanisms...**Both low and high estrogen levels promote enhanced expression and transcriptional activity of ERs aiming to maintain the crucial cellular estrogen surveillance**. (21)

The Pitfalls of Anti-Estrogen Treatment

In 2015, Dr. Suba reviewed the pitfalls of antiestrogen treatment and the mechanism of antiestrogen resistance. The treatment of breast cancer cells by either estrogen or antiestrogen drug results in estrogen receptor overexpression making the tumor cells more sensitive to estrogen-induced apoptosis. In the 30 percent of breast cancer tumors that respond to tamoxifen administration, up-regulation of estrogen receptors is sufficient to supply the cancer cells with estrogen signaling, causing tumor regression from the apoptotic effects. However, if estrogen signaling is completely blocked, the tumor cell's capacity to compensate becomes exhausted. This is called tamoxifen resistance, leading to unrestrained tumor proliferation and metastatic spread. Even at this stage, all is not lost. The re-introduction of high-dose estrogen restores signaling and tumor regression. In 2015, Dr. Zsuzsanna Suba writes:

> ER overexpression in tumor cells treated with either estrogens or antiestrogens may be explained by the fundamental regulatory capacity of estrogens. These apparently contradictory treatments can similarly upregulate abundant ER expressions and transcriptional activities. Estrogen-induced upregulation of estrogen signaling is a physiologic process, **while in case of antiestrogen administration, it may be regarded as a counteraction for the defense of endangered cellular estrogen surveillance**...Increased estrogen signaling displays a unique dichotomy effect: it safeguards the survival and proliferative activity of healthy cells, **while induces apoptotic death of malignant tumor cells**... In patients with breast cancer, estrogen administration is capable of exerting

self-generating, increased ER expression and estrogen synthesis as well so as to achieve the strong upregulation of estrogen signaling and apoptotic tumor cell damage. Paradoxically, antiestrogen treatment may also provoke compensatory ER overexpression in tumor cells and extreme estrogen synthesis in the patient so as to **restore the apoptotic capacity of estrogen signaling**. When these counteractions are sufficient, tamoxifen administration seems to be deceivingly effective, resulting in transient tumor regression...By contrast, primary insufficiency or exhaustion of the defensive counteractions in patients treated with antiestrogens may lead to the complete blocking of estrogen signaling, since the patient has limited capacities for extreme ER expression and estrogen synthesis. In such cases, the result is unrestrained proliferative activity of the tumor cells, and the rapid tumor spread is mistakenly evaluated as de novo or acquired antiestrogen resistance. Nevertheless, high-dose estrogen treatment is capable of restoring the suppressed estrogen signaling even after heavy exposure to antiestrogen treatment. Exogenous and newly synthesized estrogens are in competition with tamoxifen for binding sites on newly expressed, abundant ERs, and the higher the concentration of estrogens, the higher is the probability of successful defeat of tamoxifen. **The triumph of estrogens leads to apoptotic tumor cell death and clinical regression of the disease**...In conclusion, during long-term antiestrogen treatment, the upregulation of both estrogen and ER synthesis exhibits gradual exhaustion and the development of complete antiestrogen blockade of estrogen signaling results in rapid tumor spread. Fortunately, high-dose estrogen treatment is capable of restoring the estrogen signaling even after exhaustive antiestrogen therapy with the achievement of rapid tumor regression. (22-28)

The Three Phases of Antiestrogen Administration

In 2015, Dr. Zsuzsanna Suba described the three phases of antiestrogen administration and the mechanism of antiestrogen resistance. In the first phase, preservation of estrogen signaling indicates successful antiestrogen therapy with tumor regression. In the second phase, the tumor appears stable with no progression. In the third and final phase, the tumor cell's ability to compensate by upregulating ERs and aromatase enzyme is exhausted leading to rapid proliferation, metastatic disease, and death, writing:

There are three phases of antiestrogen administration in genetically proficient breast cancer cases, treated by either ER blocker [tamoxifen] or aromatase inhibitor, which can be characterized by [first] good tumor regression, [second] stagnation of tumor growth, and [third] aggressive tumor spread...During this first period, **estrogen signaling may exhibit compensatory upregulation** in tamoxifen-treated premenopausal patients experiencing regular cycles and ovulatory activity with the potential to become pregnant...Moreover, in breast cancer biopsy specimens, active estrogen signaling characterized by intense phosphorylation at Ser(167) of ERs predicted longer disease-free survival and overall survival for the patients...[Note: When AKT phosphorylates Ser167 on the estrogen receptor (ER), this activates ER-alpha's transcriptional activity without the need for estrogen ligand binding. This massively upregulates estrogen signaling during periods of estrogen deprivation without depending upon serum estrogen levels.] In sequential biopsies of large primary breast tumors, measurement of aromatase content before and during effective treatment with aromatase inhibitor showed a surprisingly marked, counteractive **increase in [aromatase] enzyme activity [which converts testosterone into estrogen]**... These results support [the concept] that the **provoked compensatory upregulation** of estrogen signaling may be in correlation with successful tumor regression instead of an antiestrogenic effect...Tamoxifen treatment induces artificial estrogen resistance in women. These patients mimic the chaotic findings, which are characteristic of cases suffering from the genetic failures of ER expression and/or transcriptional activity coupled with counteractive defense mechanisms, such as extreme [increase in] estrogen synthesis...In tamoxifen-treated cases, very high compensatory estrogen levels and even an increased ER expression may be associated with the symptoms of ER blockade-induced estrogen resistance, such as multiple ovarian cysts or endometrial hyperplasia [which may lead to endometrial cancer]...On the other hand, exhaustive aromatase inhibitor treatment in breast cancer cases mimics the symptoms of aromatase deficiency syndrome deriving from the genetic defect of aromatase synthesis...In the case of this artificial aromatase deficiency, patients may exhibit even an increased counteractive expression in aromatase enzyme synthesis besides the overexpression of ERs. In the second phase of aromatase inhibitor treatment, apparent resistance to aromatase inhibitors is mistakenly regarded as an increased cross talk between intensified ER and GFR [growth factor receptor] signaling pathways...In the third phase of antiestrogen treatment, both extreme ER expression and aromatase synthesis are exhausted, and the completion of estrogen signal blockage results in rapid growth and metastatic spread of breast cancer leading to the death of the patient. At the same time, diverse toxic effects of estrogen deficiency and accidentally developing cancers at different sites, particularly in the endometrium [endometrial cancer], may be diagnosed. This phase of complete antiestrogen blockade of estrogen signaling is mistakenly referred to as acquired antiestrogen resistance...In the remaining half of patients [ER positive patients with no response to tamoxifen, i.e. de novo resistance], ...These cases do not have sufficient extra capacities for defensive ER overexpression

and increased estrogen synthesis against the artificial inhibition of estrogen signaling. In such patients, the failure of tumor prevention or regression by antiestrogen treatment is erroneously regarded as de novo resistance against antiestrogen treatment. (29-36)

Explaining What Went Wrong with the HABITS Study

In the previous chapter on Hormone Replacement for Breast Cancer Survivors, Part Two, we discussed the 26 studies compiled by Dr. Avrum Bluming. Of these 26, the only one showing increased breast cancer recurrence was the 2004 HABITS study by Dr. Lars Holmberg of Sweden. The HRT used in this study was estradiol (E2) combined with norethisterone, a highly carcinogenic progestin derived from a testosterone backbone. Additionally, all the patients with recurrent cancer in the HRT group were also taking tamoxifen, the antiestrogen drug that blocks estrogen receptors. The HRT users on tamoxifen had a 3-fold increase in cancer recurrence compared to non-users. Dr. Holmberg suggested the carcinogenic progestin, norethisterone was the main culprit. Now that we have learned about the three phases of antiestrogen resistance of Dr. Suba (above), we can speculate as to what went wrong with the HABITS study. Note HRT= hormone replacement therapy. (37-39)

Switching to Androgenic Signaling

Firstly, in the HABITS study patients with breast cancer recurrence were all taking tamoxifen which blocked the estrogen receptors, so giving exogenous estradiol (E2) is a futile gesture. The exogenous estrogen can not reach the blocked receptors. As mentioned above, the first phase of tamoxifen resistance is a desperate attempt by the breast cells to upregulate estrogen receptors and estrogen signaling. When this fails, the cancer cell switches from estrogenic to androgenic signaling. According to Dr. Charles Dai (2023), AR (androgen receptor) can bind to the same DNA regions as the ER (estrogen receptor). In 2021, Dr. Theresa E. Hickey did an in-vitro breast cancer study finding AR (androgen receptor protein) was detected at 42% of estrogen-stimulated ER-binding sites on chromatin. Thus, androgens are upregulated to replace deficient estrogen signaling during anti-estrogen treatment. In addition, androgen signaling, much like estrogen, can induce apoptosis within breast cancer cells. (40-45)

The Switch to Androgenic Signaling

This switch to androgenic signaling as an alternate pathway is supported by studies in 2014 by Dawn R. Cochrane PhD. She is an adjunct professor and Staff Scientist within the Department of Molecular Oncology at the University of British Columbia, and the University of Colorado in Denver. Dr. Cochrane studied ER-positive breast cancer tissue samples from 192 women in vitro and in vivo xenografts. Dr. Cochrane showed that in tamoxifen-resistant breast cancer, there is a switch from ER (estrogen receptor) signaling to AR (androgen receptor) signaling, resulting in increased nuclear AR. On the other hand, in tamoxifen-responsive breast cancers, there is less nuclear AR. Dr. Cochrane interpreted the upregulation of AR as an indicator of tamoxifen resistance. My interpretation differs. In tamoxifen-responsive cancer, estrogenic signaling is upregulated, thus compensating for the estrogen receptor-blocking drug. However, in tamoxifen-resistant cases, compensation with upregulation of estrogenic signaling has failed. Instead, the cell tries to compensate by upregulating the alternate androgen signaling pathway. Thus, accounting for the high nuclear AR present tamoxifen-resistant cancer types. In 2022, Dr. Sara Ravaioli writes:

> Cochrane et al. demonstrated that AR nuclear expression in relation with ER in primary tumors predicts the benefit from adjuvant tamoxifen, on the basis of previous findings reporting that AR expression

decreases in neo-adjuvant endocrine therapy [TAM]-responsive tumors. (46-47)

In 2014, Dr. Dawn Cochrane thought that "AR overexpression increases tamoxifen resistance in breast cancer models in vitro and in vivo." **Note:** I would view this differently. I would instead say AR overexpression is a failed compensatory mechanism and an indicator of tamoxifen resistance. I would not say androgen overexpression causes tamoxifen resistance. Hopefully, future research will resolve this difference of opinion. Dr. Dawn Cochrane writes:

> In a cohort of 192 women with ER + breast cancers, a high ratio of AR:ER (≥2.0) indicated an over **four-fold increased risk for failure** while on tamoxifen (**HR = 4.43**). The AR:ER ratio had an independent effect on risk for failure above ER % staining alone. AR:ER ratio is also an independent predictor of disease-free survival (**HR = 4.04**, 95% CI: 1.68, 9.69; p = 0.002) and disease-specific survival (**HR = 2.75**, 95% CI: 1.11, 6.86; p = 0.03)...**In ER + tumors that respond to neoadjuvant endocrine therapy, we previously observed that AR mRNA and protein expression decrease, while in tumors that fail to respond AR mRNA does not decrease. AR overexpression increases tamoxifen resistance in breast cancer models in vitro and in vivo...** In mice, **treatment with an AI markedly elevated intratumoral testosterone** [in DMBA induced] rat mammary tumors. In postmenopausal women with ER + breast cancer, particularly those being treated with AIs, circulating levels of estradiol (E2) are extremely low, while **circulating androgen levels are increased** since AIs block the conversion of androgens to estrogen. (46-48)

Androgens Provide Complete Relief of Menopausal Symptoms

In 2011, Dr. Rebecca Glaser showed testosterone therapy alone, without administering exogenous estrogen completely relieves menopausal symptoms. Her study showed excellent crosstalk between estrogen and androgen signaling, strong enough to completely relieve menopausal symptoms of estrogen deficiency. This is more completely discussed in Chapter 18 on Testosterone Prevention of Breast Cancer. (50)

Blocking Both Estrogen and Androgen Signaling

What happened when the HABITS trial patient on tamoxifen was given the synthetic progestin, norethisterone, an androgen receptor-blocking drug? Assuming the above discussion is accepted, that androgens represent an alternate estrogen signaling pathway, we can now speculate what happened. Tamoxifen blocks the estrogen receptors, and the cancer cell will then try to compensate by upregulating ER-alpha. When this fails, androgen signaling is upregulated as an alternate pathway. The norethisterone synthetic progestin drug blocks this alternate pathway. According to Dr. Suba's three phases of antiestrogen administration, I speculate that these two drugs, tamoxifen and progestin together completely blocked all ER signaling, and exhausted the cancer cells' ability to compensate, thus leading to tamoxifen resistance and aggressive cancer progression. Thus, resulting in a 300 percent increased cancer recurrence. This is a bad thing.

The Solution to the HABITS Study

What is the solution? Firstly, do not give androgen-blocking drugs such as synthetic progestins to patients on tamoxifen or aromatase inhibitors (AIs). Instead of synthetic progestins, use natural progesterone or reduce exposure to the progestin as was done in the Stockholm trial. The parallel study to the HABITS trial was the 2005 Stockholm trial by Dr. Eva von Schoultz who recognized this problem with norethisterone and switched to a less androgenic blocking progestin, medroxyprogesterone (MPA). Dr. Eva von Schoultz further reduced the patient exposure to MPA with a schedule of 10 days a month every other month, thus obtaining a more favorable result. This simple modification turned everything around! The Stockholm Study HRT users enjoyed a **50 percent reduction in both breast cancer mortality and overall mortality compared to non-users.** There were 4 deaths in the HRT group and 9 deaths in the non-user group. The recurrence rate for HRT users was not increased as it was in the HABITS trial, 11 in the menopausal hormone therapy group, and 13 in the control no-treatment group. (51)

In support of the above HABITS trial speculation are the studies by Dr. Kiyoshi Takagi (2010 and 2022) that reveal intra-tumoral androgen (DHT) levels increase after treatment with aromatase inhibitors (AI) as a desperate measure to compensate for loss of estrogen signaling. In my opinion, AI-induced dramatic upregulation of intratumoral AR signaling is a major anti-cancer benefit. The definitive studies on AI induced increased intratumoral androgen levels were done by Dr. Luciana Macedo (2006) and Dr. Niramol Chanplakorn (2011). Both authors suggest the beneficial anti-cancer effects of aromatase inhibitors are not due to estrogen depletion, rather, it is due to the upregulation of intratumoral androgens. In 2006, Dr. Luciana Macedo writes:

> In conclusion, the results suggest that **aromatase inhibitors may exert their antiproliferative effect not only by reducing the intracellular production of estrogens but also by unmasking the inhibitory effect of androgens acting via the AR [androgen receptor].** (52-54)

And in 2011, Dr. Niramol Chanplakorn writes:

> These results suggest that increased 5-alpha-Red2 [5-alpha reductase converts testosterone to its more potent form DHT] and AR [androgen receptor] following AI [aromatase inhibitor] treatment may partly contribute to reduce the tumor cell proliferation through **increasing intratumoral androgen concentrations and its receptor.** (55)

In 2019, Dr. Lanyang Gao studied the androgenic actions of aromatase inhibitors in vitro and in vivo breast cancer cell studies finding aromatase inhibitor drugs are themselves chemically modified within cancer cells into androgens, and this androgenic activity is the main benefit, writing:

> Direct treatment of ER (+) breast cancer with Formestane [aromatase inhibitor drug] diminishes the tumor [breast cancer xenograft] within weeks. **This is unlikely due to lack of estrogens alone**. We proposed that it is the negative influence of androgens on the growth of ER(+) breast cancer… **We found that breast cancer cells can metabolize Formestane and Exemestane to androgenic compounds which inhibit proliferation**…Our experiments have elucidated the antiproliferative effect of androgens on breast cancer cells. We could demonstrate that the steroidal aromatase inactivators formestane and exemestane are chemically modified within the cancer cells and that the modified sterols act predominantly as antiproliferative androgens. (56-58)

In 2024, Dr. Gemma Santacana-Font studied the aromatase inhibitor, exemestane, in breast cancer samples in-vitro finding exemestane (EXE) and its primary metabolite, 17β-hydroxyexemestane (HEXE), exhibit potent inhibitory effects on breast cancer tumor growth. Both HEXE and EXE bind to and activate the androgen receptor (AR) leading to androgen-induced apoptosis (cell death), writing:

> Exemestane, an aromatase inactivator of the third generation, plays a crucial role in breast cancer therapy by targeting the P450 aromatase enzyme and, thus, decreasing estrogen synthesis. Exemestane (Aromasin™) is currently the **only steroidal aromatase inactivator widely used in clinical routine treatment of ER+ breast cancer** in all phases of the disease in a global perspective. However, the complex mechanisms underlying its therapeutic effects, besides being an aromatase inhibitor, remain

incompletely understood. In this study, we employed a **combination of human samples and in vitro data** to unveil a compelling insight: Exemestane (EXE) and its primary metabolite, 17β-hydroxyexemestane (HEXE), **exhibit potent inhibitory effects on tumor growth** when present together in patient serum. Our biochemical analysis establishes a critical threshold—**20% HEXE metabolite of the total EXE in patient serum**—to trigger a tumor growth inhibition exceeding 90%, as evidenced by Ki67 staining. Mechanistically, our data reveals that **both HEXE and EXE bind to the Androgen Receptor (AR), triggering a synergistic activation that induces a transcriptional program leading to cell death while diminishing the intracellular signaling activated by the oncogene Ras**. Notably, patients with **tumors characterized by a minimum of 20% AR+ epithelial cancer cells stand to benefit the most from exemestane and HEXE**… Collectively, our findings elucidate a dual therapeutic role for Exemestane in selected patients: **it not only restrains estrogen-driven proliferation by estrogen suppression but also stimulates cell death by establishing a specific interactome with the AR at the genomic level.** (58)

How to Make Tamoxifen More Effective: Combine TAM with an Androgen Drug

Since the HABITS trial study showed poor outcomes by adding an anti-androgen to tamoxifen, what if we do the opposite? Instead of an anti-androgen, an androgenic drug is combined with tamoxifen. Will this combination yield a better outcome? Two studies in 1983 by Dr. Douglas. C. Tormey and in 1991 by Dr. James N. Ingle show the answer is YES! Dr. Ingle's study was a randomized trial in about 238 patients 65 or older with metastatic breast cancer randomized to tamoxifen (TAM) alone, or TAM plus fluoxymesterone (FLU), a synthetic androgen. Patients were followed for 5 years. The best results were obtained for a subset of 97 patients with ER-positive breast cancer, the study shows better outcomes with the combination of TAM plus FLU compared to TAM alone. Response rate, time to disease progression, and survival were all superior for the combination TAM plus FLU group. This study by Dr. Ingle is highly supportive of our above speculation that the anti-androgen norethisterone (a synthetic progestin) was the culprit in the HABITS trial. In 1991, Dr. Ingle writes:

> **Among 97 patients** with **estrogen receptor (ER) of 10 [fmol] or greater and 65 years of age or older**, there were highly significant advantages for treatment with TAM plus FLU in both response rate **[40 vs 68 percent]** and time to progression **[7.1 vs 18.3]** months). Of particular note is that in this patient group TAM plus FLU showed a survival advantage **[27.8 vs 42.9]** (Cox model, P = 0.06). **These findings support our original conclusion that the advantages observed are indicative of a real biologic effect of adding FLU to TAM**... (59-60)

AI Resistant Breast Cancer is Androgen Dependent instead of Estrogen Dependent

In 2014, Dr Rika Fuji finds that over time, AI resistance is acquired when ER-positive breast cancer loses the estrogen receptor and changes from ER-dependent to AR-dependent. When this happens, Dr. Fuji then suggests switching to androgen inhibitor drugs, writing:

> These results suggest that in some cases of ER-positive breast carcinoma, tumor cells possibly change from ER-dependent to AR-dependent, rendering them resistant to AI. AR inhibitors may thus be effective in a selected group of patients. (61)

Exemestane is an irreversible aromatase inhibitor and the most widely used AI. However, there is the problem of eventual acquired AI resistance. In 2020, Dr. Cristina Amaral from Portugal was concerned about acquired exemestane resistance, motivating her to study both exemestane-responsive and resistant breast cancer cell types in vitro. Dr. Cristina Amaral found initial exemestane treatment induces androgen receptor (AR) overexpression and activation in breast cancer cells. However, exemestane resistance is acquired after prolonged treatment, making the breast cancer cells AR-dependent and ER-independent. For the exemestane-resistant cell type, Dr. Amaral suggests anti-androgens as a treatment. Dr. Amaral's work supports our above speculation of what went wrong in the HARITS trial. Dr. Amaral says **the androgen receptor over-expression appears to be a response to reduced ER activity**. In other words, when an anti-estrogen AI drug is given, the breast cancer cell will switch from ER to AR signaling as a compensatory mechanism to maintain genomic integrity as discussed by Dr. Suba above. Dr. Amaral writes:

> Thus, the **role of AR [androgen receptor] overexpression in promoting the growth of AIs-acquired resistant cells appears to be a response to reduced ER activity**. Corroborating this, in ER+ breast tumours after AI therapy it was reported that AR levels were increased in AIs non-responders and decreased in AIs responder's tumours... we investigate the biological role of AR in sensitive (MCF-7aro) and resistant (LTEDaro) ER+ breast cancer cells treated with Exemestane. ...targeting AR with bicalutamide (CDX) in Exe-treated cells, enhances the efficacy of this AI in sensitive cells and re-sensitizes resistant cells to Exe treatment. Furthermore, by targeting AR in Exe-resistant cells, it is also possible to block the activation of the ERK1/2 and PI3K cell survival pathways, hamper ERα activation and increase ERβ expression. Thus, this study, highlights a new mechanism involved in Exe-acquired resistance, implicating AR as a key molecule in this setting and suggesting that Exe-resistant cells may have an AR-dependent but ER-independent mechanism. Hence, we propose AR antagonism as a potential and attractive therapeutic strategy to overcome Exe-acquired resistance or to enhance the growth inhibitory properties of Exe on ER+

breast cancer cells, improving breast cancer treatment. (62)

How to Do a Failed Clinical Trial: Block the Androgen Receptors

In vitro studies do not always translate into the clinic. When clinical trials are done with the androgen blocking drug, enzalutamide, this was a clinical failure. A failed trial in 2020 by Dr. Ian Krop gave the androgen receptor-blocking drug, enzalutamide to ER+ breast cancer patients on antiestrogen therapy (exemestane) showing no difference in progression-free survival compared to exemestane alone. The authors of this trial lacked an understanding of Dr. Suba's three phases of antiestrogen resistance. Blocking the desperate upregulation of androgen signaling in the cancer cell to compensate for estrogen depletion by antiestrogen drug (tamoxifen or AI), completely exhausts the cancer cells' ability to compensate. Thus, leading to phase three, aggressive cancer proliferation and patient demise. (63-64)

Dr. Rebecca Glaser to the Rescue: Give Testosterone!

What if we reverse course here and give testosterone? Instead of blocking testosterone with a progestin drug or enzalutamide (Krop, 2020), what if the antiestrogen-treated breast cancer patient is given testosterone? In 2014, Dr. Rebecca Glaser published a case series of 72 breast cancer survivors on letrozole, an aromatase inhibitor, also treated with a testosterone pellet, showing no breast cancer recurrence over 8 years of follow-up. This case series of Dr. Glaser is very supportive of the above speculation, that upregulation of androgen signaling in women on antiestrogen therapy is beneficial. (65)

In 2024, Dr. Amy H. Tien reviewed the role of androgen receptors in breast cancer and discussed a previous 2021 study by Dr. Theresa E. Hickey finding AR protein was found at 42 percent of ER binding sites on chromatin. This indicates androgens may act like estrogens and serve to upregulate estrogen signaling and induce apoptosis, writing:

> a study by Hickey and colleagues [2021] demonstrated that **AR [androgen receptor] behaves as a tumor suppressor in ER-alpha-positive breast cancer**...They showed that AR transactivation with androgen inhibited ER-driven cell proliferation in an ex vivo patient-derived explant model. Using breast cancer cell lines, they also showed that **AR was detected at 42% of estrogen-stimulated ER-binding sites on chromatin** when both AR and ER were activated. This suggests that AR could directly affect ER transcriptional activity by redistributing ER. Interestingly, the binding or recruitment of coactivators p300 and SRC-3, which are required for ER signaling, were both reduced and **replaced by AR upon AR activation**. (66-68)

Aromatase Inhibitors Up-Regulate Androgen Induced Genes

In 2024, Dr. Amy H. Tien found aromatase inhibitor (AI) administration up-regulates 305 of 610 androgen-induced genes, and suggests that **the main anticancer benefit is androgen signal upregulation**. Combining testosterone with an aromatase inhibitor was found synergistic in 2015 by Dr. Rong Chen working in vitro. However, not all breast cancers respond in this way. Dr. Tien also observed in a small number, one percent or so of cancer variants, androgens make things worse and stimulate cancer growth. One such variant is the luminal AR (LAR) subgroup of TNBC triple-negative breast cancer. Here, anti-androgen drugs would be more suitable, since for this rare variant, androgens act as proliferative oncogenic driver, rather than as tumor suppressor. Androgens are not suitable for ER-negative, HER2-amplified breast cancer, and for AR-positive TNBC. Dr. Amy H. Tien writes:

> The analysis of 19 published studies revealed AR expression in about 75% of ER-alpha-positive breast cancer and approximately

32% of ER-alpha-negative breast cancer. **In ER alpha-positive breast cancer, higher levels of nuclear AR protein are usually correlated with improved outcomes and better survival regardless of treatments.** The prognostic value of AR in ER-alpha-positive breast cancer has been demonstrated in many studies...**DHT concentrations are elevated (three-fold higher) in tissues of DCIS and breast carcinoma compared to non-neoplastic tissues.** The expression levels of 5α-reductase are elevated, and its activity is 4–8 times higher in breast carcinoma tissues than in non-neoplastic tissues...Aromatase is the enzyme that converts testosterone to estrogens. The gene expression analyses of breast cancer tissue from patients neoadjuvant treated with the aromatase inhibitor exemestane revealed that **approximately one-half of the 610 androgen-induced genes examined were increased in response to blocking aromatase activity**...AR expression also increases with neoadjuvant treatment with aromatase inhibitors...In ER-positive breast cancer, AR behaves as a tumor suppressor, with its function being opposite to ER...Patients with TNBC [triple negative breast cancer] usually have larger and more aggressive tumors, leading to poor clinical outcomes. Although TNBC patients respond to chemotherapy, they commonly develop distant recurrence and metastases. TNBC tumors account for 10–20% of all breast cancers...luminal AR (LAR). The LAR subgroup represents a range of 11–22% of TNBC depending on the population studied and analysis methods and is classified based on the luminal gene expression pattern. **The LAR subgroup is particularly sensitive to antiandrogens. In this sub- group, AR behaves as an oncogenic driver for tumor cell proliferation**...AR has distinct roles in different subtypes of breast cancers... **In HER2-amplified breast cancer, AR behaves as the oncogenic driver instead of ER...AR also acts as an oncogenic driver in AR-positive TNBC**...(37-38) (42-45) (50) (66-69)

Efficacy and Adverse Effects of Antiestrogen Treatment

Tumor Efficacy of Tamoxifen: Tamoxifen as a first-line therapy for ER-positive breast cancer induces a tumor regression rate of 40%–50%. Nearly all responders eventually acquired tumor resistance, leading to the rapid progression of aggressive disease. (13) (70-71)

ER Negative Breast Cancer and Acquired TAM Resistance in ER positives:

Tamoxifen is ineffective for ER-alpha-negative breast cancer. Loss of ER-alpha expression is associated with TAM resistance and poor prognosis. In 2014, Dr. Floris H Groenendijk writes:

> **Response to tamoxifen is rare in ER-alpha-negative breast cancer...**A portion of ER-alpha-positive tumors becomes independent of estrogen signaling after which they lose ER-alpha expression and, hence, are tamoxifen resistant...Loss of ER-alpha was associated with tamoxifen resistance and can be used as a predictor of poor response to subsequent endocrine therapy. (72)

Adverse Side Effects of Tamoxifen: Tamoxifen has estrogenic effects which increase blood clots, stroke, pulmonary emboli, coronary artery disease, and endometrial cancer. (73)

Dual Qualities of Tamoxifen as Dr. Jekyll and Mr. Hyde

in 2004, Dr. Daniel Hayes recognized the dual nature of tamoxifen, making the analogy to Dr. Jekyll and Mr. Hyde, if you will. As we all know, Dr. Jekyll is a good, healing doctor, while Mr. Hyde is an evil, serial killer. Overall, women on tamoxifen for 10-15 years will have a 15 percent reduction in mortality. ER-negative breast cancer cases rarely respond. And even among ER-positive breast cancer cases, only about half the women benefit. In women with the ER-positive and overexpressed HER-2 breast cancer variants, tamoxifen is harmful, with higher rates of recurrence and mortality than those not taking the drug. Dr. Daniel Hayes writes:

> Worldwide overviews of prospective clinical trials demonstrate that adjuvant tamoxifen reduces the annual odds of death for women with invasive breast cancer by approximately **15% over 10–15 years**…However, **tamoxifen is not ideal. Even among women with ER-positive breast cancer, only 40–50% of patients benefit, suggesting that a substantial fraction of ER-positive cancers are resistant to this drug**…patients with HER-2–overexpressing tumors who were assigned to adjuvant tamoxifen had **higher rates of recurrence and mortality than those who did not receive the agent**. (74)

Efficacy of Aromatase Inhibitors (AI): Tumor response rates are low, **less than 30 percent**, similar to results of other endocrine therapies such as tamoxifen, DES, and androgens. (140-141)

Estrogen Stimulates Breast Cancer Growth In-Vitro

Growth factors upregulated in the microenvironment feeding breast cancer cells include: estrogen (E2), platelet-derived growth factor (PDGF), epidermal growth factor (EGF), insulin-like growth factors (IGF-1, IGF-2), fibroblast growth factor (FGF), transforming growth factor beta (TGF-β), hepatocyte growth factor (HGF), and vascular endothelial growth factor (VEGF) as discussed in 2020 by Dr. Veronica Vella. In 1976, Dr. Marc Lippman administered estrogen to breast cancer cell cultures finding that estrogen stimulates proliferation and growth of cancer cells in-vitro, while antiestrogens inhibit proliferation and growth. These studies supported the belief that estrogen causes breast cancer and that antiestrogens should be the ideal treatment for the prevention and treatment of breast cancer. Is this true in actual medical studies using antiestrogen drugs to prevent breast cancer? Let us next look at the efficacy of antiestrogens for the primary prevention of breast cancer. (75-80)

Efficacy of Adjuvant Endocrine Therapy on DCIS

DCIS, ductal carcinoma in situ, is a premalignant form of breast cancer with an excellent prognosis, and commonly over-treated with surgery and radiation therapy. Although tamoxifen in this clinical setting is rarely used in other countries, in the US about half the ER-positive DCIS cases are treated with tamoxifen or aromatase inhibitors. In 2019, Dr. Maartje van Seijen, from Amsterdam reviewed the use of adjuvant hormonal therapy for DCIS. In the two randomized trials evaluating tamoxifen for DCIS, **neither trial influenced overall survival**, writing:

> Owing to the side effects of hormonal therapy and ambiguous results from clinical trials, **postmenopausal women with DCIS are rarely treated with endocrine therapy in many countries.** In addition, the notion of systemic treatment for a localized disease with an excellent outcome **is perceived as being counterintuitive**. Two randomized clinical trials have investigated the role of tamoxifen – a drug that inhibits the estrogen receptor (ER) – versus placebo in DCIS. The risk of subsequent invasive ipsilateral breast cancer was found to be reduced by tamoxifen in the NSABP trial; the UK, Australia and New Zealand (UK/ANZ) DCIS trial demonstrated a reduction in recurrent DCIS but not in invasive breast cancer. **Tamoxifen administration did not influence overall survival in either trial** and appeared to be more effective at reducing the incidence of new breast events in patients who did not receive radiotherapy in the NSABP trial. Yet, a non-significant reduction in the incidence of new breast events was seen in the prospective series from the UK, independent of whether the patients received radiotherapy or not. Furthermore, to prevent one recurrence, 15 patients would need to be treated (the number needed to

treat). In terms of efficacy, tamoxifen and anastrozole (an aromatase inhibitor) are comparable, and the percentage of women who reported side effects were 91% and 93% for anastrozole and tamoxifen, respectively. Although anastrozole administration more often causes side effects such as musculoskeletal pain, hypercholesterolaemia and strokes, tamoxifen is associated with muscle spasm, deep vein thrombosis and the development of gynaecological symptoms and gynaecological cancers. **In the USA, the uptake of endocrine treatment is higher than in other countries, and nearly half of all ER positive patients are treated by additional adjuvant tamoxifen treatment, indicating a lack of consensus on the added value of this treatment.** (81)

Efficacy of Tamoxifen and Aromatase Inhibitors for Primary Prevention of Breast Cancer

Primary prevention means a population of healthy women are given tamoxifen (TAM) or aromatase inhibitor (AI) to prevent breast cancer. In 2019, Dr. Simone Mocellin reviewed the numerous clinical trials evaluating antiestrogens for primary prevention of breast cancer in healthy women. Dr. Simone Mocellin found 17 RCT studies as of August, 2018 showing **TAM reduced breast cancer risk by 32 percent over a follow up period of 6-16 years**. The placebo group's risk for breast cancer was **7.2** percent, and TAM group was **4.9** percent. This means the placebo group had a **92.8** percent chance of remaining cancer-free. And, the TAM group had a **95.1** percent chance to remain cancer-free. **Note:** RCT = randomized control trial. For the aromatase inhibitor studies, the follow-up period was 3-5 years. The relative risk reduction in the AI group was **53 percent**. The cancer risk in the placebo group was **3.1** percent, and in the AI group was **1.4** percent. This means the placebo group had a **96.9** percent chance to remain cancer-free. And, the AI group had a **98.6** percent chance to remain cancer-free. The obvious question: is this small improvement in cancer-free survival from 96.9 percent to 98.6 percent worth the estrogen deficiency symptoms mimicking severe menopause? The other factor, as discussed below, is the lack of data showing any reduction in all-cause mortality in primary prevention studies using aromatase inhibitors (antiestrogen drugs). Lack of mortality benefit with AIs should not be surprising to us given many studies showing estrogen deficiency of premature ovarian failure or early menopause is associated with increased mortality. One such study in 2024 by Dr. Hilla Haapakoski from Finland studied 22,859 women with premature ovarian failure before age 40 compared to controls, finding **more than 300 percent increased mortality** in women with premature ovarian failure, writing:

> The mean follow-up time for all groups was 17.5 years. At the end of the follow-up, **9.8%** of women with spontaneous POI [premature ovarian insufficiency] and **2.9%** of controls were deceased. (82-83)

Anastrozole for Primary Prevention of Breast Cancer?

What is the reduction in overall mortality in studies of primary prevention of breast cancer using antiestrogen drugs? Although there is a 50 percent reduction in breast cancer, there is **NO Mortality Reduction**. In 2020, Dr. Jack Cuzick reported in Lancet the long-term results of the IBIS II clinical trial using the AI, anastrozole for primary prevention of breast cancer in healthy women with increased risk for breast cancer due to family or personal history. 1,920 women were randomly assigned to anastrozole one mg. per day oral tablet, and another 1920 women were assigned to placebo for 5 years of treatment. The study continued for 131 months (11 years) finding a **49%** reduction in breast cancer for anastrozole (85 vs 165 breast cancer cases, hazard ratio [HR] =**0.51**). This looks good. However, there was **no reduction in overall mortality**, (69 vs 70), AI vs. placebo, and **no significant reduction in breast cancer mortality** (two anastrozole vs three placebo). Of note, the AI-treated group had a 28 percent reduction in skin cancers (basal and squamous cell) suggesting the efficacy of antiestrogen drugs in skin cancer, likewise a hormonal sensitive cancer. (84-86)

Comparing Anastrozole to Premarin for Primary Prevention

In 2014, Dr. David A. Cameron commented on the IBIS-II study primary prevention study by Dr. Jack Cuzick in which use of the antiestrogen drug anastrozole over 5 years reduced the incidence of breast cancer by 53 percent in 1,920 women compared to placebo. However, there was no benefit in terms of overall mortality nor breast cancer mortality. Compare this to the second arm 2004 WHI study (Women's Health Initiative) using Premarin alone (CEE) in women with prior hysterectomy. The 18-year follow-up showed a 23 percent reduction in breast cancer and a 40 percent reduction in breast cancer mortality as reported in 2020 by Dr. Rowan T. Chlebowski. Which would you rather take, an aromatase inhibitor that induces estrogen deficiency with severe menopausal symptoms, yet has no mortality benefit (neither all-cause, nor breast cancer mortality)? Or would you rather take estrogen hormone replacement (Premarin, CEE) which reduces mortality from breast cancer by 40 percent without menopausal symptoms? This one is a no-brainer. The estrogen hormone replacement is the obvious choice. (87-88)

Time to Re-Examine Breast Cancer Risk Reduction and Clinical Strategy

In 2021, Dr. Rowan Chlebowski professor of medicine and chief of Medical Oncology and Hematology at UCLA Medical Center reviewed the medical literature on primary prevention of breast cancer with antiestrogens and concluded it is time for a change. Current antiestrogen strategies to prevent breast cancer do not reduce mortality from breast cancer. On the other hand, dietary intervention with a low-fat diet reduces mortality from breast cancer by **22 percent**. Citing the 2004 WHI study, second arm, Dr. Rowan Chlebowski states **giving post-menopausal women Premarin-alone (CEE) reduced breast cancer incidence by 22 percent and reduced breast cancer mortality by 40 percent**. These results are better than antiestrogen trials. Thus, Dr. Rowan Chlebowski agrees with Dr. Zsuzsanna Suba's position that **the use of antiestrogens for primary prevention of breast cancer is a medical mistake,** writing:

> endocrine-targeted agents (tamoxifen and aromatase inhibitors) reduce [the incidence of] estrogen receptor (ER)–positive, progesterone receptor (PR)–positive cancers **without reducing deaths from breast cancer**. Across three tamoxifen placebo-controlled **prevention trials** (N = 23,360) begun almost 30 years ago, **although there were 226 fewer breast cancer cases, there were nine more deaths from breast cancer in the tamoxifen groups**. Following clinical advances, currently more than half of breast cancer cases are solved problems with extremely low risk of death. As endocrine-targeted agents commonly prevent these cancers, **widespread implementation of current prevention strategies may not reduce deaths from breast cancer**... Against this background, in the Women's Health Initiative **Dietary Modification randomized trial** (N = 48,835), ER-positive, PR-negative cancers were statistically significantly reduced in the intervention group (by **23%** hazard ratio, **0.77;** 95% CI, 0.64 to 0.94) and **deaths from breast cancer were reduced 21%** (P = .02). In the **Women's Health Initiative** randomized, placebo-controlled trial evaluating **conjugated equine estrogen** (N = 10,739), **deaths from breast cancer were reduced 40%** (P = .04)...CEE [Premarin] alone compared with placebo among 10,739 women with a prior hysterectomy was associated with statistically significantly lower breast cancer incidence, 238 cases vs 296 cases, hazard ratio [HR], **0.78**), and was associated with statistically significantly

lower breast cancer mortality with 30 deaths vs 46 deaths (HR, **0.60**)...**These findings suggest that reexamination of breast cancer risk reduction strategies and clinical practice is needed**. (88-89)

Barriers to Antiestrogen Use for Primary Prevention

10 million high-risk women 35-79 years of age in the U.S. are eligible for tamoxifen for primary prevention of breast cancer, yet only one percent of eligible women are using tamoxifen for this purpose. In 2015, Dr. Andrea DeCensi lamented this low uptake for tamoxifen, citing concerns about adverse side effects and lack of demonstrated mortality reduction. Dr. Andrea DeCensi writes:

Unsurprisingly, in a recent survey conducted among over 200 breast cancer specialists... the top three important or very important reasons for low [tamoxifen] uptake were the following:

(1) the drugs may have serious side effects;

(2) **there is no evidence for a reduction in mortality** [for primary prevention]

(3) the drugs are off label in Europe.

The issue of lack of effect on mortality... is a main point of contention...Preventive therapy trials of **SERMs [tamoxifen] have not yet shown a reduction in breast cancer-specific or all-cause mortality,** and this has been one of the main arguments against their use. (90-91)

More on Aromatase Inhibitors, Loss of Estrogen Receptors Lead to AI Resistance

In 2014, Dr. William R. Miller discussed acquired resistance to aromatase inhibitor drugs after successful initiation, writing:

In postmenopausal women, drugs such as letrozole, anastrozole, and exemestane can inhibit aromatization of androgen in vivo by more than 99%, **often decrease circulating estrogens to undetectable levels**, and, in hormone-dependent breast cancers, reduce

tumor proliferation and growth. Third-generation aromatase inhibitors (AIs) are now front-line treatments for breast cancer. However, response rates range between 35 and 70% in neoadjuvant studies, and benefits may be lower in advanced disease. **Acquired resistance after initial successful treatment also occurs**. Note: estrogen is made in the body by the aromatase enzyme, which converts testosterone to estradiol. Thus, aromatase inhibitor drugs prevent the production of estrogen, leading to low serum levels of estradiol and menopausal symptoms. (92)

Aromatase Inhibitor Resistance - Mutation from ER Positive to ER Negative

In 2024, Dr. Pieter J. Westenend found that resistance to aromatase inhibitor therapy in about 30 percent of ER-positive patients initially responding, is due to loss of the estrogen receptors. In other words, breast cancer mutates from ER-positive to ER-negative status which is a poor prognostic indicator. Loss of the estrogen receptor has a negative effect on overall survival, writing:

In metastatic estrogen-receptor (ER)-positive HER2-negative breast cancer, resistance to endocrine therapy can be **caused by ER loss** and the mutation of ESR1, the gene coding for ERs...We found that, in a population of 136 patients, one of these mechanisms was responsible for endocrine resistance in 30% of the patients...Furthermore, we demonstrated that **ER loss has a negative effect on overall survival**...(93)

Antiestrogen Drugs May Not Be Ideal

According to Dr. Zsuzsanna Suba, almost all breast cancers responding to TAM will eventually acquire TAM resistance resulting in rapid progression of advanced metastatic disease. Regarding AIs, likewise the vast majority of patients with advanced (metastatic) breast cancer will either be resistant or eventually acquire resistance to aromatase inhibitor treatment. Because of the above findings, Dr. Zsuzsanna

Suba suggests estrogen-blocking drugs are not the ideal treatment for breast cancer, writing:

> Antiestrogen therapy of advanced breast cancer yielded many difficulties and failures from the onset. ...Considering the whole population of breast cancer patients, antiestrogen treatment could not surpass the **"magic" 30% of tumor response rate**, similarly to the weaknesses of other endocrine therapies; such as oophorectomy [surgical removal of ovaries to prevent estrogen production] or high doses of synthetic estrogens [DES was mainstay treatment for breast cancer in the 1940s]. The majority of even the targeted ER-positive tumors were not responsive to the endocrine treatment showing primary resistance...In addition, patients showing earlier good tumor responses to antiestrogens later experienced secondary resistance leading to metastatic disease and fatal outcome... These experiences strongly suggest that our therapeutic efforts against breast cancer [with estrogen blocking drugs] are not appropriate. Further insights into the mechanisms of tumor growth and tumor recurrence are necessary for the improvement of breast cancer care...(13) (94)

Maintaining Genome Integrity

Why is blocking estrogen problematic? Aside from the obvious adverse side effects of estrogen deficiency on all organ systems of the body, and the induction of severe menopausal symptoms, the estrogen signal is a fundamental regulator for all our cells, and estrogen blockade is toxic to our genome, provoking desperate countermeasures to maintain the estrogen signal. These counter-measures are provoked because all our cells, including breast cancer cells, need to maintain genome integrity, and estrogen [E2] signaling controls the gene network involved in maintaining genome integrity. In 2021, Dr. Sara Pescatori discussed the dual role of estradiol in balancing DNA damage with genome integrity, making this knowledge diffi-

cult to reconcile with the idea of estrogen working as a carcinogen, writing:

> The sex hormone 17β-estradiol (E2) exerts diverse pleiotropic physiological effects including the control of the reproductive system in females and the development of primary and secondary sexual characteristics in humans. E2 regulates a plethora of physiological functions in non-reproductive tissues including heart, bone, and brain systems. Accordingly, E2 can exert beneficial effects being protective against osteoporosis, cardiovascular and neurodegenerative diseases...the most frequent BCs [breast cancers] (i.e., 75%) express the ER-alpha at the diagnosis. The ER-alpha is an important prognostic factor because its expression drives the treatment (i.e., the endocrine therapy [with AI's]), which aims to block different aspects of the E2:ER-alpha proliferative signaling...Nonetheless, it is difficult to reconcile how E2, which controls crucial physiological processes both in females and males and contributes to body homeostasis, could work as a carcinogen inducing the development of BC [breast cancer]. (95)

Falsifying Estrogen as Carcinogen

1) Estrogen prevents breast cancer and reduces breast cancer mortality. The 2004 WHI study (using estrogen-alone, Premarin, CEE) showed estrogen treatment reduced the incidence of breast cancer by 22 percent, and the 18-year follow-up showed a 40 percent reduction in mortality from breast cancer. (88)

2) Estrogen deficiency, not estrogen excess is associated with oral cancer. (96)

3) Estrogen deficiency after menopause increases the risk of breast cancer. However, this risk for breast cancer is lower for premenopausal women when estrogen levels are higher. Roughly 80 percent of breast cancer cases occur in the postmenopausal age group. On the other hand, only 20 percent of breast cancer cases occur in the premenopausal age group. (97)

4) High estrogen levels of pregnancy con-

fer protection from breast cancer. Conversely, nulliparity [no pregnancy] increases the risk for breast and other female cancers. (98-101)

5) Pregnancy Protects Mice from Induction of Breast Cancer with Chemical Carcinogens.

Animal experiments using mice shows that pregnancy before or soon after exposure to a chemical carcinogen (DMBA) is strongly protective against developing breast cancer. Short term exogenous hormone exposure mimicking pregnancy is highly protective in mouse models of carcinogen induced breast cancer, reducing tumor formation by 60-70 percent. Even synthetic estrogens confer protection. (102-107)

6) Interrupting Adjuvant Hormone Therapy for Pregnancy

If high estrogen levels cause recurrence in breast cancer survivors, one would expect increased recurrence after pregnancy when estrogen levels are extremely high. This was studied and found to be false. In 2023, Dr. Ann Partridge of the Dana Farber Cancer Institute published her POSITIVE trial in which temporary interruption of adjuvant hormonal therapy for ER-positive breast cancer to carry a pregnancy does not increase breast cancer recurrence compared to controls. Among 516 women with a median age was 37 years, over three years, the incidence of recurrent breast cancer was 8.9% in the treatment-interruption-pregnancy group and 9.2% in the control group. (108)

In 2017, Dr. Javaid Iqbal found that a pregnancy 6 months or more after breast cancer is associated with dramatic improvement in outcome. Dr. Javaid Iqbal did a retrospective cohort study using a health database in Toronto, Canada including 7553 women aged 20 to 45 years with the diagnosis of invasive breast cancer. Dr. Javaid Iqbal was specifically interested in women who became pregnant 6 months or more after the diagnosis of breast cancer, finding this group had a **75 percent reduction**

in mortality compared to the nonpregnancy cohort, writing:

> compared with nonpregnant women, the risk of death significantly dropped for women who delivered [a full-term pregnancy] 6 months or more after diagnosis of breast cancer (multivariable HR, 0.25). The 5-year actuarial survival rate was 96.7% [**3.3%** mortality] for women who had pregnancy 6 months or more after diagnosis of breast cancer, vs 87.5% (95% CI, 86.5%-88.4%) [**12.5%** mortality] for women with no pregnancy) (age-adjusted HR, 0.22). (109-112)

Tamoxifen or AI Resistant Breast Cancers Regress with Estrogen Therapy

Long term tamoxifen or AI treatment eventually transforms initially responsive ER-positive breast cancers into TAM-resistant cancers. These aggressive cancers may now be treated with estrogen with a dramatic reduction in cancer-related mortality. In 2024, Dr Zsuzsanna Suba writes:

> In tumors exhaustively treated with tamoxifen, the achieved ER blockade may be regarded as an artificially created ER-negative status, as estrogen signaling suffers irreparable damage. Among patients with tamoxifen-resistant advanced breast cancer, [subsequent] estrogen treatment dramatically decreases breast cancer related mortality...In tumors, with tamoxifen-blocked ERs, estrogen treatment induces abundant new ER expression and drives genome stabilization. In patients with aromatase inhibitor-resistant tumors, estrogen treatment induces regression of metastatic cancers and extends survival. (9) (45) (113)

How to Prevent Breast Cancer?

In 2015, Dr. Zsuzsanna Suba reviewed the pitfalls of anti-estrogen treatment as a prevention and treatment of breast cancer. Both estrogen and anti-estrogen treatments result in extreme up-regulation of estrogen signaling, the key mechanism for prevention and treatment of breast cancer. Dr. Suba writes:

> whatever type of available endocrine therapies may be used, including estrogen, antiestrogen treatment, or oophorectomy, an extreme upregulation of ER signaling seems to be the crucial mechanism of successful prevention and treatment for breast cancer. (29) (114)

The Dawn of a New Era - Estrogen for Breast Cancer Treatment

Animal transfection experiments were done in 1992 by Dr. Marcel Garcia and in 1994 by Dr. Anait Levenson using mice given breast cancer xenografts. In an attempt to make the cancer responsive to TAM, breast cancer cells were transfected with the estrogen receptor genes, transforming the xenografted cancer cells into ER-positive cancer cells. In 1992, Dr. Marcel Garcia found that estrogen-negative breast cancer cells transfected with the ER became ER-positive, and produced lung metastatic disease when xenografted into mice. These lung metastatic lesions were unresponsive to tamoxifen but were **inhibited three-fold by estrogen**. Re-administration of tamoxifen reversed the benefits of estrogen. Dr. Garcia's experiment supports the above statement by Dr. Suba, that upregulation of estrogen signal prevents and treats breast cancer. Dr. Marcel Garcia writes:

> We transfected the human estrogen receptor into the estrogen receptor-negative metastatic breast cancer cell line MDA-MB-231 **in an attempt to restore their sensitivity to antiestrogens.** Two stable sublines of MDA-MB-231 cells (HC1 and HE5) expressing functional estrogen receptors were studied for their ability to grow and invade in vitro and to metastasize in athymic nude mice. The number and size of lung metastases developed by these two sublines in ovariectomized nude mice was not markedly altered by tamoxifen but **was inhibited 3-fold by estradiol**. Estradiol

also significantly inhibited in vitro cell proliferation of these sublines and their invasiveness in Matrigel, a reconstituted basement membrane, whereas the antiestrogens 4-hydroxytamoxifen and ICI 164,384 reversed these effects. These results show that estradiol inhibits the metastatic ability of estrogen receptor-negative breast cancer cells following transfection with the estrogen receptor, whereas estrogen receptor-positive breast cancers are stimulated by estrogen, **indicating that factors other than the estrogen receptor are involved in the progression toward hormone independence.** (115-116)

Adverse Effects of Aromatase Inhibitors

Since aromatase inhibitors induce artificial menopause, it is not surprising the adverse side effects of aromatase inhibitors mimic severe menopausal symptoms of hot flashes, night sweats, insomnia, "brain fog", vaginal dryness, arthralgia, decreased bone mineral density, and increased bone fracture rate. (117-118)

Harms and Perils of Prolonged Hypoestrogenism - 50% Discontinuation Rate

In 2021, Dr. Robert P. Kauffman from Amarillo, Texas discusses the perils of prolonged hypoestrogenism as treatment for breast cancer, and questions if the antiestrogen treatment is worse than breast cancer recurrence. Surgical-induced estrogen deprivation with hysterectomy is associated with increased all-cause mortality. For more on this, see Chapter 11, HRT Beneficial After Hysterectomy. Likewise, drug induced long-term estrogen deprivation [LTED] using tamoxifen, aromatase inhibitors, and ovarian suppression drugs, is associated with an increase in all-cause mortality. Profound estrogen deficiency, especially in the premenopausal group is not well tolerated leading to an anti-estrogen drug discontinuation rate of about fifty percent before the end of the 5 years of treatment. Patients experience severe menopausal symptoms, vasomotor instability, sleep disturbance, depression/anxiety, cognitive decline, cardiovascular disease and stroke, diabetes, bone loss, joint and connective tissue discomfort, genitourinary syndrome of menopause, and sexual dysfunction. The mainstream medicine solution is to convert the breast cancer survivor into a walking drug store. The primary care doctor will prescribe a list of drugs: SSRI antidepressants and benzodiazepines for depression and anxiety, sleeping pills for insomnia, bisphosphonates and denosumab for osteoporosis, celecoxib, ibuprofen and NSAIDs for joint pain, repeated antibiotics for recurrent urinary tract infections, cognitive therapy for sexual dysfunction, statin drugs for cardiovascular disease, and diabetic drugs for elevated blood sugar. The breast cancer patient is left to ponder the question: Could the anti-estrogen drug treatment be worse than the disease? Dr. Robert P Kauffman writes:

> **Long-term estrogen deprivation is associated with an increase in all-cause mortality...**Profound hypoestrogenism in the premenopausal age group **may not be well tolerated** due to a host of bothersome side effects (primarily vasomotor symptoms, musculoskeletal complaints, genitourinary syndrome of menopause, and mood disorders). Prolonged hypoestrogenism in younger women is associated with premature development of cardiovascular disease, bone loss, cognitive decline, **and all-cause mortality**...Sustained lower serum [estrogen] levels are associated with cardiovascular disease, osteoporosis, neurodegeneration, and inflammatory processes...The side-effects and potential harm of prolonged estrogen suppression are many: vasomotor instability, sleep disturbance, depression/anxiety, cognitive decline, cardiovascular disease and stroke, diabetes, bone loss, joint and connective tissue discomfort, genitourinary syndrome of menopause, and sexual dysfunction—to name the most common...The effect of low circulating estrogen levels on human physiology has been well-studied in non-cancer populations. Prolonged disruption or cessation of ovarian function before the age of natural menopause or for prevention of heritable ovarian cancer **is associated with multisystem disease** (diabetes, cardiovascular disease, stroke, depression, osteoporosis, and others). (119-125)

Our Office Protocol for the Breast Cancer Survivor

Approximately 50 percent of breast cancer survivors decide on their own to discontinue the antiestrogen drugs within the allotted 5 years of use. These patients are very motivated to seek relief from their severe menopausal symptoms with bioidentical hormone replacement. Our office HRT formula is Bi-est (20 percent estradiol E2 and 80 percent estriol E3) and progesterone as a topical cream or vaginal capsule. The HRT formula always includes testosterone which downregulates ER-alpha activity and metabolizes to 3-beta-diol which binds to and activates ER-beta, the tumor suppressor receptor. We also test for iodine levels and give iodine supplements when found low. All patients receive DIM, di-indole methane. Over the past 20 years, not a single breast cancer survivor treated in my office has had a recurrence.

Conclusion: Dr. Suba gives us an overview of antiestrogen therapies with tamoxifen and aromatase inhibitors. Although complete remission of advanced breast cancer with exemestane is quite dramatic and impressive. Anti-estrogen therapies can be disappointing because the majority of breast cancers show primary resistance, and those that do respond eventually develop acquired resistance leading to progressive metastatic disease and fatal outcomes. Dr. Suba's three stages of antiestrogen resistance tell us that good responders compensate by up-regulating estrogen signaling. These patients will enjoy tumor regression. In contrast, those patients with complete inhibition of ER signaling have exhausted their capacity to compensate, leading to stimulation of breast cancer growth, metastatic disease, and fatal outcomes. In my opinion, androgen signaling may represent an alternate pathway when estrogen receptors are blocked. When both estrogen and androgen signaling are blocked, this leads to the poor outcome seen in the HABITS trial by Lars Holmberg in Sweden. For mainstream oncology, antiestrogens remain the cornerstone of breast cancer treatment. I have seen dramatic regression of metastatic breast cancer using the aromatase inhibitor drug, exemestane, and can attest to its success. This is an efficacious drug. Is the success of exemestane due to undetectable serum estrogen levels, or is it due to the upregulation of intracellular androgens or both? This remains an unresolved question for mainstream medicine. Turning to antiestrogens as adjunctive therapy for the breast cancer survivor, mainstream medicine recognizes the dual features of Dr. Jekyll and Mr. Hyde. Dr. Robert P. Kauffman asks if such adjunctive use of antiestrogen drugs is worse than the risk for recurrence. The many severe adverse side effects of antiestrogens are the reason for a 50 percent discontinuation rate in breast cancer survivors on anti-estrogen drugs (TAM, AIs). As we have seen above, even prestigious academic oncologists such as Dr. Rowan T. Chlebowski are calling for the reappraisal of antiestrogens for breast cancer prevention. On the other hand, considerable medical literature has accumulated over the years making the case that estrogen treatment is breast cancer preventive. The 2004 second-arm WHI Premarin-alone RCT study shows estrogen (Premarin, CEE) is a good choice for breast cancer prevention. Yet, the wheels of change turn slowly in mainstream medicine, so I would not hold my breath for this to happen anytime soon. For more on this topic, see my book, *Cracking Cancer Toolkit*, the Use of Repurposed Drugs for Cancer Treatment (2020). Thanks, and credit goes to Zsuzsanna Suba MD, PhD, professor emeritus at Semmelweis University in Budapest, and board-certified pathologist. Dr. Suba's three phases of antiestrogen resistance have given us a framework for understanding what went wrong with the HABITS study by Lars Holmberg of Sweden. (126-139)

♦ References for Chapter 20 Adjuvant Endocrine Therapy Dr. Jekyll and Mr. Hyde

1) Arzanova, Evelina, and Harvey N. Mayrovitz. "The Epidemiology of Breast Cancer." Exon Publications (2022): 1-19.

2) National Cancer Institute S. Epidemiology End Result Program Cancer Statistic Facts: Female Breast Cancer. 2022. Available from: https://seer.cancer.gov/statfacts/html/breast.html [Accessed on 10 Jun 2022]

3) Andre, Fabrice, et al. "Breast Cancer with Synchronous Metastases: Trends in Survival During A 14-Year Period." Journal of Clinical Oncology 22.16 (2004): 3302-3308.

4) Suba, Zsuzsanna. "Rosetta Stone For Cancer Cure: Comparison of The Anticancer Capacity of Endogenous Estrogens, Synthetic Estrogens and Antiestrogens." Oncology Reviews 17 (2023): 10708.

5) Beatson, G. T. "On the Treatment of Inoperable Cases Of Carcinoma Of The Mamma: Suggestions For A New Method Of Treatment, With Illustrative Cases." Cancer journal (Villejuif) 2.10 (1989): 347-350.

6) Zamora-León, Pilar. "Are The Effects of DES Over? A Tragic Lesson from The Past." International Journal of Environmental Research and Public Health 18.19 (2021): 10309.

7) Rogers, Rachael E., et al. "Prenatal Exposure to Diethylstilbestrol Has Long-Lasting, Transgenerational Impacts on Fertility and Reproductive Development." Toxicological Sciences 195.1 (2023): 53-60.

8) Haddow, Alexander, et al. "Influence of Synthetic Estrogens on Advanced Malignant Disease." British Medical Journal 2.4368 (1944): 393.

9) Jordan, V. Craig. "Linking Estrogen-Induced Apoptosis with Decreases in Mortality Following Long-Term Adjuvant Tamoxifen Therapy." Journal of the National Cancer Institute 106.11 (2014): dju296.

10) Poirot, Marc. "Four Decades of Discovery in Breast Cancer Research and Treatment—An Interview with V. Craig Jordan." International Journal of Developmental Biology 55.7-8-9 (2011): 703-712.

11) Ingle, James N., et al. "Randomized Clinical Trial of Diethylstilbestrol Versus Tamoxifen in Postmenopausal Women with Advanced Breast Cancer." New England Journal of Medicine 304.1 (1981): 16-21.

12) Shete, Nivida, et al. "Revisiting Estrogen for The Treatment of Endocrine-Resistant Breast Cancer: Novel Therapeutic Approaches." Cancers 15.14 (2023): 3647.

13) Suba, Zsuzsanna. "Rosetta stone for Cancer Cure: Comparison of The Anticancer Capacity of Endogenous Estrogens, Synthetic Estrogens and Antiestrogens." Oncology Reviews 17 (2023): 10708.

14) Suba, Zsuzsanna. "DNA Damage Responses in Tumors Are Not Proliferative Stimuli, but Rather They Are DNA Repair Actions Requiring Supportive Medical Care." Cancers 16.8 (2024): 1573.

15) Wang, Li, and Li-Jun Di. "BRCA1 and Estrogen/Estrogen Receptor in Breast Cancer: Where They Interact?" International Journal of Biological Sciences 10.5 (2014): 566.

16) Yoshida, Kiyotsugu, and Yoshio Miki. "Role of BRCA1 and BRCA2 as Regulators of DNA Repair, Transcription, And Cell Cycle in Response to DNA Damage." Cancer science 95.11 (2004): 866-871.

17) Yedidia-Aryeh, Lia, and Michal Goldberg. "The Interplay between the Cellular Response to DNA Double-Strand Breaks and Estrogen." Cells 11.19 (2022): 3097.

18) Lewis-Wambi, Joan S., and V. Craig Jordan. "Estrogen Regulation of Apoptosis: How Can One Hormone Stimulate and Inhibit?" Breast cancer research 11 (2009): 1-12.

19) Zach, L.; Yedidia-Aryeh, L.; Goldberg, M. Estrogen and DNA Damage Modulate mRNA Levels of Genes Involved in Homologous Recombination Repair In Estrogen-Deprived Cells. J. Transl. Genet. Genom. 2022, 6, 266–280

20) Rajan, Arathi, et al. "Deregulated Estrogen Receptor Signaling and DNA Damage Response In Breast Tumorigenesis." Biochimica et Biophysica Acta (BBA)-Reviews on Cancer 1875.1 (2021): 188482.

21) Suba, Zsuzsanna. "DNA Stabilization by The Upregulation of Estrogen Signaling In BRCA Gene Mutation Carriers." Drug Design, Development and Therapy (2015): 2663-2675.

22) Suba, Zsuzsanna. "The Pitfall of The Transient, Inconsistent Anticancer Capacity of Antiestrogens and The Mechanism of Apparent Antiestrogen Resistance." Drug Design, Development and Therapy (2015): 4341-4353.

23) Suba Z. DNA Stabilization by The Upregulation of Estrogen Signaling In BRCA Gene Mutation Carriers. Drug Design Devel Ther. 2015;9: 2663–2675.

24) Liu S, Ruan X, Schultz S, et al. Estetrol Stimulates Proliferation and Estrogen Receptor Expression in Breast Cancer Cell Lines: Comparison of Four Estrogens. Eur J Contracept Reprod Health Care. 2015;20(1):29–35.

25) Stoica GE, Franke TF, Moroni M, et al. Effect of Estradiol on Estrogen Receptor-alpha Gene Expression and Activity Can Be Modulated By The Erbb2/PI 3-K/Akt Pathway. Oncogene. 2003;22(39):7998–8011.

26) Tolhurst RS, Thomas RS, Kyle FJ, et al. Transient Over-Expression of Estrogen Receptor-alpha in Breast Cancer Cells Promotes Cell Survival and Estrogen-Independent Growth. Breast Cancer Res Treat. 2011;128(2):357–368.

27) Kuske B, Naughton C, Moore K, et al. Endocrine Therapy Resistance Can Be Associated with High Estrogen Receptor Alpha (Eralpha) Expression And Reduced ER-Alpha Phosphorylation In Breast Cancer Models. Endocr Relat Cancer. 2006;13(4):1121–1133.

28) Iwase, H., et al. "Ethinylestradiol is Beneficial for Postmenopausal Patients with Heavily Pre-Treated Metastatic Breast Cancer After Prior Aromatase Inhibitor Treatment: A Prospective Study." British Journal of Cancer 109.6 (2013): 1537.

29) Suba, Zsuzsanna. "The Pitfall of the Transient, Inconsistent Anticancer Capacity of Antiestrogens and The Mechanism of Apparent Antiestrogen Resistance." Drug Design, Development and Therapy 9 (2015): 4341.

30) Jordan, V. Craig, et al. "Alteration of Endocrine Parameters in Premenopausal Women with Breast Cancer During Long-Term Adjuvant Therapy with Tamoxifen as The Single Agent." JNCI: Journal of the National Cancer Institute 83.20 (1991): 1488-1491.

31) Jiang, Jie, et al. "Phosphorylation of Estrogen Receptor-A At Ser167 Is Indicative of Longer Disease-Free and Overall Survival in Breast Cancer Patients." Clinical Cancer Research 13.19 (2007): 5769-5776.

32) Yamashita, Hiroko, et al. "Phosphorylation of Estrogen Receptor A Serine 167 Is Predictive of Response to Endocrine Therapy and Increases Post Relapse Survival in Metastatic Breast Cancer." Breast Cancer Research 7 (2005): 1-12.

33) Miller, W. R., and J. O'Neill. "The Importance of Local Synthesis of Estrogen Within the Breast." Steroids 50.4-6 (1987): 537-548.

34) Quaynor SD, et al. Delayed Puberty and Estrogen Resistance in A Woman with Estrogen Receptor Alpha Variant. N Engl J Med. 2013;369(2):164–171.

35) Madeddu, Clelia, et al. "Ovarian Hyperstimulation in Premenopausal Women During Adjuvant Tamoxifen Treatment for Endocrine-Dependent Breast Cancer: A Report Of Two Cases." Oncology Letters 8.3 (2014): 1279-1282.

36) Morishima, Akira, et al. "Aromatase Deficiency in Male and Female Siblings Caused by A Novel Mutation and The Physiological Role of Estrogens." The Journal of Clinical Endocrinology & Metabolism 80.12 (1995): 3689-3698.

37) Bluming, Avrum Zvi. "Hormone Replacement Therapy After Breast Cancer: It Is Time." The Cancer Journal 28.3 (2022): 183-190.

38) Holmberg, Lars, and Harald Anderson. "HABITS (Hormonal Replacement Therapy After Breast Cancer—Is It Safe?), A Randomized Comparison: Trial Stopped." The Lancet 363.9407 (2004): 453-455.

39) Holmberg, Lars, et al. "Increased Risk of Recurrence After Hormone Replacement Therapy In Breast Cancer Survivors." Journal of the National Cancer Institute 100.7 (2008): 475-482.

40) Hickey, Theresa E., et al. "Minireview: The Androgen Receptor in Breast Tissues: Growth Inhibitor, Tumor Suppressor, Oncogene?" Molecular Endocrinology 26.8 (2012): 1252-1267.

41) Hickey, Theresa E. et al. The Androgen Receptor Is a Tumor Suppressor in Estrogen Receptor-Positive Breast Cancer. Nat. Med. 2021, 27, 310–320.

42) Dai, Charles, and Leif W. Ellisen. "Revisiting Androgen Receptor Signaling in Breast Cancer." The Oncologist 28.5 (2023): 383-391.

43) Wang, Yu, et al. "Androgen Receptor-Induced Tumor Suppressor, KLLN, Inhibits Breast Cancer Growth and Transcriptionally Activates P53/P73-Mediated Apoptosis in Breast Carcinomas." Human Molecular Genetics 22.11 (2013): 2263-2272.

44) Thomas, Peter, et al. "Identification and Characterization of Membrane Androgen Receptors In The ZIP9 Zinc Transporter Subfamily: II. Role Of Human ZIP9 In Testosterone-Induced Prostate and Breast Cancer Cell Apoptosis." Endocrinology 155.11 (2014): 4250-4265.

45) Suba, Zsuzsanna. "Estrogen Regulated Genes Compel Apoptosis in Breast Cancer Cells, Whilst Stimulate Antitumor Activity in Peritumoral Immune Cells in a Janus-Faced Manner." Current Oncology 31.9 (2024): 4885-4907.

46) Ravaioli, Sara, et al. "Androgen Receptor in Breast Cancer: The "5W" questions." Frontiers in Endocrinology 13 (2022): 977331.

47) Cochrane, Dawn R., et al. "Role of the Androgen Receptor in Breast Cancer and Preclinical Analysis of Enzalutamide." Breast Cancer Research 16 (2014): 1-19.

48) Dimitrakakis, Constantine, and Carolyn Bondy. "Androgens and the Breast." Breast Cancer Research 11 (2009): 212.

50) Glaser, Rebecca, Anne E. York, and Constantine Dimitrakakis. "Beneficial Effects of Testosterone Therapy In Women Measured By The Validated Menopause Rating Scale (MRS)." Maturitas 68.4 (2011): 355-361.

51) von Schoultz, Eva, and Lars E. Rutqvist. "Menopausal Hormone Therapy After Breast Cancer: The Stockholm Randomized Trial." Journal of the National Cancer Institute 97.7 (2005): 533-535.

52) Macedo, Luciana F., et al. "Role of Androgens On MCF-7 Breast Cancer Cell Growth and on the Inhibitory Effect of Letrozole." Cancer Research 66.15 (2006): 7775-7782.

53) Takagi, Kiyoshi, et al. "Increased Intratumoral Androgens in Human Breast Carcinoma Following Aromatase Inhibitor Exemestane Treatment." Endocrine-Related Cancer 17.2 (2010): 415-430.

54) Takagi, Kiyoshi, et al. "Diverse Role of Androgen Action In Human Breast Cancer." Endocrine Oncology 2.1 (2022): R102-R111.

55) Chanplakorn, Niramol, et al. "Increased 5α-Reductase Type 2 Expression in Human Breast Carcinoma Following Aromatase Inhibitor Therapy: The Correlation with Decreased Tumor Cell Proliferation." Hormones and Cancer 2 (2011): 73-81.

56) Gao, Lanyang, et al. "The Beneficial Androgenic Action of Steroidal Aromatase Inactivators in Estrogen-Dependent Breast Cancer After Failure of Nonsteroidal Drugs." Cell death & disease 10.7 (2019): 494.

57) Ariazi, Eric A., et al. "Exemestane's 17-Hydroxylated Metabolite Exerts Biological Effects as An Androgen." Molecular Cancer Therapeutics 6.11 (2007): 2817-2827.

58) Santacana-Font, Gemma, et al. "Exemestane and Its Primary Metabolite 17-Hydroexemestane Inhibit Synergically the Tumor Growth Of ER/AR Positive Breast Cancer Tumors." Cancer Research 84.6_Supplement (2024): 7576-7576.

59) Tormey, Douglas. C., et al. "Evaluation of Tamoxifen Doses with And Without Fluoxymesterone In Advanced Breast Cancer." Annals of Internal Medicine 98 (1983): 139-144.

60) Ingle, James N., et al. "Combination Hormonal Therapy with Tamoxifen Plus Fluoxymesterone Versus Tamoxifen Alone in Postmenopausal Women with Metastatic Breast Cancer. An Updated Analysis." Cancer 67.4 (1991): 886-891.

61) Fujii, Rika, et al. "Increased Androgen Receptor Activity and Cell Proliferation in Aromatase Inhibitor-Resistant Breast Carcinoma." The Journal of Steroid Biochemistry and Molecular Biology 144 (2014): 513-522.

62) Amaral, Cristina, et al. "The Potential Clinical Benefit of Targeting Androgen Receptor (AR) In Estrogen-Receptor Positive Breast Cancer Cells Treated with Exemestane." Biochimica et Biophysica Acta (BBA)-Molecular Basis of Disease 1866.5 (2020): 165661.

63) Krop, I., et al. "A Randomized Placebo Controlled Phase II Trial Evaluating Exemestane with or without Enzalutamide in Patients with Hormone Receptor-Positive Breast Cancer." Clinical Cancer Research: 26.23 (2020): 6149-6157.

64) Leo, Javier, et al. "Stranger Things: New Roles and Opportunities for Androgen Receptor In Oncology Beyond Prostate Cancer." Endocrinology 164.6 (2023): bqad071.

65) Glaser, Rebecca L., et al. "Efficacy of Subcutaneous Testosterone on Menopausal Symptoms in Breast Cancer Survivors." J Clin Oncol 32. Suppl 2 (2014): 109.

66) Tien, Amy H., and Marianne D. Sadar. "Treatments Targeting the Androgen Receptor and Its Splice Variants in Breast Cancer." International Journal of Molecular Sciences 25.3 (2024): 1817.

67) Hickey, Theresa E., et al. "Minireview: The Androgen Receptor in Breast Tissues: Growth Inhibitor, Tumor Suppressor, Oncogene?" Molecular Endocrinology 26.8 (2012): 1252-1267.

68) Hickey, Theresa E. et al. The Androgen Receptor Is a Tumor Suppressor in Estrogen Receptor-Positive Breast Cancer. Nat. Med. 2021, 27, 310–320.

69) Chen, Rong, et al. "Antiproliferative Effects of Anastrozole On MCF-7 Human Breast Cancer Cells In Vitro Are Significantly Enhanced by Combined Treatment with Testosterone Undecanoate." Molecular Medicine Reports 12.1 (2015): 769-775.

70) Viedma-Rodríguez, Rubí, et al. "Mechanisms Associated with Resistance to Tamoxifen In Estrogen Receptor-Positive Breast Cancer." Oncology Reports 32.1 (2014): 3-15.

71) Osborne, C. Kent. "Tamoxifen in the Treatment of Breast Cancer." New England Journal of Medicine 339.22 (1998): 1609-1618.

72) Groenendijk, Floris H., and René Bernards. "Drug Resistance to Targeted Therapies: Deja Vu All Over Again." Molecular Oncology 8.6 (2014): 1067-1083.

73) Yang, Geniey, et al. "Toxicity and Adverse Effects of Tamoxifen and Other Anti-Estrogen Drugs." Pharmacology & Therapeutics 139.3 (2013): 392-404.

74) Hayes, Daniel F. "Tamoxifen: Dr. Jekyll and Mr. Hyde?" Journal of the National Cancer Institute 96.12 (2004): 895-897.

75) Vella, Veronica, et al. "Microenvironmental Determinants of Breast Cancer Metastasis: Focus on The Crucial Interplay Between Estrogen and Insulin/Insulin-Like Growth Factor Signaling." Frontiers in Cell and Developmental Biology 8 (2020): 608412.

76) Lippman, M., G. Bolan, and K. Huff. "The Effects of Estrogens and Antiestrogens on Hormone-Responsive Human Breast Cancer in Long-Term Tissue Culture." Cancer Research 36.12 (1976): 4595-4601.

77) Chalbos, Dany, et al. "Estrogens Stimulate Cell Proliferation and Induce Secretory Proteins In A Human Breast Cancer Cell Line (T47D)." The Journal of Clinical Endocrinology & Metabolism 55.2 (1982): 276-283.

78) Soto, A. M., and C. Sonnenschein. "The Role of Estrogens on The Proliferation of Human Breast Tumor Cells (MCF-7)." Journal Of Steroid Biochemistry 23.1 (1985): 87-94.

79) Osborne, C. Kent, et al. "Effect of Estrogens and Antiestrogens on Growth of Human Breast Cancer Cells in Athymic Nude Mice." Cancer Research 45.2 (1985): 584-590.

80) Huseby, R. A., et al. "Evidence for a Direct Growth-Stimulating Effect of Estradiol on Human MCF-7 Cells In Vivo." Cancer research 44.6 (1984): 2654-2659.

81) van Seijen, Maartje, et al. "Ductal Carcinoma in Situ: To Treat or Not To Treat, That Is The Question." British Journal of Cancer 121.4 (2019): 285-292.

82) Haapakoski, Hilla, et al. "Mortality Among Women With POI, Nationwide Register Based Case-Control Study." Endocrine Abstracts. Vol. 99. Bioscientifica, 2024.

83) Mocellin, Simone, et al. "Risk-Reducing Medications for Primary Breast Cancer: A Network Meta-Analysis." Cochrane Database of Systematic Reviews 4 (2019).

84) Cuzick, Jack, et al. "Use of Anastrozole for Breast Cancer Prevention (IBIS-II): Long-Term Results of A Randomised Controlled Trial." The Lancet 395.10218 (2020): 117-122.

85) Chen, Haifei, et al. "Downregulation of Estrogen-Related Receptor Alpha Inhibits Human Cutaneous Squamous Cell Carcinoma Cell Proliferation and Migration by Regulating EMT Via Fibronectin and STAT3 Signaling Pathways." European Journal of Pharmacology 825 (2018): 133-142.

86) Lan, Jing, Xing-Hua Gao, and Rashmi Kaul. "Estrogen Receptor Subtype Agonist Activation In Human Cutaneous Squamous Cell Carcinoma Cells Modulates Expression Of CD55 And Cyclin D1." EXCLI journal 18 (2019): 606.

87) Cameron David A. Breast Cancer Chemoprevention: Little Progress in Practice? Lancet. 2014;383(9922):1018–1020.

88) Chlebowski, Rowan T., et al. "Association of Menopausal Hormone Therapy with Breast Cancer Incidence and Mortality During Long-Term Follow-Up of The Women's Health Initiative Randomized Clinical Trials." JAMA 324.4 (2020): 369-380.

89) Chlebowski, Rowan T., et al. "Breast Cancer Prevention: Time for Change." JCO Oncology Practice 17.12 (2021): 709-716.

90) DeCensi, Andrea, et al. "Barriers to Preventive Therapy for Breast and Other Major Cancers and Strategies To Improve Uptake." Ecancermedicalscience 9 (2015).

91) Narod SA. Tamoxifen Chemoprevention - End of The Road? JAMA Oncol. 2015.

92) Miller, William R., and Alexey A. Larionov. "Understanding the Mechanisms of Aromatase Inhibitor Resistance." Breast Cancer Research 14 (2012): 1-11.

93) Westenend, Pieter J., et al. "Estrogen-Receptor Loss and ESR1 Mutation in Estrogen-Receptor-Positive Metastatic Breast Cancer and the Effect on Overall Survival." Cancers 16.17 (2024).

94) Jordan, V. Craig. "Tamoxifen: Toxicities and Drug Resistance During the Treatment and Prevention of Breast Cancer." Annual Review of Pharmacology and Toxicology 35 (1995): 195-211.

95) Pescatori, Sara, et al. "A Tale of Ice and Fire: The Dual Role For 17β-Estradiol In Balancing DNA Damage and Genome Integrity." Cancers 13.7 (2021): 1583.

96) Suba, Zsuzsanna. "Gender-Related Hormonal Risk Factors for Oral Cancer." Pathology & Oncology Research 13 (2007): 195-202.

97) McGuire, Andrew, et al. "Effects of Age on The Detection and Management of Breast Cancer." Cancers 7.2 (2015): 908-929.

98) Feigman, Mary J., et al. "Pregnancy Reprograms the Epigenome of Mammary Epithelial Cells and Blocks The Development of Premalignant Lesions." Nature Communications 11.1 (2020): 2649.

99) Katz, Tiffany A. "Potential Mechanisms Underlying the Protective Effect of Pregnancy Against Breast Cancer: A Focus on the IGF pathway." Frontiers in Oncology 6 (2016): 228.

100) Russo, Jose, and Irma H. Russo. "Molecular Basis of Pregnancy-Induced Breast Cancer Prevention." Hormone Molecular Biology and Clinical Investigation 9.1 (2012): 3-10.

101) Gleicher, N. Why Are Reproductive Cancers More Common in Nulliparous Women? Reprod. Biomed. Online 2013, 26, 416–419.

102) Sinha, D.K.; Pazik, J.E.; Dao, T.L. Prevention of Mammary Carcinogenesis in Rats By Pregnancy: Effect Of Full-Term And Interrupted Pregnancy. Br. J. Cancer 1988, 57, 390–394.

103) Medina, D.; Smith, G.H. Chemical Carcinogen-Induced Tumorigenesis in Parous, Involuted Mouse Mammary Glands. JNCI J. Natl. Cancer Inst. 1999, 91, 967–969.

104) Yang, J. et al. Protective Effects Of Pregnancy And Lactation Against N-Methyl-N-Nitrosourea-Induced Mammary Carcinomas In Female Lewis Rats. Carcinog 1999, 20, 623–628.

105) Guzman, R.C.; et al. Hormonal Prevention of Breast Cancer: Mimicking the Protective Effect of Pregnancy. Proc. Natl. Acad. Sci. USA 1999, 96, 2520–2525.

106) Lakshmanaswamy, Rajkumar, et al. Short-term Exposure to Pregnancy Levels of Estrogen Prevents Mammary Carcinogenesis. Proc. Natl. Acad. Sci. USA 2001, 98, 11755–11759.

107) Lakshmanaswamy, Rajkumar, et al. Prevention of Mammary Carcinogenesis by Short-Term Estrogen and Progestin Treatments. Breast Cancer Res. 2003, 6, R31–R37.

108) Partridge, Ann H., et al. "Interrupting Endocrine Therapy to Attempt Pregnancy After Breast Cancer." New England Journal of Medicine 388.18 (2023): 1645-1656.

109) Iqbal, Javaid, et al. "Association of the Timing of Pregnancy with Survival in Women With Breast Cancer." JAMA Oncology 3.5 (2017): 659-665.

110) Knabben, Laura, and Michel D. Mueller. "Breast Cancer and Pregnancy." Hormone molecular biology and clinical investigation 32.1 (2017).

111) Lambertini, Matteo, et al. "Pregnancy after breast cancer: a systematic review and meta-analysis." Journal of Clinical Oncology 39.29 (2021): 3293-3305.

112) Azim Jr, Hatem A., et al. "Safety of Pregnancy Following Breast Cancer Diagnosis: A Meta-Analysis Of 14 Studies." European journal of cancer 47.1 (2011): 74-83.

113) Jordan, V. Craig. "The New Biology of Estrogen-Induced Apoptosis Applied to Treat and Prevent Breast Cancer." Endocrine-Related Cancer 22.1 (2015): R1.

114) Mueck, A. O., and H. Seeger. "Estrogen as a New Option for Prevention and Treatment Of Breast Cancer—Does This Need A "Time Gap"?" Climacteric: The Journal of the International Menopause Society 18.4 (2015): 444-447.

115) Garcia, Marcel, et al. "Activation of Estrogen Receptor Transfected Into A Receptor-Negative Breast Cancer Cell Line Decreases The Metastatic And Invasive Potential Of The Cells." Proceedings of the National Academy of Sciences 89.23 (1992): 11538-11542.

116) Levenson, Anait S., and V. Craig Jordan. "Transfection of Human Estrogen Receptor (ER) CDNA Into ER-Negative Mammalian Cell Lines." The Journal of Steroid Biochemistry and Molecular Biology 51.5-6 (1994): 229-239.

117) Howell, Anthony, et al. "Results of the ATAC (Arimidex, Tamoxifen, Alone or in Combination) Trial After Completion Of 5 Years' Adjuvant Treatment for Breast Cancer." Lancet 365.9453 (2005): 60-62.

118) Ma, Yan, et al. "Symptom Experience In Endocrine Therapy For Breast Cancer Patients: A Qualitative Systematic Review And Meta-Synthesis." Asia-Pacific Journal of Oncology Nursing (2023): 100364.

119) Kauffman, Robert P., Christina Young, and V. Daniel Castracane. "Perils of Prolonged Ovarian Suppression and Hypoestrogenism in The Treatment of Breast Cancer: Is the Risk of Treatment Worse Than the Risk of Recurrence?" Molecular and Cellular Endocrinology 525 (2021): 111181.

120) Molinelli, Chiara, et al. "Ovarian Suppression: Early Menopause and Late Effects." Current Treatment Options in Oncology 25.4 (2024): 523-542.

121) Scharl, A., and A. Salterberg. "Significance of Ovarian Function Suppression in Endocrine Therapy For Breast Cancer In Pre-Menopausal Women." Geburtshilfe und Frauenheilkunde 76.05 (2016): 516-524.

122) Hershman DL, Shao T, Kushi LH, et al. Early Discontinuation and Non-Adherence to Adjuvant Hormonal Therapy Are Associated with Increased Mortality In Women With Breast Cancer. Breast Cancer Res Treat 2011; 126: 529–537

123) Hadji P, Ziller V, Kyvernitakis J, et al. Persistence in Patients with Breast Cancer Treated with Tamoxifen or Aromatase Inhibitors: A Retrospective Database Analysis. Breast Cancer Res Treat 2013; 138: 185–191

124) Sarrel, Philip M., et al. "The Mortality Toll of Estrogen Avoidance: An Analysis of Excess Deaths Among Hysterectomized Women Aged 50 To 59 Years." American journal of public health 103.9 (2013): 1583-1588.

125) Michelsen, Trond M., et al. "All-Cause And Cardiovascular Mortality After Hysterectomy and Oophorectomy In A Large Cohort (HUNT2)." Acta Obstetricia et Gynecologica Scandinavica 102.4 (2023): 465-472.

126) Suba, Zsuzsanna. "Causal Therapy of Breast Cancer Irrelevant of Age, Tumor Stage and ER-Status: Stimulation of Estrogen Signaling Coupled with Breast Conserving Surgery." Recent Patents on Anti-Cancer Drug Discovery 11.3 (2016): 254-266.

127) Suba, Zsuzsanna. "Turn in Breast Cancer Care: Upregulation of Estrogen Signal May Be Much More Effective than Its Inhibition." 2022. 1-17.

128) Suba, Zsuzsanna Key to Estrogen Anticancer Capacity, unpublished.

129) Onwude, J. L. "Does Unopposed Peri-menopausal or Post-menopausal Estrogen Protect against Breast Cancer." A Systematic Review. J Surg Surgical Res 7.2 (2021): 075-082.

130) Iwase, H., et al. "Ethinylestradiol is Beneficial for Postmenopausal Patients with Heavily Pre-Treated Metastatic Breast Cancer After Prior Aromatase Inhibitor Treatment: A Prospective Study." British Journal of Cancer 109.6 (2013): 1537.

131) Bennink, Herjan JT Coelingh, et al. "The Use of High-Dose Estrogens for The Treatment of Breast Cancer." Maturitas 95 (2017): 11-23.

132) Traphagen, Nicole A., et al. "High Estrogen Receptor Alpha Activation Confers Resistance To Estrogen Deprivation And Is Required For Therapeutic Response To Estrogen In Breast Cancer." Oncogene 40.19 (2021): 3408-3421.

133) Schwartz, Gary N., et al. "Alternating 17β-Estradiol and Aromatase Inhibitor Therapies Is Efficacious in Postmenopausal Women with Advanced Endocrine-Resistant ER+ Breast Cancer." Clinical Cancer Research 29.15 (2023): 2767-2773.

134) Hosford, Sarah R., et al. "Estrogen Therapy Induces an Unfolded Protein Response to Drive Cell Death In ER+ Breast Cancer." Molecular Oncology 13.8 (2019): 1778-1794.

135) Fan, Ping, and V. Craig Jordan. "Estrogen Receptor and The Unfolded Protein Response: Double-Edged Swords in Therapy for Estrogen Receptor-Positive Breast Cancer." Targeted oncology 17.2 (2022): 111-124.

136) Zhou, Zhenqi, et al. "Proteomic Analysis Reveals Major Proteins and Pathways That Mediate the Effect of 17-β-Estradiol in Cell Division and Apoptosis in Breast Cancer MCF7 Cells." Journal of Proteome Research 23.11 (2024): 4835-4848.

137) Maximov, Philipp Y., et al. "Estrogen Receptor Complex to Trigger or Delay Estrogen-Induced Apoptosis In Long-Term Estrogen Deprived Breast Cancer." Frontiers in Endocrinology 13 (2022): 869562.

138) Chimento, Adele, et al. "Estrogen Receptors-Mediated Apoptosis In Hormone-Dependent Cancers." International Journal of Molecular Sciences 23.3 (2022): 1242.

139) Thomas, Elizabeth T., et al. "Prevalence of Incidental Breast Cancer and Precursor Lesions in Autopsy Studies: A Systematic Review and Meta-Analysis." BMC cancer 17 (2017): 1-10.

140) Bonneterre, J., et al. "Anastrozole versus tamoxifen as first-line therapy for advanced breast cancer in 668 postmenopausal women: results of the Tamoxifen or Arimidex Randomized Group Efficacy and Tolerability study." Journal of clinical oncology 18.22 (2000): 3748-3757.

141) Bonneterre, Jacques, et al. "Anastrozole is superior to tamoxifen as first-line therapy in hormone receptor positive advanced breast carcinoma: Results of two randomized trials designed for combined analysis." Cancer 92.9 (2001): 2247-2258.

Chapter 21

Estrogen Metabolism, Iodine, DIM and 2MeOE2

ELLEN IS A 56-YEAR-OLD ICU nurse doing well with her bioidentical hormone replacement program. She is enjoying full relief from menopausal symptoms. The hot flashes and night sweats are gone and she is thankful for better quality sleep. Every day, she takes her iodine supplement, Iodoral, a 12.5 mg tablet form of Lugol's solution containing 5mg iodine and 7.5mg potassium iodide. During a telephone follow-up call Ellen asked a question about the iodine supplement: "Is it really needed? The dosage seems excessive." And, she said had been reading on a message board that iodine can be harmful. I explained to her that iodine is one of the safest nutrients we have. In this chapter, we will discuss further the importance of iodine for breast cancer prevention, and specifically, as iodine relates to estrogen metabolism. In short, iodine is very beneficial for the metabolism of estradiol into the "good" intermediates, and away from the "bad" ones.

Estrogen Metabolism in a Nutshell

Estradiol is Produced by Aromatase Conversion of Testosterone

The aromatase enzyme is the key that converts testosterone into estradiol. Most of the estradiol is made in the ovary from aromatase conversion of testosterone to estradiol. Estrogen is so important that many other sites in the body have aromatase activity such as the brain, retina, adrenal gland, liver, testis, blood vessels (vasculature), fat cells (adipose tissue), skin and bone, all capable of converting testosterone to estrogen, thus producing the much-needed estrogen locally. After the estrogen locks into receptors and sends its signal, its job is done. Afterwards, where does all this estrogen go? Excess estrogen is metabolized by the liver into intermediates which are then excreted by the liver into the bile and into the lumen of the GI tract, and then excreted in the stools. The study of estrogen metabolism involves a detailed look at these estrogen metabolite intermediates, some of which are good, and some bad. This is also called Estrogen Detoxification, the elimination of toxic estrogen metabolites. (1)

What are the Three Human Estrogens? E1, E2 and E3.

When we refer to estrogen, we must be specific about which type of estrogen. There are three major human estrogens, estrone (E1), estradiol (E2), and estriol (E3). A fourth type of estrogen is derived from pregnant horses called conjugated equine estrogen (CEE), trade name Premarin, a mixture of more than 50 estrogens, invented in 1942 by Wyeth Ayerst. Estrone (E1) is a weak estrogen produced by body fat, placenta, and ovaries. E1 and E2 are reversibly converted to each other by 17β-hydroxysteroid dehydrogenase in the liver. E1 primarily activates the estrogen receptor alpha (ER-alpha), rather than ER-beta. Estradiol (E2) is produced by aromatase conversion of testosterone to estradiol. E2 is the strongest estrogen with strong binding to estrogen receptors, 50% binding to ER-alpha and 50% binding to ER-beta. E2 is made in tissues with aromatase enzyme activity, the ovaries being the main source. Estriol (E3) is a weak estrogen, the main estrogen of pregnancy secreted by the placenta. Estriol is produced by conversion of estradiol to estriol, and predominantly activates ER-beta, the tumor suppressor. (2)

Estrogen Receptors (ER), ER-alpha and ER-beta

Notice the above discussion of estrogen involves describing the ability to bind to and activate the two different estrogen receptors, ER-alpha and ER-beta. This is important because ER-alpha is the proliferative receptor associated with increased breast cancer risk. At the same time, ER-beta is the tumor suppressor receptor that has the benefit of breast cancer prevention. Activation of ER-beta causes suppression of ER-alpha activity. (3-5)

The Three Phases of Estrogen Detoxification

As mentioned above, estradiol (E2) and estrone (E1) are reversibly inter-converted in the liver by the 17beta-hydroxysteroid dehydrogenase enzyme, so estrogen metabolism starts with the conversion of estradiol to estrone (E1) which enters the Phase One Detoxification pathway:

The Phase One Estrogen Metabolism Pathways

1) The **CYP1A1 enzyme** acts on E1 => making 2-OH-E1 (2-hydroxy-estrone) => and then makes 2-methoxy-estrone (2-MeOE1) and 2 methoxy-estradiol (2-MeOE2). **This is Very GOOD!** However, this pathway requires a functioning COMT enzyme. 2MEO has anti-cancer activity as discussed below. **Note:** COMT=catechol-o-methyl transferase.

2) The **CYP1B1 enzyme** acts on E1=> making 4-OH-E1 (4-hydroxy-estrone) => and then makes quinones and DNA adducts. These are carcinogenic and **BAD!** Quinones are associated with higher risk for breast and endometrial cancer. This pathway is driven by CYP1B1 and becomes dominant when the patient has SNP in COMT causing impaired COMT function. 4-hydroxy-estrogens lead to 4-hydroxy-quinones which produce carcinogenic DNA Adducts. If COMT is functional then 4-OH-E1 is methylated to 4-MeO-E1 (4 methoxy-estrone) This is **Good!** Note SNP=single nucleotide poly-morphism or mutation.

3) The CYP3a4 enzyme acts on E1 => making 16-OH-E1 (16-hydroxy-estrone) which is converted to **estriol (E3),** a weak estrogen with preferential ER-beta binding, this is a beneficial tumor suppressor receptor. However, the 16-OH-estrone metabolites are generally regarded as genotoxic with high affinity for the estrogen receptor causing proliferative effects. This is **BAD**. (6-7)

Why is Methylation Important?

Adding a methoxy group to estrogen metabolites is called methylation, requiring a functioning COMT gene and a functioning MTHFR gene. SNPS (mutations) in either of these two genes causes poor methylation and increased levels of bad metabolites called quinones which form DNA adducts, increasing breast cancer risk. In 2016, Dr Samia Shouman writes:

> Estradiol is metabolized into 2-OHE2 and 4-OHE2 by CYP1A1 and CYP1B1, respectively. These catechols undergo further oxidation into semiquinones and quinones that react with DNA to form depurinating adducts leading to mutations associated with breast cancer. (8)

Dr. Ercole L. Cavalieri from U. of Nebraska Medical Center devoted most of his career to the study of estrogen metabolites, finding that N-acetyl-cysteine (NAC) and resveratrol could prevent DNA adduct formation, thus preventing cancer. In 2016, Dr. Ercole L. Cavalieri writes:

> Estrogens can initiate cancer by reacting with DNA. Specific metabolites of endogenous estrogens, the catechol estrogen-3,4-quinones, react with DNA to form depurinating estrogen-DNA adducts. Loss of these adducts leaves apurinic sites in the DNA, generating mutations that can lead to the initiation of cancer. ..The levels of depurinating estrogen-DNA adducts are high in women diagnosed with breast cancer and those at high risk for the disease... Women with thyroid or ovarian cancer also have high levels of estrogen-DNA

adducts, as do men with prostate cancer or non-Hodgkin lymphoma. Depurinating estrogen-DNA adducts are initiators of many prevalent types of human cancer... The dietary supplements N-acetylcysteine and resveratrol inhibit formation of estrogen-DNA adducts in cultured human breast cells and in women. (9)

Phase Two Metabolism is "conjugation" which turns these estrogen metabolites into water soluble compounds easily excreted by liver into the gut or kidney into urine. This involves methylation, sulfation or glucuronidation. (10)

Phase Three Estrogen Metabolism involves excretion of conjugated estrogen metabolites into the gut, supported by prebiotics, probiotics, Calcium D Glucarate and fiber. Although most of these conjugated metabolites are excreted in the stool, there is enterohepatic recirculation. Some estrogen metabolites are deconjugated and reabsorbed back into the portal venous system, a process facilitated by gut bacteria that produce beta-glucuronidase, the enzyme that deconjugates estrogen metabolites. (11)

4-OH-Estrogens are Highly Carcinogenic

In 2019, Dr. Suyu Miao studied the effects of 4-hydroxy-estrogens (4-OH-estrogens) finding they bind more strongly to estrogen receptors than estradiol (E2). In addition to this more estrogenic effect, Dr. Miao found direct evidence of carcinogenicity of 4-hydroxy-estrogens by inducing breast cancer in athymic nude mice. In Dr. Suyu Miao's mouse study, **normal breast cells were pretreated with 4-OH-E2 and then injected into athymic nude mice. 80 percent of the injected mice breast developed breast cancer within 2 weeks.** Control mice injected with untreated breast cells did not develop breast cancer. Thus Dr. Suyu Miao showed 4-OH-E2 to reliably induce breast cancer in a mouse model, writing:

> ...although 2-hydroxy metabolite 2-OHE2 had limited estrogenic activity, 4-hydroxy

metabolite 4-OHE2 **had much stronger estrogenic activity, which was even more potent than that of E2** in MCF-7 cells... The tumorigenic effect of 4-OH-E2 was first evaluated in athymic nude mice models. As expected, while mice injected with MCF10A cells [normal breast cells] into the mammary fat pad of nude mice did not form tumors, [however] **four out of five mice (80%) injected with MCF10A-H cells [breast cells pre-treated with 4-OH-E2] formed tumors after 2 weeks.** (12)

CYP1B1 Enzyme Creates 4-OH-Estrogens

Dr. Suyu Miao also studied the transgenic female mouse model genetically engineered to express the CYP1B1 gene which converts estradiol (E2) to 4-OH-E2. The mice were supplemented with estradiol (E2) slow-release pellets, and were compared to control mice with no E2 pellet treatment. The E2 pellet-treated transgenic mice had greater breast proliferation and higher KI-67, a proliferation marker, when compared to control mice with undetectable KI-67. In addition, the E2-treated transgenic mice had multiple mammary tumors, while the controls had none. Thus Dr. Suyu Miao demonstrated the CYP1B1 enzyme converts estradiol to 4-OH estrogens which are carcinogenic, writing:

> The results indicated that although 2-hydroxy metabolite 2-OHE2 had limited estrogenic activity, **4-hydroxy metabolite 4-OHE2 had much stronger estrogenic activity, which was even more potent than that of E2** in MCF-7 cells...... Taken together, multiple mammary tumors were only present in mice of the [transgenic mice expressing CYP1B1] + E2 [estradiol] group, but not in mice of other groups... analysis revealed **4-OH-E1 to be the most important factor of breast cancer risk**... the most significant alteration was an **increase in 4-hydroxy metabolites ...It was found that 4-OH-E2 not only induced malignant transformation of breast epithelial cells in vitro but also stimulated tumor growth in the xenograft model and induced mammary carcinomas in the transgenic mice model**

expressing CYP1B1, a key enzyme of 4-hydroxy metabolites... At the molecular level, **4-OH-E2 compromised the function of SAC spindle-assembly checkpoint (SAC) and thus rendered genome instability**... Among many alterations of EMs [estrogen metabolites] in the breast cancer group, the most significant one in this study was an increase in 4-hydroxy metabolites. **Urine 4-OH-E1 in the patients with breast cancer was three times higher than that in healthy women,** while other EMs [estrogen metabolites] changed less. The best indicator that reflected the risk of breast cancer was the **ratio of 4-hydroxy metabolites to total estrogen**. Conclusions: **the ratio of 4-hydroxy metabolites to total estrogen is the best indicator reflects the risk of breast cancer**. Our study found 4-OH-E2 induced carcinogenesis by destroying the SAC [spindle-assembly checkpoint] and induced the abnormal mitosis. (12)

In 2007, Dr. Alexandra R. Belous showed direct proof CYP1B1-mediated, E2-induced DNA adduct formation as the basis for carcinogenesis. (13-14)

How to Divert Estrogen Metabolism Away from BAD (4-OH-estrogens) to GOOD Metabolites (2-OH-estrogens =>2-MeO-E1)

Given the above discussion, the next logical question is: How can metabolism be shifted away from the bad 4-OH-estrogens, and towards the beneficial 2-OH estrogen pathway? This can be accomplished by upregulating the CYP1A1 enzyme in the liver, thus preferentially converting E1 (estrone) to 2-OH-E1 (2-hydroxy-estrone). Next, we discuss using nutritional supplements such as Iodine (Iodoral), I3C/DIM, and resveratrol to manipulate estrogen metabolism in this beneficial manner. One useful dietary modification is the increased cruciferous vegetable intake such as broccoli and brussel sprouts. Or, one may supplement with Indole 3 Carbinol (I3C) or its more active dimer, DIM (Di-Indole-Methane). Of course, one must avoid Endocrine Disrupting Chemical (EDC's)

such as BPA (Bisphenol-A) and Phalates in plastics, insecticides, and others. (15-18)

Using Iodine (Iodoral) to Manipulate Estrogen Metabolism

Let us next review how iodine can be used to manipulate estrogen metabolism toward the beneficial pathways, thus preventing breast cancer. Firstly, we must give thanks and credit to Jonathan V. Wright, MD, the "Father of Bioidentical Hormone Replacement Therapy" and an early pioneer in the field. A graduate of Harvard University and the University of Michigan Medical School in 1969, Dr. Wright established Tahoma Clinic in 1973 in Washington State, and Meridian Valley Laboratory in 1976. During his career, Dr. Jonathan Wright discovered the significant effects of iodine on estrogen metabolism as discussed in his 2005 paper in Annals of Internal Medicine, writing:

> To maximize the safety and efficacy of human hormone replacement therapy, it is suggested that exact molecular copies of human hormones ("bio-identical" hormones) be administered in physiologic quantities and proportions, following physiologic timing and routes of administration. It is also suggested that physicians return to the practice of monitoring hormone therapy by precise laboratory measurement levels of the hormones administered. This paper also presents clinical and laboratory data concerning appropriate proportions of bio-identical estrogens, the physiologic and supraphysiologic nature of commonly employed doses, estrogen levels achieved by varying routes of administration, and **the significant effects of iodine on estrogen metabolism** and cobalt on estrogen excretion. (19-21)

Iodine Decreased ER-alpha mRNA Levels (Bernard Eskin Group)

In 2008, Dr. Frederick R Stoddard II of Bernard Eskin's group studied how iodine supplementation alters gene expression in MCF7 breast cancer cells, finding Lugol's iodine, (5%

I2, 10% KI) useful for the manipulation of estrogen metabolism. In 2012, Dr. Alexander Poor, a member of Bernard Eskin's group studied iodine, estrogen, and breast cancer. Dr. Poor reviews the earlier work in 2008 by Dr. Stoddard who did a gene array study. Dr. Stoddard found iodine inhibits the expression of estrogen-responsive genes TFF1 and WISP2, and up-regulates estrogen metabolism (CYP1A1, CYP1B1, and AKR1C1). Lastly, iodine decreases Cyclin D1 mRNA levels which then permits BRCA1 inhibition of estrogen-responsive transcription. **Note:** Cyclin D1 is a competitive inhibitor of the BRCA1 gene. Finally, ER-alpha mRNA levels are decreased by Lugols' iodine in MCF-7 (breast cancer) cells in-vitro. Dr. Alexander Poor writes:

> We analyzed the effects of iodine on global gene expression in estrogen responsive MCF-7 breast cancer cell line. Microarray analysis and quantitative real time polymerase chain reaction (RT-PCR) indicated that **iodine inhibits the expression of estrogen-responsive genes TFF1 and WISP2**...Our data further provided three potential mechanisms to explain the observed decrease in estrogen response. First, iodine treatment results in **decreased ER-alpha mRNA** levels; second, iodine **up-regulates genes involved in estrogen metabolism** (CYP1A1, CYP1B1, and AKR1C1), and finally, iodine **decreases Cyclin D1** (a competitive inhibitor of BRCA1) **mRNA levels which may functionally permit BRCA1 inhibition of estrogen responsive transcription**. Thus the interaction between iodine and estrogen signaling may inhibit breast cancer growth by affecting an intermediate, perhaps the estrogen receptor system. Note: the BRCA1 gene is a tumor suppressor gene. Note: TFF1 (Trefoil Factor 1) is the classical target gene of the Estrogen Receptor and the most studied within the medical literature. (22-25)

Note: Estrogen receptor alpha (ER-alpha) is proliferative, while Estrogen receptor beta (ER-beta) is a tumor suppressor. Activation of ER-alpha activates transcription of the pro-liferative oncogene, Cyclin D1, implicated in carcinogenesis. The BRCA1 gene is a tumor suppressor gene, so any impairment of BRCA1, such as elevated cyclin D1 level or a mutation in the BRCA1 gene is pro-carcinogenic. According to Dr. Chenguang Wang in 2005, upregulation of Cyclin D1 prevents the BRCA1 gene from repressing the proliferative effect of ER-alpha (Estrogen Receptor alpha). The main activity of the BRCA1 gene is repair of oxidative damage, such as double-strand breaks in DNA using the seleno-protein system. Thus, selenium supplementation has been found effective for reducing breast cancer risk associated with BRCA1 mutation. (26-28)

Iodine Increases the CYP1A1/CYP1B1 Ratio

The 2008 gene array study by Dr. Frederick R. Stoddard reveals iodine supplementation shifts the ratio in favor of CYP1A1 over CYP1B1, thus increasing 2-hydroxy estradiol (2-OH-E2) which is then converted to the favorable 2-methoxy-estradiol, a favorable metabolite which has anti-cancer effects. CYP1B1 activity leads to the formation of the bad metabolite, 4-hydroxy-estradiol (4-OH-E2) which converts to the quinones which form carcinogenic DNA adducts. Thus, even though CYP1B1 mRNA production is increased 40-fold by Iodine (Lugol's Solution), the increase in CYP1A1 is a much greater 250-fold. In 2008, Dr. Stoddard revealed Lugol's iodine increases the CYP1A1/CYP1B1 ratio which leads to formation of 2-methoxy-estradiol, a strong anti-cancer metabolite acting independently from ER-alpha or ER-beta. Secondly, Lugol's iodine prevents BRCA1 gene inhibition, freeing the BRCA1 gene to inhibit ER-alpha signaling and thus, downregulating cell proliferation. The BRCA1 gene is responsible for DNA damage repair, writing:

> The observed increase in the CYP1A1/CYP1B1 ratio may shift the direction of estrogen metabolism favoring 2-OH-E2 which may either directly affect proliferation through increasing 2-methoxyestradiol, decreasing 3, 4-estradiol quinone or

indirectly via the inactivation of E2. The importance of the CYP1A1/CYP1B1 ratio in-vivo is evident in the increased presence of 4-OH-E2 in breast cancer tissue compared to non breast cancer controls....Data presented suggests that iodine/iodide may inhibit the estrogen response through...**1) up-regulating proteins involved in estrogen metabolism (specifically through increasing the CYP1A1/1B1 ratio), and...2) decreasing BRCA1 inhibition thus permitting its inhibition of estrogen responsive transcription [ER-alpha].** (22-25)

I3C/DIM Upregulates CYP1A1 Estrogen Metabolism

Next, let us turn our attention to cancer prevention properties of two natural compounds, Indole-3-carbinol (I3C) and 3,3'-diindolyl-methane (DIM), found in cruciferous vegetables like broccoli, cauliflower, and Brussels sprouts. As mentioned above, upregulation of the CYP1A1 pathway is beneficial as this leads away from quinones and DNA adduct formation, and leads towards formation of **the natural anti-cancer agent, 2-methoxy-estradiol.** Several animal studies have shown indole-3-carbinol (I3C) treatment in mice **upregulates CYP1A1 gene expression.** In 2010, Dr. N.V. Trusov found upregulated mRNA content in the liver for CYP1A1, CYP1A2, and CYP3A1 enzymes in I3C-treated mice. This effect was accompanied by an increased activity of phase II detoxification metabolism including the quinone reductase enzymes. (29-32)

I3C Reduces ER-alpha by 60 Percent

In 2006, Dr. Thomas Wang used MCF-7 (breast cancer) cells in-vitro to show that I3C reduces ER-alpha mRNA expression by 60%. DIM is the more active dimer of I3C, showing a 20-fold greater potency than I3C, and is therefore the preferred supplement. (33)

I3C/DIM Anti-Cancer Effects - I3C Downregulates ER-alpha and Induces CYP1A1

In 2008, Dr. Jing-Ru Weng studied I3C and its more active dimer, DIM, finding suppression of proliferation in breast, colon, prostate, and endometrial cancer cell lines. DIM also inhibited spontaneous or chemical-induced cancer formation in breast, liver, lung, cervix, and gastrointestinal tract in various animal model studies. In 2008, Dr. Jing-Ru Weng found I3C/DIM downregulates ER-alpha signaling, while at the same time increasing the binding of ER-beta to ERE (estrogen response elements), resulting in strong anti-proliferative effects, writing:

> **Indole-3-carbinol is a negative regulator of ER-alpha signaling** in human tumor cells. In addition to altering estrogen metabolism through CYP1A1, indole-3-carbinol and its metabolites also affect ER signaling through two different mechanisms...indole-3-carbinol and DIM could **suppress ER-alpha expression** in breast cancer cells ... Moreover, indole-3-carbinol was reported to **increase the binding of ER-beta to the estrogen response element,** resulting in a significantly higher ER-beta/ER-alpha ratio that is associated with an **antiproliferative status** in human breast cancer cells...(34)

I3C Induces 6-fold increase in binding of ER-Beta to the Estrogen Response Element

In 2006, Dr. Shyam Sundar studied two breast cancer cell lines, in vitro, finding I3C strongly downregulated ER-alpha protein levels and transcription. Even though ER-beta protein levels remain unchanged, I3C induced 6-fold increased binding of ER-beta to its ERE [Estrogen Response Element], thus acting as a tumor suppressor, writing:

> Taken together, our results demonstrate that the expression and function of ER-alpha and ER-beta can be uncoupled by I3C with a key cellular consequence being a significantly higher ER-beta:ER-alpha ratio that is generally highly associated with

antiproliferative status of human breast cancer cells. (35)

I3C Effective for Ulcerative Colitis in Animal study

Indole-3-Carbinol (I3C) has shown benefits for ulcerative colitis in animal models. In 2021, Dr. Shunting Peng writes:

> I3C is a good candidate as a natural product to prevent and treat UC [Ulcerative Colitis]. (36-37)

DIM Increases 2-OH-Estrogen in Human Study

In 2024, Dr. Mark Newman studied the effect of DIM on estrogen metabolism in premenopausal women using dried urine collection showing DIM increases 2-OH-estrogens (2-OH-E1). DIM also increased the the 2/16 ratio (ratio of 2-OH-estrogens to 16-OH estrogens). (38)

Predicting Breast Cancer Risk with Genetic Testing

SNPs (Single Nucleotide Polymorphism) COMT gene

Mutations (SNPs) in genes for the enzymes, COMT, CYP1A1, CYP1B1, ER-alpha, and ER-beta are procarcinogenic because they impair estrogen metabolism. In 2015, Dr Hamed Samavat writes:

> It has been postulated that genetic polymorphisms [also called SNPs] in genes encoding enzymes involved in estrogen metabolism pathways and the genes encoding the ERs [estrogen receptors] are associated with breast cancer risk. Polymorphic variations in genes encoding COMT, CYP1A1, CYP1B1, estrogen receptor alpha (ERα), estrogen receptor beta (ERβ), CYP17A1, and CYP19A1 have received extensive attention within the last decade. (2)

The enzyme, COMT (Catechol O methyl Transferase gene) is involved in methylation of estrogen, and conversion of 2 hydroxy estrone to 2 methoxy estrone. Any SNP (single nucleotide polymorphism, mutation) in the COMT gene will reduce the activity of the COMT enzyme and will increase catechol estrogen accumulation, quinone-estrogens, and pro-carcinogenic DNA adducts. Normally, estradiol binds to ER-alpha and ER-beta receptors in our DNA without causing any oxidative damage to our DNA. However, catechol estrogens are highly oxidative, creating DNA adducts and oxidative DNA damage (mutations), thus increasing risk for breast cancer. Prevention involves supplements such as Iodine (Iodoral), I3C/DIM, resveratrol, and NAC. Methyl donors such as Vitamins B12, B6 and methylfolate assist the methylation pathways. As you might suggest from the above discussion, it is possible to predict breast cancer risk based on genetic testing for mutations in genes for the following enzymes: COMT, CYP1A1, and CYP1B1. In 2021, Dr. Feng Zhao did exactly this, finding good predictive ability. (39)

PolyMorphisms in Estrogen Detoxification Pathways Increase Cancer Risk

In 2021, Dr. Micaela Almeida studied polymorphisms in genes involving metabolic pathways in estrogen detoxification using samples from 157 women with breast cancer. Genes studied were GSTM1, GSTT1, CYP1B1 Val432Leu and MTHFR C677T. The genes for GSTM1 and GSTT1 code for phase II enzymes that detoxify catechol estrogen quinones through the conjugation of glutathione (GSH). The absence of these enzymes, due to the null polymorphism of GSTM1 and GSTT1, compromises detoxification and allows accumulation of catechol estrogens, leading to carcinogenic DNA adduct formation. The polymorphism Val432Leu (CYP1B1) increases CYP1B1 activity, contributing to higher levels of 4-hydroxy catechol estrogens detoxified by Phase II enzymes... The polymorphism of MTHFR C677T reduces MTHFR activity, leading to decreased activity of COMT (catechol-O-methyl-transferase) needed for prevention of catechol estrogen formation. Dr. Micaela Almeida found various combinations of polymorphisms increased breast cancer risk, writing:

> GSTM1 and GSTT1 are phase II enzymes that detoxify catechol estrogen quinones through the conjugation of GSH. The absence of these enzymes, due to the null polymorphism of GSTM1 and GSTT1, compromises the detoxification and allows the accumulation of catechol estrogens, leading to DNA adducts formation …**We suppose that prolonged exposure to estrogen levels combined with an inefficient detoxification due to GSTM1 and GSTT1 null genotype are related to breast cancer development at later ages.** This fact can be explained by the accumulation of catechol estrogens and DNA adducts formation during a lifetime, which culminate in breast cancer development. ..A two-way association of MTHFR C677T and GSTT1 null genotype was performed and we verified that the majority of women carriers of both altered T allele of MTHFR C677T and GSTT1 null genotype were 50 years old or more at the age of [breast cancer] diagnosis (p-value = 0.034). These results might be explained by the fact that **the metabolic pathway is extremely compromised due to inexistent GSTT1 and low COMT activity; low levels of Phase II enzymes highly compromise 4-OH-E2 detoxification and eventually will contribute to tumor development due to inefficient estrogens detoxification during reproductive life.** (40)

SNPs Increased Risk for Ovarian Cancer 6-Fold

In 2021, Dr. Ercole Cavalieri studied SNPs in estrogen metabolic pathways, finding SNPs in CYP1B1 and COMT increased ovarian cancer by either 300 percent, or 600 percent depending on whether there were one or two copies of the mutations, writing:

> When women had one or two copies of the SNP for a more active CYP1B1 plus two copies of the SNP for a less active COMT, they were three times more likely to have ovarian cancer, and had approximately

twice the ratio of estrogen-DNA adducts to estrogen metabolites and conjugates as women without the SNPs. When the women had two copies of both the CYP1B1 and COMT SNPs, they were six times more likely to have the disease and had even higher estrogen-DNA adduct ratios. (41)

Resveratrol and NAC Prevent Quinone and Adduct Formation

The next logical question is how can we prevent estrogen-quinone formation and thus reduce cancer risk. In 2021, Dr. Ercole Cavalieri found resveratrol and NAC (N-acetyl-cysteine) useful for preventing estrogen quinones. Resveratrol is a a natural substance found in grape skins and peanuts having anti-inflammatory, antioxidant, antihyperlipidemic, anticarcinogenic, immune-modulating, cardioprotective, hepatoprotective, and neuroprotective properties. Resveratrol can prevent estrogen quinone formation by reducing CYP1B1 activity and thus reducing 4-hydroxy-quinones. Both NAC and Resveratrol convert semi-quinones back to catechol estrogens which can then be methylated to safer metabolites. NAC also forms conjugates with quinones, preventing formation of DNA adducts. Dr. Ercole Cavalieri writes:

> Both resveratrol and NAC have been shown to inhibit the formation of depurinating estrogen-DNA adducts in cultured mammalian cells. Resveratrol was found to inhibit the malignant transformation of the human MCF-10F breast epithelial cell line. NAC was found to inhibit the malignant transformation of both MCF-10F and immortalized mouse mammary cells. The two compounds work together additively to reduce the formation of depurinating estrogen-DNA adducts in MCF-10F cells treated with 4-OHE2. These results lay the foundation for investigating the ability of resveratrol and NAC to reduce estrogen-DNA adduct formation in humans as an approach to cancer prevention…The observation of high ratios of depurinating estrogen-DNA adducts to estrogen metabolites and

conjugates in women at high risk for breast cancer, as well as women with breast cancer, is consistent with an ER [Estrogen Receptor] -independent mechanism of initiation. In addition, the presence of SNPs in CYP1B1 and COMT that increase both the formation of depurinating estrogen-DNA adducts and the likelihood of ovarian cancer (six-fold) supports an ER-independent mechanism of cancer initiation by estrogens. (41)

Resveratrol Downregulates ER-alpha

Another benefit of resveratrol is the down-regulation of ER-alpha, the proliferative estrogen receptor. In 2016, Dr. Julieta Saluzzo studied in-vitro breast cancer cells, finding resveratrol downregulated ER-alpha protein. (42)

Berberine Increases CYP1A1 over CYP1B1

In 2014, Dr. Chun-Jie Wen studied berberine, a botanical used as a nutritional, in MCF-7 breast cancer cells in-vitro, finding preferential induction of CYPA1A over CYP1B1, suggesting anti-cancer effects, writing:

Previous reports suggested that 2-OH E2 have putative protective effects, while 4-OH E2 is genotoxic and has potent carcinogenic activity. Thus, the ratio of 2-OH E2/4-OH E2 is a critical determinant of the toxicity of E2 in mammary cells. In the present study, we investigated the effects of the berberine on the expression profile of the estrogen metabolizing enzymes CYP1A1 and CYP1B1 in breast cancer MCF-7 cells. Berberine treatment produced significant induction of both forms at the level of mRNA expression, but with **increased doses produced 16~ to 52~fold greater inductions of CYP1A1 mRNA over CYP1B1 mRNA.** Furthermore, berberine dramatically increased CYP1A1 protein levels but did not influence CYP1B1 protein levels in MCF-7 cells. In conclusion, we present the first report to show that berberine may provide protection against breast cancer by altering the ratio of CYP1A1/CYP1B1, could redirect E2 metabolism in a more protective

pathway in the breast cancer MCF-7 cells. (43)

MethylFolate for Homozygous MTHFR Carriers

About two thirds (66%) of the population has at least one mutation in the MTHFR gene, either 677C or 1298A, the two predominant variants of the MTHFR mutation. About 8.5-13.5 percent of the population harbors the most clinically expressed homozygous variation with both matching chromosomes bearing a mutation, and about 2.25% of people have compound heterozygous variant, meaning mutations with two different variants which may or may not reach clinical expression. (44-45)

The MTHFR Mutation

MTHFR mutation has been associated with increased risk for breast and other cancer, coronary artery disease, neuro-psychiatric disorders, etc. Treatment involves supplementation with methyl-folate rather than folate. A good quality multivitamin usually contains methyl-folate, but always check the label to make sure, and avoid the lower quality multivitamins which contain folic acid. The active form of the vitamin is methyl folate. (46-48)

2-Methoxy-Estradiol (2MeOE2) is a Natural Anticancer Drug

What is the mechanism of 2MeOE2 acting as a natural anti-cancer drug? 2MeOE2 inhibits tubulin polymerization, and serves as a microtubule inhibitor. As such, 2MeOE2 bears a resemblance to many other anti-cancer microtubule inhibitor drugs. In this category, you will find the repurposed anti-parasitic drugs fenbendazole (veterinary use), mebendazole (human use), and albendazole. Conventional chemotherapy drugs in the taxane family are microtublule inhibitors such as paclitaxel (Taxol), docetaxel (Taxotere) and cabazitaxel (Jevtana). Vinblastine and colchicine are also microtubule inhibitors. (49-51)

2MeOE2 Synergy with Albendazole

In 2013, Dr. Anahid Ehteda studied 2MeOE2 synergy with a second microtubule inhibitor, albendazole using the colorectal cancer cells (HCT-116) xenografted into nude mice. Dr. Anahid Ehteda found the two drugs were indeed synergistic anticancer agents. Both colchicine and 2MeOE2 are ligands for the colchicine binding site on tubulin. Remember, the anticancer activity of 2MeOE2 is independent of estrogen receptors alpha and beta. (52-54)

Inhibiting tubulin formation prevents the spindle formation required for mitosis (cell division) thus causing cell cycle arrest in pro-metaphase (G2/M phase). The guardian of the genome, P53, senses the cell cycle arrest and becomes activated. This upregulation of P53 takes as little as 10 nM of 2MeOE2, resulting in apoptosis (programmed cell death) and inhibition of NFk-B transcriptional activity (NFk-B = Nuclear Factor kappa-B, is the master inflammatory controller). (55)

Anti-Cancer Activity of 2MeOE2 in Various Cancers

The potent anti-cancer activity of 2MeOE2 in various cancer cell types has been known for over 25 years and has been studied in many cancers. Here is the list:

> breast cancer (Brueggemeier, Robert, 2001), nasopharyngeal carcinoma (Zhou, Ning Ning 2004), osteosarcoma (Tang, Xiaoyan, 2020), colorectal cancer (Carothers, A. M, 2002) (Lee, Ji Young, 2014), lung cancer (Mukhopadhyay, Tapas, 1997), oral Cancer (Takata, Hidehiko, 2004), uterine sarcoma (Amant, Frederic, 2003), pancreatic cancer (Schumacher, Guido,1999), T lymphoblastic leukemia (Zhang, Xueya, 2010), prostate cancer (Kumar, Addanki, 2001), anaplastic thyroid cancer (Roswall, Pernilla, 2006), melanoma (Hua, Weitian, 2022). (56-68)

2MeOE2 Anti-Inflammatory Effects

2-methoxy-estradiol (2MeOE2) has strong anti-inflammatory effects by inhibition of Nuclear Factor Kappa B (NFk-B), the inflammatory master controller. 2MeOE2 prevents transcription of NFk-B. These anti-inflammatory effects have been extensively evaluated with preclinical studies showing benefits for the vascular system, brain, joints, and bones. 2MeOE2 reduced atherosclerosis, neuro-inflammation, arthritis, and osteoporosis in various preclinical models. (69-90)

2MeOE2 Remains Largely Unknown by Convention Doctors

2-methoxy-estradiol shows considerable promise as a safe highly effective cancer drug. One might ask why 25 years have gone by without any drug maker showing interest in making 2MeOE2 widely available as a safe oncology drug. Unfortunately, since there is no patent protection for natural substances, the drug will probably never be formally studied in human trials, will never receive FDA approval, and will remain largely unknown. I find this a sad commentary on the state of oncology drug discovery in the US. Despite its status outside of conventional oncology, 2-methoxy-estradiol is available for the brave at heart as an oral capsule from a few selected compounding pharmacies in the United States.

Back to the WHI Second Arm Study

Remember the second arm of the Women's Health Initiative which found a 23 percent reduction in breast cancer in the estrogen (Premarin, CEE) treated group? The mechanism described by V. Craig Jordan is the induction of apoptosis after long-term estrogen deprivation (LTED). The second arm of the WHI used older post-menopausal women who had been estrogen-deprived for longer than 5 years. In a 2023 study by Dr. Masayo Hirao-Suzuki, 2-methoxy-estradiol (2MeOE2) was found to be a selective inhibitor of LTED MCF-7 breast cancer cells. These are cells that had been long-term estrogen deprived. In these cells, 2MeOE2 induced cell cycle arrest at low concentrations (1 microMolar). None of the other microtubule inhibitor agents had this type of selectivity for LTED breast cancer cells. Dr. Masayo Hirao-Suzuki writes:

> To identify effective treatment modalities for breast cancer with acquired resistance, we first compared the responsiveness of estrogen receptor-positive breast cancer MCF-7 cells and long-term estrogen-deprived (LTED) cells...derived from MCF-7 cells to G-1 and 2-methoxyestradiol (2-MeO-E2), which are microtubule-destabilizing agents ... LTED cells displayed approximately 1.5-fold faster proliferation than MCF-7 cells. ..2-MeO-E2 exerted antiproliferative effects selective for LTED cells with an IC50 value of 0.93 μM (vs. 6.79 μM for MCF-7 cells) and induced G2/M cell cycle arrest. Moreover, we detected higher amounts of β-tubulin proteins in LTED cells than in MCF-7 cells... Other microtubule-targeting agents, i.e., paclitaxel, nocodazole, and colchicine, were not selective for LTED cells. Therefore, 2-MeO-E2 can be an antiproliferative agent to suppress LTED cell proliferation. (91)

The above study by Dr. Masayo Hirao-Suzuki suggests another mechanism of estrogen induction of apoptosis in LTED breast cancer cells as described by V. Craig Jordan. This mechanism involves the endogenous production of 2MeOE2 which then induces apoptosis in LTED breast cancer cells.

Estrogen Metabolism Testing

A few laboratories offer urine testing for estrogen metabolites such as the 2-hydroxy and 4-hydroxy estrogens. At the time of writing, these include Genova lab, ZRT lab, Doctor's Data Lab, and Precision Analytical (Dutch Test). Financial Disclosure: I have no financial interest in any of these laboratories.

Conclusion: The study of estrogen metabolism and detoxification pathways provides a new level of understanding of breast cancer risk, highlighting the use of Iodine, DIM, resveratrol, and other supplements to shift estrogen metabolism away from CYP1B1 and towards CYPA1A leading to 2-methoxy-estradiol, a natural anti-cancer drug. On the other hand, if CYP1B1 predominates, this leads to 4-hydroxy-estrogens and carcinogenic DNA adduct formation. Natural supplements such as Lugol's iodine, I3C/DIM, resveratrol, NAC, and methylfolate assist in manipulating estrogen detoxification pathways towards the favorable 2 methoxy-estradiol and away from 4-hydroxy quinones, thus reducing breast cancer risk.

♦ References for Chapter 21 Estrogen Metabolism, Iodine, DIM and 2-MeOE2

1) Barakat, Radwa, et al. "Extra-Gonadal Sites of Estrogen Biosynthesis and Function." BMB reports 49.9 (2016): 488.

2) Samavat, Hamed, and Mindy S. Kurzer. "Estrogen Metabolism and Breast Cancer." Cancer Letters 356.2 (2015): 231-243.

3) Mal, Rahul, et al. "Estrogen Receptor Beta (ER-Beta): A Ligand Activated Tumor Suppressor." Frontiers in Oncology 10 (2020): 587386.

4) Jia, Min, Karin Dahlman-Wright, and Jan-Åke Gustafsson. "Estrogen receptor Alpha and Beta In Health and Disease." Best Practice & Research Clinical Endocrinology & Metabolism 29.4 (2015): 557-568.

5) Williams, Cecilia, et al. "A Genome-Wide Study of The Repressive Effects of Estrogen Receptor Beta on Estrogen Receptor Alpha Signaling In Breast Cancer Cells." Oncogene 27.7 (2008): 1019-1032.

6) Starek-Swiechowicz, et al. "Endogenous Estrogens—Breast Cancer and Chemoprevention." Pharmacological Reports 73.6 (2021): 1497-1512.

7) Fuhrman BJ, et al. Estrogen Metabolism and Risk of Breast Cancer in Postmenopausal Women. J Natl Cancer Inst. 2012;104(4):326–339

8) Shouman, Samia, et al. "Leptin Influences Estrogen Metabolism and Increases DNA Adduct Formation in Breast Cancer Cells." Cancer Biology & Medicine 13.4 (2016): 505.

9) Cavalieri, Ercole L., and Eleanor G. Rogan. "Depurinating Estrogen-DNA Adducts, Generators Of Cancer Initiation: Their Minimization Leads To Cancer Prevention." Clinical and Translational Medicine 5 (2016): 1-15.

10) Jancova, Petra, and Michal Siller. "Phase II Drug Metabolism." Topics on Drug Metabolism (2012).

11) Hu, Shiwan, et al. "Gut Microbial Beta-Glucuronidase: A Vital Regulator in Female Estrogen Metabolism." Gut Microbes 15.1 (2023): 2236749.

12) Miao, Suyu, et al. "4-Hydroxy Estrogen Metabolite, Causing Genomic Instability by Attenuating the Function of Spindle-Assembly Checkpoint, Can Serve as A Biomarker for Breast Cancer." American Journal of Translational Research 11.8 (2019): 4992.

13) Belous, Alexandra R., et al. "Cytochrome P450 1B1-Mediated Estrogen Metabolism Results In Estrogen-Deoxyribonucleoside Adduct Formation." Cancer research 67.2 (2007): 812-817.

14) Liehr, Joachim G., and Mary Jo Ricci. "4-Hydroxylation of Estrogens as Marker of Human Mammary Tumors." Proceedings of the National Academy of Sciences 93.8 (1996): 3294-3296.

15) Kawa, Iram Ashaq, et al. "Bisphenol A (BPA) Acts as An Endocrine Disruptor in Women with Polycystic Ovary Syndrome: Hormonal and Metabolic Evaluation." Obesity Medicine 14 (2019): 100090.

16) Talpade, J., et al. "Bisphenol A: An Endocrine Disruptor." J Entomol Zool Stud 6.3 (2018): 394-7.

17) Hafezi, Shirin A., and Wael M. Abdel-Rahman. "The Endocrine Disruptor Bisphenol A (BPA) Exerts A Wide Range of Effects in Carcinogenesis and Response to Therapy." Current Molecular Pharmacology 12.3 (2019): 230-238.

18) La Merrill, Michele A., et al. "Consensus on the key characteristics of endocrine-disrupting chemicals as a basis for hazard identification." Nature Reviews Endocrinology 16.1 (2020): 45-57.

19) Wright, Jonathan V. "Bio-Identical Steroid Hormone Replacement: Selected Observations from 23 Years of Clinical and Laboratory Practice." Annals of the New York Academy of Sciences 1057.1 (2005): 506-524.

20) Wright, Jonathan V., et al. "Comparative Measurements of Serum Estriol, Estradiol, And Estrone in Non-Pregnant, Premenopausal Women: A Preliminary Investigation." Alternative Medicine Review 4 (1999): 266-270.

21) Friel, Patrick N., Christa Hinchcliffe, and Jonathan V. Wright. "Hormone Replacement with Estradiol: Conventional Oral Doses Result In Excessive Exposure To Estrone." Altern Med Rev 10.1 (2005): 36-41.

22) Poor, Alexander E., et al. "Urine Iodine, Estrogen, and Breast Disease." Journal of Cancer Therapy 3 (2012): 1164-1169.

23) Stoddard II, Frederick R., et al. "Iodine Alters Gene Expression in The MCF7 Breast Cancer Cell Line: Evidence for An Anti-Estrogen Effect of Iodine." International Journal of Medical Sciences 5.4 (2008): 189.

24) Saxena, Neela, et al. "Differential Expression Of WISP-1 And WISP-2 Genes in Normal and Transformed Human Breast Cell Lines." Molecular and Cellular Biochemistry 228 (2001): 99-104.

25) Banerjee, Snigdha, et al. "WISP-2 Gene in Human Breast Cancer: Estrogen and Progesterone Inducible Expression And Regulation Of Tumor Cell Proliferation." Neoplasia 5.1 (2003): 63-73.

26) Wang, Chenguang, et al. "Cyclin D1 Antagonizes BRCA1 Repression of Estrogen Receptor Alpha Activity." Cancer Research 65.15 (2005): 6557-6567.

27) Kowalska, Elzbieta, et al. "Increased Rates of Chromosome Breakage in BRCA1 Carriers Are Normalized by Oral Selenium Supplementation." Cancer Epidemiology Biomarkers & Prevention 14.5 (2005): 1302-1306.

28) Dziaman, Tomasz, et al. "Selenium Supplementation Reduced Oxidative DNA Damage in Adnexectomized BRCA1 Mutations Carriers." Cancer Epidemiology, Biomarkers & Prevention 18.11 (2009): 2923-2928.

29) Trusov, N. V., et al. "Effects of Combined Treatment with Resveratrol and Indole-3-Carbinol." Bulletin Of Experimental Biology and Medicine 149 (2010): 213-218.

30) Mohammadi, Saeed, et al. "Indole-3-carbinol Induces G1 Cell Cycle Arrest and Apoptosis Through Aryl Hydrocarbon Receptor In THP-1 Monocytic Cell Line." Journal Of Receptors and Signal Transduction 37.5 (2017): 506-514.

31) Tutelyan, V. A., et al. "Indole-3-Carbinol Induction of CYP1A1, CYP1A2, And CYP3A1 Activity and Gene Expression In Rat Liver Under Conditions Of Different Fat Content In The Diet." Bulletin Of Experimental Biology and Medicine 154 (2012): 250-254.

32) Ociepa-Zawal, Marta, et al. "The Effect of Indole-3-Carbinol on The Expression of CYP1A1, CYP1B1 And AHR Genes And Proliferation of MCF-7 cells." Acta Biochimica Polonica 54.1 (2007): 113-117.

33) Wang, Thomas TY, et al. "Estrogen Receptor Alpha as A Target for Indole-3-Carbinol." The Journal of Nutritional Biochemistry 17.10 (2006): 659-664.

34) Weng, Jing-Ru, et al. "Indole-3-Carbinol as A Chemopreventive and Anti-Cancer Agent." Cancer Letters 262.2 (2008): 153-163.

35) Sundar, Shyam N., et al. "Indole-3-Carbinol Selectively Uncouples Expression and Activity Of Estrogen Receptor Subtypes In Human Breast Cancer Cells." Molecular Endocrinology 20.12 (2006): 3070-3082.

36) Peng, Chunting, et al. "Indole-3-Carbinol Ameliorates Necroptosis and Inflammation of Intestinal Epithelial Cells in Mice with Ulcerative Colitis by Activating Aryl Hydrocarbon Receptor." Experimental Cell Research 404.2 (2021): 112638.

37) Busbee, Philip B., et al. "Indole-3-Carbinol Prevents Colitis and Associated Microbial Dysbiosis in an IL-22–Dependent Manner." JCI insight 5.1 (2020).

38) Newman, Mark, and Jaclyn Smeaton. "Exploring the Impact of 3, 3'-Diindolylmethane on the Urinary Estrogen Profile of Premenopausal Women." (2024).

39) Zhao, Feng, et al. "Discovery of Breast Cancer Risk Genes and Establishment of a Prediction Model Based on Estrogen Metabolism Regulation." BMC Cancer 21 (2021): 1-11.

40) Almeida, Micaela, et al. "Influence of Estrogenic Metabolic Pathway Genes Polymorphisms on Postmenopausal Breast Cancer Risk." Pharmaceuticals 14.2 (2021): 94.

41) Cavalieri, Ercole, and Eleanor Rogan. "The 3, 4-Quinones of Estrone and Estradiol Are the Initiators of Cancer whereas Resveratrol and N-acetylcysteine Are the Preventers." International Journal of Molecular Sciences 22.15 (2021).

42) Saluzzo, Julieta, et al. "The Regulation of Tumor Suppressor Protein, P53, And Estrogen Receptor (ER-alpha) By Resveratrol in Breast Cancer Cells." Genes & Cancer 7.11-12 (2016): 414.

43) Wen, Chun-Jie, et al. "Preferential Induction of CYP1A1 Over CYP1B1 In Human Breast Cancer MCF-7 Cells After Exposure to Berberine." Asian Pacific Journal of Cancer Prevention 15.1 (2014): 495-499.

44) Long, Sarah, and Jack Goldblatt. "MTHFR Genetic Testing: Controversy and Clinical Implications." Australian Family Physician 45.4 (2016): 237-240.

45) Graydon, James S., et al. "Ethnogeographic Prevalence and Implications of the 677C> T And 1298A> C MTHFR Polymorphisms in US Primary Care Populations." Biomarkers in Medicine 13.8 (2019): 649-661.

46) Cristalli, Carlotta Pia, et al. "Methylenetetrahydrofolate Reductase, MTHFR, Polymorphisms and Predisposition to Different Multifactorial Disorders." Genes & Genomics 39 (2017): 689-699.

47) Petrone, Igor, et al. "MTHFR C677T and A1298C Polymorphisms in Breast Cancer, Gliomas and Gastric Cancer: A Review." Genes 12.4 (2021): 587.

48) Li, Z., et al. "The Methylenetetrahydrofolate Reductase (MTHFR) C677T Gene Polymorphism Is Associated with Breast Cancer Subtype Susceptibility in Southwestern China." PLoS ONE 16.7 (2021): e0254267.

49) LaVallee, Theresa M., et al. "2-Methoxyestradiol Inhibits Proliferation and Induces Apoptosis Independently of Estrogen Receptors Alpha and Beta." Cancer research 62.13 (2002): 3691-3697.

50) Romero, Yair, et al. "Antitumor Therapy under Hypoxic Microenvironment by the Combination of 2-Methoxyestradiol and Sodium Dichloroacetate on Human Non-Small-Cell Lung Cancer." Oxidative Medicine and Cellular Longevity 2020.1 (2020): 3176375.

51) Cermak, Vladimir, et al. "Microtubule-Targeting Agents and Their Impact on Cancer Treatment." European journal of cell biology 99.4 (2020): 151075.

52) Ehteda, Anahid, et al. "Combination of Albendazole and 2-Methoxyestradiol Significantly Improves the Survival Of HCT-116 Tumor-Bearing Nude Mice." BMC cancer 13 (2013): 1-13.

53) Zefirov, Nikolai A., et al. "Adamantyl-Substituted Ligands of Colchicine Binding Site in Tubulin: Different Effects on Microtubule Network in Cancer Cells." Structural Chemistry 30 (2019): 465-471.

54) D'Amato, Robert J., et al. "2-Methoxyestradiol, an Endogenous Mammalian Metabolite, Inhibits Tubulin Polymerization by Interacting at The Colchicine Site." Proceedings of the National Academy of Sciences 91.9 (1994): 3964-3968.

55) Siebert, Amy E., et al. "Effects of Estrogen Metabolite 2-Methoxyestradiol on Tumor Suppressor Protein P53 And Proliferation of Breast Cancer Cells." Systems Biology in Reproductive Medicine 57.6 (2011): 279-287.

56) Brueggemeier, Robert W., et al. "2-Methoxymethylestradiol: a New 2-Methoxy Estrogen Analog That Exhibits Antiproliferative Activity and Alters Tubulin Dynamics." The Journal of steroid biochemistry and molecular biology 78.2 (2001): 145-156.

57) Zhou, Ning Ning, et al. "2-Methoxyestradiol Induces Cell Cycle Arrest and Apoptosis Of Nasopharyngeal Carcinoma Cells." Acta Pharmacologica Sinica 25 (2004): 1515-1520.

58) Tang, Xiaoyan, et al. "Anticancer Effects and The Mechanism Underlying 2-Methoxyestradiol In Human Osteosarcoma In Vitro And In Vivo." Oncology Letters 20.4 (2020): 1-1.

59) Carothers, A. M., et al. "2-Methoxyestradiol Induces P53-Associated Apoptosis of Colorectal Cancer Cells." Cancer Letters 187.1-2 (2002): 77-86.

60) Lee, Ji Young, et al. "Tumor Suppressor Protein P53 Promotes 2-Methoxyestradiol-Induced Activation of Bak and Bax, Leading to Mitochondria-Dependent Apoptosis In Human Colon Cancer HCT116 Cells." Journal Of Microbiology and Biotechnology 24.12 (2014): 1654-1663.

61) Mukhopadhyay, Tapas, and Jack A. Roth. "Induction of Apoptosis In Human Lung Cancer Cells After Wild-Type P53 Activation By Methoxyestradiol." Oncogene 14.3 (1997): 379-384.

62) Takata, Hidehiko, et al. "2-Methoxyestradiol Enhances P53 Protein Transduction Therapy-Associated Inhibition of The Proliferation of Oral Cancer Cells Through the Suppression Of NF-kappaB Activity." Acta Medica Okayama 58.4 (2004): 181-187.

63) Amant, Frederic, et al. "2-Methoxyestradiol Strongly Inhibits Human Uterine Sarcomatous Cell Growth." Gynecologic Oncology 91.2 (2003): 299-308.

64) Schumacher, Guido, et al. "Potent Antitumor Activity Of 2-Methoxyestradiol in Human Pancreatic Cancer Cell Lines." Clinical Cancer Research 5.3 (1999): 493-499.

65) Zhang, Xueya, et al. "2-Methoxyestradiol Blocks Cell-Cycle Progression at The G2/M Phase and Induces Apoptosis in Human Acute T Lymphoblastic Leukemia CEM cells." Acta Biochim Biophys Sin 42.9 (2010): 615-622.

66) Kumar, Addanki P., et al. "2-Methoxyestradiol Blocks Cell-Cycle Progression at G2/M Phase And Inhibits Growth Of Human Prostate Cancer Cells." Molecular Carcinogenesis. 31.3 (2001): 111-124.

67) Roswall, Pernilla, et al. "2-Methoxyestradiol Induces Apoptosis in Cultured Human Anaplastic Thyroid Carcinoma Cells." Thyroid 16.2 (2006): 143-150.

68) Hua, Weitian, et al. "2-Methoxy Estradiol Inhibits Melanoma Cell Growth By Activating Adaptive Immunity." Immunopharmacology And Immunotoxicology 44.4 (2022): 541-547.

69) Bourghardt, Johan, et al. "The Endogenous Estradiol Metabolite 2-Methoxyestradiol Reduces Atherosclerotic Lesion Formation in Female Apolipoprotein E-Deficient Mice." Endocrinology 148.9 (2007): 4128-4132.

70) Dantas, Ana Paula V., and Kathryn Sandberg. "Does 2-Methoxyestradiol Represent The New And Improved Hormone Replacement Therapy For Atherosclerosis?" Circulation Research 99.3 (2006): 234-237.

71) Chakrabarti, Subhadeep, et al. "Estrogen is a Modulator of Vascular Inflammation." IUBMB life 60.6 (2008): 376-382.

72) Stubelius, Alexandra, et al. "Role of 2-Methoxyestradiol as Inhibitor Of Arthritis And Osteoporosis In A Model Of Postmenopausal Rheumatoid Arthritis." Clinical Immunology 140.1 (2011): 37-46.

73) Shand, Francis Henry Warner, et al. "In Vitro And In Vivo Evidence for Anti-Inflammatory Properties Of 2-Methoxyestradiol." Journal of Pharmacology and Experimental Therapeutics 336.3 (2011): 962-972.

74) Sutherland, Tara E., et al. "2-Methoxyestradiol—a Unique Blend of Activities Generating A New Class Of Anti-Tumour/Anti-Inflammatory Agents." Drug discovery today 12.13-14 (2007): 577-584.

75) Huerta-Yepez, S., et al. "2-Methoxyestradiol (2-ME) Reduces the Airway Inflammation and Remodeling In An Experimental Mouse Model." Clinical immunology 129.2 (2008): 313-324.

76) Hu, Qiang, et al. "2-Methoxyestradiol Alleviates Neuroinflammation and Brain Edema In Early Brain Injury After Subarachnoid Hemorrhage In Rats." Frontiers in Cellular Neuroscience 16 (2022): 869546.

78) Chen, Ying-Yin, et al. "Anticancer Drug 2-Methoxyestradiol Protects against Renal Ischemia/Reperfusion Injury by Reducing Inflammatory Cytokines Expression." BioMed research international 2014.1 (2014): 431524.

79) Yan, Chunguang, et al. "2-Methoxyestradiol Protects Against Igg Immune Complex-Induced Acute Lung Injury by Blocking NF-Kb and CCAAT/Enhancer-Binding Protein B Activities." Molecular Immunology 85 (2017): 89-99.

80) Plum, Stacy M., et al. "Disease Modifying and Antiangiogenic Activity Of 2-Methoxyestradiol In A Murine Model Of Rheumatoid Arthritis." BMC Musculoskeletal Disorders 10 (2009): 1-13.

81) Schaufelberger, Sara A., et al. "2-Methoxyestradiol, an Endogenous 17β-Estradiol Metabolite, Inhibits Microglial Proliferation and Activation Via an Estrogen Receptor-Independent Mechanism." American Journal of Physiology-Endocrinology and Metabolism 310.5 (2016): E313-E322.

82) Singh, Purnima, et al. "Central CYP1B1 (Cytochrome P450 1B1)-Estradiol Metabolite 2-Methoxyestradiol Protects from Hypertension and Neuroinflammation In Female Mice." Hypertension 75.4 (2020): 1054-1062.

83) Song, Chi Young, et al. "2-Methoxyestradiol Ameliorates Angiotensin II–Induced Hypertension by Inhibiting Cytosolic Phospholipase A2α Activity in Female Mice." Hypertension 78.5 (2021): 1368-1381.

84) Zhang, Yong, et al. "Estrogen Metabolite 2-Methoxyestradiol Attenuates Blood Pressure in Hypertensive Rats by Downregulating Angiotensin Type 1 Receptor." Frontiers in Physiology 13 (2022): 876777.

85) Azhar, Ahmad S., et al. "2-Methoxyestradiol Inhibits Carotid Artery Intimal Hyperplasia Induced By Balloon Injury Via Inhibiting JAK/STAT Axis In Rats." Environmental Science and Pollution Research 29.39 (2022): 59524-59533.

86) Kumar, Addanki P., et al. "2-Methoxyestradiol Interferes With Nfκb Transcriptional Activity In Primitive Neuroectodermal Brain Tumors: Implications For Management." Carcinogenesis 24.2 (2003): 209-216.

87) Takata, Hidehiko, et al. "2-Methoxyestradiol Enhances P53 Protein Transduction Therapy-Associated Inhibition of The Proliferation Of Oral Cancer Cells Through The Suppression Of Nfkappab Activity." Acta Medica Okayama 58.4 (2004): 181-187.

88) Yan, Chunguang, et al. "2-Methoxyestradiol Protects Against Igg Immune Complex-Induced Acute Lung Injury By Blocking NF-Kb And CCAAT/Enhancer-Binding Protein B Activities." Molecular Immunology 85 (2017): 89-99.

89) Yeh, Ching-Hua, et al. "Anticancer Agent 2-Methoxyestradiol Improves Survival In Septic Mice By Reducing The Production Of Cytokines And Nitric Oxide." Shock 36.5 (2011): 510-516.

90) Duncan, Gordon S., et al. "2-Methoxyestradiol Inhibits Experimental Autoimmune Encephalomyelitis Through Suppression of Immune Cell Activation." Proceedings of the National Academy of Sciences 109.51 (2012): 21034-21039.

91) Hirao-Suzuki, Masayo, et al. "2-Methoxyestradiol as an Antiproliferative Agent for Long-Term Estrogen-Deprived Breast Cancer Cells." Current Issues in Molecular Biology 45.9 (2023): 7336-7351.

Chapter 22

Iodine for Breast Cancer Prevention

A CLOSE FAMILY FRIEND JUST went through an ordeal with breast cancer. Thankfully she is doing well after the surgical lumpectomy and was told the margins were clear, and there was no lymph node involvement. In other words, the lumpectomy was curative. In 2024, the incidence of breast cancer has increased to 1 in 8 women or about 6,000 new cases weekly. You might ask the question: have we overlooked a safe inexpensive and widely available preventive measure? The answer is **YES, it is iodine**. The essential mineral iodine was added to table salt in 1924 as a public health measure to prevent goiter in school children. Iodine deficiency causes a form of thyroid enlargement called goiter. In some cases, thyroid enlargement can be massive. Giving the population iodine supplements is preventive. (1-2)

Our Diet is Iodine Deficient

Iodized salt is the major source of dietary iodine in the American diet. Yet, many of us have been told by the doctor to avoid salt, because salt causes high blood pressure. Those following this advice will have very little dietary iodine. Sadly, despite the iodized salt, we still have iodine deficiency in our population. In 2012, 38% of the U.S. population was iodine deficient. In 2021, 23 percent of pregnant women were iodine deficient. Maternal iodine deficiency causes cretinism, mental retardation, and developmental delay in the baby. Although all processed foods contain a large quantity of salt, none of this added salt is iodized. Iodine is one of the most common nutrient deficiencies and is estimated to affect 35–45% of the world's population. Iodine deficiency is the most common cause of goiter and is estimated to affect 2.2 billion people globally. (3-5)

The RDA for Iodine is too Low for Optimal Health

Guy Abraham MD was an iodine expert, and professor of obstetrics, gynecology, and endocrinology at UCLA Medical School. In 1978, Guy Abraham, MD started the Optimox company which produced Iodoral, a 12.5 mg iodine tablet based on the same formula as Lugol's solution, 5% I2 (molecular iodine) and 10% KI (potassium iodide). Sadly, Dr. Abraham passed away in 2013. Iodoral has been rebranded and is offered by Allergy Research Group. **Note:** I have no financial interest in Allergy Research Group. The RDA (recommended daily allowance) for iodine in the U.S. is 150 mcg per day. According to Guy Abraham MD, our iodine intake is insufficient, and he recommended a higher iodine intake of 12.5 mg per day, based on the seaweed consumption in the Japanese diet which contains kelp (wakame, kombu, and nori). Some think higher dietary iodine is the main reason for the lower incidence of breast, prostate, and thyroid cancer in Japan. (6-8)

How Safe is Iodine Supplementation?

Iodine in the form of I2 or KI, is the only trace element that can be ingested safely in amounts up to 100,000 times the RDA. Of course, this is not true for medications that contain iodine such as amiodarone and iodinated contrast material which have toxicity due to the surrounding organic molecule, and not the iodine within it as discussed in 2004 by Dr. Guy Abraham who writes:

> Of all the elements known so far to be essential for health, iodine is the most misunderstood and the most feared. Yet, **it is by far the safest of all the trace elements** known to be essential for human health. **It is**

the only trace element that can be ingested safely in amounts up to 100,000 times the RDA. For example, potassium iodide has been prescribed safely to pulmonary patients in daily amounts of up to 6.0 gm/day, in large groups of such patients for several years. It is important, however, to emphasize that this safety record only applies to **inorganic, non-radioactive iodine/iodide, not to organic iodine-containing drugs and to radioiodides.** Unfortunately, the severe side effects of iodine-containing drugs have been attributed to inorganic iodine/iodide, even though published studies clearly demonstrate that it is the whole organic molecule that is cytotoxic, not the iodine covalently bound to this molecule. (13)

Pulmonary Medicine: Historically, SSKI, supersaturated potassium iodide was commonly prescribed safely to chronic obstructive pulmonary disease (COPD) patients in large amounts. Recommended by Dr. Jonathan V. Wright, SSKI is an older treatment for COPD which helps mobilize lung secretions. The recommended dosage is 3 to 6 drops of SSKI daily. At 50 mg per drop SSKI, this amounts to 150-300 mg of potassium iodide daily. (9-11)

Cardiology: Another commonly used drug with high iodine content is amiodarone, widely used in cardiology as an anti-arrhythmia drug. A 200 mg tablet of amiodarone contains about 75 mg of organic iodide, and 10 percent or 7.5 mg is released freely into circulation as iodide. The typical dosage is one to three tablets daily, meaning 7 to 21 mg of free iodide daily. (12)

Endocrinology: Graves' Hyperthyroidism: Use of Lugol's solution for Graves' disease was first described in 1862 by Armand Trousseau in Paris, France, and is still used today before thyroidectomy for the Graves' disease patient. In Japan, long-term oral potassium iodide 100 mg per day is an accepted treatment for Graves' Disease. (13-19)

Dermatology: Iodine is used for various dermatologic disorders. In 2013, Dr. Rosane Orofino Costa from Brazil reviewed the use of potassium iodide for dermatologic conditions, writing:

> Despite being used in Medicine for over a century, potassium iodide remains a good therapeutic option for the treatment of several dermatoses as a drug of first or second choice and may be part of the dermatologist's therapeutic arsenal. (20)

The FDA Says 130 mg of Iodine is Safe

The FDA has officially stated that iodine supplementation is safe and recommends 130 mg of potassium iodide for adults in case of a radiation emergency such as a nuclear reactor accident. Considered a high dose, 130 mg of potassium iodide acts as a thyroid blocking drug which prevents thyroid uptake of radioactive iodine from inhaled radioactive gas accidentally released during a nuclear accident. Radioactive iodine uptake by the thyroid increases the risk of thyroid cancer, so blocking this uptake is a good strategy. Examples of nuclear reactor accidents are Three Mile Island in the United States, Chernobyl in Ukraine, and Fukushima in Japan. (21-22)

Iodine Allergy?

"Iodine allergy" is a misnomer since this name applies to an allergy to iodinated radiographic contrast agents, and not to elemental iodine (I2) or potassium iodine (KI) which is quite different chemically. Elemental iodine is an essential mineral and is required for life. Maternal iodine deficiency causes infantile cretinism. Iodine deficiency in the developing child causes goiter and thyroid enlargement. As such, any baby that does not receive iodine in utero and the diet would be unable to survive. Thus, there can be no allergy to elemental iodine, the same as there can no allergy to oxygen or water. All are essential nutrients needed to sustain life. Iodine, a well-known topical antiseptic, and antimicrobial agent, is also an anti-cancer agent. (23-25)

Iodine Treats Fibrocystic Breast Disease

Dr. Bernard A. Eskin published 80 papers over 30 years on iodine as a breast cancer preventive, reporting iodine deficiency increases the risk for breast and thyroid cancer in humans and animals. Iodine deficiency is also known to cause a pre-cancerous condition called fibrocystic breast disease. In 1993, Drs. William R. Ghent and Bernard Eskin showed iodine supplementation is effective for treatment of fibrocystic breast disease. Most clinicians are unaware of this. Despite its obvious potential, not much has been done with iodine treatment over the past 40 years in the United States. Since iodine is not patentable and therefore unlikely to be profitable to market, there is no money to fund studies for FDA approval. However, FDA approval is not required since iodine is widely available over-the-counter (OTC) as a supplement at the health food store. (26-30)

Iodine Deficiency Causes Thyroid Cysts and Nodules and Fibrocystic Disease

As an interventional radiologist working in the hospital for 25 years, a part of my job was evaluating thyroid abnormalities, nodules, and cysts with ultrasound, radionuclide scans, and needle biopsy. Although it was obvious these common thyroid abnormalities were due to iodine deficiency, I often wondered why none of the patients received iodine supplementation. The obvious answer is they should have been. The protective role of iodine is ignored by mainstream medicine. Part of my day as a radiologist was spent reading mammograms and breast ultrasound studies. Fibrocystic breast disease was quite common, and these women would return for needle aspiration procedure of the many breast cysts, and needle biopsy for the benign solid nodules. Many of these women returned multiple times over the years for these procedures because the medical system had no other useful treatment to offer them. We now know iodine is the medical treatment. Iodine supplementation not only resolves breast cysts and fibrocystic breast disease, but it also resolves ovarian cysts and thyroid cysts. Iodine supplementation has always been available, but again this is ignored by mainstream medicine, and hospital-based physicians are mostly unaware of it. (26-27)

Which Iodine Supplement to Use? Povidone, Lugol's and Iodoral

There are many iodine supplements on the market. Introduced in 1955, topical povidone-iodine, has antibacterial, antifungal and antiviral activity, and is still used in the operating room to prepare the skin prior to surgery, and is widely available over the counter. Although not for internal use, dilute 0.5 percent povidone-iodine mouth gargle and nasal spray is commonly used and highly effective for prevention of various upper respiratory viruses. (31-36)

In 1829, Dr. Jean Lugol from France developed Lugol's solution, an oral iodine solution for treatment of tuberculosis containing 5 percent elemental iodine (I2) and 10 percent potassium iodide (KI) in distilled water. An oral tablet form of Lugol's solution is called Iodoral from Optimox, a company founded by Dr. Guy Abraham who started "The Iodine Project" in 1997, and engaged two family practice physicians, Dr. Jorge Flechas and Dr. David Brownstein to collaborate on clinical studies to examine the health benefits of consuming 12.5 mg of dietary iodine per day, equivalent to the seaweed in the Japanese diet. More than 4,000 patients in this project consumed iodine supplements from 12 to 50 mg per day, and in those with diabetes, up to 100 mg a day. They reported their findings that iodine does indeed reverse fibrocystic disease. Diabetic patients require less insulin. Hypothyroid patients require less thyroid medication. The fibromyalgia symptoms resolve. Patients with migraine headaches stop having them. Both Lugol's solution and Iodoral tablets are widely available without a prescription. Sadly, Dr. Abraham passed away on February 13, 2013. (37-46)

One Gram Standard Dose

The Nobel laureate, Dr. Albert Szent-Györgyi (1893–1986), the physician who discovered vitamin C, used large doses of potassium iodide freely in his medical practice starting in 1917. The standard dose of potassium iodide given in 1917 was one gram (1,000 milligrams, containing 770 mg of iodine). One might speculate spontaneous remission of breast cancer reported in 1906 by Sir William Osler could have been due to large doses of iodine commonly prescribed at that time, as discussed in the next Chapter 23. Dr. Albert Szent Györgi writes:

> When I was a medical student, iodine in the form of KI (Potassium Iodide) was the universal medicine. Nobody knew what it did, but it did something and did something good. We students used to sum up the situation in this little rhyme: If ye don't know where, what, and why Prescribe ye then K and I. (47-49) (65-69)

Iodine and High Selenium Prevent Breast Cancer

Dietary selenium is important for cancer prevention because selenium is incorporated into selenoproteins essential for our cancer surveillance and DNA repair system. In 2020, Dr. Jonas Manjer studied serum levels of both iodine and selenium from patients in the Malmo Diet and Cancer Study which provided serum samples for 1,159 breast cancer cases and 1,136 controls. Dr. Manjer found the higher selenium and iodine levels associated with a **25 percent reduction in breast cancer**, writing:

> Among women with high selenium levels (above the median), high iodine levels were associated with a lower risk of breast cancer; the OR for above versus below the median was **0.75** (0.57-0.99) ... Conclusions: The combination of high serum iodine levels and high selenium levels was associated with a lower risk of breast cancer. (50-53)

In 2017, Dr. Jay Rappaport suggested that a dietary iodine deficiency explains the increasing incidence of advanced breast cancer in young women. Iodine deficiency is causative for breast cancer, and iodine supplementation is preventive. Increasing environmental exposure to halogens, fluoride, and bromide interferes with iodine uptake, and exacerbates iodine deficiency. The removal of iodine from bread in the 1970s and substitution with bromine may have caused a decline in iodine levels, explaining the increased incidence of metastatic breast cancer in younger women. Japanese women have a higher dietary intake of iodine from seaweed consumption and an exceptionally low incidence of breast cancer. However, upon moving to the US and adopting a low-iodine western diet, the incidence of breast cancer increases. Iodine is taken up by breast cells by active transport, the sodium iodide symporter, a molecular pump also present in the gastric mucosa, thyroid, and salivary glands. Dr. Jay Rappaport suggests all clinicians should be testing for iodine levels in their patients and giving iodine supplements when needed, writing:

> Iodine deficiency has been proposed to play a causative role in the development of breast cancer. Dietary iodine has also been previously proposed to play a protective role in breast cancer to a large degree based on the increased iodine consumption of dietary iodine in Japanese women, having and exceptionally low incidence of breast cancer. Furthermore, emigration of Japanese women and adopting a western diet is associated with higher breast cancer rates. Iodine is taken up by the sodium/iodide symporter in the breast and its role is important in promoting the development of normal versus neoplastic breast tissue development. In animal models of breast cancer, iodine in supplement or seaweed form, has demonstrated beneficial effects in suppressing breast cancer cell and tumor growth. The mechanism of action of iodine's anticancer effect may be complex, and roles as an antioxidant, promoting differentiation and apoptosis related to breast cancer have been proposed .... **Iodine deficiency**

is associated with fibrocystic breast disease, which can be effectively treated or prevented with iodine supplementation. Fibrocystic breast disease affects at least 50% of women of child-bearing age and is associated with an increased risk of developing breast cancer ...The observed drop in urinary iodine in young women as well as in the general population, since the 1970s, is presumably due to **removal of iodine from bread and substitution with bromine as flour conditioner during this period of time**, due in a large part to previous concerns about excess iodine as well as the preferences of commercial bakers for brominated flour. **Bromine, a suspected carcinogen, may further exacerbate iodine insufficiency since bromine competes for iodine uptake by the thyroid gland** and potentially other tissues (i.e. breast).... In conclusion, **dietary iodine insufficiency represents a plausible explanation for the increasing incidence of breast cancer in young women with distant metastasis.** In view of the established reduction in iodine levels in US women of childbearing age since the mid-70s, this group would be most vulnerable to increased breast cancer risk... Based on the importance of iodine in **thyroid and breast health, fetal brain development**, as well as deficits in nutritional trends among younger women, **iodine testing and management may be considered as a potentially important aspect for clinical practice.** (54-57)

In 2024, Dr. Yulia Kurniawati studied urinary iodine excretion finding significantly lower urine iodine concentrations in breast cancer patients when compared to controls (80 vs. 144).

In BC [breast cancer] patients, regardless of subtypes, breast cancer subjects showed a **significantly lower iodine excretion level**. The median of UIC [urinary iodine concentration] [breast cancer] patients and controls were 80.05 μg/L and 144.25 μg/L, respectively. (58-59)

Conclusion: Ignoring the important role of iodine in health is one of the tragic errors of modern medicine. Iodine is our major breast cancer preventive and an excellent treatment for fibrocystic disease. I would rank iodine alongside testosterone as our two main breast cancer preventives. Testosterone is discussed in Chapter 18. There is by now, overwhelming evidence iodine deficiency is a causative factor for breast cancer, and iodine supplementation is preventive. I agree with Dr. Jay Rappaport who advises all physicians to test their patients for iodine levels, easily done with the random spot urine for iodine, available at all laboratories. Iodine supplements may be purchased without a prescription. Although iodine in large doses is safe for healthy people, iodine should be used with caution under the care of a knowledgeable physician in people with auto-immune thyroid disorders such as Hashimoto's thyroiditis. Iodine is contraindicated for those with toxic nodular goiter or autonomous thyroid nodule as this may cause thyroid storm, a medical emergency. For more on iodine and the thyroid, see my book, *Natural Thyroid Toolkit.* (60-64)

♦ **References for Chapter 22 Iodine for Breast Cancer Prevention**

1) Zimmermann, Michael B., and Kristien Boelaert. "Iodine Deficiency and Thyroid Disorders." The Lancet Diabetes & Endocrinology 3.4 (2015): 286-295.

2) Leung, Angela M., et al. "History of US Iodine Fortification and Supplementation." Nutrients 4.11 (2012): 1740-1746.

3) Zimmermann, Michael B. "Iodine Deficiency in Pregnancy and The Effects of Maternal Iodine Supplementation on The Offspring: A Review." The American Journal of Clinical Nutrition 89.2 (2009): 668S-672S.

4) Graedon, Terry. Is the Salt in Prepared Foods Iodized? (2019) Peoples Pharmacy. https://www.peoplespharmacy.com/articles/is-the-salt-in-prepared-foods-iodized

5) Pennington, J. A., and S. A. Schoen. "Contributions of Food Groups to Estimated Intakes of Nutritional Elements: Results from The FDA Total Diet Studies, 1982-1991." International Journal for Vitamin and Nutrition Research. 66.4 (1996): 342-349.

6) Abraham, Guy E., Jorge D. Flechas, and J. C. Hakala. "Orthoiodosupplementation: Iodine Sufficiency Of The Whole Human Body." The Original Internist 9.4 (2002): 30-41.

7) Smyth, Peter PA. "The Thyroid, Iodine and Breast Cancer." Breast Cancer Research 5 (2003): 1-4.

8) Zava, Theodore T., and David T. Zava. "Assessment of Japanese Iodine Intake Based on Seaweed Consumption in Japan: A Literature-Based Analysis." Thyroid Research 4.1 (2011): 14.

9) Wright, Jonathan V. "Nutrition and Healing. SSKI for COPD" (2007). https://jeffreydachmd.com/wp-content/uploads/2015/03/COPD-Natural-Treatments-Jonathan-V-Wright-Aug-2002.pdf

10) Petty, Thomas L. "The History Of COPD." International Journal of Chronic Obstructive Pulmonary Disease 1.1 (2006): 3-14.

11) Bernecker, C. "Intermittent Therapy with Potassium Iodide in Chronic Obstructive Disease Of The Airways. A Review Of 10 Years' Experience." Acta Allergologica 24.3 (1969): 216-225.

12) Basaria, Shehzad, and David S. Cooper. "Amiodarone and the Thyroid." The American Journal Of Medicine 118.7 (2005): 706-714.

13) Abraham, Guy E. "The Safe and Effective Implementation of Orthoiodosupplementation In Medical Practice." The Original Internist 11.1 (2004): 17-36.

14) Fujikawa, Megumi, and Ken Okamura. "Graves' Hyperthyroidism Treated with Potassium Iodide: Early Response and After 2 Years of Follow-Up." European Thyroid Journal 13.6 (2024).

15) Trousseau, Armand. Clinical Medicine: Lectures Delivered at the Hôtel-Dieu, Paris. Vol. 2. P. Blakiston, Son, 1882.

16) Kopp, Peter A. "Iodine in the Therapy of Graves' Disease: A Century After Henry S. Plummer." Thyroid 33.3 (2023): 273-275.

17) Schiavone, Donatella, et al. "Role of Lugol solution Before Total Thyroidectomy for Graves' Disease: Randomized Clinical Trial." British Journal of Surgery 111.8 (2024): znae196.

18) Bloomfield, Arthur L. "The History of the Use of Iodine in Toxic Diffuse Goiter (Graves' Disease)." AMA Archives of Internal Medicine 100.4 (1957): 678-683.

19) Elaraj, Dina M., and Thomas J. Fahey. "The Use of Potassium Iodide in Graves' Disease." Surgery 163.1 (2018): 73-74.

20) Costa, Rosane Orofino, et al. "Use of Potassium Iodide in Dermatology: Updates on An Old Drug." Anais Brasileiros De Dermatologia 88.3 (2013): 396-402.

21) Leung, Angela M., et al. "American Thyroid Association Scientific Statement on The Use of Potassium Iodide Ingestion in a Nuclear Emergency." Thyroid 27.7 (2017): 865-877.

22) Guidance: Potassium Iodide as a Thyroid Blocking Agent in Radiation Emergencies. U.S. Department of Health and Human Services. Food and Drug Administration. Center for Drug Evaluation and Research (CDER) December 2001. Procedural.

23) Abraham, Guy E., Jorge D. Flechas, and J. C. Hakala. "Optimum Levels of Iodine for Greatest Mental and Physical Health." The Original Internist 9.3 (2002): 5-20.

24) Winder, Mateusz, et al. "The impact of Iodine Concentration Disorders on Health and Cancer." Nutrients 14.11 (2022): 2209.

25) Patrick, Lyn. "Iodine: Deficiency and Therapeutic Considerations." Alternative Medicine Review 13.2 (2008).

26) Gaby, Alan R. "Iodine Treatment of Fibrocystic Breast Disease." Townsend Letter for Doctors and Patients 256 (2004): 24-25.

27) Ghent, William R., et al. "Iodine Replacement in Fibrocystic Disease of The Breast." Canadian Journal of Surgery 36.5 (1993): 453-460.

28) Eskin, Bernard A. "Iodine and Breast Cancer a 1982 Update." Biological Trace Element Research 5 (1983): 399-412.

29) Bernard A. Eskin et al. Rat Mammary Gland Atypia Produced by Iodine Blockade with Perchlorate. Cancer Res. 1975 Sep;35(9):2332-9

30) Eskin, Bernard A, and Bruce V Stadel. "Dietary Iodine and Cancer Risk." The Lancet 308.7989 (1976): 807-808.

31) Fuse, Yozen, et al. "High Ingestion Rate of Iodine from Povidone-Iodine Mouthwash." Biological Trace Element Research (2022): 1-8.

32) Seikai, T., et al. "Gargling with Povidone Iodine Has a Short-Term Inhibitory Effect On SARS-Cov-2 In Patients with COVID-19." Journal of Hospital Infection 123 (2022): 179-181.

33) Lim, Nicole-Ann, et al. "Repurposing Povidone-Iodine to Reduce the Risk Of SARS-Cov-2 Infection and Transmission: A Narrative Review." Annals of Medicine 54.1 (2022): 1488-1499.

34) Sundar, Shyam. "Repurposing 0.5% Povidone Iodine Solution in Otorhinolaryngology Practice in Covid 19 Pandemic." International Journal of Otolaryngology and Surgery 1.2 (2024): 17-19.

35) Ang, W. X., et al. "Antiviral Activity of Povidone-Iodine Gargle and Mouthwash Solution Against Enterovirus A71, Coxsackieviruses A16, A10, and A6." Tropical Biomedicine 41.3 (2024): 241-250.

36) Friedland, Peter L., and Simon Tucker. "Phase II Trial of The Impact 0.5% Povidone-Iodine Nasal Spray (Nasodine®) On Shedding of SARS-CoV-2." The Laryngoscope (2024).

37) Durani, Piyush, and David Leaper. "Povidone–iodine: Use in Hand Disinfection, Skin Preparation and Antiseptic Irrigation." International Wound Journal 5.3 (2008): 376-387.

38) Calissendorff, Jan, and Henrik Falhammar. "Lugol's Solution and Other Iodide Preparations: Perspectives and Research Directions in Graves' Disease." Endocrine 58 (2017): 467-473.

39) Buchanan, J. Arthur. "Lugol, his Work and His Solution." Annals of Medical History 10.2 (1928): 202.

40) Powell, John L. "Powell's Pearls: Jean Guillaume Auguste Lugol, MD (1788–1851)." Obstetrical & Gynecological Survey 61.1 (2006): 1.

41) Neuzil, Eugène. "Jean Guillaume Auguste Lugol (1788-1851): His Life and His Works: A Brief Encounter, 150 Years After His Death." Histoire des Sciences Medicales 36.4 (2002): 451-464.

42) Abraham, Guy E. "The Historical Background of The Iodine Project." The Original Internist 12.2 (2005): 57-66.

43) Abraham, Guy E. "Iodine: The Universal Nutrient." Townsend Letter for Doctors and Patients 269 (2005): 85.

44) Abraham, Guy E., Jorge D. Flechas, and J. C. Hakala. "Optimum Levels of Iodine for Greatest Mental and Physical Health." The Original Internist 9.3 (2002): 5-20.

45) Flechas, Jorge D. "Orthoiodosupplementation in A Primary Care Practice." The Original Internist 12.2 (2005): 89-96.

46) Brownstein, David. "Clinical Experience with Inorganic, Non-Radioactive Iodine/Iodide." The Original Internist 12.3 (2005): 105-108.

47) Morell, Sally Fallon. "The Great Iodine Debate." Wise Traditions in Food, Farming, And the Healing Arts, Summer (2009). https://www.westonaprice.org/health-topics/modern-diseases/the-great-iodine-debate/

48) Miller, D. W. "Extrathyroidal Benefits of Iodine." Journal of American Physicians and Surgeons 11.4 (2006): 106.

49) Abraham, Guy E. "The History of Iodine in Medicine Part I: From Discovery to Essentiality." The Original Internist 13.1 (2006): 29-36.

50) Manjer, Jonas, et al. "Serum Iodine and Breast Cancer Risk: A Prospective Nested Case–Control Study Stratified for Selenium Levels." Cancer Epidemiology, Biomarkers & Prevention 29.7 (2020): 1335-1340.

51) Longtin, Robert. "Selenium for Prevention: Eating Your Way to Better DNA Repair?" Journal of the National Cancer Institute 95.2 (2003): 98-100.

52) Lubinski, J., et al. "Serum Selenium Levels Predict Survival After Breast Cancer." Breast Cancer Research and Treatment 167.2 (2018): 591-598.

53) Harris, Holly R., et al. Selenium Intake and Breast Cancer Mortality in A Cohort of Swedish Women." Breast Cancer Research and Treatment 134 (2012): 1269-1277.

54) Rappaport, Jay. "Changes in Dietary Iodine Explains Increasing Incidence of Breast Cancer with Distant Involvement in Young Women." Journal of Cancer 8.2 (2017): 174.

55) Cann, Stephen A., Johannes P. Van Netten, and Christiaan van Netten. "Hypothesis: Iodine, Selenium and The Development of Breast Cancer." Cancer Causes & Control 11 (2000): 121-127.

56) Waugh, Declan Timothy. "Fluoride Exposure Induces Inhibition of Sodium/Iodide Symporter (NIS) Contributing to Impaired Iodine Absorption and Iodine Deficiency: Molecular Mechanisms of Inhibition and Implications for Public Health." International Journal of Environmental Research and Public Health 16.6 (2019): 1086.

57) Vobecký, Miloslav, et al. "Interaction of Bromine with Iodine in The Rat Thyroid Gland at Enhanced Bromide Intake." Biological Trace Element Research 54 (1996): 207-212.

58) Kurniawati, Yulia, et al. "Analysis of Urinary Iodine Concentration in Differentiated Thyroid Cancer and Breast Cancer Cases." Asian Pacific Journal of Cancer Prevention: APJCP 25.6 (2024): 1869.

59) Kargar, Saeed, et al. "Urinary Iodine Concentrations in Cancer Patients." Asian Pacific journal of cancer prevention: APJCP 18.3 (2017): 819.

60) Livadas, D. P., et al. "The Toxic Effects of Small Iodine Supplements in Patients with Autonomous Thyroid Nodules." Clinical endocrinology 7.2 (1977): 121-127.

61) Hedberg, Fredric, et al. "Assessing the Impact of Short-Term Lugol's Solution On Toxic Nodular Thyroid Disease: A Pre-Post-Intervention Study." Frontiers in Endocrinology 15 (2024): 1420154.

62) Derry, David M. Breast Cancer, and Iodine: How to Prevent and How to Survive Breast Cancer. Trafford Publishing; 2nd edition (2001)

63) Brownstein, David. Iodine: Why You Need It, Why You Can't Live Without It. Medical Alternatives Press; (2008)

64) Farrow, Lynn. The Iodine Crisis: What You Don't Know About Iodine Can Wreck Your Life Devon Press; (2018)

65) Onuigbo, Wilson IB. "Spontaneous Regression of Breast Carcinoma: Review of English publications from 1753 to 1897." Oncology Reviews 6.2 (2012): e22.

66) Osler, William. "An Address on the Medical Aspects of Carcinoma of The Breast." British Medical Journal 1.2349 (1906): 1.

67) D'Alessandris, Nicoletta, et al. "What Can Trigger Spontaneous Regression of Breast Cancer?" Diagnostics 13.7 (2023): 1224.

68) Sasamoto, Mahato, et al. "Breast Carcinoma with Spontaneous Regression After Needle Biopsy: A Case Report and Literature Review." Gland Surgery 12.6 (2023): 853.

69) Zahl, Per-Henrik, and Jan Mæhlen. "Model of Outcomes of Screening Mammography: Spontaneous Regression of Breast Cancer May Not Be Uncommon." BMJ: British Medical Journal 331.7512 (2005): 350.

Chapter 23

Iodine for Treatment of Breast Cancer

Spontaneous Regression of Breast Cancer

DAVID BROWNSTEIN MD REPORTS THREE cases of spontaneous regression of breast cancer after women take iodine supplementation in his book, on page 63, *Iodine, Why You Need It, Why You Can't Live Without It*. The first patient, Joan a 63-year-old English teacher, was diagnosed with breast cancer in 1989, declined conventional treatment, and took 50 mg per day of Iodoral, 20mg of molecular iodine I2, and 30mg of potassium iodide in each 50 mg tablet. Six weeks later, Joan's PET scan showed, "all of the existing tumors were disintegrating". The second patient, 73-year-old Delores, was diagnosed with breast cancer in 2003. She declined conventional treatment with radiation and chemotherapy. Instead, Dolores took 50 mg of Iodoral daily. A follow-up breast ultrasound 18 months later showed, "It appears that these malignancies have diminished in size since the last examination. Interval improvement is definitely seen." Two years later a follow-up mammogram and ultrasound failed to show any abnormality and were read by the radiologist as normal. The third patient, 52-year-old Joyce was diagnosed with breast cancer two years prior and started on Iodoral, 50 mg per day. Three years after starting Iodoral, her follow-up mammograms and ultrasound exams show a decreasing size of the tumor with no progression. (1)

Iodine Deficiency Causes Breast Cancer - The Overwhelming Evidence

Iodine deficiency is associated with a higher rate of goiter and breast cancer. Similarly, higher dietary Iodine intake is associated with less goiter and breast cancer. For example, Japan has the highest dietary intake of iodine, 13 mg per day, from seaweed consumption, and the lowest rates for goiter and breast cancer. However, when Japanese women immigrate to the U.S. and change their dietary intake of iodine to the lower 150 mcg/day in the U.S., breast cancer rates increase. Iceland is another country with high iodine intake depending on fish consumption. Fish are naturally high in iodine. In 2017, Dr. Alfheidur Haraldsdottir from Reykjavik, Iceland studied 9,340 Icelandic women over 27 years of follow-up, during which time 744 women were diagnosed with breast cancer. Women who grew up in coastal villages with higher iodine intake from fish consumption had **54 percent reduction in breast cancer**, compared to women who grew up in the city with lower iodine intake from lower fish consumption. (2-5)

Iodine Suggested as Adjuvant Treatment for Breast Cancer

In previous chapters, we discussed the DMBA mouse model of breast cancer as the most studied animal model. DMBA is a chemical carcinogen that induces breast cancer reliably when injected into mice. Over the years, hundreds of independent researchers have confirmed molecular iodine suppresses DMBA induction of breast cancer. In 2005, Dr. Carmen Aceves from Mexico found in both animal models and human studies that molecular iodine (I2) suppresses the development of breast cancer. Dr. Aceves proposed iodine (I2) as adjuvant breast cancer therapy. The anti-cancer effect of molecular iodine (I2) is from intra-cellular iodolactones, writing:

> Seaweed is an important dietary component in Asian communities and a rich source of iodine in several chemical forms. The high

consumption of this element (25 times more than in Occident) has been associated with the low incidence of benign and cancer breast disease in Japanese women. **In animal and human studies, molecular iodine (I2) supplementation exerts a suppressive effect on the development and size of both benign and cancer neoplasias.** This effect is accompanied by a significant reduction in cellular lipoperoxidation. Iodine, in addition to its incorporation into thyroid hormones, is bound into antiproliferative iodolipids in the thyroid called **iodolactones**, which may also play a role in the proliferative control of mammary gland. We propose that an I2 supplement should be considered as an adjuvant in breast cancer therapy... (6-8)

In 2009, the Carmen Aceves Velasco Group in Mexico reported iodine as safe, with no harmful effects on thyroid function, and an anti-proliferative effect on human breast cancer cell cultures. Iodine binds to membrane lipids called lactones forming **iodo-lactones which regulate apoptosis, programmed cell death. Iodine causes apoptosis, forcing cancer cells to undergo programmed cell death**. Dr. Aceves concluded that continuous molecular iodine treatment has a "potent antineoplastic effect" on the progression of mammary cancer. In 2011, Dr. Ofelia Soriano and Carmen Acevedo from Mexico discussed the beneficial anticancer effect of 0.1 percent iodine in the DMBA mouse model of breast cancer, triggering caspase-mediated apoptosis in breast cancer cells without harming normal breast cells. Dr. Ofelia Soriano proposes iodine as an adjuvant treatment for premenopausal breast cancer, writing:

> we analyzed the effect of various concentrations of iodine and/or iodide in the dimethylbenz[a] anthracene **(DMBA) mammary cancer model in rats.** The results show that **0.1% iodine or iodide increases the expression of peroxisome proliferator-activated receptor type g (PPARg), triggering caspase-mediated apoptosis pathways in damaged mammary tissue (DMBA-treated mammary gland)** as

well as in frank mammary tumors, but not in normal mammary gland... we propose that it [iodine] be considered as an adjuvant treatment for premenopausal mammary cancer. (9-15)

Apoptosis and The Thrill of Discovery

Apoptosis, or programmed cell death, is important for the development of the embryo and in the treatment of cancer. Do you remember the old spy movies? Upon capture the spy takes a cyanide pill to commit suicide, a fate far better than falling into the hands of the enemy. All of our cells can commit suicide. For example, the P53 gene, "the guardian of the genome" will instruct the cell to commit programmed cell death, apoptosis, if the cell has genetic damage that cannot be repaired. The exact mechanism is described in 2002 by Dr. Bruce Alberts in *Molecular Biology of the Cell*:

> When cells are damaged or stressed, they can also kill themselves by triggering procaspase aggregation and activation from within the cell. In the best understood pathway, mitochondria are induced to release the electron carrier protein cytochrome c into the cytosol, where it binds and activates an adaptor protein called Apaf-1...DNA damage, for example, as discussed earlier, can trigger apoptosis. This response usually requires p53, which can activate the transcription of genes that encode proteins that promote the release of cytochrome c from mitochondria. These proteins belong to the Bcl-2 family. (16)

In 2002, three biologists were awarded the Nobel Prize for their discovery of the mechanism of apoptosis, Sydney Brenner, H. Robert Horvitz and John E. Sulston. The Brenner laboratory in Cambridge made their discovery using a roundworm, the microscopic soil nematode, Caenorhabditis elegans. After news of the award, Dr. Sidney Brenner's wife was told by a lady in a shop, "Someone with the same name as your husband appears to have gotten the Nobel Prize!" In 2002, Inder Verma, the

Editor-in-Chief of Molecular Therapy Journal wrote the following:

> Sydney Brenner, Distinguished Professor at The Salk Institute in La Jolla, CA, Robert Horvitz, Professor of Biology at MIT, Cambridge, MA, and Sir John Sulston... have been awarded the 2002 Nobel Prize in Physiology and Medicine for their pioneering work on "genetic regulation of organ development and programmed cell death"...Sydney Brenner is well known to the scientific community not only for his vision, but also his biting wit. Although he is a Distinguished Professor at The Salk Institute, he prefers to refer to himself as the "Extinguished Professor." At a ceremony honoring and celebrating Sydney's Nobel Prize, he retold to the audience what his wife told him about a lady in a shop who said to her "Someone with the same name as your husband appears to have gotten the Nobel Prize!" (17)

During his Nobel acceptance speech, Dr. Robert Horvitz recounted the "Eureka" moment, the day in 1992 when they discovered humans share the same apoptosis genes found in the roundworm, writing:

> People sometimes ask, when does a scientist feel that "aha!" thrill of discovery? In the case of our studies of programmed cell death, my biggest thrill was probably on February 12, 1992. That was the day that Michael Hengartner obtained the CED-9 sequence, searched the database and found Bcl-2 at the top of the similarity list. We immediately realized that the pathway of cell-death genes we were studying in C. elegans was very likely to be similar to the pathway that controls apoptosis/programmed cell death in humans. (18)

Dr. Ashutosh Shrivastava from India

What if we could turn on apoptosis in cancer cells without harming normal cells? Would this be the cure for cancer? Yes, this is exactly what iodine does. In 2006, Dr. Ashutosh Shrivastava from India studied how iodine induces apopto-sis in breast cancer cells (MCF-7 cells) in vitro. Dr. Shrivastava found the apoptosis mechanism was caspase-independent and caused depletion of intracellular thiols, writing:

> The iodine-induced apoptotic mechanism was studied in MCF-7 cells... iodine-induced apoptosis in a time- and dose-dependent manner in MCF-7 cells. Iodine-induced apoptosis was independent of caspases. **Iodine dissipated mitochondrial membrane potential, exhibited antioxidant activity, and caused depletion in total cellular thiol content [thiols are intracellular antioxidants].** Western blot results showed a **decrease in Bcl-2 and up-regulation of Bax [involved in apoptosis control].** Immunofluorescence studies confirmed the activation and mitochondrial membrane localization of Bax [triggering apoptosis]. Ectopic Bcl-2 overexpression did not rescue iodine-induced cell death. Iodine treatment induces the translocation of apoptosis-inducing factor from mitochondria to the nucleus, and treatment of N-acetyl-l-cysteine prior to iodine exposure restored basal thiol content [antioxidant capability], ROS levels, and completely inhibited nuclear translocation of apoptosis-inducing factor and subsequently cell death, indicating that **thiol depletion may play an important role in iodine-induced cell death**. These results demonstrate that iodine treatment **activates a caspase-independent and mitochondria-mediated apoptotic pathway...ability of iodine to cause significant reduction in free radical generation and glutathione levels supports mitochondria-mediated apoptotic cell death.** Note: the primary intracellular thiol is the antioxidant, glutathione, synthesized from cysteine, glycine, and glutamate. Glutathione readily neutralized ROS, reactive oxygen species. Note: Bcl-2 protein is anti-apoptotic, and Bax protein is pro-apoptotic. (19)

Dr. Hiroomi Funahashi from Japan

In 2001, Dr. Hiroomi Funahashi from Japan reported a common seaweed food (wakame and mekabu), containing high iodine content is more beneficial than chemotherapy for breast cancer. Seaweed also contains Fucoxanthin (Fucoidan) a type of carotenoid having considerable anti-cancer activity as discussed in 2022 by Dr. Tsz-Ying Lau. Dr. Hiroomi Funahashi writes:

> It [seaweed] showed an extremely strong suppressive effect on rat mammary carcinogenesis when used in daily drinking water, without toxicity. In vitro, mekabu [seaweed] solution strongly induced apoptosis in 3 kinds of human breast cancer cells. **These effects were stronger than those of a chemotherapeutic agent widely used to treat human breast cancer.** Furthermore, no apoptosis induction was observed in normal human mammary cells. In Japan, mekabu is widely consumed as a safe, inexpensive food. Our results suggest that mekabu has potential for chemoprevention of human breast cancer... administration of Lugol's iodine or iodine-rich Wakame seaweed to rats treated with the carcinogen dimethyl benzanthracene [DMBA] suppressed the development of mammary tumors. **Note: The same group demonstrated that seaweed induced apoptosis in human breast cancer cells with greater potency than that of fluorouracil, a chemotherapeutic agent used to treat breast cancer.** (20-27)

Iodine Alters Gene Expression in Breast Cancer Cells

The study of estrogen metabolism reveals the presence of both favorable and unfavorable metabolic pathways. A 2008 paper by Drs. Frederick Stoddard and Bernard A. Eskin showed that iodine alters gene expression in breast cancer cells, upregulating beneficial metabolic pathways and downregulating unfavorable pathways. For more on this, see Chapter 21 on Estrogen Metabolism, Iodine and 2MeOE2. (28)

Iodine for Lung Cancer

A 2003 Dr. Ling Zhang genetically modified small cell lung cancer cells to increase iodine uptake by adding genes for the NIS, sodium iodide symporter, and TPO, thyroperoxidase. These genetically modified lung cancer cells were then xenografted into mice which were later treated with intra-peritoneal potassium iodide injections. Dr. Ling Zhang found iodine causes lung cancer cells to undergo apoptosis, programmed cell death. Another similar case report in 1993 by Dr. Aleck Hercbergs describes spontaneous remission of lung cancer in a patient incidentally treated with amiodarone which releases about 9 mg of iodine into circulation for each 200 mg tablet. **Note:** amiodarone is a commonly used cardiology drug used to treat cardiac arrythmias. (29-30)

Using Adjunctive Iodine Combined with Chemotherapy

In 2013, Dr. Yunuen Alfaro studied breast cancer xenografts in mice, finding a good anti-cancer synergy of molecular iodine combined with conventional chemotherapy (doxorubicin). Dr. Yunuen Alfaro writes:

> The DOX-I2 (Doxorubicin / Iodine I2) combination exerts antineoplastic, chemosensitivity, and cardioprotective effects and could be a promising strategy against breast cancer progression. Note: Doxorubicin (Adriamycin) is a commonly used chemotherapy drug. (31)

In 2018, Dr Zambrano-Estrada studied the Iodine/Doxorubicin combination in dogs with canine mammary carcinoma finding excellent anti-cancer synergy, writing:

> The mDOX+I2 [Doxorubicin plus molecular iodine] scheme **improves the therapeutic outcome,** diminishes the invasive capacity, attenuates the adverse events, and **increases disease-free survival.** These data led us to propose mDOX+I2 as an effective treatment for canine mammary cancer. (32)

Iodine for Triple Negative Breast Cancer

In 2019, Dr. Irasema Mendieta studied molecular iodine (I2) treatment for breast cancer both in vitro in vivo xenografts in athymic nude mice, using two different cell lines. The first cell line is estrogen-receptor positive, and the second is **triple-negative** for estrogen, progesterone and HER2 receptors. **Note:** triple negative means lacking receptors for estrogen, progesterone, and HER (human epithelial receptor). The iodine treatment prevented cachexia in the xenografted mice and decreased the invasive nature of the triple-negative breast cancer xenografts. This is significant since triple-negative breast cancer rarely responds to endocrine adjuvant therapy with tamoxifen or aromatase inhibitors, and is usually treated with chemotherapy with poor prognosis. Thus, there is a need for safer and more effective treatments. Dr. Mendieta proposed Iodine as an adjuvant treatment for breast cancer, writing:

> we analyze the effect of iodine on two types of mammary tumor cell lines that represent epithelial breast cancers. The first is laminal estrogen-responsive, which exhibits low invasive potential (MCF-7), and the second is a basal triple negative with elevated invasive potentiality (MDA-MB231)...iodine supplementation prevents cachexia to hinder cancer progression by maintaining a small tumor size and decreases markers like TNF-alpha in tumorous cells...I2, Molecular Iodine, decreases the invasive potential of a **triple negative basal cancer cell line**, and under in vivo conditions the oral supplement of this halogen activates the antitumor immune response, preventing progression of xenografts from laminal and basal mammary cancer cells. These effects allow us to **propose iodine supplementation as a possible adjuvant in breast cancer therapy.** Note: other natural treatments for triple-negative breast cancer include fucoidan found in brown algae seaweed, hesperidin, paeonol, naringenin, matrine, Korean red ginseng, and actein, as discussed in 2023 by Dr. Lea Ling-Yu Kan. (33-36)

Iodine as Adjuvant Treatment in 30 Breast Cancer Patients

In 2019 Dr. Moreno-Vega did a randomized human trial on adjuvant use of iodine alone or combined with chemotherapy for breast cancer. Thirty women had early-stage, and thirty women had advanced-stage breast cancer. Five-year disease-free survival was significantly higher (86 percent) for the patients given iodine (I2) supplements before and after surgery compared to I2 supplements only after surgery (46 percent). I2-treated tumors exhibit less invasive potential, and significant increases in apoptosis, estrogen receptor expression, and immune cell infiltration. Transcriptome analysis of I2-supplemented tumors showed activation of the immune response. Dr. Moreno-Vega writes:

> This study analyzes an oral supplement of molecular iodine (I2), alone and in combination with the neoadjuvant therapy 5-fluorouracil/epirubicin/cyclophosphamide or taxotere/epirubicin (FEC/TE) in women with **Early (stage II) and Advanced (stage III) breast cancer**. In the Early group, 30 women were treated with I2 (5 mg/day) or placebo (colored water) for 7–35 days before surgery. For the Advanced group, 30 patients received I2 or placebo, along with FEC/TE treatment. After surgery, all patients received FEC/TE + I2 for 170 days. **I2 supplementation showed a significant attenuation of the side effects and an absence of tumor chemoresistance. The control, I2, FEC/TE, and FEC/TE + I2 groups exhibited response rates of 0, 33%, 73%, and 100%, respectively, and a pathologic complete response of 18%, and 36% in the last two groups. Five-year disease-free survival rate was significantly higher in patients treated with the I2 supplement before and after surgery compared to those receiving the supplement only after surgery (82% versus 46%). I2-treated tumors exhibit less invasive potential, and significant increases in apoptosis, estrogen receptor expression, and immune cell infiltration.**

Transcriptomic analysis indicated **activation of the antitumoral immune response.** (37)

In 2021, Dr. Olga Cuenca-Micó from Mexico studied the in vitro effect of 5 mg/day I2 supplementation alone or together with conventional chemotherapy on breast cancer tumor samples obtained from 30 patients removed at time of surgery. Dr. Olga Cuenca-Micó found that I2 supplementation induces beneficial activation of the immune response, writing:

two pilot study groups were established based on the stage of cancer diagnosed: Early (stage II) and Advanced (stage III) breast cancer groups. Thirty patients were randomly assigned (double-blind) to receive either molecular iodine (I2, 5 mg/day) or a placebo (vegetable-colored water) for 7–35 days (as determined by the preoperative oncologist's protocol). In the Advanced group, 30 patients were randomly (double-blind) divided into the I2 or placebo groups, and both groups received 4–6 cycles of neoadjuvant chemotherapy (Cht) using either 5-fluorouracil/ epirubicin/ cyclophosphamide or taxotere/epirubicin. Daily, after breakfast, I2 or placebo was diluted in drinking water. During the surgical procedure, the tumor sample was kept in dry ice to avoid degradation and stored at −80 °C until further analysis...Molecular iodine (I2) induces apoptotic, antiangiogenic, and antiproliferative effects in breast cancer cells. ... I2 and Cht [chemotherapy]+I2 samples showed **significant increases in the expression of Th1 and Th17 pathways**. Tumor immune composition...revealed **significant increases in M0 macrophages and B lymphocytes in both I2 groups**. Real-time PCR [genetic testing] showed that I2 tumors overexpress T-BET (p = 0.019) and interferon-gamma (IFN-gamma; p = 0.020) and **silence tumor growth factor-beta (TGF-beta**; p = 0.049), whereas in Cht+I2 tumors, GATA3 is silenced (p = 0.014). Preliminary methylation analysis shows that **I2 activates IFN-gamma gene promoter (by increasing its unmethylated form) and silences TGF-beta in Cht+I2.** In conclusion, our data showed that **I2 supplements induce the activation of the immune response and that when combined with Cht, the Th1 pathways are stimulated.** Note: TH1 and TH17 pathways produce interferon-gamma (IFN-gamma) which activates cytotoxic CD8+ T cells to directly kill cancer cells. The TH1 pathway drives interleukin (IL)-2 production which stimulates the production of cytotoxic T cells and natural killer (NK) cells which attack and kill cancer cells. (38)

Iodine Combined with Zoledronic Acid for Triple Negative Breast Cancer

About 15-20 percent of all breast cancers are classified as triple-negative, meaning the cells lack receptors for estrogen, progesterone, and human epidermal growth factor (HER-2). Triple-negative breast cancer is one of the most aggressive cell types and is unresponsive to endocrine adjuvant therapy with tamoxifen and aromatase inhibition. Aggressive breast cancer has a predilection for metastatic spread to the bone, and frequently to the ribs. This is routinely treated by mainstream oncology with an IV bisphosphonate drug called Zoledronate. In 2016, Dr. Ranu Tripathi used a mouse xenograft model to study triple negative breast cancer treated with the combination of molecular Iodine (I2) with Zoledronate, finding good synergy, and enhanced apoptosis when given in combination, writing:

We analyzed the effect of combination of I2 (molecular iodine) with Zol (Zolendronate) as a potent adjuvant therapeutic agent **for triple negative breast cancer cells** (MDA-MBA-231) and in the mice model of breast cancer. ...We report that Zol potentiates the efficacy of I2 by inducing non-mitochondrial intrinsic apoptosis by increasing intracellular calcium and ER stress. We show that MDA-MB-231cells register minimal hypodiploidy in response to individual treatment with either I2 or Zol, but **synergistically enhances apoptosis when given in combination**. Similar potentiating effect as reflected by enhanced apoptotic index on I2-mediated cell death was also reported in these cells by **addition of chloroquine** and by addition of **doxorubicin** in other animal tumor models and cancer cells. Note: cordycepin (cordyceps) a parasitic fungus natural product has anti-cancer activity against triple-negative breast cancer as discussed in 2022 by Dr. Chunli Wei. (39-40)

In Conclusion: In 2023, Drs. Ibrahim and Elliyanti from Indonesia reviewed all previous research on iodine(I2) as a treatment for breast cancer, concluding:

the consumption of large amounts of iodine will help cure breast cancer gradually. The Japanese people have the lowest incidence of thyroid cancer, prostate cancer, and breast cancer. This, of course, can be evidence that **iodine can be used as an alternative ingredient to treat breast cancer**... **Iodine has potential as a substance with anticancer activity through antiproliferative, apoptotic, and immune system activating mechanisms**. (41)

Current research suggests molecular iodine (I2) as adjuvant treatment for breast cancer. Entry of iodine into normal and malignant breast epithelial cells is facilitated by the sodium iodide symporter (NIS). This NIS molecular pump machine is present in about 80 percent of breast cancer cell types. Other cancers such as lung and prostate may also benefit. Iodine has good synergy when used in combination with conventional chemotherapy and with I.V. Zoledronate. Given the poor prognosis and limited treatment options for triple-negative breast cancer, molecular iodine emerges as an important adjuvant therapy. In addition, we have uncovered a list of natural supplements with anticancer activity particularly useful for triple negative breast cancer. While most patients with triple-negative breast cancer are treated with chemotherapy, very few if any oncologists will recommend adding iodine for the triple-negative patient. This is one of the errors of mainstream oncology. In my opinion, combining molecular iodine with chemotherapy in triple-negative breast cancer is benefi-

cial and strongly indicated, with the potential to improve outcomes. Further research on molecular iodine for breast cancer should receive top priority for NIH funding. However, we may never see randomized controlled trials for iodine or any other natural substances. How many breast cancer patients buy iodine supplements over the counter without the knowledge of their oncologist? We may never know. For information on the use of repurposed drugs for cancer treatment, see my book, *Cracking Cancer Toolkit.* For more information on the health benefits of iodine, see my book, *Natural Thyroid Toolkit.* (42-51)

♦ **References for Chapter 23 Iodine for Treatment of Breast Cancer**

1) Brownstein, David. Iodine, Why You Need It, Why You Can't Live Without It, Fourth Edition 2009, Medical Alternatives Press.

2) Derry, David M. Breast cancer and iodine. Trafford on Demand Pub, 2001.

3) Smyth, Peter PA. "The Thyroid, Iodine and Breast Cancer." Breast Cancer Research 5 (2003): 1-4.

4) Haraldsdottir, Alfheidur, et al. "Early Life Residence, Fish Consumption, And Risk of Breast Cancer." Cancer Epidemiology, Biomarkers & Prevention 26.3 (2017): 346-354.

5) Nyström, Helena Filipsson, et al. "Iodine Status in The Nordic Countries–Past and Present." Food & Nutrition Research 60.1 (2016): 31969.

6) Aceves, Carmen, Brenda Anguiano, and Guadalupe Delgado. "Is Iodine A Gatekeeper of The Integrity of The Mammary Gland?" Journal Of Mammary Gland Biology and Neoplasia 10 (2005): 189-196.

7) Arroyo-Helguera, O., et al. "Uptake and Antiproliferative Effect of Molecular Iodine in the MCF-7 Breast Cancer Cell Line." Endocrine-Related Cancer 13.4 (2006): 1147-1158.

8) Arroyo-Helguera, O., et al. "Signaling Pathways Involved in The Antiproliferative Effect of Molecular Iodine in Normal and Tumoral Breast Cells: Evidence That 6-Iodolactone Mediates Apoptotic Effects." Endocrine-Related Cancer 15.4 (2008): 1003-1011.

9) Aceves, Carmen, et al. "Antineoplastic Effect of Iodine in Mammary Cancer: Participation Of 6-Iodolactone (6-IL) And Peroxisome Proliferator-Activated Receptors (PPAR)." Molecular Cancer 8.1 (2009): 1-9.

10) Soriano, Ofelia, et al. "Antineoplastic Effect of Iodine and Iodide in Dimethylbenz [A] Anthracene-Induced Mammary Tumors: Association Between Lactoperoxidase and Estrogen-Adduct Production." Endocrine-Related Cancer 18.4 (2011): 529-539.

11) Anguiano, B., et al. "Uptake and Gene Expression with Antitumoral Doses of Iodine in Thyroid and Mammary Gland: Evidence That Chronic Administration Has No Harmful Effects." Thyroid: Official Journal of the American Thyroid Association 17.9 (2007): 851.

12) Aceves, Carmen, et al. "Molecular Iodine Has Extrathyroidal Effects as An Antioxidant, Differentiator, And Immunomodulator." International Journal of Molecular Sciences 22.3 (2021): 1228.

13) García-Solís, Pablo, et al. "Inhibition of N-methyl-N-nitrosourea-Induced Mammary Carcinogenesis by Molecular Iodine (I2) But Not by Iodide (I–) Treatment: Evidence That I2 Prevents Cancer Promotion." Molecular And Cellular Endocrinology 236.1-2 (2005): 49-57.

14) Rösner, Harald, et al. "Antiproliferative/Cytotoxic Activity of Molecular Iodine and Iodolactones In Various Human Carcinoma Cell Lines. No Interfering With EGF-Signaling, But Evidence for Apoptosis." Experimental And Clinical Endocrinology & Diabetes 118.07 (2010): 410-419.

15) Rösner, Harald, et al. "Antiproliferative/Cytotoxic Effects of Molecular Iodine, Povidone-Iodine and Lugol's Solution in Different Human Carcinoma Cell Lines." Oncology letters 12.3 (2016): 2159-2162.

16) Alberts, Bruce, et al. "Programmed Cell Death (Apoptosis)." Molecular Biology of the Cell. 4th edition. Garland Science, 2002.

17) Verma, Inder. "2002 Nobel Prize for Physiology and Medicine." Molecular Therapy 6.6 (2002): 698.

18) Horvitz, H. Robert. "Worms, Life, And Death (Nobel Lecture)." Chembiochem 4.8 (2003): 697-711.

19) Shrivastava, Ashutosh, et al. "Molecular Iodine Induces Caspase-Independent Apoptosis in Human Breast Carcinoma Cells Involving the Mitochondria-Mediated Pathway." Journal of Biological Chemistry 281.28 (2006): 19762-19771.

20) Lau, Tsz-Ying, and Hiu-Yee Kwan. "Fucoxanthin is a Potential Therapeutic Agent for The Treatment of Breast Cancer." Marine drugs 20.6 (2022): 370.

21) Zhang, Zhongyuan, et al. "Fucoidan Extract Induces Apoptosis In MCF-7 Cells Via a Mechanism Involving The ROS-Dependent JNK Activation And Mitochondria-Mediated Pathways." PloS one 6.11 (2011): e27441.

22) Kato N, Funahashi H, Ando K, Takagi H. Suppressive Effect of Iodine Preparations On Proliferation Of DMBA-Induced Breast Cancer In Rat. J Jpn Soc Cancer Ther 1994; 29:582–588.

23) Funahashi, Hiroomi, et al. Suppressive Effect of Iodine On DMBA-Induced Breast Tumor Growth in The Rat. J Surg Oncol 1996; 61:209–213.

24) Funahashi, Hiroomi, et al. "Wakame Seaweed Suppresses the Proliferation Of 7, 12-Dimethylbenz (A)-Anthracene-Induced Mammary Tumors in Rats." Japanese Journal of Cancer Research 90.9 (1999): 922-927.

25) Funahashi, Hiroomi, et al. "Seaweed Prevents Breast Cancer?" Japanese Journal of Cancer Research 92.5 (2001): 483-487.

26) Teas, Jane, et al. "The Consumption of Seaweed as A Protective Factor In The Etiology Of Breast Cancer: Proof Of Principle." Journal Of Applied Phycology 25 (2013): 771-779.

27) Chan, Eric Wei Chiang, et al. "The Health-Promoting Properties of Seaweeds: Clinical Evidence Based on Wakame and Kombu." Journal of Natural Remedies (2023): 687-698.

28) Stoddard II, Frederick R., et al. "Iodine Alters Gene Expression in The MCF7 Breast Cancer Cell Line: Evidence for An Anti-Estrogen Effect of Iodine." International Journal of Medical Sciences 5.4 (2008): 189.

29) Zhang, Ling, et al. "Nonradioactive iodide effectively induces apoptosis in genetically modified lung cancer cells." Cancer research 63.16 (2003): 5065-5072.

30) Hercbergs, Aleck, and John T. Leith. "Spontaneous Remission of Metastatic Lung Cancer Following Myxedema Coma—An Apoptosis-Related Phenomenon?" JNCI: Journal of the National Cancer Institute 85.16 (1993): 1342-1343.

31) Alfaro, Yunuen, et al. "Iodine and Doxorubicin, A Good Combination for Mammary Cancer Treatment: Antineoplastic Adjuvancy, Chemoresistance Inhibition, And Cardioprotection." Molecular cancer 12.1 (2013): 1-11.

32) Zambrano-Estrada, Xóchitl, et al. "Molecular Iodine/Doxorubicin Neoadjuvant Treatment Impair Invasive Capacity and Attenuate Side Effect in Canine Mammary Cancer." BMC veterinary research 14.1 (2018): 1-14.

33) Mendieta, Irasema, et al. "Molecular Iodine Exerts Antineoplastic Effects by Diminishing Proliferation and Invasive Potential and Activating the Immune Response In Mammary Cancer Xenografts." BMC cancer 19.1 (2019): 261.

34) Mahtani, Reshma, et al. "Advances in Therapeutic Approaches For Triple-Negative Breast Cancer." Clinical Breast Cancer 21.5 (2021): 383-390.

35) Hsu, Wen-Jing, et al. "Fucoidan from Laminaria Japonica Exerts Antitumor Effects on Angiogenesis and Micrometastasis in Triple-Negative Breast Cancer Cells." International Journal of Biological Macromolecules 149 (2020): 600-608.

36) Kan, Lea Ling-Yu, et al. "Natural-Product-Derived Adjunctive Treatments to Conventional Therapy and Their Immunoregulatory Activities in Triple-Negative Breast Cancer." Molecules 28.15 (2023): 5804.

37) Moreno-Vega, Aura, et al. "Adjuvant Effect of Molecular Iodine in Conventional Chemotherapy for Breast Cancer. Randomized Pilot Study." Nutrients 11.7 (2019): 1623.

38) Cuenca-Micó, Olga, et al. "Effects of Molecular Iodine/Chemotherapy in the Immune Component of Breast Cancer Tumoral Microenvironment." Biomolecules 11.10 (2021): 1501.

39) Tripathi, Ranu, et al. "Zoledronate and Molecular Iodine Cause Synergistic Cell Death In Triple Negative Breast Cancer Through Endoplasmic Reticulum Stress." Nutrition and Cancer 68.4 (2016): 679-688.

40) Wei, Chunli, et al. "Cordycepin Inhibits Triple-Negative Breast Cancer Cell Migration and Invasion By Regulating EMT-TFs SLUG, TWIST1, SNAIL1, and ZEB1." Frontiers in Oncology 12 (2022): 898583.

41) Ibrahim, Raihan Syah, and Aisyah Elliyanti. "The Potential of Iodine as a Treatment for Breast Cancer: A Narrative Review." Jurnal Kesehatan Manarang 9.3 (2023): 159-165.

42) Cann, Stephen A., et. al. "Hypothesis: Iodine, Selenium and The Development of Breast Cancer." Cancer causes & control 11.2 (2000): 121-127.

43) Venturi, Sebastiano, et al. "Role of Iodine in Evolution and Carcinogenesis of Thyroid, Breast and Stomach." Advances In Clinical Pathology 4 (2000): 11-18.

44) Venturi, Sebastiano. "Is There a Role for Iodine in Breast Diseases?" The Breast 10.5 (2001): 379-382.

45) Renier, Corinne, et al. "Endogenous NIS Expression in Triple-Negative Breast Cancers." Annals of Surgical Oncology 16 (2009): 962-968.

46) Elliyanti, Aisyah, et al. "Analysis Natrium Iodide Symporter Expression in Breast Cancer Subtypes For Radioiodine Therapy Response." Nuclear Medicine and Molecular Imaging 54 (2020): 35-42.

47) Kogai, Takahiko, and Gregory A. Brent. "The Sodium Iodide Symporter (NIS): Regulation and Approaches to Targeting for Cancer Therapeutics." Pharmacology & therapeutics 135.3 (2012): 355-370.

48) Renier, Corinne, et al. "Breast Cancer Brain Metastases Express the Sodium Iodide Symporter." Journal of Neuro-Oncology 96 (2010): 331-336.

49) Willhauck, Michael J., et al. "Functional Sodium Iodide Symporter Expression in Breast Cancer Xenografts In Vivo After Systemic Treatment with Retinoic Acid And Dexamethasone." Breast Cancer Research and Treatment 109 (2008): 263-272.

50) Dach, Jeffrey. Cracking Cancer Toolkit: Using Repurposed Drugs for Cancer Treatment. Medical Muse Press (2020).

51) Dach, Jeffrey. Natural Thyroid Toolkit: Hashimoto's, Graves,' Iodine and Natural Desiccated Thyroid. Medical Muse Press (2023)

Chapter 24

The Folly of Non-Hormonal Menopausal Drugs

MARY IS A 55-YEAR-OLD BREAST cancer survivor, having breast cancer at age 47 treated with lumpectomy alone. Mary is in remission with no recurrence after 8 years of follow-up. Recently, Mary has been experiencing menopausal symptoms of night sweats and hot flashes, so she asked her primary care doctor about hormone replacement. Since she has a history of breast cancer, her primary care doctor said hormone replacement is contraindicated, and instead, advised Mary to take a new non-hormonal drug called Veozah for relief of the night sweats and hot flashes. Mary started the drug and thankfully, the night sweats and hot flashes have disappeared.

Brisdell, SSRI Antidepressant Called Paxil

The first FDA-approved non-hormonal treatment for night sweats and hot flashes was Brisdelle, FDA-approved in 2013. Brisdelle is an SSRI antidepressant, originally called Paxil (paroxetine). The problem is all SSRI drugs are ineffective for menopausal symptoms of hot flashes. That is why Brisdelle never caught on for treatment of hot flashes. Adverse effects of Brisdelle and all SSRI antidepressants include sexual dysfunction, emotional blunting, and increased risk for suicide, homicide, and violent behavior. SSRI drugs are addictive and discontinuation carries side effects, so they must be tapered gradually under supervision. Needless to say, this is not a good thing. SSRI antidepressants are discussed more completely in Chapter 26, the Depressing State of Antidepressants.

The New Drug is Veozah

Veozah (fezolinetant) is the new non-hormonal menopause drug FDA approved May 2023 for the treatment of night sweats and hot flashes. Unlike the failed Brisdelle SSRI antidepressant drug, Veozah has efficacy for reducing hot flashes by blocking the Neurokinin3 receptors (NK3r) in the brain. The NK3 receptors are the neuron receptors in the hypothalamus that radiate to the autonomic centers controlling thermoregulation. A 12-week study showed Veozah's efficacy for the reduction of hot flashes is about the same as estrogen hormone replacement. This study was funded by Astella Pharmaceuticals and was extended for a 40-week blinded observation period, so the entire study lasted one year. (1-2)

Gastrointestinal Adverse Effects

Neurokinin 3 receptors are also found in the gastrointestinal tract in the vagus nerve efferent fibers which innervate the parasympathetic nervous system, thus accounting for the gastrointestinal side effects of abdominal pain and diarrhea reported in the clinical trials for Veozah. In 2002, Dr. C. Blondeau writes:

> neurokinin-1 and neurokinin-3 receptors are involved in the parasympathetic control of digestive functions. (1-3)

Neurokinin 3 receptors have also been implicated in various auto-immune diseases. Whether or not Veozah will increase auto-immune disease remains unknown. Our knowledge of neurokinin receptors is still in its infancy, and it may take years of research to uncover the adverse effects of blocking NK3 receptors in the brain and gastrointestinal tract. On December 16, 2024 the FDA added a boxed warning, of serious liver injury associated with the use of Veozah (fezolinetant). (3-7)

Drug Marketing to the Hormone Averse Woman

Veozah was FDA approved May 12, 2023 in menopausal women who have contra-indications to estrogen hormone replacement such as history of breast cancer, endometrial cancer, or venous thromboembolic disease. However, this drug market segment is too small to generate enough profit to recoup drug development costs. A much larger and more lucrative market segment is women who have been brainwashed by the drug industry to fear estrogen. Veozah print and television marketing targets this larger demographic. Off-label marketing of the Veozah drug to "hormone averse" women creates a windfall for the drug industry, and a catastrophe for the patient. Menopause is an estrogen deficiency disease and requires estrogen, not an NK3 receptor blocker that addresses only thermoregulation while ignoring the benefits of estrogen for all other organ systems. To review the benefits of hormone replacement, see Chapter 1, Health Benefits of Menopausal Hormone Replacement.

An Even Larger Secondary Drug Market

Postmenopausal estrogen deficiency is associated with various degenerative diseases. Each disease category has corresponding pharmaceutical drug treatments. These estrogen deficiency disease categories are: osteoarthritis, osteoporosis, depression, coronary artery disease, dementia, vaginal atrophy, and genito-urinary symptoms, to name a few, all treated with their category of pharmaceutical drugs as seen in this list, below. Call me a conspiracy theorist, but this looks like a plan to create another windfall using pharmaceuticals to treat the estrogen deficiency diseases of menopause. Follow the money. Nothing happens in medicine unless it generates profits for the drug industry. What is the marketing formula that never fails? If only we could persuade postmenopausal women to fear estrogen and decline treatment, this would further the chronic disease epidemic and create a windfall for the drug industry. (8-13)

Disease Entity	Drug
Osteoporosis:	Fosamax, Bisphosphonates, Denosumab
Osteoarthritis:	Celebrex, NSAIDS
Depression:	SSRI antidepressants
Heart Disease:	Statin Drugs
Dementia:	Donepezil (Aricept)
Weight Gain:	Phentermine, Ozempic
Cognitive Dysfunction:	Ritalin, Adderal, Amphetamines
Chronic Fatigue:	Amphetamines
Vaginal Atrophy:	Prasterone, laser treatments
Recurrent UTI's:	Antibiotics, Bactrim, Macrobid
Insomnia:	Benzodiazepines, Z-drug Sleeping Pills

How to Make Your Patient Hormone Averse

Imagine if every primary care and OB/GYNE doctor handed out off-label Veozah to every "hormone averse" post-menopausal patient. For patients walking into the office who are not "hormone averse", a short conversation with the doctor explaining the "evils" of estrogen fixes this. If the patients were not "hormone averse" before, they are now, and they have become good customers for the Veozah drug. This is an error and a tragedy. The correct treatment for menopausal estrogen deficiency is estrogen hormone replacement.

Everything Seems OK

Once starting Veozah drug, the night sweats and hot flashes resolve, and the patient is thinking everything is going OK. However, a few years later, after it is too late, the patient will realize that their menopausal symptoms are getting worse. They will realize the osteoarthritis, osteoporosis, cognitive dysfunction, genitourinary symptoms, vaginal atrophy, depression and chronic fatigue are all due to estrogen deficiency. And, they will realize they have been deceived and lied to. They should have been taking bioidentical hormone replacement all along. They may then seek a physician willing to give them the real thing, estrogen.

Withdrawal Effects of Blocking Neurotransmitter Receptors

What happens when women develop tolerance or stop the Veozah drug? The night sweats and hot flashes return with a vengeance. This is a drug withdrawal effect commonly observed for all neurotransmitter receptor-blocking drugs. The neurons compensate for the Veozah drug by upregulating the receptors. This is the same mechanism as other neurotransmitter-blocking drugs such as amphetamines, benzodiazepines and opiates. Patients on these drugs develop tolerance requiring higher doses for effect. When the patient stops taking the Veozah drug, the withdrawal effects of debilitating night sweats and hot flashes force the patient to go back on the drug. It is impossible to get off. The patient discovers she must be on Veozah for life or suffer the withdrawal effects of even worse debilitating night sweats and hot flashes.

Liver Injury Black Box Warning

On December 16, 2024, the FDA added a black box warning label to Veozah (fezolinetant) for the risk of elevated liver enzymes and severe liver injury. The FDA added recommendations to perform monthly liver function testing for 2 months after starting the drug, then at 3, 6 and 9 months after starting.

No Long-Term Studies - Drug Discontinuation Effects?

The Veozah hot flash study lasted 12 weeks, with another 40 weeks of blinded observation. Astellas Pharmaceuticals, from Tokyo, Japan is the drug manufacturer who funded the studies, and they did not report what happens when the drug is discontinued. The FDA approved the drug without considering long-term effects of

drug discontinuation, or any long-term effects of blocking Neurokinin receptors of neurons of the hypothalamus of the brain. Neurokinin receptor blockers have been studied for use as anti-psychotic drugs, and for drug and alcohol addiction. Do women really want to take an antipsychotic drug for menopause? This drug is a disaster waiting to happen. Neurokinin receptors exist throughout the body in various organ systems, mainly the gastrointestinal tract. What happens when these are blocked? We do not know. (14-15)

Dr. Felice Gersh on Veozah

In 2024, Dr. Felice Gersh, an integrative OB/Gyne doctor from Irvine, California, expresses many of the same concerns mentioned above. The drug company advertising targets all "hormone averse" post-menopausal women, potentially leading to a catastrophe for the menopausal patient not receiving bioidentical hormone replacement with estrogen, the correct treatment. Here is a partial transcript of Dr. Gersh discussing Veozah:

> Many women are living in fear of what should be a beloved hormone of the body, ovarian produced estradiol. At menopause, we lose our ovarian production of estradiol. So, because there are so many women who are fearful of hormones [because of] the Women's Health Initiative [Study] from over 20 years ago... But because the negativity is so prevalent, drug manufacturers have been trying to find alternatives to estrogen to treat night sweats and hot flashes. Estrogen is involved with regulating temperature. So, when we go through menopause and have estrogen deficiency, hot flashes and night sweats are triggered by this particular peptide neurokinin3 (NK3). This peptide is made in the hypothalamus of the brain where the thermoregulatory centers reside. So, the drug industry created a drug to block the receptor for NK3. How brilliant is that? This drug blocks the receptor and is pretty effective at blocking night sweats and hot flashes. And I am totally for it for

women who can not take estradiol. It is a small segment of the female population, but it does exist. What kind of a patient would that be? A woman who has an ongoing estrogen-positive cancer, like breast cancer and endometrial cancer... That is not a large enough audience to target this drug... They are growing that market by making women think it is good to be "hormone averse" and to think it is better to use something that is non-hormonal. So, that is how they are promoting this. **What they are doing in their advertising breaks my heart.** And I want you to know, please don't listen to them unless you have breast or uterine cancer. Because otherwise, you are a candidate, in almost every case, for going on real estrogen, estradiol. Why would you take a drug (Veozah) that... only treats the hot flashes and night sweats. It is not treating every other organ system that is benefited by estrogen. Let us name a few of these systems: The cardiovascular, musculoskeletal, neurological, gastrointestinal, skin, and genitourinary system.... Estradiol has a direct benefit on all those organ systems. So, if you have a choice and there is no contraindication to take real estradiol, that is what should take. (16)

Treating Vasomotor Symptoms While Ignoring They Are a Marker for Chronic Disease

As mentioned above, menopausal hormone deficiency is a harbinger of chronic disease, completely preventable with estrogen hormone replacement therapy. The idea of treating the vasomotor symptoms of menopause, hot flashes, and night sweats with a psych drug blocking the neurokinin receptors in the brain is complete folly. This strategy ignores menopausal estrogen deficiency as the harbinger of chronic disease. In 2017, Dr. Nicoletta Biglia MD, PhD, Full Professor of Gynaecology and Obstetrics at the University of Torino Medical School makes this same observation writing:

there is a growing body of evidence demonstrating that VMS [vasomotor

symptoms] may be a biomarker for chronic disease. In this review, the association between VMS and a range of chronic post-menopausal conditions including CVD [cardiovascular disease], osteoporosis, and cognitive decline is discussed. Prevention of CVD in women, as for men, should be started early, and effective management of chronic disease in postmenopausal women has to start with the awareness that **VMS during menopause are harbingers of things to come and should be treated accordingly**. (17)

In 2014, Dr. Rogerio Arnaldo Lobo, MD, Professor of Obstetrics and Gynecology and Fellowship Director in the Division of Reproductive Endocrinology, Columbia University says the onset of menopause heralds many chronic diseases such as obesity, metabolic syndrome and diabetes, cardiovascular disease, osteoporosis and osteoarthritis, cognitive decline, dementia and depression, and cancer. Dr. Rogerio Lobo says the Women's Health Initiative author's recommendation to avoid estrogen therapy is no longer valid based on current data. Dr. Lobo recommends estrogen therapy as part of a "comprehensive strategy to prevent chronic disease after menopause, menopausal hormone therapy, particularly estrogen therapy may be considered as part of the armamentarium," writing:

Women may expect to spend more than a third of their lives after menopause. Beginning in the sixth decade, many chronic diseases will begin to emerge, which will affect both the quality and quantity of a woman's life. Thus, the onset of menopause heralds an opportunity for prevention strategies to improve the quality of life and enhance longevity. **Obesity, metabolic syndrome and diabetes, cardiovascular disease, osteoporosis and osteoarthritis, cognitive decline, dementia and depression, and cancer are the major diseases of concern.** ... Although the most recent publications from the follow-up studies of the Women's Health Initiative do not recommend menopause hormonal therapy as a prevention strategy, **these conclusions may not be fully valid for midlife women, on the basis of the existing data.** For healthy women aged 50–59 years, estrogen therapy decreases coronary heart disease and all-cause mortality; this interpretation is entirely consistent with results from other randomized, controlled trials and observational studies. **Thus, as part of a comprehensive strategy to prevent chronic disease after menopause, menopausal hormone therapy, particularly estrogen therapy may be considered as part of the armamentarium.** (18)

In 2020, Dr. Micheline McCarthy, Professor of Neurology, the University of Miami Miller School of Medicine says menopause is the onset of a systemic inflammatory state that "sets the stage for late-life neurodegenerative/neurovascular disease with co-morbid cognitive dysfunction or decline". The idea of treating vasomotor symptoms with a psych drug while completely ignoring estrogen deficiency-induced systemic inflammation and neurologic degeneration is complete folly. Dr. Micheline McCarthy feels that ER-beta activation is protective of mitochondria in the brain and prevents neurologic decline, writing:

> Emerging evidence is showing that peri-menopause is pro-inflammatory and disrupts estrogen-regulated neurological systems. Estrogen is a master regulator that functions through a network of estrogen receptors subtypes alpha (ER-α) and beta (ER-β). Estrogen receptor-beta has been shown to regulate a key component of the innate immune response known as the inflammasome, and it also is involved in regulation of neuronal mitochondrial function....There is increasing and compelling evidence showing that estrogen decline during the menopausal transition drives a systemic inflammatory state. This state is characterized by systemic pro-inflammatory cytokines derived from reproductive tissues, alteration in the cellular immune profile, increased availability of inflammasome proteins in the CNS [central nervous system],

and a pro-inflammatory microenvironment which makes the brain more susceptible to ischemic and other stressors. These pro-inflammatory processes appear to compromise ER-beta's role in protecting the brain from ischemic damage and to compromise **mitochondrial functions that modulate inflammasome activation. This state sets the stage for late life neurodegenerative/neurovascular disease with co-morbid cognitive dysfunction or decline. The use of ER-beta-selective agonists may constitute a safer and more effective target for future therapeutic research than an ER-alpha agonist or E2 [estradiol]. ER-beta activation in the brain confers ischemic protection, stimulates mitochondrial functions, and inhibits inflammasome activation.** ER-beta agonists may be safer in that ER-beta lacks the ability to stimulate the proliferation of breast or endometrial tissue. The ER-beta agonist may be able to act both on the cerebro- and cardiovascular system to reduce the ischemic burden...the model of **reproductive senescence as a systemic inflammatory phase of life is crucial to understanding neurological changes that can occur in menopausal women.**

Conclusion: Who profits from a deceptive drug marketing campaign to convince the female population to fear estrogen? Follow the money. Nothing happens in medicine without the blessing and support of the drug industry. For every estrogen deficiency degenerative disease, the drug industry has a profitable drug for you. Drug marketing persuades women to believe a non-hormonal drug is a sufficient treatment. Menopause is an estrogen deficiency state, and a non-hormonal menopausal drug is not a sufficient treatment. The tragedy is that after many years of being deceived by the false drug industry marketing to convince them to take a non-hormonal menopause drug, they will eventually discover the deception. The awakened patient may then decide to try bioidentical hormone therapy. Yet, it may be too late because severe withdrawal symptoms may make it impossible to stop the neurokinin receptor blocker drug. (20-32)

◆ References for Chapter 24 The Folly of Non-Hormonal Menopausal Drugs

1) Lederman, Samuel, et al. "Fezolinetant for Treatment of Moderate-To-Severe Vasomotor Symptoms Associated With Menopause (SKYLIGHT 1): A Phase 3 Randomised Controlled Study." The lancet 401.10382 (2023): 1091-1102.

2) Shaukat, Ayesha, et al. "Veozah (Fezolinetant): A Promising Non-Hormonal Treatment for Vasomotor Symptoms in Menopause." Health Science Reports 6.10 (2023): e1610.

3) Blondeau, C., N. Clerc, and A. Baude. "Neurokinin-1 and Neurokinin-3 Receptors Are Expressed in Vagal Efferent Neurons That Innervate Different Parts of The Gastro-Intestinal Tract." Neuroscience 110.2 (2002): 339-349.

4) Sanger, Gareth J. "Neurokinin NK1 and NK3 Receptors as Targets for Drugs to Treat Gastrointestinal Motility Disorders and Pain." British journal of pharmacology 141.8 (2004): 1303-1312.

5) Sanger, Gareth J., et al. "Defensive and Pathological Functions of The Gastrointestinal NK3 Receptor." Vascular pharmacology 45.4 (2006): 215-220.

6) Poole, Daniel Philip, et al. "Stimulation of the Neurokinin 3 Receptor Activates Protein Kinase Cε And Protein Kinase D In Enteric Neurons." American Journal of Physiology-Gastrointestinal and Liver Physiology 294.5 (2008): G1245-G1256.

7) Mishra, Amrita, and Girdhari Lal. "Neurokinin Receptors and Their Implications In Various Autoimmune Diseases." Current Research in Immunology 2 (2021): 66-78.

8) Holman, Halsted R. "The relation of the chronic disease epidemic to the health care crisis." ACR open rheumatology 2.3 (2020): 167-173.

9) Cifuentes, Mariana, et al. "Low-grade chronic inflammation: a shared mechanism for chronic diseases." Physiology 40.1 (2025): 4-25.

10) Meetoo, Danny. "Chronic diseases: the silent global epidemic." British journal of nursing 17.21 (2008): 1320-1325.

11) Horton, Richard. "The neglected epidemic of chronic disease." The Lancet 366.9496 (2005): 1514.

12) Bennett, Jeanette M., et al. "Inflammation—nature's way to efficiently respond to all types of challenges: implications for understanding and managing "the epidemic" of chronic diseases." Frontiers in medicine 5 (2018): 316.

13) Mills, Dora Anne. "Chronic disease: the epidemic of the twentieth century." Maine Policy Review 9.1 (2000): 50-65.

14) Schank, Jesse R. "Neurokinin Receptors in Drug and Alcohol Addiction." Brain research 1734 (2020): 146729.

15) Griebel, Guy, and Sandra Beeské. "Is There Still a Future for Neurokinin 3 Receptor Antagonists as Potential Drugs for The Treatment of Psychiatric Diseases?" Pharmacology & Therapeutics 133 (2012): 116-123.

16) Gersh, Felice. The New Nonhormonal Drug for Hot Flashes. 2024. https://youtu.be/qst2V7n_9tw

17) Biglia, Nicoletta, et al. "Vasomotor Symptoms in Menopause: A Biomarker of Cardiovascular Disease Risk And Other Chronic Diseases?" Climacteric 20.4 (2017): 306-312.

18) Lobo, Rogerio A., et al. "Prevention of Diseases After Menopause." Climacteric 17.5 (2014): 540-556.

19) McCarthy, Micheline, and Ami P. Raval. "The Peri-Menopause In A Woman's Life: A Systemic Inflammatory Phase That Enables Later Neurodegenerative Disease." Journal of Neuroinflammation 17 (2020): 1-14.

20) Jett S, Schelbaum E, Jang G, et al. Ovarian steroid Hormones: A Long Overlooked But Critical Contributor To Brain Aging And Alzheimer's Disease. Front Aging Neurosci. 2022;14:948219.

21) Peters KJ. What Is Genitourinary Syndrome of Menopause and Why Should We Care? Perm J. May 2021;25.

22) Kodoth V, Scaccia S, Aggarwal B. Adverse Changes in Body Composition During the Menopausal Transition and Relation to Cardiovascular Risk: A Contemporary Review. Women's Health Rep (New Rochelle). 2022;3(1):573-581.

23) Maki PM, Henderson VW. Cognition and the Menopause Transition. Menopause. Jul 2016;23(7):803-5.

24) Mosconi L, Berti V, Dyke J, et al. Menopause Impacts Human Brain Structure, Connectivity, Energy Metabolism, And Amyloid-Beta Deposition. Sci Rep. Jun 9 2021;11(1):10867.

25) Giannini, Andrea, et al. "Neuroendocrine Changes During Menopausal Transition." Endocrines 2.4 (2021): 405-416.

26) Rinaldi, Fabio, et al. "The Menopausal Transition: Is the Hair Follicle "Going through Menopause"?." Biomedicines 11.11 (2023): 3041.

27) Zouboulis, C. C., et al. "Skin, Hair and Beyond: The Impact Of Menopause." Climacteric 25.5 (2022): 434-442.

28) Bravo, Bruna, et al. "Dermatological Changes during Menopause and HRT: What to Expect?" Cosmetics 11.1 (2024): 9.

29) Watt FE. Musculoskeletal Pain and Menopause. Post Reprod Health. Mar 2018;24(1):34-43.

30) Thurston RC. Vasomotor Symptoms: Natural History, Physiology, And Links with Cardiovascular Health. Climacteric: The Journal of the International Menopause Society. Apr 2018;21(2):96-100.

31) Veozah Is a New Non-Hormonal Drug for Hot Flashes. Peoples Pharmacy. Joe Graedon. May 30, 2023.

32) Aschenbrenner, Diane S. "Second Nonhormonal Drug for Menopausal Hot Flashes." AJN The American Journal of Nursing 123.11 (2023): 22-23.

Chapter 25

Adverse Effects of Birth Control Pills

Depression, Anxiety, Mood Disturbance

AMY, A 21-YEAR-OLD COLLEGE STUDENT came into the office with her mother because of depression, anxiety, and severe mood disorder. Her symptoms included forgetfulness, insomnia, and alternating depression and euphoria. Her medical history was unremarkable except for the past two years she had been on birth control pills (BCPs). Her laboratory studies showed a severely low B12 level of 217pg/mL. In 2005, Dr. Lawrence Solomen found patients with serum B12 (cobalamin) values less than 300 pg/mL may experience unexplained neuropsychiatric or hematologic disorders due to occult B12 deficiency. My office uses a serum B12 cutoff level of 450 pg/mL. Below this level, we give B12 supplements, usually 5,000 to 10,000 mcg per day as a sublingual tablet. Patients with gastric atrophy or pernicious anemia cannot absorb B12 and may need injections. The patient was advised to discontinue the BCPs since they were likely causing her symptoms. Six weeks later, Amy reported all symptoms had resolved. Oral contraceptive pills (OCPs) cause vitamin B6, B12, folate and tryptophan deficiency. These vitamin deficiencies can cause mood disorders such as depression, anxiety, and even frank psychosis. In addition, B12 and folate deficiency can cause peripheral neuropathy and megaloblastic anemia. For women contemplating going off BCPs to carry a pregnancy, prenatal vitamins containing methylfolate is advised pre-conception. The best version of folate is the 5-methyl-tetra-hydro-folate. Maternal folate deficiency causes neural tube defects in the developing fetus, prevented by folate supplements. In the 1990s folate fortification of flour and grain products was mandated, arguably one of the most successful public health measures in history. (1-7)

Avoid the Copper T IUD

Amy was advised to avoid synthetic hormone-impregnated IUDs (intrauterine device), the Mirena , Kyleena, Liletta, and Skyla. The synthetic hormones cause adverse effects very similar to the oral BCPs What about the ParaGard Copper T-IUD? The Copper T-IUD is no longer recommended and is currently in litigation for breakage of the IUD arm during removal with retention of fragments in the uterine cavity. Other issues include heavy bleeding, disruption of vaginal microbiome with recurrent vaginal infection, copper toxicity or sensitivity, etc. Instead of the Copper T-IUD, some form of barrier method is recommended such as the diaphragm, cervical cap and/or condom. Thanks and Credit goes to Lindsey Berkson and David Brownstein for alerting me to this information in their online course entitled: "Everything Hormones". (8-11)

Adverse Effects of OC's Are Well Documented in Medical Literature

The medical literature is full of reports of various nutritional deficiencies caused by oral contraceptives (OCs) also called Birth Control Pills (BCPs). In 1971, Dr. Brenda Herzberg reported that 25% of her patients stopped the BCPs because of headaches, depression, weight gain, and loss of libido. In 2007, Dr. Chris D. Melitis reported that BCPs deplete the body of nutrients such as B6, B12, folate, magnesium. riboflavin (vitamin B2), thiamine, ascorbic acid (vitamin C), and zinc. Both B12 and folate levels decrease by 40% with oral contraceptive use which may cause megaloblastic anemia. Lower folic acid levels correlate with increased prevalence of abnormal Papanicolaou (Pap) smear results. Increased coagulation leads to an increased risk of venous thrombosis and stroke. BCPs disturb tryptophan metabolism, a precursor of serotonin, which may cause depression. (12-13)

Vitamin and Mineral Deficiencies Caused by Oral Contraceptives

1) Folate Depletion from BCPs causes increased risk of cervical dysplasia with megaloblastic changes in the cervical epithelium, vascular thrombosis, megaloblastic anemia, platelet hyperactivity and stroke: The authors recommend supplementation with methylfolate at doses of 400-800 mcg per day, especially in women contemplating pregnancy after stopping the BCPs. Folate deficiency increases the risk of neural tube defect in the developing fetus. (14-16)

2) B6 Depletion: BCPs cause vitamin B6 depletion and clinical depression, most likely associated with interference in the role that vitamin B6 plays as a cofactor in conversion of tryptophan to serotonin and in biosynthesis of dopamine, and gamma-aminobutyric acid (GABA). B6 deficiency disruption of tryptophan pathways cause clinical depression. Thankfully this can be promptly relieved with B6 supplements. Pyridoxal-5-Phosphate is the active form of Vitamin B-6. Since pyridoxine in high doses can cause neurotoxicity, the Pyridoxal-5-Phosphate version should be used instead. (17-22)

3) BCP-induced B12 deficiency causes a long list of neuropsychiatric disorders. Supplementation with B12 is recommended. (23-27)

4) Vitamin C Deficiency: BCPs cause decreased Vitamin C levels- supplementation is recommended. (28)

5) Venous Thrombosis and Stroke: BCPs increase the risk of venous thrombosis and stroke. (29-30)

Oral Birth Control Pills Increase Strokes and Heart Attacks

In 2005 Dr. Jean-Patrice Baillargeon from Quebec did a meta-analysis of studies examining the risks associated with the use of low-dose BCPs, finding risk for myocardial infarctions (heart attacks) and ischemic strokes doubled compared to non-users. That is a 200% increase! For more on the coagulation and clotting dangers of oral estrogen pills, see Chapters 4 and 5. (31-32)

Use of Oral Contraceptive as Menopausal Hormone Replacement

One of the errors in modern medicine is to give birth control pills as a form of menopausal hormone replacement. In 2021, Dr. Felice Gersh reviewed the issue, stating clearly that oral contraceptives should not be used as menopausal hormone replacement because they are thrombogenic and predispose to hypertension. BCPs also raise serum CRP and cortisol levels. The correct treatment is bioidentical hormone replacement, not synthetic endocrine-disrupting chemicals found in oral birth control pills. Dr. Gersh writes:

> Of note, oral contraceptives contain ethinyl estradiol and various progestins, are thrombophilic [cause blood clots] and **predispose to hypertension**. Oral contraceptives are **not recommended as HT [Hormone Therapy] in PM [Post Menopausal] women** and must be used cautiously, with individualized risk management, during the menopausal transition. (33-35)

Birth Control Pills Raise Serum CRP and Cortisol

In my office, laboratory studies of women on birth control pills commonly demonstrate a pronounced elevation of serum cortisol. This finding is caused by elevated cortisol binding protein induced by oral estrogen pills. Transdermal estrogen does not cause this effect. We also typically find marked elevation of serum CRP in BCP users. **Note:** CRP is C-reactive protein, a nonspecific inflammatory marker. In 2018, Dr. Joao Victor Guedes from Brazil found BCP users had elevated serum CRP and D-Dimer levels, indicating chronic subclinical inflammation with thrombogenic and atherogenic potential. (36-39)

In 2024, Dr. Summer Mengelkoch studied 153 women from a private university in the southern United States between August 2021 and April 2022. Half the women used BCPs and the other half were non-users of BCPs (NC). Dr. Summer Mengelkoch found that BCP use causes hypothalamic-pituitary-adrenal axis (HPA axis) changes which mimic chronic stress, writing:

> A growing body of research has found that HC [hormonal contraceptive, BCP] users exhibit more chronic inflammation oxidative stress, and inflammation-related health conditions such as cardiovascular disease, autoimmune disorders, and depression compared to NC [BCP nonuser] women... we examined whether HC use predicted exaggerated inflammatory reactivity to an acute psychosocial stressor...we found that women using HCs exhibited a more robust, rather than blunted, cortisol response to the laboratory-based stressor than NC women... research has found that HC use predicts changes in HPA activity that mimic those found for women exposed to chronic stress in early life. Note: HPA is hypothalamic-pituitary-adrenal axis. (40-42)

Testosterone and Binding Globulin

Another disturbing finding is that BCP suppression of ovarian function reduces testosterone levels by 50-60 percent, causing loss of libido. The low testosterone effect is sometimes used to treat acne and hirsutism, both associated with elevated testosterone. The low testosterone with loss of libido may continue many years later after discontinuing the BCPs. Serum-binding globulin is increased dramatically by BCPs. The increased thyroid-binding globulin makes thyroid testing inaccurate. (43-47)

BCPs are Endocrine Disrupting Chemicals

One might ask, why are oral contraceptives so problematic? The answer is that they contain synthetic, chemically altered hormones that are endocrine-disrupting chemicals (EDCs), foreign to the human body. Any slight alteration in a hormone structure creates a "Frankenstein monster" that can wreak havoc on the fine-tuned biochemistry of the human body. In 2022, Dr. Murphy Lam Yim Wan from the University of Hong Kong considers BCPs as toxic EDCs, playing a role in breast cancer, writing:

> Oral contraceptives, which contain both levonorgestrel (a progestin) and ethinyl estradiol (an estrogen), **are considered as toxic EDCs** because they alter normal reproductive function by mimicking the action of estrogen and thus preventing normal hormone production. A large body of evidence showed that the ethinyl estradiol in oral contraceptive pills not only affect the fertility and sexual behavior of mice but also affects fetal growth and survival, as well as causes disruption of the fetal development of the prostate, urethra, especially when the mice were exposed to ethinyl estradiol during early pregnancy... **seven prospective and retrospective studies identified an increased breast cancer risk** associated with contraceptive pills use, regardless of contraceptive pills type. (48-53)

In 2015, Dr. Kim Strifert and others suggested that EDCs in oral contraceptives impair brain estrogen signaling in our progeny. DES was the first EDC to induce transgenerational cancer effects. Like DES, BCPs could be inducing adverse effects in progeny. Thus, BCPs could be contributing to the autism epidemic, writing:

> Oral contraceptives, synthetic hormones created to imitate natural human hormones and disrupt endogenous endocrine function to inhibit pregnancy, **may be causing the harmful neurodevelopmental effects that result in the increased prevalence of ASD [autism spectrum disorder]**. It is conceivable that the synthetic hormones

repeatedly assault the oocyte [egg] causing persistent changes in expression of the estrogen receptor beta gene. **Ethinylestradiol, a known endocrine disruptor, may trigger DNA methylation of the estrogen receptor beta gene causing decreased mRNA resulting in impaired brain estrogen signaling in progeny**. In addition, it is possible **the deleterious effects are transgenerational** as the estrogen receptor gene and many of its targets may be imprinted and the methylation marks protected from global demethylation and preserved through fertilization and beyond to progeny generations. (54-57)

◆ References Chapter 25 Adverse Effects of Birth Control Pills

1) Hjelt, K., et al. "Oral Contraceptives and The Cobalamin (Vitamin B12) Metabolism." Acta Obstetricia et Gynecologica Scandinavica 64.1 (1985): 59-63.

2) Shojania, A. Majid. "Oral Contraceptives: Effect of Folate and Vitamin B12 Metabolism." Canadian Medical Association Journal 126.3 (1982): 244.

3) Issac, Thomas Gregor, et al. "Vitamin B12 Deficiency: An Important Reversible Co-Morbidity in Neuropsychiatric Manifestations." Indian Journal of Psychological Medicine 37.1 (2015): 26-29.

4) Jha, Roshan Kumar, et al. "Vitamin B12 Deficiency and Psychiatric Manifestations-A Concise Review." Indian Journal of Forensic Medicine & Toxicology 15.3 (2021): 2664-2667.

5) Sahu, Prashant, et al. "Neuropsychiatric Manifestations in Vitamin B12 Deficiency." Vitamins and Hormones. Vol. 119. Academic Press, 2022. 457-470.

6) Crider, Krista S., et al. "Folic acid and The Prevention of Birth Defects: 30 Years of Opportunity and Controversies." Annual Review of Nutrition 42.1 (2022): 423-452.

7) Crider, Krista S., Lynn B. Bailey, and Robert J. Berry. "Folic Acid Food Fortification—Its History, Effect, Concerns, And Future Directions." Nutrients 3.3 (2011): 370-384.

8) Kaunitz, Andrew M. "Patient Education: Birth Control, Which Method Is Right for Me? (Beyond the Basics)." (2024). UptoDate.

9) Dubovis, Marina, and Naglaa Rizk. "Retained Copper Fragments Following Removal of A Copper Intrauterine Device: Two Case Reports." Case Reports in Women's Health 27 (2020): e00208.

10) Sarver, Jordan, et al. "Fractured Copper Intrauterine Device (IUD) Retained in The Uterine Wall Leading to Hysterectomy: A Case Report." Case Reports in Women's Health 29 (2021): e00287.

11) Dolman, Matthew. Paragard IUD Lawsuit: Updates and Settlements. Feb 21, 2025 https://lawsuitlegal-news.com/paragard-iud-lawsuit/

12) Herzberg, Brenda N., et al. "Oral Contraceptives, Depression, and Libido." Br Med J 3.5773 (1971): 495-500.

13) Meletis, Chris D., and Nieske Zabriskie. "Common Nutrient Depletions Caused by Pharmaceuticals." Alternative & Complementary Therapies 13.1 (2007): 10-17.

14) Wald, Nicholas J. "Folic acid and Neural Tube Defects: Discovery, Debate and The Need for Policy Change." Journal of Medical Screening 29.3 (2022): 138-146.

15) Sütterlin, Marc W., et al. "Serum Folate and Vitamin B12 Levels in Women Using Modern Oral Contraceptives (OC) Containing 20 µg Ethinyl Estradiol." European Journal of Obstetrics & Gynecology and Reproductive Biology 107.1 (2003): 57-61.

16) Hvas, Anne-Mette, et al. "Vitamin B6 level is associated with symptoms of depression." Psychotherapy and psychosomatics 73.6 (2004): 340-343.

17) C. Curtin, Anne, and Carol S. Johnston. "Vitamin B6 Supplementation Reduces Symptoms of Depression In College Women Taking Oral Contraceptives: A Randomized, Double-Blind Crossover Trial." Journal of Dietary Supplements 20.4 (2023): 550-562.

18) Hemminger, Adam, and Brandon K. Wills. "Vitamin B6 Toxicity." (2020).

19) Hellmann, Hanjo, and Sutton Mooney. "Vitamin B6: a Molecule for Human Health?" Molecules 15.1 (2010): 442-459.

20) Hvas, Anne-Mette, et al. "Vitamin B6 Level Is Associated with Symptoms of Depression." Psychotherapy and Psychosomatics 73.6 (2004): 340-343.

21) Leeton, John. "Depression Induced by Oral Contraception and The Role of Vitamin B6 In Its Management." Australian & New Zealand Journal of Psychiatry 8.2 (1974): 85-88.

22) Adams, P. W., et al. "Effect of pyridoxine Hydrochloride (Vitamin B6) Upon Depression Associated with Oral Contraception." The Lancet 301.7809 (1973): 897-904.

23) Wertalik, Louis F., et al. "Decreased serum B12 levels with oral contraceptive use." Jama 221.12 (1972): 1371-1374.

24) Lussana, Federico, et al. "Blood levels of homocysteine, folate, vitamin B6 and B12 in women using oral contraceptives compared to non-users." Thrombosis research 112.1-2 (2003): 37-41.

25) Issac, Thomas Gregor, et al. "Vitamin B12 Deficiency: An Important Reversible Co-Morbidity in Neuropsychiatric Manifestations." Indian Journal of Psychological Medicine 37.1 (2015): 26-29.

26) Jha, Roshan Kumar, et al. "Vitamin B12 Deficiency and Psychiatric Manifestations-A Concise Review." Indian Journal of Forensic Medicine & Toxicology 15.3 (2021): 2664-2667.

27) Sahu, Prashant, Harish Thippeswamy, and Santosh K. Chaturvedi. "Neuropsychiatric Manifestations in Vitamin B12 Deficiency." Vitamins and Hormones. Vol. 119. Academic Press, 2022. 457-470.

28) Rivers, J. M. "Oral Contraceptives and Ascorbic Acid." The American Journal of Clinical Nutrition 28.5 (1975): 550-554.

29) Dinger, Jürgen, et al. "Risk of Venous Thromboembolism and The Use of Dienogest-And Drospirenone-Containing Oral Contraceptives: Results from A German Case-Control Study." BMJ Sexual & Reproductive Health 36.3 (2010): 123-129.

30) Baillargeon, Jean-Patrice, et al. "Association Between the Current Use of Low-Dose Oral Contraceptives and Cardiovascular Arterial Disease: A Meta-Analysis." The Journal of Clinical Endocrinology & Metabolism 90.7 (2005): 3863-3870.

31) Letnar, Gasper, et al. "Ischemic Stroke in Users of Combined Hormonal Contraceptives: A Danish Registry Study." Stroke (2024).

32) Yonis, H., et al. "Association of Contemporary Hormonal Contraception and The Risk of Arterial Thrombosis." European Heart Journal 45. Supplement 1 (2024): ehae666-1535.

33) Gersh, Felice L., James H. O'Keefe, and Carl J. Lavie. "Postmenopausal Hormone Therapy for Cardiovascular Health: The Evolving Data." Heart 107.14 (2021): 1115-1122.

34) Liu, Hui, et al. "Association Between Duration of Oral Contraceptive Use and Risk of Hypertension: A Meta-Analysis." The Journal of Clinical Hypertension 19.10 (2017): 1032-1041.

35) Cameron, Natalie A., et al. "Oral Contraceptive Pills and Hypertension: A Review of Current Evidence and Recommendations." Hypertension 80.5 (2023): 924-935.

36) Guedes, João Victor M., et al. "Evaluation of lipid profile, high-sensitivity C-reactive protein and D-dimer in users of oral contraceptives of different types." Jornal Brasileiro de Patologia e Medicina Laboratorial 54.1 (2018): 14-20.

37) Yu, Run. "Exaggerated Increases in the Serum Cortisol Level in a Woman Following Oral Contraceptive Treatment." AACE Clinical Case Reports 10.5 (2024): 206-209.

38) Qureshi, Ayesha C., et al. "The Influence of The Route of Oestrogen Administration on Serum Levels of Cortisol-Binding Globulin and Total Cortisol." Clinical Endocrinology 66.5 (2007): 632-635.

39) Masama, Coleka, et al. "Hormone Contraceptive Use in Young Women: Altered Mood States, Neuroendocrine and Inflammatory Biomarkers." Hormones and Behavior 144 (2022): 105229.

40) Mengelkoch, Summer, et al. "Hormonal Contraceptive Use Is Associated with Differences In Women's Inflammatory And Psychological Reactivity To An Acute Social Stressor." Brain, Behavior, and Immunity 115 (2024): 747-757.

41) Hertel, Johannes, et al. "Evidence for Stress-Like Alterations In The HPA-Axis In Women Taking Oral Contraceptives." Scientific Reports 7.1 (2017): 14111.

42) Jentsch, Valerie L., et al. "Hormonal Contraceptive Usage Influences Stress Hormone Effects On Cognition And Emotion." Frontiers In Neuroendocrinology 67 (2022): 101012.

43) Sänger, Nicole, et al. "Effects of an Oral Contraceptive Containing 30 Mcg Ethinyl Estradiol And 2 Mg Dienogest on Thyroid Hormones and Androgen Parameters: Conventional Vs. Extended-Cycle Use." Contraception 77.6 (2008): 420-425.

44) Palatsi, Ret al, et al. "Serum Total and Unbound Testosterone and Sex Hormone Binding Globulin (SHBG) In Female Acne Patients Treated With Two Different Oral Contraceptives." Acta Dermato-Venereologica 64.6 (1984): 517-523.

45) Woznicki, Katarina. Birth Control Pills May Produce Protracted Effects on Testosterone Levels. January 03, 2006. MedPage Today. www.medpagetoday.com/OBGYN/HRT/2423

46) Panzer, Claudia, et al. "Impact of Oral Contraceptives on Sex Hormone-Binding Globulin and Androgen Levels: A Retrospective Study in Women with Sexual Dysfunction." The Journal of Sexual Medicine 3.1 (2006): 104-113.

47) Toldy, E., et al. "Comparative Analytical Evaluation of Thyroid Hormone Levels In Pregnancy And In Women Taking Oral Contraceptives: A Study From An Iodine Deficient Area." Gynecological endocrinology 18.4 (2004): 219-226.

48) Wan, Murphy Lam Yim, et al. "Endocrine Disrupting Chemicals and Breast Cancer: A Systematic Review of Epidemiological Studies." Critical Reviews in Food Science and Nutrition 62.24 (2022): 6549-6576.

49) Clouzot, Ludiwine, et al. "17α-Ethinylestradiol: An Endocrine Disrupter of Great Concern. Analytical Methods and Removal Processes Applied to Water Purification. A Review." Environmental Progress 27.3 (2008): 383-396.

50) Timms, Barry G., et al. "Estrogenic Chemicals in Plastic and Oral Contraceptives Disrupt Development of The Fetal Mouse Prostate and Urethra." Proceedings of the National Academy of Sciences 102.19 (2005): 7014-7019.

51) Dante, Giulia, et al. "Vitamin and Mineral Needs During the Oral Contraceptive Therapy: A Systematic Review." International Journal of Reproduction, Contraception, Obstetrics and Gynecology 3.1 (2014): 1-10.

52) Westhoff, Carolyn L. et al., the Quick Start Study Group. "Oral Contraceptive Discontinuation: Do Side Effects Matter?" American Journal of Obstetrics and Gynecology 196.4 (2007): 412.e1–412.e7. PMC. Web. 12 Jan. 2015.

53) Schneider-Kamp, Anna, and Jennifer Takhar. "Interrogating The Pill: Rising Distrust and The Reshaping of Health Risk Perceptions in The Social Media Age." Social Science & Medicine 331 (2023): 116081.

54) Strifert, Kim. "An Epigenetic Basis for Autism Spectrum Disorder Risk and Oral Contraceptive Use." Medical Hypotheses 85.6 (2015): 1006-1011.

55) Li, Ling, et al. "Prenatal Progestin Exposure Is Associated with Autism Spectrum Disorders." Frontiers in Psychiatry 9 (2018): 611.

56) Donhauser, Justin. "Hormonal Contraceptives and Autism Epidemics." Medical Hypotheses 141 (2020): 109729.

57) Hargreave, Marie, et al. "Maternal Use of Hormonal Contraception and Risk Of Childhood Autism Spectrum Disorders: A Parental Exposures And Child Health (PECH) Cohort Study." Psychiatry Research 332 (2024): 115695.

Chapter 26

The Depressing State of Antidepressants

MARY, A 65-YEAR-OLD RETIRED ACCOUNTANT has been a patient in my office for 15 years. For the past 20 years, she has been taking two different SSRI antidepressant drugs prescribed by her primary care doctor. Mary's physical examination showed hyperactive reflexes, dilated pupils, and fine hand tremors, all from the SSRI drugs. One day, Mary told me she wanted to get off the SSRI drugs. Since her doctor would not help, would I help her to get off the SSRI drugs? I replied, yes of course, I am always happy to provide an SSRI tapering schedule. Without the gradual tapering, withdrawal effects can be quite severe, so we try to gradually taper down the dosage over 12 weeks or more before discontinuing them altogether. In Mary's case, since she is on two separate SSRI drugs, we taper them one at a time. **Caution: DO NOT STOP YOUR SSRI DRUG without gradual tapering under the supervision of your physician**. (1-6)

In 2023, Dr. James Davies, PhD in medical anthropology from the University of Oxford and practicing psychotherapist reviewed SSRI drug use between the years 2011 to 2023 in the United Kingdom. During this time, SSRI prescriptions doubled to 85 million, and nearly 20 percent of the adult population is now taking an SSRI antidepressant drug. How can we explain the disconnect between widespread use despite the lack of meaningful clinical benefit? The medical literature on SSRI antidepressants reveals these drugs have no meaningful clinical benefit, except for the most severe depression where any treatment is better than a placebo. Dr. James Davies writes:

> Over the past decade, antidepressant prescriptions have almost doubled in England, rising from 47.3 million in 2011 to 85.6 million in 2022-23. Over 8.6 million adults in England are now prescribed them annually (nearly 20% of adults) ...Multiple meta-analyses have shown **antidepressants to have no clinically meaningful benefit beyond placebo for all patients but those with the most severe depression**. (7)

SSRI Antidepressants Are Dangerous

SSRI antidepressant drugs are dangerous for two reasons. Firstly, they are addictive drugs causing morphological changes in the terminal ends of the neurons. The SSRI drug is a serotonin reuptake inhibitor which means the drug blocks the serotonin transporters which are microscopic pumps, protein machines located at the outer membrane of the pre-synaptic neuron. Their job is to take up serotonin from the synaptic cleft and return it to the pre-synaptic neuron. Blocking the transporter inhibits the reuptake of serotonin from the synaptic cleft, increasing the amount of neurotransmitter available in the synaptic space, thus increasing the neurotransmitter signal. There are many classes of such reuptake inhibitors. Cocaine inhibits the reuptake of dopamine, serotonin, and norepinephrine by blocking all three transporters. Amphetamines block the dopamine transporters. Tiagabine (Gabitril) is a GABA reuptake inhibitor. Desvenlafaxine (Pristiq), duloxetine (Cymbalta) and venlafaxine (Effexor) are SNRI drugs, both serotonin and norepinephrine reuptake inhibitor drug. **Note:** SNRI is Serotonin and Norepinephrine Reuptake Inhibitor. What happens after prolonged use of the SSRI drug? The brain compensates by upregulating and increasing the number of transporters, thus creating drug tolerance. As in all addictive drugs, once drug tolerance is achieved, higher doses are needed for the same effect. What happens if the drug is stopped? This will mimic serotonin deficiency syndrome and will trigger drug withdrawal symptoms. Withdrawal symptoms are both psychological and physical. Psychological withdrawal symptoms are worsening anxiety and depression, insomnia, despair, panic, suicidal thoughts, and possible psychotic breakdown. Physical symptoms are autonomic nervous system dysfunction, sweating, flushing, fatigue, irritable bowel syndrome, vertigo, loss of balance, greater sensitivity to pain, and binge eating. This is not good. Recovery from drug withdrawal may take a few months, the time it takes for the neurons to down-regulate the transporters. Eventually, withdrawal symptoms abate, meaning the brain has restored normal serotonin concentrations at the synaptic cleft. For the stronger SNRI drugs such as Effexor (venlafaxine), a drug that inhibits both serotonin and norepinephrine reuptake, its withdrawal effects include electric shock-like "brain zaps" or "shivers" which may take months to resolve. (8)

Akathisia and Psychosis

The second reason SSRI drugs are dangerous is an adverse effect called akathisia; a form of agitation often described as a feeling like scraping your fingernails across the blackboard at the front of the classroom. Akathisia and manic psychosis are known side effects of SSRI drugs, cocaine, methamphetamine, ADHD stimulants (methylphenidate), and antipsychotic drugs. Akathisia is a form of psychomotor psychosis that drives the drug user to violence, suicide, homicide, and or other bizarre behaviors. Many of the mass shootings reported in the media are the result of SSRI-induced akathisia. The link between SSRI drugs and mass shootings, suicide, and homicide has been known for decades. Yet this information is largely ignored or intentionally suppressed by mainstream media to preserve massive drug company profits from this class of drugs. (9-18)

One Percent of SSRI Users Become Violent

Most people seem to tolerate SSRI drugs without becoming psychotic. Why do SSRI drugs cause a small number of people to exhibit suicidal thoughts, hostility, and violent behavior? In 2021, Dr. Eikelenboom-Schieveld examined this question, finding genetic mutations in drug metabolism in one percent of the population, leading to the accumulation of excess SSRI drug levels. This could account for aggressive, hostile, violent, and psychotic behavior of people on SSRI drugs, and other neuro-psychiatric drugs such as amphetamines, ADHD stimulants, anti-psychotics, etc. This mutation involves the CYP450 enzyme system in the liver which is responsible for metabolizing and removing drugs and chemicals from the body. If the CYP450 is mutated and not working properly, then the SSRI drug accumulates, reaching toxic blood levels. **Note:** ADHD= attention deficit hyperactivity disorder. Dr. Eikelenboom-Schieveld says violence is a known side effect of psychoactive medication, writing:

> There is an association between prescription drugs, most notably **antidepressants and other psychoactive medication**; having variant alleles for CYP2B6, CYP2C8, CYP2C9, CYP2C19, CYP2D6 and CYP3A4; and **the occurrence of an altered emotional state or acts of violence**. Based on these results, genotyping [genetic testing] patients for these six CYP450s would provide information as to who might be susceptible to adverse drug reactions, e.g., the development of an altered emotional state or **assault/suicide/homicide**. This would be an improvement to personalized medicine.... **Violence is a known side effect of psychoactive medication**, as is recognized in the literature. Psychoactive medication is mainly metabolized by enzymes generated by CYP450 genes. Reduced or non-functional alleles will have an effect on the blood levels of drugs and can cause side effects, e.g., **acts of violence**...This might explain why millions of people take prescription drugs and **only around 1% commit acts of violence**, a

number the FDA nevertheless considers "frequent". When either the variant alleles or the amount of medication increase, one might develop an **altered emotional state**. This should be taken as a warning sign. From the medical histories, such emotional states are often considered as a sign that the medication is not working enough. **A typical result is to add more or different medication, elevating a patient to a level with an increased risk of acts of violence.** (19-20)

Depression is More Complicated

Amphetamines are performance-enhancing drugs. In the 1940s, amphetamines were handed out freely to WWII soldiers on both sides. After the war in the 1950s, amphetamines made their way into civilian life. They were the first drugs marketed for the treatment of depression. Amphetamines do work quite well as a pick-me-up brain stimulant. However, it was discovered that people on amphetamines eventually had psychotic episodes, addiction, and withdrawal effects. This eventually led to the 1965 FDA restrictions on amphetamine use. For SSRI and amphetamine drugs that block neurotransmitter re-uptake, the neurons eventually adapt to the drug requiring higher doses to reach the same effect. This is called drug tolerance in which the brain has reached an equilibrium state. The drug dosage is no longer effective. Many SSRI users will continue the drug to avoid the uncomfortable withdrawal effects of insomnia and anxiety. A similar scenario is found with many amphetamine users who eventually find themselves in a rehabilitation clinic undergoing drug withdrawal. SSRI drugs (Serotonin Reuptake Inhibitor Drugs) are based on the theory that depression is due to a deficiency of serotonin, a brain neurotransmitter. Recent studies show that depression is much more complicated, and not caused by lack of serotonin. Causes of depression include menopausal hormone deficiency, endocrine disorders, HPA dysfunction, low thyroid, low testosterone, low estrogen, inflammatory conditions such as leaky gut, diabetes, etc. as described by Dr. Angelos Halaris below. Note HPA= Hypothalamic Pituitary axis. (21)

Serotonin Theory of Depression Falsified

In 2023, Dr. Joanna Moncrieff, a psychiatrist from London did a systematic review of the medical literature finding no basis for the serotonin theory of depression. Even worse, long-term use of SSRI antidepressants reduces brain serotonin, a compensating effect associated with drug tolerance as mentioned above. Dr. Joanna Moncrieff writes:

> The main areas of serotonin research provide **no consistent evidence of there being an association between serotonin and depression**, and no support for the hypothesis that depression is caused by lowered serotonin activity or concentrations. Some evidence was consistent with the possibility that **long-term antidepressant use reduces serotonin concentration.** (22)

In a second article in 2023, Dr. Joanna Moncrieff rebuts criticism by her psychiatric colleagues saying the clinical trials of "SSRI antidepressants show marginal differences from placebo and do not fulfill criteria for clinical relevance. SSRI drugs perturb brain chemistry in unpredictable ways, and emotional blunting has emerged as a clear drug effect," writing:

> whether antidepressants produce a genuine and useful pharmacological effect that is independent of the placebo effect, has not been established. **Antidepressants show marginal differences from placebo, which do not fulfil criteria for clinical relevance**, and may represent amplified placebo effects due to unblinding. **It is hard to reconcile even the most generous appraisal of their efficacy with the vast numbers of people now taking them**...Antidepressants produce varied and more or less subtle effects on arousal, sensations, thoughts and feelings, commonly including numbing of emotions, now demonstrated even in healthy volunteers ...From the public's point

of view, taking a drug that is believed to reverse an underlying chemical imbalance or other brain abnormality is quite a different prospect from **taking a drug that perturbs brain chemistry in incompletely known and potentially unpredictable ways,** with poorly researched effects on mood and behavior, with **emotional numbing** emerging as a clear effect. Yet, this approach to marketing drugs by drawing on unproven, implausible single neurotransmitter hypotheses to provide biological justifications for their use continues apace. (23)

Sexual Dysfunction, Emotional Blunting

In addition to increased violence and suicide, adverse side effects of SSRI antidepressants include sexual dysfunction and emotional blunting, two inconvenient truths that have been largely ignored. A medical literature search for SSRI antidepressant sexual dysfunction returns 40,000 articles. In 2024, Dr. Judith J. Stephenson reported sexual dysfunction from SSRI drug use in 19 percent of patients which may be irreversible after discontinuing the drug. (24)

In 2024, David Cox discusses the sexual consequences of SSRI drugs, writing: "Long-term sexual dysfunction is a recognized side-effect for some patients who take these widely prescribed antidepressants, and can leave sufferers devastated." In this same article appearing in the Guardian, Rosie Tilli says her SSRI drug use has eliminated her ability to have a sexual response:

> "I reassured myself that I would be fine as soon as I fully ceased the [SSRI] medication, but I wasn't," she says. "Now nearly four years on, I've learned to put on a sunny disposition, but internally I am riddled with psychological grief and anguish. **I can't experience any physiological sexual response**. **No arousal even when physically touched.** It's as if the entire electrical hardwiring of the sexual system has been short-circuited. My clitoris feels like my

elbow now, and there's nothing I can do to reverse it." (25)

Emotional Blunting and Personality Changes

One might speculate that emotional blunting may represent the main clinical benefit of SSRI drugs for people with emotional disturbances or difficult-to-control emotional responses. In my personal experience as a clinician, I find that people on long-term SSRI drugs have characteristic personality changes. They tend to be more aggressive, hostile, talkative, and self-centered than the control population. People are unaware that the SSRI drug transforms them into an unpleasant, obnoxious, argumentative personality type. Surprisingly, the medical literature supports this viewpoint. (26-28)

What Causes Depression and What Are the Treatments?

Current pharmacological interventions with SSRI antidepressants provide remission in only 30 percent of patients. With the failure of the drug industry serotonin hypothesis, you might ask the next logical question. What is the real cause of depression and what are the treatments? In 2021, Angelos Halaris, MD, a board-certified psychiatrist, and Professor of Psychiatry at Loyola University Medical School addresses this question. Dr. Halaris finds endocrinological aberrations, notably hypothalamic-pituitary-adrenal (HPA) axis dysregulation, thyroid disorders, and menopausal hormone deficiency most common. Dr. Angelos Halaris discusses how low Vitamin D contributes to depression. He suggests the medical evaluation of depression should include genetic testing for Single Nucleotide Polymorphisms (SNPs) in Cytochrome P450, Serotonin Transporter, COMT, and folate conversion to methyl-folate (the MTHFR test). He also mentions the role of immune system dysregulation and generalized inflammation, and the value of testing for CRP (C reactive Protein) an inflammatory marker. He mentions salivary cortisol testing, and the dexamethasone suppression test useful for adrenal dysfunction. He acknowledges that methyl-folate, estrogen and testosterone can serve as highly effective antidepressant treatments. **Note:** COMT= catechol-O-methyl transferase enzyme. MTHFR=methylenetetrahydrofolate reductase. Dr. Angelos Halaris writes:

> Major Depressive Disorder (MDD) is a highly prevalent psychiatric disorder worldwide. … Current pharmacologic interventions fail to produce at least partial response to approximately **one third** of these patients, and **remission is obtained in approximately 30% of patients**. This is known as Treatment-Resistant Depression (TRD)… We discuss **endocrinological aberrations, notably, hypothalamic-pituitary-adrenal (HPA) axis dysregulation and thyroid and gonadal dysfunction.** We address the role of **Vitamin D** in contributing to depression. Pharmacogenomic testing is being increasingly used to determine **Single Nucleotide Polymorphisms in Cytochrome P450, Serotonin Transporter, COMT, folic acid conversion (MTHFR)**. As the role of **immune system dysregulation** is being recognized as potentially a major contributory factor to TRD, the measurement of **C-reactive protein (CRP)** and select immune biomarkers, where testing is available, can guide combination treatments with anti-inflammatory agents (e.g., selective COX-2 inhibitors) reversing treatment resistance…As mentioned above, among the most consistent biological changes in MDD [Major Depressive Disorder] patients is **increased plasma cortisol and overall dysregulation of the HPA axis** …**Salivary cortisol** appears to be the most accurate measurement of cortisol because it reflects non-protein-bound cortisol in blood…but how do we measure HPA axis function? HPA axis function can be measured using challenge tests. One such challenge test is the **dexamethasone-suppression test (DEX),** which was among the first tests used to assess stress-related psychiatric disorders…**Testosterone and Estrogen act as antidepressants…**Vitamin D Studies

of Vitamin D supplementation show no adverse effects even at high doses of up to 10,000 IU daily, and doses of 800 IU are generally sufficient to reach a 25(OH) D level of at least 50 nmol/L (or 20 ng/mL)…**Folic Acid, L-methylfolate, MTHFR…** Numerous studies have demonstrated the association between depression and folate deficiency. L-methylfolate, the active form of vitamin B9, is the only form that can cross the blood–brain barrier. **Low levels of L-methylfolate are associated with multiple neuropsychiatric diseases, including MDD, schizophrenia and Alzheimer's…** Studies have shown the effectiveness of both folic acid and L-methylfolate as both a monotherapy and as an adjunctive therapy, suggesting the importance of assessing folate status in patients that appear to be "treatment resistant". In fact, **L-methylfolate is among the only medical foods licensed by the FDA for the treatment of depression** …it is also essential to consider the **MTHFR gene**; individuals may have genetic polymorphisms that affect the conversion of synthetic or dietary folate to L-methylfolate, the biochemically active form. If someone is homozygous for the T variant (TT), studies suggest that they have about 30% of the enzyme activity of people with the wild-type (CC) variant [100,101]. Heterozygous (CT) individuals have about 65% of the enzyme activity of CC individuals. In Caucasian North Americans, 8–20% of the population has the TT genotype…**Recommendation: perform measurement of blood level folic acid and genomic test for MTHFR**, preferably as part of a more comprehensive pharmacogenomic profile as available on the market in your country…**Anti-Inflammatory Treatments in Affective Disorders**…Recommendations: Investigate the presence of an inflammatory process anywhere in body and take corrective action to reduce or eliminate this source of inflammation. **Measurement of hsCRP** may be a useful marker of an inflammatory process in the body. (21) (29-32)

In my opinion, depression is not a disease. Rather, depression is a symptom of an underlying disorder, most related to low thyroid disorders, menopausal hormone deficiency, inflammatory disorders, and mitochondrial dysfunction disorders. Note an overlap between two disorders, chronic fatigue, and depression. For all practical purposes in the clinic, they frequently overlap and can sometimes be regarded as identical syndromes. In 2011, Dr. Michael Maes, Professor, Department of Neuropsychiatry, University of Ghent, Belgium says depression and chronic fatigue "should be regarded as 'co-associated disorders' that are clinical manifestations of shared pathways." Dr. Michael Maes writes:

There is a significant 'comorbidity' between **depression** and myalgic encephalomyelitis/**chronic fatigue syndrome** (ME/CFS). Depressive symptoms frequently occur during the course of ME/CFS. **Fatigue and somatic symptoms (F&S), like pain, muscle tension, and a flu-like malaise, are key components of depression.** At the same time, depression and ME/CFS show major clinical differences, which allow to discriminate them with a 100% accuracy... Numerous studies have shown that depression and ME/CFS are characterized by **shared aberrations in inflammatory, oxidative and nitrosative (IO&NS) pathways, like systemic inflammation ... dysfunctional mitochondria; lowered antioxidant levels, like zinc and coenzyme Q10; autoimmune responses** to neoepitopes formed by O&NS; ...and **increased translocation of gram-negative bacteria. ...Depression and ME/CFS are not 'comorbid' disorders, but should be regarded as 'co-associated disorders' that are clinical manifestations of shared pathways**. (32-33)

Estrogen Treats Menopausal Depression

In 2001, Dr. Claudio de Novaes Soares from Brazil ran a double-blind placebo-controlled trial in 50 women with perimenopausal depression treated with an estradiol skin patch (100 mcg), finding excellent efficacy, writing:

Perimenopausal women (aged 40-55 years, with irregular menstrual periods and serum concentrations of follicle-stimulating hormone >25 IU/L), meeting criteria for major depressive disorder, dysthymic disorder, or minor depressive disorder, according to DSM-IV, were randomized to receive transdermal patches of 17beta-estradiol (100 μg) or placebo in a 12-week, double-blind, placebo-controlled study... Fifty women were enrolled in the study; 26 met DSM-IV criteria for **major depressive disorder,** 11 for dysthymic disorder, and 13 for minor depressive disorder... **Remission of depression was observed in 17 (68%) women treated with 17beta-estradiol compared with 5 (20%) in the placebo group (P = .001). Conclusion: Transdermal estradiol replacement is an effective treatment of depression for perimenopausal women.** (34)

In 2001, Dr. Uriel Halbreich from Buffalo, New York found that estrogen was effective for treatment of depression in postmenopausal women, writing:

There is growing evidence suggesting that estrogen may be efficacious as a sole antidepressant for depressed perimenopausal women. (35-39)

In 2024, Dr. Louise Newsome, an OB/Gyne practitioner in the U.K. suggests that psychiatrists are missing the opportunity to prescribe HRT (hormone replacement therapy) instead of SSRI antidepressants for their depressed post-menopausal patients. However, the typical psychiatrist lacks the training and expertise to prescribe HRT to menopausal women. It would be unrealistic to expect them to do so. It would also be unrealistic to expect the psychiatrist to refer these patients out to another doctor who has expertise in HRT. This would mean losing a lucrative source of income, as the patient on an SSRI drug becomes a patient for life. At the end of the day, the patient must vote with their feet, flee the psychiatrist's office, and seek out a doctor with expertise in bioidentical hormone replacement. A doctor with expertise in thyroid disorders is an added plus. Dr. Louise Newsome says SSRI antidepressants are unnecessary and inappropriate, writing:

> However, far too many women are being offered or prescribed antidepressants instead of HRT [hormone replacement therapy] which is **often unnecessary and inappropriate**. Most psychiatrists do not prescribe HRT which is a massive missed opportunity for women. (40)

On November 5, 2024, in her weekly podcast, Dr. Louise Newsome was joined by Dr. Marty Makary, a pancreatic cancer surgeon at Johns Hopkins Medical Center and newly appointed commissioner for the U.S. Food and Drug Administration (FDA). Dr. Makary is the author of the best-selling book, *Blind Spots in Medicine*. In this podcast, Dr. Marty Makary discusses how menopause is ignored in medical education. Women seeking menopausal hormone replacement are more likely to be prescribed an SSRI antidepressant. This is an error and tragedy of modern medicine. Dr. Marty Makary speaks:

> [00:21:29] You know, tragically because of this Women's Health initiative dogma 22 years ago, medical schools just kind of concluded, well, there's nothing you can do for menopause, so why teach it? ...And so we ignored menopause altogether in medical education. So today, tragically, at least in the United States, a woman is more likely to be prescribed an antidepressant for menopause than they are hormone replacement. (41)

My Clinical Experience

In my office, I frequently see post-menopausal women with depression caused by estrogen deficiency. These patients typically report dramatic improvement after treatment with bio-identical hormone replacement with a combination of estrogen, progesterone, testosterone, and DHEA. We also use nutritional supplements such as cordyceps, lithium orotate, 5-HTP, and theanine. Many of these women arrive in my office currently taking multiple anti-depressant drugs for their post-menopausal symptoms of depression. Typically, we try to taper these women off the SSRI anti-depressants, once they start treatment with bio-identical hormone replacement and are feeling better. The correct treatment for the post-menopausal female with depressive symptoms is bioidentical hormone replacement, not an SSRI drug. Progesterone is excellent for relief of anxiety-type mood disturbance. I would venture to say testosterone therapy could very well be the most effective anti-depressant known to mankind. (42-47)

Estrogen Enhances Mitochondrial Respiration in the Brain

Estrogen controls mitochondrial energy production working in synergy with thyroid and testosterone hormones, all three binding to receptors in nuclear and mitochondrial DNA, regulating the structure and function of the mitochondrial respiratory chain and generating new mitochondria. Estrogen upregulates energy production and is required by the high energy demands of brain and heart. Thus, it is not difficult to understand menopausal estrogen deficiency as a major cause of clinical depression, and menopausal estrogen replacement therapy as a highly effective treatment. It is important to remember that if clinical depression is caused by a deficiency in mitochondrial energy production, SSRI drugs are ineffective, since they do not improve mitochondrial energy production. In 2023 by Dr. Matej Luptak from Prague studied pig brain-isolated mitochondria showing **all SSRI antidepressants harm mitochondrial function** by inhibiting the mitochondrial electron transport chain. Quite the opposite, in 2023, Dr. Jing Zhu found **estrogen upregulates energy production** by enhancing the respiratory function of the mitochondrial electron transport chain, writing:

Furthermore, the E2 [estradiol] signal has been identified as one of the major signals that converge upon mitochondria to exert its **neuroprotective effect**. Mitochondria malfunction may cause many neurocognitive and neurodegenerative disorders, such as AD [alzheimer's disease], depression, and anxiety, which show a sex-specific prevalence. Proteomic analysis of brain mitochondria of female rats indicates that E2 regulates the expression of pyruvate dehydrogenase (PDH), a pivotal enzyme that transforms the pyruvate to acetyl CoA, provides substrate in the citric acid cycle, **concomitantly increases oxidative phosphorylation and ATP synthase, and decreases beta-oxidation**. According to another in vivo data, **E2 [estradiol] and progesterone treated rat brain mitochondria display enhanced respiratory function coupled with increased expression of the electron transport chain complex IV (cytochrome C oxidase).** (48-54)

Depression Caused by Leaky Gut, LPS, Microglia Activation in the Brain

A major cause of depression/chronic fatigue is "leaky gut" that releases inflammatory cytokines into the blood stream which then cross the blood-brain barrier entering the brain causing inflammation and immune activation. This immune-inflammatory activation in the brain causes activation of microglia and clinical depression/chronic fatigue. Dr. Allesio Fasano, chief of Pediatric Gastroenterology at Mass General Children's Hospital has revealed that, in susceptible individuals, the ingestion of wheat gluten triggers the release of Zonulin, a hormone that opens the "tight junctions" between epithelial cells of the GI mucosa. This is called "leaky gut." For people with gluten sensitivity, the prolonged opening of channels between the epithelial cells makes the gut lining permeable to undigested food particles and gut bacteria which "leak" into the bloodstream, thus we have a "leaky gut." This is also called low-level endotoxemia, or LPS, short for Lipo-Poly-Saccharide, the outer membrane of enteric gram-negative bacteria which activates macrophages and immune cells, releasing inflammatory cytokines into the bloodstream. This slurry of LPS and inflammatory mediators eventually reaches the cerebral circulation causing inflammation in the brain with activation of microglia. This inflammatory response disturbs the autonomic nervous system, and neurotransmitter production, resulting in autonomic dysfunction, depression, and chronic fatigue. In 2021, Dr. Michael Maes reviews this immune-inflammatory response as fundamental in the pathophysiology of depression and chronic fatigue, writing:

> In the last three decades, the robust scientific data emerged, demonstrating that the **immune-inflammatory response is a fundamental component of the pathophysiology of major depressive disorder (MDD)**.... The gastrointestinal (GI) tract, along with gut-associated lymphoid tissue (GALT), constitutes the largest lymphatic organ in the human body and forms the biggest surface of contact with the external environment. **It is also the most significant source of bacterial and food-derived antigenic material**. ...A broad range of factors, including psychological stress, inflammation, dysbiosis and other, may compromise the permeability of this barrier. This leads to **excessive bacterial translocation [leaky gut]** and **the excessive influx of food-derived antigenic material that contributes** to **activation of the immune-inflammatory response and depressive psychopathology**. (55-62)

Depression Caused by Mitochondrial Dysfunction

Let us go to the next category of Depression/Chronic Fatigue. This category is caused by mitochondrial dysfunction with deficient cellular energy production. These are very treatable disorders which include hypothyroidism, post-menopausal estrogen deficiency, low testosterone in males, HPA dysfunction with low cortisol, and genetic mutations such as MTHFR.

Mitochondrial toxins such as NSAIDS, certain antibiotics, statins, anti-diabetics and SSRI anti-depressants (themselves) cause mitochondrial dysfunction resulting in deficient energy production. Antibiotics that impair mitochondrial function include fluoroquinolones, erythromycin, doxycycline, etc. Statin drugs deplete CoQ10, a key intermediate in the electron transport chain, thus acting as a mitochondrial toxin. Mitochondrial dysfunction may be caused by deficiencies in vitamin B12, folate, alpha lipoic acid, thiamine, Co-Q10, iron, selenium, and other co-factors involved in the electron transport chain (ETC). In 2018 Dr. Josh Allen says that the serotonin theory of depression has fallen out of favor, and instead suggests mitochondrial dysfunction should be regarded as the true etiology of depression, writing:

> For more than 50 years, the dominant theory for the pathogenesis of depression was the monoamine hypothesis (Schildkraut, 1965), which arose from observations that [SSRI] antidepressant drugs work by inhibiting the reuptake of monoamines such as serotonin and norepinephrine. However, **this theory has largely fallen out of favor due to a number of discrepancies**, such as the fact that the therapeutic effects of antidepressants take weeks to develop even though monoamine levels are elevated within hours of administration, and the fact that only about 40% of patients respond satisfactorily to treatment (Trivedi et al., 2006)...Human and animal studies suggest an **intriguing link between mitochondrial diseases and depression**... Mitochondria are the cellular powerhouse of eukaryotic cells, and they also regulate brain function through oxidative stress and apoptosis. In this paper, **we make the case that mitochondrial dysfunction could play an important role in the pathophysiology of depression.** Alterations in mitochondrial functions such as oxidative phosphorylation (OXPHOS) and membrane polarity, which increase oxidative stress and apoptosis, may precede the development of depressive symptoms. However, the data in relation to **[SSRI] antidepressant drug effects are contradictory**: some studies reveal they have no effect on mitochondrial function **or even potentiate dysfunction...** Overall, the data suggest an intriguing link between mitochondrial function and depression that warrants further investigation. (63-68)

Natural Products for Depression

We have found natural products useful in depression. These include berberine, cordyceps, maca, lithium and 5-HTP. Testosterone is perhaps nature's best antidepressant. Also, do not forget caffeine in coffee and tea serve as brain stimulant and excellent antidepressant. Caffeine and methylxanthines in chocolate have antidepressant effects. For assisting mitochondrial energy production, we have Coenzyme Q10, D-ribose and L-carnitine, alpha lipoic acid, and benfothiamine. Low B12 and low iron may cause fatigue and depression, which resolves upon taking B12 and iron supplements. A leaky gut with elevated inflammatory cytokines has been linked to depression. Leaky gut can be addressed with berberine, food sensitivity testing, gluten-free diet, probiotics, and glutamine to heal the gut. (69-79)

SSRI Withdrawal and Return to Normalcy

Long-term use of SSRI drugs induces adaptive, morphological changes in the serotonergic neurons to compensate for the inhibition of serotonin transporters and resulting excess serotonin within the synaptic cleft. In the SSRI drug user, their brain quickly compensates for the SSRI drug effect by increasing the number of serotonin transporters, thus removing the excess serotonin from the synaptic cleft. Once serotonin levels return to equilibrium, this is called drug tolerance. What happens when the patient suddenly stops the SSRI drug? With no SSRI drug to maintain serotonin levels in the synaptic cleft, the upregulated serotonin transporters quickly deplete serotonin from the synaptic cleft, leading to withdrawal symptoms identical to serotonin deficiency syndrome.

Withdrawal effects include insomnia, anxiety, dizziness, vertigo, electric shock sensations in the head, and flu-like symptoms. Gradually over time, the SSRI neurons decrease the number of serotonin transporters at the terminal synapse, and withdrawal symptoms abate as normal brain chemistry is restored. (80-82)

Many people can withdraw from SSRI drugs uneventfully if the drug is tapered down gradually over 3-6 months. This is true for the younger patient on the drug for less than 1-2 years. However, it is not unusual to see patients on two or three different SSRI drugs for many years, as well as taking a benzodiazepine drug for sleep. These patients on higher doses or multiple SSRI drugs over 10 to 20 years may not be able to taper off their SRRI drug, even with a gradual taper, because of severe withdrawal effects. Also, SSRI drugs vary in ease or difficulty to withdraw from them. And, there is variation in severity of withdrawal symptoms. Perhaps the most severe withdrawal symptoms are found with the combined serotonin/norepinephrine reuptake inhibitors (SNRIs). These are notorious for causing electric shock-like "brain zaps" during withdrawal. To help ameliorate the withdrawal effects, progesterone is especially useful helping to reduce symptoms of anxiety, depression, and insomnia associated with SSRI drug withdrawal. In 2024, Dr. Julia Stimpfl discussed deprescribing SSRI antidepressants in adolescents, writing:

> Antidepressant withdrawal symptoms are related to the pharmacokinetics of the medication, which vary across antidepressants and may include irritability, palpitations, anxiety, nausea, sweating, headaches, insomnia, paresthesia, and dizziness. These symptoms putatively involve changes in serotonin transporter expression and receptor sensitivity, impacting the serotonin, dopamine, and norepinephrine pathways. (83-85)

Dr. James Greenblatt's Adjunctive SSRI Withdrawal Program

In the above discussion, oral micronized progesterone was mentioned as adjunct to SSRI tapering. In 2022, Dr. James Greenblatt, a Johns Hopkins-trained child and adolescent psychiatrist suggests a basket of supplements helpful for SSRI tapering: B12, methyl-folate, vitamin D3, B6, Zinc, 5-HTP, lithium orotate, B-complex, magnesium, curcumin, N-acetylcysteine (NAC), and cannabidiol (CBD). Lithium is especially useful because it increases the volume of the hippocampus, the brain area involved in neurogenesis, memory, and spatial learning. Lithium is a mineral ubiquitous in our water supply. Lithium was added to popular soft drinks from the 1930s to the 1950s, such as lithiated 7-Up. In 2014, Anna Fels writes in the New York Times, "Should we all take a bit of lithium?" (86-94)

Do Antidepressants Cause More Harm Than Good?

In 2012, Dr. Paul Andrews asked the question: do antidepressants do more harm than good? After an exhaustive study of the medical literature, Dr. Andrews concludes that SSRI antidepressants are neither safe nor effective, and do more harm than good:

> it is widely believed that antidepressant medications are both safe and effective; however, this belief was formed **in the absence of adequate scientific verification. The weight of current evidence suggests that, in general, antidepressants are neither safe nor effective; they appear to do more harm than good.** (95)

Dr. Paul Andrews also states the evidence of harm is greatest in the elderly where SSRI use is associated with increased risks of falling, hyponatremia, bleeding, stroke, and death, writing:

> Patients should be informed that current research suggests that unless they have very severe depression, the **symptom reducing effects of antidepressants are modest and**

are not considered clinically significant. Unless there are rapid-onset adverse side effects, antidepressant therapy usually lasts for months. Patients should be advised that prolonged use might cause mild cognitive impairment and interfere with tasks that require highly focused concentration, such as driving, which may increase the risk of accidents. Patients should also be advised that antidepressants might trigger even more severe depressive episodes when they are discontinued. All patients should be advised of the possible bleeding risks, and physicians should exercise particular caution in prescribing these drugs in conjunction with other diuretic or anti-thrombotic medications. The evidence of harm is strongest in the elderly, who should be advised of the risks of falling, hyponatremia, bleeding, stroke, and death... **In fact, antidepressants cause neuronal damage and mature neurons to revert to an immature state, both of which may explain why antidepressants also cause neurons to undergo apoptosis (programmed death**). Antidepressants can also cause developmental problems, they have **adverse effects on sexual and romantic life, and they increase the risk of hyponatremia (low sodium in the blood plasma), bleeding, stroke, and death in the elderly**. Our review supports the conclusion that antidepressants **generally do more harm than good by disrupting a number of adaptive processes regulated by serotonin**...Antidepressants perturb monoamine levels through a variety of mechanisms, the most common of which is by **binding to monoamine transporters.** In the normally functioning rodent brain, transporter blockade prevents the reuptake of monoamines into the presynaptic neuron, which causes extracellular monoamines to increase from equilibrium levels in forebrain regions within minutes to hours of administration. With prolonged antidepressant use, however, the brain's homeostatic mechanisms buffer this effect by making **a number of compensatory changes, including an inhibition of synthesis that causes the entire pool of serotonin in the forebrain (intracellular

plus extracellular) to decline. Consequently, extracellular levels in the forebrain return to equilibrium levels with prolonged treatment.** There are other changes that take place with chronic antidepressant use to maintain homeostasis, including **alterations in the density and functioning of serotonin receptors, transporters, and enzymes** ...Effects of Prolonged Antidepressant Treatment...Even among those who respond to antidepressant treatment, **longer-term use is associated with a loss of symptom reducing efficacy** – sometimes causing a full-blown relapse. This is also consistent with the brain pushing back against the symptom reducing effect of antidepressants. In an early review, studies showed that 9–57% of long-term antidepressant users met formal criteria for a relapse or a recurrence. More recent studies have found similarly **high rates of relapse** among those who initially remitted on the drug. In one study of fluoxetine, 35.2% met relapse criteria after 6 months of continuous treatment, increasing to 45.9% after 12 months. In another study, 68% of patients who initially met remission criteria, and were exposed only to continuous antidepressant treatment, had a relapse over a 2 year period. Of course, these studies only report increases in symptoms that meet formal criteria for a relapse. **A more general loss of efficacy with prolonged antidepressant use must be substantially higher**...Three recent, large, prospective epidemiological studies have found that, even after controlling for depressive symptoms, **antidepressant use is associated with an increased risk of death in the elderly** ....out of 1000 elderly people taking antidepressants, the number of deaths per year caused by antidepressants was estimated to be **10.8** (for TCAs), **35.7** (for SSRIs), and **43.9** (for other antidepressants). (95)

Spiritual Depression, What is It?

So far, we have considered depression from the standpoint of medical practice, biology, and physiology. Mainstream treatments are based on molecular biology and reductionist science. However, this model goes only so far. We are not robots. We have emotions and spiritual connections with each other and to our creator. Consider the patient who is grieving over the loss of a loved one, mother, father, or spouse. Consider the emotionally battered patient whose boss at work is a bully constantly berating them. Consider the father who has lost his job and is financially unable to feed his wife and children. Consider the drug addict who resorts to petty crime to support a drug habit. Consider the depressed cancer patient undergoing harsh chemotherapy treatments and brutal surgical procedures. Where do these categories of depression fit into our treatment algorithm? These people are trapped in a bad situation they cannot escape. Is an SSRI drug the answer for them? Or is the answer a deeper connection with family and friends serving as a support group? Can we achieve better outcomes with a deeper connection with our creator through meditation, reading sacred texts, appreciating nature, and attending religious services? Institutional medicine has recognized that spirituality as a medical treatment is a protective factor, and improves outcomes and all-cause mortality. In 2025, Kate Fiona Jones, a social worker from the University of Notre Dame, Sydney, Australia ponders some of these questions, writing:

> Spirituality has been defined as "a dynamic and intrinsic aspect of humanity through which persons seek ultimate meaning, purpose, and transcendence, and experience relationship to self, family, others, community, society, nature, and the significant or sacred" (Puchalski et al., 2014, p. 5). While such a definition may include religious belief, it encompasses other sources of connection and ultimate meaning...spirituality was identified as a **protective factor, reducing the likelihood or severity of hopelessness, suicidality and depression** (Malviya, 2023). This protective aspect of spirituality or religion within the context of mental illness has been identified among a range of populations...Spiritual

practices identified to be helpful included prayer, meditation, attending religious services, reading sacred texts or spiritual self-help books or spending time in nature (Yamada et al., 2020). (107-108)

In 2024, Dr. Katelyn N.G. Long, a doctorate in public health and post-doctorate fellow at Harvard, says spirituality is a determinant of health leading to improved outcomes and lower all-cause mortality, writing:

> The growing body of robust, empirical research strongly links spiritual beliefs, states of being, communal practices, and private rituals to a range of beneficial health outcomes including lower all-cause mortality. This has led some public health scholars to call spirituality and religion determinants of health. (109-110)

Conclusion: One of the tragic errors of modern medicine is the refusal to prescribe bioidentical hormone replacement for menopausal depression. Some patients may need thyroid medication in addition to HRT. Instead, the postmenopausal patient is given an ineffective SSRI antidepressant causing loss of libido, drug tolerance, and addiction. Rather than improving mitochondrial energy production, studies show SSRI drugs harm the mitochondria. (103-106)

Estrogen, testosterone and thyroid hormones all increase mitochondrial biogenesis and function, thus are superior to SSRI drugs in depression when caused by mitochondrial dysfunction. A small percentage of patients on SSRIs exhibit akathisia, a form of psychomotor psychosis that drives the patients to perform aggressive, homicidal, suicidal, and violent activities, such as mass shootings. The falsification and abandonment of the serotonin theory of depression has been largely ignored by mainstream medicine which considers the SSRI antidepressant drug "the standard of care" for depression, anxiety, and a wide variety of neuropsychiatric symptoms. The SSRI drug market in the U.S. is lucrative with an annual growth rate of 3.1 percent, with sales of 8.6 billion dollars in 2024. The sad reality is that SSRI and SNRI drugs are dangerous, addictive, and mostly ineffective for depression. Depression is more complicated than handing out an SSRI drug, as described by Dr. Angelos Halaris, who reminds us the diagnosis and treatment of depression, much like the practice of medicine, requires a complex diagnostic thought process to uncover the underlying cause and devise a treatment program. Perhaps that is why the busy doctor will quickly prescribe an SSRI antidepressant drug and move on to the next patient, a simplification of medicine required by the health insurance business model. In my office, we do not use SSRI drugs. We offer a tapering schedule for those who wish to get off them. This drug tapering schedule is accompanied by an adjunct program that includes progesterone, 5HTP, lithium orotate, and theanine. For new patients presenting with depression/chronic fatigue, instead of mindlessly dispensing an SSRI drug, we use diagnostic testing to uncover the underlying cause of depression and treat it appropriately as described above.

What is the mechanism of estrogen as antidepressant, in eliminating anxiety and depression? Estrogen receptors have been found in the brain, and estrogen increases the expression of an enzyme in the brain called tryptophan hydroxylase-2 (TPH2). This enzyme's job is to convert tryptophan to serotonin, and other neurotransmitters responsible for anti-anxiety and calming effects in the brain. In 2024, Dr. Peyton Christine Bendis provides evidence that estradiol enhances serotoninergic, dopaminergic and glutamatergic neurotransmission, writing:

> In conclusion, we provide a comprehensive review of the many effects of E2 [estradiol] on neurotransmitter systems and more specifically we provide evidence supporting the hypothesis that E2 may enhance serotoninergic, dopaminergic and glutamatergic neurotransmission. Investigating animal, human, and cellular data has proven beneficial in understanding how, where, and why E2 exerts its effects on both the male and female brain. (96-102)

◆ References for Chapter 26 The Depressing State of Antidepressants

1) Badar, Ahmed. "Serotonin Syndrome: An Often-Neglected Medical Emergency." Journal of Family and Community Medicine 31.1 (2024): 1-8.

2) Scotton, William J., et al. "Serotonin Syndrome: Pathophysiology, Clinical Features, Management, And Potential Future Directions." International Journal of Tryptophan Research 12 (2019): 1178646919873925.

3) Horowitz, Mark Abie, and David Taylor. "Tapering of SSRI Treatment to Mitigate Withdrawal Symptoms." The Lancet Psychiatry 6.6 (2019): 538-546.

4) Palmer, Emilia G., et al. "Withdrawing from SSRI Antidepressants: Advice for Primary Care." British Journal of General Practice 73.728 (2023): 138-140.

5) McLaren, Niall. "Clarifying Deprescribing." Australasian Psychiatry (2024): 10398562241303398.

6) Looi, Jeffrey CL, et al. "Deprescribing Antidepressants for Depression–What Is the Evidence for And Against?" Australasian Psychiatry (2024): 10398562241282377.

7) Davies, James, et al. "Politicians, Experts, and Patient Representatives Call for The UK Government to Reverse the Rate of Antidepressant Prescribing." BMJ 383 (2023).

8) Cortes, Jose A., and Rajiv Radhakrishnan. "A Case of Amelioration of Venlafaxine-Discontinuation" Brain Shivers" With Atomoxetine." The Primary Care Companion for CNS Disorders 15.2 (2013): 26751.

9) Fox News: SSRI Antidepressants Causing School Shootings? With Dr. Peter Breggin. https://www.youtube.com/watch?v=WAO5_Hk06Mc

10) Healy, David, Andrew Herxheimer, and David B. Menkes. "Antidepressants and Violence: Problems at The Interface of Medicine and Law." PLoS Medicine 3.9 (2006): e372.

11) Lagerberg, Tyra, et al. "Associations Between Selective Serotonin Reuptake Inhibitors and Violent Crime in Adolescents, Young, And Older Adults–A Swedish Register-Based Study." European Neuropsychopharmacology 36 (2020): 1-9.

12) The Decades of Evidence That Antidepressants Cause Mass Shootings As we have seen with the vaccines, almost no social cost can keep a lucrative pharmaceutical off the market. A Midwestern Doctor. Mar 29, 2023. https://www.midwesterndoctor.com/p/the-decades-of-evidence-that-antidepressants

13) Moran, Lauren V., et al. "Risk of Incident Psychosis and Mania with Prescription Amphetamines." American Journal of Psychiatry 181.10 (2024): 901-909.

14) Lucire, Yolande, and Christopher Crotty. "Antidepressant-Induced Akathisia-Related Homicides Associated with Diminishing Mutations in Metabolizing Genes of The CYP450 Family." Pharmacogenomics and Personalized Medicine (2011): 65-81.

15) Kalniunas, Arturas, et al. "The Relationship Between Antipsychotic-Induced Akathisia and Suicidal Behavior: A Systematic Review." Neuropsychiatric Disease and Treatment (2021): 3489-3497.

16) Knoll IV, James L. "Warning: Antidepressants May Cause Bank Robbery." Psychiatric Times, September 27 (2013).

17) Adshead, Gwen. "Antidepressants and Murder: Case Not Closed." BMJ 358 (2017).

18) Reeves, Roy R., and Mark E. Ladner. "Antidepressant-Induced Suicidality: An Update." CNS Neuroscience & Therapeutics 16.4 (2010): 227-234.

19) Eikelenboom-Schieveld, Selma JM, and James C. Fogleman. "Cytochrome P450 genes: Their Role in Drug Metabolism and Violence." Handbook of Anger, Aggression, and Violence. Cham: Springer International Publishing, 2022. 1-29.

20) Eikelenboom-Schieveld, Selma JM, and James C. Fogleman. "Psychoactive Medication, Violence, And Variant Alleles for Cytochrome P450 Genes." Journal of personalized medicine 11.5 (2021): 426.

21) Halaris, Angelos, Emilie Sohl, and Elizabeth A. Whitham. "Treatment-Resistant Depression Revisited: A Glimmer of Hope." Journal of Personalized Medicine 11.2 (2021): 155.

22) Moncrieff, Joanna, et al. "The Serotonin Theory of Depression: A Systematic Umbrella Review of The Evidence." Molecular psychiatry 28.8 (2023): 3243-3256.

23) Moncrieff, Joanna, et al. "The Serotonin Hypothesis of Depression: Both Long Discarded and Still Supported?" Molecular Psychiatry 28.8 (2023): 3160-3163.

24) Stephenson, Judith J., et al. "Antidepressant Use and Treatment-Emergent Sexual Dysfunction Among Patients with Major Depressive Disorder: Results from an Internet-Based Survey" Journal of Affective Disorders Reports (2024): 100750.

25) The Observer Mental Health: 'It Feels Like We've Been Lobotomised': The Possible Sexual Consequences of SSRIs by David Cox the Guardian. Mar 2, 2024. https://www.theguardian.com/society/2024/mar/02/ssri-antidepressants-sexual-dysfunction-side-effects-consequences-libido

26) Knutson, Brian, et al. "Selective Alteration of Personality and Social Behavior by Serotonergic Intervention." American Journal of Psychiatry 155.3 (1998): 373-379.

27) Harmer, C. J., et al. "Acute SSRI Administration Affects the Processing Of Social Cues In Healthy Volunteers." Neuropsychopharmacology 28.1 (2003): 148-152.

28) Tang, Tony Z., et al. "Personality Change During Depression Treatment: A Placebo-Controlled Trial." Archives Of General Psychiatry 66.12 (2009): 1322-1330.

29) Jain, Rakesh, et al. "Good, Better, Best: Clinical Scenarios For The Use of L-Methylfolate In Patients With MDD." CNS spectrums 25.6 (2020): 750-764.

30) Altaf, Rabail, et al. "Folate as Adjunct Therapy To SSRI/SNRI For Major Depressive Disorder: Systematic Review & Meta-Analysis." Complementary therapies in medicine 61 (2021): 102770.

31) Tunio, Ali Gul, et al. "Efficacy of Sertraline with L. Methyl Folate and Without L. Methyl Folate in The Treatment of Major Depressive Disorder: A Comparative Study." Journal of Population Therapeutics and Clinical Pharmacology 31.2 (2024): 313-321.

32) Maes, Michael. "An Intriguing and Hitherto Unexplained Co-Occurrence: Depression And Chronic Fatigue Syndrome Are Manifestations Of Shared Inflammatory, Oxidative And Nitrosative (IO&NS) Pathways." Progress in Neuro-Psychopharmacology and Biological Psychiatry 35.3 (2011): 784-794.

33) Harvey, Samuel B., et al. "The Relationship Between Fatigue and Psychiatric Disorders: Evidence For The Concept Of Neurasthenia." Journal of psychosomatic research 66.5 (2009): 445-454.

34) de Novaes Soares, Cláudio, et al. "Efficacy of Estradiol for The Treatment of Depressive Disorders in Perimenopausal Women: A Double-Blind, Randomized, Placebo-Controlled Trial." Archives Of General Psychiatry 58.6 (2001): 529-534

35) Halbreich, Uriel, and Linda S. Kahn. "Role of Estrogen in The Aetiology and Treatment of Mood Disorders." CNS drugs 15 (2001): 797-817.

36) Gnanasegar, Rahavi, et al. "Does Menopause Hormone Therapy Improve Symptoms of Depression? Findings From a Specialized Menopause Clinic." Menopause 31(4) (2024):320-325

37) Herson, Megan, and Jayashri Kulkarni. "Hormonal Agents for The Treatment of Depression Associated with The Menopause." Drugs & Aging 39.8 (2022): 607-618.

38) Gordon, Jennifer L., and Susan S. Girdler. "Hormone Replacement Therapy in The Treatment of Perimenopausal Depression." Current Psychiatry Reports 16 (2014): 1-7.

39) Rasgon, Natalie L., et al. "Estrogen Replacement Therapy in The Treatment of Major Depressive Disorder in Perimenopausal Women." The Journal of Clinical Psychiatry 63 (2002): 45-48.

40) Newsome, Louise, Oct 26, 2024. https://x.com/drlouisenewson/status/18500507632857415835)

41) The Dr. Louise Newson Podcast 281 - Blind spots in modern medicine, with Dr. Marty Makary. Nov. 5, 2024. https://www.balance-menopause.com/menopause-library/blind-spots-in-modern-medicine-with-dr-marty-makary/

42) Wieland, Scott, et al. "Anxiolytic Activity of The Progesterone Metabolite 5α-Pregnan-3α-Ol-20-One." Brain Research 565.2 (1991): 263-268.

43) Picazo, O., and A. Ferna. "Anti-Anxiety Effects of Progesterone and Some of Its Reduced Metabolites: An Evaluation Using the Burying Behavior Test." Brain Research 680.1-2 (1995): 135-141.

44) Studd, John, and N. Panay. "Hormones and Depression in Women." Climacteric 7.4 (2004): 338-346.

45) Walther, Andreas, et al. "Association of Testosterone Treatment with Alleviation of Depressive Symptoms in Men: A Systematic Review and Meta-analysis." JAMA psychiatry 76.1 (2019): 31-40.

46) Sun, Qihan, et al. "Role of Estrogen in Treatment of Female Depression." Aging (Albany NY) 16.3 (2024): 3021.

47) Kulkarni, Jayashri. "Estrogen—a Key Neurosteroid in The Understanding and Treatment of Mental Illness in Women." Psychiatry Research 319 (2023): 114991.

48) Zhu, Jing, et al. "Role of Estrogen in The Regulation of Central and Peripheral Energy Homeostasis: From A Menopausal Perspective." Therapeutic Advances in Endocrinology and Metabolism 14 (2023): 20420188231199359.

49) Luptak, Matej, et al. "Different Effects of SSRIs, Bupropion, And Trazodone on Mitochondrial Functions and Monoamine Oxidase Isoform Activity." Antioxidants 12.6 (2023): 1208.

50) Irwin, Ronald W., et al. "Progesterone and Estrogen Regulate Oxidative Metabolism in Brain Mitochondria." Endocrinology 149.6 (2008): 3167-3175.

51) Klinge, Carolyn M. "Estrogens Regulate Life and Death in Mitochondria." Journal of Bioenergetics and Biomembranes 49 (2017): 307-324.

52) Klinge, Carolyn M. "Estrogenic Control of Mitochondrial Function." Redox biology 31 (2020): 101435.

53) Ahmad, Irshad, and Annie E. Newell-Fugate. "Role of Androgens and Androgen Receptor in Control of Mitochondrial Function." American Journal of Physiology-Cell Physiology (2022).

54) Sagliocchi, Serena, et al. "The Key Roles of Thyroid Hormone in Mitochondrial Regulation, At Interface of Human Health and Disease." Journal of Basic and Clinical Physiology and Pharmacology 35.4-5 (2024): 231-240.

55) Rudzki, Leszek, and Michael Maes. "From "Leaky Gut" To Impaired Glia-Neuron Communication in Depression." Major Depressive Disorder: Rethinking and Understanding Recent Discoveries. Singapore: Springer Singapore, 2021. 129-155.

56) Maes, Michael. "Leaky Gut In Chronic Fatigue Syndrome: A Review." Activitas Nervosa Superior Rediviva 51.1-2 (2009): 21-8.

57) Fasano, Alessio. "Zonulin, Regulation of Tight Junctions, And Autoimmune Diseases." Annals of the New York Academy of Sciences 1258.1 (2012): 25-33.

58) Batey, L., et al. "Lipopolysaccharide Effects on Neurotransmission: Understanding Implications for Depression." ACS Chemical Neuroscience (2024).

59) Troubat, Romain, et al. "Neuroinflammation and Depression: A Review." European Journal of Neuroscience 53.1 (2021): 151-171.

60) Kim, Yong-Ku, et al. "The Role of Pro-Inflammatory Cytokines in Neuroinflammation, Neurogenesis and The Neuroendocrine System in Major Depression." Progress in Neuro-Psychopharmacology and Biological Psychiatry 64 (2016): 277-284.

61) Hurley, Laura L., and Yousef Tizabi. "Neuroinflammation, Neurodegeneration, And Depression." Neurotoxicity Research 23 (2013): 131-144.

62) Qin L, Wu X, Block ML, et. al. Systemic LPS Causes Chronic Neuroinflammation and Progressive Neurodegeneration. Glia. 2007;55(5):453-62.

63) Allen, Josh, et al. "Mitochondria and Mood: Mitochondrial Dysfunction as A Key Player in The Manifestation of Depression." Frontiers in Neuroscience 12 (2018): 386.

64) Gardner, Ann, and Richard G. Boles. "Beyond the Serotonin Hypothesis: Mitochondria, Inflammation and Neurodegeneration in Major Depression and Affective Spectrum Disorders." Progress in Neuro-Psychopharmacology and Biological Psychiatry 35.3 (2011): 730-743.

65) Klinedinst, N. Jennifer, and William T. Regenold. "A Mitochondrial Bioenergetic Basis of Depression." Journal of Bioenergetics and Biomembranes 47 (2015): 155-171.

66) Bansal, Yashika, and Anurag Kuhad. "Mitochondrial Dysfunction in Depression." Current neuropharmacology 14.6 (2016): 610-618.

67) Zielińska, Magdalena, et al. "Dietary Nutrient Deficiencies and Risk of Depression (Review Article 2018–2023)." Nutrients 15.11 (2023): 2433.

68) Du, Jing, et al. "The Role of Nutrients in Protecting Mitochondrial Function and Neurotransmitter Signaling: Implications for The Treatment of Depression, PTSD, And Suicidal Behaviors." Critical Reviews in Food Science and Nutrition 56.15 (2016): 2560-2578.

69) Noori, Tayebeh, et al. "The Role of Natural Products in Treatment of Depressive Disorder." Current Neuropharmacology 20.5 (2022): 929-949.

70) Liu, Jiawen, et al. "Natural Products for The Treatment of Depression: Insights into Signal Pathways Influencing the Hypothalamic–Pituitary–Adrenal Axis." Medicine 102.44 (2023): e35862.

71) Levine, Bruce. Psychiatry's Discredited Theories and Drugs Versus a Sane Model and Approach. Mad in America (2024). https://www.madinamerica.com/2024/02/psychiatrys-discredited-theories/

72) Stone, Marc B., et al. "Response to Acute Monotherapy For Major Depressive Disorder In Randomized, Placebo Controlled Trials Submitted To The US Food And Drug Administration: Individual Participant Data Analysis." BMJ 378 (2022).

73) Chen, Guang, et al. "Enhancement of Hippocampal Neurogenesis by Lithium." Journal of neurochemistry 75.4 (2000): 1729-1734.

74) Kim, Jin Seuk, et al. "Lithium Selectively Increases Neuronal Differentiation of Hippocampal Neural Progenitor Cells Both In Vitro And In Vivo." Journal of neurochemistry 89.2 (2004): 324-336.

75) Yan, Xue-Bo, et al. "Lithium Regulates Hippocampal Neurogenesis by ERK Pathway and Facilitates Recovery of Spatial Learning and Memory in Rats After Transient Global Cerebral Ischemia." Neuropharmacology 4.53 (2007): 487-495.

76) Quiroz, Jorge A., et al. "Novel Insights into Lithium's Mechanism of Action: Neurotrophic and Neuroprotective Effects." Neuropsychobiology 62.1 (2010): 50.

77) Dach, Jeffrey. "Gut–Brain: Major Depressive Disorder, Hypothalamic Dysfunction, and High Calcium Score Associated with Leaky Gut." Alternative Therapies in Health and Medicine 21 (2015): 10-15.

78) Twardowska, Agata, et al. "Preventing Bacterial Translocation in Patients with Leaky Gut Syndrome: Nutrition and Pharmacological Treatment Options." International Journal of Molecular Sciences 23.6 (2022): 3204.

79) Li, Ning, et al. "Berberine Attenuates Pro-Inflammatory Cytokine-Induced Tight Junction Disruption in An In Vitro Model of Intestinal Epithelial Cells." European Journal of Pharmaceutical Sciences 40.1 (2010): 1-8.

80) Sghendo, Lino, and Janet Mifsud. "Understanding the Molecular Pharmacology of The Serotonergic System: Using Fluoxetine as A Model." Journal of Pharmacy and Pharmacology 64.3 (2012): 317-325.

81) Kalia, Madhu, et al. "Comparative Study of Fluoxetine, Sibutramine, Sertraline and Dexfenfluramine on The Morphology of Serotonergic Nerve Terminals Using Serotonin Immunohistochemistry." Brain research 858.1 (2000): 92-105.

82) Collins, Helen M., et al. "Rebound Activation of 5-HT Neurons Following SSRI Discontinuation." Neuropsychopharmacology (2024): 1-10.

83) Stimpfl, Julia N., et al. "Deprescribing Antidepressants in Children and Adolescents: A Systematic Review of Discontinuation Approaches, Cross-Titration, and Withdrawal Symptoms." Journal of Child and Adolescent Psychopharmacology (2024).

84) Cosci, Fiammetta, and Guy Chouinard. "Acute and Persistent Withdrawal Syndromes Following Discontinuation of Psychotropic Medications." Psychotherapy and psychosomatics 89.5 (2020): 283-306.

85) Sharp, Trevor, and Helen Collins. "Mechanisms of SSRI Therapy and Discontinuation." Emerging Neurobiology of Antidepressant Treatments (2023): 21-47.

86) Greenblatt, James, and Winnie Lee. Integrative Medicine for Depression: A Breakthrough Treatment Plan that Eliminates Depression Naturally. Friesen Press, 2019.

86) Greenblatt, James, Functional Medicine for Antidepressant Withdrawal: An Integrative and Functional Medicine Approach to The Treatment and Prevention of Antidepressant Withdrawal, Friesen Press, June 27, 2022

87) Li, Hongfu, et al. "Lithium-Mediated Long-Term Neuroprotection in Neonatal Rat Hypoxia-Ischemia Is Associated with Antiinflammatory Effects and Enhanced Proliferation and Survival of Neural Stem/ Progenitor Cells." Journal of Cerebral Blood Flow & Metabolism 31.10 (2011): 2106-2115.

88) Diniz, Breno Satler, et al. "Lithium and Neuroprotection: Translational Evidence and Implications For The Treatment Of Neuropsychiatric Disorders."Neuropsychiatric disease and treatment 9 (2013): 493.

89) Gray, Jason D., and Bruce S. McEwen. "Lithium's Role In Neural Plasticity And Its Implications For Mood Disorders." Acta Psychiatrica Scandinavica 128.5 (2013): 347-361.

90) Hajek, T., et al. "Neuroprotective Effect of Lithium on Hippocampal Volumes in Bipolar Disorder Independent of Long-Term Treatment Response." Psychological Medicine 44.3 (2014): 507-517.

91) Kim, Jin Seuk, et al. "Lithium Selectively Increases Neuronal Differentiation of Hippocampal Neural Progenitor Cells Both In Vitro And In Vivo." Journal of Neurochemistry 89.2 (2004): 324-336.

92) Yan, Xue-Bo, et al. "Lithium Regulates Hippocampal Neurogenesis by ERK Pathway and Facilitates Recovery of Spatial Learning and Memory in Rats After Transient Global Cerebral Ischemia." Neuropharmacology 4.53 (2007): 487-495.

93) Quiroz, Jorge A., et al. "Novel Insights into Lithium's Mechanism of Action: Neurotrophic and Neuroprotective Effects." Neuropsychobiology 62.1 (2010): 50

94) Fels, Anna. "Should we all take a bit of lithium." New York Times 14 (2014). https://www.openphilanthropy.org/wp-content/uploads/Opinion-_-Should-We-All-Take-a-Bit-of-Lithium_-The-New-York-Times.pdf

95) Andrews, Paul W., et al. "Primum Non Nocere: An Evolutionary Analysis of Whether Antidepressants Do More Harm Than Good." Frontiers in Psychology 3 (2012): 117.

96) Bendis, Peyton Christine, et al. "The impact Of Estradiol on Serotonin, Glutamate, And Dopamine Systems." Frontiers in Neuroscience 18 (2024): 1348551.

97) Plöderl, Martin, et al. "Observational Studies of Antidepressant Use and Suicide Risk Are Selectively Published in Psychiatric Journals." Journal Of Clinical Epidemiology 162 (2023): 10-18.

98) Turner, Erick H., et al. "Selective publication of antidepressant trials and its influence on apparent efficacy: Updated comparisons and meta-analyses of newer versus older trials." PLoS medicine 19.1 (2022): e1003886.

99) Paludan-Müller, Asger Sand, et al. "Extensive Selective Reporting of Quality of Life In Clinical Study Reports And Publications Of Placebo-Controlled Trials Of Antidepressants." International Journal of Risk & Safety in Medicine 32.2 (2021): 87-99.

100) Mnie-Filali, Ouissame, et al. "Long-Term Adaptive Changes Induced by Antidepressants: From Conventional to Novel Therapies." Mood Disorders (2013).

101) Pan, Lisa A., et al. "Neurometabolic Disorders: Potentially Treatable Abnormalities in Patients with Treatment-Refractory Depression and Suicidal Behavior." American Journal of Psychiatry 174.1 (2017): 42-50.

102) Why Are Antidepressants So Harmful? Exploring The Common Side Effects of SSRIS And the Nightmare of Quitting Them. A Midwestern Doctor. Nov 26, 2023, the Forgotten Side of Medicine. https://www.midwesterndoctor.com/p/why-are-antidepressants-so-harmful

103) Then, Chee-Kin, et al. "Antidepressants, Sertraline, And Paroxetine, Increase Calcium Influx And Induce Mitochondrial Damage-Mediated Apoptosis Of Astrocytes." Oncotarget 8.70 (2017): 115490.

104) Li, Yan, et al. "Mitochondrial Dysfunction Induced by Sertraline, An Antidepressant Agent." Toxicological Sciences 127.2 (2012): 582-591.

105) Hroudová, Jana, and Zdeněk Fišar. "In Vitro Inhibition of Mitochondrial Respiratory Rate By Antidepressants." Toxicology letters 213.3 (2012): 345-352.

106) Charles, Emilie, et al. "The Antidepressant Fluoxetine Induces Necrosis by Energy Depletion And Mitochondrial Calcium Overload." Oncotarget 8.2 (2016): 3181.

107) Jones, Kate Fiona, and Megan C. Best. "How are the Spiritual Resources and Needs of Mental Health Consumers Identified and Documented by Staff upon Admission to an Australian Mental Health Service? A Mixed Methods Study." Journal of Religion and Health (2025): 1-18.

108) Pececnik, Tatjana Markelj, and Christian Gostecnik. "Use of Spirituality in The Treatment of Depression: Systematic Literature Review." Psychiatric Quarterly 93.1 (2022): 255-269.

109) Long, Katelyn NG, et al. "Spirituality as a Determinant of Health: Emerging Policies, Practices, And Systems: Article Examines Spirituality as A Social Determinant of Health." Health Affairs 43.6 (2024): 783-790.

110) Orr, Robert D. "Incorporating Spirituality Into Patient Care." AMA Journal of Ethics 17.5 (2015): 409-415.

Chapter 27

The Failure of Cholesterol Lowering Drugs

ELLEN IS A 56-YEAR-OLD LAWYER doing well on her post-menopausal bioidentical hormone program. Ellen had a routine visit with her primary care doctor at the Mayo Clinic and was prescribed a statin anti-cholesterol drug for slightly elevated cholesterol (224 mg/DL). Ellen runs 3 miles in the morning twice a week. Soon after starting the statin drug, she noticed cramping in her leg muscles with each run. She called me at the office to discuss the new myalgia symptoms. I explained to Ellen that muscle pain, and muscle damage are well-known adverse effects of statin drugs, and I suggested she take a break from the drug. Sure enough, off the statin drug, the muscle pain promptly resolved. About 6 months later, Ellen was due for another routine visit with her primary care doctor at the Mayo Clinic, who again prescribed a new statin drug, stating this one is safer and does not cause muscle pain. Ellen obediently started the new statin drug, and just like clockwork, muscle pain came back again. At Ellen's next follow-up call with me at the office, I explained to Ellen she is statin drug intolerant, and should not be taking statins. I explained to Ellen that doing the same thing over and over again and expecting a different result is the definition of insanity. My plea to stop the statin drug went unheeded. Ellen replied no, she is not crazy at all, and she plans to stay on the statin drug as prescribed by the Mayo Clinic. Part of my job as a physician is to "reverse brainwash" patients deceived by drug industry marketing. Unfortunately, in Ellen's case, my attempted reverse brainwashing failed.

Statin Drugs for Marathon Runners

For statin drug users, the most common complaint is myalgia, muscle soreness, weakness, cramping, and aching in the legs, all of which resolve after stopping the statin drug. Because of the muscle pain, people on statin drugs usually reduce their physical activity. Statins are mitochondrial toxins that deplete Coenzyme Q10 (ubiquinone) and reduce mitochondrial energy production by 30 percent. At the same time, reactive oxygen species (ROS) increase, producing cellular damage. In 2020, Dr. Allyson M. Schweitzer studied the impact of statin drugs on physical activity finding that up to 29% of statin users develop statin-associated muscle symptoms (SAMS) including muscle pain (myalgia), inflammation (myositis), muscle damage (myopathy), and in extreme cases, rhabdomyolysis (necrosis of muscle cells). Statins may also cause neuropathy (nerve pain), and the patient may have a severely painful arm keeping them awake at night. Dr. Allyson M Schweitzer writes:

> Statins in particular have been demonstrated to **reduce mitochondrial respiration and increase ROS production.** Allard et al. (2018) demonstrated that muscle from symptomatic statin users had reductions in complex II and IV activity [of the electron transport chain] and an approximate 28% decrease in ATP [energy] production capacity…compared to muscle from non-statin users. Additionally, Dohlmann et al. (2019) demonstrated reduced complex II respiration in both symptomatic and asymptomatic statin users…Due to its roles within the mitochondria, it was hypothesized that this lack of CoQ10 may be contributing to the reduced respiration and increased ROS production demonstrated in the mitochondria of statin users…Unsurprisingly, CoQ10 supplementation has been proposed as a treatment for SAMS. (1)

Vitamin D deficiency makes statin-induced muscle toxicity worse and supplementation with vitamin D may ameliorate symptoms. Statin users have a 40% reduction in the sarcolemma membrane protein, called CLC-1 (chloride channel proteins) causing muscle cramping, muscle spasm, and weakness. Dr. Allyson M Schweitzer writes:

> reduced vitamin D levels can reduce statin metabolism and clearance, resulting in a greater chance for myotoxicity...statin users with SAMS [Statin Associated Muscle Symptoms] had significantly lower plasma vitamin D levels than asymptomatic statin users...Khayznikov et al. (2015) found previously statin-intolerant patients to be free of SAMS after 24 months of vitamin D supplementation...In addition, statin users with SAMS demonstrate a 40% reduction in CLC-1 chloride channel proteins when compared to non-statin users. This channel is responsible for resting sarcolemma membrane stability, and thus reductions may lead to membrane hyperexcitability, and a decreased threshold for subsequent muscle cramping, spasm, and weakness....It has been suggested that statins alter mPTP [mitochondrial permeability transition pore] permeability and increase mitochondrial calcium uptake. If enough calcium is taken up by the mitochondria, respiration is impeded, swelling may occur, and apoptotic/necrotic pathways may be initiated. This could explain the deficits in mitochondrial respiration demonstrated in statin users as well as the reduction in ATP production capacity. In support of this concept, Busanello et al. (2018) showed that blocking the mPTP, mitochondrial permeability transition pore in statin-treated mice restored normal mitochondrial respiration... Compared to non-users, Parker et al. (2013) and Bouitbir et al. (2016) found statin users to have elevated CK [creatine kinase] levels at rest, indicating a greater baseline level of muscle damage. A recent case study also describes the development of acute rhabdomyolysis in a marathon runner treated with rosuvastatin, characterized by severe pain and elevated CK levels. Note: rhabdomyolysis is a severe form of muscle necrosis. (1)

Statins Impair Muscle Repair and Regeneration

Statin drugs block the mevalonate (MVA) pathway responsible for cholesterol biosynthesis. Other downstream products such as geranylgeraniol are blocked as well. This leads to deficient muscle repair and regeneration, and difficulty healing muscle injuries. Dr. Allyson M Schweitzer writes:

> Geranylgeraniol, downstream of mevalonate in the cholesterol biosynthesis pathway, has been shown to induce myogenic differentiation of murine myoblasts derived from muscle satellite cells... Similarly, statin treatment has been shown to affect satellite cell differentiation and reduce myoblast fusion, attenuating the production of multinucleated entities. These impairments were **rescued with mevalonate co-treatment**, validating the importance of downstream products of the cholesterol biosynthesis pathway in muscle maturation. As satellite cells are primary mediators of **muscle regeneration**, these results are indeed concerning for muscle repair and regeneration in the presence of statin therapy...Taken together, the above preclinical studies suggest that **statins delay muscle regeneration** and impede satellite cell function. (1)

Statin-induced muscle pain may be aggravated by physical activity, and statin users have reduced muscle strength and higher CK (creatine kinase) serum levels compared to non-users indicating ongoing muscle damage. Dr. Allyson M Schweitzer writes:

> A 2005 study by Bruckert et al. is commonly cited as evidence that SAMS [statin-associated muscle symptoms] are **aggravated by physical activity**. In this study, 40% of the population reporting myalgia indicated that certain factors triggered their symptoms. Of this 40%, over half indicated 'unusual physical exertion' as a trigger. Bruckert et al. (2005) also report the prevalence of myalgia to be greater in physically active individuals...Thompson et

al. (1997) demonstrated serum CK levels [creatine kinase, a muscle breakdown product indicating muscle damage] to be elevated in statin users compared to non-users following treadmill exercise, indicating a **greater presence of muscle damage**. These serum CK levels returned to baseline within 3 days...Established literature has demonstrated that statins **reduce muscle strength** during strength testing, as well as elevate serum CK both at rest and after exercise, raising the concern that physical exertion may exacerbate muscle damage... The current review demonstrates that statins do alter skeletal muscle metabolism, in particular, mitochondrial metabolism. (1)

Statin Drugs Cause Mitochondrial Dysfunction

In 2023, Dr. Lize Kroon from South Africa wrote her PhD thesis on muscle injuries in runners, finding statin drugs cause mitochondrial dysfunction in skeletal muscle, thus producing muscle pain, compromising aerobic exercise capacity, and increasing muscle injury. Because of the above side effects, 80% of top sports performers cannot tolerate statin drugs. Dr. Lize Kroon writes:

New evidence suggests that statin use can directly compromise aerobic exercise capacity and increase risk of muscle injuries; however, it is just the beginning of comprehension between the interactions of statins with exercise training. The use of statins may result in **mitochondrial dysfunction in skeletal muscle cells**, which in turn may interfere with aerobic capacity and training adaptations...Research has demonstrated that statins induce high levels of myalgia [muscle pain] during moderate physical exertion, resulting in a decrease in participation in physical activity, consequently affecting exercise performance. A growing body of evidence in both human and animal models indicates that physical exercise, especially when it involves eccentric contractions, can aggravate statin-induced myopathy, while

statin therapy may predispose runners to exercise-induced muscle damage. Therefore, **adults on statin therapy avoid moderate physical exertion during everyday life**...A cross-sectional survey of amateur runners in the Netherlands found an association between statin use and the prevalence of **exercise-related injuries**. Research indicated that among statin users, 41% reported an injury in the previous year: tendon- or ligament-related sport injuries 22%, muscle-related injuries 15% and other injuries 13%. The majority (30%) of the statin users reported one injury and 10% reported two injuries. Simvastatin 57% and atorvastatin 25% were the most frequently used statins. Findings indicate that **in top sports performers only about 20% tolerate statin treatment**, involving atorvastatin, fluvastatin, lovastatin, pravastatin and simvastatin, demonstrating no side-effects. Research suggested that older athletes are at increased risk of falling and decreased muscle strength and muscle quality are present in adults receiving statin therapy. (2-8)

Cardiotoxicity of Statin Drugs

In addition to skeletal muscle toxicity, statin drugs are toxic to cardiac muscle cells, a possible etiology for the increasing epidemic of congestive heart failure related to Coenzyme Q10 (CoQ10) depletion, an important co-factor in the mitochondrial transport chain. In 2020, Dr. Hans-Ulrich Kloer advises all patients on statin drugs with depressed cardiac function should be given CoQ-10, writing:

Heart failure (HF) is one of the most common causes of death in Western society. Recent results underscore the utility of coenzyme Q10 (CoQ10) addition to standard medications in order to reduce mortality and to improve quality of life and functional capacity in chronic heart failure (CHF)... Previous reports have shown that **CoQ10 concentration is decreased in myocardial tissue in CHF and by statin therapy, and the greater the CoQ10 deficiency the**

more severe is the cardiocirculatory impairment. In patients with CHF and hypercholesterolaemia being treated with statins, the combination of CoQ10 with a statin may be useful for two reasons: decreasing skeletal muscle injury and improving myocardial function...However, particular caution is advisable with the use of strategies of extreme lowering of cholesterol that may negatively impact on myocardial function. All in all there is a strong case for considering co-administration of ubiquinol [CoQ-10] with statin therapy in patients with depressed or borderline myocardial function. (9-12)

Statins Are Causing an Epidemic of Congestive Heart Failure

In 2015, Dr. Harumi Okuyama reviewed the effects of statin drugs on the mevalonate (MVA) pathway inhibition of prenyl intermediates causing vitamin K depletion which increases coronary calcification. Increasing coronary calcification is the definition of progressive coronary artery disease as demonstrated by the coronary calcium score test. Statin inhibition of prenyl intermediates also causes selenoprotein depletion which decreases anti-oxidant protection in mitochondria. Similarly, statins cause Coenzyme Q10 depletion and Heme A depletion which inhibits mitochondrial energy production causing congestive heart failure. Dr. Okuyama believes the statin-induced heart damage is permanent, representing a direct cause of the epidemic of congestive heart failure with more than one million cases annually. This is the epidemic of congestive heart failure. Sadly, cardiologists deny the culpability of the statin drug, and attribute the congestive heart failure to other causes. Dr. Harumi Okuyama writes:

Physicians in general are not aware that statins can cause heart failure **and are clearly not recognizing it.** Although vast majority of physicians readily recognize and diagnose heart failure in patients taking statins, the heart failure is almost always attributed to other non-statin related factors, such as aging, hypertension and coronary artery disease...The mechanism for the impairment in heart muscle function appears to be related to impaired mitochondrial function, which in turn is related to **statin depletion of CoQ10, selenoproteins and 'heme A'**, all required for normal mitochondrial function...**Statin-induced impairment in heart muscle function appears to be permanent,** and even though patients may clinically benefit from discontinuation of the statin along with supplemental CoQ10, we believe that many years of statin drug therapy result in the gradual accumulation of mitochondrial DNA damage. A prolonged decrease in mitochondrial CoQ10 would diminish the ability to protect mitochondrial DNA from free radical damage. After a critical percentage of mitochondrial DNA is mutated, offspring mitochondria will progressively lose their efficiency to produce ATP and simultaneously can generate more free radicals and result in a self-perpetuating vicious cycle. The negative consequences of statin-induced increase in coronary artery disease, coupled with a direct statin toxicity upon the myocardium, can be expected to be additive with enormous clinical implications...**With more than one million heart failure hospitalizations every year in the USA, the rapidly increasing prevalence of congestive heart failure is now described as an epidemic and it is likely that statin drug therapy is a major contributing factor**. (13)

Low Cholesterol, Violent Crime and Increased Mortality in Elderly

If cholesterol was the evil substance we are led to believe, one would expect increased mortality from higher cholesterol levels and decreased mortality from lower cholesterol levels. Medical research shows the exact opposite. Low cholesterol is associated with increased mortality in the elderly, and increased suicide, violent crime, and psychosis for all other ages. (14-25)

Why Are Statin Drugs So Toxic?

Statin drugs were originally derived from fungal toxins, representing a fungal adaptation to drive away predators in the wild. There are many poisonous mushrooms, and statin drugs are another poison by blocking the mammalian mevalonate (MVA) pathway. Notice statins strongly inhibit the HMGR enzyme (hydroxy-methyl-glutaryl-reductase) which converts Acetyl-Co-A into cholesterol and other end-products required for proper functioning of the mammalian cell. Statin drugs are poisons that throw a monkey wrench into mammalian cell function. In 2019, Dr. Anna Fracassi writes:

> This [mevalonate, MVA] pathway mainly produces cholesterol, but it is also responsible for the generation of other important end-products that play several physiological roles: Coenzyme Q-10 (mitochondrial respiratory chain), farnesyl and geranylgeranyl moieties (protein post-translational modifications), isopentenyl tRNAs (RNA transcription involved in selenoprotein production), and dolichol (protein N-glycosylation)....The MVA pathway, also known as cholesterol/isoprenoid biosynthetic pathway, is a pivotal metabolic pathway expressed in all mammalian cells. It leads to the production of several end-products, required for the proper functioning of cell physiology. HMGR represents the key and rate-limiting enzyme of the whole pathway, and is responsible for the conversion of HMG-CoA into MVA. HMGR is strongly inhibited by statins. (26)

Statins Block Hormone Production

Statin drugs block HMG reductase, thus inhibiting the production of cholesterol, leading to low serum cholesterol levels, an intervention thought to prevent atherosclerotic coronary artery disease. Whether or not this is true will be discussed later. For now, remember that cholesterol is the precursor of all hormones in the body. **Blocking cholesterol blocks all hormone production.** Needless to say, we need our hormones, and doing this makes people sicker, not healthier.

Neurotoxicity of Statin Drugs - Memory Loss

In 2012, the FDA added a warning label for possible memory loss caused by statin drugs. Although the brain is only 2 percent of the body weight, the brain contains about 25% of the total body cholesterol. Since the blood-brain barrier (BBB) prevents cholesterol uptake from the bloodstream, the neurons in the brain make their own cholesterol, called de novo synthesis. Unfortunately, statin drugs easily cross the blood-brain-barrier and block production of cholesterol in the brain. **Note:** lipophilic statins such as lovastatin, simvastatin, atorvastatin, fluvastatin, pitavastatin, and cerivastatin easily cross the blood-brain barrier (BBB). However, for hydrophilic statins such as pravastatin and rosuvastatin, this ability is less, yet there is still entry into the cerebral cortex. In 2010, Dr. Gibson Wood cites two studies in mice showing detection of hydrophilic pravastatin in the cerebral cortex. About 70% of the brain cholesterol is found in myelin sheaths of oligodendrocytes, and 30 percent in cell membranes of neurons and brain cells called astrocytes. Would you take a drug that prevents your brain from making essential cholesterol? Cholesterol depletion in the hippocampus "leads to progressive loss of dendritic spines and synapses". In 2019, Dr. Anna Fracassi writes:

> Synaptic vesicle biogenesis depends on high cholesterol levels. The remarkable need of cholesterol for vesicle membranes seems to be fundamental for maintaining a proper vesicle curvature and for the assembly of vesicle-specific proteins and lipids. For instance, intracellular cholesterol levels are essential … for the release of synaptic vesicles. Experimental evidence has demonstrated that cholesterol is not only crucial in presynaptic terminals, but also for postsynaptic functions. For instance, a number of neurotransmitter receptors and other postsynaptic components are closely associated to cholesterol-rich lipid rafts, suggesting that an optimal cholesterol concentration is imperative for the structural and the functional organization of post-

synaptic terminals. Both reduction and enrichment of cholesterol hamper the activity of gamma aminobutyric acid A (GABAA) receptor. Similarly, cholesterol depletion in hippocampal neurons destabilizes surface...(AMPA) receptors and leads to **progressive loss of dendritic spines and synapses.** (26)

Brain Derived Estrogen

In 2023, Dr. Jiewei Hu studied aged female mice, finding brain-derived estrogen critical for maintaining cognitive health. Statin drugs cross the blood-brain barrier and block the production of brain-derived estrogen. In 2025, Dr. Sara Taylor from Chapel Hill, North Carolina studied the in-vitro effect of statin drug, pitavastatin, on local hippocampal estrogen biosynthesis, finding statins reduce local brain estrogen production and BDNF (brain-derived neurotrophic factor) needed for neurogenesis and synaptic density, thus causing impairment of memory and cognition. Dr. Sara Taylor writes:

> The aim of this study is to investigate the effects of pitavastatin on hippocampal synaptogenesis because the hippocampus is crucial for memory formation. We also evaluated the effects of pitavastatin on local hippocampal estrogen synthesized in the hippocampus itself and its effect on Brain-Derived Neurotrophic Factor (BDNF). Using a hippocampal cell line, H19-7, we found that hippocampal neurons exposed to pitavastatin demonstrate a **significant reduction in the synaptic marker postsynaptic density protein 95 (psd-95).** The pitavastatin treated neurons also exhibited **decreased production of local estrogen and their expression of BDNF mRNA was decreased. These results suggest that statins reduce the ability of hippocampal neurons to form synapses by restricting the production of local estrogen.** Because neural connections in the hippocampus are crucial for memory formation, our findings implicate statins as

medications that **may compromise cognitive function.** (27-28)

Lipophilic Statin Drugs Accelerate Early Dementia

In 2023, Dr P. Padmanabham did a prospective study of 299 patients with early cognitive impairment enrolled in the Alzheimer's Disease Neuroimaging Initiative database. The patient's statin use was categorized as: none (nonS), lipophilic (LS) or hydrophilic (HS), along with recorded baseline serum cholesterol values. The rate of conversion to dementia was determined with global cognition (ADAS13 and MoCA), and memory domain testing. Metabolic brain imaging was also performed. Use of lipophilic statins was associated with **triple the risk of being demented within 8 years** compared to statin non-users. This decline into dementia was preceded by metabolic brain imaging changes associated with early Alzheimer's Dementia. **Note:** the lipophilic statins and not the hydrophilic statins are implicated above. This is probably related to the ease of crossing the BBB for lipophilic statins. **Note:** Lipophilic statins are atorvastatin (lipitor), simvastatin, lovastatin, fluvastatin, cerivastatin and pitavastatin. Hydrophilic statins are rosuvastatin (Crestor) and pravastatin (Pravachol). Dr. Prasanna Padmanabham writes:

> Conclusions: Use of **lipophilic, but not hydrophilic**, statins by subjects with early cognitive impairment and normal cholesterol levels at baseline was associated with nearly double the risk of becoming demented within 4 years, and **risk of becoming demented within 8 years was triple that for non-statin users.** Moreover, this was preceded during the first 2 years after baseline by metabolism in brain regions associated with early Alzheimer's demonstrating significantly greater decline, and correlating with magnitudes of clinically measurable loss of cognitive and general function. (29-32)

Coenzyme Q, Prenylated Proteins and Dolichol

Other than cholesterol, another end-product of the MVA pathway essential for the brain is CoQ10, responsible for electron transport in the mitochondria. A deficiency in CoQ10 may result in cerebellar ataxia, encephalomyopathy and brain atrophy. Yet another end-product is involved in protein prenylation, needed for correct localization of proteins in the cell membrane which is required for intracellular signaling cascades. Production of isoprenoids, such as farnesylpyrophosphate (FPP) and geranylgeranylpyrophosphate (GGPP) are blocked by statin drugs. In 2019, Dr. Anna Fracassi writes:

> For instance, isoprenoids constitute the side chain of coenzyme Q (CoQ), which assures ATP [energy] production in all mammalian cells including neurons. In mitochondria, CoQ is responsible for the electron transport during the oxidative phosphorylation. CoQ also preserves brain cells from central neurotoxic damages, acting as a powerful anti-oxidant and neuroprotective compound. In addition, clinical evidence indicates that CoQ10 deficiency often results in **neuropathological conditions, such as cerebellar ataxia, encephalomyopathy and multiple system atrophy**....Protein prenylation consists in the covalent binding of farnesyl pyrophosphate (FPP) or geranylgeranyl pyrophosphate (GGPP) moieties to proteins. The attachment of a prenyl group is an essential prerequisite for the regulation of protein localization on cell membranes and, in turn, for key signaling cascades... It was reported that HMGR inhibition by statins significantly **decreases cholesterol biosynthesis in rodent brains** and, more recently...statins also lead to a significant reduction in protein prenylation in this organ. From these observations, it is not surprising that clinically relevant doses of statins can induce important biological effects in the brain....Several reports indicated that statins may alter synaptic transmission by modulating the function of neurotransmitter receptors. (26) (33-37)

Statins and New Onset Diabetes

In 2015, Dr. Lorenzo Arnaboldi studied new onset diabetes in statin users, noting that in 2012, the FDA added a warning label that statins may raise blood sugar and could cause memory loss, writing:

> Large meta-analyses, posthoc and genetic studies showed that **statins might increase the risk of new-onset diabetes (NOD), particularly in insulin-resistant, obese, old patients**. ...Based on this evidence, to warn against the possibility of statin-induced NOD or worsening glycemic control in patients with already established diabetes, FDA and EMA changed the labels of all the available statins in the USA and Europe. [In 2012, the FDA added warning label: statins may raise levels of blood sugar and could cause memory loss]. Recent meta-analyses and retrospective studies demonstrated that statins' diabetogenicity is a dose-related class effect, but the mechanism(s) is not understood. Among statins, only pravastatin and pitavastatin do not deteriorate glycemic parameters in patients with and without type 2 diabetes mellitus. (38)

Statins and Neuropathy

Statin drugs are known to cause painful neuropathies in the extremities. Symptoms include numbness, tingling, or pain in the hands and feet. I have seen this in clinical practice. When the statin drug is stopped, the neuropathy pain disappears. Sadly, the American Heart Association has been captured and serves as a mouthpiece for the drug industry, and claims there is no evidence statin drugs cause peripheral neuropathy. This is a blatantly false statement based on flawed research, published in 2019 by Dr. Connie Newman, writing:

> There is no convincing evidence for a causal relationship between statins and cancer, cataracts, cognitive dysfunction, **peripheral**

neuropathy, erectile dysfunction, or tendonitis. [**Note**: this statement is blatantly false]. (39)

In 2014, Dr. Brenton West studied sensory perception among 30 statin-treated patients compared to controls finding statin users had decreased vibration sensation indicating peripheral sensory neuropathy. In 2002, Dr. David Gaist studied the risk of polyneuropathy in a population-based patient registry study in Denmark, finding for patients on long-term statin drugs, for 2 or more years, their risk of polyneuropathy **increased 26-fold**, writing:

> The authors verified a diagnosis of idiopathic polyneuropathy in 166 cases. The cases were classified as definite (35), probable (54), or possible (77). The odds ratio linking idiopathic polyneuropathy with statin use was 3.7 (95% CI 1.8 to 7.6) for all cases and 14.2 (5.3 to 38.0) for definite cases. The corresponding odds ratios in current users were 4.6 (2.1 to 10.0) for all cases and 16.1 (5.7 to 45.4) for definite cases... For patients treated with statins for 2 or more years the odds ratio of definite idiopathic polyneuropathy was **26.4** (7.8 to 45.4). Conclusions: Long-term exposure to statins may substantially increase the risk of polyneuropathy. (40-41)

In 2020, Dr. Neha Gurha from New Delhi, India studied CoQ10 and HMGCo-A levels, as well as nerve conduction studies in 50 statin users compared to controls, finding statin-induced lower CoQ10 and HMGCR levels are associated with nerve conduction deficits, writing:

> Statin users had lower serum CoQ10 and HMGCR levels associated with nerve conduction deficits suggesting a role of CoQ10 in the occurrence of the neurological adverse effects. (42-43)

The Failure of Cholesterol-Lowering Drugs

Now that we have reviewed the toxicity and adverse effects of statin drugs which are broad-spectrum poisons of the mammalian cell,

let us now turn our attention to the question of the efficacy of lowering cholesterol with a drug. This is called the cholesterol theory of heart disease and is the same question asked by Dr. Mikael Rabaeus in 2019 who reviewed all the studies on the new cholesterol-lowering drugs called CETP and PCSK9 Inhibitors. Firstly, Dr. Rabaeus tells us we have no access to the raw data of these studies, and we should. Why is the raw data kept secret? In the absence of the raw data, Dr. Mikael Rabaeus suggests we should rely on mortality outcomes as the most reliable data. Dr. Rabaeus was disappointed to find, despite a very significant lowering of cholesterol, these new drugs were a complete failure, writing: **"neither anti-CETP nor anti-PCSK9 treatment can significantly reduce the risk of cardiovascular death,".** This new data overturns the theory that reducing cholesterol levels prevents heart disease. Dr. Mikael Rabaeus writes:

> In the absence of open access to the raw data for independent scientists, it is clear that the **mortality endpoint** should be the main criterion to test efficacy...If cholesterol-lowering treatments other than statins reduce the risk of cardiovascular complications, it would confirm that the cholesterol-heart theory is correct. **However, if these substances fail to reduce the risk, the cholesterol-heart theory should be rejected**...The two groups of medications are (i) the cholesteryl ester transfer protein **(CETP) inhibitors**, which we will call anti-CETP in the present study, and (ii) the proprotein convertase subtilisin/kexin type 9 serine protease **(PCSK9) inhibitors**, which we will call anti-PCSK9 in this study...**The review did show that neither anti-CETP nor anti-PCSK9 treatment can significantly reduce the risk of cardiovascular death, thereby giving credit to the questioning of the cholesterol-heart theory**...despite a very significant effect on cholesterol levels, the CETP and PCSK9 inhibitors have not been shown to diminish the frequency of clinical events in high-risk patients, especially not the important ones represented by total and cardiovascular deaths...Another

consequence of these findings is that they speak strongly against the cholesterol-heart theory, confirming the doubts that have already been raised by a large group of scientists all over the world. As this theory leads to millions of people taking statin drugs, it appears highly necessary that access to raw data of all statin trials be allowed so as to re-appreciate them. This is an important aspect considering the **very strong conflicts of interest that the majority of scientists present**, all the more concerning as many of these scientists exercise official activities in Association boards and guidelines committees and in medical journals. Therefore, we continue to maintain that the cholesterol-heart theory should be seriously challenged. (44)

In 2023, Dr. Jahnavi Grover from Blacktown Hospital, Sydney, Australia did a systematic review and meta-analysis of 10 PCSK9 inhibitor clinical trials. The authors declared no industry funding for their study. Ten randomized controlled trials of PCSK9 inhibitors with 57,890 patients were reviewed. Dr. Grover found: **"no statistically significant effect was observed... for all-cause mortality."** In other words, a placebo gave the same mortality benefit as the PCSK9 drug, indicating the failure of cholesterol-lowering drugs. Think of it this way. Approximately 680,000 deaths every year in the US are attributed to cardiovascular disease. An effective drug for the prevention of cardiovascular disease should reduce this mortality number by 50-90 percent. That is what we are looking for. Why in the world would anyone take a drug to prevent heart disease that has no mortality benefit? Such a drug is a fraud. (45)

CETP Inhibitor Drugs All Abandoned

After disappointing clinical trials, all four CETP inhibitor drugs, anacetrapib, dalcetrapib, evacetrapib, and torcetrapib were abandoned by their drug company sponsor. They were never FDA approved, and never brought to the market. This is the definition of a failed drug. Torcetrapib (Pfizer) was abandoned in 2006. Evacetrapib (Eli Lilly) failed a clinical trial in 2014 and was abandoned. Dalcetrapib (Hoffmann–La Roche) drug development was abandoned in May 7, 2012 "due to a lack of clinically meaningful efficacy." Anacetrapib (Merck) halted drug development in 2017. Obicetrapib (Amgen, AMG-899) abandoned in 2017. In 2020, Amgen licensed the drug to NewAmsterdam Pharma which is in Phase III trial as of 2023. (46-48)

44 Randomized Controlled Trials (RCTs)

As pointed out in 2017 by Maryanne Demasi, there are 44 randomized controlled trials of drug or dietary interventions that lower LDL cholesterol for primary or secondary prevention of coronary artery disease. None of the 44 show any mortality benefit. (49-51)

Making things even worse, a 2019 meta-analysis by Dr. Morten Hansen found that taking a statin drug would lengthen a person's life by only 3.1 days for primary and 4.2 days for secondary prevention. (52)

In 2021, Dr. Robert Dubroff reviewed the latest cholesterol targets and analyzed the prevailing medical studies on cholesterol-lowering drugs. Over 75 percent of cholesterol-lowering drug trials reported no mortality benefit, and 50 percent no cardiovascular benefit. This contradictory evidence has been largely ignored by mainstream medicine because it rejects the cholesterol theory of heart disease. Dr. Robert Dubroff writes:

> In this analysis **over three-quarters of the cholesterol lowering trials reported no mortality benefit and nearly half reported no cardiovascular benefit at all**... In most fields of science the existence of contradictory evidence usually leads to a paradigm shift or modification of the theory in question, but **in this case the contradictory evidence has been largely ignored simply because it doesn't fit the prevailing paradigm**. (53)

Statin Drugs Increase Coronary Calcium Score

One of the glaring problems is the cholesterol level is not predictive of which patients will have heart attacks and which patients will not. The cholesterol panel has been replaced by a new technology called the coronary calcium score, highly predictive of future heart attack risk. There is no correlation between cholesterol level and calcium score as demonstrated in 2001 by Dr. Harvey Hecht, clinical professor and cardiologist at Mount Sinai Hospital in New York. A greater than 15 percent annual increase in calcium score is highly predictive of active coronary artery disease and impending heart attack. Under 15 percent annual increase indicates dormant disease with a low risk for impending heart attack. Interventions that decrease the annual progression of calcium score prevent heart attack as demonstrated in 2004 by Dr. Paolo Raggi. Unfortunately for the statin drug industry, statin drugs do not decrease calcium score progression. It is now widely accepted by mainstream cardiology that statin drugs tend to increase calcium score, or have an effect indistinguishable from placebo. The current explanation is the statin induced increase in calcium score is a benign stabilization of the atherosclerotic plaque, since "we know how beneficial statin drugs are". If you believe this, I have a bridge to nowhere to sell you. In a 2004 study by Dr. Paolo Raggi, over 15 percent annual progression of calcium score while on a statin drug predicted impending myocardial infarction, not stabilization. In 2018, Dr. Joshua Mitchel did a retrospective registry study at the Walter Reed Hospital reviewing the impact of statin drugs on cardiovascular outcomes following calcium score testing. Dr. Mitchell found that statin drugs have no clinical benefit for a calcium score of 100 or lower, regardless of cholesterol level. This indicates a paradigm shift in mainstream cardiology. The calcium score has replaced the cholesterol level for determining which patient should be treated with a statin drug. On a practical basis for the practicing clinician who wishes to get their patient off the statin drugs without incurring the wrath of the cardiologist, a calcium score test is indicated. If the calcium score is under 100, this justifies stopping the statin drug. In my opinion, the adverse side effects of statin drugs are so hideous that nobody should be forced to take them. For people who decide to decline statin drugs, a reasonable alternative is the calcium score protocol listed below. (54-57)

Our Calcium Score Protocol

In my office, our calcium score protocol for preventing calcium score progression, and thereby preventing and reversing coronary artery disease is based on a more comprehensive theory of heart disease described in my book, *Heart Book* (2018) and is as follows: Plant-based diet, vitamin C, tocotrienol vitamin E, aged garlic, magnesium, vitamin K (MK-7), berberine, curcumin, proteolytic enzymes (lumbrokinase, serrapeptidase, nattokinase), dental hygiene and dental cleaning, the control of metabolic syndrome, insulin sensitivity and fasting blood sugar with dietary modification and metformin. For diabetics and others with leaky gut, a probiotic program from Microbiome Labs called Mega-spore-biotic Total Gut Restore is useful. I suggest this calcium score protocol is safer and more effective than cholesterol-lowering drugs. The plant-based diet alone is safer and more effective than statin drugs, as discussed in 2017 by Dr. Caldwell Esselstyn. Aged garlic is another supplement with 4 of 6 randomized controlled trials showing attenuation of calcium score progression, a result superior to statins which fail to have any beneficial effect on calcium score progression. I would like to see a randomized controlled trial comparing the progression of calcium score on statin drugs versus our protocol: plant-based diet, aged garlic, MK7 vitamin K, tocotrienol vitamin E, vitamin C, and lumbrokinase, berberine, probiotic. However, this will never happen. For more on this topic, see my book entitled, *Heart Book* (2018). (58-66)

Statins for the Menopausal Female?

The single most effective treatment in menopausal women for the prevention of coronary artery disease is not statin drugs. It is estrogen hormone replacement therapy (HRT) within the first 5 years after menopause. In 2022, Dr. Howard Hodis discusses the benefits of HRT in the Danish Osteoporosis study showing a 52 percent reduction in cardiovascular disease and a 30 percent reduction in all-cause mortality. This benefit is far superior to anything seen in statin drug studies:

> Survival curve from the Danish Osteoporosis Study showing **a statistically significant reduction of cardiovascular disease by 52%** (HR, 0.48; 95% CI, 0.27–0.89) after 10 years of randomized hormone replacement therapy (HRT) (estrogen with or without progestogen) relative to no HRT. **Note:** For more on this, see Chapters 9 and 10, Preventing Coronary Artery Disease with Bioidentical Hormones. (68-69)

Conclusion: When prescribing a drug for the prevention or treatment of a disease, one must weigh the benefits against the adverse effects. Since statins are mitochondrial and cellular poisons with extensive adverse side effects on muscle and brain, one would expect that high efficacy outweighs the negatives. Quite to the contrary, we find cholesterol lowering is a complete failure for preventing cardiovascular disease. There is no mortality reduction, and half the RCTs have no cardiovascular benefit. Why in the world would any rational-thinking human being take such a drug? This brings us back to my original question: why do we have so many failed randomized drug trials for cholesterol-lowering drugs? Why keep repeating the same drug experiment repeatedly expecting a different result? This is the definition of insanity. In 2016, Dr. David Diamond studied the entire statin drug medical literature, finding the small benefit in cholesterol-lowering trials is independent of cholesterol-lowering. This is exactly what we are seeing with the failed clinical trials for the newer drugs, both anti-CETP and anti-PCSK9 drugs. They reduce cholesterol levels even lower than statins, yet have no mortality benefit. The small benefit seen in statin drug clinical trials is due to pleiotropic effects, the anti-inflammatory and antimicrobial effects. The same clinical benefit can be obtained without statin drugs by using our calcium score protocol listed above, thus avoiding statin drug adverse effects on muscle, heart, and brain. (67)

♦ References for Chapter 27 The Failure of Cholesterol Lowering Drugs

1) Schweitzer, Allyson M., et al. "The Impact of Statins on Physical Activity and Exercise Capacity: An Overview of The Evidence, Mechanisms, And Recommendations." European Journal of Applied Physiology 120 (2020): 1205-1225.

2) Kroon, Lize. The Epidemiology and Associated Risk Factors for Muscle Strain Injuries in Endurance Running Athletes. Diss. University of Pretoria, South Africa, 2023.

3) Murlasits Z, Radak Z. The Effects of Statin Medications on Aerobic Exercise Capacity and Training Adaptations. Sports Medicine. 2014;44(11):1519-30.

4) Mikus CR, et al. Simvastatin Impairs Exercise Training Adaptations. Journal of the American College of Cardiology. 2013;62(8):709-14.

5) Sinzinger H, O'Grady J. Professional Athletes Suffering from Familial Hypercholesterolaemia Rarely Tolerate Statin Treatment Because Of Muscular Problems. British Journal of Clinical Pharmacology. 2004;57(4):525-8.

6) Scott D, Blizzard L, Fell J, Jones G. Statin Therapy, Muscle Function and Falls Risk in Community-Dwelling Older Adults. QJM: An International Journal of Medicine. 2009;102(9):625-33.

7) Muraki A, et al. Coenzyme Q10 Reverses Mitochondrial Dysfunction in Atorvastatin-Treated Mice and Increases Exercise Endurance. Journal of Applied Physiology. 2012;113(3):479-86.

8) Bouitbir J, et al. Atorvastatin Treatment Reduces Exercise Capacities in Rats: Involvement of Mitochondrial Impairments and Oxidative Stress. Journal of Applied Physiology. 2011;111(5):1477-83.

9) Kloer, Hans-Ulrich, et al. "Combining Ubiquinol with A Statin May Benefit Hypercholesterolaemic Patients with Chronic Heart Failure." Heart, Lung and Circulation 29.2 (2020): 188-195.

10) Chaulin, Aleksey. "Cardiotoxicity as a Possible Side Effect of Statins." Reviews in Cardiovascular Medicine 24.1 (2023): 22.

11) Chaulin, Aleksey M. "The Negative Effects of Statin Drugs on Cardiomyocytes: Current Review of Laboratory and Experimental Data (Mini-Review)." Cardiovascular & Hematological Agents in Medicinal Chemistry 22.1 (2024): 7-16.

12) Langsjoen, Peter H., et al. "Statin-Associated Cardiomyopathy Responds to Statin Withdrawal and Administration of Coenzyme Q10." The Permanente Journal 23 (2019).

13) Okuyama, Harumi, et al. "Statins Stimulate Atherosclerosis and Heart Failure: Pharmacological Mechanisms." Expert Review of Clinical Pharmacology 8.2 (2015): 189-199.

14) Turusheva, Anna, et al. "Low Cholesterol Levels Are Associated with A High Mortality Risk in Older Adults Without Statins Therapy: An Externally Validated Cohort Study." Archives of Gerontology and Geriatrics 90 (2020): 104180.

15) Zhou, Zhen, et al. "Low-Density-Lipoprotein Cholesterol and Mortality Outcomes Among Healthy Older Adults Not Taking Lipid-Lowering Agents: A Cohort Study With 12,334 Participants." Circulation 146.Suppl_1 (2022): A12463-A12463.

16) Golomb, Beatrice A., Håkan Stattin, and Sarnoff Mednick. "Low Cholesterol and Violent Crime." Journal Of Psychiatric Research 34.4-5 (2000): 301-309.

17) Leppien, Emily, et al. "Effects of Statins and Cholesterol on Patient Aggression: Is There A Connection?" Innovations in clinical neuroscience 15.3-4 (2018): 24.

18) Virkkunen, Matti. "Serum Cholesterol Levels in Homicidal Offenders: A Low Cholesterol Level Is Connected with A Habitually Violent Tendency Under the Influence of Alcohol." Neuropsychobiology 10.2-3 (1983): 65-69.

19) Zhang, Jian, et al. "Low HDL Cholesterol Is Associated with Suicide Attempt Among Young Healthy Women: The Third National Health and Nutrition Examination Survey." Journal of Affective Disorders 89.1-3 (2005): 25-33.

20) Vevera, J., et al. "Cholesterol Concentrations in Violent and Non-Violent Women Suicide Attempters." European Psychiatry 18.1 (2003): 23-27.

21) Apter, Alan, et al. "Serum Cholesterol, Suicidal Tendencies, Impulsivity, Aggression, And Depression in Adolescent Psychiatric Inpatients." Biological psychiatry 46.4 (1999): 532-541.

22) Çetin, Mesut, et al. "Low Cholesterol Level in Patients with Antisocial Personality Disorder: The Association with Homicidal Behavior." Bull Clin Psychopharmacol 9 (1999): 185-188.

23) Sen, Piyal, et al. "How Do Lipids Influence Risk of Violence, Self-Harm and Suicidality In People With Psychosis? A Systematic Review." Australian & New Zealand Journal of Psychiatry 56.5 (2022): 451-488.

24) Edgar, P. F., et al. "Violent Behavior Associated with Hypocholesterolemia Due to A Novel APOB Gene Mutation." Molecular psychiatry 12.3 (2007): 258-263.

25) Suneson, Klara, et al. "Low Total Cholesterol and Low-Density Lipoprotein Associated with Aggression and Hostility in Recent Suicide Attempters." Psychiatry Research 273 (2019): 430-434.

26) Fracassi, Anna, et al. "Statins and The Brain: More Than Lipid Lowering Agents?" Current Neuropharmacology 17.1 (2019): 59-83.

27) Hu, Jiewei, et al. "Brain-Derived Estrogen: A Critical Player in Maintaining Cognitive Health of Aged Female Rats, Possibly Involving GPR30." Neurobiology of Aging 129 (2023): 15-27.

28) Taylor, Sara, and Rabin Adhikari. "The Effect of Statin Treatment on Synaptogenesis in the Hippocampus." Biological research for nursing 27.1 (2025): 71-80.

29) Padmanabham, Prasanna, et al. "Cognitively Impaired Subjects with Normal Total Cholesterol Using Lipophilic Statins Undergo Accelerated Decline to Dementia and Loss of Cognition and General Function Correlating with Loss of Regional Brain Metabolism in a Multi-Center Longitudinal Study." (2023): P1445-P1445.

30) Padmanabham, Prasanna, Stephen Liu, and Daniel Silverman. "Lipophilic Statin Users without Dementia Undergo Accelerated Decline in Regions Associated with Dementia and Those Regions are Predictive of Further Decline over Subsequent Two-Year Period in a Multi-Center Longitudinal Study." (2024): 242555-242555.

31) Sahebzamani, Frances M., et al. "Examination of the FDA Warning for Statins and Cognitive Dysfunction." J. Pharmacovigil 2.4 (2014): 1000141-100015.

32) Padala KP, Padala PR, McNeilly DP, et al. The Effect Of HMG-Coa Reductase Inhibitors on Cognition in Patients with Alzheimer's Dementia: A Prospective Withdrawal and Rechallenge Pilot Study. Am J Geriatr Pharmacother 2012; 10:296-302

33) Mailman, Tiffany, Manoj Hariharan, and Barbara Karten. "Inhibition of Neuronal Cholesterol Biosynthesis with Lovastatin Leads to Impaired Synaptic Vesicle Release Even in The Presence of Lipoproteins or Geranylgeraniol." Journal of Neurochemistry 119.5 (2011): 1002-1015.

34) Miron, Veronique E., et al. "Statin Therapy Inhibits Remyelination in The Central Nervous System." The American Journal of Pathology 174.5 (2009): 1880-1890.

35) Klopfleisch, Steve, et al. "Negative Impact of Statins on Oligodendrocytes and Myelin Formation In Vitro And In Vivo." Journal of Neuroscience 28.50 (2008): 13609-13614.

36) Xiang, Zhongmin, and Steven A. Reeves. "Simvastatin Induces Cell Death in A Mouse Cerebellar Slice Culture (CSC) Model of Developmental Myelination." Experimental neurology 215.1 (2009): 41-47.

37) Smolders, Inge, et al. "Simvastatin Interferes with Process Outgrowth and Branching of Oligodendrocytes." Journal of Neuroscience Research 88.15 (2010): 3361-3375.

38) Arnaboldi, Lorenzo, and Alberto Corsini. "Could Changes in Adiponectin Drive the Effect Of Statins On The Risk Of New-Onset Diabetes? The Case of Pitavastatin." Atherosclerosis Supplements 16 (2015): 1-27.

39) Newman, Connie B., et al. "Statin Safety and Associated Adverse Events: A Scientific Statement from the American Heart Association." Arteriosclerosis, thrombosis, and vascular biology 39.2 (2019): e38-e81.

40) West, Brenton, et al. "Statin Use and Peripheral Sensory Perception: A Pilot Study." Somatosensory & Motor Research 31.2 (2014): 57-61.

41) Gaist, Davd., et al. "Statins and Risk of Polyneuropathy: A Case-Control Study." Neurology 58.9 (2002): 1333-1337.

42) Gurha, Neha, et al. "Association of Statin Induced Reduction in Serum Coenzyme Q10 Level and Conduction Deficits in Motor and Sensory Nerves: An Observational Cross-Sectional Study." Clinical Neurology and Neurosurgery 196 (2020): 106046.

43) Golomb, Beatrice A., and Marcella A. Evans. "Statin adverse effects: a review of the literature and evidence for a mitochondrial mechanism." American Journal of Cardiovascular Drugs 8 (2008): 373-418.

44) Rabaeus, Mikael, and Michel de Lorgeril. "A Systematic Review of Clinical Trials Testing CETP and PCSK9 Inhibitors: The Cholesterol-Heart Theory—Time for a Requiem?" Journal of Controversies in Biomedical Research 5.1 (2019): 4-11.

45) Grover, Jahnavi, et al. "Longer-term Impact of PCSK9 Inhibitors on Major Adverse Cardiovascular Events and All-Cause Mortality: A Systematic Review and Meta-Analysis of Randomised Controlled Trials." European Heart Journal 44. Supplement 2 (2023): ehad655-2808.

46) Hey, Spencer Phillips, et al. "Success, Failure, And Transparency in Biomarker-Based Drug Development: A Case Study of Cholesteryl Ester Transfer Protein Inhibitors." Circulation: Cardiovascular Quality and Outcomes 10.6 (2017): e003121.

47) Doggrell, Sheila A. "No Cardiovascular Benefit with Evacetrapib–Is This the End of The Road for the 'Cetrapibs'?" Expert Opinion on Pharmacotherapy 18.14 (2017): 1439-1442.

48) Eyvazian, Vaughn A., and William H. Frishman. "Evacetrapib: Another CETP Inhibitor for Dyslipidemia with No Clinical Benefit." Cardiology In Review 25.2 (2017): 43-52.

49) Demasi, Maryanne, Robert H. Lustig, and Aseem Malhotra. "The Cholesterol and Calorie Hypotheses Are Both Dead—It Is Time to Focus on The Real Culprit: Insulin Resistance." Pharmaceutical Journal (2017).

50) DuBroff, Robert. "Cholesterol Paradox: A Correlate Does Not a Surrogate Make." BMJ Evidence-Based Medicine 22.1 (2017): 15-19.

51) DuBroff, Robert. "A Reappraisal of The Lipid Hypothesis." The American Journal of Medicine 131.9 (2018): 993-997.

52) Hansen, Morten Rix, et al. "Postponement of Death by Statin Use: A Systematic Review and Meta-Analysis of Randomized Clinical Trials." Journal of General Internal Medicine 34 (2019): 1607-1614.

53) DuBroff, Robert, Aseem Malhotra, and Michel de Lorgeril. "Hit or miss: the new cholesterol targets." BMJ Evidence-Based Medicine 26.6 (2021): 271-278.

54) Hecht, Harvey S., et al. "Relation of Coronary Artery Calcium Identified by Electron Beam Tomography to Serum Lipoprotein Levels and Implications for Treatment." The American journal of cardiology 87.4 (2001): 406-412.

55) Raggi, Paolo, Tracy Q. Callister, and Leslee J. Shaw. "Progression of Coronary Artery Calcium and Risk of First Myocardial Infarction in Patients Receiving Cholesterol-Lowering Therapy." Arteriosclerosis, thrombosis, and vascular biology 24.7 (2004): 1272-1277.

56) Mitchell, Joshua D., et al. "Impact of Statins on Cardiovascular Outcomes Following Coronary Artery Calcium Scoring." Journal of the American College of Cardiology 72.25 (2018): 3233-3242.

57) Kaushik, Atul, Nilashish Dey, and Peeyush Jain. "Clinical utility of computed tomography coronary calcium scoring in real-life practice." Journal of Current Cardiology 2.3 (2024): 146-149.

58) Esselstyn, Caldwell B. "A Plant-Based Diet and Coronary Artery Disease: A Mandate for Effective Therapy." Journal Of Geriatric Cardiology: JGC 14.5 (2017): 317.

59) Esselstyn, Caldwell, and Mladen Golubic. "The Nutritional Reversal of Cardiovascular Disease, Fact or Fiction? Three Case Reports." Exp Clin Cardiol 20 (2014): 1901-8.

60) Massera, Daniele, et al. "A Whole-Food Plant-Based Diet Reversed Angina without Medications or Procedures." Case reports in cardiology 2015.1 (2015): 978906.

61) Massera, Daniele, et al. "Angina Rapidly Improved with A Plant-Based Diet and Returned After Resuming a Western Diet." Journal Of Geriatric Cardiology: JGC 13.4 (2016): 364.

62) Liu, Xiaohui, et al. "Plant-Based Diet and All-Cause and Cause-Specific Mortality among Patients with Cardiovascular Disease: A Population-Based Cohort Study." Circulation 150.Suppl_1 (2024): A4139937-A4139937.

63) Budoff, Matthew. "Aged Garlic Extract Retards Progression of Coronary Artery Calcification." The Journal of nutrition 136.3 (2006): 741S-744S.

64) Wlosinska, Martiné, et al. "The Effect of Aged Garlic Extract on The Atherosclerotic Process—A Randomized Double-Blind Placebo-Controlled Trial." BMC Complementary Medicine and Therapies 20 (2020).

65) Murali, Shashank, et al. "Interventions to Attenuate Cardiovascular Calcification Progression: A Systematic Review of Randomized Clinical Trials." Journal of the American Heart Association 12.23 (2023): e031676.

66) Kasim, Manoefris, et al. "Improved Myocardial Perfusion in Stable Angina Pectoris by Oral Lumbrokinase: A Pilot Study." The Journal of Alternative and Complementary Medicine 15.5 (2009): 539-544.

67) Diamond, David M., and Uffe Ravnskov. "How Statistical Deception Created the Appearance That Statins Are Safe and Effective in Primary and Secondary Prevention of Cardiovascular Disease." Expert Review of Clinical Pharmacology 8.2 (2015): 201-210.

68) Hodis, Howard N., and Wendy J. Mack. "Menopausal Hormone Replacement Therapy and Reduction of All-Cause Mortality and Cardiovascular Disease: It Is About Time and Timing." The Cancer Journal 28.3 (2022): 208-223.

69) Schierbeck, Louise Lind, et al. "Effect of Hormone Replacement Therapy on Cardiovascular Events In Recently Postmenopausal Women: Randomised Trial." BMJ 345 (2012).

70) Saheki, Akira, et al. "In Vivo And In Vitro Blood-Brain Barrier Transport Of 3-Hydroxy-3-Methylglutaryl Coenzyme A (HMG-Coa) Reductase Inhibitors." Pharmaceutical Research 11 (1994): 305-311.

71) Wood, W. Gibson, et al. "Statins and Neuroprotection: A Prescription to Move the Field Forward." Annals Of The New York Academy of Sciences 1199.1 (2010): 69-76.

Chapter 28

Waking Up from the Sleeping Pill Nightmare

MARY IS A 52-YEAR-OLD HOUSEWIFE from Wyoming who has taken an Ambien (zolpidem) sleeping pill every night for the last three years. Because of worrisome menopausal symptoms of vaginal dryness, hot flashes, and night sweats, Mary was starting on hormone replacement therapy (HRT) with bioidentical estrogen and progesterone. A problem was apparent early on. Mary required three times the standard dosage of hormone replacement to relieve her menopausal symptoms. Perhaps the Ambien sleeping pills are interfering with or blocking the progesterone and estrogen receptors in the brain?

Sleeping Pills Mechanism of Action

The most prescribed sleeping pills in the U.S. are the "Z drugs" (zolpidem, zopiclone, and zalephon). How do they work? The Z drugs have a mechanism very similar to the benzodiazepine (BZD) class of drugs. The drug binds to the Type A- GABA neurons in the brain, thus increasing sensitivity to GABA (gamma-aminobutyric acid). The GABA neurons are inhibitory while the glutamate neurons are excitatory. Z-drugs are thought to be safer and less addictive than the benzodiazepine class of drugs. However, this is false, as the Z-drugs carry the same risks of addiction and withdrawal. Dr. Daniela F Curado's 2021 study in Brazil showed chronic users self-reported chemical dependency about the same for both Z-drugs (70%) and BZD drugs (77%). In 2018, Dr. Aasha Agravat compared Z-drug hypnotics to benzodiazepines, writing:

Current perception that Z-drugs are safer than benzodiazepines is inaccurate; they tend to have a quicker onset and shorter duration of action but **produce similar side-effects, so are not suitable substitutions**. There is a variation between the different half-lives; hypnotics with the shorter half-lives tend to be safer producing less residual problems and adverse effects in most patients, but have higher risks of withdrawal so **should only be used short-term. Benzodiazepines still carry the greatest risk of tolerance and abuse potential when compared with Z-drugs**. Consideration of patient characteristics as well as pharmacokinetic differences between drugs before offering a hypnotic is vital. **All hypnotics should be used short-term or intermittently and be reviewed regularly**. This **caution should be explained to patients** along with information about the hypnotic prescribed to limit abuse potential. (1-2)

In 2023, Dr. Justyna Cabaj compared benzodiazepines (BZDs) and Z-drugs and found that although prescribing trends suggest Z-drugs are safer than BZD's, this is a medical myth. Z-drugs are not safer and carry the same risks and addiction potential. Dr. Justyna Cabaj writes:

Despite their effectiveness, **both drug classes carry the risk of addiction, physical and psychological dependence, and withdrawal symptoms**. Side effects, such as **drowsiness, dizziness, and cognitive impairment,** are also associated with their use. Recent studies indicate that chronic use of BZDs and Z-drugs may lead to **cognitive impairment and an increased risk of dementia in older adults**. Furthermore, individual factors, dosage, duration of use, and drug interactions can affect their efficacy. Prescribing trends show a **decline in benzodiazepine prescriptions** and **an increase in Z-drug use due to perceived safety advantages**. However, **evidence suggests that Z-drugs carry similar risks of adverse effects and addiction potential as benzodiazepines. (3)

Benzodiazepines for Sleep

Although benzodiazepines are commonly prescribed short-term for insomnia and anxiety, when used long-term, benzodiazepines create drug tolerance and chemical addiction. Withdrawal effects are quite severe, with worsening insomnia and anxiety. Thus, long-term use is hazardous and not recommended. It is best to avoid this class of drugs altogether. The most used benzodiazepines for sleep are: diazepam (Valium), lorazepam (Ativan), alprazolam (Xanax), temazepam (Restoril), and triazolam (Halcion). Benzodiazepine addiction is the most notorious and most difficult to withdraw, requiring medical supervision or a drug rehabilitation facility. Benzodiazepines can be dangerous in combination with alcohol or other sedative drugs, a potentially lethal combination causing death from respiratory depression. (4)

Z-Drugs Interfere with Estrogen and Progesterone Receptors in the GABA Neurons

Insomnia or sleep disturbance is a common menopausal symptom. HRT provides complete relief of insomnia with good quality sleep. However in menopausal women on long-term Z-drug sleeping pills, this can cause problems. The sleeping pill drugs (z-drugs) bind to the GABA neurons. The GABA neurons also have estrogen and progesterone receptors. However the sleeping pill "Z-drug" prevents binding of estrogen and progesterone to their receptors on the GABA neuron, and requires higher doses of HRT to be effective for the relief of menopausal symptoms. Progesterone acts through the metabolite allopregnanolone, a neurosteroid that binds to GABA receptors. Progesterone has well-documented neuroprotective effects. (5-10)

Menopausal Insomnia is a Symptom of Estrogen and Progesterone Deficiency

Because of the "fear of estrogen," menopausal women seeking relief from insomnia will be given a hypnotic Z-drug sleeping pill or BZD drug, rather than estrogen hormone replacement (HRT), one of the tragic errors of modern medicine. HRT should be the first-line treatment for insomnia in menopausal women. HRT (estradiol/ progesterone) is more effective and treats all menopausal symptoms, not just insomnia.

Progesterone Induces Sleep via GABA Receptors

How does progesterone help with sleep? What is the mechanism of action? In 1996, Dr. Marike Lancel found that the mechanism of sleep induction with the progesterone metabolite, allopregnanolone, is very similar to the mechanism of Z-drugs that bind to the GABA receptor. In 2021, Dr. Brendan J. Nolan from Australia studied the efficacy of micronized progesterone for sleep, finding improvement in sleep outcomes in nine randomized controlled trials in postmenopausal women. Preclinical studies in mice show the mechanism of action of progesterone resembles benzodiazepine drugs, with increased metabolites of allopregnanolone binding to and activating GABA receptors. **Note:** GABA is the inhibitory neurotransmitter system that is required for sleep. The major difference is that progesterone is a safe neuroprotective, natural hormone with no adverse side effects. On the other hand, hypnotic sleeping pills are associated with chemical addiction, tolerance, increased mortality, and a long list of adverse effects. Dr. Brendan J. Nolan writes:

> Preclinical data has shown progesterone metabolites improve sleep parameters through positive allosteric modulation of the gamma-aminobutyric acid type A receptor [GABA Type A Receptor].... (11-14)

In 2024, Dr. Gyun-Ho Jeon discusses how to treat menopausal insomnia, suggesting hormone replacement with estrogen and progesterone should be the treatment of choice because of the favorable effects of estrogen and progesterone on the brain. Both hormones are neurosteroids made in the brain. Dr. Gyun-Ho Jeon also recommends 5-HTP (5-hydroxytryptophan) and melatonin. **Note:** 5-HTP is the precursor to serotonin synthesis in the brain, thus increasing serotonin and melatonin levels. 5HTP is a good alternative to SSRI antidepressants (selective serotonin reuptake inhibitors) which increase serotonin at the synaptic cleft. I have found the combination of lithium orotate, theanine and 5-HTP useful for sleep. Dr. Gyun-Ho Jeon writes:

> **menopausal hormone therapy (MHT) should be considered as the treatment of choice** among pharmacological treatments, following **cognitive behavioral therapy**, which is suggested as the first-line treatment in the general population insomnia treatment guidelines. Additionally, **melatonin and 5HT-based drugs [theanine, lithium orotate, and melatonin]**, which have fewer side effects, along with MHT should be preferentially recommended in menopausal women...Overall, estrogen seems to increase the rapid eye movement sleep [REM sleep] and total sleep time and decrease sleep latency and awakenings after sleep. Estrogen may also exert an antidepressant effect by regulating 5HT [5-hydroxy-tryptophan, a precursor to serotonin]. Progesterone stimulates benzodiazepine receptors, causing the release of gamma-aminobutyric acid (GABA), a sedating neurotransmitter, and thus induces sleep favoring non-rapid eye movement sleep. Progesterone is also known to exert an anxiolytic and respiratory stimulant effect, which may also help promote good sleep. (15-21)

In 2020 and 2023, Dr. Annika Haufe reviewed the role of hormones in postmenopausal sleep disturbances, finding low estrogen and low progesterone levels correlate with poor sleep, writing:

Results from good-quality studies demonstrated that the **postmenopausal decline in estrogen and progesterone contributes to sleep disturbances in women and that timely treatment with estrogen and/or progesterone therapy improved overall sleep quality**...Estrogen: Polysomnographic measurements demonstrated a positive relationship between estrogen and sleep, **specifically low estradiol was associated with lower sleep efficiency**, sleep-disordered breathing, and a high frequency of movement arousals. Subjective sleep measures from questionnaires demonstrating a positive association between estrogen and sleep found significantly **lower levels of estradiol in women with insomnia**. In addition, **decreasing estradiol levels were associated with poor sleep and trouble falling and staying asleep**. A higher estrogen level was significantly associated with decreased awakening during the night as well as early morning awakenings... **the results clearly demonstrate a positive effect of estrogen on sleep in perimenopausal women... Progesterone:** In subjective questionnaires, **lower concentrations of progesterone were associated with increased frequency of sleep disturbances and insomnia**. Likewise, allopregnanolone [a progesterone metabolite] levels were negatively correlated with shallow sleep and sleep disturbances. (22-24)

Z- Drug Sleeping Pills Increase Mortality and Cancer

Multiple studies show that Z drug sleeping pills are associated with increased mortality, increased cancer, and a host of other medical problems. In addition, sleeping pills become ineffective through tolerance for chronic users, and reduce functionality the following day. In 2012, Dr. Daniel F. Kripke Professor of Psychiatry, Emeritus at UCSD, in La Jolla did a matched cohort study of hypnotic, Z-sleeping pill use in 10,531 patients, using the Geisinger Health System database, a large rural integrated health system serving 2.5 million people in 41 counties in Pennsylvania. Zolpidem (Z-drug) and Tamazepam (benzodiazepine) were the most frequently used. Dr. Daniel F. Kripke found that hypnotic sleeping pills are associated with a **three-to-five-fold increase in mortality** and a 35 percent increase in cancer. Dr. Daniel F. Kripke writes:

> As predicted, patients prescribed any hypnotic had substantially **elevated hazards of dying compared to those prescribed no hypnotics**. For groups prescribed 1–18, 18–132 and >132 doses/year, HRs [hazard ratios] were **3.60** (2.92 to 4.44), **4.43** (3.67 to 5.36) and **5.32** (4.50 to 6.30), respectively, demonstrating a dose–response association. HRs were elevated in separate analyses for several common hypnotics, including **zolpidem, temazepam, eszopiclone, zaleplon, other benzodiazepines, barbiturates and sedative antihistamines**. Hypnotic use in the upper third was associated with a significant elevation of incident cancer; HR=**1.35.** (25)

In 2016, Dr. Daniel Kripke reported hypnotic sleeping pills increase the risk for excess mortality (especially overdose deaths, quiet deaths at night, and suicides), infections, cancer, depression, automobile crashes, falls, other accidents, and hypnotic-withdrawal insomnia. Contrary to advertising claims, sleeping pills have little or no health benefit, and daytime performance is often made worse, not better. Rather than use sleeping pills, Dr. Kripke advises cognitive therapy and phototherapy are safer and more effective than Z-sleeping pills. Dr. Daniel Kripke writes:

> The most important risks of hypnotics include excess mortality (especially overdose deaths, quiet deaths at night, and suicides), infections, cancer, depression, automobile crashes, falls, other accidents, and hypnotic-withdrawal insomnia... Hypnotics have usually been prescribed without approved indication, most often with specific contraindications, **but even when indicated, there is little or no benefit**. The recommended doses objectively increase

sleep little if at all, daytime performance is often made worse (not better) and the lack of general health benefits is commonly misrepresented in advertising. Treatments such as the cognitive behavioral treatment of insomnia and bright light treatment of circadian rhythm disorders offer safer and more effective alternative approaches to insomnia. (26-30)

What is Cognitive Behavioral Therapy for Insomnia (CBT-I)?

Cognitive behavioral therapy is a form of therapy that retrain the patient to have healthy sleeping habits. The health care provider or psychotherapist teaches the patient how to modify unwanted thoughts and behaviors creating insomnia. In other words, cognitive behavioral therapy is a form of psychotherapy that challenges and modifies our irrational or distorted thoughts about sleep. It also includes behavioral techniques such as sleep restriction, stimulus control therapy, relaxation techniques, and education about sleep hygiene. (31)

What is Phototherapy for Insomnia?

In 2020, Dr. Marc Dalmau Rodríguez asks the question: is bright light an effective treatment for insomnia? Insomnia has been associated with disruption of the circadian rhythm, the 24-hour wake-sleep cycle, our internal clock which can be reset by exposure to light. The first-line treatment of insomnia should be cognitive therapy and phototherapy, both proven to be effective in treating insomnia. Dr. Marc Dalmau Rodríguez writes:

> Light therapy or phototherapy consists of exposing the eyes to **bright light from the sun or from artificial lamps...** Sleep onset insomnia and early-morning-awakening insomnia have been associated, respectively, **to delays and advances in circadian rhythms. Circadian rhythms determine the intervals of sleep and wakefulness and depend mainly on an internal clock: the suprachiasmatic nucleus, a group of**

neurons receiving direct information from the retinas and controlling the synthesis and release of melatonin based on the incoming light information. Darkness stimulates melatonin secretion, while light conditions inhibit it, leading to awakening (Lovato & Lack, 2013). **With light being such an important factor in sleep regulation, it is not surprising that some types of insomnia can be treated by using exposure to bright light…Morning light stimulation has proven to be more effective in the treatment of sleep-onset insomnia (Lack & Wright, 2007). When waking up, light is applied for a full hour, and every subsequent morning the waking hour is advanced 15 minutes, while the session is reduced for the same time, until the desired waking up time has been achieved.** The treatment is then continued for, at least, 14 extra mornings to ensure the results (Lovato & Lack, 2013). Lack, Wright & Paynter (2007) observed that, when treating sleep-onset insomnia, **phototherapy was able to decrease the latency of the onset of sleep and increase its duration, decrease anxiety, fatigue and sleepiness before going to bed, thus providing an overall improvement in daytime functioning.** On the other hand, when treating early-morning-awakening insomnia, it has been observed that it is better to apply light stimulation at night (Lack & Wright, 2007). It is enough to apply light therapy four hours before going to bed during two nights to appreciate an improvement in patients (Lovato & Lack, 2013). Furthermore, other variables such as light wavelength have an impact on the results of the treatment. Wright, Lack & Kennaway (2004) applied phototherapy with five different types of wavelengths and a control situation without light: they discovered that **those lights with shorter wavelengths (blue and green) were more effective in treating insomnia than lights with longer wavelengths (red, orange, and/or yellow)…In conclusion, the treatment of insomnia should follow, first and foremost, a non-pharmacological approach.** Although **cognitive therapy is the most effective treatment to date**, its limitations in terms of involvement and abandonment must be considered. **Phototherapy, on the other hand, has proven to be effective in treating insomnia**, although its efficacy must be investigated in greater depth. (32-34)

First Line Therapy for Insomnia

For the general population, cognitive behavior therapy and bright light (phototherapy) are considered first-line treatments for sleep. For post-menopausal women, hormone replacement with estrogen and progesterone is highly effective and should be the first-line therapy.

Conclusion: Z-drug and BZD drug hypnotic sleeping pills are unsafe, addictive, and associated with increased mortality. These Z-drug sleeping pills interfere with estrogen and progesterone receptors in GABA neurons in the brain, requiring much larger hormone dosage to have efficacy for menopausal symptoms. Post-menopausal women should get off the sleeping pills, and instead use progesterone and estrogen HRT for sleep, a much safer and more effective choice. Non-drug sleep solutions such as cognitive therapy and phototherapy have shown efficacy and should be more generally adopted.

♦ **References for Chapter 28 Waking Up from the Sleeping Pill Nightmare**

1) Agravat, Aasha. "'Z'-hypnotics versus benzodiazepines for the treatment of insomnia." Progress in Neurology and Psychiatry 22.2 (2018): 26-29.

2) Curado, Daniela F., et al. "Dependence on hypnotics: a comparative study between chronic users of benzodiazepines and Z-drugs." Brazilian Journal of Psychiatry 44.3 (2021): 248-256.

3) Cabaj, Justyna, Julia Bargieł, and Ewelina Soroka. "Benzodiazepines and z-drugs-between treatment effectiveness and the risk of addiction." Journal of Education, Health and Sport 46.1 (2023): 468-480.

4) Edinoff, Amber N., et al. "Benzodiazepines: Uses, Dangers, And Clinical Considerations." Neurology international 13.4 (2021): 594-607.

5) Maki, Pauline M., Nick Panay, and James A. Simon. "Sleep disturbance associated with the menopause." Menopause (2024): 10-1097.

6) Almey, Anne, Teresa A. Milner, and Wayne G. Brake. "Estrogen receptor α and G-protein coupled estrogen receptor 1 are localized to GABAergic neurons in the dorsal striatum." Neuroscience letters 622 (2016): 118-123.

7) Thind, Khushdev K., and Paul C. Goldsmith. "Expression of Estrogen and Progesterone Receptors In Glutamate And GABA Neurons Of The Pubertal Female Monkey Hypothalamus." Neuroendocrinology 65.5 (1997): 314-324.

8) Kapur, Jaideep, and Suchitra Joshi. "Progesterone Modulates Neuronal Excitability Bidirectionally." Neuroscience Letters 744 (2021): 135619.

9) Guennoun, Rachida. "Progesterone in the Brain: Hormone, Neurosteroid and Neuroprotectant." International Journal Of Molecular Sciences 21.15 (2020): 5271.

10) Cheng, Yu-Shian, et al. "Pharmacologic and Hormonal Treatments for Menopausal Sleep Disturbances: A Network Meta-Analysis Of 43 Randomized Controlled Trials And 32,271 Menopausal Women." Sleep Medicine Reviews 57 (2021): 101469.

(11) Nolan, Brendan J., B. Liang, and A. S. Cheung. "Efficacy of Micronized Progesterone for Sleep: A Systematic Review and Meta-analysis of Randomized Controlled Trial Data." The Journal of Clinical Endocrinology and Metabolism 106.4 (2021): 942-951.

12) Lancel, Marike, et al. "Progesterone induces changes in sleep comparable to those of agonistic GABA-A receptor modulators." American Journal of Physiology-Endocrinology and Metabolism 271.4 (1996): E763-E772.

13) Carrasco-Nuñes, Nayely, Marta Romano, and Marisa Cabeza. "Sex hormone dose escalation for treating abnormal sleep in ovariectomized rats: in vitro GABA synthesis in sleep-related brain areas." Canadian Journal of Physiology and Pharmacology 101.10 (2023): 529-538.

14) Carrasco-Nuñez, Nayely, Martha Romano, and Marisa Cabeza. "17β-Estradiol and Progesterone Role on Glutamate Decarboxylase Activity in Brain Regions Sleeping-Process Related." Available at SSRN 3989807.

15) Jeon, Gyun-Ho. "Insomnia in Postmenopausal Women: How to Approach and Treat It?" Journal of Clinical Medicine 13.2 (2024): 428.

16) Cheng, Yu-Shian, et al. "Pharmacologic and Hormonal Treatments For Menopausal Sleep Disturbances: A Network Meta-Analysis Of 43 Randomized Controlled Trials And 32,271 Menopausal Women." Sleep Medicine Reviews 57 (2021): 101469.

17) Brann, Darrell W., et al. "Brain-derived estrogen and neural function." Neuroscience & Biobehavioral Reviews 132 (2022): 793-817.

18) Guennoun, Rachida. "Progesterone in the brain: hormone, neurosteroid and neuroprotectant." International journal of molecular sciences 21.15 (2020): 5271.

19) Sutanto, Clarinda Nataria, et al. "The impact of 5-hydroxytryptophan supplementation on sleep quality and gut microbiota composition in older adults: A randomized controlled trial." Clinical Nutrition 43.3 (2024): 593-602.

20) Jacobsen, Jacob PR, et al. "Adjunctive 5-hydroxy-tryptophan slow-release for treatment-resistant depression: clinical and preclinical rationale." Trends in pharmacological sciences 37.11 (2016): 933-944.

21) Turner, Erick H., Jennifer M. Loftis, and Aaron D. Blackwell. "Serotonin a la carte: supplementation with the serotonin precursor 5-hydroxytryptophan." Pharmacology & therapeutics 109.3 (2006): 325-338.

22) Baker, and Brigitte Leeners. "The role of ovarian hormones in the pathophysiology of perimenopausal sleep disturbances: A systematic review." Sleep medicine reviews 66 (2022): 101710.

23) Haufe, Annika, and Brigitte Leeners. "Sleep Disturbances Across a Woman's Lifespan: What Is the Role of Reproductive Hormones?" Journal of the Endocrine Society 7.5 (2023): bvad036.

24) Haufe, Annika, Fiona C. Baker, and Brigitte Leeners. "The role of ovarian hormones in the pathophysiology of perimenopausal sleep disturbances: A systematic review." Sleep medicine reviews 66 (2022): 101710.

25) Kripke, Daniel F., Robert D. Langer, and Lawrence E. Kline. "Hypnotics' association with mortality or cancer: a matched cohort study." BMJ open 2.1 (2012): e000850.

26) Kripke, Daniel F. "Hypnotic drug risks of mortality, infection, depression, and cancer: but lack of benefit." F1000Research 5 (2016).

27) Sun, Yu, et al. "Association between zolpidem and suicide: a nationwide population-based case-control study." Mayo Clinic Proceedings. Vol. 91. No. 3. Elsevier, 2016.

28) Choi JW, Lee J, Jung SJ, Shin A, Lee YJ. Use of sedative-hypnotics and mortality: a population-based retrospective cohort study. J Clin Sleep Med. 2018;14(10):1669–1677.

29) Haines, Adam, et al. "The association of hypnotics with incident cardiovascular disease and mortality in older women with sleep disturbances." Sleep medicine 83 (2021): 304-310.

30) Umbricht, Annie, and Martha L. Velez. "Benzodiazepine and nonbenzodiazepine hypnotics (z-drugs): the other epidemic." Textbook of Addiction Treatment: International Perspectives (2021): 141-156.

31) Tal J.Z., Suh S.A., Dowdle C.L., Nowakowski S. Treatment of Insomnia, Insomnia Symptoms, and Obstructive Sleep Apnea During and After Menopause: Therapeutic Approaches. Curr. Psychiatry Rev. 2015; 11:63–83.

32) Rodríguez, Marc Dalmau. "Phototherapy: Is Bright Light An Effective Alternative For The Treatment Of Insomnia Disorders?." 2020, Ciencia Cognitiva. https://www.cienciacognitiva.org/?p=2014

33) Xerfan, Ellen MS, et al. "The Influence Of Phototherapy On Circadian Melatonin And Sleep Regulation And Potential Benefits Of These Pathways In The Management Of Vitiligo: A Narrative Review: Vitiligo, Phototherapy, Sleep And Melatonin." Archives of Dermatological Research 316.9 (2024): 632.

34) Lu, Xinlian, Chengyu Liu, and Feng Shao. "Phototherapy Improves Cognitive Function In Dementia: A Systematic Review And Meta-Analysis." Brain And Behavior 13.5 (2023): e2952.

Chapter 29

Screening Mammography and Overdiagnosis

WHAT IS SCREENING MAMMOGRAPHY? HOW does it differ from diagnostic mammography? Screening mammography means we submit an entire population of healthy women to an annual mammogram. Diagnostic mammography means we only do mammograms on symptomatic women who seek medical attention such as a palpable lump. Diagnostic mammography is a useful tool in the medical work-up of a breast lump or mass, or architectural distortion, all clinical signs of breast cancer. The medical evaluation will usually include ultrasound imaging to gain more information about the size and interior of the mass, whether cyst or solid. A very different project is screening mammography in which the entire healthy female population goes for annual or semi-annual mammogram. When first introduced in 1983, the medical community had high hopes screening mammography would reduce national mortality from breast cancer. Unfortunately, this hope was never realized. Dr. Laura Esserman is a professor of Surgery and Radiology and director of the UCSF Breast Care Clinic in San Fransisco. In 2009, Dr. Esserman startled the medical community by questioning screening mammography with an article in the Journal of the American Medical Association (JAMA). After reviewing the national mortality data on breast cancer for the last 20 years, from 1983 to 2003, Dr. Laura Esserman writes the expected reduction in national mortality from breast cancer never materialized:

> Mammography screening for breast cancer has significant drawbacks, and expected survival benefits have not materialized... While the incidence of early-stage breast cancer has decreased due to mammography, the incidence rates for the killer cancers, (the advanced cancers) have remained

stable. While it is true that overall mortality rates have declined slightly, this is attributed to better treatment rather than increased detection. (1)

Reviewing the national data, beginning in 1983 with the introduction of mammography screening, the annual incidence of breast cancer increased dramatically. When this total incidence is separated into localized, regional, and metastatic cases and the data from 1983 to 2003 reviewed, the killer cancers, the regional and metastatic cancers have remained stable with little change, despite the detection of massive numbers of localized cancer cases by screening mammography. When reviewing national mortality data from 1930 to 2006, the incidence of breast cancer dramatically increased with the introduction of screening mammography in the 1980s. However, the annual mortality from breast cancer shows only a modest decline from 30 cases per 100,000 women in 1930 to 25 cases for 100,000 women in 2006. Dr. Esserman suggests this rather modest decline in mortality is not due to increased detection with mammography, rather it is due to improvement in treatment.

Annual Breast Cancer Mortality - Where's the Benefit?

While the incidence of early-stage breast cancer has decreased by 2.8% per year since 2001, incidence rates of advanced (distant-stage) disease have remained stable. In 2009, 192,370 women were diagnosed with breast cancer and 40,170 women died of breast cancer. Mammography has increased the detection of early-stage cancer, called DCIS, with 60,000 cases of DCIS detected annually. In 2023, 297,790 women were diagnosed with breast cancer and 43,170 women died of breast cancer. (2-3)

Dr. Barnett Kramer in 2002

In 2002, Dr. Barnett Kramer, the Director of Disease Prevention at the National Institutes of Health, was interviewed by Gina Kolata of the New York Times on screening mammography. Dr. Barnett Kramer pointed out the same findings observed by Dr. Laura Esserman, although screening mammography found a huge population of early breast cancers, this failed to materialize into a corresponding decline in advanced cancers, the cancers with higher mortality, writing:

> The number of women with breast cancers with the worst prognosis, those that spread to other organs, had been fairly constant in the years before mammography was introduced, and **that trend did not change after the introduction of mammography...** If screening worked perfectly, every cancer found early would correspond to one fewer cancer found later. That did not happen. **Mammography, instead has resulted in a huge new population of women with early stage cancer but without a corresponding decline in the numbers of women with advanced cancer**. (4)

Weighing the Pluses and Minuses of Screening Mammography

In 2009, Dr. Gilbert Welch, Professor of Medicine and Public Policy at Dartmouth Medical School, wrote a BMJ editorial reviewing benefits and adverse effects of screening mammography. Dr. Gilbert Welch says about 0.1 percent of women who undergo a screening mammogram will avoid dying from breast cancer. However, 2-10 women will be overdiagnosed and treated needlessly. 10 to 15 women will be told they have early breast cancer, but this will not affect their prognosis. 50-250 will undergo a breast biopsy for a "false alarm". (5)

In 2012, Drs. Bleyer and Welch summarized three decades of screening mammography, esti-

mating that **1.3 million women were overdiagnosed** with breast cancer, meaning tumors were detected that never would have led to clinical symptoms. In 2008, about **one-third** of breast cancer cases found by screening represented overdiagnosis, writing:

> we estimated that breast cancer was overdiagnosed (i.e., tumors were detected on screening that would never have led to clinical symptoms) in **1.3 million U.S. women in the past 30 years**. We estimated that in 2008, breast cancer was overdiagnosed in more than **70,000 women**; this **accounted for 31% of all breast cancers diagnosed**. (6)

Mammography Finds the Reservoir of DCIS Lesions

What is the explanation for all this overdiagnosis? The explanation is ductal carcinoma in situ (DCIS) considered a premalignant lesion, however pathologists call it cancer. DCIS is treated with surgery and radiation therapy when found on screening mammography and confirmed with biopsy. How is it possible to find so much DCIS with screening mammography? The mammography is an xray technique with the exquisite ability to find small calcifications, the tell-tale signs of DCIS. In 2024, Dr. Mark McArthur a radiologist at UCLA Health reviewed DCIS calcifications writing:

> On mammography, the most suggestive feature of DCIS is calcification, which is seen in the majority of cases. The calcification morphologies frequently seen with DCIS are fine linear, fine pleomorphic, coarse heterogeneous, and amorphous. In particular, fine linear morphology directly corresponds to disease along a duct. The distribution of these calcifications also lends credence to a diagnosis of DCIS. Linear and segmental distribution patterns are associated with DCIS; these patterns reflect the spread of disease through the ducts and TDLUs [terminal ductal lobular unit]. In fact, linear and segmental distributions may be related to malignancy up to 80% of the time. (29)

Typically, DCIS lesions are small, indolent, pre-cancerous, and found in 18 percent of the women on autopsy studies. DCIS is a reservoir of silent disease. Since DCIS is so common, and is usually indolent, treating DCIS with surgery and radiation therapy leads to overdiagnosis and overtreatment. For the truly invasive cancers found in 1-2% of the population at autopsy, screening mammography has not reduced the number of advanced cancers causing about 40,000 deaths from breast cancer annually. Dr. Gilbert Welch sums it up with the following sage advice:

> doctors who recommend less-aggressive mammography (less frequently, waiting until you are age 50, or stopping it when you are older) or are less quick to biopsy may not be bad doctors but good ones. (7)(37)

Just Stop Calling It Cancer - DCIS

One glaring problem with screening mammography is the detection of DCIS at a rate of 60,000 cases per year. DCIS is ductal carcinoma in situ, a pathology diagnosis that carries a good prognosis, a nearly 100 percent 5-year survival, even with no treatment. Despite the rather benign natural history of DCIS, mainstream medicine treats these lesions aggressively with surgery and radiation. One solution is to change the terminology and have pathologists stop calling it "cancer." In 2010, Dr. Carmen Allegra did exactly this, with an NIH consensus statement calling for a change in terminology, asking pathologists to stop calling it cancer, writing:

> Because of the noninvasive nature of DCIS, coupled with its favorable prognosis, strong consideration should be given to **elimination of the use of the anxiety-producing term "carcinoma" from the description of DCIS."** (8-9)

A Large Reservoir of Silent Disease - For All Three

All three types of cancer, breast, thyroid, and prostate have a large reservoir of indolent or biologically insignificant disease that remains silent during the patient's lifetime. We know this from autopsy studies. In 1987, Dr. Maja Nielsen from Copenhagen, Denmark did a study of 110 medico-legal autopsies in women (20-54 years) dying in auto accidents using specimen radiographs of thin sections and extensive histopathology, finding breast cancer in 20% of the cases. 2% had invasive cancer, and 18% of the 110 cases had in-situ cancer (14% had DCIS). (10-11)(37)

In 2019, Dr. Maartje van Seijen from Amsterdam reviewed the harms of overdiagnosis and overtreatment of DCIS which arise with screening mammography. How many women harbor a reservoir of undetected DCIS that never becomes symptomatic? Screening mammography detects the small calcifications of DCIS, an indolent lesion, thus triggering overdiagnosis and overtreatment. Dr. Maartje van Seijen writes:

> Ductal carcinoma in situ (DCIS) now represents 20–25% of all 'breast cancers' consequent upon detection by population-based breast cancer screening programmes. Currently, **all DCIS lesions are treated, and treatment comprises either mastectomy or breast-conserving surgery supplemented with radiotherapy. However, most DCIS lesions remain indolent.** Difficulty in discerning harmless lesions from potentially invasive ones can lead to overtreatment of this condition in many patients. The Marmot Report in 2012 recognised the burden of overtreatment to women's wellbeing. In effect**, women with DCIS are labelled as 'cancer patients', with concomitant anxiety and negative impact on their lives, despite the fact that most DCIS lesions will probably never progress to invasive breast cancer**. Owing to the uncertainty regarding which lesions run the risk of progression to invasive cancer, current risk perceptions are misleading and consequently bias the dialogue between clinicians and women diagnosed with DCIS, **resulting in overtreatment for some, and potentially many, women**...The diagnosis of DCIS labels

women as being at risk for invasive breast cancer. **Despite the good prognosis and normal life-expectancy, women diagnosed with DCIS may experience substantial psychological distress and overestimate the implications of a DCIS diagnosis**. Comorbidity of surgery and prior depression have been reported as important factors related to worse quality of life in these women...The number of women diagnosed with DCIS over the past few decades largely follows the introduction of population-based breast cancer screening... **the incidence of mortality from early-stage breast cancer has not decreased concurrently with DCIS detection and treatment, indicating that managing DCIS does not reduce breast-cancer-specific mortality and therefore could be considered as overtreatment**. A review of autopsies in women of all ages revealed a median prevalence of **8.9%** (range 0–14.7%). **For woman aged >40 years, this prevalence was 7–39%, whereas breast cancer is diagnosed in only 1% of women in the same age range. These data suggest that a large number of women might have an undetected source of DCIS that will never become symptomatic**...DCIS is usually straightforward to detect by mammography because of its association with **calcifications**...Generally, patients diagnosed with **DCIS have an excellent long-term breast-cancer-specific survival of around 98% after 10 years of follow-up and a normal life expectancy**. (12)

In 2023, Dr. Ismail Jatoi, professor of Surgical Oncology at UT Health San Antonio, reviews the problems with screening mammography and detection of DCIS. Screening mammography has increased the incidence of DCIS by 6-fold. Contrary to expectations, the greater detection and treatment of DCIS has **not led to a reduction in the incidence of invasive breast cancer**. "Screening programs may be tapping into a reservoir of indolent nonprogressive DCIS within a target population where the disease is prevalent," writing:

With the wider use of screening mammography, **DCIS has become a major public health issue.** There are now more than 60,000 women diagnosed with DCIS each year in the United States... Although DCIS was very rarely diagnosed prior to the advent of screening mammography, it now constitutes up to **25% of all breast tumors in countries that have implemented breast cancer screening programs**...There was a surge in rates of DCIS detection following the implementation of mammography screening. For example, mammography screening was introduced in the United States in the late 1970s, and rates of DCIS rose from **5.8 per 100,000 women in 1973 to 1975 to 32.5 per 100,000 women in 2004**. The incidence of DCIS relative to **1975 increased nearly 6-fold within approximately 25 years following the introduction of mammography** screening in the United States...Yet, despite the dramatic increase in rates of DCIS [ductal carcinoma in situ] detection and treatment following the widespread use of mammography screening in many parts of the world, **rates of IBC [invasive breast cancer] have continued to climb. Contrary to expectations, the detection and treatment of greater numbers of DCIS cases has not led to any reduction in the incidence of IBC**. For example, in the United States, the age-adjusted incidence of IBC increased from **100.0 to 124.3 cases per 100,000** women **between the years 1975 and 2004, despite the sharp increase in the detection and treatment of DCIS during the same time period**...Additionally, **screening mammography detects forms of IBC that might never have become clinically apparent in the absence of screening (ie, it results in overdiagnosis,** discussed further in the paragraphs below)...Analysis of the results of the Canadian National Breast Screening Study (CNBSS) suggests that **although women with DCIS have higher odds of developing IBC [invasive breast cancer] , more than 80% would be expected to remain free of invasive disease**. Results of epidemiological studies seem to suggest that **most DCIS cases do not progress to IBC**. Thus, if we accept the premise that DCIS is a precursor of IBC, then it should be viewed as nonobligate in nature with data from the Canadian study suggesting that **only 1 out of every 4 or 5 cases of DCIS would progress to an invasive breast malignancy. Autopsy studies also provide corroboration for the notion that mammography screening taps into a sizeable reservoir of occult, nonprogressive DCIS,** and further supports the hypothesis that DCIS is a nonobligate precursor of IBC. **Nielsen and colleagues reported results of 110 medicolegal autopsies undertaken at the Fredericksburg Hospital in Copenhagen, Denmark.** These autopsies were undertaken in women between 20 and 54 years of age who had **died of accidents,** and with no known history of breast cancer at the time of death. **Occult DCIS was an unexpected finding in 15% of these women, a prevalence of 4 to 5 times greater than the number of overt cancers expected to develop during a 20-year period. In autopsies of women with a previous breast cancer diagnosis, Alpers and Wellings found DCIS in 48% of the contralateral breast that was not previously known to harbor malignancy.** Yet, **only approximately 12.5% of all patients who present with unilateral breast cancer would have been expected to develop a contralateral breast cancer after a 20-year follow-up.** In 1989, Peeters and colleagues introduced the concept of **"overdiagnosis."** These authors defined **"overdiagnosis" as "a histologically established diagnosis of invasive or intraductal breast cancer that would never have developed into a clinically manifest tumor during the patient's normal life expectancy if no screening examination had been carried out."** Thus, **the detection of DCIS with mammography screening is often used as a good example of overdiagnosis and may illustrate the potential harms of overdiagnosis.** Many cases of DCIS that are detected with mammography screening would likely never have been detected in the absence of screening and **pose no threat to life**. Hence, the **overdiagnosis of DCIS leads to unnecessary treatments, and patients may incur a small excess risk of morbidity and mortality from those unnecessary**

treatments. Also, overdiagnosis may have adverse effects on patients' quality of life. For example, it often leads to not only unnecessary treatments, but also unnecessary anxiety, possible increases in health insurance premiums because of the diagnosis, potential difficulties obtaining life insurance, and possibly even difficulties finding employment (ie, some employers might be reluctant to hire an individual with a previous diagnosis of cancer). Thus, increases in DCIS detection rates because of screening likely results in substantial overdiagnosis and overtreatment, and potentially severe adverse effects on quality of life...Most DCIS lesions manifest as mammographic calcifications and are therefore not discernible macroscopically. The calcification is often linear or casting. **DCIS became a major public health issue only after the introduction of mammography screening, and this is particularly evident in the United States, where there was nearly a 6-fold increase in the incidence of DCIS within approximately a 25-year period following the introduction of mammography screening**...Autopsy studies have consistently revealed a finite incidence of DCIS, ranging from less than 1% to as much as 15%, suggesting that some forms of DCIS that are detected with mammography and subsequently excised with either complete or partial mastectomy **would have carried no clinical consequence to the patient during their lifetime.** This raises the issue of "non-obligate progression" and resulting uncertainties in terms of potential overdiagnosis, innate biological behavior, and clinical management...**Screening programs may be tapping into a reservoir of indolent nonprogressive DCIS within a target population where the disease is prevalent.** DCIS is a heterogeneous disease with a variable natural history and prediction of recurrence after treatment remains a significant challenge; not all DCIS will progress to invasive disease, with estimates for this proportion ranging from 25% to 50%, depending partly on the grade of the lesion. **It is now acknowledged that some cases**

of DCIS would never have become invasive during a woman's lifetime and are clinically nonconsequential. Indeed, some cases of low-grade DCIS are more appropriately termed pseudo-disease and are linked to a degree of overdiagnosis within screening programs. It has been estimated that overdiagnosis represents 19% of cases within the National Health Service Breast Screening Program and **this can potentially undermine the benefits of screening when 3 cancers are overdiagnosed for each life saved**. It should be noted that the risk of death after any treatment for DCIS is less than 2% and life expectancy within the older age group is more likely to be determined by competing causes of death such as atherosclerosis and ischemic heart disease. **Nonetheless, once DCIS has been detected, treatment must be offered and this involves mastectomy in up to 30% of cases and often irradiation to the breast when BCS [breast conservation surgery] is carried out**. (13)

In 2015, Dr. John D. Keen, a radiologist from Chicago discusses the 4 principles used to advise women on screening mammography, suggesting **overdiagnosis can be avoided by abstaining from screening mammography**.

> **The first principle is that screening may help, hurt, or have no effect. In order to reduce mortality and mastectomy rates, screening must reduce the rate of advanced disease, which likely has not happened.** Through overdiagnosis, screening produces substantial harm by increasing both lumpectomy and mastectomy rates, which offsets the often-promised benefit of less invasive therapy. Next, **all-cause mortality is the most reliable way to measure the efficacy of a screening intervention. Disease-specific mortality is biased due to difficulties in attribution of cause of death and to increased mortality due to overdiagnosis and the resulting overtreatment with radiotherapy and chemotherapy.** To enhance participation, the benefit from screening is often presented in relative instead of absolute terms. Third, some screening statistics must be interpreted with caution. Increased survival time and the percentage of early-stage tumors at detection sound plausible, but are affected by lead-time and length biases. In addition, analyses that only include women who attend screening cannot reliably correct for selection bias. The final principle is that accounting for tumor biology is important for accurate estimates of lead time, and the potential benefit from screening. Since "early detection" is actually late in a tumor's lifetime, the time window when screen detection might extend a woman's life is narrow, as many tumors that can form metastases will already have done so. **Instead of encouraging screening mammography, physicians should help women make an informed decision as with any medical intervention... Once a cancer is detected, it is currently not possible to distinguish life-threatening from indolent cases. Therefore, overdiagnosis can only be avoided by abstaining from breast screening.** (14)

Spontaneous Remission of Breast Cancer?

In 2008 by Drs. Per-Henrik Zahl and Gilbert Welch reviewed mammography screening in 4 Norwegian counties. The cumulative incidence of invasive breast cancer was 22% higher in the screened group compared to non-screened controls. The authors concluded that many small breast cancers spontaneously regress, writing:

> Because the cumulative incidence among controls never reached that of the screened group, it appears that some breast cancers detected by repeated mammographic screening would not persist to be detectable by a single mammogram at the end of 6 years. **This raises the possibility that the natural course of some screen-detected invasive breast cancers is to spontaneously regress.** (35-36)

Spontaneous regression of breast cancer has been reported many times in the medical literature. Sir William Osler, a legendary and revered doctor reported 14 cases himself. Perhaps breast cancer remission was more common during his lifetime. I have seen a case of spontaneous regression of breast cancer documented by a follow-up MRI scan. This patient was self-treating with Lugol's solution (iodine). How many spontaneous remissions of breast cancer are due to high-dose iodine treatment, a common treatment between 1840 to 1940? In 1901 when Dr. Osler wrote his report on spontaneous regression of breast cancer, his patients may have been taking high-dose iodine. For more on Iodine as Prevention and Treatment of Breast Cancer, see Chapters 22 and 23. (15-18) (35-36)

The Problem of Overdiagnosis

What does overdiagnosis mean? Overdiagnosis means the patient is given a diagnosis of cancer which leads to invasive treatments such as surgery, and radiation therapy for a lesion which would have never come to clinical attention if not for the screening mammogram. The diagnosis of DCIS is largely due to finding the tell-tale calcifications on a screening mammogram. Many of these DCIS cases are clinically insignificant. We know this from autopsy studies from Dr. Maja Nielsen from Denmark. In 2018, Ujas Parikh discusses this:

> the increased frequency of DCIS diagnosis is **largely due to screening**. Given that **DCIS may never reach clinical significance over a woman's lifetime,** critics argue that **screening mammography may induce more harm than good, citing overdiagnosis, overtreatment, and increased radiation.** On the other hand, because there is a lack of understanding of DCIS and its natural progression, proponents argue that limiting screening because of overdiagnosis of DCIS will cause undue harm. Numerous randomized controlled trials have found reduced [breast cancer] mortality with screening, citing decreases in [breast cancer] mortality in up to 32% of patients. The Gothenburg Breast Screening Trial, for

example, found a significant 21% reduction in breast cancer–related mortality among 51,611 women. (19)

Note: Many randomized trials show a 20-30 percent reduction in breast cancer mortality with screening mammography. However, this does not translate into a reduction in all-cause mortality because of the harms of overdiagnosis.

What to Measure: Mortality from Breast Cancer or All-Cause Mortality?

In 2003, Dr. Nils Bjurstam reported on the Gothenberg breast screening trial showing a 21-24 percent reduction in breast cancer mortality. However, there was no significant reduction in all-cause mortality, due to the harms from overdiagnosis.

> The results show a reduction of 21–24% in breast carcinoma mortality associated with invitation to screening. These results approach statistical significance and are consistent with expectations among women in this age range according to other trials women ages 40 –59 years...**We observed a nonsignificant reduction (5%) in all-cause mortality in the study group** (RR, 0.95; 95% CI, 0.89 –1.02). (20-21)

Increased All-Cause Mortality from Over-Diagnosis

In 2021, Dr. Amanda E Kowalski, Professor of Applied Economics and Public Policy, the University of Michigan, reviewed the evolving evidence on screening mammography and mortality. A review of the data finds no statistically significant reduction in all-cause mortality for women in any age group. Overdiagnosis harms increase with each year of enrollment in screening mammography causing an additional 7 excess deaths per 100,000 women. After two years, screening causes an additional 14 deaths per 100,000 women which is greater than the annual death from auto accidents in the US (12.4 per 100,000). Dr. Amanda E Kowalski writes:

The rationale for widespread mammography is that early detection of potentially fatal breast cancers enables earlier and more effective treatment. But there is a potential drawback: mammography can detect some early-stage cancers [DCIS] that will never progress to cause symptoms—a phenomenon often referred to as overdiagnosis. In such cases, the emotional, financial, and physical costs of a cancer diagnosis and any subsequent treatments occur without any corresponding health benefit. Because it is hard to tell which women will be harmed by their cancers, there is a tendency to treat all women as if their cancers will be lethal (Mukherjee 2017). Even if the initial cancer would have never proven life-threatening, exposure to **chemotherapy, radiotherapy, and surgery can potentially lead to new conditions, even to new fatal cancers**...The possibility of overdiagnosis turns out to be central to guidelines for mammography screening...**The US Preventive Services Task Force identifies overdiagnosis as the most important harm that mammograms pose**. Though false positives can also pose harm, overdiagnosis is a separate phenomenon... According to the task force definition, overdiagnosis refers to "the diagnosis and treatment of noninvasive and invasive breast cancer that would otherwise not have become a threat to their health, or even apparent, during their lifetime" overdiagnosis has become apparent in two studies of mammography trials based on at least 15 years of follow-up data, which imply overdiagnosis rates of **5 to 55 percent** depending on the subgroup and base rate (Zackrisson et al. 2006; Baines, To, and Miller 2016)...**findings suggest that decreases in breast cancer mortality are due to factors other than screening, such as decreased use of menopausal hormone therapy (Ravdin et al. 2007)** [referring to the sudden decline in women using combined Premarin and medroxyprogesterone after the 2002 WHI study] **and improved treatments**...In all age groups, mammography has increased dramatically over time in the United States...By 2015, **58.3** percent of women aged 40–49, **71.3** percent of women aged

50–65, and **63.3 percent of women aged 65+ reported receiving a mammogram within the past two years.** Widespread mammography seems embedded in the US health care system, both as a matter of the acculturation of patients and health care providers, and also as a matter of **financial incentives: the aggregate annual cost of mammography has been estimated to be $2.1 billion just among US women in their 40s with private health insurance** (Kunst et al. 2020)...**It is hard to understate the controversy surrounding the results from the mammography trials.** The idea that finding small, treatable cancers will save lives by stopping them from growing into larger malignant cancers is appealingly simple...**For advocates of mammography, it was thus disappointing when the evidence from the randomized controlled trials was underwhelming.** Some began to update their thinking on the value of mammography...Overall, the meta-analysis finds **no statistically significant reduction in all-cause mortality for women in any age group**... some trials even show imprecise increases in all-cause mortality across all age groups or within an age group. Results focused only on breast cancer mortality are slightly more promising. The meta-analysis finds **statistically significant but small reductions in breast cancer mortality for women in their 50s and 60s, and it finds imprecise reductions for women aged 39 to 49 and women aged 70 to 74**...Life-saving benefits of mammograms outweigh their collateral harms at first, **but as more time passes, collateral harms rise**...Examination of the trend in all-cause mortality reveals that the tradeoff between the harms and benefits of mammography has been shifting toward harms over time...A pronounced upward-sloping and statistically significant trend is visible in the all-cause mortality results. **With each additional year that passes after enrollment, an additional seven excess deaths per 100,000 women become apparent among intervention-arm participants relative to control-arm participants.** To put this trend in perspective, annual road traffic deaths in the United

States are 12.4 per 100,000 (World Health Organization 2018)...**The underwhelming evidence on benefits of mammography** should not motivate a wholesale rejection of the practice, but rather it should motivate research aimed at uncovering the contexts in which mammograms may provide benefits. (22-24)

Breast Cancer Self Detection is Key

The breast, much like the skin, is a superficial organ, and cancer is readily visible and palpable for both. Except for the smaller masses visible only by imaging, most invasive breast cancers are self-detected and will come to clinical attention. Masses or lumps are readily palpable in the breast, unlike deeper organs such as the lung, kidney, pancreas, or liver where a cancer may lurk and grow quite large to an advanced stage before making itself known to the patient. Thus, the self-detection of breast cancer plays an important role in 30-60 percent of cases despite the availability of screening mammography. (25)

Diagnosis is Not Screening

We must be careful about the difference between screening, and diagnosis. Screening pertains to mass screening of a healthy population. As mentioned above, screening mammography may lead to overdiagnosis and overtreatment, while failing to reduce the incidence of advanced breast cancers. Diagnostic mammography pertains to the evaluation and workup of a symptomatic patient, and in this context provides benefits. Breast mammography and ultrasound remain excellent diagnostic tools for workup and evaluation of the palpable breast lump.

How to Prevent Breast Cancer

Although screening mammography can save one out of a thousand screened patients from dying of breast cancer, the cost is over-diagnosis and over-treatment for many other patients which results in cancelling out the perceived benefits. One must realize that screening mammography, although marketed as breast cancer prevention, is a form of cancer detection, and is not preventive of anything. Breast cancer prevention as described in the previous chapters involves an understanding of estrogen metabolism, estrogen receptors, and hormone formulations. Breast cancer prevention involves avoiding carcinogenic chemicals such as progestins, bisphenol A (BPA), DES, and other Endocrine Disrupting Chemicals (EDCs). My office breast cancer prevention program involves vitamin D, selenium, and DIM (di-indole-methane). My office recommendation is to avoid hormone formulas that contain synthetic progestins such as medroxyprogesterone and norethisterone. These are known to be carcinogenic, endocrine disrupting chemicals (EDCs). Thus, the title of this book is bioidentical hormones which means using only exact replicas of natural hormones made by the human body. These include the two human estrogens in bi-est, originally formulated by Jonathan V. Wright, containing 20 percent estradiol (E2), and 80 percent estriol (E3). Also included are progesterone and testosterone as described in previous chapters. Both estriol (E3) and testosterone, or their metabolites, target ER-beta, the tumor suppressor.

Advocating for Screening Mammography

To be fair, I should present the other side of the argument advocating for screening mammography. Perhaps this is best represented by Dr. Elaine Schattner a former oncologist at Weil Cornell Medical Center in NYC, and breast cancer survivor. Dr. Schattner disagrees with Dr. Esserman and says the US death rate from breast cancer fell by 40 percent from 1989 to 2017. Dr. Schattner attributes this 40 percent decline to both screening mammography and better treatment. Dr. Schattner says the latest mammogram technology is much improved, and feels that the risks of overdiagnosis and overtreatment have been exaggerated while the clinical benefits of screening mammography have been dismissed, writing:

> Overall, the U.S. death rate from breast cancer fell by 40% from 1989 to 2017—an astonishing drop attributed to screening and better treatments. This progress occurred despite a concomitant rise in disease incidence. Invasive cases are on the rise, climbing by 0.3% per year, on average, since 2004. One in eight women are affected, and the decline in mortality has slowed since 2011. Breast cancer remains a leading threat to women's health and longevity. More than 42,000 American women will die of breast cancer this year, a tragic figure...I have described the role of traditional and social media in amplifying outdated, misogynistic, and agenda-driven assessments that selectively exaggerate mammography risks, "overdiagnosis," and costs, while dismissing its clinical benefits. Current facts about the potential risks and benefits should be provided in a clear and unbiased manner. (30-31)

We find another rebuttal to Dr. Laura Esserman in 2020 by Drs. Stephen Duffy and Laszlo Tabar. They reviewed mammogram screening data in Sweden, finding women who participated in screening had a 40 percent reduction in breast cancer mortality over 10 years, writing:

> Women who participated in mammography screening had a statistically significant 41% reduction in their risk of dying of breast cancer within 10 years (relative risk, 0.59; 95% CI, 0.51-0.68 [P < .001]) and a 25% reduction in the rate of advanced breast cancers (relative risk, 0.75; 95% CI, 0.66-0.84 [P < .001]). **Note:** This sounds impressive. However, this data does not include total mortality. Measuring breast cancer mortality does not include harm from overdiagnosis and overtreatment. Overall mortality includes deaths from overtreatment and should be included in the data. However, as you will see below, overall mortality is the problem. (32)

No Evidence of Overall Mortality Reduction

In 2024, Dr. Arn Migowski from Brazil says there is no evidence screening mammography has any impact on overall mortality, and to date there are no published studies on the increased death from harms of overdiagnosis and overtreatment, writing:

> **In the absence of evidence on the impact of mammographic screening on overall mortality**, many attempts to determine the balance between the harms and benefits of screening compare different outcomes, generating a search for an arbitrary and questionable ideal balance value between these outcomes.... Some authors argue that if all the harms associated with screening are considered, the probable conclusion is that **more screened women than unscreened women would die**. However, to date, no published study on the balance between the harms and benefits of this intervention has covered all the harms known to be associated with screening and a potential increase in deaths. (33)

Lack of Effect on Overall Mortality

In 2013, Dr. Sylvie Erpeldinger from France says the lack of an effect on overall mortality, and the modesty of the benefit size raises questions about the relevance of mass screening mammography, writing:

> **The lack of effect of screening on overall mortality** could be explained by a balance between benefit on breast cancer deaths and an increase in other death causes, but also by the inability of these trials to observe significant change on mortality, due to the small proportion of breast cancer deaths (less than 10%) in overall mortality. **The modesty of the benefit size, which was estimated at 1 breast cancer death prevented in 10 years for every 2,000 women screened**, put into question the relevance of mass screening, and highlights the need for clear and complete information for the concerned patients. (34)

Screening Mammography is Big Business

Annual expenditures for screening mammography in the US is estimated between 8 and 30 billion dollars per year. This represents a considerable amount of money for institutional medicine, health care workers, and equipment manufacturers. With this amount of money at stake for an entrenched industry, there is a need for favorable studies to justify screening mammography such as the above study by Drs. Duffy and Tabar. However, there is an obvious element of bias. One must maintain a high index of suspicion and keep an open mind. The screening mammography debate is not over by any means, and will go on. Will the introduction of new improved mammogram technology come to the rescue? We will have to wait and see.

Conclusion: I agree with Dr. John Keen (above) who advises women on the four principles of screening mammography, and then lets them make an informed decision and decide which path to follow. Screening may help, hurt, or have no effect. Screening has been a disappointment because it does not reduce the rate of advanced breast cancer, total mortality or U.S. annual mortality from breast cancer. Screening mammography leads to harm from overdiagnosis and overtreatment inherent in the diagnosis of DCIS. Dr. John D. Keen writes: "Once a cancer [DCIS] is detected, it is currently not possible to distinguish life-threatening from indolent cases. Therefore, overdiagnosis can only be avoided by abstaining from breast screening." About half of the female population wishes to avoid the overdiagnosis and overtreatment related to the detection of DCIS inherent in screening mammography. About another half to two-thirds of the female population is not concerned and obediently undergoes the annual mammogram. This is a personal preference, and we leave this decision up to the patient to decide. (26-28)

◆ References for Chapter 29 Screening Mammography Harms from Overdiagnosis

1) Esserman, Laura, Yiwey Shieh, and Ian Thompson. "Rethinking Screening for Breast Cancer and Prostate Cancer." Jama 302.15 (2009): 1685-1692.

2) Siegel, Rebecca L., et al. "Cancer Statistics, 2024." CA: a Cancer Journal for Clinicians 74.1 (2024): 12-49.

3) Dizon, Don S., and Arif H. Kamal. "Cancer Statistics 2024: All Hands on Deck." CA: A Cancer Journal For Clinicians 74.1 (2024).

4) Confronting Cancer; Breast Cancer: Mammography Finds More Tumors. Then the Debate Begins. by Gina Kolata, April 9, 2002 New York Times. https://www.nytimes.com/2002/04/09/science/confronting-cancer-breast-cancer-mammography-finds-more-tumors-then-debate.html

5) Welch, H. Gilbert. "Overdiagnosis and Mammography Screening." BMJ: British Medical Journal (Online) 339 (2009).

6) Bleyer, Archie, and H. Gilbert Welch. "Effect of Three Decades of Screening Mammography on Breast-Cancer Incidence." New England Journal of Medicine 367.21 (2012): 1998-2005.

7) Welch, H. Gilbert, and William C. Black. "Using Autopsy Series to Estimate the Disease "Reservoir" For Ductal Carcinoma in-Situ of The Breast: How Much More Breast Cancer Can We Find?" Annals of Internal Medicine 127.11 (1997): 1023-1028.

8) Allegra, Carmen J., et al. "National Institutes of Health State-of-the-Science Conference Statement: Diagnosis and Management of Ductal Carcinoma in Situ September 22–24, 2009." Journal of the National Cancer Institute 102.3 (2010): 161-169.

9) van Seijen, Maartje, et al. "Ductal Carcinoma in Situ: To Treat or Not To Treat, That Is The Question." British journal of cancer 121.4 (2019): 285-292.

10) Nielsen, Maja, et al. "Breast Cancer and Atypia Among Young and Middle-Aged Women: A Study Of 110 Medicolegal Autopsies." British Journal of Cancer 56.6 (1987): 814-819.

11) Welch, H. Gilbert, and William C. Black. "Using Autopsy Series to Estimate the Disease "Reservoir" For Ductal Carcinoma in Situ of The Breast: How Much More Breast Cancer Can We Find?" Annals of Internal Medicine 127.11 (1997): 1023-1028.

12) van Seijen, Maartje, et al. "Ductal carcinoma in situ: to treat or not to treat, that is the question." British journal of cancer 121.4 (2019): 285-292.

13) Jatoi, Ismail, et al. "The Biology and Management of Ductal Carcinoma in Situ of the Breast." Current Problems in Surgery 60.8 (2023): 101361-101361.

14) Keen, John D., and Karsten J. Jørgensen. "Four Principles to Consider Before Advising Women On Screening Mammography." Journal of Women's Health 24.11 (2015): 867-874.

15) Abraham, Guy E. "The History of Iodine In Medicine Part I: From Discovery To Essentiality." The Original Internist 13.1 (2006): 29-36.

16) Oler, William. The Medical Aspects of Carcinoma of the Breast, with a Note on the Spontaneous Disappearance of Secondary Growths, American Medicine, (1901). 17-19; 63-66.

17) Cole, Warren H. "Efforts to explain spontaneous regression of cancer." The Guthrie Journal 50.3 (1981): 99-112.

18) Ross, Michael B., et al. "Spontaneous Regression Of Breast Carcinoma: Follow-Up Report And Literature Review." Journal Of Surgical Oncology 19.1 (1982): 22-24.

19) Parikh, Ujas, Chloe M. Chhor, and Cecilia L. Mercado. "Ductal Carcinoma In Situ: The Whole Truth." American Journal of Roentgenology 210.2 (2018): 246-255.

20) Bjurstam, Nils, et al. "The Gothenburg Breast Screening Trial." Cancer: Interdisciplinary International Journal of the American Cancer Society 97.10 (2003): 2387-2396.

21) Farmer, Chris, and Kevin Y. Kane. "Screening Decreases Breast Cancer-Specific Deaths but Not All-Cause Mortality. (Patient-Oriented Evidence that Matters)." Journal of Family Practice 51.6 (2002): 513-514.

22) Kowalski, Amanda E. "Mammograms and Mortality: How Has the Evidence Evolved?" Journal of Economic Perspectives 35.2 (2021): 119-140.

23) Ravdin, Peter M., et al. "The Decrease in Breast-Cancer Incidence In 2003 In the United States." New England Journal of Medicine 356.16 (2007): 1670-1674.

24) Ravdin, Peter M. "Hormone Replacement Therapy and The Increase in The Incidence of Invasive Lobular Cancer." Breast Disease 30.1 (2009): 3-8.

25) Malmgren, Judith A., et al. "Persistence of Patient-Detected Breast Cancer Over Time: 1990–2019." Cancer 129.24 (2023): 3862-3872.

26) Keen, John D., and Karsten J. Jørgensen. "Four Principles to Consider Before Advising Women on Screening Mammography." Journal of Women's Health (15409996) 24.11 (2015).

27) Autier, Philippe, and Mathieu Boniol. "Mammography Screening: A Major Issue in Medicine." European Journal of Cancer 90 (2018): 34-62.

28) Autier, Philippe, et al. "Effect of Screening Mammography on The Risk of Breast Cancer Death and Of All-Cause Death: A Systematic Review with Meta-Analysis of Cohort Studies." Journal of Clinical Epidemiology (2024): 111426.

29) McArthur, Mark, and James Chalfant. "Case: Ductal Carcinoma In Situ." UCLA Health. https://www.uclahealth. org/departments/radiology/education/breast-imagingteaching-resources/cases/case-ductal-carcinoma-situ.

30) Schattner, Elaine. "Correcting a Decade of Negative News About Mammography." Clinical Imaging 60.2 (2020): 265-270.

31) Schattner, Elaine. "High-Quality Mammography: A Step Forward for Women's Health." Radiology 298.2 (2021): 306-307.

32) Duffy, Stephen W., Laszlo Tabar, et al. "Mammography Screening Reduces Rates of Advanced and Fatal Breast Cancers: Results In 549,091 Women." Cancer 126, no. 13 (2020): 2971.

33) Migowski, Arn, et al. "Harms and Benefits of Mammographic Screening for Breast Cancer In Brazil." Plos one 19.1 (2024): e0297048.

34) Erpeldinger, Sylvie, et al. "Is There Excess Mortality in Women Screened with Mammography: A Meta-Analysis of Non-Breast Cancer Mortality." Trials 14 (2013): 1-9.

35) Zahl, Per-Henrik, Jan Maehlen, and H. Gilbert Welch. "The Natural History of Invasive Breast Cancers Detected by Screening Mammography." Archives of Internal Medicine 168.21 (2008): 2311-2316.

36) Zahl, Per-Henrik, Peter C. Gøtzsche, and Jan Mæhlen. "Natural History of Breast Cancers Detected In The Swedish Mammography Screening Programme: A Cohort Study." The Lancet Oncology 12.12 (2011): 1118-1124.

37) Thomas, Elizabeth T., et al. "Prevalence of Incidental Breast Cancer and Precursor Lesions in Autopsy Studies: A Systematic Review and Meta-Analysis." BMC cancer 17 (2017): 1-10

Chapter 30

Fine Tuning the Universe

WE LIVE IN A UNIVERSE that is fine-tuned for the emergence of life. In 2002, Lord Martin Rees, Emeritus Professor of Cosmology and Astrophysics at the University of Cambridge published his book, *Just Six Numbers,* referring to six numerical constants in the physical world. A few examples of these constants are: the mass of the proton (Mp), the mass of the neutron (Mn), the speed of the light (c) and the Newtonian gravitational constant (G). There are many more. These are all calculated based on measurement of the physical world. If any of these six constants changed even slightly at the beginning of the "Big Bang," then the stars, galaxies, and ultimately, life would not exist. We would not be here. The amazing thing is that this fine-tuning of the universe is also expressed in all living things. Next, we explore the fine-tuning in the molecular biology of the cell. **Note:** The "Big Bang" is the modern physics explanation for the creation of the universe. The Big Bang theory is based on red shift measurements of light from distant galaxies revealing an expanding universe, first described by Edwin Hubble in 1929. For more on the Big Bang, see the 2021 article by Dr. E. Nihal Ercan from Turkey. (1) (39)

Fine Tuning in Cell Biology

In 2020, Dr. Steinar Thorvaldsen, a Professor of Information Science in Norway, used statistical methods to model the fine-tuning of molecular machines and biological systems. Dr. Thorvaldsen says that fine-tuning is a clear feature of biological systems, and is even more extreme and complicated than the fine-tuning of the large-scale universe. Protein structures in our cells are examples of nano-engineering that surpass anything human engineers have created. Each one of the millions of protein machines in our cells is more complex and sophisticated than the space shuttle, writing:

> In both physics and molecular biology, fine-tuning emerges as a uniting principle and synthesis – an interesting observation by itself...A major conclusion of our work is that **fine-tuning is a clear feature of biological systems. Indeed, fine-tuning is even more extreme in biological systems than in inorganic systems**. It is detectable within the realm of scientific methodology. **Biology is inherently more complicated than the large-scale universe and so fine-tuning is even more a feature...**Living forms exhibit structures and functions that can best be understood as **nano-level engineering.** In 1998 Bruce Alberts published an important paper...**The Cell as a Collection of Protein Machines** (Alberts, 1998)...The strong synergy within the protein complex makes it irreducible to an incremental process. They are rather to be acknowledged as **fine-tuned initial conditions of the constituting protein sequences. These structures are biological examples of nano-engineering that surpass anything human engineers have created**. Such systems pose a serious challenge to a Darwinian account of evolution, since irreducibly complex systems have no direct series of selectable intermediates...As Denis Noble states, biological systems function as a full orchestra with its different elements playing ensemble the score of life (Noble, 2006)...One of the surprising discoveries of modern biology has been that the cell operates in a manner similar to modern technology, while biological information is organized in a manner similar to plain text. (2)

Intelligence in the Cell and All Life

The claim that there is intelligent life in the universe has been present in human thought for millennia and can be found in ancient documents over five thousand years old. We can find popular novels and books that examine the idea, is there extra-terrestrial intelligence in the universe? For example, in Woody Allen's 1973 science fiction comedy, *Sleeper,* the character Miles Monroe was asked: Do you believe in God? He answers:

> I'm what you would call a teleological, existential atheist. I believe that there's an intelligence to the universe, *with the exception of certain parts of New Jersey.* (5)

The truth is all life is permeated with intelligence, from the lowly one-celled organism to the trillion cells in our bodies that support our own lives. Consider the intelligence in the green leaf rising towards the sun, converting electromagnetic radiation into usable energy for building carbon molecules into more leaves and trees. Consider the intelligence of the lowly ant hill in your backyard. Each ant works together as a community towards a common goal, building and feeding the ant hill. As you will read below, the universe is fine-tuned for the emergence of life. The logical extension of this realization means there must be life on planets of other solar systems. Eventually over millennia, if human civilization does not self-destruct, we will develop the tools to detect it or even share information with it. In 2023, Brian J. Ford a biologist from Cambridge and London, compared the cell to a secret agent imbued with intelligence and autonomy, writing:

> Unambiguous signs of intelligence can be seen in the manner in which living cells recognize a serious situation and devise remedial responses that will rectify the damage...**Current philosophy admits that cells exhibit sentience and even consciousness** [Baluska, 2019]...I have demonstrated that cells are more than biochemistry and physics: **they are intelligent**...Living cells perform acts of unimaginable intricacy and we can evince ingenuity in so many simple micro-organisms...Cell intelligence will provide a revolution in understanding how the brain might function and shows us much of the unfathomable realms of biology with which we have yet to engage...Living organisms are themselves communities of curious, adaptable, and thoughtful living cells... will not resolve the existence of organisms in communities until we can comprehend **the cellular community that lies within** [our own bodies] ...**For decades I have held that the living cell is intelligent. This is what we need to understand.** (6-7)

Fine-Tuning Human Cells

One of the expressions of this fine-tuned universe is a trillion cells in our body. Each cell contains millions of molecular machines all working together in harmony, like an orchestra. The ribosomes in our cells make proteins specified by instructions encoded in our DNA. Energy production to power the cell comes from mitochondria, millions of times more complex than the space shuttle. The mitochondria are microscopic generators converting electrical energy stored in carbon bonds into usable ATP [adenine tri-phosphate]. During cell replication, microtubules in the cell arrange in a spindle formation to pull apart the 23 pairs of chromosomes and create two cells from one. The photoreceptors in our retina can detect a single photon of light. The electrical pulse from the photoreceptor travels through the optic nerve to the visual cortex where an image in our mind is created. These are only a few examples of fine-tuning in cell biology. If the cells of our body fail us, then we get sick and die.

Why Use Cheap Gas?

Would you put cheap gas into the space shuttle? Of course not. Then, why would anyone put fake hormones into their bodies? Synthetic chemically altered hormones are fake, endocrine-disrupting chemicals, a monkey wrench thrown into the fine-tuning of our cell biology and the universe. Why do we have a medical system that contradicts the fine-tuning of cell biology and the universe? The answer is the drug patent system requires it. The drug industry must obtain a patent to protect drug profits. Only chemically altered substances can be patented. When we chemically alter a natural substance found in cell biology, we have changed the fine-tuning and the consequences can be devastating. Unfortunately, that is exactly what the practice of medicine is based on. We have a medical system that is not sustainable because it alters the fine-tuning of cell biology, eventually leading to the extinction of mankind. It may take 50 years, 100 years, or 300 years, but it will eventually lead to our extinction. The chronic disease epidemic is a wake-up call telling us where we are headed. What Robert F. Kennedy Jr. is trying to do is to save humanity from this extinction event. (34-36)

The Chronic Disease Epidemic

The drug industry has deceived us into thinking synthetic hormones are hormone replacement. They are not hormone replacement. They are endocrine-disrupting chemicals. This is a basic flaw in the house of medicine, an edifice dominated by a drug industry whose business model contradicts the fine-tuning in our cells. Why has the drug industry created the chronic disease epidemic? The obvious answer is to maximize profits for the drug industry. Is it starting to make sense now? Our greatest public health crisis is the chronic disease epidemic brought to public attention by Robert F. Kennedy Jr.'s presidential campaign and ascendancy to Secretary of Human Health Services (HHS). His stated goal is to eliminate the chronic disease epidemic. Its most glaring feature is the exponential rise in autism. In 1980, autism was 1 in 10,000 children. Today, autism is 1 in 36 children. What is causing it? Mainstream medicine throws up its hands, shrugs its shoulders, and says we do not know. Our hallowed halls of academic medicine, legislators, and government agencies refuse to discuss the CDC vaccine schedule as a possibility. In 1986, Congress granted the drug industry immunity from liability for their vaccine products. If vaccines are so safe and effective, then why does the drug industry need a liability shield? If their product injures or kills anyone, they have no financial responsibility. This eliminates any incentive to make a safe product and creates incentives to add new lucrative vaccines to the CDC schedule. They have a captive market of school-age children forced to take their products to attend school. In the 1950s, the CDC childhood schedule listed four vaccines: diphtheria, tetanus, pertussis, and smallpox. The 2024 CDC schedule is now 18 vaccines and 76 doses. This combination of 18 vaccines and 76 doses has never been tested for safety. At what point do we reach the number of vaccines that degrade the immune system and cause chronic disease? We reached that point long ago.

The Fear of Estrogen and Chronic Disease

Treating our chronic disease epidemic accounts for 90% of our annual 4.1 trillion-dollar healthcare budget for coronary artery disease, cancer, diabetes, osteoporosis, osteoarthritis, depression, autoimmune disease, obesity, and any chronic non-communicable condition. Notice many of these conditions are post-menopausal chronic diseases prevented by hormone replacement. Why are we heading towards extinction as a human race? Follow the money. More chronic disease means more drug sales and greater profits for the drug industry. If the business plan of the drug industry is to create more chronic disease and more customers for its drugs, then a perfect set-up is fear of estrogen and elimination of menopausal hormone replacement (HRT). 1.3 million women enter menopause annually, viewed by the drug industry as new customers for drugs to treat chronic diseases of estrogen deficiency. Throughout this book, we have laid out the basic concept of estrogen deficiency as harbinger of chronic disease. If we are serious about making America healthy again, this means eliminating the irrational fear of estrogen and opening the door to menopausal hormone replacement for the 1.3 million women entering menopause annually. The hormone formula matters. Bioidentical estrogen, progesterone, and testosterone are preferred. Synthetic chemically altered hormones are endocrine-disrupting chemicals and should be avoided. Examples of HRT formulas are listed in the appendix for your reading enjoyment.

Errors in Modern Medicine

Hopefully, this book will open the eyes of menopausal women to why the US medical system fails to provide them with the single most important medical intervention, bioidentical menopausal hormone replacement. This is only one of many errors in modern medicine. Our medical schools, training programs, medical research, medical journals, and the entire medical system are captured by the drug industry which colludes with health insurance companies, government agencies, and mass media spewing out nonstop propaganda. Nothing happens in medicine without the blessing and support of the drug industry, a racketeering criminal enterprise. In 2020, Dr. Denis Arnold reviewed the criminal and civil penalties paid out by the drug industry from 2003 to 2016 totaling 33 billion dollars, writing:

> Among 26 firms in our sample, 22 (85%) had **financial penalties for illegal activities**. The combined dollar value of financial penalties **totaled $33 billion for 2003 to 2016**... Given the scope and nature of the illegal activities involving financial penalties, **physicians and regulators should exhibit vigilance over the activities of large pharmaceutical firms.** (33) (10-13)

Instilling fear of estrogen into the hearts of post-menopausal women is only a tiny fraction of the criminal activities of the drug industry. Hopefully, this book will open your eyes and expel the irrational fear of estrogen, concocted by the drug industry in their endless drive for more and more profits. The main goal of this book is to provide you with the massive medical literature revealing we have been deceived and lied to by the media, drug industry, and government agencies. This book is reverse brainwashing opening the eyes of the reader and removing the fear of estrogen. To think that medical evidence rests solely on the randomized controlled trial (RCT) is another drug industry lie. Medical evidence includes preclinical, in vitro, and in vivo animal studies, observational clinical studies, epidemiological studies, as well as RCTs. This book provides all the medical evidence supporting the safety and efficacy of bioidentical hormone replacement for menopause, not just the RCT required for FDA drug approval. (8-34)

The Miracle of Life

What is this miracle called life? Think about our own life that began as two cells, an egg and a sperm containing paired instruction sets in their DNA. The underlying intelligence required for the origin of this instruction set in our DNA is something the ancients understood well. Whether we use the most powerful instruments to examine the galaxies in space or the smallest particles of the atom, or whether we use the naked eye and the five senses to examine our immediate natural world, we are confronted with the obvious conclusion that the world is permeated with intelligence. This intelligence makes all life possible, including our individual lives. This realization leads to a sense of gratitude to the Creator and a sense of wonder and awe. Yes, we are faced with the eventual prospect of aging, disease, and death. But until that time comes, we celebrate life. (37-38)

A Recurring Question

A recurring question my patients ask every day in my office is: why won't my doctor give me bioidentical hormone replacement? Why is my doctor like that? My answer is always the same: Yes, they are all like that. They are never going to change. And that is a good thing because **otherwise, I would have nothing to do!**

♦ References for Chapter 30 Fine Tuning of the Universe

1) Rees, Martin. Just Six Numbers: The Deep Forces That Shape the Universe. Basic Books, 2008.

2) Thorvaldsen, Steinar, and Ola Hössjer. "Using statistical methods to model the fine-tuning of molecular machines and systems." Journal of Theoretical Biology 501 (2020): 110352.

3) Alberts, Bruce. "The cell as a collection of protein machines: preparing the next generation of molecular biologists." cell 92.3 (1998): 291-294.

4) Noble, Denis. "Systems Biology and The Heart." Biosystems 83.2-3 (2006): 75-80.

5) IMDB Quotes, Sleeper, Woody Allen: Miles Monroe. https://www.imdb.com/title/tt0070707/characters/nm0000095

6) Ford Brian J. The Cell as Secret Agent—Autonomy and Intelligence of The Living Cell: Driving Force of Development. Academia Biology 2023;1

7) Baluska, Frantisek, and Arthur Reber. "Sentience and Consciousness in Single Cells: How the First Minds Emerged In Unicellular Species." BioEssays 41.3 (2019): 1800229.

8) Angell M. The Truth About the Drug Companies: How They Deceive Us and What To Do About It. New York, NY: Random House; 2004.

9) Smith R. The Trouble with Medical Journals. London, UK: Royal Society of Medicine Press; 2006

10) Kassirer JP. On The Take: How Medicine's Complicity with Big Business Can Endanger Your Health. New York, NY: Oxford University Press; 2005.

11) Gotzsche, Peter. Deadly Medicines and Organized Crime: How Big Pharma Has Corrupted Healthcare. CRC Press, 2019.

12) McCarthy, Eugene. "A Call To Prosecute Drug Company Fraud As Organized Crime." Syracuse L. Rev. 69 (2019): 439.

13) Goldacre, Ben. Bad Pharma: How Drug Companies Mislead Doctors and Harm Patients. Macmillan, 2014.

14) Mogensen, Søren Wengel, et al. "The Introduction of Diphtheria-Tetanus-Pertussis and Oral Polio Vaccine Among Young Infants in An Urban African Community: A Natural Experiment." EBioMedicine 17 (2017): 192-198.

15) Aaby, Peter, et al. "Evidence of Increase in Mortality After the Introduction of Diphtheria–Tetanus–Pertussis Vaccine to Children Aged 6–35 Months in Guinea-Bissau: A Time for Reflection?" Frontiers In Public Health 6 (2018): 79.

16) "Turtles All the Way Down - Vaccine Science and Myth" Anonymous (Author), Zoey O'Toole (Editor), Mary Holland J.D. (Editor, Foreword)

17) Humphries, Suzanne, and Roman Bystrianyk. "Dissolving illusions: Disease, Vaccines and The Forgotten History." (2013). CreateSpace.

18) White, William. The Story of a Great Delusion in A Series of Matter-Of-Fact Chapters. EW Allen, 1885.

19) Holman, Halsted R. "The Relation of The Chronic Disease Epidemic to The Health Care Crisis." ACR open rheumatology 2.3 (2020): 167-173.

20) Mills, Dora Anne. "Chronic Disease: The Epidemic of The Twentieth Century." Maine Policy Review 9.1 (2000): 50-65.

21) Yang, Jane L., et al. "Estrogen Deficiency In The Menopause And The Role Of Hormone Therapy: Integrating The Findings Of Basic Science Research With Clinical Trials." Menopause (2024): 10-1097.

22) Panevin, Taras S., et al. "Endogenous Estrogen Deficiency and The Development Of Chronic Musculoskeletal Pain: A Review." Terapevticheskii arkhiv 94.5 (2022): 683-688.

23) El Mohtadi, Mohamed, et al. "Estrogen Deficiency—A Central Paradigm In Age-Related Impaired Healing?" EXCLI journal 20 (2021): 99.

24) Roman-Blas, Jorge A., et al. "Osteoarthritis Associated with Estrogen Deficiency." Arthritis research & therapy 11 (2009): 1-14.

25) Emmerson, Elaine, and Matthew J. Hardman. "The Role of Estrogen Deficiency In Skin Ageing And Wound Healing." Biogerontology 13 (2012): 3-20.

26) Isola, José VV, et al. "Role of Estrogen Receptor Alpha in Aging And Chronic Disease." Advances In Geriatric Medicine and Research 5.2 (2023): e230005.

27) Sowers, M. R., and Margaret T. La Pietra. "Menopause: its Epidemiology and Potential Association With Chronic Diseases." Epidemiologic Reviews 17.2 (1995): 287-302.

28) Lobo, Rogerio A., et al. "Prevention of Diseases After Menopause." Climacteric 17.5 (2014): 540-556.

29) Ramezani Tehrani, Fahimeh, and Mina Amiri. "The Association Between Chronic Diseases And The Age At Natural Menopause: A Systematic Review." Women & health 61.10 (2021): 917-936.

30) Geraghty, Patricia. "The Interaction of Menopause and Chronic Disease." Each Woman's Menopause: An Evidence Based Resource: For Nurse Practitioners, Advanced Practice Nurses, and Allied Health Professionals. Cham: Springer International Publishing, 2021. 91-120.

31) Chun, Sungwook. "The Relationship of Hot Flush To Other Menopausal Symptoms And Chronic Disease Related To Menopause." The Journal of Korean Society of Menopause 19.2 (2013): 54.

32) Mishra, Gita D., et al. "Optimising Health After Early Menopause." The Lancet 403.10430 (2024): 958-968.

33) Arnold, Denis G., et al. "Financial Penalties Imposed on Large Pharmaceutical Firms For Illegal Activities." JAMA 324.19 (2020): 1995-1997.

34) DeVoe, Jennifer. "The Unsustainable US Health Care System: A Blueprint For Change." The Annals of Family Medicine 6.3 (2008): 263-266.

35) Lockwood, Charles J. "Our Unsustainable Health-Care System." Contemporary Ob/Gyn 53.5 (2008).

36) Williams, Jackson. "Unsustainable Health Care Spending In The USA: How Will It End?." Futures 126 (2021): 102674.

37) Meyer, Stephen C. Signature in the Cell: DNA and the Evidence for Intelligent Design. Zondervan, 2009.

38) Meyer, Stephen C. Return of the God Hypothesis: Three Scientific Discoveries That Reveal the Mind Behind the Universe. HarperOne (April 4, 2023)

39) Ercan, E. Nihal. "The Big Bang." Research & Reviews in Science and Mathematics (2021): 15.

Appendix

Hormone Replacement Therapy (HRT) Formulas

Post-Menopausal Hormone Replacement Therapy (HRT):

Topical Application

Item #1 Progesterone/ Bi-Est Topical Cream (Topiclick Dispenser), Progesterone 50 mg per gram and Bi-Est 5.0 mg per gram.

Apply Half Gram Topically BID, 28 days on, 2 days off. Rotate application site. Do not apply to breast tissue.

Item #2 Progesterone Caps, 100 mg caps.

One cap PO QHS before sleep.

Item #3 Testosterone and Progesterone Topical Combination Cream, 1.2% testosterone (12 mg per gram), and 1% progesterone (10 mg per gram).

Dispense QHS One Click (0.25 ml.), and apply a tiny dot to eyelids. Apply excess topically to the Face.

Item #4 (Optional) Estriol / Testosterone Vaginal Capsule (clear capsule w/olive oil base) 2.0 mg/gm Estriol and 4.0 mg/gm Testosterone.

Insert ONE capsule, 2-3 times per week at bedtime.
(clear capsule w/olive oil base)
30 (Thirty) capsules - with applicator.

Full Program Vaginal Capsule or Suppository Delivery

Item #1 Progesterone / Bi-Est Vaginal Capsule (clear capsule w/olive oil base), Progesterone 20 mg per gram and Bi-Est 2.0 mg per gram.

Insert 1 gram Capsule QHS, 90 (Ninety) capsules- with applicator.

Item #2 Progesterone Caps, 100 mg caps.

One cap PO QHS before sleep.

Item #3 Testosterone and Progesterone Topical Cream, 1.2% testosterone (12 mg per gram), and 1% progesterone (10 mg per gram).

Dispense QHS One Click (0.25 ml.), and apply a tiny dot to eyelids. Apply excess topically to the Face.

Premenopausal Progesterone Schedule

Item #1 Progesterone Caps, 100 mg caps.

One cap PO BID for days 12-26 of menstrual cycle.

Note: These are suggested starting formulas. Dosage adjustments are made based on estrogen excess symptoms, or testosterone excess symptoms. The vaginal application provides much higher hormone levels compared to topical application due to better absorption by vaginal mucosa.

Index

B

brain fog 16, 29, 269, 311

brainwash 383

brain zaps 366, 374

BRCA1 83, 88, 89, 90, 98, 118, 314, 324, 325, 331

BRCA gene 88, 89, 90, 92, 295

breast cancer 5-9, 15, 16, 17, 19, 22-25, 28, 32,
       33, 44, 45, 47- 61, 63, 64, 74, 77-95, 97, 98,
       100-109-122, 126, 127, 128, 130, 131, 137,
       140, 144, 145, 161-164, 166, 171, 174, 178,
       180-184, 186, 190, 196-199, 219-234, 236, 240,
       241, 248-264, 267, 269-289, 291, 293-313,
       318-330, 335, 337, 338, 339, 343-349, 352,
       353, 355, 361, 404-414

breast cancer protective 6, 16, 48, 49, 54, 56, 58, 59,
       74, 84, 114, 226, 273, 274, 282

breast cancer survivors 25, 32, 33, 81, 82, 85, 86, 87,
       88, 91, 92, 100, 115, 116, 174, 233, 252, 253,
       261, 264, 269-273, 275, 276, 278, 279, 280,
       281, 286-289, 303, 310, 312, 313

Breast Cancer Survivors 3, 24, 25, 36, 77, 81, 87, 91,
       93, 95, 97, 98, 116, 183, 253, 262, 266, 268,
       269, 288, 289, 292, 298, 315, 316

breast mass 56, 248, 249, 250

breast tenderness 176, 188, 189, 204

brexanolone 190, 195, 196, 211

B-ring 16, 50, 87, 103, 107, 224, 226, 233, 273, 274,
       282, 289

Brisdelle 352

bromine 338, 339

bromocriptine 185, 209

burden of suffering 20, 164

C

cabergoline 209

Cache County 26, 37

cachexia 347

calcium score 138, 146, 151, 164, 225, 226, 386, 392,
       393

calcium score protocol 392, 393

Caldwell Esselstyn 392

California Teachers Study 163

canine mammary carcinoma 346

Cannabidiol 212, 216

cannabis 196, 212, 213

captured 177, 241, 272, 285, 389, 419

carcinogen 23, 25, 48, 103, 228, 229, 231, 240, 255,
       309, 310, 339, 343, 346

Cardio-Protective 136

Cardiotoxicity 385, 394

cardiovascular disease 21, 22, 24, 29, 33, 44, 72, 111,
       135, 136, 137, 143, 144, 147, 155, 157, 159,
       160, 165, 182, 225, 228, 241, 270, 312, 355,
       361, 391, 393, 403

Carol Petersen 3, 9

cartilage 25, 130, 169, 170, 171, 172, 173

cartilage regeneration 25, 171, 172

catechol 78, 84, 85, 110, 321, 326, 327, 369

Cathleen Rivera 162

CBD 212, 213, 374

CEE 16, 23, 25, 31, 33, 49, 50, 51, 57, 66, 67, 71, 72,
       73, 77, 79, 87, 88, 100, 101, 102, 104-108, 111,
       116, 126, 136-140, 144, 145, 157, 162, 163,
       166, 169, 181, 182, 190, 217, 221, 223-227,
       233, 241, 269, 274, 281-289, 307, 309, 313,
       320, 329

Celecoxib 194

celiac disease 125

cell cycle control 295

CeMCOR 10

cerebellar ataxia 389

cervical cancer 295

cervical mucus 188

CETP 390, 391, 393, 395

cGMP 264

Charles Conner 30

Charles Darwin 230, 233

Charles E. Wood 49, 56, 107

Chasteberry 185, 215

Chaste Tree 185, 208, 215

chemical castration 65, 232

chemotherapy 50, 51, 91, 100, 108, 248, 249, 250,
        255, 278, 286, 293, 304, 328, 343, 346, 347,
        348, 349, 376, 409, 411

Chernobyl 336

chimpanzees 179, 208

Chlebowski 23, 35, 61, 79, 80, 85, 87, 94, 97, 98, 102,
        104, 117, 163, 223, 224, 236, 267, 287, 292,
        293, 307, 313, 317

cholesterol 43, 45, 135, 138, 144, 147, 155, 157, 195,
        224, 383, 384, 386, 387, 388, 389, 390, 391,
        392, 393, 396

cholesterol-heart theory 390, 391

chondrocytes 169, 170

chondroitin 127, 172

Chris D. Melitis 359

chronic disease 17, 126, 241, 243, 244, 245, 353, 355,
        356, 357, 418

chronic disease epidemic 17, 241, 243, 244, 245, 353,
        357, 418

chronic psychological stress 153

chronic stress 153, 195, 196, 361

CIMT 145

Claudia Lanari 48, 81, 164, 178

clear cell adenocarcinoma 224

Climara 52, 124, 132, 264

clomiphene 31

c-Myc 57

cMYC oncogene 230

Cocaine 365

cocaine craving 196

Co-Enzyme Q10 144

cognition 7, 29, 33, 242, 252, 261, 388

cognitive 26, 28, 29, 31, 32, 47, 143, 144, 154, 157,
        165, 196, 213, 228, 273, 312, 353, 355, 356,
        375, 388, 389, 397, 399, 400, 401

colchicine 328, 329, 330

colitis 153, 158, 194, 326

collagen 29, 30, 127, 130, 142, 143, 170, 171, 172,
        242

colon cancers 24, 182, 199

colorectal cancer 5, 30, 101, 110, 329

compounded 53, 60, 69, 74, 92, 124, 250, 253, 263,
        264

compounding pharmacies 213, 264, 272, 329

COMT 53, 85, 276, 321, 326, 327, 328, 368, 369

congestive heart failure 144, 252, 385, 386

conjugation 322, 326, 327

COPD 244, 336, 340

copper T-IUD 232

CoQ10 373, 383, 385, 386, 389, 390

cordyceps 349, 371, 373

coronary artery disease 16, 22, 135, 136, 137, 143,
        144, 145, 146, 151, 152, 157, 159, 162, 166,
        177, 226, 304, 328, 353, 386, 387, 391, 392,
        393, 418

coronary calcification 386

corpus luteum 80, 82, 111, 161, 179, 187, 188, 196,
        209

corticosteroids 192

cortisol 71, 199, 360, 361, 368, 369, 372

cortisone 44

coumestrol 31

COX 180, 194, 369

C-reactive protein 71, 72, 73, 361, 363, 369

Creator 419

cretinism 335, 336

criminal 177, 245, 419

criminal penalties 177

CRP 71, 72, 73, 74, 360, 361, 368, 369

C-telopeptide 170

Curcumin 172, 175

Current Good Manufacturing Practice 264

CVD 22, 135, 136, 355

cyclin D1 50, 51, 118, 274, 277, 324

cyclin D1 mRNA 277

Cynomolgus 139, 173, 292

Cynthia L. Bethea 16, 58, 103, 104, 136

CYP1A1 276, 321, 323, 324, 325, 326, 328, 331, 332

CYP1B1 116, 276, 321, 322, 323, 324, 325, 326, 327, 328, 330, 331, 332, 334

CYP450 enzyme system 366

cytokines 146, 152, 153, 154, 156, 194, 195, 242, 356, 372, 373

D

daidzein 31, 44, 51, 61

Daniel F. Kripke 399, 400

Danish Osteoporosis Study 21, 22, 128, 393

Darwin 230, 231, 233

Darwinian 217, 416

David Brownstein 3, 6, 8, 337, 343, 359

David C. Slawson 177

DCIS 90, 110, 304, 305, 404, 405, 406, 407, 408, 409, 410, 411, 414

death signal 51, 52, 93, 111, 112, 258, 273, 281

Deborah Grady 117, 136, 144

declined by 90 percent 102

deep vein thrombosis 9, 306

definition of insanity 383, 393

dementia 5, 6, 16, 20, 21, 26, 29, 32, 33, 144, 177, 353, 355, 388, 397

dendritic spines 387, 388

denosumab 126, 132, 312

dental plaque 155

depression 27, 29, 33, 58, 84, 127, 165, 176, 177, 179, 190, 195, 196, 199, 204, 205, 206, 207, 211, 212, 213, 243, 244, 263, 272, 311, 312, 353, 355, 359, 360, 361, 362, 365, 366, 367, 368, 369, 370, 371, 372, 373, 374, 375, 376, 377, 398, 400, 402, 407, 418

dermatology 257

DES 6, 7, 54, 93, 224, 225, 236, 255, 256, 273, 294, 295, 305, 309, 313, 361, 412

DES daughter 6

DEXA 18, 123, 129, 262

dexamethasone 368, 369

dexamethasone suppression 368

DHT 80, 109, 220, 254, 255, 256, 277, 300, 304

diabetes 6, 24, 29, 31, 32, 44, 65, 68, 144, 152, 165, 172, 228, 244, 312, 337, 355, 367, 389, 418

diazepam 398

diclofenac 172

diethylstilbestrol 6, 93, 113, 231, 295

dihydrotestosterone 51, 80, 108, 254, 256, 277

di-indole methane 19, 312

dilated pupils 365

DIM 3, 19, 53, 55, 84, 85, 109, 110, 111, 225, 256, 285, 288, 289, 312, 320, 323, 325, 326, 330, 412

Dioscorea 43, 45

diosgenin 43, 44, 45, 51

DMBA 48, 167, 180, 185, 225, 236, 299, 310, 343, 344, 346, 350

DNA 78, 83, 84, 85, 88, 89, 90, 96, 98, 110, 116, 139, 148, 154, 155, 156, 159, 225, 230, 236, 241, 242, 243, 246, 252, 253, 255, 260, 276, 277, 295, 296, 299, 309, 314, 317, 321, 322, 323,

epidemic 17, 57, 241, 243, 244, 245, 247, 353, 357, 361, 385, 386, 403, 418

epidermal growth factor 57, 58, 89, 251, 261, 305, 348

epiphany 8

ER alpha 51, 55, 107, 221, 224, 260, 304

ER-alpha-knockout mice 141

ER beta 16, 51, 107, 224, 233

ER-beta 6, 30, 31, 44, 50, 51, 54, 55, 58, 59, 61, 74, 78, 80, 84, 87, 91, 103, 106-110, 118, 119, 127, 141, 152, 153, 154, 182, 220, 221, 224, 226, 242, 257, 258, 273, 274, 275, 280, 282, 283, 285, 287, 288, 289, 292, 312, 320, 321, 324, 325, 326, 356, 412

Ercole Cavalier 276

Ercole Cavalieri 84, 109, 110, 115, 225, 327

Ercole L Cavalieri 78

Erika Schwartz 240

ER loss 308

ER-negative HERS 260

Errors in Modern Medicine 3, 77, 100, 117, 130, 419

errors of modern medicine 33, 126, 131, 213, 273, 339, 376, 398

estradiol 16, 22, 24-28, 32, 42, 47-60, 63, 67, 71-74, 77, 78, 80, 81, 84, 87, 88, 92, 93, 104, 106-109, 111, 113-116, 118, 120-126, 128, 130, 132, 136, 139, 140, 141, 143, 144, 153, 154, 157, 158, 159, 162, 163, 165, 170, 178, 180, 181, 182, 187, 189, 190, 191, 195, 199, 204, 205, 207, 209, 218, 220-225, 229, 231, 248, 250, 251, 253, 256, 257, 258, 261-264, 269, 270, 272-289, 298, 299, 308, 309, 311, 312, 320, 321, 322, 324, 325, 326, 329, 330, 354, 355, 356, 360, 361, 370, 372, 377, 398, 399, 412

estriol 11, 16, 28, 31, 50, 51, 53, 54, 55, 59, 60, 73, 74, 84, 92, 104, 107, 108, 109, 110, 119, 120, 123, 124, 125, 132, 159, 162, 218, 224, 233, 257, 264, 272, 273, 275, 278, 279, 280, 285, 287, 288, 289, 312, 320, 321, 412

estrogen avoidance 162, 166, 239

estrogen-blocking drugs 30, 31, 44, 79, 111, 116, 262, 269, 271, 280, 281, 286, 287, 294, 295, 309

estrogen deficiency 17, 18, 21, 26, 28-32, 41, 51, 81, 125, 126, 127, 130, 131, 136, 140, 141, 142, 145, 152, 153, 156, 158, 165, 166, 168, 170, 171, 172, 177, 217, 218, 243, 245, 252, 261, 269, 276, 281, 298, 300, 306, 307, 309, 311, 353, 354, 355, 356, 371, 372, 418

estrogen dominance 188, 189, 204, 205

estrogen excess 123, 124, 125, 257, 309, 422

Estrogen Metabolism 3, 55, 83, 84, 85, 97, 110, 169, 276, 320, 321, 322, 323, 325, 330, 331, 332, 346

Estrogen Paradox 51, 52, 106, 111

estrogen receptor alpha 55, 63, 93, 112, 119, 154, 182, 198, 253, 320, 326

estrogen receptor beta 6, 30, 31, 93, 119, 152, 158, 257, 326, 362

Estrogen Response Element 325

estrogen-responsive element 253, 277

estrogen signaling 132, 198, 254, 295, 296, 297, 298, 299, 300, 303, 304, 310, 311, 312, 324, 361, 362

estrogen signaling in progeny 362

Estrone 54, 62, 96, 120, 236, 257, 291, 292, 320, 331, 332

Eureka 345

Eva von Schoultz 87, 283, 284, 300

Evenity 128

Everything Hormones 6, 7, 289, 359

evolution 217, 416

Excess Estrogen 218

genistein 31, 44

genito-urinary syndrome 28, 261, 272, 273

genome integrity 296, 309

genome stability 295, 296

George Beatson 294

Gilbert Welch 405, 406, 409, 414, 415

glucosamine 172

glucose 29, 31, 74, 199, 218, 243

glutamate 193, 345, 397

glutathione peroxidase 83

gluten sensitivity 125, 152, 372

goiter 335, 336, 339, 343

Gombe National Park 179

goserelin 269

Gothenburg Breast Screening Trial 410, 415

GPER1 153

G-protein coupled receptor 153, 158

growth factor 16, 29, 49, 51, 57, 58, 89, 106, 114, 116, 141, 223, 233, 250, 251, 256, 261, 271, 273, 275, 298, 305, 348

Growth factors 305

growth signal 52, 93, 112, 258, 271, 273, 281

guardian of the genome 232, 329, 344

Gustafsson 7, 62, 106, 118, 273, 290, 330

gut barrier 152, 153, 157, 193

gut dysbiosis 85, 155

gut permeability 152, 158, 193

Guy Abraham 335, 337

gynecologist 125, 189, 204, 217

H

HABITS 25, 81, 82, 86, 87, 88, 95, 97, 248, 250, 270, 276, 279, 282, 283, 284, 285, 288, 292, 298, 300, 301, 302, 313, 315

hair 30, 32, 196, 219, 242, 257, 261

hair loss 30, 257

hair regrowth 196, 257

Hanahan 49, 60, 78, 94

hand tremors 365

Harry Genant 168, 173

Harumi Okuyama 386

Harvey Hecht 392

headache 176, 204

heart attack 135, 138, 164, 392

Heart Book 137, 147, 151, 157, 392

heart failure 22, 143, 144, 157, 172, 252, 385, 386

hematocrit 161

Heme A 386

hemoglobin 161, 192

Henry Lemon 54, 108, 109

hepatocyte growth factor 305

HERS 89, 136, 145, 146, 260

HEXE 301

hijacking of PMS by psychologists 205

hip fracture 101

hippocampus 5, 29, 374, 387, 388

hirsutism 234, 261, 361

Hisham Mohammed 53, 108, 182, 198, 221, 274

hives 192

HMGR 387, 389

homicide 352, 366

Hopi Indians 7

hormone averse 353, 354, 355

Hormone Deception 7, 225, 236

hormone-impregnated IUDs 359

Hormones After Breast Cancer: Is it Safe? 270

hot flashes 16, 18, 26, 33, 71, 124, 165, 179, 180, 269, 294, 311, 320, 352, 353, 354, 355, 397

hot flushes 29, 33, 179, 189, 206

Howard Hodis 20, 22, 48, 51, 77, 100, 101, 104, 116, 143, 145, 151, 157, 163, 226, 275, 286, 393

HPA 179, 211, 361, 363, 367, 368, 369, 372

HPLC 188

I

mortality 16, 20, 21, 22, 23, 25, 31, 32, 33, 44, 49, 51, 52, 61, 77, 78, 82, 83, 85, 86, 88, 90, 91, 101, 102, 110, 112, 115, 116, 121, 129, 135, 136, 137, 138, 141, 143, 144, 145, 151, 157, 161, 162, 163, 164, 165, 166, 172, 183, 194, 227, 229, 230, 232, 233, 239, 240, 253, 262, 270, 271, 276, 278, 279, 280, 282, 283, 286, 287, 300, 304, 305, 306, 307, 308, 309, 310, 311, 312, 356, 376, 385, 386, 390, 391, 393, 398, 399, 400, 401, 402, 403, 404, 405, 407, 408, 409, 410, 411, 413, 414

MPA 6, 16, 17, 19, 21, 22, 23, 24, 26, 28, 47, 48, 49, 53, 54, 55, 56, 57, 58, 59, 60, 61, 64, 66, 67, 68, 71, 73, 74, 79, 80, 81, 92, 93, 94, 101, 102, 103, 104, 105, 109, 111, 112, 113, 116, 117, 128, 133, 136, 138, 139, 140, 144, 145, 146, 149, 162, 163, 164, 166, 167, 169, 176, 177, 178, 180, 181, 182, 183, 184, 190, 191, 198, 212, 220, 221, 222, 225, 226, 227, 229, 233, 236, 240, 241, 253, 254, 255, 256, 266, 269, 272, 274, 275, 283, 284, 285, 286, 288, 290, 300

MPA mouse breast cancer model 81, 166, 184

MPA mouse model 49, 93, 128, 164, 233

MSM 172

MTHFR 53, 68, 85, 110, 120, 257, 267, 321, 326, 327, 328, 332, 368, 369, 372

muscle 23, 31, 33, 130, 144, 157, 193, 194, 218, 243, 251, 252, 261, 263, 306, 370, 383, 384, 385, 386, 393

myalgia 383, 384, 385

MYC 55, 58, 59, 63, 118, 220, 230, 235, 261, 274, 290

myelin sheaths 387

myocardial infarction 22, 140, 141, 143, 145, 147, 159, 172, 226, 392

myoinositol 65, 191, 192, 212, 213

myopathy 144, 383, 385

myositis 383

N

NAC 84, 85, 110, 321, 326, 327, 330, 374

N-acetyl-cysteine 321, 327

nano-engineering 416

natural treatments for triple-negative breast cancer 347

nausea 205, 206, 374

neoplastic transformation 78

neural tube defects 359

neurogenesis 84, 194, 211, 242, 374, 388

Neurokinin3 receptors 352

neuropathy 144, 157, 359, 383, 389, 390

neuroprotection 194, 211

neuroprotective 29, 58, 84, 182, 193, 194, 199, 327, 372, 389, 398

neurosteroid 29, 56, 189, 190, 194, 195, 398, 402

neurotransmitters 27, 242, 377

newsletter 14, 17, 206

New Zealand White rabbits 140

NF-KappaB 154

NF-kB 142, 194

NF-κB 146, 154, 158, 159, 194

night sweats 16, 18, 26, 33, 71, 124, 165, 180, 189, 269, 294, 311, 320, 352, 353, 354, 355, 397

NK3 352, 353, 354, 357

Noble 78, 114, 416, 420

non-human primates 47

norepinephrine 27, 365, 366, 373, 374

norethindrone 44, 83, 103, 109, 111, 229

norethisterone 22, 24, 25, 55, 67, 81, 83, 87, 88, 95, 126, 127, 143, 178, 180, 223, 270, 274, 276, 281, 282, 283, 285, 288, 298, 300, 302, 412

NSAID 171, 174, 177, 194, 220

NSAIDs 171, 172, 173, 174, 220, 239, 312

nuclear factor kappa-B 142, 152

nuclear reactor accident 336

null effect 104, 105, 143

nulliparity 310

nulliparous 230, 285

Nurses' Health Study 138, 147, 162, 166

O

obesity 24, 31, 65, 68, 85, 158, 171, 355, 418

OB/GYN 5, 26, 28, 33, 40, 52, 101, 204, 263

OCPs 65, 66, 71, 74, 204, 213, 229, 359

old dogmas 157

oncogene 50, 55, 57, 58, 230, 235, 274, 275, 277, 301, 324

oncologists 79, 82, 262, 285, 313, 349

oocyte 187, 362

oophorectomy 88, 90, 135, 141, 145, 162, 163, 294, 295, 309, 311

Optimox 335, 337

oral cancer 309

oral cavity 155

oral CEE 72, 73, 140

oral contraceptive pills 65, 71, 204, 361

Oral Contraceptives 45, 63, 70, 75, 235, 360, 362, 363, 364

oral estrogen 16, 21, 27, 65, 66, 67, 68, 71, 72, 73, 74, 84, 100, 101, 126, 128, 139, 144, 146, 165, 168, 251, 271, 360

orgasm 28, 196, 272

orthomolecular medicine 207

osteoarthritis 16, 25, 33, 137, 143, 168, 169, 170, 171, 172, 174, 269, 353, 355, 418

osteoblasts 126, 127

osteoclasts 125, 126, 127, 130

osteoid 126, 127

osteonecrosis 18, 127, 132

osteonecrosis of the jaw 18, 127, 132

osteopenia 18

osteoporosis 16, 18, 19, 20, 21, 27, 33, 34, 44, 111, 123, 124, 125, 126, 127, 128, 129, 130, 131, 132, 133, 134, 143, 177, 194, 270, 309, 312, 329, 353, 355, 418

Our Stolen Future 225, 234, 236

ovarian ablation 90, 91

ovarian carcinoma 199

ovarian cysts 298, 337

ovary 14, 69, 80, 82, 111, 176, 187, 189, 320

overdiagnosis 110, 405, 406, 407, 408, 409, 410, 411, 412, 413, 414

overtreatment 110, 406, 407, 408, 409, 410, 412, 413, 414

ovulation 60, 65, 161, 176, 178, 179, 180, 187, 188, 189, 191, 192, 196, 204, 205, 209, 212, 219, 232

oxidized LDL 142

P

P5P 212, 213

P53 114, 230, 238, 258, 267, 315, 329, 332, 333, 334, 344

P53-induced apoptosis 258

P53 protein 230

p65 154, 159

paclitaxel 328, 330

Palbiciclib 259

Paolo Raggi 138, 392

paradigm shift 16, 115, 138, 157, 227, 391, 392

paroxetine 352

patent 43, 44, 65, 144, 177, 188, 190, 229, 232, 239, 329, 418

PCOS 69, 234

PCSK9 390, 391, 393, 395

peaks and troughs 124

pelvic exam 125, 189

pelvic examination 161

pelvic sonogram 125, 161, 189, 191, 209

PEMF 172

perimenopausal 33, 158, 180, 183, 189, 190, 199, 370, 399, 402

Perimenopause 3, 41, 96, 185, 187, 189, 200

peripheral neuropathy 359, 389

pernicious anemia 359

Phase One Detoxification 321

Philip J. DiSaia 85, 86, 88, 97

Phillip Sarrel 104, 162

phototherapy 400, 401

Phthalates 224, 225

Phyllis Bronson 3, 14, 206, 207

PIBF 194, 195

pitting edema 205

placebo 18, 23, 31, 48, 49, 56, 57, 58, 73, 79, 80, 81, 87, 90, 91, 100, 101, 102, 104, 105, 107, 111, 116, 128, 129, 130, 131, 137, 138, 143, 151, 162, 163, 164, 169, 181, 205, 208, 209, 210, 212, 222, 224, 226, 227, 240, 252, 253, 275, 276, 277, 286, 287, 288, 305, 306, 307, 347, 348, 365, 367, 370, 391, 392

platelet-derived growth factor 305

Platelet-Rich Plasma 172, 175

pleiotropic effects 157, 393

PMS 14, 176, 178, 179, 183, 184, 188, 189, 190, 194, 199, 204, 205, 206, 207, 208, 209, 210, 212, 213, 214, 216, 218, 219

POF 135

POI 22, 165, 166, 306, 316

Polycystic Ovary Syndrome 331

polymorphisms 85, 110, 120, 326, 327, 369

post-finasteride syndrome 196

post-menopausal 19, 21, 26, 28, 30, 50, 53, 54, 56, 60, 66, 67, 71, 72, 73, 74, 84, 86, 88, 91, 100, 101, 107, 113, 115, 116, 123, 125, 126, 127, 128, 129, 131, 132, 135, 136, 138, 139, 141, 142, 143, 144, 145, 146, 151, 154, 157, 164, 168, 169, 170, 171, 172, 173, 181, 183, 188, 191, 219, 221, 223, 233, 240, 243, 245, 248, 252, 261, 262, 278, 307, 329, 353, 354, 355, 370, 371, 372, 383, 401, 418, 419

postmenopausal 16, 18, 19, 20, 21, 22, 23, 24, 27, 28, 29, 30, 33, 48, 56, 57, 59, 60, 62, 63, 67, 72, 73, 74, 79, 84, 86, 92, 94, 104, 109, 115, 118, 120, 126, 127, 128, 129, 131, 133, 134, 136, 137, 140, 141, 142, 143, 144, 145, 151, 158, 160, 163, 167, 169, 171, 181, 182, 199, 209, 220, 221, 222, 223, 227, 233, 240, 248, 250, 254, 255, 262, 263, 269, 276, 277, 278, 279, 284, 291, 295, 299, 305, 308, 309, 319, 353, 355, 370, 376, 398, 399

postpartum depression 190, 195, 196, 211, 263

post-partum state 127

post-traumatic stress disorder 195, 196

povidone-iodine 337

pravastatin 139, 385, 387, 388, 389

pregnancy 16, 48, 49, 54, 114, 119, 122, 153, 187, 194, 195, 199, 208, 214, 218, 224, 230, 231, 232, 233, 271, 275, 280, 281, 287, 288, 295, 309, 310, 320, 359, 360, 361

pregnancy reprograms the epigenome 230

pregnanediol 178, 179, 188

pregnenolone 195, 196

Premarin 16, 23, 24, 25, 28, 31, 33, 48, 49, 50, 51, 52, 53, 54, 56, 57, 58, 66, 67, 72, 73, 79, 87, 88, 100, 101, 102, 103, 104, 106, 107, 116, 126, 136, 137, 138, 139, 140, 144, 145, 157, 162, 163, 166, 169, 181, 190, 217, 221, 223, 224, 225, 226, 227, 229, 233, 240, 241, 269, 271, 272, 273, 274, 275, 279, 282, 283, 285, 286,

pseudo-disease 409

psychosis 359, 366, 376, 386

PTEN 258, 259

pterostilbene 44

PTSD 195, 196, 380

pulmonary embolus 65, 71, 101

pyridoxal 5-phosphate 212

Pyridoxine 191, 202, 216

Q

quinone 78, 85, 109, 110, 116, 120, 324, 325, 326,
327

R

RAAS 22, 23, 35

radiation 231, 251, 294, 305, 336, 343, 405, 406, 410,
417

Rahul Mal 50, 108, 109

Rajkumar 48, 60, 79, 94, 113, 114, 115, 121, 122,
180, 186, 231, 237, 261, 262, 268, 317

Rajkumar Lakshmanaswamy 94, 114, 122, 180, 231,
262, 268

randomized clinical trials 23, 77, 120, 121, 284, 305

RCT 22, 48, 80, 82, 86, 87, 92, 137, 143, 197, 205,
210, 222, 223, 264, 288, 306, 313, 419

RDA 335, 336

Rebecca Glaser 30, 84, 108, 219, 248, 249, 257, 263,
264, 276, 299, 303

Reclast 18

recurrent cancer 269, 298

regional cerebral blood flow 28

REM sleep 399

renin-angiotensin-aldosterone 23

reprogram the epigenome 275

reservoir of silent disease 406

resveratrol 44, 84, 85, 111, 225, 321, 322, 323, 326,
327, 328, 330

rhabdomyolysis 383, 384

Romosozumab 128

ROS 23, 61, 345, 350, 383

Rosalind Franklin 241, 246

roundworm 344, 345

Royal Osteoporosis Society 129, 134

RU-486 187, 200

Russell Jaffe 125, 132

Russell Marker 3, 43, 44, 45, 51, 80

S

safety 6, 17, 19, 34, 47, 67, 72, 74, 100, 102, 118, 119,
120, 124, 145, 182, 208, 217, 218, 219, 220,
226, 263, 264, 276, 281, 288, 323, 336, 397,
418, 419

Safety of Progesterone 218

salivary cortisol 368

salivary glands 338

Sally Fields 18

SCID 49, 78

screening mammography 110, 404, 405, 406, 407,
409, 410, 412, 413, 414

seaweed 335, 337, 338, 343, 345, 346, 347

Sebastián Giulianelli 55, 220, 274

selenium 19, 53, 55, 83, 89, 90, 110, 111, 125, 225,
256, 288, 324, 338, 373, 412

selenoprotein 83, 89, 386, 387

selenoprotein repair system 89

serotonin 27, 29, 58, 210, 211, 212, 360, 365, 366,
367, 368, 373, 374, 375, 376, 377, 399, 402

serotonin theory of depression 367, 373, 376

serotonin transporters 365, 373, 374

Serum-binding globulin 361

serum estrogen levels 124, 152, 294, 297, 313

sexual dysfunction 165, 263, 312, 352, 368

silibinin 44

simvastatin 139, 385, 387, 388

T

tuberculosis 337

tumor suppressor 6, 16, 44, 50, 51, 54, 80, 84, 88, 93,
    107, 108, 109, 114, 182, 220, 221, 224, 226,
    233, 257, 258, 259, 260, 273, 275, 280, 282,
    283, 288, 289, 295, 303, 304, 312, 320, 321,
    324, 325, 412

type II collagen 170, 171

U

ubiquinone 383

ubiquitin-proteasome 142, 143

ulcerative colitis 326

Unopposed estrogen 271

unsaturated B-ring steroids 16, 50, 87, 103, 107, 226,
    233, 282, 289

urinary tract infection 261

urticaria 192, 193

uterine bleeding 125, 177, 180, 183, 189, 191, 192,
    199, 212, 233

uterine fibroids 161, 191

V

vaginal atrophy 26, 257, 353

vaginal capsule 28, 257, 312

vaginal dryness 26, 29, 165, 252, 261, 294, 311, 397

Vaginal Estrogen 38, 39, 278, 288, 291, 292

vagus nerve 352

Valerie A. Flores 107

vascular endothelial growth factor 141, 305

vasomotor symptoms 26, 27, 33, 71, 84, 95, 124, 180,
    272, 312, 355, 356

V. Craig Jordan 51, 52, 93, 99, 106, 111, 112, 113,
    118, 121, 223, 258, 273, 290, 295, 296, 313,
    314, 318, 319, 329, 330

venlafaxine 365, 366

venous thromboembolism 22, 24, 65, 67, 68, 71, 72,
    73, 74, 84, 100, 101, 107, 126, 143, 165, 222,
    233

Veozah 33, 352, 353, 354, 355, 357, 358

violent behavior 352, 366

Violent Crime 377, 386, 394

Vioxx 194, 203, 243

visceral fat 28

vitamin A 191

Vitamin B6 212, 216, 362, 363

vitamin C 338, 359, 392

Vitamin D 82, 83, 95, 132, 134, 234, 267, 368, 369,
    383

Vitamin D3 53, 55, 125

vitamin K 201, 386, 392

Vitex 69, 179, 180, 185, 208, 209, 210, 212, 213, 214,
    215

Vivelle 52, 264

VTE 65, 67, 68, 71, 72, 73, 74, 101, 107, 222

W

WADA 82, 95

walking drug store 312

water retention 176, 179, 205, 206, 219

Watson and Crick 241

weight 7, 28, 29, 33, 65, 85, 125, 165, 171, 192, 204,
    205, 250, 359, 374, 387

weight control 29

weight gain 28, 65, 204, 250, 359

Weinberg 49, 60, 78, 94, 215, 225, 237

Wellness by Design Project 9

wheat gluten sensitivity 152

WHI 5, 16, 19, 20, 21, 23, 25, 28, 33, 48, 49, 50, 51,
    52, 66, 67, 73, 74, 77, 79, 80, 85, 87, 88, 92,
    97, 100, 101, 102, 103, 104, 105, 106, 107, 111,
    112, 116, 117, 122, 126, 128, 129, 136, 137,
    138, 140, 144, 145, 151, 162, 163, 164, 166,

Z

Printed in Dunstable, United Kingdom

68801886R00252